INTRODUCTION TO NEARSHORE HYDRODYNAMICS

ADVANCED SERIES ON OCEAN ENGINEERING
ISSN: 1793-074X

Series Editor-in-Chief
Philip L- F Liu (*Cornell University*)

---

*Published**

Vol. 17 The Mechanics of Scour in the Marine Environment
by B Mutlu Sumer (Technical University of Denmark) and
Jørgen Fredsøe (Technical University of Denmark)

Vol. 18 Beach Nourishment: Theory and Practice
by Robert G Dean (University of Florida, USA)

Vol. 19 Saving America's Beaches: The Causes of and Solutions to Beach Erosion
by Scott L Douglass (University of South Alabama, USA)

Vol. 20 The Theory and Practice of Hydrodynamics and Vibration
by Subrata K Chakrabarti (Offshore Structure Analysis, Inc, Illinois, USA)

Vol. 21 Waves and Wave Forces on Coastal and Ocean Structures
by Robert T Hudspeth (Oregon State University, USA)

Vol. 22 The Dynamics of Marine Craft: Maneuvering and Seakeeping
by Edward M Lewandowski (BMT Designers & Planners, VA, Arlington, USA)

Vol. 23 Theory and Applications of Ocean Surface Waves
Part 1: Linear Aspects
Part 2: Nonlinear Aspects
by Chiang C Mei (Massachusetts Institute of Technology, USA),
Michael Stiassnie (Technion-Israel Institute of Technology, Israel) and
Dick K-P Yue (Massachusetts Institute of Technology, USA)

Vol. 24 Introduction to Nearshore Hydrodynamics
by Ib A Svendsen (University of Delaware, USA)

Vol. 25 Dynamics of Coastal Systems
by Job Dronkers (Rijkswaterstaat, The Netherlands)

Vol. 26 Hydrodynamics Around Cylindrical Structures (Revised Edition)
by B Mutlu Sumer (Technical University of Denmark) and
Jørgen Fredsøe (Technical University of Denmark)

*To view the complete list of the published volumes in the series, please visit:
http://www.worldscientific.com/series/asoe

Advanced Series on Ocean Engineering — Volume 24

INTRODUCTION TO NEARSHORE HYDRODYNAMICS

Ib A. Svendsen

University of Delaware, USA

NEW JERSEY • LONDON • SINGAPORE • BEIJING • SHANGHAI • HONG KONG • TAIPEI • CHENNAI

Published by

World Scientific Publishing Co. Pte. Ltd.
5 Toh Tuck Link, Singapore 596224
USA office: 27 Warren Street, Suite 401-402, Hackensack, NJ 07601
UK office: 57 Shelton Street, Covent Garden, London WC2H 9HE

British Library Cataloguing-in-Publication Data
A catalogue record for this book is available from the British Library.

INTRODUCTION TO NEARSHORE HYDRODYNAMICS

ISBN-13 978-981-256-142-8
ISBN-10 981-256-142-0
ISBN-13 978-981-256-204-3 (pbk)
ISBN-10 981-256-204-4 (pbk)

Printed in Singapore

To my wife

Karin

Prologue

My husband, distinguished professor emeritus, Dr. Ib A. Svendsen, died on Sunday, December 19, 2004.

He had been working intensely to finish this book for the past several months following his retirement from University of Delaware on September 1, 2004. It seemed to be both a rewarding, but on the same time highly frustrating job, as anybody who has given birth to a book will probably recognize.

As he mentions in his preface, one of the problems he was facing was deciding what to include in the book. He knew that some topics might have been included or covered in more details, and he was considering the possibility of an additional book exploring these subjects, and embodying the response from this edition.

On December 10 Ib sent the manuscript to World Scientific Publisher. On Tuesday December 14 he made the last organizational changes to his files on the book, and inquired of the publisher how much longer he would have for changes and additions. He was looking forward to discussions with colleagues and students about the contents of the book.

But early on December 15 he collapsed with cardiac arrest at the fitness center at University of Delaware. He died without regaining consciousness.

It is my hope that this book will become the means of learning and inspiration for future graduate students and others within coastal engineering as was Ib's sincere wish.

The royalties from this and Ib's other publications will be used to finance a memorial fund in his honour: **Ib A. Svendsen Endowment**, c/o Department of Civil Engineering, University of Delaware, Newark, DE 19716. This fund will benefit University of Delaware civil engineering students in their international studies.

Karin Orngreen-Svendsen
Landenberg, March 21, 2005

Preface

The objective of this book is to provide an introduction for graduate students and other newcomers to the field of nearshore hydrodynamics that describes the basics and helps de-mystify some of the many research results only found in journals, reports and conference proceedings.

When I decided to write this book I thought this would be a fairly easy task. From many years of teaching and research in the field of nearshore hydrodynamics I had extensive notes about the major topics and I thought it would be a straight forward exercise to expand the notes into a text that would meet that objective.

Not so. From being a task of considering how to expand the notes - which I found enjoyable - the work rapidly turned into the more stressful task of deciding what to omit from the book and how to cut. I had completely underestimated the number of relevant topics in modern nearshore hydrodynamics, the amount of important research results produced over the last decades, and the complexity of many of those results.

In the end I came up with a compromise that became this book. I have considered some topics are so fundamental that they have to be covered in substantial detail. Otherwise one could not claim this to be a textbook. On the other hand, for reasons of space, sections describing further developments have been written in a less detailed, almost review style and supplemented with a selection of references to the literature. The list of references is not exhaustive but rather meant to give the author's modest suggestions for what may be the most helpful introductory reading for a newcomer to the field. This unfortunately means that many excellent papers are not included which in no way should be taken as an indication of lesser quality. The transition between the two styles may be gradual within each subject. It is hoped that the detailed coverage of the fundamental topics such as linear wave theory, the basics of nonlinear Stokes and Boussinesq wave theory, the nearshore circulation equations, etc., will bring the reader's insight and understanding to a point where he/she is able to benefit from the sections that discuss the latest developments, is able to read the current literature, and perhaps to start their own research.

Though no computational models are described in detail and the presentation focuses on the hydrodynamical aspects of the nearshore the choice of topics and the presentation is oriented toward including the hydrodynamical basis in a wider sense for some of the most common model equations. The principles that form the basis of good modelling can perhaps be simplified as the following:

If you want to model nature you must copy nature
It you want to copy nature you must understand nature

This has been the motto behind the writing of this book.

The purpose of hydrodynamics is the mathematical description of what is happening in nature, and the basic equations such as the Navier-Stokes equations are as close to an exact copy of nature as we can come. Therefore misrepresentation of nature only comes in through the simplifications and approximations that we introduce to be able to solve the particular problem we consider. No model/equation is more accurate than the underlying assumptions or approximations. An important task in providing the background for responsible applications of the equations of nearshore hydrodynamics is therefore to carefully monitor and discuss the physical implications of the assumptions and approximations we introduce. I have tried to do just that throughout the text.

Todays models are becoming more and more sophisticated and complex. Usually this also means more and more accurate and the use of them is becoming part of everyday life. Mostly this also means they become more and more demanding of computer time and of man power to use and interpret them. So in many applications there will be a decision about which accuracy is needed. Is linear wave theory good enough? Are we outside the range of validity of a particular Boussinesq model? Nobody can prevent users from deciding to use model equations/theories for situations where they are insufficient or do not properly apply. Sometimes the results are acceptable sometimes they are misleading. One parameter, such as for example the wave height, may be accurately predicted for the conditions considered while another, say the particle velocity, is not. It is generally outside the scope of this book to provide estimates of errors for particular theories.

It may be easy to run a computer model. One has to remember, however, that all that comes out of that is numbers. An enormous amount of numbers. They may be interpreted and plotted in diagrams to look like nearshore flow properties. But knowing/understanding the powers and the limitations of models requires understanding the basis for the equations. Which features are represented in the equations, which not, why this or that effect is important, and when, etc., is a first condition for generating confidence in the results. It is the hope that the content of the book will help serving that purpose and thereby promote the prudent and constructive use of models.

For reasons of space many important topics and aspects of nearshore hydrodynamics have been left out.

One such is the testing of the theories using laboratory measurements. A major reason is of course lack of space, but there are some important concerns too. In a moment of outrageous provocation and frustration I once wrote about laboratory experiments: "If there is a discrepancy between the theory and the measurements it is likely to be due to errors in the experiments". The reason is that, while it is fairly easy to create good theories, it is so difficult to conduct good experiments, in particular with waves. Anybody who has tried can testify to all the many unwanted - and often unanticipated - side effects and disturbances that occur even in a simple wave experiment in a wave flume. And often those are the major reasons for the deviations between the theory and the experiment designed to test it. Therefore we have to be careful before we use an experimental result to deem a well documented theory inaccurate or poor as long as we are within the range of validity of the assumptions. This is also why I prefer to replace the commonly used term "verification" of a model against experimental data with the term "testing". So though comparisons with measurements can be found many places a systematic testing of theories against laboratory measurements has not been one of the main objectives of the book. In fact comparison of the simpler theories to more advanced and accurate ones is often more revealing.

In a different role experimental results have been quoted extensively to gain physical insight into areas where theoretical understanding is lacking. This particularly applies to the hydrodynamics of waves in the surfzone.

Extensive field experiments have been conducted in particular over the last two decades. The comprehensive and careful data analysis of those experiments has provided insights and ideas for further study. However, those results are described in the book only to the extent it is needed to understand the hydrodynamical phenomena and the theories covering them. It should be noted, though, that the way nearshore modelling is developing today the direct comparison of model results with the complex conditions on natural beaches will be one of the promising research areas in the coming years. Unfortunately we can only just touch upon this subject in an introductory book like this.

Wind wave spectra is an area that is more related to data analysis than to hydrodynamics. Only the concept of energy spectra is explained as an example of wave superposition and a brief overview of the ideas is given. This also applies to nonlinear spectral Boussinesq models. The reader is referred to the relevant literature.

Again for space reasons, evolution equations and concepts for time and space varying waves based on Stokes' wave theory have not been explored at all. This applies to topics such as the side band instability, which mainly occurs in deeper water, to the theory of slowly varying Stokes waves and the nonlinear Schrödinger equation. An important reason for this choice is that the Stokes' wave theory has an uncanny habit of not working well in the shallow water regions nearshore. Instead the Boussinesq wave theory, which leads to nonlinear evolution equations for waves in shallow water, has been covered in great detail. This theory has over the last one or two decades been developed into an extremely useful and accurate tool for nearshore applications. In fact it has even been extended to depths that approach the deep water limit of the nearshore region which further adds to its relevance.

Acknowledgements

A book like this is really influenced by a great number of contributions over many years, often from people who do not even realize they have contributed. I cannot here mention them all but I do want to thank my colleague through many years Ivar G. Jonsson for numerous discussions that helped develop my insight into the topics described in this book. He was also co-author on an earlier book on The Hydrodynamics of Coastal Regions, which has been the starting point for the description of linear waves in this book.

Also a special thank to Howell Peregrine. Our extensive scientific discussions have been ongoing for decades and he more than anybody helped open my eyes to the fascinations of fluid mechanics.

Over the years I have also received many comments from graduate students to the notes I have used in my courses. Those notes have formed the initial basis for the book.

More focused on the present book has been valuable comments and suggestions from Mick Haller and Ap Van Dongeren who reviewed early versions of the first chapters and helped improving their content and form. Also thanks to Jack Puleo for input to the chapter on swash, to Kevin Haas for his many suggestion for the chapter on breaking waves and surfzone dynamics, and to Francis Ting for generously providing his unpublished data shown in that chapter.

Also sincere thanks to Qun Zhao who throughout the work has been assisting in many ways, including in preparing the many drawings and patiently responded to my many requests for changes. And to my secretary Rosalie Kirlan for taking care of scanning figures from papers and reports.

I also want express my gratitude to Per Madsen for numerous discussions and extensive help with and insight into many subjects, particularly on the more advanced topics on linear and nonlinear waves.

And special thanks to Jurjen Battjes who undertook the task to read a late version of the entire manuscript. His meticulous comments and suggestions have been invaluable as they helped not only to remove typos but to improve many unclear or ambiguous passages in the manuscript.

More than anybody, however, I am indebted to my wife Karin. It is an understatement to say that without her patient and caring support this book would not have been finished.

However, inspite of all efforts to avoid it, it is inevitable that a book like this will have misprints and errors and, even worse, reflect my misunderstandings of other scientists work. I apologize to the authors for any such mistakes and hope that they will have time and patience to point them out to me.

The work on this book has been partially funded by the National Oceanographic Partnerships Program (NOPP) under the ONR grant N0014-99-1-1051, which has lead to the development of the open-source Nearshore Community Model (NearCoM) briefly described in the last chapter. It is hoped that this book will help more students, engineers and coming scientists to understand the basic theories of nearshore hydrodynamics which such models are based on and thereby be able to use them wisely.

IAS

Landenberg, PA, December 2004

Contents

Chapter 1

Introduction

1.1 A brief historical overview

The nearshore coastal region is the region between the shoreline and a fictive offshore limit which usually is defined as the limit where the depth becomes so large that it no longer influences the waves. This depth depends on the wave motion itself and in simple terms it can be identified as a depth of approximately half the wave length. Thus in storms with larger and longer waves the offshore limit moves further out to sea. This definition is practical because the influence of the bottom on the waves is one of the most important mechanisms in nearshore hydrodynamics.

Nearshore hydrodynamics could probably be said to have been founded by G. G. Stokes, who in 1847 developed the first linear and nonlinear wave theory. Today this theory is often referred to as *Stokes waves* (see also Stokes, 1880). Over the following century various wave phenomena were analysed and a great number of results, remarkable from a mathematical point of view, were obtained. Of particular importance from todays perspective was the development by Boussinesq (1872) of the consistent approximation for nonlinear waves in shallow water, a situation for which Stokes himself recognized that his theory was failing. Korteweg and DeVries (1895) added to this result by finding analytical solutions to the Boussinesq equations. These solutions are known as *cnoidal* and *solitary waves.* Interestingly the infinitely long solitary waves had already been observed in real channels by Russell (1844). Finally even this ultra brief historical review would be incomplete without mentioning the pioneering discovery of the wave radiation stress by Longuet-Higgins and Stewart (1962). This established the insight that forms an essential element in all later research related to currents and long wave generation in the nearshore.

The advent of computers has radically changed the perspective of what is relevant hydrodynamics in todays world. Equations or theories that, when developed before the computer age, were merely of theoretical interest have become central to modern engineering applications while many of the remarkable mathematical results that helped the understanding of how waves behave have become mainly of academical interest. The content of this book partially reflects that in the choice of which subjects and results are pursued in detail.

1.2 Summary of content

As an introduction Fig. 1.2.1 from Svendsen and Jonsson (1976) shows a schematic of most of the major wave phenomena that occur in the nearshore. These, and some more that are not visible in such a picture, are the phenomena that are analysed further in the following chapters.

Chapter 2

The first chapter (Chapter 2) is meant as a reference chapter that essentially presents the main hydrodynamical results used later in the book. For most sections there are no derivations in this chapter. If the reader needs further explanation reference is made to the textbooks quoted in the list of references at the end of the chapter. Exceptions are the sections on boundary conditions, turbulence and energy flux which contain material not so easily found in standard books.

Chapter 3

Remaining central to the understanding of nearshore wave and current motion is the Stokes theory, which in its simplest linear form represents the most important theoretical background for nearshore hydrodynamics. Chapter 3 therefore gives a thorough analysis, not only of the linear wave theory itself but also of the most important of the results that have been derived on the basis of that theory.

The main objective of the linear theory is to establish a first approximation for all the flow details of small amplitude waves on a constant depth. This is done in Section 3.2.

The characteristic surface profile of such waves is described by the sine function, whence they are also called **sinusoidal** waves. It turns out that even though the average over a wave period of such wave profiles is zero

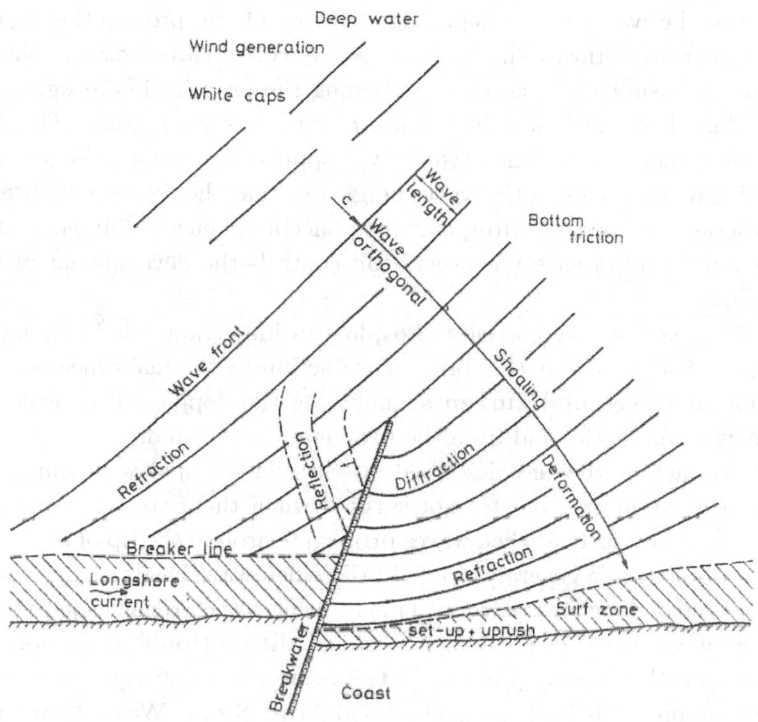

Fig. 1.2.1 Nearshore wave processes (Svendsen and Jonsson, 1976).

they still have properties that in average over a period are non-zero (Section 3.3). Linear waves transport energy (the socalled **energy flux**) which is the mechanism that causes waves generated in an area to spread forward in the direction of wave propagation. Waves also represent both a **mass (or volume) flux** of fluid and they exert a net force on the surroundings that is called the **radiation stress.**

Section 3.4 explores what happens when we utilize the freedom of linear theory to form new wave solutions by adding solutions of waves. Thus two waves added can form **standing waves** or **wave groups**. And in particular the results of adding arbitrarily many waves leads to the concept of **wave spectra** which can be used in the analysis and description of random seas.

A particularly important element in the hydrodynamics of waves on a coast is the variation of the water depth. The effect of wave propagation over a varying depth is analysed in Section 3.5. The major effect causing changes of the waves is the depth dependence of the propagation velocity for the wave forms. This induces **wave refraction** which is shown to follow laws similar to the laws controling the propagation of light and sound. The depth variations also cause change in wave heights. This becomes particularly important as the waves approach the shore, because the decreasing depth increases the waves heights so that they eventually break. The process is termed **shoaling** and the concepts of energy flux generated by the waves and of energy conservation controls the development of the wave height.

In the nearshore the currents also play an important role in changing the waves. Section 3.6 gives a brief introduction to the main mechanisms of combined **waves and currents**, including the doppler effect which is also known from optics and from the propagation of sound.

The refraction theories described in Section 3.5 makes assumptions about the wave motion theatre not satified when there are rapid changes along wave fronts such as when waves propagate around the tip of breakwaters and also out in a general wave field the when wave height changes over short distances along a wave front. This influences the propagation pattern even for linear waves, a phenomenon called **diffraction**. In Section 3.7 we develop a theoretical approach to the combination of depth refraction and diffraction. This leads to the socalled **Mild Slope Wave Equation (MSE)** which describes the variation over a domain of the wave height and wave pattern. The derivation and properties of this equation is discussed in detail.

Chapter 4

Chapter 4 is dedicated to a closer look at the energy balance in waves both before and after breaking. This expands the analysis in Section 3.5 and involves discussion of the various types of energy present in the nearshore and derivation of the **energy equation** which is an equation that describes the transformation and propagation of energy in areas with varying depth and currents.

Chapter 5

One of the most improtant physical processes in the nearshore region is the **wave breaking** that occurs close to the shore of beaches. As mentioned this is caused by the (gradual) decrease in depth closest to the shore. On sufficiently gently sloping beaches such as most littoral beaches the breaking process destroys or dissipates almost all the incoming wave energy in the nearshore region called the **surfzone**. This causes rapid changes in the waves with violent particle velocities that highly contribute to the movement of sediment material and beach erosion. The rapid changes in wave height also imply rapid changes in radiation stresses for the waves. This create the most important forcing mechanism for nearshore currents. Our knowledge about the wave breaking is still limited and in Chapter 5 we use both measured data and theoretical analysis in an attempt to describe and understand the details of the wave motion.

Chapter 6

In Chapter 6 a brief overview is given of the *types* of wave models based on of the results described in the previous chapters that are frequently used today.

Chapter 7

The following three chapters are focusing on the effect of finite wave heights. In the linear wave theory discussed and utilized in Chapter 3 we assumed the wave height was infinitely small (which is the reason the equations become linear). Real waves, however, can be quite steep and the finite wave height has profound effects on the motion. In Chapter 7 we first derive the different forms of the model equations that properly describe the flow in different parameter ranges for the waves. The key parameters here are the water depth, the wave length (or wave period) and the wave height. It is shown how that this leads to the governing equations for the classical theories of **Stokes waves**, **Boussinesq waves** and **nonlinear shallow water waves** to mention the most important.

Chapter 8

Chapter 8 is then presenting a detailed derivation of the **second order stokes theory** and briefly outlines third and fifth order versions of the theory. The chapter also includes a description of the special computer version of the stokes wave theory called the **stream function theory**, which makes it possible to calculate the properties of stokes waves to very high order. This theory has been proven a very effective numerical tool.

Chapter 9

In recent years the **Boussinesq wave theory** has become one of the most effective ways of analysing nearshore wave motion computationally. Chapter 9 derives the basic equations. It also gives a relatively detailed account of the constant form solution for Boussinesq waves called **cnoidal waves** which is the Boussinesq wave equivalent to the sinusoidal waves of linear wave theory, and the infinitely long version of those waves waves called **solitary waves**.

A strength of the Boussinesq equations is that when solved computationally they provide the development in time and space of the entire wave motion in a coastal domain, which makes it possible also to analyze irregular waves such as wind generated storm waves. One of the weak points of Boussinesq wave theory is its limitation to relatively shallow water. However, numerous recent results have modified the equations to forms that extend the validity of the theory almost to the limt of what we have defined as the nearshore region. From the perspective of practical applications this has tremendous importance by making the method viable. These developments are also presented in the chapter. Another problem is that solving the Boussinesq equations in a realistically large domain over a sufficiently long time period of time for practical applications still requires very substantial computational efforts.

Chapter 10

When waves propagate over a domain with depth small enough that the depth influences the waves (as in the nearshore region) a **boundary layer** develops at the bottom. In the traditional approach to wave motion analysis (such as described in the chapters above) the effect of this boundary layer is disregarded: the motion is considered irrotational described by a velocity potential and at the bottom we essentially have a **slip velocity**. However, the boundary layer is real and it does produce both a local disturbance of the flow near the bottom and a shear stress (or **bottom friction**) acting on the fluid above. This stress dissipates energy and when waves propagate over longer distances the accumulative effect of the energy dissipation due to the bottom friction causes the wave height to decrease slowly but significantly. Chapter 10 presents the classical theory of viscous wave boundary layers for stokes waves to first and second order in the wave amplitude. It also derives and discusses the expressions for the socalled **steady streaming** in the boundary layer which is a net current generated by nonlinear mechanisms active inside the boundary layer. The chapter then proceeds with analysis of turbulent boundary layers giving results based on the empirical concept of a fricion coefficient. The general case of combined wave-current motion is presented in detail.

Chapter 11

The wave averaged properties are important parts of the mechanisms resposible for the wave generated currents, such as longshore and cross-shore currents, socalled **nearshore circulation**. These currents are important in the nearhsore environments where they contribute significantly to the morphodynamic changes of beaches. Because the currents are esentially wave averaged flows those currents are governed by wave averaged equations which are also depth integrated. In Chapter 11 we describe the derivation of those equations which also reveals the exact definitions of the wave mass (or rather volume) flux and the radiation stress. Those concepts are analyzed in detail for linear waves which is the form most frequently used in applications and also put into context of waves in two horizontal dimensions.

The chapter then goes through two important special ("canonical") cases of nearshore circulation: the cross-shore momentum balance on a long straight coast with shorenormal wave incidence, and the wave generated longshore current on such a coast with oblique wave incidence. Because the equations are wave averaged they require as input information about the volume flux and the radiation stresses at all points of the domain which correspnds to demanding the wave motion known. This information is usually provided by wave models of the type described in earlier chapters, particularly linear models.

Finally the use of boundary conditions along the free boundaries of nearshore models is described. Such boundaries are artificial in the sense that they only exist because we limit the computations to a section of a coast. The demand along such boundaries is that they form no obstacle to the wave motion. In particular the waves that would want to propagate out of the domain - either because they were generated inside or were reflected from the beach or engineering structues inside the computational domain - should be able to do so freely.

Chapter 12

The nearshore currents covered by the equations discussed in the previous chapter are essentially depth averaged and therefore no resolution is obtained for the vertical variation of those currents. In the case of a long straight beach with shorenormal wave incidence it is clear, however, that since the waves have shoreward net volume flux then there must also be a seaward going current. This is called the **undertow** and the mechanisms governing this flow are analyzed in Section 12.2.

Chapter 13

The results for the undertow can be supplemented with a similar analysis of the longshore currents and this reveals that the nearshore currents in general vary over depth both in magnitude and direction. It turns out that this feature is important for the way in which the currents interact and change in the horizontal direction (Chapter 13). This chapter analyses general 3-dimensional currents by expanding the depth integrated and time averaged equations for nearshore circulation derived in Chapter 11 to depth varying currents and gives analytical results for the vertical profiles of the currents. The resulting equations are called **quasi-3D equations** because the depth varying currents is represented in the modified depth integrated equations as coefficients that account for the horizontal effects of the depth variations, much like the momentum correction factor in engineering hydraulics equations account for the depth variation of the flow in a river.

Chapter 14

Finally the variation in height and period of the irregular wind waves approacing a beach leads to variation in the radiation stresses which ends up generating new, much longer waves. These **infragravity or IG** waves become particular important in the inner part of the nearhsore region where the wind waves are breaking while the IG-waves usually are not (Section 14.1). Canonical examples of IG waves are the socalled **edge waves** which anre waves propagating along the shore with their strongest motion closest to the shoreline and decreasing seaward.

The chapter also analyses the fact that the simple models of nearshore currents, developed under the assumption of steady flow, turn out to be unstable - socalled **shear instabilities**. A consequence is that many (or most) longshore currents show fluctuations in time and space that again have profound influence also on the mean currents. The initial linear instability theory is developed and numerical computations of what happens as the instabilites grow into complex longshore flows are discussed.

The advanced present day nearshore models are now opening such situations from natural beaches to realistic computational analysis. A food for thought discussion is offered at the end of the Chapter 14 about these and other complex flow situations found on natural beaches and how models can help improve our understanding.

1.3 References - Chapter 1

Boussinesq, J. (1872). Theorie des onde et des resous qui se propagent le long d'un canal rectangulaire horizontal, en communiquant au liquide contenu dans ce canal des vitesses sensiblement pareilles de la surface au fond. Journal de Math. Pures et Appl., Deuxieme Serie, **17**, 55–108.

Korteweg, D. J. and G. DeVries (1895). On the change of form of long waves advancing in a canal, and on a new type of long stationary waves. Phil. Mag., Ser. 5, **39**, 422 – 443.

Longuet-Higgins, M. S. and R. W. Stewart (1962). Radiation stress and mass transport in gravity waves with application to 'surf-beats'. J. Fluid Mech., **8**, 565 – 583.

Russell, J. S. (1844). Report on waves. Brit. Ass. Adv. Sci. Rep.

Stokes, G. G. (1847). On the theory of oscillatory waves. Trans. Cambridge Phil. Soc., **8**, 441 – 473.

Stokes, G. G. (1880). Mathematical and Physical Papers, Vol 1. Cambridge University Press.

Svendsen, I. A. and I. G. Jonsson (1976). Hydrodynamics of Coastal Regions. Den Private Ingeniørfond. Copenhagen, 285 pp.

Chapter 2

Hydrodynamic Background

2.1 Introduction

As an introduction, this chapter summarizes the basic hydrodynamic principles and laws on which we base the entire analysis of nearshore hydrodynamics and modelling. The presentation largely assumes that the reader is familiar with basic vector and tensor analysis which will be the language used for the rest of the book, and with the principles of basic fluid mechanics. The reference list at the end of the chapter includes a list of fluid mechanics texts that can be used as reference backup for the details of the basic parts of this chapter. Many of those will also contain useful lists of vector relationships. The book by Kundu (1990, 2001) is particularly designed for ocean fluid mechanics applications.

The review includes only results needed for the rest of the text. The first two sections on the kinematics and the dynamics of fluid flow only contain well-known equations and principles. They are just reviewed **without proof or detailed explanations**. The subjects of boundary conditions covered in Section 2.4, of turbulence in Section 2.5, and the general expression for energy flux in Section 2.6 are less commonly known, and therefore covered in more detail.

The assumptions made will be included in the review, since they form the fundamental limitations for the validity of the results. This is important and attempts will be made throughout the book to keep track of this and emphasize it. The reason is that some of the model equations to be discussed later turn out to be almost as general as the basic hydrodynamic equations of motion, which highly increases the credibility of the results. Other model equations, on the other hand, turn out to be based on more empirical assumptions that potentially limit their generality and accuracy.

In any case, understanding the assumptions underlying each result and how they limit the validity of those results is one of the most powerful tools in the process of building and using models.

2.2 Kinematics of fluid flow

Modern fluid mechanics is based on the concept that the fluid is a continuum. This allows all fluid properties to be described by mathematical functions that, except for isolated times and locations, are continuous and differentiable.

2.2.1 *Eulerian versus Lagrangian description*

Lagrangian description

There are two fundamental ways to describe the fluid flow. One is the **Lagrangian** description that identifies the position $\mathbf{r}$ of each fluid particle at all times. This requires the particles are initially marked, e.g., by their position $\mathbf{r_0}$ at time t_0. Therefore $\mathbf{r} = \mathbf{r}(\mathbf{r_0}, t)$. In this description, the fluid velocity $\mathbf{v}$ and acceleration $\mathbf{a}$ are given by

$$\mathbf{v} = \left(\frac{d\mathbf{r}}{dt}\right)_{r_0=\text{const}} \tag{2.2.1}$$

$$\mathbf{a} = \left(\frac{d^2\mathbf{r}}{dt^2}\right)_{r_0=\text{const}} \tag{2.2.2}$$

This description of particle motion is rarely used though it can be useful in certain cases, such as the description of the particle motion in waves.

Eulerian description

In the Eulerian description, the fluid motion is described by specifying the velocity $\mathbf{v} = u,\ v,\ w$ in $x,\ y,\ z$ coordinates at all (fixed) points $\mathbf{r}$.

Unfortunately, the basic physical laws of hydrodynamics are formulated for bodies that essentially consist of a fixed set of particles or molecules. In other words the basic laws such as Newton's 2nd law are formulated in Lagrangian systems. Thus we need to transform the time derivatives for quantities f following the particles, i.e., $f(\mathbf{r}(t), t)$ into Eulerian coordinates. Using the chain rule, we therefore define the **material** or **total** derivative

$\frac{d}{dt}$ by

$$\frac{d}{dt} = \frac{\partial}{\partial t} + \mathbf{v} \cdot \frac{\partial}{\partial \mathbf{r}} = \frac{\partial}{\partial t} + \mathbf{v} \cdot \nabla$$

$$= \frac{\partial}{\partial t} + v_i \frac{\partial}{\partial x_i} = \frac{\partial}{\partial t} + u \frac{\partial}{\partial x} + v \frac{\partial}{\partial y} + w \frac{\partial}{\partial z} \tag{2.2.3}$$

where

$$\nabla = \frac{\partial}{\partial x}, \frac{\partial}{\partial y}, \frac{\partial}{\partial z}$$

is the gradient operator.

Here, $\frac{\partial}{\partial t}$ is called the **local**, $v_i \frac{\partial}{\partial x_i}$ the **convective** (also called "advective") derivative. Applied to $\mathbf{v}$ itself, we get the acceleration

$$\mathbf{a} = \frac{d\mathbf{v}}{dt} = \frac{\partial \mathbf{v}}{\partial t} + \mathbf{v} \cdot \nabla \mathbf{v} \tag{2.2.4}$$

2.2.2 *Streamlines, pathlines, streaklines*

Overall impressions of the flow field can be obtained by three different types of lines linking points in the flow domain.

Streamlines

Streamlines are defined so that they at all points in the flow have the velocity vector as tangent. Since there at any time is one velocity vector $\mathbf{v}$ at each point $\mathbf{r}$ there is one streamline going through each point at a given time. The definition expresses that along the streamline, $\mathbf{v}$ is parallel to $d\mathbf{r}$, i.e.,

$$\mathbf{v} \times d\mathbf{r} = 0 \tag{2.2.5}$$

from which it may be infered that the streamlines must satisfy

$$\frac{dx}{u} = \frac{dy}{w} = \frac{dz}{w} \tag{2.2.6}$$

The streamlines give an **instantaneous** illustration of the flow field.

Pathlines

Pathlines show the paths of fluid particles. Thus, along a pathline, the position vector $\mathbf{r}$ changes so that

$$\frac{d\mathbf{r}(t)}{dt} = \mathbf{v}(\mathbf{r}(t),\ t) \tag{2.2.7}$$

Solution of (2.2.7) gives the pathlines which essentially represents a Lagrangian description of the flow. The pathlines can be said to give the **history** of the flow.

Streaklines

A Streakline connects all particles that at some time have passed or will pass through a chosen point in the fluid. Thus, streaklines show the **trace of dye** continuously injected at points in the flow domain. For details see Currie (1974).

In steady flow streamlines, pathlines and streaklines coincide.

2.2.3 *Vorticity* ω_i *and deformation tensor* e_{ij}

The vorticity **vector** of the flow is defined as

$$\vec{\omega} = \nabla \times \mathbf{v} \tag{2.2.8}$$

or

$$\omega_i = \epsilon_{ijk} \frac{\partial}{\partial x_j} v_k \tag{2.2.9}$$

where ϵ_{ijk} is the alternating unit tensor. In x, y, z coordinates

$$\begin{Bmatrix} \omega_x \\ \omega_y \\ \omega_z \end{Bmatrix} = \begin{Bmatrix} \frac{\partial w}{\partial y} - \frac{\partial v}{\partial z} \\ \frac{\partial w}{\partial x} - \frac{\partial u}{\partial z} \\ \frac{\partial v}{\partial x} - \frac{\partial u}{\partial y} \end{Bmatrix} \tag{2.2.10}$$

The **deformation tensor** e_{ij} is defined by

$$e_{ij} = \frac{1}{2}\left(\frac{\partial v_i}{\partial x_j} + \frac{\partial v_j}{\partial x_i}\right) \tag{2.2.11}$$

In some texts e_{ij} is called the "rate of strain tensor" e.g., Kundu (1990, 2001).

2.2.4 *Gauss' theorem, Green's theorems*

Gauss' theorem

The rule called **Gauss' theorem** applies to a vector field with the following characteristics:

- A differentiable vector field **V**.
- defined at all points in space.

For this vector field, we consider a **closed, simply connected** region Ω, bounded by the surface S. Then **V** satisfies

$$\int_\Omega \nabla \cdot \mathbf{V} d\Omega = \int_S \mathbf{V} \cdot \mathbf{n} dS \tag{2.2.12}$$

where **n** is the outward normal to the surface S. In tensor form (2.2.12) reads

$$\int_\Omega \nabla_j \mathbf{V}_j \, d\Omega = \int_S \mathbf{V}_j \, \mathbf{n}_j \, dS \tag{2.2.13}$$

The definition of "simply connected" region Ω is as follows: **any** closed curve inside Ω surrounding a point P in Ω must be reducible, which means the curve can be continuously shrunk to the point P, without any point of the curve leaving the region.

An example of a region which is **not** simply connected is the doughnut (a "torus"): a curve inside the doughnut which initially circumscribes the hole in the torus cannot be reduced to a point P in the region without part of the curve passing through the hole.

Gauss' theorem also applies to higher order tensors. Thus, for a second order tensor, such as a stress σ_{ij}, the theorem takes the form

$$\int_\Omega \nabla_i \sigma_{ij} d\Omega = \int_S \sigma_{ij} n_i dS \tag{2.2.14}$$

Similarly, for a second order tensor $\rho v_i v_j$, which occurs if the flux of momentum is integrated over a larger domain, we get

$$\int_\Omega \frac{\partial}{\partial x_j}(\rho v_i v_j) d\Omega = \int_S \rho v_i v_j n_j dS \tag{2.2.15}$$

The Gauss theorem essentially reduces the 3-dimensional volume integral to a 2-dimensional surface integral. It can also be used in 2- and

1-dimensional domains. As an illustration, in the case of a 1-dimensional domain, Ω becomes an interval $x \in (a|b)$. Then (2.2.12) reduces to

$$\int_{x=a}^{x=b} \frac{\partial \mathbf{V}}{\partial x} dx = \mathbf{V}(b) - \mathbf{V}(a) \tag{2.2.16}$$

which is the well known integration from a to b of a function.

Green's theorems

Green's theorems are derived from Gauss' theorem by considering a special vector field defined as

$$\mathbf{V} = \phi_1 \nabla \phi_2 \tag{2.2.17}$$

where ϕ_1 and ϕ_2 are arbitrary differentiable scalar functions. For this field, the Gauss theorem can be written

$$\int_\Omega \left[\phi_1 \nabla^2 \phi_2 + \nabla \phi_1 \cdot \nabla \phi_2\right] d\Omega = \int_S \phi_1 \nabla \phi_2 \cdot \mathbf{n} dS \tag{2.2.18}$$

which is known as the **First form of Green's theorem.**

Interchanging ϕ_1 and ϕ_2 and subtracting from (2.2.18) gives

$$\int_\Omega \left(\phi_1 \nabla^2 \phi_2 - \phi_2 \nabla^2 \phi_1\right) d\Omega = \int_S \left(\phi_1 \nabla \phi_2 - \phi_2 \nabla \phi_1\right) \cdot \mathbf{n} dS \tag{2.2.19}$$

which is the **Second form of Green's theorem.**

As for the Gauss theorem, the Green's theorems can also be applied to 2- and 1-dimensional domains. For a 1-dimensional domain, $x \; \epsilon(a|b)$, we get

$$\int_a^b \left(\phi_1 \nabla^2 \phi_2 - \phi_2 \nabla^2 \phi_1\right) dx = \left[\phi_1 \nabla \phi_2 - \phi_2 \nabla \phi_1\right]_a^b \tag{2.2.20}$$

2.2.5 *The kinematic transport theorem, Leibniz rule*

The kinematic transport theorem describes the rate of change in time of the content of some quantity $F(x, y, z, t)$ inside a volume $\Omega(t)$. Thus, we are seeking

$$\frac{d}{dt} \int_{\Omega(t)} F(x, y, z, t) d\Omega \tag{2.2.21}$$

The volume $\Omega(t)$ has a surface $S(t)$ that at all points is assumed to move with the fluid velocity $\mathbf{v}(x, y, z, t)$. Thus, the volume Ω will contain the same fluid particles at all times (it is a **material volume**).

The rate of change (2.2.21) can then be written

$$\frac{d}{dt}\int_{\Omega(t)} F(x,y,z,t)d\Omega = \int_{\Omega(t)} \frac{\partial F(t)}{\partial t} d\Omega + \int_{S(t)} F(t)\mathbf{v}_S \cdot \mathbf{n} dS \qquad (2.2.22)$$

where $\Omega(t)$ and $S(t)$ refer to the instantaneous positions of the volume Ω and its surface S. $\mathbf{v_s}$ is the particle velocity along S.

Applying Gauss' theorem to the last integral in (2.2.22) brings this on the form

$$\frac{d}{dt}\int_{\Omega(t)} F\, d\Omega = \int_{\Omega(t)} \left[\frac{\partial F}{\partial t} + \nabla \cdot (\mathbf{v}F)\right] d\Omega \qquad (2.2.23)$$

which is the kinematic transport theorem.

The 1-dimensional form of (2.2.22) in the interval $x \in (a(t)|b(t))$ becomes

$$\frac{d}{dt}\int_{a(t)}^{b(t)} f(x,t)dx = \int_{a(t)}^{b(t)} \frac{\partial f}{\partial t} dx + \frac{\partial b}{\partial t} f(t, b(t)) - \frac{\partial a}{\partial t} f(t, a(t)) \qquad (2.2.24)$$

which is known as **Leibniz rule.**

2.3 Dynamics of fluid flow

The dynamics of fluid flow describes the three conservation principles: the conservation of mass, momentum and energy.[1]

2.3.1 *Conservation of mass*

In its general form (varying fluid density, ρ), the conservation of mass is expressed by

$$\frac{d\rho}{dt} + \rho\nabla \cdot \mathbf{v} = 0 \qquad (2.3.1)$$

where by (2.2.3)

$$\frac{d\rho}{dt} = \frac{\partial \rho}{\partial t} + \mathbf{v} \cdot \nabla\rho \qquad (2.3.2)$$

[1]Strictly speaking, the conservation of mass is also a purely kinematic principle.

Combining (2.2.3) and (2.3.2) we can also write (2.2.3) as

$$\frac{\partial \rho}{\partial t} + \nabla \cdot (\rho \mathbf{v}) = 0 \tag{2.3.3}$$

If the flow is incompressible – as most flows in water are – and ρ is constant – which is often a realistic assumption in nearshore flows – then we get

$$\nabla \cdot \mathbf{v} = 0 \qquad \text{or} \qquad \frac{\partial u}{\partial x} + \frac{\partial v}{\partial y} + \frac{\partial w}{\partial z} = 0 \tag{2.3.4}$$

This is usually termed **the continuity equation.**

2.3.2 *Conservation of momentum*

Stress components

Stresses essentially are second order tensors, σ_{ij}. This reflects that

- A stress is a force per unit area
- Acting on an (infinitesimal) surface area with normal vector $\mathbf{n}$. The magnitude and direction of the stress depends not only on the position of the infinitesimal surface, but also on the direction of the normal vector $\mathbf{n}$: changing the direction of $\mathbf{n}$ changes the stress.
- The stresses on a particle represent the effect of the fluid surrounding the particle.

In order to properly account for the effect of the stresses, it is necessary to **define positive directions for the stress components in the direction of the coordinate axes.**

Thus, we consider the cubic particle in Fig. 2.3.1. It is recalled that the elements of a tensor represent the components in the coordinate directions. For σ_{ij},

- the first index (i) indicates the direction of the normal vector to the surface element considered
- the second index (j) indicates the direction of the stress on the surface.

The 9 components of σ_{ij} can therefore be labeled

$$\sigma_{ij} = \begin{Bmatrix} \sigma_{xx} & \sigma_{xy} & \sigma_{xz} \\ \sigma_{yx} & \sigma_{yy} & \sigma_{yz} \\ \sigma_{zx} & \sigma_{zy} & \sigma_{zz} \end{Bmatrix} \tag{2.3.5}$$

The directions of the stress components shown in Fig. 2.3.1 are the **positive** directions of the coordinate components. It is noticed that the sign-conventions are as follows[2]

- On surface elements where the outward normal vector is in the **positive** coordinate axis direction, all stress components are positive in the axis directions.
- on surface elements where the outward normal vector is in the direction **opposite** the coordinate axis, stresses are positive in directions opposite the coordinate axes.

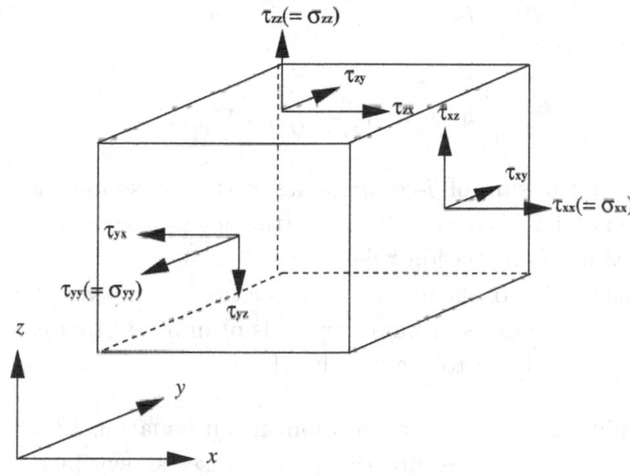

Fig. 2.3.1 This figure shows two things: (a) Symbols for stress components. (b) Definition of positive directions.

Conservation of momentum in terms of stresses

The equation for conservation of momentum is the fluid mechanic

[2] This is the classical system of definitions for *indices*.
It is **not** uniformly used in the literature, however. Some older, but still valuable, references use the opposite definitions.

Examples:			
	White (1991) uses	$\tau_{\text{normal, stress dir.}}$	as here
	C.C. Mei (1983) uses	$\tau_{\text{normal, stress dir.}}$	as here
	Batchelor (1968) uses	$\tau_{\text{stress dir., normal}}$	opposite of here
	Hinze (1975) uses	$\tau_{\text{stress dir., normal}}$	opposite of here

version of Newton's second law for a solid body saying

$$\text{Mass} \cdot \text{acceleration} = \sum \text{all forces on the body} \tag{2.3.6}$$

This is usually called **the momentum equation**.

The analogy of the body is a unit volume material fluid particle with mass $\rho \cdot 1$. Since the particle follows the flow, its acceleration is given by (2.2.4). The forces on the particle are volume forces (which we here will limit to gravity with acceleration **g**), and surface forces, which are the stresses σ_{ij} (or $\overline{\overline{\sigma}}$) in the fluid.

The momentum equation in terms of the stresses then is

$$\frac{d\mathbf{v}}{dt} = \frac{\partial \mathbf{v}}{\partial t} + (\mathbf{v} \cdot \nabla)\mathbf{v} = \mathbf{g} + \frac{1}{\rho}\nabla \cdot \overline{\overline{\sigma}} \tag{2.3.7}$$

or

$$\frac{\partial v_j}{\partial t} + (v_i \nabla_i) v_j = g_j + \frac{1}{\rho} \nabla_i \sigma_{ij} \tag{2.3.8}$$

Here $\nabla_i \sigma_{ij}$ is the sum of j-components of the stress increases over all surfaces of the particle; i.e., all values of i (the normal vector direction) are considered in $\nabla_i \sigma_{ij}$ (summation rule).

Though this version of the momentum equation gives a very transparent description of the dynamics of fluid flow, it is of limited practical use until the stresses σ_{ij} are related to the flow itself.

Exercise 2.3-1 Use the momentum equation (2.3.8) to verify the positive directions for stresses shown in Fig. 2.3.1 assuming that velocities are counted positive in the positive direction of the coordinate axis.

(Hint: notice that the index j in $\frac{\partial v_j}{\partial t}$ indicates tat (2.3.8) describes the j-th component of the vector equation).

2.3.3 *Stokes' viscosity law, the Navier-Stokes equations*

The stresses σ_{ij} are essentially due to intermolecular forces. In fluid flow, these are modelled by defining the **dynamic viscosity** μ for the fluid. To conveniently introduce this, the average value of the normal components of the stresses is subtracted from the total stress. This is called the **pressure** p defined by

$$p = -\frac{1}{3}(\sigma_{11} + \sigma_{22} + \sigma_{33}) \tag{2.3.9}$$

Thus σ_{ij} is written

$$\sigma_{ij} = -p\delta_{ij} + \tau_{ij} \tag{2.3.10}$$

which defines the residual or **deviatoric stress** τ_{ij}. Then, (2.3.8) becomes

$$\frac{\partial v}{\partial t} + v_i \nabla_i v_j = -\frac{1}{\rho}\nabla_j p + g_j + \frac{1}{\rho}\nabla_i \tau_{ij} \tag{2.3.11}$$

Through extensive analysis (for details see Aris (1962)), it has then been found that in so-called Newtonian fluid (of which water is one) we have for incompressible flows (i.e. (2.3.4) applies)

$$\sigma_{ij} = -p\delta_{ij} + 2\mu e_{ij} \tag{2.3.12}$$

where e_{ij} is the deformation tensor given by (2.2.11) and μ is is the above mentioned proportionality factor called the **dynamic viscosity** that simply indicates the magnitude of the stresses caused by the deformations described by e_{ij}.[3] μ cannot be measured directly. In practice it is determined by measuring combined values of stresses and deformations in well defined flows and calcualting μ as the ratio.

This is called **Stokes' viscosity law**. Though introducing the additional unknown p, this relation essentially links the fluid stresses σ_{ij} to the fluid motion v_j (through e_{ij}).

Substituting (2.3.12) for σ_{ij} into (2.3.8) then gives the momentum equation in terms of the velocity v_j and the pressure p. Utilizing again the incompressibility of the flow, we then get (2.3.8) as

$$\frac{\partial v_j}{\partial t} + v_i \frac{\partial}{\partial x_i} v_j = -\frac{1}{\rho}\frac{\partial p}{\partial x_j} + g_j + \frac{\mu}{\rho}\frac{\partial^2 v_j}{\partial x_i \partial x_i} \tag{2.3.13}$$

This is know as the **Navier-Stokes equation(s)**. Since v_j is a vector, (2.3.13) has three components in 3-dimensional space.

In vector form (2.3.11) reads

$$\frac{\partial \mathbf{v}}{\partial t} + (\mathbf{v} \cdot \nabla)\mathbf{v} = -\frac{1}{\rho}\nabla p + \mathbf{g} + \nu \nabla^2 \mathbf{v} \tag{2.3.14}$$

[3] Actually μ is a "fudge factor" which establishes the connection between two quantities, the deformation tensor and the viscous stresses, which is due to physical processes (the Brownian motions of the molecules in the fluid) that are not modelled.

The quantity

$$\nu = \frac{\mu}{\rho}$$

is called the **kinematic viscosity**.

Usually, the positive z direction is taken in the direction opposite gravity. If we then write the pressure p as

$$p = \rho g z + p_D$$

where p_D is the so-called **dynamic pressure**, then the momentum equation simplifies to

$$\frac{\partial \mathbf{v}}{\partial t} + (\mathbf{v} \cdot \nabla)\mathbf{v} = -\frac{1}{\rho}\nabla p_D + \nu \nabla^2 \mathbf{v} \tag{2.3.15}$$

Along with the continuity equation, the N-S equations represent a complete set of equations for the fluid flow with ($\mathbf{v}$, p) or ($\mathbf{v}$, p_D) as unknowns. As mentioned the assumptions behind those equations are incompressible flows and constant (ρ, μ).

2.3.4 *The boundary layer approximation*

At solid walls, a viscous fluid is forced to adhere to the wall, making $\mathbf{v} = 0$ at the wall. Except for very small reynolds numbers this creates large velocity gradients in the direction perpendicular to the wall, and hence large stresses in a layer along the wall. This is called a **boundary layer**. Often, boundary layers are thin, and for plane walls this makes it possible to simplify the Navier-Stokes equations for the flow in the boundary layer. The result is, for a simple 2-dimensional (x, z) boundary layer along a wall in the x-direction.

$$\frac{\partial u}{\partial x} + \frac{\partial w}{\partial z} = 0 \tag{2.3.16}$$

Momentum equation, x-component

$$\frac{\partial u}{\partial t} + u\frac{\partial u}{\partial x} + w\frac{\partial u}{\partial z} = -\frac{1}{\rho}\frac{\partial p_D}{\partial x} + \nu\frac{\partial^2 u}{\partial z^2} \tag{2.3.17}$$

Momentum equation z-component

$$\frac{\partial p_D}{\partial z} = 0 \tag{2.3.18}$$

This is called the boundary layer approximation, and (2.3.17) is termed the **boundary layer equation.**

2.3.5 *Energy dissipation in viscous flow*

The internal viscous forces transform some of the mechanical energy of the flow into heat. The process is irreversible so that heat energy cannot be turned back into mechanical enegy. This is called **energy dissipation.**

The energy dissipation ϵ per unit volume in a viscous flow is given by

$$\epsilon = \mu \frac{\partial v_i}{\partial x_j} \left(\frac{\partial v_i}{\partial x_j} + \frac{\partial v_j}{\partial x_i} \right) \tag{2.3.19}$$

which can also be written

$$\epsilon = \tau_{ij} \frac{\partial v_i}{\partial x_j} \tag{2.3.20}$$

An alternative form of (2.3.19) is

$$\epsilon = \frac{1}{2} \mu \left(\frac{\partial v_i}{\partial x_j} + \frac{\partial v_j}{\partial x_i} \right)^2 = 2\mu e_{ij}^2 \tag{2.3.21}$$

2.3.6 *The Euler equations, irrotational flow*

Away from boundary layers and turbulence, the viscous stresses τ_{ij} will be small. The Navier-Stokes equations then reduce to the inviscid form

$$\frac{\partial \mathbf{v}}{\partial t} + (\mathbf{v} \cdot \nabla)\mathbf{v} = -\frac{1}{\rho} \nabla p + \mathbf{g} \tag{2.3.22}$$

or the equivalent with p_D. This is called the **Euler equation(s)**. Along with the continuity equation, this represents a complete set of equations for the flow with unknown $(\mathbf{v}, p)$.

The so-called **Kelvin's theorem** states that in inviscid flows the vorticity (following a particle) does not change. Therefore, if to start with the vorticity is zero everywhere, it remains so, i.e., $\omega \equiv 0$ at all times. This is called **irrotational flow.**

The velocity potential

For irrotational flow, we can determine a scalar function ϕ which describes the entire flow field. ϕ is defined in such a way that

$$\mathbf{v} = \nabla \phi \tag{2.3.23}$$

Introducing this definition of $\mathbf{v}$ into the continuity equation gives that ϕ must satisfy

$$\nabla^2\phi = \frac{\partial^2\phi}{\partial x^2} + \frac{\partial^2\phi}{\partial y^2} + \frac{\partial^2\phi}{\partial z^2} = 0 \qquad (2.3.24)$$

which is the **Laplace equation**.

The Bernoulli equation

Similarly, the three components of the Euler equations collapse into one equation which can be integrated in space to give

$$\frac{\partial\phi}{\partial t} + \frac{1}{2}(\nabla_j\phi)^2 + \frac{1}{\rho}p + g(z) = C(t) \qquad (2.3.25)$$

where $C(t)$ is an arbitrary integration function and $(\nabla_j\phi)^2)$ stands for $\nabla_j\phi\nabla_j\phi$. (2.3.25) is the **Bernoulli equation**.

Exercise 2.3-2

Verify that for $\mathbf{v}$ defined by (2.3.23), the flow automatically becomes irrotational.

2.4 Conditions at fixed and moving boundaries

In general, boundary conditions are mathematical expressions for the physical conditions at the boundary which specify how the flow behaves there. When solving the flow equations, enough boundary conditions are needed to determine the integration constants and introduce the effects of external forces on the fluid such as wind stresses on the surface or shear stresses at the bottom.

In this section we analyze the boundary conditions at fixed and moving boundaries such as the free surface and fixed bottoms. The conditions are for potential and viscous flows.

For general turbulent flows with stresses at free and fixed boundaries the kinematic conditions are the same. The dynamic conditions, however, are more complicated. A detailed derivation of those conditions is given in Chapter 11 where they are needed for the depth integration of the equations of motion.

2.4.1 *Kinematic conditions*

One of the physical conditions imposed at boundaries leads to the kinematic boundary conditions.

The basic assumption is that *a fluid particle at a boundary stays on that boundary.* At a fixed surface, such as an impermeable sea bottom, this is equivalent to assuming that the fluid velocity is parallel to the bottom at all points. At a free surface it means the fluid particles must follow the motion of the free surface.

To formulate this principle mathematically, we let the position of the surface be given by

$$z = \zeta(x_\alpha,\ t)\ (= \zeta(x, y,\ t)) \tag{2.4.1}$$

where index α is a tensor notation representing horizontal (x, y) coordinates.

This relationship applies to all points on the surface at all times. Therefore, whether we move from one point to another on the surface at a fixed time, or we move in time so that the surface changes position, such changes must be related by the differential of Eq. (2.4.1):

$$dz = \frac{\partial \zeta}{\partial x_\alpha} dx_\alpha + \frac{\partial \zeta}{\partial t} dt \tag{2.4.2}$$

i.e., if we want to stay on the surface they must satisfy Eq. (2.4.2)

This simply states that if we follow the surface we can freely change all variables except one. We can, for example, move horizontally at will by choosing (infinitely small but arbitrary) values of $dx_\alpha = (dx,\ dy)$ and we can consider the surface at time t and $t + dt$. Eq. (2.4.2) then tells us that if point $(x_\alpha,\ z,\ t)$ was on the surface at time t and we move to $x_\alpha + dx_\alpha$, $t + dt$, then we must change z to $z + dz$ where dz is given by Eq. (2.4.2) if we want the point $(x_\alpha + dx_\alpha,\ z + dz,\ t + dt)$ to be on the surface too.

The second step in the derivation is to implement into Eq. (2.4.2) our physical assumption that a fluid particle at the surface with velocity components (u_α, w) follows exactly those rules as the particle is moving around and the surface is changing position. This means requiring that in (2.4.2)

$$dz = w dt \ ; \qquad dx_\alpha = u_\alpha dt \tag{2.4.3}$$

which, after division by dt gives

$$\boxed{\frac{\partial \zeta}{\partial t} - w + u_\alpha \frac{\partial \zeta}{\partial x_\alpha} = 0 \quad \text{at} \ z = \zeta} \tag{2.4.4}$$

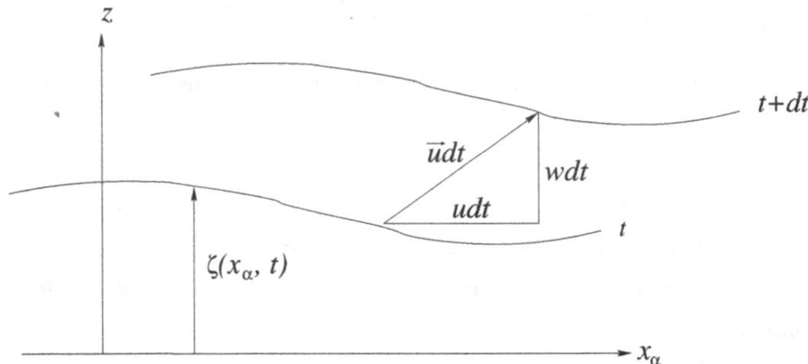

Fig. 2.4.1 Graphical illustration of the kinematic free surface boundary condition (KFSBC).

This is the general form of the kinematic boundary condition at a surface on which fluid particles stay.

In x, y, z coordinates, this becomes

$$\frac{\partial \zeta}{\partial t} - w + u\frac{\partial \zeta}{\partial x} + v\frac{\partial \zeta}{\partial y} = 0 \qquad \text{at } z = \zeta \tag{2.4.5}$$

Thus for the free surface $z = \zeta$, (2.4.4) or (2.4.5) apply directly.

The bottom condition

In the case the flow can be assumed inviscid it will exhibit a finite velocity at the bottom. Thus the proper boundary condition is simply that the velocity u_n normal to the bottom is zero

$$u_n = 0 \tag{2.4.6}$$

In viscous flows and flows in boundary layers the proper boundary condition at a solid bottom will be

$$\mathbf{u} = 0 \tag{2.4.7}$$

which clearly automatically satisfies the condition that a particle at the bottom stays there.

In many practical flow models the flow inside the bottom boundary layer is not resolved directly. In stead the models allow a socalled **slip velocity** which then represents the velocity immediately outside the boundary layer similar to the conditions in a potential flow.

In the case of potential flow or a slip velocity being used the equation (2.4.4) also applies as the kinematic condition. Thus for a *fixed bottom* (no time dependence) at $z = -h(x_\alpha)$, we get the kinematic condition

$$\boxed{w + u_\alpha \frac{\partial h}{\partial x_\alpha} = 0 \quad \text{at} \quad z = -h(x_\alpha)} \tag{2.4.8}$$

where (u_α, w) is then the finite velocity.

However, with a slip velocity the dynamic effect of the bottom must also be included in the sense that the shear stresses (if any) actually developed by the bottom friction in the boundary layer must be included as a dynamical boundary condition at the bottom (see Chapter 10).

2.4.2 *Dynamic conditions*

Kinematic conditions relate the motion of the particles in time – i.e., the velocity field – to the development in time of the position of the surface.

In contrast, dynamic conditions are **instantaneous conditions** expressing that at all times the external stresses on a boundary surface must be balanced by equivalent internal stresses immediately inside the fluid.

The dynamic free surface condition in potential flows

The dynamic condition at the free surface in a potential flow can be written in much simpler form than for general turbulent flows.

At the free surface we assume the potential flows are inviscid so that no stresses other than pressures acting on a boundary can be transfered onto the fluid. On the free surface the relevant stress is the atmospheric pressure p_a which varies little over the length scales we consider. Since a constant pressure on the surface does not contribute to the flow, p_a is often just put equal to zero. In potential flow the Bernoulli equation is valid. The dynamic surface condition at a free surface position at $z = \eta$ then follows directly by setting the pressure equal to zero in that equation. We get

$$\phi_t + g\zeta + \frac{1}{2}(\nabla_i \phi)^2 = \frac{C(t)}{\rho} \quad ; \quad z = \zeta \tag{2.4.9}$$

where $C(t)$ is the Bernoulli constant. The proper specification of $C(t)$ is not quite trivial and will be discussed for each of the flows analyzed later.

For the derivation of the dynamic free surface condition in general turbulent flows reference is made to Chapter 11.

2.5 Basic ideas for turbulent flow

Most real flows are *turbulent*. This particularly applies to surf zone waves and currents. The Fig. 2.5.1 shows an example of a time series for the velocity component u in a (steady) turbulent flow.

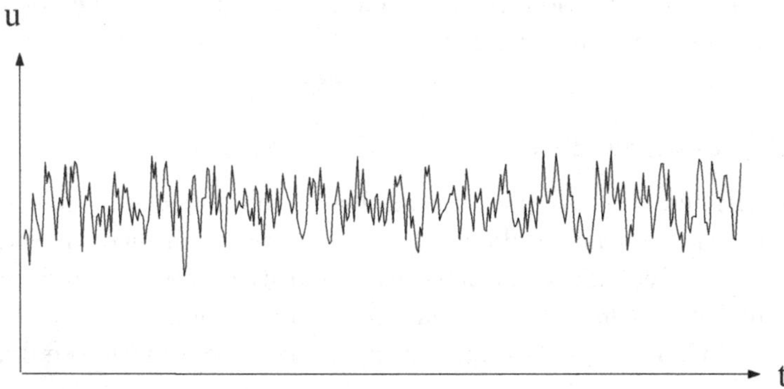

Fig. 2.5.1 Velocity u in steady turbulent flow.

The flow is in principle determined by the Navier-Stokes equations. The solution of the full equations, even on large computers, is still not possible for problems of engineering relevance, and certainly not for any problems appearing in nearshore hydrodynamics.

However, even if we could solve those equations, we would get an answer which, due to the immense amount of data and richness in small details, would be of limited practical interest. To get practical information, we would need to compress the data significantly by defining and extracting information about convenient parameters.

From a technical point of view, such a data compression is done by separating each flow property (such as u, v, p, etc.) into a "mean" component $\widehat{f}$ and a turbulent fluctuation f'.

2.5.1 *Reynolds' decomposition of physical quantities*

The separation of the flow into a mean and a fluctuation part is done by observing that *any* flow-related quantity can be written

$$\begin{aligned} f &= \widehat{f} + f' \\ \widehat{f} &= \text{mean value} \\ f' &= \text{fluctuation} \end{aligned} \tag{2.5.1}$$

(without loss of generality).

This is called a **Reynolds decomposition**, and by definition we have

$$\widehat{f'} = 0 \tag{2.5.2}$$

Using (2.5.1) and (2.5.2), we then get the following rules:

$$\widehat{\widehat{f}} = \widehat{f} \tag{2.5.3}$$

$$\widehat{f + g} = \widehat{f} + \widehat{g} \tag{2.5.4}$$

$$\widehat{\widehat{f}\, g} = \widehat{f} \cdot \widehat{g} \quad ; \quad \text{espec.} \ \widehat{\widehat{f}\, g'} = \widehat{f}\ \widehat{g'} = 0 \tag{2.5.5}$$

$$\widehat{\frac{\partial f}{\partial s}} = \frac{\partial \widehat{f}}{\partial s} \quad ; \quad \widehat{\int_{\text{fixed}} f ds} = \int_{\text{fixed}} \widehat{f}\, ds \tag{2.5.6}$$

These relationships are useful to keep in mind when working with turbulent flow equations. They apply as long as (2.5.2) is valid.

The philosophy behind splitting each total flow quantity (velocity components, pressure, etc.) into a mean and a fluctuating part can be justified quite rigorously, but that is beyond the scope of the present course (for reference see, e.g., Tennekes and Lumley (1972) or Hinze (1975)).

In popular terms, the philosophy is the following: we can never predict the detailed development of a particular realization of a turbulent flow. What we may (and should) expect is that the *effects*, which the turbulence has on the *mean* flow properties, are predictable so that—by properly modeling those effects—we will be able to predict the *mean* flow (i.e. $\widehat{f}$).

Clearly, the Reynolds decomposition depends on our capability to define the average $\widehat{f}$. It turns out that this is not always easy. This is discussed in the following, using the horizontal velocity component u as an example.

2.5.2 *Determination of the turbulent mean flow*

Ensemble averaging

One way of defining the mean value of u is through the **ensemble average** :

> **The ensemble average of a quantity at a (fixed) point is defined as the mean of the values of the quantity at that point at the same time in many identical repetitions of the flow experiment.**

We use $\widehat{\ }$ to indicate ensemble averaged quantities. Any flow property (velocity, pressure, temperature) can be ensemble averaged. Fig. 2.5.2 shows an example for steady mean flow.

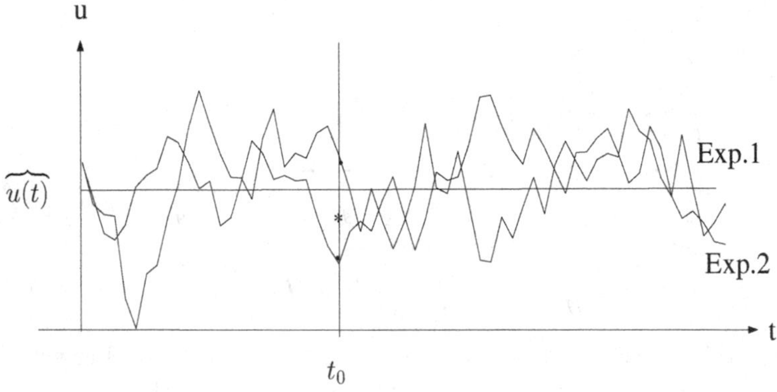

Fig. 2.5.2 Ensemble averaging in steady turbulent flow. This figure shows the time series of two independent realizations of the same (steady) flow. The ensemble average for these two realizations at t_0 is the mean value of the two observed values at t_0.

The advantage of the concept of ensemble averaging is that it is also defined in flows with time varying mean values. An example is shown in Fig. 2.5.3.

In mathematical terms, the ensemble average at $t = t_0$ is defined by

$$\widehat{u}(t_0) = \frac{1}{n} \sum_{i=1}^{n} u_i(t_0) \qquad i = \text{ experiment no.} \tag{2.5.7}$$

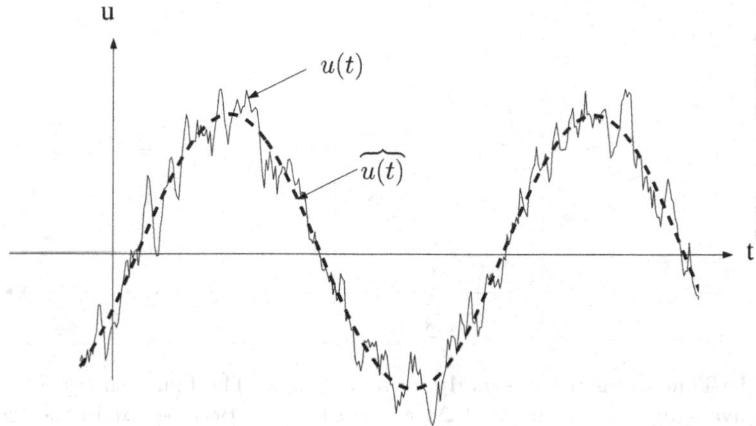

Fig. 2.5.3 Ensemble average for time varying mean flow (for clarity only one time series is shown) u.

In practice ensemble averaging is almost imposssible to generate. For laboratory experiments with strictly periodic waves, however, the ensemble averaging can credibly be replaced with averaging of the measurements obtained at the same phase in each of the waves in the wave experiment. This works very well outside the breaking point where the waves all propagate at the same speed so that the constant wave period is kept throughout the wave flume. However, inside breaking the nonlinear effects in the waves have a tendency make the (slightly) higher waves move faster and catch up on the (slightly) smaller ones. The result is changing wave periods, which makes it difficult to identify which phase over the period of each wave to use in the averaging. And obviously in irregular waves such as on a real coast this surrogat for the ensemble average is not possible.

Time averaging

As mentioned it is very difficult to repeat the same experiment sufficiently many times to obtain a reasonably accurate assessment of the ensemble average. In contrast the time series collected from gages offers a natural basis for determining a time averaged mean value. Therefore many texts and scientists use **time averaging** instead of ensemble averaging.

In steady flow, ensemble and time averaging are the same because the time average $\widehat{u_{(t)}}$ over time Δt is defined as

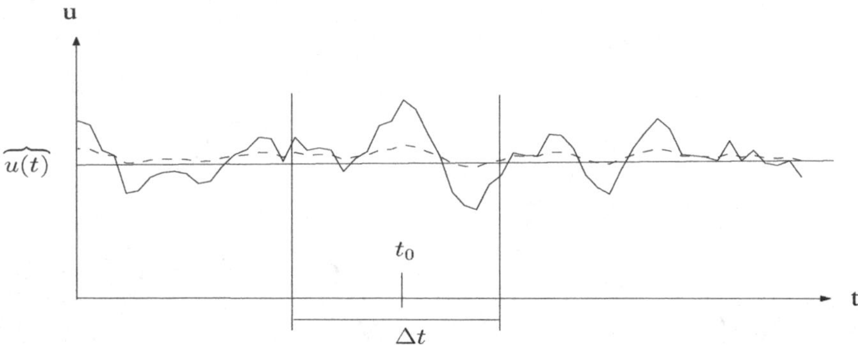

Fig. 2.5.4 Time averaging in **steady** turbulent flow. The figure shows the idea of time averaging over an interval Δt around $t = t_0$. Because Δt in the figure is not much longer than the characteristic time scale of the turbulence, the time average value shows some time variation even though the mean flow is steadwe introduce y.

$$\widehat{u_{(t)}} = \frac{1}{\Delta t} \int_t^{t+\Delta t} u(t)dt; \quad \Delta t \to \infty \tag{2.5.8}$$

which is similar to (2.5.7).

The time averaging can be turned into a sliding average by continuously sliding the finite-length interval Δt along the t axis, which makes $\widehat{u_{(t)}}$ a continuous function of t.

Discussion

We see that in steady mean flow

$$\widehat{u_{(t)}} \to \widehat{u}$$
$$\Delta t \to \infty$$

As Fig. 2.5.4 shows, however, for this to be valid even in flows with a steady long term mean value Δt needs to be much larger than the typical timescale of the turbulence.

When it comes to a time varying mean flow, the situation is even more complicated as the example in Fig. 2.5.5 shows. Though $\widehat{u_{(t)}}$ may be approximated by the value at t_0 in the middle of the time interval we see

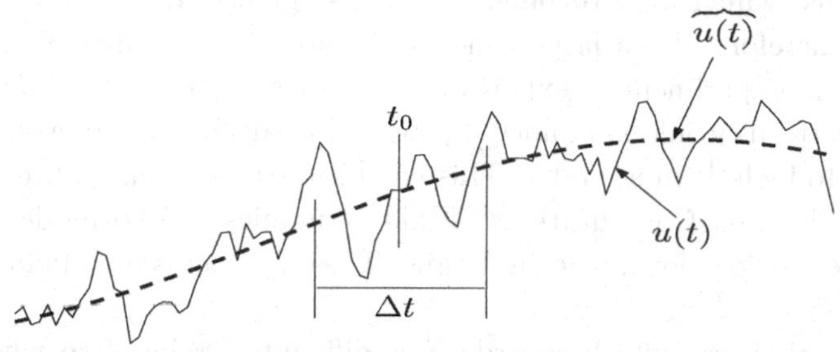

Fig. 2.5.5 Time averaging in **time-varying** mean flow.

that $\widehat{u_{(t)}}$ is not a well defined quantity. In the first place, it is not clear what t_0 is. Secondly, it is also obvious that we must make the assumption that the mean flow does not change significantly within the time Δt.

Hence we see that to obtain a reliable result from time averaging time varying mean flows we must require that the time scale T_t of the turbulence is so much shorter than time scale T_m of the mean flow that we can select a Δt which satisfies $T_t \ll \Delta t \ll T_m$.

This requirement can be written in terms of the turbulent fluctuations $\widehat{u'_{(t)}}$ which are determined as:

$$u'_{(t)} = u - \widehat{u'_{(t)}} \qquad (2.5.9)$$

The requirement is that

$$\Delta t \text{ must be long enough to give } \quad \widehat{u'_{(t)}} = \frac{1}{\Delta t} \int_t^{t+\Delta t} u' dt = 0 \qquad (2.5.10)$$

This can of course not always be satisfied. Only if the time scale T_m of the variation of the mean motion (e.g., the wave period) is large in comparison to the turbulent time scale T_t, can we choose Δt so large that

$$\widehat{u} \approx \widehat{u_{(t)}} \qquad (2.5.11)$$

which would be ideal.

On the other hand, the ensemble average $\widehat{u}(t_o)$ remains a well defined quantity, also in time varying mean flows.

It is finally pointed out that in many types of turbulence (including the early stages of wave breaking), there is an additional problem: Certain vortex patterns, which look turbulent at a first glance and have a large spatial (and therefore also a large time) scale, are likely to almost repeat themselves from experiment to experiment. According to both (2.5.7) and (2.5.8) such patterns would fall under the mean flow although they would be very difficult to include in a model. This complicates modeling of breaking waves. Only detailed Computational Fluid Dynamics (CFD)-models are able to resolve the flow down to such details. These problems remain largely unsolved.

In the following we will disregard these difficulties related to what is turbulence and what is mean flow and simply assume that we can perform a Reynolds decomposition (meaning the $\widehat{\cdot}$ can be determined), and we then discuss how to derive and use the resulting equations.

2.5.3 *The Reynolds equations*

When we introduce the Reynolds decomposition (2.5.1) into the equations for conservation of mass (the continuity equation), and momentum (the Navier-Stokes equations), an equation system results which is called the **Reynolds equations**. The following gives a brief outline of the derivation for the mass and momentum equations.

The Navier-Stokes equations can be written
Continuity:

$$\frac{\partial u}{\partial x} + \frac{\partial v}{\partial y} + \frac{\partial w}{\partial z} = 0 \tag{2.5.12}$$

Momentum:

$$\begin{aligned}
x: \frac{\partial u}{\partial t} + u\frac{\partial u}{\partial x} + v\frac{\partial u}{\partial y} + w\frac{\partial u}{\partial z} &= \frac{1}{\rho}\left(-\frac{\partial p}{\partial x} + \frac{\partial \tau^v_{xx}}{\partial x} + \frac{\partial \tau^v_{yx}}{\partial y} + \frac{\partial \tau^v_{zx}}{\partial z}\right) \\
y: \frac{\partial v}{\partial t} + u\frac{\partial v}{\partial x} + v\frac{\partial v}{\partial y} + w\frac{\partial v}{\partial z} &= \frac{1}{\rho}\left(-\frac{\partial p}{\partial y} + \frac{\partial \tau^v_{xy}}{\partial x} + \frac{\partial \tau^v_{yy}}{\partial y} + \frac{\partial \tau^v_{zy}}{\partial z}\right) \\
z: \frac{\partial w}{\partial t} + u\frac{\partial w}{\partial x} + v\frac{\partial w}{\partial y} + w\frac{\partial w}{\partial z} &= -g + \frac{1}{\rho}\left(-\frac{\partial p}{\partial z} + \frac{\partial \tau^v_{xz}}{\partial x} + \frac{\partial \tau^v_{yz}}{\partial y} + \frac{\partial \tau^v_{zz}}{\partial z}\right)
\end{aligned} \tag{2.5.13}$$

where τ^v_{ij} represents the viscous stresses given by

$$\tau^v_{ij} = \rho\nu\left(\frac{\partial u_i}{\partial x_j} + \frac{\partial u_j}{\partial x_i}\right) \tag{2.5.14}$$

with ν being the kinematic viscosity.

We observe that the structure of the three components of the momentum equations above is the same. In order to reduce the number of equations to be examined, it is convenient (and customary when dealing with turbulence) to use tensor notation instead of the equivalent (but more cumbersome) x, y, z-coordinates used in (2.5.12) and (2.5.13).

Derivation of the Reynolds equations

In tensor notation (using the summation rule), we can write (2.5.12) and (2.5.13) in the form

$$\text{Continuity}: \quad \frac{\partial u_i}{\partial x_i} = 0 \qquad i = 1, 2, 3 \tag{2.5.15}$$

(incompressibility has already been assumed in the Navier-Stokes equations). Momentum, j-th component:

$$\frac{\partial u_j}{\partial t} + u_i \frac{\partial u_j}{\partial x_i} = g_j + \frac{1}{\rho}\frac{\partial}{\partial x_i}\left(-p\delta_{ij} + \tau_{ij}^v\right) \qquad i, j = 1, 2, 3 \tag{2.5.16}$$

where $g_j = (0, 0, -g)$.

Before embarking on the Reynolds decomposition we change the convective acceleration term $u_i\, \partial u_j / \partial x_i$ to a more convenient form. We write

$$u_i \frac{\partial u_j}{\partial x_i} = u_i \frac{\partial u_j}{\partial x_i} + u_j \frac{\partial u_i}{\partial x_i} \tag{2.5.17}$$

which, according to (2.5.15), for incompressible flow means adding 0.

Hence

$$u_i \frac{\partial u_j}{\partial x_i} = \frac{\partial u_i u_j}{\partial x_i} \tag{2.5.18}$$

Thus (2.5.16) can be written

$$\frac{\partial u_j}{\partial t} + \frac{\partial u_i u_j}{\partial x_i} = g_j + \frac{1}{\rho}\frac{\partial}{\partial x_i}\left(-p\delta_{ij} + \tau_{ij}^v\right) \tag{2.5.19}$$

We now introduce the Reynolds Decomposition given by

$$u_i = \overbrace{u_i} + u_i' \quad ; \quad p = \overbrace{p} + p_j' \tag{2.5.20}$$

In the continuity equation, this yields

$$\frac{\partial \overbrace{u_i}}{\partial x_i} + \frac{\partial u_i'}{\partial x_i} = 0 \tag{2.5.21}$$

This entire equation can be ensemble (or "turbulent") averaged which gives

$$\frac{\widehat{\partial \widehat{u_i}}}{\partial x_i} + \frac{\widehat{\partial u_i'}}{\partial x_i} = 0 \tag{2.5.22}$$

or since

$$\frac{\widehat{\partial u_i'}}{\partial x_i} = \frac{\partial \widehat{u_i'}}{\partial x_i} = 0 \tag{2.5.23}$$

we get

$$\boxed{\frac{\partial \widehat{u_i}}{\partial x_i} = 0} \tag{2.5.24}$$

Hence both the *mean* and the *turbulent* fluctuation satisfy the continuity equation independently as (2.5.23) and (2.5.24) show.

For the momentum equation (2.5.19), we get

$$\begin{aligned}
&\frac{\partial \widehat{u_j}}{\partial t} + \frac{\partial u_j'}{\partial t} + \frac{\partial (\widehat{u_i} + u_i')(\widehat{u_j} + u_j')}{\partial x_i} \\
&\quad = -\frac{1}{\rho}\frac{\partial}{\partial x_i}\left(-(\widehat{p} + p')\delta_{ij}\right) + g_j \\
&\quad\quad + \frac{1}{\rho}\frac{\partial}{\partial x_i}\nu\left(\frac{\partial(\widehat{u_i} + u_i')}{\partial x_j} + \frac{\partial(\widehat{u_j} + u_j')}{\partial x_i}\right)
\end{aligned} \tag{2.5.25}$$

This equation is ensemble averaged using:

$$\frac{\widehat{\partial u_j'}}{\partial t} = 0 \quad ; \quad \frac{\partial \widehat{u_j}}{\partial x_j} = 0 \quad ; \quad \widehat{\widehat{u_i}\,\widehat{u_j}} = \widehat{u_i}\,\widehat{u_j} \quad ; \quad \widehat{\;'\;} = 0 \tag{2.5.26}$$

(2.5.25) then becomes

$$\begin{aligned}
&\frac{\partial \widehat{u_j}}{\partial t} + \frac{\partial \widehat{u_i}\,\widehat{u_j}}{\partial x_i} + \frac{\partial \widehat{\widehat{u_i}\, u_j'}}{\partial x_i} + \frac{\partial \widehat{u_i'\,\widehat{u_j}}}{\partial x_i} + \frac{\partial \widehat{u_i' u_j'}}{\partial x_i} \\
&\quad = -\frac{1}{\rho}\frac{\partial}{\partial x_i}\left(-\widehat{p}\,\delta_{ij} + \nu\left(\frac{\partial \widehat{u_i}}{\partial x_j} + \frac{\partial \widehat{u_j}}{\partial x_i}\right)\right) + g_j
\end{aligned} \tag{2.5.27}$$

and hence we get

$$\boxed{\frac{\partial \widehat{u_j}}{\partial t} + \frac{\partial \widehat{u_i}\,\widehat{u_j}}{\partial x_i} = -\frac{1}{\rho}\frac{\partial \widehat{p}}{\partial x_j} + \frac{1}{\rho}\frac{\partial}{\partial x_i}\left(\nu\left(\frac{\partial \widehat{u_i}}{\partial x_j} + \frac{\partial \widehat{u_j}}{\partial x_i}\right) - \rho\,\widehat{u_i'\,u_j'}\right) + g_j} \tag{2.5.28}$$

Eqs. (2.5.24) and (2.5.28) are the **Reynolds equations** for the conservation of mass and momentum in the mean flow. A similar equation can be derived for the energy of the mean flow, for the turbulent (kinetic) energy, and in fact for other turbulent quantities. For further reference see e.g. Tennekees and Lumley (1972), Hinze (1975), or Wilcox (1998).

The equations we have derived here describe the mean flow. We see from (2.5.28) that the turbulence affects the mean flow *only* through the terms

$$-\frac{\partial}{\partial x_j}\,\rho\,\widehat{u_i'\,u_j'} \tag{2.5.29}$$

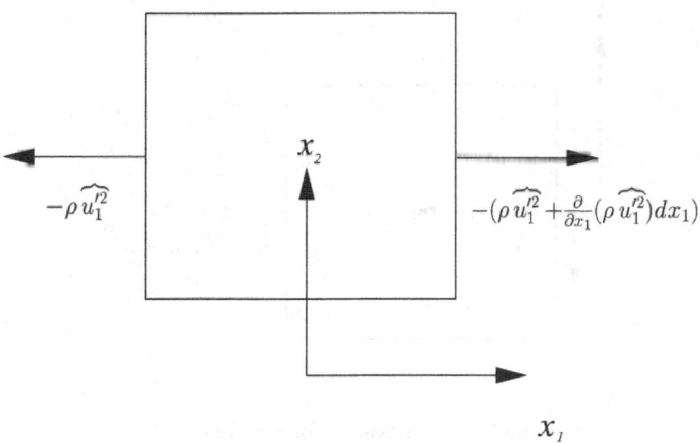

Fig. 2.5.6 The stresses for $i = 1, j = 1$.

They are equivalent to additional surface stresses.

Example 1

We consider the case $i = 1$; $j = 1$. This is illustrated in Fig. 2.5.6. The net contribution on the fluid element is

$$\frac{\partial}{\partial x_j} \rho \widehat{u'_i\, u'_j} = \rho \frac{\partial}{\partial x_1} \widehat{u_1'^{\,2}} \tag{2.5.30}$$

Using the rules for stress components we see that $-\rho \frac{\partial \widehat{u_1'^{\,2}}}{\partial x_1}$ is a normal stress.

Example 2

$i = 1$; $j = 2$, See Fig. 2.5.7 for illustration.

On each side of the fluid element in Example 2, we have contributions of the type $-\rho \widehat{u'_1\, u'_2}$ but the net effect on fluid element is a shear stress of the size

$$-\rho \frac{\partial}{\partial x_2} \left(\widehat{u'_1\, u'_2} \right) dx_2 \tag{2.5.31}$$

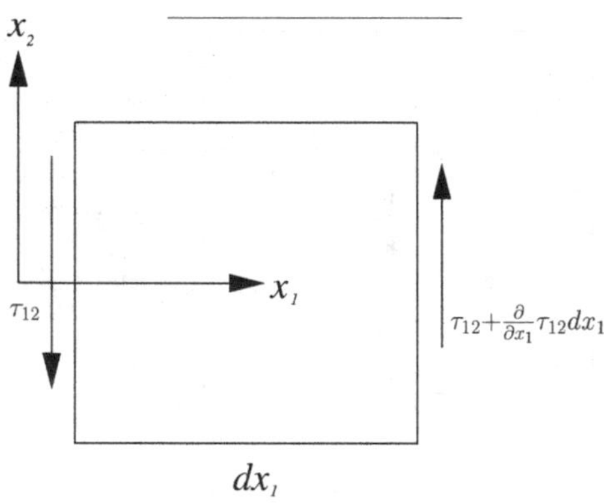

Fig. 2.5.7 The stresses for $i = 1, j = 2$.

Thus, realizing that $-\rho \widehat{u'_i\, u'_j}$ has the nature of a stress distribution we introduce the definition

$$\tau'_{ij} = -\rho \widehat{u'_i\, u'_j} \qquad i, j = 1, 2, 3 \tag{2.5.32}$$

which are called the **Reynolds stresses**. We see that τ'_{ij} has the components

$$\tau'_{ij} = -\rho\,\widehat{u'_i u'_j} = \begin{Bmatrix} -\rho\,\widehat{u'^2_1} & -\rho\,\widehat{u'_1 u'_2} & -\rho\,\widehat{u'_1 u'_3} \\ -\rho\,\widehat{u'_2 u'_1} & -\rho\,\widehat{u'^2_2} & -\rho\,\widehat{u'_2 u'_3} \\ -\rho\,\widehat{u'_3 u'_1} & -\rho\,\widehat{u'_3 u'_2} & -\rho\,\widehat{u'^2_3} \end{Bmatrix} \tag{2.5.33}$$

Hence the momentum equation(s) may be written

$$\frac{\partial \widehat{u_j}}{\partial t} + \frac{\partial \widehat{u_i}\,\widehat{u_j}}{\partial x_i} = -\frac{1}{\rho}\frac{\partial \widehat{p}}{\partial x_j} + \frac{1}{\rho}\frac{\partial}{\partial x_i}\left(\tau^v_{ij} + \tau'_{ij}\right) + g_j \tag{2.5.34}$$

or, if we, for convenience, introduce the definition

$$\underset{\text{total}}{\tau_{ij}} = \underset{\text{visc.}}{\tau^v_{ij}} + \underset{\text{turb.}}{\tau'_{ij}} \tag{2.5.35}$$

we get

$$\frac{\partial \widehat{u_i}}{\partial t} + \frac{\partial \widehat{u_i}\,\widehat{u_j}}{\partial x_j} = -\frac{1}{\rho}\frac{\partial \widehat{p}}{\partial x_i} + \frac{1}{\rho}\frac{\partial \tau_{ij}}{\partial x_j} + g_i \tag{2.5.36}$$

In view of the fact that

$$\tau^v_{ij} \ll \tau'_{ij} \tag{2.5.37}$$

the viscous stresses are usually neglected altogether in turbulent flows, which means we assume

$$\tau_{ij} \simeq \tau'_{ij} \tag{2.5.38}$$

In conclusion, by comparing (2.5.19) and (2.5.36) we see that if we replace the viscous shear stresses τ^v_{ij} with the total (or just the turbulent) shear stresses then the mean velocity $\widehat{u_i}$ satisfies the same equations as the total (instantaneous) velocity u_i.

2.5.4 *Modelling of turbulent stresses*

When deriving the Navier-Stokes equations we found that the equations in terms of the stresses had too many unknowns. We resolved that problem by developing Stokes viscosity law which relates the stresses to the flow velocities. This lead to introduction of the fluid viscosity μ (or the kinematic viscosity $\nu = \mu/\rho$) which is a **fluid property** independent of the flow.

In turbulence we are faced with the same task: to express the turbulent stresses $\tau_{ij} = -\rho \overbrace{u_i' \, v_j'}$ in terms of the mean flow.

Unfortunately, as is readily understood, an analogous turbulent viscosity ν_t can be expected to depend on the turbulence which means on the flow itself. The problem of finding an equivalent to the Stokes viscosity is therefore far more complicated. This is called the **turbulent closure problem** because we need at least one extra relationship to close the equations so the number of unknowns equals the number of equations. **This problem has not been solved in a satisfactory way yet.** It constitutes the core of all turbulence research.

In the following some of the simplest closure models are described very briefly.

Prandtl's mixing length

The concept of a mixing length to model the turbulent stresses was introuced by Prandtl (see Prandtl, 1952). The basic idea is that due to the turbulent fluctuations fluid particles with velocity u in the x-direction travel a certain distance l in the cross-flow direction y before acquiring the velocity of the new region with velocity $u + \Delta u$. Omitting for simplicity the overbrace the momentum difference between the two regions is

$$m \, \Delta u = m \, l \, \frac{\partial u}{\partial y} \tag{2.5.39}$$

The mass of fluid actually moving from one level y_0 to level $y_0 + l$ is proportional to $\rho|v'|$, which leads to the expression for τ

$$\tau = m \, \Delta u = \rho |v'| \, l \, \frac{\partial u}{\partial y} \tag{2.5.40}$$

For continuity reasons we must have

$$v' \sim u' \sim |l| \, \frac{\partial u}{\partial y} \tag{2.5.41}$$

This means that

$$\tau = \rho \, l^2 \, |\frac{\partial u}{\partial y}| \frac{\partial u}{\partial y} \tag{2.5.42}$$

which expresses τ in terms of the mean flow.

The eddy viscosity model

The mixing length hypothesis is mainly seen used in older literature. It has the disadvantage that it makes the equations non-linear because the shear stress is proportional to the velocity gradient squared.

Today a more popular shear stress model is the simple concept of an eddy viscosity ν_t which assumes that the shear stress can be expressed as

$$\tau = \rho \, \nu_t \, \frac{\partial u}{\partial y} \tag{2.5.43}$$

This was first introduced by Boussinesq. The usefulness of this closure model lies of course in its simpliciy. It turns out to work relatively well for a number of simple flows provided ν_t is properly modelled.

It is also worth to mention the suggestion that ν_t can be expressed as

$$\nu_t = l_t \, \sqrt{k} \tag{2.5.44}$$

where l_t is a characteristic length scale for the turbulence, and k the turbulent kinetic energy defined by

$$k = \frac{1}{2}(\widehat{u'^2} + \widehat{v'^2} + \widehat{w'^2}) \tag{2.5.45}$$

represents a characteristic velcity. This can sometimes help judging the relevance of a ν_t value as in Chapter 13.

As can be seen from the above both the mixing length and the eddy viscosity hypotheses are introduced on the basis of simple one dimensional shear flow. However, the eddy viscosity concept can directly be expanded to include randoml three dimensional flows the same way as the ordinary viscosity, meaning defining the shear stresses in a flow by

$$\tau_{ij} = \rho \, \nu_t \left(\frac{\partial u_i}{\partial x_j} + \frac{\partial u_j}{\partial x_i} \right) \tag{2.5.46}$$

In the present text the turbulent closure modelling is entirely limited to the simple concept of an eddy viscosity. This kind of closure is also termed a simple algebraic closure model.

Advanced turbulence modelling

It is important to realize, however, that far more advanced closure models are available. These models require numerical computations and are based on developing evolution equations for the turbulent quantities. The first such quantity modelled is usually the turbulent kinetic energy k. The equation for k is based on the energy equation for the turbulent flow. k is often modelled along with the dissipation ϵ of k in a so called $k-\epsilon$-model, but this is only one type of turbulent closure model. Such advanced models of course have the advantage of being able to represent more realistically more complex flows. In the $k-\epsilon$-models k and ϵ are used to determine a generalized (time and space varying) eddy viscosity ν_t by the expression

$$\nu_t = C_\nu \, \frac{k}{\epsilon^{3/2}} \tag{2.5.47}$$

where C_ν is an emppirical constant. This ν_t is then normally used to determine the turbulent stresses by means of (2.5.46), by which the closure of the equations is achieved.

However, all closure models have the same problem: in the evolution equations for each turbulent quantity such as k or ϵ will always emerge other, more complicated turbulent quantitites such as triple correlations between turbulent fluctuations, that need to be estimated on the basis of physical reasoning, experimental results etc. Thus the closure problem is always just transferred to other unknowns.

The spectrum of modern turbulence methods also includes the Large Eddy Simulation (LES) approach in which the larger eddies in the flow are included in the numerical results as part of the mean flow, and only the small scale turbulence is covered by empirical model equations. Modern computers are also becoming so powerful that Direct Numerical Simulation (DNS) by direct solution of the (viscous) Navier-Stokes equations, where all turbulent scales are computed, can give answers to small scale flow problems.

For more information reference is made to recent literature. White (1991) gives a description of the $k-\epsilon$ model concept, and Wilcox (1998) gives an extensive recent overview of a wider selection of turbulence techniques. The review of turbulence Reynolds-stress closure models by Speziale (1991) still remains useful in spite of the rapid developments in this area, and there is a large paper by Vreman et al. (1997) covering the LES method, to mention a few.

2.6 Energy flux in a flow

To analyze the energy flux in a flow across an arbitrary surface, we first consider a small element dS of the surface, see Fig. 2.6.1.

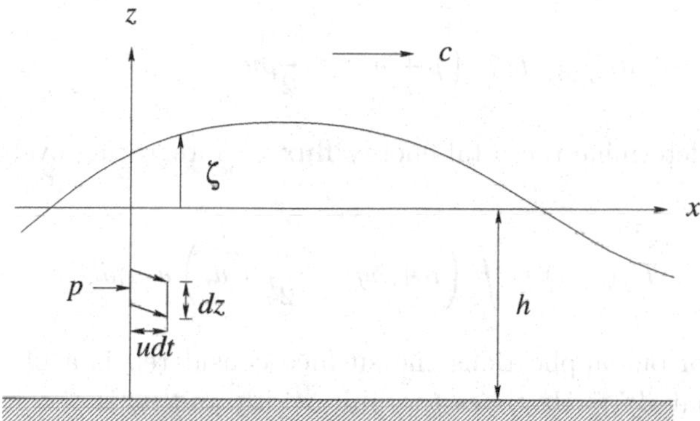

Fig. 2.6.1 Definition sketch for analysis of the energy flux through a vertical section.

The energy flux through this element has two contributions:

(i) the actual flow of potential and kinetic energy carried by the water particles passing through the section.
(ii) the work done by the pressure on the moving particles.

The energy density (energy per unit volume) is therefore given by

$$\rho g z + \frac{1}{2}\rho\left(u^2 + v^2 + w^2\right) = \rho g z + \frac{1}{2}\rho u_i u_i \tag{2.6.1}$$

where the first term is the potential energy relative to some (arbitrarily chosen) **horizontal** reference level, the second the kinetic energy.

Thus, the instantaneous energy flux through a surface element dS from transport of potential and kinetic energy is, per unit time

$$\left(\rho g z + \frac{1}{2}\rho u_i u_i\right) u_i \, n_i dS \tag{2.6.2}$$

where

$$u_i n_i = u_n \tag{2.6.3}$$

is the velocity component normal to the surface element.

The work done by the pressure is, per unit time

$$p u_i \, n_i dS = p u_n dS \tag{2.6.4}$$

so that the total energy flux through dS is

$$dE_f(x_i, t) = \left(p + \rho g z + \frac{1}{2} \rho u_i u_i \right) u_n dS \tag{2.6.5}$$

and we can determine the total energy flux by integrating over the entire surface

$$E_f(x_i, t) = \int_S \left(p + \rho g z + \frac{1}{2} \rho u_i u_i \right) u_i n_i dS \tag{2.6.6}$$

In most of our applications the surface considered is a plane vertical section so that $dS = dz$ times the unit width and n_i is the same for all z (see Figure 2.6.1). This implies that $u_i \; n_i = u_\alpha \; n_\alpha$, where index α refers to a vector in the (x, y)-plane. Since we assume a plane surface here we can bring the constant normal vector n_α outside the integral. The total instantaneous energy flux through such a vertical section therefore becomes

$$E_f(x_\alpha, t) = \int_{-h_0}^{\eta} \left(p + \rho g z + \frac{1}{2} \rho u_i u_i \right) u_\alpha dz \; n_\alpha \tag{2.6.7}$$

At this point, however, we have not yet assumed that the velocity vector u_α has the same direction at all points over the section. In fact if we are considering a flow that is a combination of waves (that may be plane so that $u_{w\alpha}$ has the same direction over the vertical) and depth varying currents $U_\alpha(z)$, then obviously the total velocity u_α will vary in direction over depth. The total integral will then be a vector in a direction that depends on that depth variation, given by

$$\boxed{E_{f\alpha}(x_\alpha, t) = \int_{-h_0}^{\eta} \left(p + \rho g z + \frac{1}{2} \rho u_i u_i \right) u_\alpha dz} \tag{2.6.8}$$

valid for an arbitrary flow sending energy through a plane vertical surface extending from the bottom to the instantaneous free surface. The flux through the plane surface with normal vector n_α is then

$$E_f = E_{f\alpha} n_\alpha \tag{2.6.9}$$

In vector notation this can be written

$$\mathbf{E_f}(x, y, t) = \int_{-h_0}^{\eta} \left(p + \rho g z + \frac{1}{2}\rho \mathbf{u} \cdot \mathbf{u}\right) \mathbf{u}\, dz \tag{2.6.10}$$

For completeness the x, y components of this are

$$E_{fx}(x, y, t) = \int_{-h_0}^{\eta} \left(p + \rho g z + \frac{1}{2}\rho \mathbf{u} \cdot \mathbf{u}\right) u\, dz \tag{2.6.11}$$

$$E_{fy}(x, y, t) = \int_{-h_0}^{\eta} \left(p + \rho g z + \frac{1}{2}\rho \mathbf{u} \cdot \mathbf{u}\right) v\, dz \tag{2.6.12}$$

2.7 Appendix: Tensor notation

Tensor notation (or "index notation") utilizes the fact that many equations look the same in all three coordinate directions. Thus instead of writing out the x, y and z equations, we use the symbols i or j as indexes. Each of those can then be x, y or z.

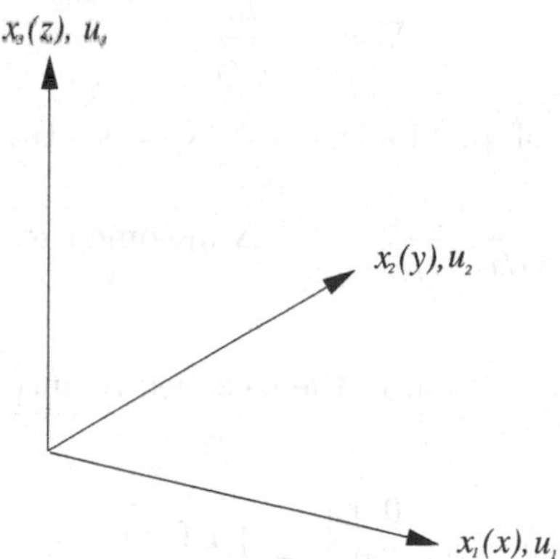

Fig. 2.7.1 Tensor-coordinate system.

Example: u_j can be u, v or w depending on whether j corresponds to x, y or z.

In addition, it is customary in tensor notation to talk about x_1, x_2, x_3 instead of x, y, z. This is illustrated in Fig. 2.7.1 showing the cartesian coordinate system normally used in tensor notation.

For that reason, we also normally say that i (or j) is 1, 2, 3 instead of x, y, z.

Finally, tensor notation becomes really elegant because of the summation rule: If the same index is repeated as index within the same term, it means the term should be considered the sum of all the three terms we get by letting the repeated index be 1, 2 and 3. Thus we have as examples

$$i) \quad \tau_{ii} = \tau_{11} + \tau_{22} + \tau_{33} \tag{2.7.1}$$

$$ii) \quad \frac{\partial u_i}{\partial x_i} = \frac{\partial u_1}{\partial x_1} + \frac{\partial u_2}{\partial x_2} + \frac{\partial u_3}{\partial x_3} \tag{2.7.2}$$

$$iii) \quad u_i \frac{\partial u_j}{\partial x_i} = u_1 \frac{\partial u_j}{\partial x_1} + u_2 \frac{\partial u_j}{\partial x_2} + u_3 \frac{\partial u_j}{\partial x_3} \tag{2.7.3}$$

$$iv) \quad \frac{\partial^2 \phi}{\partial x_i \partial x_i} = \frac{\partial^2 \phi}{\partial x_1^2} + \frac{\partial^2 \phi}{\partial x_2^2} + \frac{\partial^2 \phi}{\partial x_3^2} \tag{2.7.4}$$

Note that $ii)$ is the same as

$$\nabla_i u_i = \frac{\partial u_i}{\partial x_i} \tag{2.7.5}$$

which is the divergence of u_i. Similarly $iv)$ expresses the Laplacian operator

$$\nabla_i^2 = \frac{\partial}{\partial x_i \partial x_i} \qquad (= \Delta \text{ in some texts}) \tag{2.7.6}$$

Kronecker's δ_{ij}

In tensor notation we also meet the so called Kronecker δ_{ij}. It is defined as

$$\delta_{ij} = \begin{Bmatrix} 1\,0\,0 \\ 0\,1\,0 \\ 0\,0\,1 \end{Bmatrix} = \begin{cases} 1 \text{ for } i = j \\ 0 \text{ for } i \neq j \end{cases} \tag{2.7.7}$$

Hence δ_{ij} is really the unit matrix.

Essentially, when δ_{ij} occurs in a term in an equation, it has the effect of changing i to j (or j to i) in that term.

Example 1

$$\delta_{ij} u_i = u_j \tag{2.7.8}$$

because by the summation rule

$$\delta_{ij}u_i = \delta_{1j}u_1 + \delta_{2j}u_2 + \delta_{3j}u_3 \tag{2.7.9}$$

However by the definition of δ_{ij}, only the term which has j equal to the first index is 1, the others are 0. So

$$\left.\begin{array}{l} j = 1 \rightarrow \delta_{i1}u_i = u_1, \\ j = 2 \rightarrow \delta_{i2}u_i = u_2, \\ j = 3 \rightarrow \delta_{i3}u_i = u_3 \end{array}\right\} \quad \text{or} \quad \delta_{ij}u_i = u_j \tag{2.7.10}$$

Example 2

$$\delta_{ij}\tau_{ij} = \tau_{ii} \tag{2.7.11}$$

(or τ_{jj}, which makes no difference because of the summation rule) and

$$\delta_{ij}\frac{\partial p}{\partial x_i} = \frac{\partial p}{\partial x_j} \tag{2.7.12}$$

2.8 References - Chapter 2

Useful references on fluid mechanics:

Aris, Rutherford (1962). Vectors, tensors and the basic equations of fluid mechanics. Prentice Hall, Englewood Cliffs, NJ.

Batchelor, G.K. (1968). An Introduction to Fluid Dynamics, Cambridge Univ. Press.

Currie, I.G. (1974). Fundamental mechanics of fluids. McGraw-Hill, New York.

Hinze (1975). Turbulence, McGraw Hill.

Kundu, P.K. (1990, 2001). Fluid Mechanics. Academic Press, New York.

Mei, C.C. (1983). The Applied Dynamics of Ocean Surface Waves, Wiley Interscience.

Panton, R. (1984). Incompressible flow. Wiley, New York.

Prandtl, L. (1952). Essentials of fluid dynamics. Hafner Publishing Comp., New York.

Speziale, C. G. (1991) Analytical methods for the development of Reynolds-stress closures in turbulence. Ann. Rev. Fluid Mech., **23**, 107–137.

Tennekes, H. and J.L. Lumley (1972). A First Course in Turbulence, The MIT Press.

Vreman, B., G. Bernard, and H. Kuerten (1997). Large eddy simulation of the turbulent mixing layer. J. Fluid Mech., **339**, 357–390.

White, F. M. (1974, 1991). Viscous fluid flow. McGraw-Hill.

Wilcox, D. C. (1998). Turbulence modeling for CFD. DCW Industries, Inc., La Cañada, Calif 91011.

Yuan, (1967). Foundations of fluid mechanics, Prentice Hall,

Chapter 3

Linear Waves

Introduction

The simplest possible water wave theory is that of linear waves, which emerge as solution to a simplified version of the general equations of motion. In this chapter, we first discuss the assumptions leading to the simplified equations, then derive the solution for linear waves. We then analyze the physical characteristics of that solution such as velocity and pressure fields, and special cases of the theory such as waves in very deep and very shallow water. We also analyze wave averaged quantities such as mean wave energy density, and the mean flux of mass (or volume), momentum and energy caused by a linear wave motion. We also consider the superposition of linear waves, both some simple canonical examples and the most general form described by wave spectra. We then go on to consider the propagation of linear waves in regions of varying depth and thereby derive the basic equations for many of the modern wave models.[1]

Throughout we seek to clarify the physical properties and the limitations of the approximate solutions found.

3.1 Assumptions and the simplified equations

It is a fundamental scientific principle to start with the simplest possible formulation that contains the principal features of the problem.

So we first restrict the analysis to the following situation

a) Constant depth, h.

[1]Part of the derivation in this chapter is a slightly modified version of the presentation by Svendsen and Jonsson (1976), and many of the figures are taken from that book.

b) Periodic waves with period T.
c) 2 dimensional motion in the vertical x, z-plane (2DV), i.e. one dimensional motion in the horizontal direction (1DH).

These assumptions merely restrict the physical situation considered and do not represent approximations about the wave motion.

However, as a necessary approximation, we neglect the effect of the boundary layer that develops at the bottom. This includes neglecting the viscous or turbulent stresses that develop in the boundary layer and also neglecting the disturbances the boundary layer causes in the flow above. There are two major reasons for this: one is that the shear stresses inside the boundary layer are small in comparison to the inertia and pressure forces that are the primary forces in the wave motion outside the boundary layer; the other that under waves with wave periods like wind waves the boundary layer is usually very thin in comparison to the water depth h. Both reasons imply that the *local* disturbances from the bottom boundary layer are really small. This will be discussed further in Chapter 10.[2]

The internal viscous stresses in the motion are also extremely small.[3] Notice, however, that away from boundary layers these stresses do not create vorticity so the flow remains irrotational inspite of the existence of the very small viscous stresses.

This means the flow can effectively be considered frictionless and, according to Kelvin's theorem, it will remain irrotational if it is assumed initially to be so.

The assumption of irrotational flow is necessary for a solution. It implies we can introduce a velocity potential ϕ defined so that (see chapter 2.3.6)

$$u,\ w = \nabla\phi \tag{3.1.1}$$

where u, w are the horizontal and vertical velocity components, respectively, and ∇ is the gradient operator defined by

$$\nabla = \frac{\partial}{\partial x}\ ,\ \frac{\partial}{\partial z} \tag{3.1.2}$$

[2]Notice the emphasis on *local* because over longer distances the energy dissipation due to bottom friction accumulates and over longer distances (many wave lengths) can dissipate substantial fractions of the wave energy.

[3]Calculations show for example that a wave in deep water could move all the way around the earth loosing only half of its energy due to internal friction.

See Figure 3.1.1 for definitions. The velocity potential satisfies the Laplace equation

$$\nabla^2 \phi = \phi_{xx} + \phi_{zz} = 0; \quad -h < z < \eta(x,t) \tag{3.1.3}$$

where index $_x$ or $_z$ represent partial differentiation with respect to x and z, respectively.

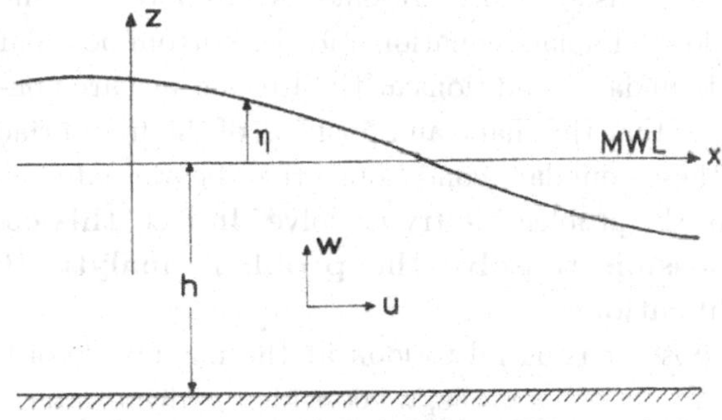

Fig. 3.1.1 Definition sketch.

Notice that we have chosen to place the x-axis at the mean water level (MWL) which is determined so that

$$\int_0^L \eta dx = 0 \tag{3.1.4}$$

The boundary conditions for the problem are the bottom condition (see Chapter 2.4)

$$\phi_z = 0 \qquad \text{at} \qquad z = -h \tag{3.1.5}$$

and the two free surface conditions: the kinematic condition

$$\phi_z - \eta_t - \phi_x \eta_x = 0 \qquad \text{at} \qquad z = \eta \tag{3.1.6}$$

and the dynamic condition

$$\eta + \frac{1}{2g}\left(\phi_x^2 + \phi_y^2\right) + \frac{1}{g}\phi_t = 0 \qquad \text{at} \qquad z = \eta \tag{3.1.7}$$

Finally, we need a boundary condition in the x direction. For this we utilize the assumption that the motion is periodic with a wave length L which is the horizontal distance between two consequtive wave crests (see

Fig. 3.2.1). This means the conditions at an arbitrary x value will repeat itself for each L. Hence, we only need to consider the section $0 < x < L$. This applies to all variables but is conveniently expressed in terms of the horizontal velocity as

$$\phi_x(0, z, t) = \phi_x(L, z, t) \qquad (-h < z < 0) \tag{3.1.8}$$

Mathematically this system represents two important difficulties. The first is that, while the Laplace equation and the bottom boundary condition are linear both boundary conditions at the free surface are non-linear. The second difficulty is that the shape and position of the free surface boundary $z = \eta$ at which these boundary conditions are to be evaluated is itself one of the unknowns of the problem we try to solve. In fact, **this combination makes it impossible to solve the problem analytically without further simplifications.**

For this purpose, it is useful to look at the magnitude of the terms in (3.1.6) and (3.1.7). As an example, we consider a "deep water wave" which is a wave in water deeper than half the wave length. It will be found later that in such a wave the fluid particles at the free surface move in circular paths, each orbit taking one wave period T (which is the time elapsed between the passage at a fixed point of two consecutive wave crests). Since a fluid particle at the surface stays there, the diameter of the circular paths must be the wave height H defined as the vertical distance between the crest and trough levels in the wave (Fig. 3.2.1). Hence, the speed of such a particle can be approximated by $V = \pi H/T$, which means that

$$(\phi_x)_{\max} = (\phi_z)_{\max} = \frac{\pi H}{T} = O\left(\frac{H}{T}\right) \tag{3.1.9}$$

For η_t, a typical value must be

$$\eta_t = O\left(\frac{H}{T}\right) \tag{3.1.10}$$

whereas, we for η_x have

$$\eta_x = O\left(\frac{H}{L}\right) \tag{3.1.11}$$

These estimates, how crude they are, suffice for the present purpose if we further realize that the propagation velocity c, the wave length L, and wave period T are related by

$$c = \frac{L}{T} \tag{3.1.12}$$

Then we get the magnitude of the terms in (3.1.6) as follows

$$\begin{aligned} \phi_z &= O\left(\frac{H}{T}\right) = O\left(c\frac{H}{L}\right) \\ \eta_t &= O\left(c\frac{H}{L}\right) = O(\phi_z) \\ \phi_x\, \eta_x &= O\left(c\frac{H^2}{L^2}\right) = O\left(\frac{H}{L}\phi_z\right) \end{aligned} \tag{3.1.13}$$

by which we have, of course, said nothing about the phase differences of the maximum values for each term.

Equation (3.1.13) shows that the order of magnitude of the last, non-linear term in (3.1.6) is H/L times the linear terms

Hence if we assume that H/L – which also is called the wave steepness – is a small quantity (i.e. $\ll 1$), we thereby obtain the result that the non-linear term in (3.1.6) becomes much smaller than the two linear terms in (3.1.6). (In nature $H/L = 0.05$ represents a quite steep wave which justifies the assumption.)

Exercise 3.1-1 Show by analogous arguments that the same will be the result in (3.1.7): the nonlinear terms are $O(H/L)$ times smaller than the linear terms.

We can now obtain a first approximation to the problem simply by assuming $H/L \ll 1$ and so, **neglecting the non-linear terms**, (3.1.6) and (3.1.7) then simplify to

$$\phi_z - \eta_t = 0 \text{ at } z = \eta \tag{3.1.14}$$

$$\eta + \frac{1}{g}\phi_t = 0 \text{ at } z = \eta \tag{3.1.15}$$

which are linear.

In fact, the second problem, that of the unknown position of the free surface, can be attacked by the same approach. Thus we can write by a

Taylor expansion

$$\phi_z(x,\eta,t) = \phi_z(x,0,t) + \eta\,\phi_{zz}(x,0,t) + \cdots$$

$$= \phi_z(x,0,t) - \eta\,\phi_{xx}(x,0,t) + \cdots \tag{3.1.16}$$

where the Laplace equation has been invoked in the second term. Since we have

$$\eta = O(H) \tag{3.1.17}$$

we get

$$\phi_{xx} = O\left(\frac{\phi_x}{L}\right) \tag{3.1.18}$$

and (3.1.16) becomes by virtue of (3.1.9) and (3.1.15)

$$\phi_z(x,\eta,t) = \phi_z(x,0,t) + O\left(c\,\frac{H^2}{L^2}\right) = \phi_z(x,0,t) + O\left(\frac{H}{L}\phi_z\right) \tag{3.1.19}$$

This means that if we in (3.1.14) evaluate ϕ_z at $z = 0$ instead of at $z = \eta$, it corresponds to neglecting terms of the same order of magnitude as the non-linear terms we decided above to neglect.

Similar results can, of course, be obtained for ϕ_t in (3.1.15) so that these two boundary conditions at the free surface to the first approximation both become linear, **and** can be evaluated at the fixed horizontal level $z = 0$.

Thus, we can summarize the list of simplifications and approximations as follows:

We

a) consider waves on constant depth h
b) periodic waves with period T
c) restrict consideration to the 2D vertical plane (x, z)
d) neglect viscous (and turbulent) stresses so the motion becomes irrotational
e) assume the wave height H is much smaller than the wave length L

As described above, each of these assumptions contributes to simplification of the problem, but only the last two are also approximations.

The resulting wave solution has many names. Often it is called "small waves," "small amplitude waves" or "infinitesimal waves" because of the assumption of $H/L \ll 1$. More frequently "linear waves" or "sinusoidal waves" is used, the latter, of course, referring to the sine function which turns ou to describe the phase motion of these waves. In some contexts, they are also referred to as "Airy waves" because G. B. Airy (1845) was the first to derive the expressions describing these waves, or "first order Stokes waves" because G. G. Stokes (1847) was the first to derive a higher order theory.

3.2 Basic solution for linear waves

3.2.1 *Solution for ϕ and η*

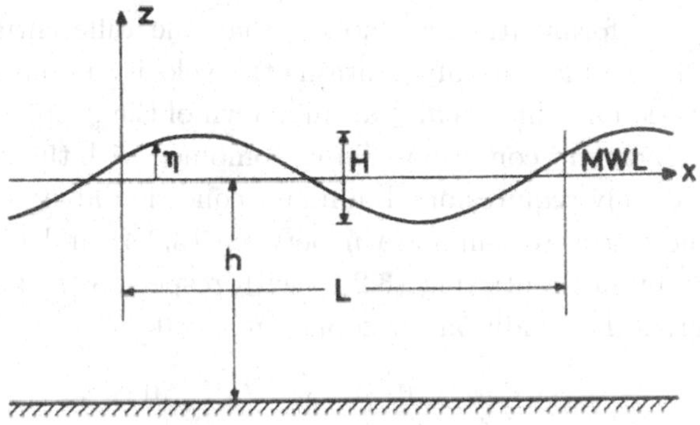

Fig. 3.2.1 Periodic waves on constant depth (SJ76).

The mathematical problem we are now facing consists of solving the Laplace equation

$$\phi_{xx} + \phi_{zz} = 0 \tag{3.2.1}$$

in the rectangular domain

$$0 \le x \le L \quad ; \quad -h \le z \le 0 \tag{3.2.2}$$

The boundary conditions are

a) The (kinematic) condition at the bottom

$$\phi_z = 0 \quad \text{at} \quad z = -h \tag{3.2.3}$$

b) The (linearized) kinematic condition at the mean water surface (MWL)

$$\phi_z - \eta_t = 0 \qquad \text{at} \qquad z = 0 \tag{3.2.4}$$

c) The (linearized) dynamic condition at the MWL

$$g\eta + \phi_t = 0 \qquad \text{at} \qquad z = 0 \tag{3.2.5}$$

d) The periodicity condition giving the boundary condition in the x-direction. As mentioned, we have chosen to impose this condition on the horizontal velocity.

$$\phi_x(0, z, t) = \phi_x(L, z, t) \qquad\qquad -h \leq z \leq 0 \tag{3.2.6}$$

To finalize the formulation we notice that the differential equation (3.2.1) governing the motion only contains the velocity potential ϕ, while the surface elevation η, which is also an unknown of the problem, occurs in the two surface boundary conditions. This, combined with the fact that the Laplace equation only requires one boundary condition along each boundary, makes it necessary to eliminate η between (3.2.4) and (3.2.5). This is accomplished by differentiating (3.2.5) with respect to t and adding to (3.2.4) which gives the condition for ϕ only at $z = 0$.

$$g\,\phi_z + \phi_{tt} = 0 \qquad \text{at} \qquad z = 0 \tag{3.2.7}$$

The physical situation is illustrated in Fig. 3.2.1, the mathematical formulation described above in Fig. 3.2.2.

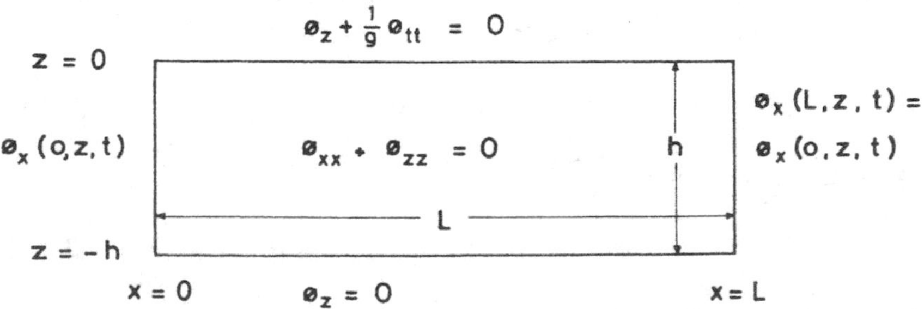

Fig. 3.2.2 The mathematical formulation of the problem (SJ76).

We see that the formulation now represents a regular boundary value problem for the Laplace equation with ϕ as unknown. We also note that this problem is linear (linear equation with linear boundary condition).

It turns out that solution to this problem can be expressed in different mathematical ways that also represent different types of wave motion. One is waves that propagate in either the positive or the negative x-direction without change of form ("progressive" waves), and at this point we choose to consider only that case. We, therefore, choose to further limit the analysis to

e) consider propagating waves of constant form.

The case of standing waves will be considered in Section 3.4.[4]

Waves of constant form

To utilize assumption f), we first examine the implications this has on the physical wave motion.

Fig. 3.2.3 shows the surface profiles of a wave of constant form at two different times, $t = 0$ and $t = t_0$. The wave is assumed to propagate in the positive x-direction with the speed c, which means that the horizontal distance in the picture between points with the same η is $x_0 = ct_0$. In particular, the value of η at point (x_0, t_0) will be the same as the value of η at $(0, 0)$ on

$$\eta(0,0) = \eta(x_0, t_0) \tag{3.2.8}$$

only if x_0 and t_0 are related by

$$x_0 - ct_0 = 0 \tag{3.2.9}$$

Recalling that $c = L/T$ this means that (3.2.8) is satisfied for all (x_0, t_0) provided

$$\frac{t_0}{T} - \frac{x_0}{L} = 0 \tag{3.2.10}$$

which therefore is the only possible combination of x and t that can occur in the description of constant form waves. In more general form, we define

[4]In fact, it is unnecessary at this point to introduce "progressive waves of constant form" as an assumption. The equations can be solved without further assumptions (see Exercise 3.2-7). It is, however, illustrative to see how the progressive, constant form assumption can be utilized to simplify the solution.

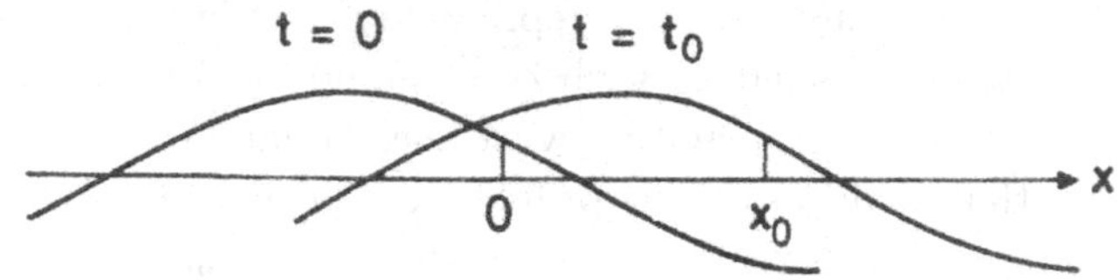

Fig. 3.2.3 Waves of constant form (SJ76).

the variable

$$\theta = 2\pi\left(\frac{t}{T} - \frac{x}{L}\right) \tag{3.2.11}$$

where the factor 2π is introduced to streamline the later results. θ is called the **phase angle**.

Exercise 3.2-1

Show that if we consider a wave that propagates in the negative x-direction, the combination of x and t becomes $t/T + x/L$.

We can therefore simplify the Laplace equation and the boundary conditions by replacing (x,t) with θ. In the Laplace equation, this means

$$\frac{\partial^2}{\partial x^2} = k^2 \frac{\partial^2}{\partial \theta^2} \tag{3.2.12}$$

where the constant

$$k = \frac{2\pi}{L} \tag{3.2.13}$$

is called **the wave number**. Thus, we get

$$k^2 \phi_{\theta\theta} + \phi_{zz} = 0 \tag{3.2.14}$$

Similarly the free surface condition (3.2.7) becomes

$$g\phi_z + \frac{\omega^2}{g}\phi_{\theta\theta} = 0 \tag{3.2.15}$$

where we have defined the wave frequency ω as

$$\omega = \frac{2\pi}{T} \tag{3.2.16}$$

Finally, the periodicity condition (3.2.6) can be written

$$\phi_\theta(0, z) = \phi_\theta(2\pi, z) \tag{3.2.17}$$

The bottom condition does not change. Thus, we have reduced the number of independent variables from three (x, z, t) to two (θ, z). We also see that we have

$$\theta = \omega t - kx \tag{3.2.18}$$

and

$$c = \frac{\omega}{k} \tag{3.2.19}$$

Exercise 3.2-2

Show by considering the original problem in (x, z, t) that this corresponds to saying: we can find a coordinate system (x', y, z) that moves in the horizontal direction with velocity c in which the wave does not change with time.

Solution for the velocity potential

It is now straightforward to determine the solution for ϕ. This can be done by means of the separation method which implies seeking solutions of the form

$$\phi(\theta, z) = f(\theta)\ Z(z) \tag{3.2.20}$$

Substituted into (3.2.14), this gives

$$k^2 f'' Z + f Z'' = 0 \tag{3.2.21}$$

where the $'$ implies ordinary differentiation. The standard approach is to divide (3.2.10) by fZ and write the result as

$$-k^2 \frac{f''}{f} = \frac{Z''}{Z} \tag{3.2.22}$$

In this equation, the left hand side is a function of θ only, the right hand side of z only. Thus, both must be equal to a constant, λ^2 say, which we here assume > 0. This means that the single equation (3.2.22) splits into two equations

$$f'' + \frac{\lambda^2}{k^2}\ f = 0 \tag{3.2.23}$$

and

$$Z'' - \lambda^2 Z = 0 \tag{3.2.24}$$

which have the complete solutions (assuming for simplicity that λ is the positive root of λ^2)

$$f = A_1 \sin\left(\frac{\lambda}{k}\,\theta\right) + A_2 \cos\left(\frac{\lambda}{k}\,\theta\right) \tag{3.2.25}$$

$$Z = B_1 e^{\lambda z} + C_1 e^{-\lambda z} \tag{3.2.26}$$

This form of f can be simplified if we realize that a sum of a sine and a cosine of the same variable can be written as a single sine (or cosine) with a different amplitude A and phase angle δ, i.e.,

$$f = A \sin\left(\frac{\lambda}{k}\;\theta + \delta\right) \tag{3.2.27}$$

However, since we can freely choose the origin of x and t (and therefore θ), we can put $\delta = 0$ without loss of generality.

Similarly, we can choose $B_1 = (B + C)/2$ and $C_1 = (B - C)/2$ where B and C are two other constants. This changes Z to the more convenient form

$$Z = B \cosh \lambda z + C \sinh \lambda z \tag{3.2.28}$$

In addition to λ, the constants A, B and C are then the integration constants that need to be determined by means of the boundary conditions.

The periodicity condition (3.2.17) becomes $f'(0) = f'(2\pi)$ which we see from (3.2.27) will only be satisfied if we require

$$\lambda = k \tag{3.2.29}$$

Exercise 3.2-3

Show that since by definition one wave length corresponds to $\theta = 2\pi$, solutions like $\lambda = 2k,\ 3k, \cdots$ are not allowed.

The bottom condition (3.2.3) applied to (3.2.20) gives

$$Z'(-h) = 0 \tag{3.2.30}$$

which with (3.2.28) for Z gives

$$B = C \coth kh \tag{3.2.31}$$

Substituted into (3.2.28), that equation can then be written

$$\begin{aligned} Z &= \frac{C}{\sinh kh}(\cosh kh \cosh kz + \sinh kh \sinh kz) \\ &= C\,\frac{\cosh k(z+h)}{\sinh kh} \end{aligned} \tag{3.2.32}$$

At this point, the solution (3.2.20) for ϕ can then be written as

$$\phi = AC\,\frac{\cosh k(z+h)}{\sinh kh}\sin\theta \tag{3.2.33}$$

Phase velocity, surface profile and velocity potential

The only boundary condition left for ϕ is the combined kinematic and dynamic condition (3.2.15). When we substitute the z and t derivatives of (3.2.33) into this, we get (after division by $AC\sin\theta$)

$$k - \frac{\omega^2}{g}\coth kh = 0 \tag{3.2.34}$$

which can also be written

$$\boxed{\omega^2 = gk \tanh kh} \tag{3.2.35}$$

This is called the **dispersion relation** for reasons which will become clear later. It essentially specifies the correlation between ω and k (or between L and T). Since we also have that $c = \omega/k$, (3.2.35) can also be written

$$\boxed{c^2 = \frac{g}{k}\tanh kh} \tag{3.2.36}$$

which gives the value of the phase velocity (or propagation velocity or "phase speed") c for given water depth h and wave length $L(= 2\pi/k)$.

Exercise 3.2-4

What will the expression (3.2.36) be for a wave moving in the negative x-direction?

Exercise 3.2-5 Why does c not depend on the wave height H?

The surface elevation η was initially eliminated from the two free surface boundary conditions. Therefore, η can now be determined from either of those and we see the dynamic condition (3.2.6) is the simplest to use. Using (3.2.33), we get directly

$$\eta = -\frac{1}{g}\phi_t = -AC\,\frac{\omega}{g}\coth kh \cos\theta \tag{3.2.37}$$

If we substitute (3.2.34) and (3.2.19), this can also be written

$$\eta = -\frac{AC}{c}\cos\theta \tag{3.2.38}$$

Obviously this means that the parameter combination $-AC/c$ represents the wave amplitude a or half the wave height H.

This brings us to the question, which will be discussed further in the next section: What is required to specify the wave? It is in accordance with intuitive physical understanding of the problem that the result describes the motion of a wave of any amplitude a or height H, provided we stay within the limits of $H/L \ll 1$ which was a basic assumption for the mathematical formulation. Hence, we are free to choose this combination of constants. We define

$$a = \frac{H}{2} = -\frac{AC}{c} \tag{3.2.39}$$

Then the surface elevation can be written

$$\boxed{\eta = a\cos\theta = \frac{H}{2}\cos(\omega t - kx)} \tag{3.2.40}$$

Substituting (3.2.39) into (3.2.33) then also gives directly the final expression for the velocity potential ϕ:

$$\boxed{\phi = -\frac{Hc}{2}\,\frac{\cosh k(z+h)}{\sinh kh}\,\sin(\omega t - kx)} \tag{3.2.41}$$

With η and ϕ determined we have essentially determined the solution for linear (or sinusoidal) water waves. As expected, the parameters are

$$h : \text{the water depth}$$
$$H : \text{the wave height}$$
$$L : \text{the wave length}$$

However, in addition both the propagation speed c and the wave period T occur in the solution. It is therefore obvious that some further analysis is required to clarify how to properly specify the wave motion for practical applications. This is done in Chapter 3.2.2.

It may be recalled that we initially obtained the simplified formulation of the equations by assuming $H/L << 1$ and consequently evaluated the two free surface boundary conditions at the MWL $z = 0$ instead of at $z = \eta$. This implies, however, that the waves described by the solution have (at least theoretically) an infinitely small wave height, and that the results we have found for ϕ and all the variables that can be derived from that are only valid up to $z = 0$.

Exercise 3.2-6

Show that the solution for ϕ can also be written as

$$\phi = -\frac{H}{2}\frac{g}{\omega}\frac{\cosh\, k(z+h)}{\cosh\, kh}\sin\theta. \tag{3.2.42}$$

This form will be used in e.g. Chapters 3.5.4 and 3.7.

Exercise 3.2-7 To illustrate the effect of the assumption of constant form waves, this exercise briefly outlines the solution of the equations without making that assumption.

Thus, assume in the Laplace (3.2.1) that

$$\phi = T(t)\;\; X(x)\;\; Z(z) \tag{3.2.43}$$

a) Show that his leads to the equivalent to (3.2.22)

$$\frac{X''}{X} = -\frac{Z''}{Z} = -\lambda^2 \tag{3.2.44}$$

with the general solutions

$$X = A_1 \sin \lambda x + A_2 \cos \lambda x \tag{3.2.45}$$

$$Z = B_1 e^{\lambda z} + B_2 e^{-\lambda z} \tag{3.2.46}$$

b) Show that with suitable definitions of constants, the bottom condition (3.2.3) brings Z on the form

$$Z = B' \frac{\cosh \lambda(z+h)}{\sinh \lambda h} \tag{3.2.47}$$

and that the periodicity condition again leads to

$$\lambda = k \tag{3.2.48}$$

so that

$$Z = B' \frac{\cosh k(z+h)}{\sinh kh} \tag{3.2.49}$$

c) Show that the combined free surface condition then yields

$$g \frac{Z'}{Z} = -\frac{T''}{T} \tag{3.2.50}$$

where

$$g \frac{Z'}{Z'} = gk \tanh kh \equiv \omega^2 \tag{3.2.51}$$

in which the second equality defines the constant ω. Thus, show that this implies that

$$T = C_1 \sin \omega t + C_2 \cos \omega t \tag{3.2.52}$$

d) Define constants A, B, C and D so that the total ϕ can be written

$$\phi = \{A \sin \omega t \sin kx + B \cos \omega t \sin kx\} \frac{\cosh k(z+h)}{\sinh kh} + \{C \sin \omega t \cos kx + D \cos \omega t \cos kx\} \frac{\cosh k(z+h)}{\sinh kh} \tag{3.2.53}$$

which is the general solution.

e) Show that any one of the four constants $\neq 0$ together with the others $= 0$ will give one of the standing waves considered in Section 3.4.1.

f) Show that $C = -B = a, \quad A = D = 0$ will give the progressive wave

$$\phi_a = a \frac{\cosh k(z+h)}{\sinh kh} \sin(\omega t - kx) \tag{3.2.54}$$

and $C = B = b, \; A = D = 0$ gives

$$\phi_b = b \frac{\cosh k(z+h)}{\sinh kh} \sin(\omega t + kx) \tag{3.2.55}$$

which are progressive waves in the + and − x-directions, respectively.

Thus by suitable choice of the four integration constants the general solution contains all these options.

As can also be seen, however, the solution procedure is more complicated without the constant form assumption.

The solution presented above is based on utilizing that a frictionless flow that initially is irrotational will remain irrotational (Kelvin's law), and hence the flow can be described by a velocity potential. However, such a flow will also satisfy the Euler equations in combination with the continuity equation.

The linearized form of the Euler equations are (introducing the dynamic pressure $p_D = p + \rho g z$)

$$\frac{\partial u}{\partial x} + \frac{\partial w}{\partial z} = 0 \tag{3.2.56}$$

$$\frac{\partial u}{\partial t} + \frac{1}{\rho}\frac{\partial p_D}{\partial x} = 0 \tag{3.2.57}$$

$$\frac{\partial w}{\partial t} + \frac{1}{\rho}\frac{\partial p_D}{\partial z} = 0 \tag{3.2.58}$$

In addition, we want the flow to be irrotational so we must also require

$$\frac{\partial u}{\partial z} - \frac{\partial w}{\partial x} = 0 \tag{3.2.59}$$

It can be shown that derivation of the linear wave results from this set of equations is also possible.

Exercise 3.2-8

Frequently the formulas for the wave motion (surface elevation, velocity potential or the velocity components themselves) are written as the Real value of a complex expression. For example

$$\eta = \Re\{a\, e^{i\theta}\} \tag{3.2.60}$$

Show that this is equivalent to (3.2.40). Show also that if ve write

$$\eta = a\, e^{i\theta} + c.c \tag{3.2.61}$$

where *c.c.* stands for the complex conjugate of the $e^{i\theta}$, then that also sorresponds to (3.2.40).

Often the $\Re$ or the *c.c* is omitted in the derivations for brievity, but is understood throughout. In particular it is meant that we take the Real value of the final complex function results to get the actual physical result.

This way of writing the wave motion makes it possible to use the powerful methods of complex variable theory, which often simplifies the derivations (see e.g. Chapters 3.5.2 and 3.7).

3.2.2 *Evaluation of linear waves*

As (3.2.41) shows, in addition to the water depth h, we need H, L, c and T to specify the wave motion – a total of 5 parameters. However, these parameters are linked by two important relations which must be satisfied. One is the definition of c

$$c = \frac{L}{T} \tag{3.2.62}$$

the other is the dispersion relation (3.2.35). Thus, we can only choose three of the five parameters.

Therefore, the question becomes how to determine the necessary parameters when the wave is specified by various combinations of the five. This choice, however, is not completely free. H must always be specified in addition to two other parameters and in practical applications the depth h is usually known as well. Here we discuss only the two most common combinations

a) The wave specified by h, H and L
b) The wave specified by h, H and T.

The simplest is case a), which literally provides the frame within which the surface profile varies. In the mathematical formulas, the problem also is very simple: when L is known, so is the wave number k, and along with h this means ω and hence the wave period T follow directly from the dispersion relation (3.2.35). c is then obtained from (3.2.62) and thereby all parameters are known.

Exercise 3.2-9 List the reasons why the wave scannot be specified uniquely by the combination h, L and T?

Case b) is more difficult because we now need to determine L. This is done from the dispersion relation (3.2.35). However, with ω and h known, this is a transcendental equation for L, which can only be solved iteratively. The most convenient way is to write (3.2.35) in the dimensionless form

$$kh = \frac{\omega^2 h}{g} \coth kh \tag{3.2.63}$$

which is then solved for kh. We see that this equation has one parameter only, namely $\omega^2 h/g$ which can also be written as

$$\frac{\omega^2 h}{g} = (2\pi)^2 \left(T\sqrt{g/h}\right)^{-2} \tag{3.2.64}$$

Hence $T\sqrt{g/h}$ is a convenient dimensionless wave period.

Exercise 3.2-10 Eq. (3.2.63) yields to direct iteration which is also straightforward to program on a computer.

To demonstrate the procedure, perform a simple hand calculation based on the iteration

$$(kh)_{n+1} = \frac{\omega^2 h}{g} \coth(kh)_n \tag{3.2.65}$$

in which the previous (n'th) approximation for kh is used on the RHS to obtain the next $(n+1)'th$ approximation. Start with the approximation

$$(kh)_1 = \frac{\omega^2 h}{g} \tag{3.2.66}$$

(corresponding to $\coth kh = 1$ or $kh = \infty$), substitute into the right hand side and calculate $(kh)_2$ by (3.2.65), etc.

This method converges rapidly for all values of $\omega^2 h/g (> 0)$ (see Fig. 3.2.4) and it is much simpler to program than for example a Newton-Raphson iteration which is sometimes used.

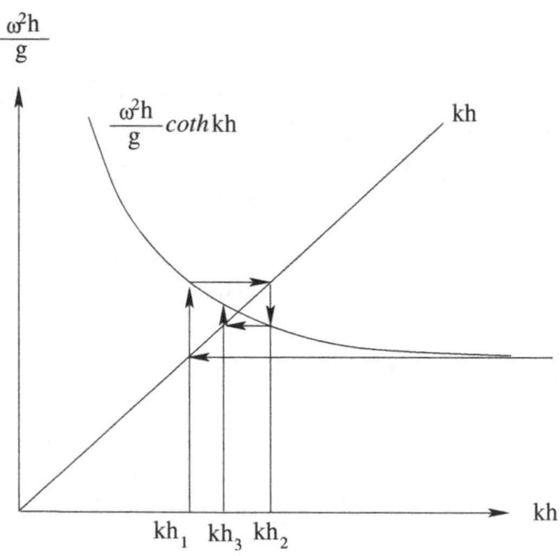

Fig. 3.2.4 The iterative solution of (3.2.64). The solution is at the crossing of the two curves, the straight line representing the left hand side of the equation, the coth-curve the right hand side. The figure shows graphically how the iteration procedure progresses. The initial estimate of $\coth kh = 1$ leads to a new estimate kh_2, etc.

Variation of L and c

Interesting questions are: how do L and c vary with wave period T and with depth h? This is discussed below.

We see that the dispersion relation (3.2.35) can also be written

$$L = \frac{gT^2}{2\pi} \tanh kh \tag{3.2.67}$$

which is useful for analyzing the variation of the wave length L versus T and h. Note that since $L = cT$ the equivalent formula for c simply reads

$$c = \frac{gT}{2\pi} \tanh kh = \frac{g}{\omega} \tanh kh \tag{3.2.68}$$

Deep and shallow water approximations for L and c

We first consider the extreme case of very large depth h (relative to L). Then kh is large, $\tanh 2\pi h/L \sim 1$ and the wave length L is given by

$$L(h \to \infty) \equiv L_0 = \frac{gT^2}{2\pi} \tag{3.2.69}$$

This is called 'deep water,' L_0 the deep water wave length, and we see that 'deep water' means h large enough relative to L to make $\tanh kh \sim 1$ a good approximation. For the phase velocity c, we similarly get

$$c(h \to \infty) \equiv c_0 = \frac{gT}{2\pi} \tag{3.2.70}$$

which shows that in deep water, the wave length and the phase speed only depends on T.

Notice that for general kh, we can write

$$L = L_0 \tanh kh \tag{3.2.71}$$

and similarly

$$c = c_0 \tanh kh \tag{3.2.72}$$

The other extreme is the situation where $h \ll L$. This is termed "shallow water" and we see it implies that $\tanh kh \sim kh$ so that

$$L = T\sqrt{gh} \tag{3.2.73}$$

The equivalent expression for c is

$$c = \sqrt{gh} \tag{3.2.74}$$

which shows that in shallow water, the phase speed depends only on the depth.

> **Exercise 3.2-11** Prove that (3.2.73) and (3.2.74) correspond to kh is so small that $\tanh kh$ is given by the first term in its Taylor expansion.

Variation of L and c with T, h constant

Since both L (and c) are given as transcendental functions of T, assessing the variation with T can only be done by first solving (3.2.64)) and then determine L and c from (3.2.67) and (3.2.68). The result is shown in Fig. 3.2.5 for L/h in terms of $T\sqrt{g/h}$. The figure also shows the deep and shallow water approximations for L/h.

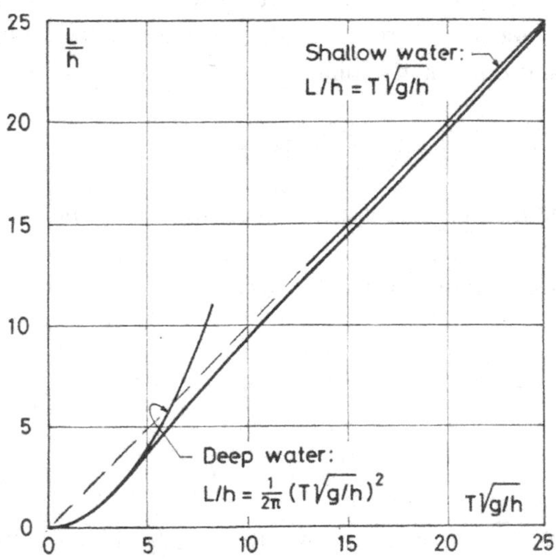

Fig. 3.2.5 L/h versus $T/\sqrt{gh}$. The figure also shows the deep and shallow water approximations (SJ76).

Frequency dispersion

The relation (3.2.68) shows that in general the propagation speed c depends on the wave period T and it turns out that c grows monotonically with T.

The most direct way to see this is to determine $\partial c/\partial\omega$ and show it is ≤ 0 meaning for given h c increases when ω decreases, that is it increases with T increasing. Eliminating k from (3.2.68) gives

$$c = \frac{g}{\omega} \tanh \frac{\omega h}{c} \tag{3.2.75}$$

Differentiating this with respect to ω while keeping $h = const$ we then get

$$\frac{\partial c}{\partial \omega} = -\frac{c}{\omega} \frac{1-G}{1+G} \tag{3.2.76}$$

where G is given by

$$G = \frac{2kh}{\sinh 2kh} \tag{3.2.77}$$

Since $0 < G \leq 1$ we see that $\partial c/\partial\omega \leq 0$ for all kh, meaning c increases momotonically with T

Hence, waves with different periods will propagate with different speeds. In fact, the longer the wave period, the faster the propagation speed. This applies until the wave gets so long that it becomes a shallow water wave ($=> G = 1$). Then, $c = \sqrt{gh}$ is independent of the wave period.

The property that cc varies with ω is called **frequency dispersion**, and, as we just saw, shallow water waves are **not** frequency dispersive.

Exercise 3.2-12

Derive (3.2.76)

Variation of L and c with h; T constant

As (3.2.67) and (3.2.68) show, the variation with h is seemingly straightforward. For T constant, both c and L vary as $\tanh kh$. However, $kh = 2\pi h/L$ is in itself a function of h/L. Hence, we again need to resort to a full numerical solution of these equations to get the full picture.

By using (3.2.71), we see that h/L in kh is given by

$$\frac{h}{L_0} = \frac{h}{L} \tanh kh \tag{3.2.78}$$

where for given T, the left hand side is an explicit function of h that varies monotonously with kh. Hence, dimensionless plots of L and c can be generated as functions of h/L_0. Fig. 3.2.6 shows the variation of L/L_0 and c/c_0 versus h/L_0. It confirms that for given wave period T, phase speed as well as wave length grows monotonously with h.

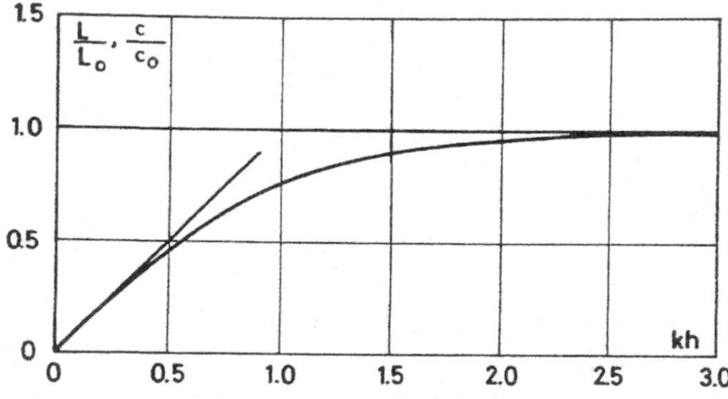

Fig. 3.2.6 L/L_0 and c/c_0 versus h/L_0 (SJ76).

3.2.3 *Particle motion*

Once the velocity potential ϕ is known, we in principle know everything about the wave motion. Thus, we can determine the complete velocity field, the particle motion and the pressure variation created by the wave motion.

The velocity field

The horizontal and vertical velocity components (u, w) are determined directly by partial differentiation of ϕ with respect to (x, z). The results can be written in different forms, but from (3.2.41), we get directly

$$u = \phi_x = c\frac{kH}{2}\,\frac{\cosh k(z+h)}{\sinh kh}\,\cos(\omega t - kx) \tag{3.2.79}$$

$$w = \phi_z = -c\frac{kH}{2}\,\frac{\sinh k(z+h)}{\sinh kh}\,\sin(\omega t - kx) \tag{3.2.80}$$

We see that (apart from the z an θ variations) $kH/2 = ka$ represents u/c. Other forms are obtained by replacing $ckH/2$ with $\pi H/T$ (which refers to the deep water value) or (less illustratively) with $a\omega$.

Exercise 3.2-13

Show that if we instead consider a wave that propagates in the negative x-direction (i.e., $\theta = \omega t + kx$), then u changes sign while w does not.

Exercise 3.2-14

Show that in linear waves, the instantaneous local volume flux Q over a vertical is given by

$$Q = \int_{-h}^{\eta} u dz = \int_{-h}^{0} u dz + O(H^2)$$

$$= c\eta + O(H^2) \tag{3.2.81}$$

Fig. 3.2.7 shows the velocity variations in a linear progressive wave illustrated by the vertical profiles of the horizontal velocities under the crest of the wave and by the vertical profiles at the phase angle where $\eta = 0$. Similarly Fig. 3.2.8 shows an instantaneous picture of the velocity field. For illustrative purposes, the surface elevation has been exaggerated.

Exercise 3.2-15

Show that u can also be written as

$$u = c\frac{\eta}{h} \cdot \frac{kh \cosh k(z+h)}{\sinh kh} \tag{3.2.82}$$

where η is the surface elevation given by (3.2.40). Thus u is in phase with η and consequently the largest particle velocities in the direction of wave propagation occur under the wave crest, the largest in the opposite direction are found under the wave trough.

Similarly, show that w can be written as

$$w = -c\eta_x \frac{\sinh k(z+h)}{\sinh kh} \tag{3.2.83}$$

which shows the vertical velocity is largest where the slope on the surface is largest.

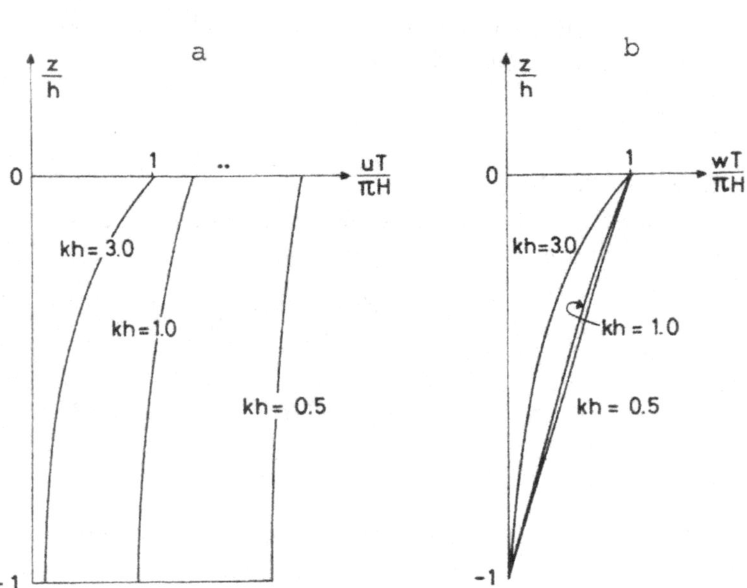

Fig. 3.2.7 The vertical velocity profiles in a linear progressive wave illustrated by the vertical profiles of the horizontal velocities under the crest (a) and profiles of the vertical velocities at the phase angle where $\eta = 0$ (b) (SJ76).

The particle paths

The description of the velocity field given above provides information on the velocity at any fixed point below MWL at any time (**Eulerian** description). This, however, does not indicate how the individual fluid particles move, though it may be sensed from Fig. 3.2.8 that the particles describe some sort of orbital motion.

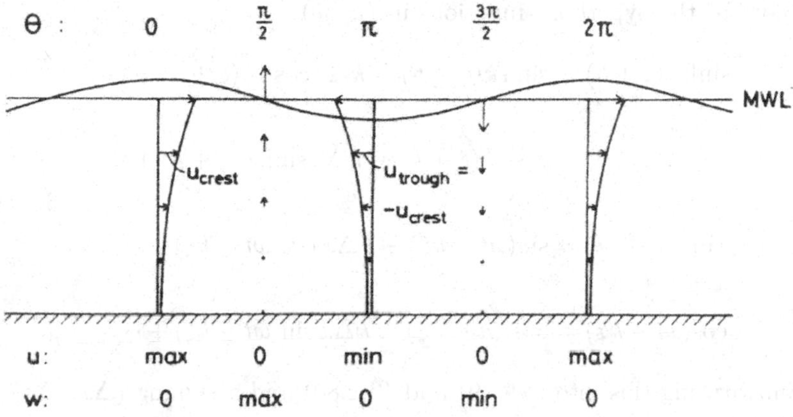

Fig. 3.2.8 The instantaneous velocity vectors in a linear progressive wave motion. For illustrative purposes, the surface elevation has been exaggerated (SJ76).

As described in Section 2, the particle paths are described by solving the equations

$$\frac{dx}{dt} = u(x(t), z(t), t) \qquad\qquad \frac{dz}{dt} = w(x(t), z(t), t) \tag{3.2.84}$$

with u, w given by (3.2.79) and (3.2.80), and where the RHS depends on t both directly and through $x(t)$, $z(t)$. This gives a **Lagrangian** description of the motion.

These equations cannot be solved exactly, because when we sustitute the results for (u, w) we find tht the unknowns (x, z) appear both outside and inside the transcendental functions of cos, cosh etc. On the other hand, we have already made assumptions about small wave height which suggests that the orbits indicated above let the particles move in small excursions Δx, Δz around a mean particle position (ξ, ζ). We can utilize this by writing

$$\begin{aligned} x &= \xi + \Delta x(t) \\ z &= \zeta + \Delta z(t) \end{aligned} \tag{3.2.85}$$

and further introduce the Taylor expansion of $(u,\ w)$ from the point $(\xi,\ \zeta)$.

We have for the hyperbolic functions in $(u,\ w)$

$$\begin{aligned}
\sinh k(z+h) &= \sinh k(\zeta+h) + k\Delta z \cosh k(\zeta+h) + \cdots \\
\cosh k(z+h) &= \cosh k(\zeta+h) + k\Delta z \sinh k(\zeta+h) + \cdots \\
\sin(\omega t - kx) &= \sin(\omega t - k\xi) - k\Delta x \cos(\omega t - k\xi) + \cdots \\
\cos(\omega t - kx) &= \cos(\omega t - k\xi) + k\Delta x \sin(\omega t - k\xi) + \cdots
\end{aligned} \tag{3.2.86}$$

Substituting this into (3.2.79) and (3.2.80) and assuming $(\Delta x,\ \Delta z) = O(H)$ we get

$$u(x,z,t) = \frac{Hc}{2}\,k\,\frac{\cosh k(\zeta+h)}{\sinh kh}\,\cos(\omega t - k\xi) + O(H^2) \tag{3.2.87}$$

$$w(x,z,t) = -\frac{Hc}{2}\,k\,\frac{\sinh k(\zeta+h)}{\sinh kh}\,\sin(\omega t - k\xi) + O(H^2) \tag{3.2.88}$$

This shows that to $O(H^2)$ the velocity for a particle at point (x,z) can be approximated by the velocity at the mean position (ζ,ξ) of that particle. Thus, (3.2.84) can be written

$$\frac{d\Delta x}{dt} = u(\xi,\zeta,t) + O(H^2) \quad ; \quad \frac{d\Delta z}{dt} = w(\xi,\zeta,t) + O(H^2) \tag{3.2.89}$$

where the RHS's now only depend on t explicitly. Neglecting terms $O(H^2)$ the equations can then be solved by simple integration to give (using also (3.2.85))

$$x = \xi + \frac{H}{2}\,\frac{\cosh k(\zeta+h)}{\sinh kh}\,\sin(\omega t - k\xi) \tag{3.2.90}$$

$$z = \zeta + \frac{H}{2}\,\frac{\sinh k(z+h)}{\sinh kh}\,\cos(\omega t - k\xi) \tag{3.2.91}$$

Exercise 3.2-16 Show that the particle path given by (3.2.90) and (3.2.91) is an ellipse with center in the mean particle position (ξ,ζ) and the major semi-axis.

$$\alpha = \frac{H}{2}\,\frac{\cosh k(\zeta+h)}{\sinh kh} \tag{3.2.92}$$

which is also the horizontal particle excursion or displacement amplitude. Show, similarly, the minor axis or vertical particle amplitude is

$$\beta = \frac{H}{2} \frac{\sinh k(\zeta + h)}{\sinh kh} \tag{3.2.93}$$

Show also that the horizontal distance between the focal points of the ellipse is constant over the depth and equals $\frac{H}{\sinh kh}$.

The particle paths along a vertical line under the wave are shown in Fig. 3.2.9. We see that at the surface ($\zeta = 0$), the horizontal amplitude is $\frac{H}{2} \coth kh$ (which is always $\geqq \frac{H}{2}$) while the vertical amplitude is $\frac{H}{2}$ for all kh (as the kinematic surface condition would dictate).

At the bottom, we of course have $\beta = 0$ but also $\alpha_b = \frac{1}{2 \sinh kh}$. Thus, at the bottom the ellipse degenerates into a horizontal line: the particles move back and forth along the bottom as dictated by the bottom boundary condition.

3.2.4 *The pressure variation*

As always in potential flow problems, the pressure can be determined from the Bernoulli equation. Neglecting again terms of $O(H^2)$ and introducing that $p = 0$ at $z = 0$, we get

$$gz + \frac{p}{\rho} + \phi_t = 0 \tag{3.2.94}$$

or

$$p = -\rho g z - \rho \phi_t \tag{3.2.95}$$

We see that p (as could be expected) consists of a hydrostatic component ($-\rho g z$) and a component generated by the wave motion ($-\rho \phi_t$). Since the hydrostatic component does not contribute to the motion, it is convenient to define the **dynamic pressure** p_D as the part generated by the wave motion

$$p_D = p + \rho g z \tag{3.2.96}$$

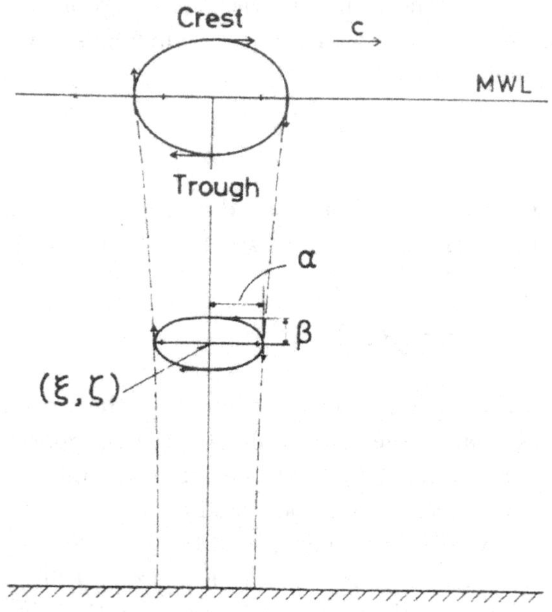

Fig. 3.2.9 The particle paths in a progressive linear wave motion (SJ76).

which shows that

$$p_D = -\rho\phi_t \tag{3.2.97}$$

Substituting for ϕ we then get

$$p_D = \rho g \frac{H}{2} \frac{\cosh k(z+h)}{\cosh kh} \cos(\omega t - kx) \tag{3.2.98}$$

We see that p_D can also be written

$$p_D = \rho g \eta \frac{\cosh k(z+h)}{\cosh kh} \tag{3.2.99}$$

Thus, the pressure generated at a point by the wave motion is proportional to the surface elevation vertically above that point. In fact, at the MWL we have

$$p_D(z=0) = \rho g \eta \tag{3.2.100}$$

which shows that at the MWL the pressure simply corresponds to the weight of the water above. As (3.2.99) also shows, the dynamic pressure is largest at the surface and decreases downward with the factor $\cosh k(z+h)/\cosh kh$ which is ≤ 1 for all relevant values of z. At the bottom, the value of p_D is $\rho g\eta/\cosh kh$.

Exercise 3.2-17 Show that the vertical projection of the Euler equation can be written (in linearized form)

$$\frac{\partial w}{\partial t} = -\frac{1}{\rho}\frac{\partial p_D}{\partial z} \tag{3.2.101}$$

Substitute (3.2.80) for w and use the equation to derive (3.2.98) for p_D.

It is emphasized that the result of Exercise 3.2-17 shows that in linear waves **the dynamic pressure is caused entirely by the vertical accelerations in the wave motion.**

Exercise 3.2-18

Use (3.2.101) to determine $\partial p_D/\partial z$ at $z = -h$. Compare the result to 3.2.99.

Fig. 3.2.10 illustrates the pressure variations along a vertical. The two straight lines indicate the hydrostatic pressure corresponding to $z = 0$ and $z = \eta$, and between them we find the dynamic pressure which decreases from $\rho g\eta$ at the surface to a value at the bottom that depends on kh.

As with the velocities, we notice that the theory provides no information about the pressure between $z = 0$ and $z = \eta$. However the pressure at the MWL corresponds to the estimate that the pressure above that level is hydrostatic.

A frequent way of measuring wave heights in the field is to place a pressure sensor at or near the bottom. That will measure pressure variations that can be transformed into wave height variations using (3.2.98) if the value of k is known. Usually k is found by determining the wave period T from the time series of the recording and solving the dispersion relation for k.

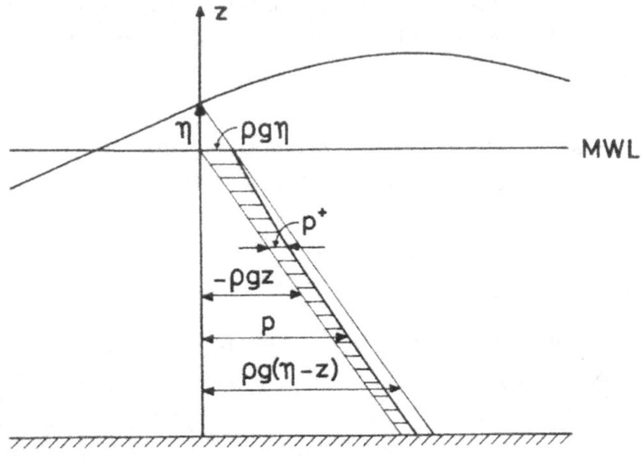

Fig. 3.2.10 The pressure variations along a vertical (SJ76).

3.2.5 *Deep water and shallow water approximations*

In section 3.2.2 we briefly discussed the deep and shallow water approximations for the parameters L and c. Here we analyze the details of the motion in the case of very large, or infinitely large water depth, and the case of very small water depth (both relative to the wave length). In both these cases, the expressions found for sinusoidal properties such as ϕ, c, u, w, p_D, etc. will simplify somewhat. However, in both cases there are also special features and pitfalls worth emphasizing. And in all cases, the surface profile remains sinusoidal with η given by (3.2.40).

Deep water waves

The deep water approximation for linear water waves is obtained by assuming that the water depth becomes very large (ideally infinitely large) relative to the wave length L, that is, $kh \gg 1$.

In deep water the hyperbolic functions in the results change as follows:

$$\sinh kh = \frac{e^{kh} - e^{-kh}}{2} \to \frac{1}{2}e^{kh} \tag{3.2.102}$$

$$\cosh kh = \frac{e^{kh} + e^{-kh}}{2} \to \frac{1}{2}e^{kh} \tag{3.2.103}$$

$$\tanh kh = \frac{e^{kh} - e^{-kh}}{e^{kh} + e^{-kh}} \to 1 \tag{3.2.104}$$

$$G = \frac{2kh}{\sinh 2kh} \to 0 \tag{3.2.105}$$

Fig. 3.2.11 shows the variation of the hyperbolic functions.

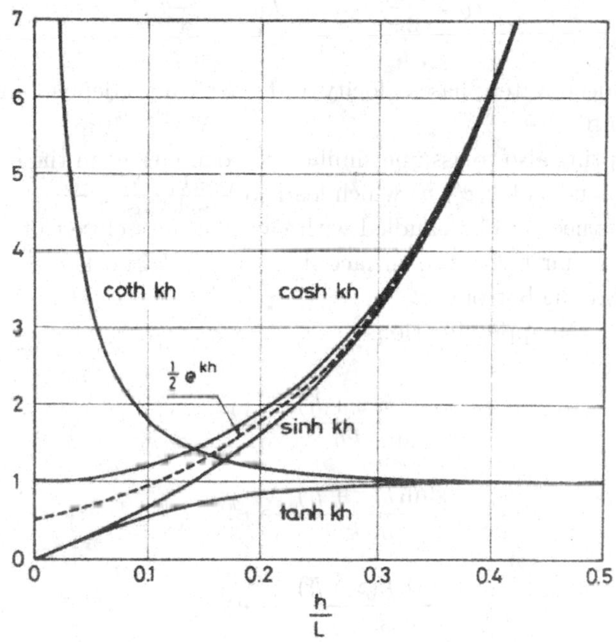

Fig. 3.2.11 Hyperbolic functions versus h/L (SJ76).

We already saw that in deep water, the dispersion relation reduces to

$$\omega^2 = gk_0 \quad \text{or} \quad k_0 = \omega^2/g \tag{3.2.106}$$

where we follow traditional convention and mark deep water values by index $_0$.

Similarly, for the phase velocity, we get

$$c_0^2 = \frac{g}{k_0} \tag{3.2.107}$$

and as mentioned in this case we can also get an explicit expression for $c(T)$ and $L(T)$ by eliminating k_0 from (3.2.106) and (3.2.107).

$$c_0 = \frac{gT}{2\pi} \quad ; \quad L_0 = \frac{gT^2}{2\pi} \tag{3.2.108}$$

Thus, in deep water phase velocity and wave length depend on the wave period, T, only.

It is tempting also to assume similar approximations in the expressions $\cosh k(z+h)$ and $\sinh k(z+h)$ which lead to $\frac{\cosh k(z+h)}{\sinh kh}$, $\frac{\sinh k(z+h)}{\sinh kh} \to e^{kz}$. However, this needs to be handled with care, because close to the bottom, no matter how far below the surface it is, $z \to -h$ and hence $k(z+h)$ is **not** large near the bottom.

Therefore, the approximations

$$\frac{\cosh k(z+h)}{\sinh kh} \to e^{kz} \tag{3.2.109}$$

$$\frac{\sinh k(z+h)}{\sinh kh} \to e^{kz} \tag{3.2.110}$$

$$\frac{\cosh k(z+h)}{\cosh kh} \to e^{kz} \tag{3.2.111}$$

are only valid in the upper part of the water column. Though the errors from using these approximations near the bottom are small in absolute value, they become as large as 50% of the correct values which can be important. This implies for example that the expression (3.2.109) should never be used for p_D when determining wave heights from a pressure cell placed near the bottom.

Exercise 3.2-19

Show that the uniformly valid deep water approximations are

$$\frac{\cosh k(z+h)}{\sinh kh} \to e^{kz}\left(1+e^{-2k(z+h)}\right) \quad (3.2.112)$$

$$\frac{\sinh k(z+h)}{\sinh kh} \to e^{kz}\left(1-e^{-2k(z+h)}\right) \quad (3.2.113)$$

$$\frac{\cosh k(z+h)}{\cosh kh} \to e^{kz}\left(1+e^{-2k(z+h)}\right) \quad (3.2.114)$$

which can then be used to simplify the z=dependent variables such as ϕ, u, w and p_D. Show also that for reasonaly large values of $kh(\gtrsim\frac{1}{2})$ these expressions are very close approximations to the full hyperbolic functions for the entire depth $-h < z < 0$.

The particle paths also become simpler in deep water waves. From the results presented in Section 3.2.3 for the elliptic shape with axes α and β, we realize that in deep water $\alpha = \beta$ so the ellipses degenerate to circles. Fig. 3.2.12 shows an example of the variation of the particle paths in a deep water wave, and the values of α. The figure also illustrates the basic rule that at a depth of $z = -L/2$, the effect of the wave motion is negligible.

Exercise 3.2-20 Show that in a deep water wave the instantaneous position $x(t)$, $z(t)$ of a particle with mean position $(\xi,\ \zeta)$ is given by

$$x(t) = \xi + \frac{H}{2}\ e^{k\zeta}\left(1+e^{-2k(\zeta+h)}\right)\sin\theta \quad (3.2.115)$$

$$z(t) = \zeta + \frac{H}{2}e^{k\zeta}\left(1-e^{2k(\zeta+h)}\right)\cos\theta \quad (3.2.116)$$

The deep water limit

The limit for the region in which the deep water approximations are usually considered valid is $h/L > 1/2$. Use of the deep water approximations at that limit causes errors relative to the general formulas that depend on the wave parameter considered. For the phase speed c, the error at $h/L = 1/2$ is $+0.4\%$ (c_0 is larger).

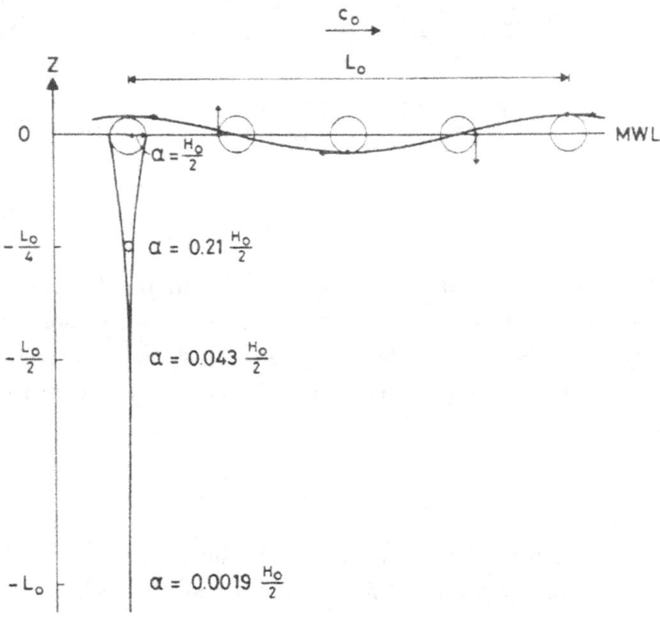

Fig. 3.2.12 Particle paths in deep water (SJ76).

The deep water limit for typical short waves can be said to be the limit, shoreward of which the waves begin to "feel the bottom." Thus, a typical storm wave of $T = 8$ s will have a deep water wave length of $L_0 \simeq 100m$, and hence a deep water limit of $h = 50$ m. This could also be a reasonable definition of the region termed "Nearshore" in the title of this book, though obviously this can change with the application.

The shallow water approximation

In the shallow water approximation, the waves are assumed to be much longer than the water depth that is $L/h \gg 1$.

As mentioned earlier this simplifies the dispersion relation to

$$\omega^2 = gk^2h \tag{3.2.117}$$

and the phase velocity to

$$c = \sqrt{gh} \qquad\qquad L = T\sqrt{gh} \tag{3.2.118}$$

Here we see right away that the phase speed is only a function of the depth, independent of T. The waves are nondispersive as already shown in section 3.2.2.

The shallow water assumption leads to straightforward approximations for the hyperbolic functions in the results

$$\sinh kh \sim kh \tag{3.2.119}$$

$$\cosh kh \sim 1 \tag{3.2.120}$$

$$\tanh kh \sim kh \tag{3.2.121}$$

$$G = \frac{2kh}{\sinh 2kh} \sim 1 \tag{3.2.122}$$

Since now $k(z+h)$ must also be small when kh is, we can introduce similar approximations for the z dependent functions in the results for ϕ, u, w, etc.

$$\frac{\cosh k(z+h)}{\sinh kh} \sim \frac{1}{kh} \quad (\to \infty \text{ for } kh \to 0) \tag{3.2.123}$$

$$\frac{\sinh k(z+h)}{\sinh kh} \sim 1 + \frac{z}{h} \tag{3.2.124}$$

$$\frac{\cosh k(z+h)}{\cosh kh} \sim 1 \tag{3.2.125}$$

For the z-direction variables, we get that the approximation is

$$u = c\frac{H}{2h}\cos\theta = \frac{\pi H}{T}\frac{1}{kh}\cos\theta \tag{3.2.126}$$

$$w = -\frac{\pi H}{T}\left(1 + \frac{z}{h}\right)\sin\theta \tag{3.2.127}$$

$$p_D = \rho g\frac{H}{2}\cos\theta \tag{3.2.128}$$

As for general waves, both u and p_D are in phase with and proportional to the surface elevation η. For u, this gives the useful relation

$$u = c\frac{\eta}{h} \tag{3.2.129}$$

and for p_D, we get

$$p_D = \rho g \eta \qquad \text{or} \qquad p = \rho g(\eta - z) \tag{3.2.130}$$

at all points over the vertical.

The result for p_D (or p) also shows that in linear long waves, the pressure due to the waves is hydrostatic. In more physical terms, it means that the phase motion is so slow that the vertical accelerations are negligible so that locally the pressure corresponds just to the weight of the water above the point considered.

It is also worth to notice that, while the expressions for u and p_D are independent of z, the appropriate expression for the vertical velocity (3.2.127) includes a linear variation with z. Since for long waves $kh \ll 1$, the result means that $w \ll u$, and in fact one often sees the approximation $w \sim 0$ in shallow water waves. However, this obviously is in conflict with the kinematic free surface condition (3.2.4) which states that $w = \eta_t$, and $w = w(z) \neq 0$ is also required for satifaction of the continuity equation $u_x + w_z = 0$.

More controversially, this also means that if we calculate the velocities from the shallow water aproximation for the velocity potential ϕ, then we need to use two different approximations for ϕ depending on whether we determine u or w. For w, we need the first two terms in the Taylor expansion of $\cosh k(z+h)$ in (3.2.41).

$$\phi = -\frac{Hc}{2}\left(1 + \frac{1}{2}\left(\frac{z}{h}\right)^2\right)\sin\phi \tag{3.2.131}$$

to obtain (3.2.127), while (3.2.126) corresponds to using only the first term in (3.2.131). This paradox is caused by the use of the velocity potential for shallow water waves and is illustrated again in Chapter 9 about the Boussinesq equations.

The particle paths for shallow water waves can be determined the same way as the velocities by the lowest order Taylor expansions with respect to $z + h$ of the general expressions.

Exercise 3.2-21

Show that the particle paths $x(t)$, $z(t)$ are given by

$$x(t) = \xi + \frac{H}{2}\frac{1}{kh}\sin\theta \tag{3.2.132}$$

$$z(t) = \zeta + \frac{H}{2}\left(1 + \frac{\zeta}{h}\right)\cos\theta \tag{3.2.133}$$

where (ξ, ζ) is the mean position of the particle.

Fig. 3.2.13 shows the variation of the particle paths along a vertical in a shallow water wave motion.

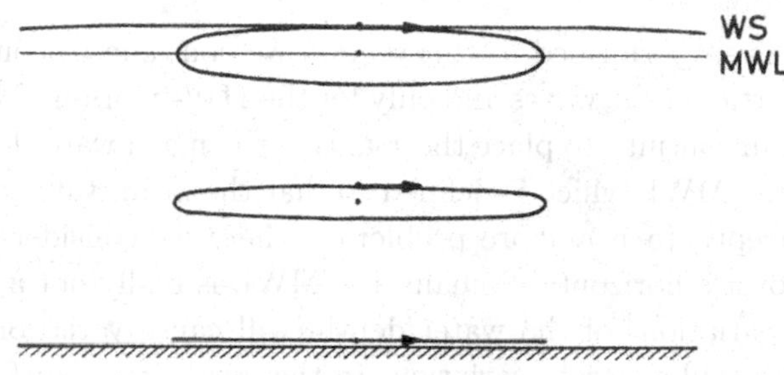

Fig. 3.2.13 The particle paths in a linear shallow water motion (SJ76).

The shallow water limit

The limit for the region in which the shallow water approximation is usually considered valid is $h/L < 1/20$. Use of the shallow water approximations at that limit causes errors relative to the general formulas that depend on which parameter is considered. For the phase velocity c, the error at $h/L = 1/20$ is 0.6%: $c = \sqrt{gh}$ is larger (in accordance with the fact that on any depth, the longest wave (shallow water wave) is the fastest possible wave).

Exercise 3.2-22

Show that for a fixed period T, the wave length L_0 in deep water is related to the shallow water wave length L by the expression

$$L = \sqrt{2\pi h L_0} \tag{3.2.134}$$

and therefore in shallow water

$$\frac{h}{L} = \sqrt{\frac{h}{2\pi L_0}} \tag{3.2.135}$$

These expression provide direct values of the shallow water L and h/L in terms of the deep water L_0 which is often known because it only depends on T.

3.3 Time averaged properties of linear waves in one horizontal dimension (1DH)

3.3.1 *Introduction*

In the analysis performed in this section we continue to consider only the local properties of the waves and only for the 1DV- motion. This implies that here we can continue to place the x-axis in the mean water level MWL.

However, the MWL which is defined so that the mean value of η is zero is a local concept. In nearshore problems, where we consider the waves and currents over a horizontal domain, the MWL is really not a horizontal surface. The variations of the water depth, will cause variations in wave height, and this will generate variations in the mean water surface. Hence the mean water **level** MWL should rather be called the mean water **surface** MWS. Furthermore we will need to distinguish between the MWS and the x-axis (or in 2DH the x, y-plane) (i.e. the level for $z = 0$), which of course **is** horizontal. In Chapter 11 we will choose the vertical position of the x-axis at a level corresponding to no waves. Therefore this level is also called the **still water level, SWL**.

All this will be discussed in the chapters on nonlinear waves and on circulation modelling, but it is pointed out that the definitions of vertical distances used here will have to be somewhat modified in the more general presentation later.

For linear (or "sine") waves, we have found the following basic results

Surface elevation

$$\eta = \frac{H}{2}\cos\theta \qquad \theta = \omega t - kx \tag{3.3.1}$$

Velocity potential

$$\phi = -\frac{H}{2}c\frac{\cosh k(z+h)}{\sinh kh}\sin\theta \tag{3.3.2}$$

When we start considering the properties of the waves averaged over a wave period ("wave averaging") definition

$$\bar{\cdot} = \frac{1}{T}\int_0^T \cdot \, dt \tag{3.3.3}$$

that from (3.1) and (3.2) we get

$$\bar{\eta} = 0 \quad , \quad \bar{\phi}_x = 0 \quad \text{etc.} \tag{3.3.4}$$

On the other hand, we clearly have $\overline{\phi_x^2} \neq 0$ and similar results apply for other nonlinear contributions. These constitute time or "wave" averaged properties of the waves which have important physical meaning.

The four wave averaged quantities we will analyze for linear waves in this section are the mass flux, the momentum flux, the energy density, and the energy flux caused by linear waves. Essentially, these are the contributions due to the waves that will appear in the wave averaged equations that will be derived later for the conservation of mass, momentum and energy for the combined waves and currents that characterize the nearshore circulation flows.

3.3.2 *Mass and volume flux*

The instantaneous mass flux $M(t)$ per unit width[5] through a vertical is given by

$$M(t) = \int_{-h}^{\eta} \rho u(x, z, t) dz \tag{3.3.5}$$

Thus the general wave averaged mass flux is the defined as

$$M = \overline{\int_{-h}^{\eta} \rho u(x, z, t) dz} \tag{3.3.6}$$

The particle velocities (u, w) in the sine wave motion is obtained as usual by partial differentiation of ϕ. In particular for u we found

$$u = \frac{H}{2} c\, k \, \frac{\cosh k(z+h)}{\sinh kh} \cos\theta \tag{3.3.7}$$

In (3.3.5) the integral can be divided into

$$= \int_{-h}^{0} \rho u \; dz + \int_{0}^{\eta} \rho u \; dz \tag{3.3.8}$$

This division of the integral is motivated by the fact that linear waves are assumed to be infinitesimal in height, and the linear solution (strictly speaking) valid only in the interval $-h \geq z \geq 0$.

[5] It is understood everywhere in the following that the wave averaged quantities are "per unit width".

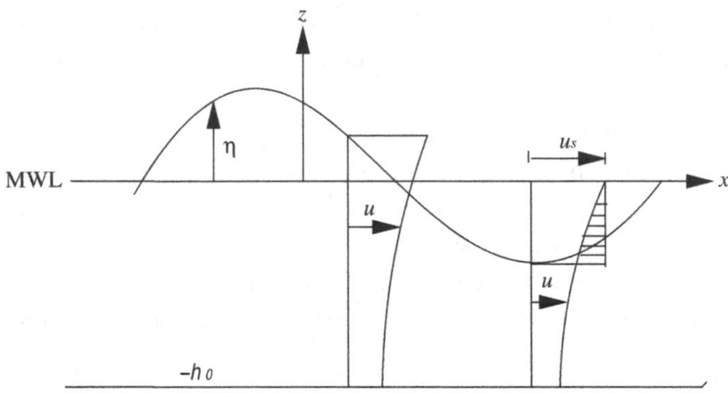

Fig. 3.3.1 The velocity profiles predicted by the sine wave velocity potential (3.3.2). Note that strictly speaking the theory only predicts the velocity up to the MWS. This also implies, however, that according to the theory there is a velocity all the way up to the MWS also over the wave trough even though there is not water all the way to the MWS there. This is consistent because the wave height is assumed infinitely small.

Averaging over a wave period (i.e., using the definition (3.3.3)) gives the wave averaged mass flux M

$$
\begin{aligned}
M &= \overline{\int_{-h}^{0} \rho u \, dz} + \overline{\int_{0}^{\eta} \rho u \, dz} = \int_{-h}^{0} \rho \bar{u} \, dz + \overline{\int_{0}^{\eta} \rho u \, dz} \\
&= \overline{\int_{0}^{\eta} \rho u \, dz}
\end{aligned} \tag{3.3.9}
$$

In obtaining a zero for the first integral, we have *formally* assumed that ϕ is defined by (3.3.2) also between wave trough and mean water level, although for real waves, which do not have mathematically infinitely small height, there is no water in that region. It may be verified, however, that this additional contribution is cancelled by the similar contribution of opposite sign from the second integral, so we have not in the total result of (3.3.9) assumed that there is water in the trough.

In (3.3.9), however, linear wave theory does not define u above MWL for $\eta > 0$. We circumvent that problem by replacing u in this integral by its Taylor expansion from $z = 0$:

$$
u(x, z, t) = u(x, 0, t) + z u_z(x, 0, t) + \cdots \tag{3.3.10}
$$

When evaluating (3.3.9), we use this expression in the entire region between trough and crest. We then get

$$\begin{aligned} M &= \rho \overline{\int_0^\eta \{u(x,0,t) + z u_z(x,0,t) + \cdots\} \, dz} \\ &= \rho \overline{u(x,0,t)\eta} + \frac{\rho}{2} \overline{\eta^2 u_z(x,0,t)} + \cdots \\ &= \rho \overline{u(x,0,t)\eta} + 0(H/L)^3. \end{aligned} \tag{3.3.11}$$

(Actually the $O(H/L)^3$-term will also be 0 so the next term is $O(H/L)^4$). For sine waves we have

$$u(z=0) = \frac{\pi H}{T} \coth kh \cos\theta = c\pi \frac{H}{L} \coth kh \cos\theta \tag{3.3.12}$$

We therefore see that the first term in (3.3.11) is $O((H/L)^2)$ so that we can use linear theory to evaluate this term. We get to leading order

$$M = \rho \overline{u\eta} + O(H/L)^3 \tag{3.3.13}$$

Hence we have

$$M = \rho \frac{\pi H^2}{2T} \coth kh \, \overline{\cos^2\theta} = \rho \frac{\pi H^2}{2T} \cdot \frac{1}{2} \frac{1}{\tanh kh} \tag{3.3.14}$$

The dispersion relation $\omega^2 = gk \tanh kh$ can be cast in the form

$$\tanh kh = 2\pi c/gT \tag{3.3.15}$$

which substituted into (3.3.14) yields

$$M = \frac{1}{8} \rho g H^2 / c \tag{3.3.16}$$

Notice again that M given by (3.3.16) is $O(H^2)$ and that this represents the lowest order of approximation for the mass flux (we have neglected all terms $O(H^3)$ and higher).

In surf zone dynamics, we are generally more interested in the volume flux $Q_{wx} = M/\rho$ so that for linear waves we have

$$Q_{wx} = \frac{1}{8} g H^2 / c \tag{3.3.17}$$

It is important to emphasize here that M (and Q_w) is associated with the motion between trough and crest.[6]

[6]The expression for Q_w in (3.3.17) can also be written $Q_w = \frac{E}{\rho c}$ where $E = 1/8 \rho g H^2$ is the energy density per unit bottom area. This expression is often seen used. However,

Sometimes one sees Q_w transformed into an equivalent (fictive) "mass transport velocity" U_s by division with the depth h:

$$U_s = \frac{Q_w}{h} \tag{3.3.18}$$

Given, as the derivation shows, that Q_w is caused by nonlinear effects between wave trough and crest, this is a physically confusing concept. It would suggest that such a mean velocity $\overline{u}$ should occur at all points over a vertical with the value $\overline{u} = U_s$. Of course this is not true and in fact we used $\overline{u} = 0$ in the derivation of the result (3.3.16).

The problem of the actual mean velocity below trough level is quite complicated and a key problem in nearshore hydrodynamics. For further details see Chapter 12.2. However, in the simplest case of waves generated in a closed region (such as a wave flume) it is obvious that there cannot over an extended period of time be a net mass flux Q_w. Once the wave motion in the flume is established, a return current U will develop which compensates for the mass flux between trough and crest. Disregarding viscous or turbulent effects, we must have

$$0 = Q_w + Uh \qquad \text{or} \qquad U = -\frac{Q_w}{h} \tag{3.3.19}$$

(which is entirely different from what (3.3.18) suggests).

The situation is sketched in Fig. 3.3.2.

3.3.3 *Momentum flux—radiation stress*

Consider again a progressive linear wave motion on a horizontal bottom The total flux of x-momentum through a vertical plane x = const for such a wave is

$$F(t) = \int_{-h}^{\eta} \left(\rho\, u^2 + p\right)\, dz \tag{3.3.20}$$

A detailed derivation of this expression will be given later (see Sect. 11).

Essentially, this is the instantaneous force which the water to the left of the section exerts on the water to the right of the section. And vice versa, see Fig. 3.3.3.

---

it is important to emphasize that this relationship is a mathematical coincidence which is only valid for linear sine waves. It does for example not apply for nonlinear waves in general, or for the wave motion inside the surfzone. The volume flux Q_w is a purely kinematical quantity while the energy E which is a dynamical quantity. We will therefore avoid using this relationship here.

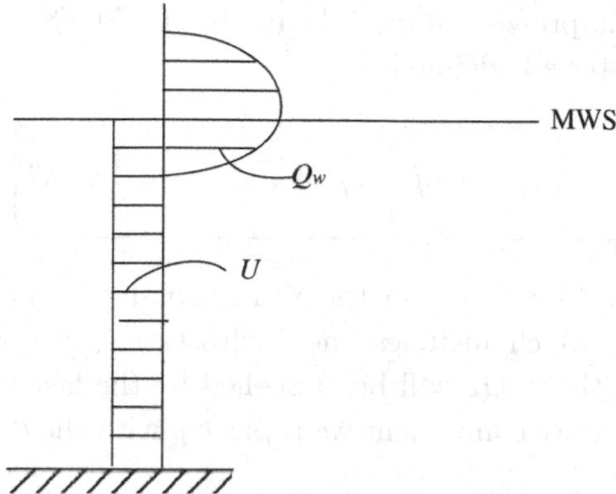

Fig. 3.3.2 Illustration of the volume flux Q_w and the return flow U as it occurs in a closed wave flume.

The time average over a wave period of $F(t)$ is

$$F = \overline{\int_{-h}^{\eta} (\rho\, u^2 + p)\, dz} \tag{3.3.21}$$

Part of that force, however, is due to the hydrostatic pressure which would be there even without the waves. We now define the **Radiation Stress**, S_{xx} for the wave motion as the **mean momentum flux (or mean force) caused by the waves only**. According to this definition S_{xx} is obtained by subtracting the hydrostatic part of the pressure from F.

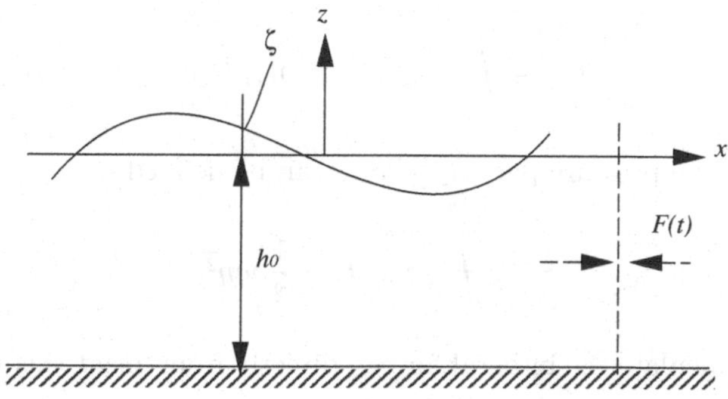

Fig. 3.3.3 $F(t)$ acting on the two sides of a vertical section.

The hydrostatic pressure force relative to the MWS is $\frac{1}{2}\rho g h^2$ so that S_{xx} as described above is defined by

$$\boxed{S_{xx} = F - \frac{1}{2}\rho g h^2 = \overline{\int_{-h}^{\eta} (\rho u^2 + p)\ dz} - \frac{1}{2}\rho\ g h^2} \tag{3.3.22}$$

which is the general definition of the radiation stress. Furthermore, p is the total pressure which includes the hydrostatic pressure contribution. Therefore, part of the $\int p dz$ will be cancelled by the last term in (3.3.22). To extract the net wave component we replace p with the dynamic pressure p_D defined by

$$p_D = p + \rho g z \tag{3.3.23}$$

i.e., p_D is the (additional) pressure generated by the waves.

If p_D is introduced into (3.3.22), it turns out, after some algebra which we return to later (see Chapter 11), that S_{xx} can be written as

$$S_{xx} = \overline{\int_{-h}^{\eta} (\rho u^2 + p_D)\ dz} - \frac{1}{2}\ \rho g \overline{\eta^2} \tag{3.3.24}$$

It is convenient to define the momentum part S_m of the radiation stress by

$$S_m = \overline{\int_{-h}^{\eta} \rho u^2\ dz} \tag{3.3.25}$$

or

$$S_m = \int_{-h}^{0} \rho \overline{u^2}\ dz + O(H/L)^3 \tag{3.3.26}$$

Similarly the pressure part S_p of S_{xx} can be defined by

$$S_p \equiv \overline{\int_{-h}^{\eta} p_D\ dz} - \frac{1}{2}\rho g \overline{\eta^2} \tag{3.3.27}$$

We can calculate the integral for S_m directly from linear theory, and we get

$$S_m = \rho \frac{\pi^2 H^2}{T^2 sh^2} \cdot \frac{1}{2} \int_{-h}^{0} \cosh^2 k(z+h)\ dz \tag{3.3.28}$$

where $sh = \sinh kh$. Since $\omega = 2\pi/T$, this can further be transformed into

$$= \frac{\rho}{8} \frac{\omega^2 H^2}{sh^2} \left[\frac{sh\, 2k(z+h)}{4} + \frac{k(z+h)}{2} \right]_{-h}^{0}$$

$$= \frac{\rho}{32} \frac{gH^2}{sh\; ch} (sh2kh + 2kh) \tag{3.3.29}$$

or

$$S_m = \frac{1}{16} \rho\, g\, H^2 \left[1 + \frac{2kh}{\sinh 2kh} \right] + O(H/L)^3$$

$$\equiv \frac{1}{16} \rho\, gH^2 (1+G) + O(H/L)^3 \tag{3.3.30}$$

where we have introduced the convenient definition

$$G \equiv \frac{2kh}{\sinh 2kh} \tag{3.3.31}$$

We notice that in parallel with what was found for the mass flux, we have $S_m = O(H^2)$ (or $(H/L)^2$) and clearly that is the lowest order approximation for S_m.

If we try to calculate S_p as a repeat of the success from S_m by substituting the linear expression for p_D

$$p_D = \rho\, g\, \eta \, \frac{\cosh k(h+z)}{\cosh kh} \tag{3.3.32}$$

into (3.3.27), we see that $\overline{\int p_D \, dz}$ is zero.

Unfortunately, this result is *not* correct. There is in fact an $O(H^2)$ contribution from p_D. This contribution clearly cannot be found directly. It appears that the correct value of $\overline{p_D \, dz}$ can be obtained if the second order Stokes approximation is used for p_D. On the other hand, since it is a contribution present even in linear waves one would also expect that linear wave theory should be able to provide the magnitude of this contribution, and this is indeed the case though it requires a more detailed analysis than we want to embark on here.

It will be shown in Chapter 11 that for linear waves p_D can be written as

$$p_D = -\rho w^2 - \frac{\partial}{\partial t} \int_z^{\eta} w dz \tag{3.3.33}$$

When this expression is averaged over the wave period the integral becomes zero because of the periodicity of the motion. Hence, using that above trough level we can use the approximation $p_D = \rho g \eta$, $\overline{\int p_D \, dz}$ can be written as

$$\overline{\int_{-h}^{\eta} p_D \, dz} = -\overline{\int_{-h}^{\eta} \rho w^2 \, dz} + \rho g \overline{\eta^2} \tag{3.3.34}$$

and we see that this expression does give a non-zero result $O(H^2)$ for the integral of p_D when we substitute linear wave results for w and η. We can therefore evaluate (3.3.34), and hence (3.3.27), from linear wave theory to get

$$S_p = \frac{1}{16} \rho \, g \, H^2 \, G + O(HL/L)^3 \tag{3.3.35}$$

Exercise 3.3-2

Derive (3.3.35) from (3.3.34).

Therefore, for linear waves the total S_{xx} becomes

$$S_{xx} = S_m + S_p = \frac{1}{16} \rho g \, H^2 (1 + 2G) + O(H/L)^3 \tag{3.3.36}$$

which is the result for the radiation stress in linear waves.[7] Notice that using (3.3.34), we can write S_{xx} as

$$S_{xx} = \overline{\int_{-h}^{\eta} \rho (u^2 - w^2) \, dz} + \rho g \frac{1}{2} \overline{\eta^2} + O(H/L)^3 \tag{3.3.38}$$

which is another form for radiation stress valid for all wave descriptions.

We finally emphasize that again it has turned out that the time mean (or wave averaged) value of a wave quantity (here the momentum flux) is nonzero and $O(H^2)$ in the lowest approximation.

[7] As was the case with the mass flux, for linear waves **only** S_{xx} can be expressed in terms of the energy density E. This yields the following alternative expressions for S_{xx}

$$S_{xx} = \frac{1}{2} E (1 + 2G) \tag{3.3.37}$$

which is fequently used in the literature.

3.3.4 *Energy density*

The wave motion also represents both potential and kinetic energy.

Potential energy

The potential energy must be measured relative to a horizontal level, which we naturally choose as $z = 0$. Hence the instantaneous potential energy of the wave motion $E_p(t)$, per unit area of the horizontal plane is given by

$$E_p(t) = \int_{-h}^{\eta} \rho g z \, dz - \int_{-h}^{0} \rho g z \, dz \tag{3.3.39}$$

where the last term represents the potential energy of the fluid at rest. Thus we have

$$E_p(t) = \int_{0}^{\eta} \rho g z \, dz = \frac{1}{2} \rho g \eta^2 \tag{3.3.40}$$

The mean value E_p over a wave period then becomes

$$E_p = \frac{1}{2} \rho g \overline{\eta^2} \tag{3.3.41}$$

which is exact. For sine waves this becomes

$$E_p = \frac{1}{16} \rho g H^2 + O(H^3) \tag{3.3.42}$$

which is correct to second order in H.

Kinetic energy

For the kinetic energy we first consider the energy density $e_k(t)$ per unit volume of fluid, which is (for the 2DV wave motion we are looking at)

$$e_k = \frac{1}{2} \rho (u^2 + w^2) \tag{3.3.43}$$

which again is exact, and, of course, always positive. For sine waves we get, using (3.2.79) and (3.2.80)

$$e_k(t) = \frac{1}{2} \rho \left(\frac{H\omega}{2 \sinh kh} \right)^2 (\cosh^2 k(z+h) \, \cos^2 \theta + \sinh^2 k(z+h) \, \sin^2 \theta) \tag{3.3.44}$$

which is correct to second order. Introducing further the dispersion relation into the first factor in brackets and using trigonometric and hyperbolic relations this can be written

$$e_k(t) = \frac{1}{4}\rho\,\frac{gkH^2}{\sinh 2kh}\,(\cos^2\theta + \sinh^2 k(z+h)) \tag{3.3.45}$$

The instantaneous kinetic energy $E_k(t)$ per unit area of the horizontal plane can then be obtained by integrating $e_k(t)$ over depth

$$\begin{aligned} E_k(t) &= \int_{-h}^{\eta} e_k(t)dz \\ &= \frac{1}{4}\rho\,\frac{gkH^2}{\sinh\,2kh}\left[h\,\cos^2\theta + \frac{1}{2}\int_{-h}^{0}(\cosh 2k(z+h) - 1)\,dz\right] \\ &= \frac{1}{16}\rho gH^2 + \frac{1}{8}\rho gH^2\frac{2kh}{\sinh 2kh}\left(\cos^2\theta - \frac{1}{2}\right) \end{aligned} \tag{3.3.46}$$

Averaging this over a wave period we then get the average kinetic energy E_k per unit area

$$E_k = \frac{1}{16}\rho gH^2 \tag{3.3.47}$$

which is the complete expression to second order.

Exercise 3.3-3

Derive equation (3.3.47)

Notice that

$$E_p = E_k \tag{3.3.48}$$

which is in fact the case for any conservative system unergoing small oscillations.

The total energy density then be comes

$$\boxed{E = E_p + E_k = \frac{1}{8}\rho gH^2} \tag{3.3.49}$$

This is also sometimes called the specific energy. It is noted that E is a function of the wave height only, not the wave period or wave length.

3.3.5 *Energy flux*

In Section 2.6 we found that the energy flux through a plane vertical section from the bottom to the surface with normal n_α is given by

$$E_f(x_\alpha, t) = \int_{-h_0}^{\eta} \left(p + \rho g z + \frac{1}{2}\rho u_i u_i\right) u_\alpha dz \; n_\alpha \tag{3.3.50}$$

where u_i is the total flow velocity (waves plus currents), and u_α is the horizontal component of u_i.

To determine the energy flux in waves we let u_i be the wave particle velocities u, v, w and p the pressure generated by the waves. Since u_α in plane waves has the same direction for all points over the vertical the energy flux is a vector $E_{f\alpha}$ in the direction of u_α, that is in the direction of propagation of the wave. Thus the general expression for the simple case of a pure wave motion we simply get for the instantaneous flux

$$E_{f\alpha}(x_\alpha, t) = \int_{-h}^{\eta} \left(\rho g z + p + \frac{1}{2}\rho\left(u^2 + v^2 + w^2\right)\right) u_\alpha \, dz \tag{3.3.51}$$

If we average this over a short wave period we get the total energy flux in the flow averaged over T:

$$\boxed{E_{f\alpha}(x_\alpha) = \overline{\int_{-h}^{\eta} \left(p + \rho g z + \frac{1}{2}\rho\left(u^2 + v^2 + w^2\right)\right) u_\alpha dz}} \tag{3.3.52}$$

For a plane 2DV wave in the x direction and a section perpendicular to the wave direction we have $u_\alpha n_\alpha = u$ and $v = 0$ so the energy flux through the section (and in the wave direction) is

$$E_{f,x} = \overline{\int_{-h}^{\eta} \left(p + \rho g z + \frac{1}{2}\rho\left(u^2 + w^2\right)\right) u dz} \tag{3.3.53}$$

Exercise 3.3-4

Show that for a linear wave train propagating in the x-direction the first approximation to the energy flux is $O(H/L)^2$ and that $E_f(t)$ can be written as

$$E_{fx}(t) \sim \int_{-h}^{0} p_D \, u dz + O(H/L)^3 \tag{3.3.54}$$

Then we get that for linear waves propagating in the x-direction the mean energy flux E_f can be determined as

$$\boxed{E_{fx} = \frac{1}{16}\rho g\, H^2\, c(1+G) + O(H/L)^3} \tag{3.3.55}$$

If we define a velocity c_g by

$$c_g = \frac{1}{2}\, c(1+G) \tag{3.3.56}$$

the energy flux can also be written

$$E_f = c_g E \tag{3.3.57}$$

which can be interpreted as that the energy flux occurs as if the energy density in the wave motion is traveling with the speed c_g. We shall later (Chapter 3.4.2) see that c_g (the so-called group velocity) is also the speed with which variations in the wave amplitude are traveling.

3.3.6 *Dimensionless functions for wave averaged quantities*

Thus we have found that even in linear waves the wave averaged mass, momentum, and energy fluxes are non-zero. Less surprising perhaps the waves also represent a non-zero energy density. This means that in such a wave motion the waves leave mean signatures for these quantities. The waves move a certain amount of water Q_w per second along in the direction of wave propagation, they excert a net force S_{xx} on a vertical section in the fluid (in addition to the hydrostatic pressure), and they transport an amount of energy E_f per second across such a vertical section.

It is convenient for later discussions to introduce dimensionless functions for the three fundamental fluxes and the energy density generated by the waves. Using the definitions for the quantities presented above we define

Dimensionless mass flux

$$B_{Qx} = \frac{Q_{wx}c}{gH^2} = \frac{c}{gH^2}\overline{\int_{-h}^{\eta} u\, dz} \tag{3.3.58}$$

Dimensionless radiation stress

$$P_{xx} = \frac{S_{xx}}{\rho g H^2} = \frac{1}{gH^2}\overline{\int_{-h}^{\eta}\left(u^2 + \frac{p_D}{\rho}\right) dz} + \frac{1}{2}\frac{\overline{\eta^2}}{H^2} \tag{3.3.59}$$

Dimensionless energy density

$$B_E = \frac{E}{\rho g H^2} = \frac{1}{\rho g H^2}\overline{\int_{-h}^{\eta}\left(\rho g z + \frac{1}{2}(u^2 + v^2 + w^2)\right) dz} \tag{3.3.60}$$

Dimensionless energy flux

$$B_x = \frac{E_{fx}}{\rho g H^2 c} = \frac{1}{gH^2 c}\overline{\int_{-h}^{\eta}\left(\frac{p_D}{\rho} + \frac{1}{2}(u^2 + v^2 + w^2)\right) u dz} \tag{3.3.61}$$

These expressions may readily be generalized to situations in the 2DH where x is replaced by x_α. This generalization will be introduced in the individual applications as needed.

It is important to realize that since the original integral expressions for E, Q_w, S_{xx} and E_{fx} are *exact*, the equivalent dimensionless parameters in (3.3.58), (3.3.59), (3.3.60) and (3.3.61) can also be made exact provided the description for the wave motion we substitute into these formulas in terms of p_D, u, etc. are exact. What we have done above is to calculate the results for Qx, S_{xx} E, and $E_{f,x}$ under the assumption that the wave motion was adequately described by the linear sine wave theory. The results we found were apparently, that for linear waves we have

$$B_{Qx} = \frac{1}{8} \tag{3.3.62}$$

$$P_{xx} = \frac{1}{16}(1 + 2G) \tag{3.3.63}$$

$$B_E = \frac{1}{8} \tag{3.3.64}$$

$$B_x = \frac{1}{16}(1 + G) \tag{3.3.65}$$

Later we will discuss the actual value of these dimensionless functions also for non-sinusoidal waves, e.g. for the very steep waves just before breaking and for the broken waves in the surf zone.

3.4 Superposition of linear waves

One of the most important features of linear problems is that solutions can be superimposed: If A and B each are solutions then $A + B$ and $A - B$ are also solutions to the problem. This is explored further in this section. However, adding solutions that are waves of simple sinusoidal shape can lead to solutions of any shape (as in Fourier series). Therefore, a corrolary to this is that the equations for linear waves we have solved supports waves of any shape. Or in other words, **the linear wave theory cannot predict the shape of the waves**.

This fundamental fact is often overlooked. In order to extract information about the wave motion that includes results about the wave shape (that is the phase variation), it is necessary to include the nonlinear terms we disregarded to obtain the linear form of the equations. This is discussed extensively in Chapters 5 – 8.

Adding even just two sinusoidal wave motions (not to speak of several) of different amplitude and frequency very quickly leads to complicated solutions in terms of the phase variation for the waves. Usually, a computer is required to analyze e.g. the time or space variation of such general solutions. In the following, we only focus on two important examples: standing waves and wave groups. In addition, a brief overview is given of the problem formulation and theory for arbitrarily many wave components (wave spectra).

Exercise 3.4-1

The simplest sum of two sine waves is obtained when we add two waves with the **same** frequency that propagate in the same direction but are phase shifted the angle μ.

$$\eta_1 = a_1 \cos(\omega t - kx)$$
$$\eta_2 = a_2 \cos(\omega t - kx + \mu) \tag{3.4.1}$$

Show that this leads to a new wave

$$\eta_3 = a_3 \cos(\omega t - kx + \delta) \tag{3.4.2}$$

with

$$a_3 = \sqrt{a_1^2 + a_2^2 + 2a_1 a_2 \cos\mu} \tag{3.4.3}$$

and

$$\tan\delta = \frac{a_2 \sin\mu}{a_1 + a_2 \cos\mu} \tag{3.4.4}$$

3.4.1 *Standing waves*

Standing waves develop in the 2D vertical plane when a propagating wave normally incident against a vertical wall is reflected from the wall. The resulting wave system is a superposition of two waves of the same height and period but propagating in opposite directions.

Since the wall represents a vertical plane along which the horizontal velocity must be zero, the problem can be solved by replacing the periodicity condition in Section 3.2 with the condition

$$\phi_x = 0 \qquad \text{at} \qquad x = x_{\text{wall}} \tag{3.4.5}$$

and go through the entire solution procedure again, omitting the assumption of constant form, progressive waves. This was in fact done in Exercise 3.2-6, except that the boundary condition at the wall now requires a different selection of integration constants.

However, a simpler approach is to follow the physical lead and form the solution as the sum of two, opposite moving progressive waves. Thus, we consider the two components

$$\eta_1 = a\cos(\omega t - kx)$$
$$\eta_2 = a\cos(\omega t + kx) \qquad (3.4.6)$$

It is seen that the sum of these two waves can be written

$$\eta = \eta_2 + \eta_2 = 2a\cos\omega t \cos kx \qquad (3.4.7)$$

The velocity potential for the sum can be obtained the same way by adding two contributions similar to (3.2.41), one with $\theta_- = \omega t - kx$, one with $\theta_+ = \omega t + kx$. The result is

$$\phi = -2ac\frac{\cosh k(z+h)}{\sinh kh}\sin\omega t\cos kx \qquad (3.4.8)$$

This wave motion is called a **standing wave.**

As before, from this expression we can obtain all other properties of the motion, such as the velocity field (u, w), the particle paths, the pressure distribution, etc. We also note that since each of the original wave components satisfy the dispersion relation

$$\omega^2 = gk\tanh kh \qquad (3.4.9)$$

the sum does too.

> **Exercise 3.4-2** Determine which constants in the general solution found in Exercise 3.2-6 will satisfy the boundary condition (3.4.5) and show that this results in (3.4.8).

Notice that the result can also be interpreted as a standing wave with total height $2H = 4a$.

Analysis of the surface variation and the velocity field

Analysis of the expression (3.4.7) shows that the motion has points (nodes) where the surface shows no vertical motion, at

$$kx = \frac{\pi}{2} + p\pi \qquad \text{or} \qquad x = \frac{L}{4} + p\frac{L}{2} \qquad (3.4.10)$$

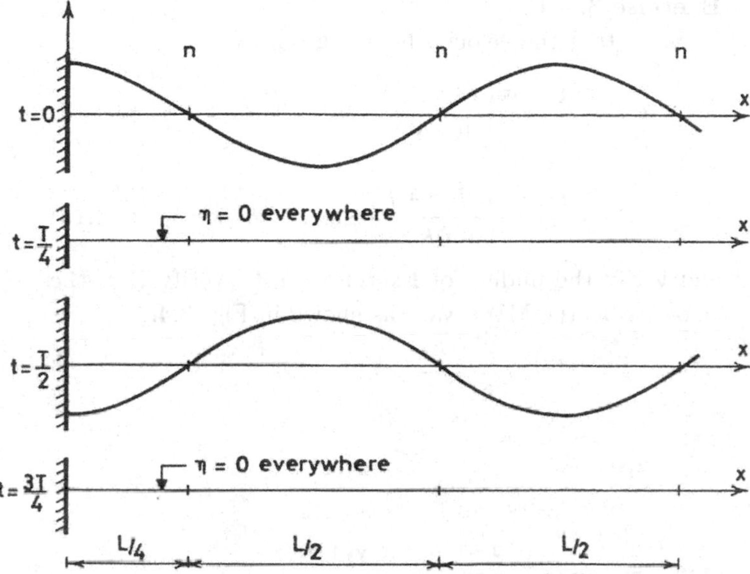

Fig. 3.4.1 Variation of the surface elevation in a standing wave (SJ76).

where p is an integer and points (antinodes) where the vertical motion is $\pm 2a$ at

$$kx = 0 + p\pi \qquad \text{or} \qquad x = 0 + p\frac{L}{2} \tag{3.4.11}$$

As Fig. 3.4.1 shows, the surface oscillates between two extreme positions which are attained at $t = 0 + pT$ and $t = \frac{T}{2} + pT$. At those positions in time, the surface is stagnant everywhere at the extreme position ($\frac{\partial \eta}{\partial t} = 0$). On the other hand at the instants $t = \frac{T}{4} + p\frac{T}{2}$ ($\cos \omega t = 0$) there is no surface elevation anywhere, but $\partial \eta / \partial t$ is maximum at all points in x.

Exercise 3.4-3

Determine all the positions where a wall can be placed without disturbing the motion.

Exercise 3.4-4

Show that the velocity field is given by

$$u = 4\,\frac{\pi a}{T}\,\frac{\cosh\,k(z+h)}{\sinh\,kh}\,\sin\,\omega t\,\sin\,kx \qquad (3.4.12)$$

$$w = -4\,\frac{\pi a}{T}\,\frac{\sinh\,k(z+h)}{\sinh\,kh}\,\sin\,\omega t\,\cos\,kx \qquad (3.4.13)$$

and verify the finding of Exercise 3.4-3. Verify that the velocities at the MWL vary as shown in Fig. 3.4.2.

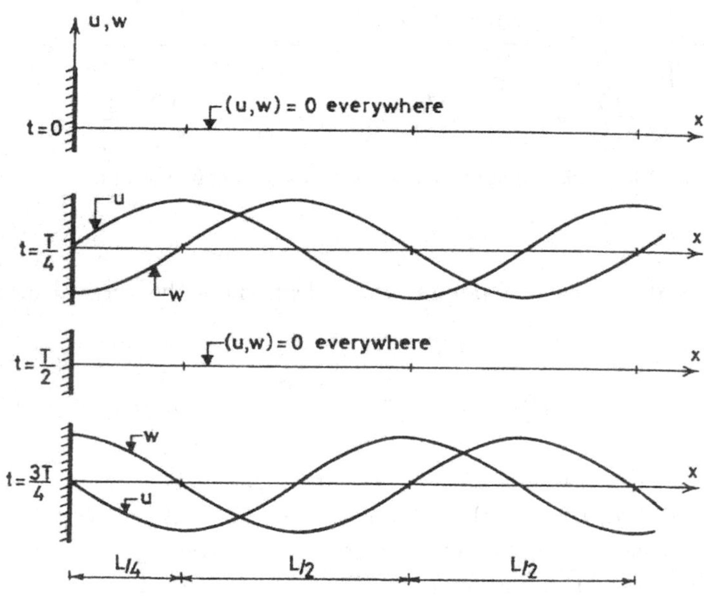

Fig. 3.4.2 Particle velocities at the free surface in a standing wave (SJ76).

Exercise 3.4-5

Determine the mathematical expression for the streamlines in a standing wave. Verify that the velocities shown at $T/4$ and $3T/4$ in Fig. 3.4.2 are consistent with the streamline pictures you can draw from the mathematical expressions.

The pressure variation

The pressure variation in standing waves is found the same way as for progressive waves from the linearized Bernoulli equation.

Exercise 3.4-6

Show that in a standing wave with node at $x = 0$ and amplitude $2a$ the pressure variation becomes

$$p_D = \rho g \, 2a \, \frac{\cosh \, k(z+h)}{\cosh \, kh} \cos \, \omega t \, \cos \, kx \qquad (3.4.14)$$

It is of interest (though not surprsing) to notice that, as in progressive waves, this can also be written

$$p_D = \rho g \eta \, \frac{\cosh \, k(z+h)}{\cosh \, kh} \qquad (3.4.15)$$

because this shows that p_D is proportional to η and hence shows the same pattern of variation with x and t as η.

We als note that the pressure decreases downward from the surface with the same attenuation factor as found for the pressure in progressive waves.

The variation of p_D at antinodes such as $x = 0$ where the horizontal velocity is zero is of particular interest. Those are the points where a vertical wall could be placed without disturbing the flow. Hence $p_D(x = 0 + p\frac{L}{2})$ represents the (first approximation to) the pressure on a vertical wall generated by the waves. The variation is illustrated in Fig. 3.4.3.

The arguments used in section 3.2 about the pressure at the MWL also apply here.

Exercise 3.4-7 Consider partially standing waves: Two opposite waves with the same period but different wave height are given by

$$\eta_1 = a \cos \, \theta_-$$

$$\eta_2 = a \cos \, \theta_+ \qquad (3.4.16)$$

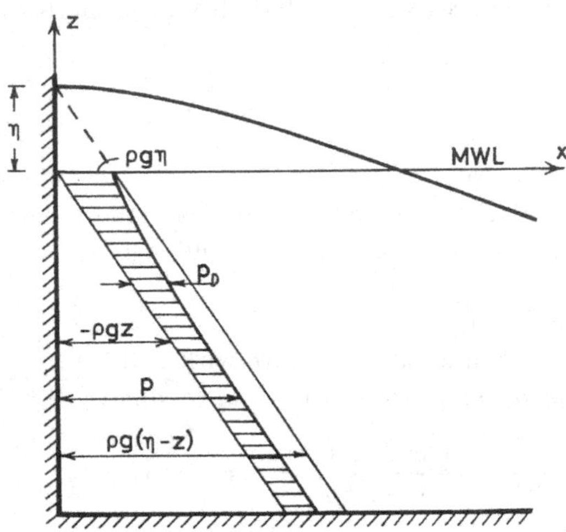

Fig. 3.4.3 The wave pressure along a vertical wall generated by a standing wave (SJ76).

Show this gives a combination of a standing and a progressive wave and determine the amplitudes of those two wave components.

3.4.2 *Wave groups*

While standing waves were the outcome of adding waves in opposite directions with the same wave period, the addition of two waves propagating in the same direction but with (slightly) different wave frequency will generate another canonical type of wave motion. If the two wave components have the same amplitude, the total wave motion will form what is called a wave group, which is a progressive wave with a (slowly) varying wave amplitude. We first consider this simple case.

Thus, we consider the two wave components propagating toward $+x$ and given by:

$$\eta_1 = a\cos(\omega_1 t - k_1 x) \qquad ; \qquad \eta_2 = a\cos(\omega_2 t - k_2 x) \tag{3.4.17}$$

To define the problem, we consider $T_1 < T_2$ (i.e., $\omega_1 > \omega_2$), which also implies that $L_1 < L_2$ and $k_1 > k_2$ since L decreases with T for constant water depth.

Using a simple trigonometric relation, the sum η_s of η_1 and η_2 can be written

$$\eta_s = 2a\cos\left(\frac{\omega_1-\omega_2}{2}t - \frac{k_1-k_2}{2}x\right)\cos\left(\frac{\omega_1+\omega_2}{2}t - \frac{k_1+k_2}{2}x\right) \quad (3.4.18)$$

The clearest illustration of this expression is obtained if we consider a case where T_1 is only slightly smaller than T_2. Then $\omega_1 + \omega_2 \gg \omega_1 - \omega_2$. Thus, the last factor in (3.4.18) will vary much faster with x and t than the first factor. In fact, we will get a wave motion which can be interpreted as a wave motion with a space and time varying amplitude written as

$$\eta_s(x,t) = A(x,t)\cos(\omega_m t - k_m x) \quad (3.4.19)$$

where the frequency ω_m and the wavenumber k_m are given by

$$\omega_m = \frac{\omega_1+\omega_2}{2} \quad ; \quad k_m = \frac{k_1+k_2}{2} \quad (3.4.20)$$

The wave amplitude $A(x,t)$ is given by

$$A(x,t) = 2a\cos(\omega_g' t - k_g' x) \quad (3.4.21)$$

where ω_g' and k_g' are defined by

$$\omega_g' = \frac{\omega_1-\omega_2}{2} \quad ; \quad k_g' = \frac{k_2-k_2}{2} \quad (3.4.22)$$

Fig. 3.4.4 shows this wave motion which is called a wave group because where and when $\cos(\omega_g t - k_g x) = 0$ the amplitude $A(x,t) = 0$. We see from (3.4.19) that the instantaneous surface (wave) profile propagates with the

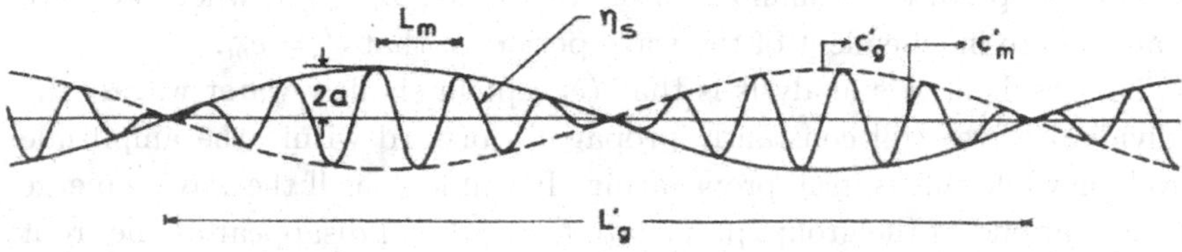

Fig. 3.4.4 The surface elevation in a wave group. The instantaneous surface profile given by (3.4.19) is shown as a thick line. This motion is oscillating within the amplitude envelop given by (3.4.21) and shown as a thinner line above and a dashed line below in the figure (SJ76).

phase velocity

$$c_m = \frac{\omega_m}{k_m} \tag{3.4.23}$$

and has wave period and wave length given by

$$T_m = \frac{2\pi}{\omega_m} = \frac{2T_1T_2}{T_1 + T_2} \tag{3.4.24}$$

$$L_m = \frac{2\pi}{L_m} = \frac{2L_1L_2}{L_1 + L_2} \tag{3.4.25}$$

However, the amplitude variation, or **amplitude modulation**, of this wave is also propagating, but with the phase speed

$$c_g' = \frac{\omega_g'}{k_g'} = \frac{\omega_1 - \omega_2}{k_1 - k_2} \tag{3.4.26}$$

While it is evident from (3.4.22) that $T_g' = \frac{2\pi}{\omega_g'} \gg T_m$ and $L_g' = \frac{2\pi}{k_g'} \gg L_m$, it is not a priori clear whether c_g' or c_m is the larger of the two. If we form the difference $c_m' - c_g$, we see that this can be written

$$c_m - c_g' = \frac{\omega_1 + \omega_2}{k_1 + k_2} - \frac{\omega_1 - \omega_2}{k_1 - k_2} = \frac{2k_1k_2}{{k_1}^2 - {k_2}^2}(c_2 - c_1) \tag{3.4.27}$$

The assumption $T_1 < T_2$ will imply $c_1 < c_2$ because the waves are frequency dispersive and $k_1 > k_2$. Hence, we see that in general $c_g' < c_m$, i.e., the amplitude envelope in Fig. 5 propagates **slower** than the instantaneous surface profile.

The exception is for shallow water waves, i.e., $L_m \gg h$, where we have c_1 and c_2 are independent of the wave period so that $c_g' = c_m$.

The result of this analysis is that (except in shallow water waves) the individual waves will constantly propagate forward within the amplitude envelope while this is itself propagating. It will look as if the waves emerge at the rear end of the group, propagate forward and disappear at the front of each group.

It is important to notice that the apparent group length between two successive nodes is only $L_g'/2$. This is illustrated in Fig. 3.4.4 by the distinction between the full and the dashed envelope curves.

Infinitely long wave groups and group velocity

If we assume that $\omega_1 \to \omega_2$ so that $k_1 \to k_2$, we see from (3.4.22) that

$$L'_g = \frac{2\pi}{k'_g} = \frac{2\pi}{k'_1 - k'_2} \to \infty \qquad (3.4.28)$$

and we have infinitely long wave groups. From the definition (3.4.27) of c'_g, we see that this implies

$$c'_g = \frac{\omega_1 - \omega_2}{k_1 - k_2} \quad \underset{k_1 \to k_2}{\longrightarrow} \quad c_g = \frac{\partial \omega}{\partial k} \qquad (3.4.29)$$

Since here ω and k are connected by the dispersion relation $\omega^2 = gk \tanh kh$, we can in the limit obtain an expression for c_g by differentiation of this relation. We get, keeping h = const.

$$2\omega \frac{\partial \omega}{\partial k} = g \tanh kh + \frac{gk}{\cosh 2kh} \qquad (3.4.30)$$

Using the relation $c = \omega/k$, this can after some algebra be reduced to

$$c_g = \frac{\partial \omega}{\partial k} = \frac{1}{2} c(1 + G) \qquad (3.4.31)$$

where G is again defined as $G \equiv \frac{2kh}{\sinh 2kh}$.

This expression is usually quoted as the "group velocity" for the waves. It is worth to notice, however, that this is strictly speaking a misnomer, because the expression (3.4.31) is only valid for infinitely long wave groups where the wave height variation is so slow that there is no wave height variation. In other words the waves look like a train of uniform waves. The correct definition of the velocity of a wave group is given by (3.4.26) and this expression clearly depends on the length of the wave group.

Exercise 3.4-8

Derive the expression (3.4.31). Analyze the deviation $c'_g - c_g$ from this expression for characteristic values of L'_g/L_m and $k_m h$.

Exercise 3.4-9

Show that $\partial\omega/\partial k$ can also be written

$$c_g = \frac{\partial \omega}{\partial k} = c + k\frac{dc}{dk} \qquad (3.4.32)$$

and use this to deduce that in shallow water waves $c = c_g$.

The variation of G versus h/L_0 (where $L_0 = gT^2/2\pi$ is the deep water wave length of a (linear) wave with period T) is shown in the Fig. 3.4.5 along with the variation of c/c_0 and $c_g/c = n$. We see that n varies from 1 in shallow water to 1/2 in deep water.

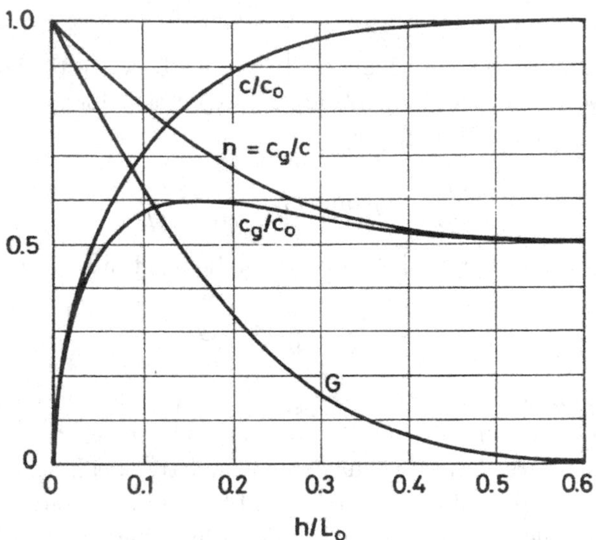

Fig. 3.4.5 The variation of n, the phase velocity c over the deep water value c_0, the ratio of group velocity c_g over c_0 and G with h/L_0 where L_0 is the deep water wave length.

It was shown in Section 3.3 that the energy flux E_f is given by (3.3.55)

$$E_{fx} = \frac{1}{16}\rho g\, H^2\, c(1+G) \tag{3.4.33}$$

As mentioned earlier with the expression above for c_g the energy flux E_f can be written as

$$\boxed{E_f = c_g E} \tag{3.4.34}$$

where $E = \frac{1}{8}\rho g H^2$ is the wave energy density. Thus c_g can actually be interpreted as the speed at which energy travels in steady uniform wave motion.

3.4.3 *Wave spectra*

Introduction

As mentioned earlier the superposition of just a few sine waves with different amplitude and frequency can create a highly complicated signal, with amplitudes of the total signal that changes both in time and space. This on the other hand leads to the natural question: would it be possible to represent a real, irregular wave motion in the ocean by simple superposition of a large number of sine wave components?

The answer to this question is: yes. However, it means we need to determine the amplitudes and phases of each of the (very) many wave components, and that turns out to be a nontrivial task.

This leads to the introduction of wave spectra as a theoretical concept. This will eventually make it possible to determine the amplitudes of the wave components that describe e.g. the surface variation in a given (measured) time series for a natural wave motion. However, it turns out not to be the simple spetrum of the amplitudes that works but the socalled energy spectrum, which is the spectrum of the energy of each wave component.

It also leads to development for practical applications of parameterized (standard) wave (energy) spectra that are based on results from numerous practical wave records.

Finally it leads to use of statistical methods for analysing wave records to determine the gross parameters used to get an overall picture of a wave situation and as parameters in the standard spectra.

There is a rich literature and a large number of spectral and statistical wave models that draw from these, mostly empirical, results. However, it is beyond the scope of the present text to go into the details of this complex area. Therefore we will only look at a simple formulation of the problems involved and point to some literature for further information.

Fourier representation of a complex wave motion

A registration at a fixed point of the surface elevation of a wave motion can be thought of as consisting of many wave components, each of which can be written

$$\eta_n(t) = A_n \cos(\omega_n t - \delta_n) \; ; \qquad n = 1, 2, ..., n \tag{3.4.35}$$

These wave components have different amplitude A_n, frequency ω_n, and phase δ_n. Hence the total wave motion at the point will be given by

$$\eta(t) = \sum_{n=1}^{\infty} A_n \cos(\omega_n t - \delta_n) \tag{3.4.36}$$

For our purpose a more convenient way of writing this is to introduce the amplitudes a_n and b_n given so that

$$A_n^2 = a_n^2 + b_n^2 \ ; \qquad \tan \delta_n = \frac{b_n}{a_n} \tag{3.4.37}$$

whereby $\eta(t)$ becomes

$$\eta(t) = \sum_{n=0}^{\infty} (a_n \cos \omega_n t + b_n \sin \omega_n t) \tag{3.4.38}$$

We see that this corresponds to the Fourier series for a record in time of finite length T_M provided we place three additional constraints on the inividual wave components

(1) We assume the longest wave has the period

$$T_M = \frac{2\pi}{\omega_1} = \frac{1}{f_1} \tag{3.4.39}$$

which defines f_1.

(2) For the rest of the wave components we choose periods so that

$$T_1 (= T_M) = 2T_2 = ... = nT_n \tag{3.4.40}$$

or

$$f_n = n f_1 \tag{3.4.41}$$

(3) We assume that $a_0 = b_0 = 0$ corresponding to

$$\overline{\eta}(t) = \frac{1}{T_M} \int_0^{T_M} \eta(t) dt = 0 \tag{3.4.42}$$

This means that (3.4.38) can be written

$$\eta(t) = \sum_{n=1}^{\infty} a_n \cos\left(\frac{2\pi n}{T_M}\right) t + b_n \sin\left(\frac{2\pi n}{T_M}\right) t \tag{3.4.43}$$

We then know from the theory for Fourier series that the sum (3.4.43) can represent an arbitrary continuous function as long as it is periodic with the period T_M. We also know that the coefficients a_n and b_n are given by

$$a_n = \frac{2}{T_M} \int_{-0}^{T_M} \eta(t) \, \cos \frac{2\pi n}{T_M} t dt \tag{3.4.44}$$

$$b_n = \frac{2}{T_M} \int_{0}^{T_M} \eta(t) \, \sin \frac{2\pi n}{T_M} t dt \tag{3.4.45}$$

Thus (3.4.38) with the above constraints satisfies our requirement that it can represent an arbitrary signal recorded over the period T_M.

The assumption that the wave record is periodic with period T_M implies that if the record were continued beyond $t = T_M$ then the already obtained record would repeat itself.

Of course no real wave motion will actually exhibit this periodicity and unfortunatley this deviation from our assumptions is far from trivial.

This can be expressed in a different way: the dominating difference from a real wind wave motion is that from our knowledge about the time series of $\eta(t)$ over a time interval T_M we will **not** be able to predict the variation of $\eta(t)$ at any time beyond T_M. A phenomenon with this property is called a **stochastic process**.

For the Fourier series the consequence of this is: there is nothing that prevents us from determining the Fourier coefficients. So consider a record (a time series) for $\eta(t)$ for T_M so large that we would be tempted to assume that the time series was representative for the whole stochastic process. Then determine the Fourier coeffients a_n, b_n or (A_n, δ_n) assuming periodicity with period T_M even though it is not correct. The consequence of the stochastic property of the motion is then that **any** change in T_M, even by an arbitrary small amount, will generate arbitrary large changes in the Fourier coefficients. In other words **the process of determing A_n, δ_n by letting $T_M \to \infty$ does not converge**. The Fourier spectrum for a stochastic process is not defined.[8]

The energy spectrum

However, it turns out that for a stochastic process it is possible to give a unique definition of another quantity, the **energy (or "power") spectrum**.

[8] That is unless $\eta(t)$ goes to zero for $t \to \pm\infty$.

To define the energy spectrum we first assume that we - for some T_M - have determined the set of coefficients a_n, b_n in (3.4.43). For this Fourier series Parceval's thorem states that

$$\frac{1}{T_M}\int_0^{T_M} \eta^2(t)dt = \overline{\eta^2}$$

$$= \frac{1}{2}\sum_{n=1}^{\infty}(a_n^2 + b_n^2) = \frac{1}{2}\sum_{n=1}^{\infty} A_n^2 \qquad (3.4.46)$$

However, $\eta^2(t)$ can be compared to the instantaneous potential energy $E_p(t)$ for which we have

$$E_p(t) = \frac{1}{2}\rho g \eta^2 \qquad (3.4.47)$$

so that over a long period of time (T_M) we have

$$\frac{E_p}{\rho g} = \frac{1}{2}\overline{\eta^2} \qquad (3.4.48)$$

At this point we utilize the result from section 3.3 that in small amplitude (i.e. sinusoidal) waves the total energy E_n of the n'th component is $2E_{p,n}$ so that

$$E_n = 2\ E_{p,n} = \rho g \overline{\eta_n^2} \qquad (3.4.49)$$

Hence we get for the entire wave motion

$$\frac{E}{\rho g} = \sum_{n=1}^{\infty}\overline{\eta_n^2} = \frac{1}{2}\sum_{N=1}^{\infty}(a_n^2 + b_n^2) \qquad (3.4.50)$$

where E is the total energy in the wave motion per m^2 bottom and averaged over long time.

We can then define an energy spectrum $S_{\eta\eta}(f_n)$ in the interval $f_n - \frac{\Delta f}{2} < f < f_n + \frac{\Delta f}{2}$ by

$$S_{\eta\eta}(f_n)\Delta f = \frac{1}{2}(a_n^2 + b_n^2) = \frac{1}{2}A_n^2 \qquad (3.4.51)$$

with

$$\Delta f = \frac{1}{T_M} \qquad (3.4.52)$$

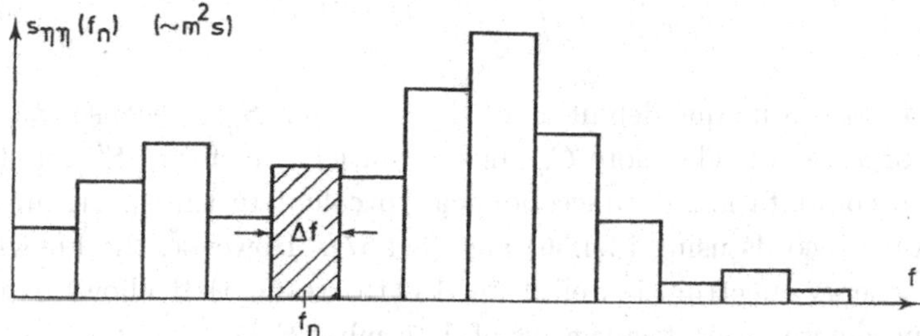

Fig. 3.4.6 The raw spectrum.

Thus we have with $\overline{\eta^2}$ defined as the variance of the total wave motion

$$\overline{\eta^2} = \frac{1}{2}\sum_{n=1}^{\infty} A_n^2 = \sum_{n=1}^{\infty} S_{\eta\eta}\Delta f \tag{3.4.53}$$

$S_{\eta\eta}$ is a step or discrete spectrum as shown in Fig. 3.4.6. It is also called the sample spectrum or the **raw spectrum**, and it represents an estimate of the true spectrum. It is a one-sided spectrum in the sense that it is defined only for positive values of f. If we go to the limmit this can be generalized to

$$var(\eta) = \overline{\eta^2} = \int_0^{\infty} S_{\eta\eta} df \tag{3.4.54}$$

A direct computation of this spectrum shows the same lack of convergence as the Fourier spectrum for a_n, b_n. However, the energy spectrum can be defined uniquely by a different approach, which is based on the **autocovariance function** $C_{\eta\eta}$ defined as the autocorrelation for $\eta(t)$

$$C_{\eta\eta} = \lim_{T_M \to \infty} \frac{1}{T_M}\int_0^{T_M} \eta(t)\eta(t+\tau)dt \tag{3.4.55}$$

It has been shown (see e.g. Jenkins and Watts, (1968)) that the correct energy spectrum $S'_{\eta\eta}$ and $C_{\eta\eta}$ are the Fourier transfoms of each other, that is

$$S'_{\eta\eta}(f) = \int_{-\infty}^{\infty} C_{\eta\eta}(\tau)e^{-i2\pi f\tau} d\tau \tag{3.4.56}$$

$$C_{\eta\eta}(\tau) = \int_{-\infty}^{\infty} S'_{\eta\eta}(f) e^{i2\pi f\tau} df \tag{3.4.57}$$

Here (3.4.56) is a unique definition of the spectrum $S'_{\eta\eta}$, because $C_{\eta\eta}$ goes to zero for $\tau \to \infty$. Therefore $C_{\eta\eta}$ has a Fourier transform, $S'_{\eta\eta}$. It turns out to be computationally uneconomical to calculate the spectrum from actual wave records using (3.4.56) and (3.4.57). However, the knowledge that the energy spectrum is well defined is the result that allows us to go ahead and devise other, faster ways of determing $S_{\eta\eta}$.

Since both $S_{\eta\eta}$ and $S'_{\eta\eta}$ must correspond to the same $var(\eta)$ we have

$$\overline{\eta^2} = \int_0^{\infty} S_{\eta\eta} df = \int_{\infty}^{\infty} S'_{\eta\eta} df \tag{3.4.58}$$

which shows that the raw spectrum has twice the intensity as the spectrum $S'_{\eta\eta}$ defined by (3.4.56).[9]

The details of how to calculate $S_{\eta\eta}$ for a given finite length time series $\eta(t)$ is essentially a problem of signal processing and beyond the scope of the present text. In brief it is done by dividing the time series into segments, each of which provides a raw spectrum, which represents an estimate of the true spectrum. The Fast Fourier Transform (FFT) technique is used for this. In accordance with our findings above each of these raw spectra will give different estimates of $S_{\eta\eta}$ at each frequency f, so the ensemble of raw spectra estimates will show significant scatter around the (unknown) true value for $S_{\eta\eta}$. However, the mean value over all the raw spectra will be closer to the true value and have smaller standard deviation. For more details of the techniques involved see e.g. Jenkins and Watts (1968).

Reproduction of the wave motion specified by a spectrum

We now consider the inverse problem of generating a time series $\eta(t)$ that represents a given energy spectrum. Thus we assume that $S_{\eta\eta}$ is given. Based on (3.4.51) we then have for the amplitude of a wave component in a small interval Δf around the frequency f

$$A^2(f) = 2\, S_{\eta\eta}(f) \cdot \Delta f \tag{3.4.59}$$

However, the spectrum only defines the distribution of energy with frequency. The total time series of length T_1 for the wave motion will consist

[9] To compensate for this the raw spectrum is sometimes defined based on $\frac{1}{4}A_n^2$ instead of $\frac{1}{2}A_n^2$ as in (3.4.53).

of N components where $N = \frac{T_1}{\Delta f}$ and can be written as by

$$\eta(t) = \sum_{n=1}^{N} \sqrt{2S_{\eta\eta}(f_n)\Delta f} \cos(2\pi f_n t - \delta_n) \tag{3.4.60}$$

where δ_n is the phase angle of each wave component. This phase angle δ_n **cannot** be determined from $S_{\eta\eta}(f)$. It is customary as a remedy to assume that δ_n is a stochastic variable evenly distributed over the interval $0 < \delta_n \leq 2\pi$. This is also called a random phase realization of the spectrum. This procedure for generating a wave signal is often used in laboratory experiments with random waves.

It is noticed that by this method we do not represent the true phase conditions in the original wave motion from which the spectrum was determined - the information was lost in the construction of the energy spectrum. Thus when constructing a time series for $\eta(t)$ from $S_{\eta\eta}(f)$ we are not able to reproduce deterministic phenomena such as groupiness of the waves which depend on the phases between the individual wave components. The random choice of phases influence the water surface elevation but it does not change the spectrum. To retain the phase information we need to extend the concept of the amplitude spectrum $S_{\eta\eta}$. For details see e.g. Kirby (2005).

It is important to emphasize that the entire analysis so far of the time series $\eta(t)$ is only considering the conditions at a fixed point. In fact the only wave dynamics involved is in the knowledge that small amplitude linear waves can have sinusoidal shape. Accordingly the wave components in the Fourier series are not propagating waves. No assumptions have so far been made as to how the components may propagate to or from the point considered.

This is also the situation with the time series for $\eta(t)$ constructed above from a spectrum: it is strictly a time series describing the variation at a point. This variation could of course represent the wave motion generated by a wave maker in a laboratory facility, and then the water in the facility would by itself take care of propagating the wave. Our theory so far does not cover that part.

However, if we make the **additional** assumption that the individual components in the spectrum are progressive linear waves, that propagate as constant form waves down the tank without interacting with the other components, then this correponds to replacing $2\pi f_n t = \omega_n t$ with $\omega_n t - k_n x$

in (3.4.60), which then becomes

$$\eta(x,t) = \sum_{n=1}^{N} \sqrt{2S_{\eta\eta}(f_n)\Delta f} \cos(\omega_n t - k_n x - \delta_n) \tag{3.4.61}$$

where k_n is determined by the dispersion relation

$$\omega_n^2 = g k_n \tanh\, k_n h \tag{3.4.62}$$

It is emphasized, however, that because of nonlinear interactions between the wave components in real irregular waves the assumption of independent propagation of the individual components will only give reasonable results for a few wave lengths from the point where spectrum $S_{\eta\eta}(f)$ is determined. See e.g. Sand (1979) who investigated over how long distances linear theory would give acceptable predictions.

Directional wave spectra

In real wave situations the waves usually do not have long crests and constant wave heights along the crests. They are "short crested" and look 3-dimensional. In terms of wave components this can be approximated by assuming the different components come from different directions. That means we could consider dividing the energy in the spectrum (which in principle is an exact quantity) into parts associated with the direction of the wave components. The directional spectrum defines the distribution of wave energy with respect to frequencies and directions.

Since the energy spectrum $S_{\eta\eta}(f)$ (which is not able to distinguish between wave directions) contains all the energy in the wave record obtained at a given point the total energy in a directional spectrum will remain the same as in $S_{\eta\eta}(f)$. Therefore we can formally spread the energy in $S_{\eta\eta}(f)$ by multiplying $S_{\eta\eta}(f)$ with the spreading function $F(f, \alpha_w)$, which satisfies that

$$\int_0^\infty F(f, \alpha_w) d\alpha_w = 1 \tag{3.4.63}$$

Thus for each frequency we have the frequency directional spectrum $S_{\eta\eta}(f, \alpha_w)$ given by

$$S_{\eta\eta}(f, \alpha_w) = S_{\eta\eta}(f) F(f, \alpha_w) \tag{3.4.64}$$

Since the spectral wave component has an amplitude A_n and an energy

$E_n(f, \alpha_w)$ given by

$$\frac{1}{\rho g} E_n(f, \alpha_w) = \frac{1}{2} A_n^2(f, \alpha_w) = S(f, \alpha_w) \tag{3.4.65}$$

we get the total energy in the spectrum given by

$$\begin{aligned}\overline{\eta^2} &= \int_0^{2\pi} \int_0^{\infty} S(f, \alpha_w) d\alpha_w df \\ &= \int_0^{2\pi} \int_0^{\infty} S_{\eta\eta}(f) F(f, \alpha_w) d\alpha_w df\end{aligned} \tag{3.4.66}$$

which shows the realationship between the total spectrum $S_{\eta\eta}(f)$ and the directional spectrum $S(f, \alpha_w)$.

To actually determine the spreading function $F(f, \alpha_w)$ at a location requires a number of wave sensors, not just the one gage needed for the frequency spectrum $S_{\eta\eta}(f)$, and for a limited number of wave gages only an approximation for F can be determined. Both one (i.e. line) and two dimensional arrays have been used. The theory behind this has been discussed in the literature and reference is made to Longuet-Higgins et al. (1963), Barber (1963), Seymour and Higgins (1977), Mitsuyasu et al. (1975) to mention a few. A detailed analysis and review will be given in Kirby (2005).

Parameterized wave spectra for wind generated waves

The analysis of numerous actual wave records from many different locations has lead to the development of parameterized wave spectra, which are mathematical formulas for the the variation of $S_{\eta\eta}(f)$ with f.

One of the frequently used spectra for fully developed seas is the **Pierson-Moskowitz-spectrum**, Pierson and Moskowitz (1964) which is represented by

$$S_{\eta\eta}(f) = \alpha\, g^2 (2\pi)^{-4} f^{-5} e^{-\frac{5}{4}(f_0^4/f^4)} \tag{3.4.67}$$

where α is an empirical constant, and $f_0 = g/2\pi U$ where U is the wind velocity 19.5 m above the sea surface. A frequently used number for α is $\alpha = 8.1 \cdot 10^{-3}$

A further development of the Pierson- Moskowitz-spectrum is the **JONSWAP-spectrum** (Joint North Sea Wave Project) Hasselmann et al. (1973), which is given by multiplying the Pierson-Moskowitz-spectrum by an additional factor. It has the form

$$S_{\eta\eta}(f) = \alpha\, g^2 (2\pi)^{-4} f^{-5} e^{-\frac{5}{4}(f_0^4/f^4)} \gamma^{e^{[-\frac{(f/f_p-1)^2}{2\sigma^2}]}} \tag{3.4.68}$$

Here f_P is the peak frequency of the spectrum while α, γ and σ are further empirical parameters that determine the shape of the spectrum. Values for σ are

$$\begin{aligned} \sigma = \sigma_a = 0.07 \quad &; \quad f \leq f_p \\ \sigma = \sigma_b = 0.09 \quad &; \quad f > f_p \end{aligned} \tag{3.4.69}$$

The other parameters depend on the wave situation, though frequently the values used are $\alpha = 0.0081$ and $\gamma = 3.3$. The latter means that the JONSWAP-spectrum is more peaked than the Pierson-Moskowitz-spectrum.

Today there are many different parameterized wave spectra available. For further details see e.g. Young (1999).

Statistical analysis of waves

Another, less detailed, way of analysing wave records is through the gross parameters for the wave motion such as wave heights and periods. This is useful, also because many of the empirical constants appearing in the parameterized spectra are linked to the gross parameters.

The wave height for the individual wave in a record is usually defined as the **zero-upcrossing-wave height**. As Fig. 3.4.7 shows this is defined as the vertical distance between maximum and minimum surface elevation in the time interval between two succesive times at which the surface passes zero in the upward direction.

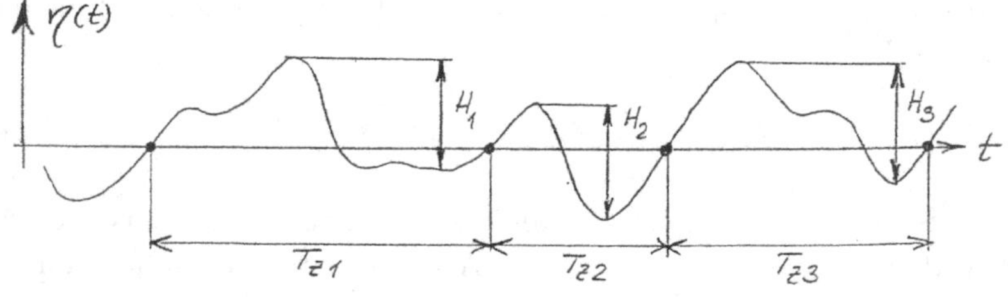

Fig. 3.4.7 Definition sketch for the zero-upcrossing wave height.

This sample of wave heights form a discrete distibution from which we can define a probability density function $p(H)$ that indicates the probability of the wave height having a certain value. Theory shows measurements have confirmed that that under certain simplifying conditions $p(H)$ is close to a **Rayleigh-distribution**, which is given by

$$p(H) = 2\frac{H}{H_{rms}}e^{-(H/H_{rms})^2} \tag{3.4.70}$$

where $H_{rms}^2 = \overline{H^2}$, the "root-mean-squared-wave-height" is the mean value of H^2

Similarly one can define a zero-upcrossing wave period T_z.

The equivalent distribution function $P(H < H_0)$ is then

$$P(H < H_0) = 1 - e^{-(H_0/H_{rms})^2} \tag{3.4.71}$$

Fig. 3.4.8 shows the Rayleigh distribution, which only has the single parameter H_{rms}.

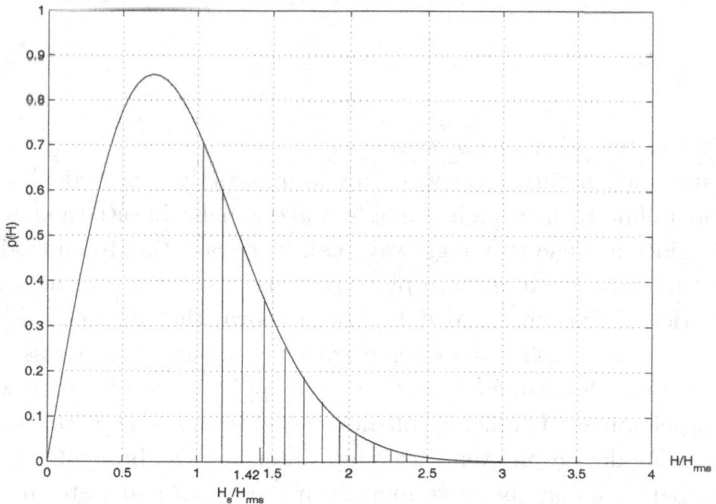

Fig. 3.4.8 The Rayleigh dsitribution for wave heights.

From the wave height distribution (3.4.70) we can of course define a **mean wave height** $\overline{H}$, and the **significant wave height** H_s is defined as the mean height of the highest one third of the waves.

Exercise 3.4-10

Show using (3.4.70) that

$$\overline{H} = \sqrt{\frac{\pi}{4}}\, H_{rms} \sim 0.886\, H_{rms} \tag{3.4.72}$$

$$H_s = H_{1/3} = \sqrt{2.0011}\, H_{rms} = 1.60\, \overline{H} \tag{3.4.73}$$

Thus we also have

$$\overline{H}^2 = 0.785\overline{H^2} \tag{3.4.74}$$

Show also that the most probable wave height H_p (max for the frequency function $p(H)$) is given by

$$H_p = \sqrt{\frac{2}{\pi}}\, \overline{H} = 0.798\overline{H} \tag{3.4.75}$$

and that (in average) $1/10th$ of the waves will have a height larger than

$$H_{1/10} = 1.71\, \overline{H} \tag{3.4.76}$$

Waves of extreme height, freak waves

The statistical nature of the wave height implies that theoretically there is no upper limt to how high a single wave can be in situation with a given $\overline{H}$. Since a single very high wave can be devastating to an engineering structure such as an offshore platform this has lead to a field termed the Statistics of Extremes and it has been shown that certain statistical distributions characterise the extreme values of stochastic processes. The Rayleigh distribution cannot be expected to apply to the very high waves.

The development of numerous offshore oilfields served by platforms with wave recording devices in even very deep water has also brought extensive new information about the wave motion in the ocean under storm conditions. From those records has appeared a (rare but critically important) phenomenon called a "Freak Wave" or a "Rogue Wave", which is a single (or a 2-3) wave(s) with heights is more than $3\overline{H}$. The mechanisms causing this phenomenon are not clear but it is suspected that a focusing effect involving waves of different frequency (longer waves overtaking shorter waves) combined with strong nonlinear interactions between the waves can create these enourmous mountains of wave crests. This also explains why the freak waves seem to emerge out of nowhere, only exist for a very short time, and then disappear.[10]

Connection to the spectrum

It turns out that under certain simplifiying assumptions the gross wave parameters described above can also be connected to the wave energy spectrum. For this purpose we consider the moments m_n of nth order of the spectrum defined as

$$m_n = \int_0^\infty f^n \, S_{\eta\eta}(f) \, df \qquad (3.4.77)$$

We see that the zero'th moment m_0 is

$$m_0 = \int_0^\infty S_{\eta\eta}(f) \, df = \overline{\eta^2} \qquad (3.4.78)$$

It turns out that an important parmeter is ϵ defined by

$$\epsilon^2 = 1 - \frac{m_2^2}{m_0 \, m_4} \qquad (3.4.79)$$

which is called the **spectral width.** It can be shown (Cartwright and Longuet-Higgins, 1956) that for $\epsilon = 0$ the wave *amplitudes* a will be Rayleigh distributed. If in addition we assume the wave heights are $H = 2\,a$ as in linear waves, then the wave heights are Rayleigh distributed too.

$\epsilon = 0$ corresponds to the wave period T being constant. This is of course not the case for real waves in which the wave heights are also not $2a$. However, since measurements show that the actual zero-upcrossing wave heights are close to being Rayleigh distributed the relations which can be derived between the spectrum and the wave heights for the limit of $\epsilon = 0$ are of significant interest.

Exercise 3.4-11

Show that this leads to

$$\overline{H^2} = 8 \, m_0 = 8 \, \overline{\eta^2} \qquad (3.4.80)$$

and that this means that the Reyleigh distribution can also

[10]The recordings of those waves confirm numerous reports from sailors about the sudden occurrence of monstrous waves that for many years were regarded with great suspicion.

be written in terms of the spectral moments as

$$p(H) = \frac{H}{4m_0} e^{-\frac{H^2}{8m0}} \tag{3.4.81}$$

This also implies that

$$\begin{aligned} H_s &= (\sqrt{8\,ln\,3} + 3\sqrt{2\pi}(1 - erf(\sqrt{ln\,3})))\sqrt{m_0} \\ &= 4.0083\sqrt{m_0} \sim 4\sqrt{m_0} \end{aligned} \tag{3.4.82}$$

where erf is the error function.

More generally it is found that for $\epsilon \neq 0$ the more complicated Rice-distribution applies. If this is used we get a relation for H_s which can be approximated with

$$H_s = 4(1 - 0.092\epsilon^2)\sqrt{m_0} \tag{3.4.83}$$

For additional information see also Battjes (1978), Ochi (1982) and Komen et al. (1994).

3.5 Linear wave propagation over uneven bottom

3.5.1 *Introduction*

Up to this point, we have explored the solution to the linearized wave problem on a constant depth. This has lead to determination of essentially all the mechanical properties of the linear wave solution, from propagation speed to velocity and pressure field, and including instantaneous properties as well as depth integrated, time averaged properties.

However, an important limitation of all investigations and results so far has been the assumption of constant water depth and one horizontal dimension (also called 1-D horizontal (1DH) motion or 2-D vertical (2DV) motion).

In the situations encountered in nature, these simplifications are hardly ever satisfied: In practice the depth varies from point to point and the wave motion changes accordingly.

Therefore, it is important to examine to which extent the linear wave theory we have developed can be extended to the more general situations of varying depth and the propagation of the waves in a 2DH domain. This is the objective of the next three sections. The linear wave models described later in Chapter 6 are essentially also based on the results which we will develop in these sections. Therefore they are also subject to the same idealizations and limitations.

Description

In short, we are planning to look at the following generalizations

- Refraction of waves in the horizontal plane (2DH) in regions with varying water depth (depth refraction) (section 3.5).
- The modification of waves by currents (section 3.6).
- Wave propagation in 2DH regions with varying depth and variation of wave heights along wave crests (combined refraction-diffraction) (section 3.7).

It is emphasized from the outset that the objective is to examine how the existing constant depth, linear wave theory can be modified to provide information about the wave motion in the more general situations, not to develop new, more general wave theories.

Thus in mathematical terms the equations we are analyzing are the 3D Laplace equation

$$\phi_{xx} + \phi_{yy} + \phi_{zz} = 0 \tag{3.5.1}$$

with the linearized, combined kinematic and dynamic free surface boundary condition given by

$$\phi_{tt} + g\phi_z = 0 \quad ; \qquad z = 0 \tag{3.5.2}$$

The new feature relative to Chapter 3.2 is the boundary condition at the bottom which is given by

$$\phi_z + \phi_x\, h_x + \phi_y\, h_y = 0 \quad ; \qquad z = -h \tag{3.5.3}$$

The processes that will be examined in this seection are almost exclusively due to the changes in phase velocity that follows as a consequence of the changing depth. Similar phenomena occur in other areas of physics such as optics. They are called **refraction**. In the cases we discuss here where the changes are caused by changing depths the term **depth refraction** is also used.

Fig. 3.5.1 gives a quantitative illustration of the process. When a wave is propagating at an angle to the depth contours one part of the wave front, defined as a curve along a wave crest, is at a smaller depth than another

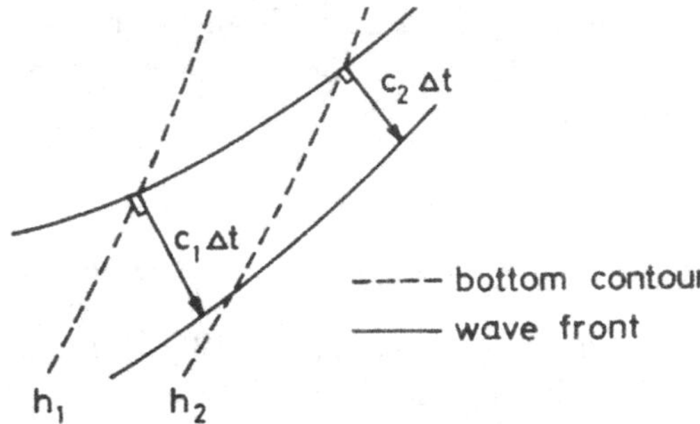

Fig. 3.5.1 The local depth refraction of a wave over a sloping topography (SJ76).

and therefore moves with a smaller speed. This causes the front to change direction as it moves on, as shown in the figure.

Over larger distances, and considering the entire wave pattern it is more illustrative to consider lines perpendicular to the wave fronts, so-called orthogonals. As Fig. 3.5.2 show orthogonals can then be spread or focused depending on the geometry of the depth variation. The similarity to the effect optical lenses have on light beams is clear.

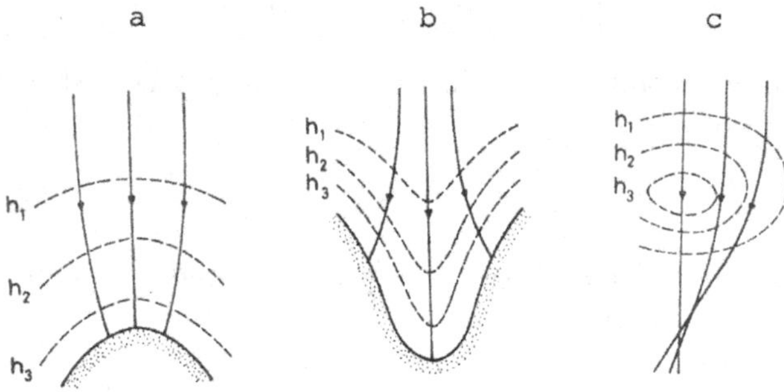

Fig. 3.5.2 Examples of changes in the pattern of wave orthogonals as result of various topograpical variations (SJ76).

In the last frame of Fig. 3.5.2 the wave orthogonals end up crossing each other on the back side of a region with smaller depth (a socalled shoal) in the bottom. If the waves are moving from shallower to deeper water this may also develop to the pattern illustrated in Fig. 3.5.3 where a line develops (called a caustic curve) beyond which the waves orthogonals do not penetrate. The dynamics of the wave motion in the region near the caustic will not be well described by the theories described in the following sections since there is a rapid variation particularly of the wave height around the caustic.

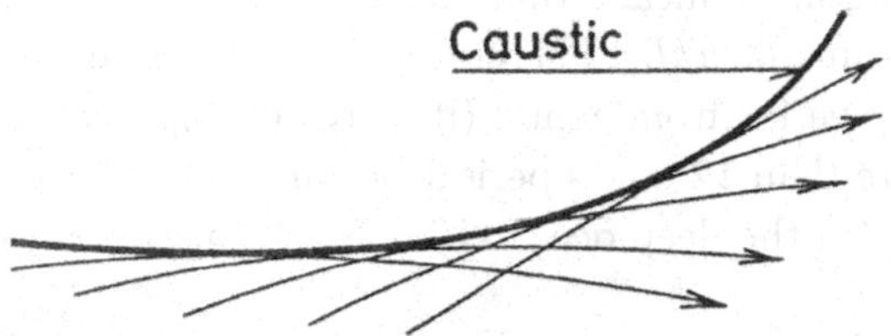

Fig. 3.5.3 Wave orthogonals forming a caustic (SJ76).

Slowly varying depth

The first assumption made to stretch the application of the theory is that the depth is only varying slowly. This is also called "gently varying water depth". Gentle enough that **the constant depth theory derived in the previous section can be applied locally**. In the following this assumption is termed the **locally constant depth assumption**. In a more physical sence it can be expressed as: gently enough that the waves constantly have time to adjust to the new depth as they propagate. This, however, implies that, as we follow the wave from one depth to a slightly different depth, the phase speed c will change, and that is the effect that leads to changes in all other wave properties. Hence, the task becomes to establish ways of tracking these changes.

We can only determine what "sufficiently gentle" means by developing a wave theory that provides information about how a sine wave is modified by a small but noticeable bottom slope h_x. It then turns out that such modifications are proportional to the slope parameter S given by

$$S = \frac{h_x L}{h} \tag{3.5.4}$$

where L is the local wave length $L = cT$ with c determined by the local depth using the constant depth formula (3.2.36).[11] The concept of "gentle

[11]The parameter S is often seen defined as

$$S_k = \frac{h_x}{kh} = \frac{1}{2\pi} S \tag{3.5.5}$$

This of course only changes the numerical value of the parameter.

slope" is then taken to mean that the terms proportional to S^2 (i.e. h_x^2) can be neglected.

We see that S can be interpreted as the relative change in water depth h over a wave length. It means that the gentleness $S \ll 1$ which we seek depends on the value of h/L. Therefore, what is gentle for a storm wave of $T = 8$ sec may be far from gentle (it may act like a vertical wall) for a tidal wave of more than 12 hours period because the tidal wave has a huge value of L/h even in the deep ocean. Fig. 3.5.4 illustrates the idea.

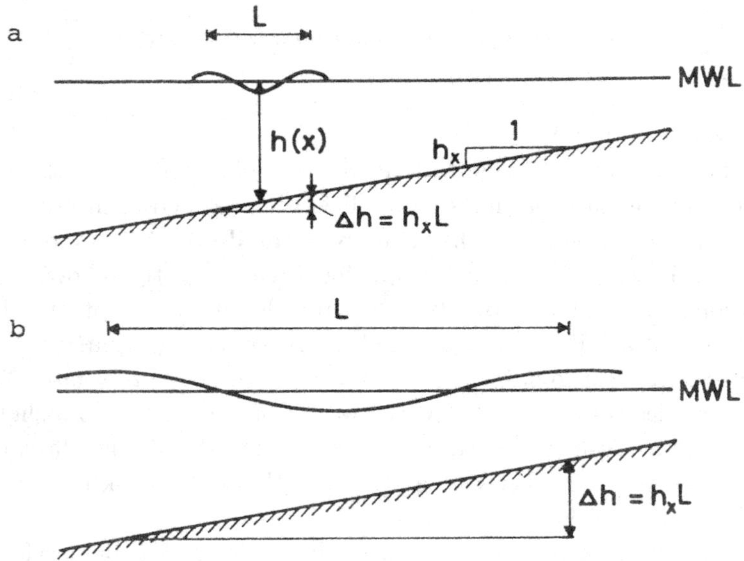

Fig. 3.5.4 The slope parameter S for two different waves on the same slope (SJ76).

While some wave properties are more sensitive to the effects of bottom slope than others, it is generally considered that overall parameters such as wave height H and phase speed c can within reason be determined from the constant depth expressions for S up to 0.5–1. This is also termed the **locally constant depth** assumption. This is discussed further in section 3.7 for the mild slope equation where we also look at some of the effects of the terms proportional to h_x^2 and also h_{xx} toward the end of the section.

Also notice that if we consider a beach with a constant value of S, this will correspond to $h_x L/h$ = constant. Using simple shallow water theory whereby $h/L \propto h^{1/2}$ we see this corresponds to the depth variation

$$h(x) \propto x^2 \tag{3.5.6}$$

x being the distance from the shoreline. We see this coastal profile has a slope decreasing toward the shoreline, which differs from most natural beaches where the slope is increasing toward the shore. Therefore, for a given wave motion, the **requirement of S small generally becomes more restrictive as we approach the shoreline.** However, as long as S is small enough everywhere, there is of course no further constraint on the slope.

3.5.2 *Shoaling and refraction*

Conservation of wave period T

It is implicit in the linear wave theory derived above that each wave component has one wave period T, that is the waves are monochromatic.

When the waves propagate from one point to another, T is conserved. This can be seen by realizing that the number of waves that per unit time pass a point is $1/T$. Since linear waves do not emerge or disappear that must also be the number of waves that pass any other point, i.e., T is the same at all points. This is sometimes called "conservation of wave crests".

Within the framework of the linear wave theory considered, the monochromatic waves will also have long continuous wave fronts.

The problem of tracking the waves as they propagate is then reduced to determining the parameters required to describe the wave motion at one depth h_2 (typically by determining L_2 and H_2) from a full description of the motion at another (reference) depth h_1.

Determination of the wave length and phase velocity variation

Once we know the wave period T, and can assume locally constant depth we can in principle determine the wave length at any depth by solving the dispersion relation (3.2.63) with respect to kh. This was discussed in Section 3.2.2. Thus if we know the wave solution at some depth h_1, we have for a different depth h_2 that

$$\omega^2 = k_1 \tanh k_1 h_1 = k_2 \tanh k_2 h_2 \tag{3.5.7}$$

This equation can of course be solved directly by the iterative procedure described earlier. However, for illustration purposes, it is useful to let the reference depth h_1, say, be infinite. Then (dropping index 2) $\tanh\, k_1 h_1 = 1$ and (3.5.7) can be written in the form

$$\frac{h}{L_0} = \frac{h}{L} \tanh 2\pi \frac{h}{L} \tag{3.5.8}$$

which links h/L at an arbitrary depth h to h/L_0, where $L_0 = gT^2/2\pi$ is the deep water wave length. Table 3.5.1 below shows the solution to this relation. The columns h/L_0 and h/L in the table represent the corresponding values in the equation.

Table 3.5.1 Solutions to (3.5.8) (SJ76).

$\frac{h}{L_0}$	tanh kh	$\frac{h}{L}$	kh	sinh kh	cosh kh	G	$\frac{H}{H_0}$
0.000	0.000	0.0000	0.000	0.000	1.00	1.000	∞
002	112	0179	112	113	01	0.992	2.12
004	158	0253	159	160	01	983	1.79
006	193	0311	195	197	02	975	62
008	222	0360	226	228	03	967	51
0.010	0.248	0.0403	0.253	0.256	1.03	0.958	1.43
015	302	0496	312	317	05	938	31
020	347	0576	362	370	07	918	23
025	386	0648	407	418	08	898	17
0.030	0.420	0.0713	0.448	0.463	1.10	0.878	1.13
035	452	0775	487	506	12	858	09
040	480	0833	523	548	14	838	06
045	507	0888	558	588	16	819	04
0.050	0.531	0.0942	0.592	0.627	1.18	0.800	1.02
055	554	0993	624	665	20	781	1.01
060	575	104	655	703	22	762	0.993
065	595	109	686	741	24	744	981
070	614	114	716	779	27	725	971
0.075	0.632	0.119	0.745	0.816	1.29	0.707	0.962
080	649	123	774	854	31	690	955
085	665	128	803	892	34	672	948
090	681	132	831	929	37	655	942
095	695	137	858	0.968	39	637	937
0.10	0.709	0.141	0.886	1.01	1.42	0.620	0.933
11	735	150	940	08	48	587	926
12	759	158	0.994	17	54	555	920
13	780	167	1.05	25	60	524	917
14	800	175	10	33	67	494	915
0.15	0.818	0.183	1.15	1.42	1.74	0.465	0.913
16	835	192	20	52	82	437	913
17	850	200	26	61	90	410	913
18	864	208	31	72	1.99	384	914
19	877	217	36	82	2.08	359	916
0.20	0.888	0.225	1.41	1.94	2.18	0.335	0.918
0.20	0.888	0.225	1.41	1.94	2.18	0.335	0.918
21	899	234	47	2.05	28	313	920
22	909	242	52	18	40	291	923
23	918	251	57	31	52	271	926
24	926	259	63	45	65	251	929
0.25	0.933	0.268	1.68	2.60	2.78	0.233	0.932
26	940	277	74	75	2.93	215	936
27	946	285	79	2.92	3.09	199	939
28	952	294	85	3.10	25	183	942
29	957	303	90	28	43	169	946
0.30	0.961	0.312	1.96	3.48	3.62	0.155	0.949
31	965	321	2.02	69	3.83	143	952
32	969	330	08	3.92	4.05	131	955
33	972	339	13	4.16	28	120	958
34	975	349	19	41	53	110	961
0.35	0.978	0.358	2.25	4.68	4.79	0.100	0.964
36	980	367	31	4.97	5.07	091	967
37	983	377	37	5.28	37	083	969
38	984	386	43	61	5.70	076	972
39	986	395	48	5.96	6.04	069	974
0.40	0.988	0.405	2.54	6.33	6.41	0.063	0.976
41	989	415	60	6.72	6.80	057	978
42	990	424	66	7.15	7.22	052	980
43	991	434	73	7.60	7.66	047	982
44	992	443	79	8.07	8.14	042	983
0.45	0.993	0.453	2.85	8.59	8.64	0.038	0.985
46	994	463	91	9.13	9.18	035	986
47	995	472	2.97	9.71	9.76	031	987
48	995	482	3.03	10.3	10.4	028	988
49	996	492	09	11.0	11.0	026	990
0.50	0.996	0.502	3.15	11.7	11.7	0.023	0.990
∞	1.000	∞	∞	∞	∞	0.000	1.000

The table can be used in the following ways:

- If T and h are known, calculate L_0 and hence h/L_0. Use the column for h/L to get that quantity. This corresponds to solving (3.5.8) with h/L_0 known.
- If the wave length L_1 is known at one depth h_1 and sought at another depth h_2 (the problem described above), calculate h_1/L_1 and use the table to get h_1/L_0, and hence L_0. This corresponds to calculating h/L_0 from (3.5.8) with h/L known. Then calculate h_2/L_0 and determine h_2/L_2 from the table.

As can be seen, the table also provides values of other functions and parameters encountered in the linear wave theory using any of the columns as input.

Thus, we have determined the wave length at an arbitrary depth from either the wave period T or from the wave length at a reference depth.

Determination of the wave height

The second problem is to deterine the waveheight H at arbitrary points from the reference information. This, however, turns out to be more complicated than determining the wave length, phase speed, etc. While L and c are (assumed to be) strictly local quantities that depend only on the local depth, the refraction process will cause the wave height at a point to depend on the entire propagation pattern leading the wave orthogonal to that point.

Here we first present the classical refraction theory with determination of the wave heights. This approach is well suited to provide insight into the mechanisms and it was for many years used through manual calculations. Later, we examine the more advanced theories that also include the effect of diffraction, but which can only be solved by computational methods.

The physical mechanism responsible for the change in wave height is the change in local energy density in the wave motion. In the slowly varying waves of refraction theory, energy is not moving across wave orthogonals. Therefore, when the wave refraction changes the distance between adjacent wave orthogonals, the energy density changes too and hence the wave heights. To determine the wave height variation we therefore need to determine the refraction pattern. This will be discussed later.

However, when the refraction pattern has been determined, we can consider the wave motion betwen two adjacent wave orthogonals as shown in Fig. 3.5.5. We imagine the reference point is at h_1 where the distance between the orthogonals is b_1 and wave height H_1, and want to determine the wave height H_2 at h_2. The energy balance for the control volume between 1 and 2 then states that the energy flowing into the volume at 1, $b_1 E_{f,1}$ equals the energy flowing out at 2, $b_2 E_{f,2}$, plus the energy $\Delta E_{1,2}$ added or subtracted between 1 and 2. In mathematical terms

$$b_2\, E_{f,2} = b_1\, E_{f,1} + \Delta\, E_{1,2} \tag{3.5.9}$$

Notice that the energy equation (3.5.9) relates **fluxes** of energy, not local energy densities.

The change in energy $\Delta E_{1,2}$ can be due to bottom friction, or energy dissipation due to wave breaking between 1 and 2, but it could also represent energy added due to wind generation. However, even if the dissipation is only due to the (seemingly small) bottom friction, the term $\Delta E_{1,2}$ can be substantial if the distance between 1 and 2 is large enough. In the following we will at first neglect this energy dissipation. However, a warning needs to be issued. The concern is that **energy dissipation accumulates**. It is therefore not always realistic to neglect the energy dissipation in a wave propagation problem, even for non-breaking waves.

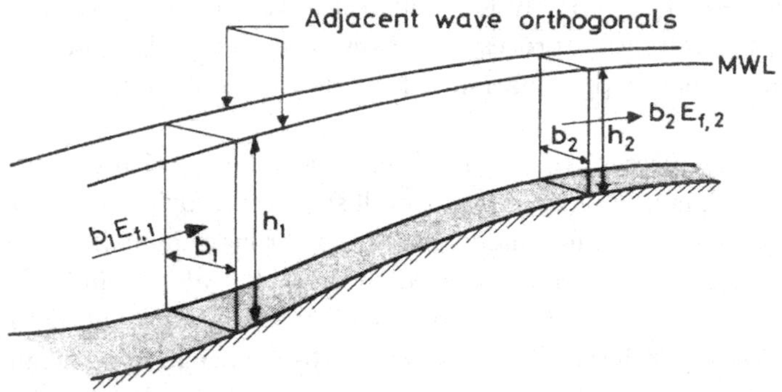

Fig. 3.5.5 Energy flux through sections between adjacent wave orthogonals (SJ76).

For now we ignore the energy change between the two section. Then the wave height at section 2 can be determined by substituting the expressions (3.3.55) for the energy flux. We get

$$b_2 \, H_2{}^2 \, c_2(1 + G_2) = b_1 \, H_1{}^2 \, c_1(1 + G_1) \tag{3.5.10}$$

where G is (again) defined by $2kh/\sinh 2kh$. Using that $c = c_0 \tanh kh$, we can then eliminate the phase velocities, and solving with respect to H_2 then gives

$$\frac{H_2}{H_1} = \left(\frac{b_1}{b_2} \, \frac{\tanh k_1 h_1 (1 + G_1)}{\tanh k_2 h_2 \; 1 + G_2} \right)^{1/2} \tag{3.5.11}$$

It may be noticed that since c_g is given by $c_g = \frac{1}{2}c(1+G)$ we can also write (3.5.10) as

$$b_2 \, {H_2}^2 \, c_{g2} = b_1 \, {H_1}^2 \, c_{g1} \tag{3.5.12}$$

or

$$\frac{H_2}{H_1} = \left(\frac{b_1}{b_2} \frac{c_{g1}}{c_{g2}} \right)^{1/2} \tag{3.5.13}$$

To determine b_1/b_2 we clearly need to determine the propagation pattern for the wave motion in the x, y domain considered.

(3.5.12) may be modified to a form which is mathematically more useful. Consider the closed volume Ω in x, y in Fig. 3.5.2 which has the surface S consisting of sections 1 and 2 and the two orthogonals. Since there is only flux of energy through the two sections 1 and 2 and (3.5.12) has already been integrated over depth we can write (3.5.12) as

$$\int_{b_1, b_2} H^2 \, \mathbf{c_g} \cdot \mathbf{n} \, dS = 0 \tag{3.5.14}$$

where $\mathbf{c_g} = c_g \frac{\mathbf{k}}{k}$. We can then apply Gauss' theorem to (3.5.14) in Ω which gives that

$$\int_{\Omega} \nabla_h \cdot (\mathbf{c_g} H^2) \, d\Omega = 0 \tag{3.5.15}$$

And since this must be true at all points we see that we must have

$$\nabla_h \cdot (\mathbf{c_g} H^2) = 0 \tag{3.5.16}$$

or

$$\nabla_h \cdot \mathbf{E_f} = 0 \tag{3.5.17}$$

where $\mathbf{E_f}$ is the energy flux in the direction of the wave number vector $\mathbf{k}$. This essentially is the wave version of the conservation of energy equation.

3.5.2.1 *Simple shoaling*

In the special case of a long straight coast, the bottom contours are straight and parallel. For waves perpendicularly incident on the ocast, this represents the simple case also reproduced in a 2DV wave flume. Then,

$b_1 = b_2$ everywhere. If we continue to neglect the energy dissipation between 1 and 2, (3.5.11) reduces to

$$H_2 = H_1 \left(\frac{\tanh k_1 h_1}{\tanh k_2 h_2} \frac{1+G_1}{1+G_2} \right)^{1/2} = H_1 \left(\frac{c_{g1}}{c_{g2}} \right)^{1/2} \tag{3.5.18}$$

As for L and c, it is convenient to consider deep water ($k_1 h_1 \to \infty$) as the reference section 1 by which (3.5.18) can be written (neglecting index 2).

$$\frac{H}{H_0} = (\tanh kh(1+G))^{-1/2} = \left(\frac{c_g}{c_0} \right)^{-1/2} \equiv K_s \tag{3.5.19}$$

The ratio H/H_0 is often termed the **shoaling coefficient** K_s and the case is called **simple shoaling.**

Fig. 3.5.6 shows the variation of H/H_0 as a function of h/L_0. It illustrates how the shoaling process causes a significant increase in wave height as the depth is decreasing towards the shore.

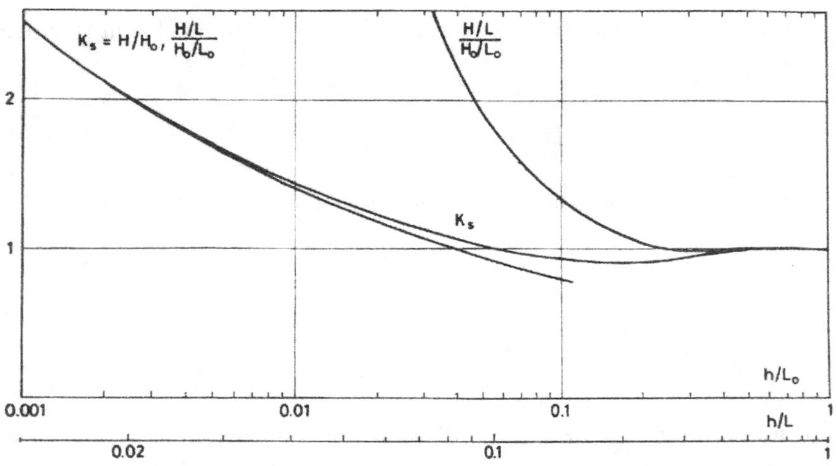

Fig. 3.5.6 The variation of H/H_0, and $\frac{H}{L}/\frac{H_0}{L_0}$ versus h/L_0. The lowest curve represents Green's law (3.5.21) (SJ76).

An interesting feature shown in Fig. 3.5.6 is the minimum for H/H_0 which turns out to occur at $h/L_0 = 1/2\pi = 0.16$. The minimum value of H/H_0 is 0.913.

Exercise 3.5-1

Show that the minimum for H/H_0 occurs where c_g/c_0, the ratio of the group velocity to the deep water phase velocity has a maximum (see Fig. 3.5.6) and verify the minimum value of $H/H_0 = 0.913$.

Exercise 3.5-2

For $h/L < 1/20$, we can use the shallow water approximation for the functions in (3.5.19). Show that this gives the approximation

$$\frac{H}{H_0} = \left(4\pi\frac{h}{L}\right)^{-1/2} \quad \text{for} \quad \frac{h}{L} < 0.05 \tag{3.5.20}$$

or, using (3.2.114)

$$\frac{H}{H_0} = \left(8\pi\frac{h}{L_0}\right)^{-1/4} \quad \text{for} \quad \frac{h}{L_0} < 0.015 \tag{3.5.21}$$

Eq. (3.5.21) is called **Green's law** (Green, 1837), and it is one of the oldest results in classical wave theory. It shows that in shallow water, the wave height changes proportional to $h^{-1/4}$. The curve for (3.5.21) is also shown in Fig. 3.5.6.

Notice also that at the shallow water limit of $h/L = 0.05$, $L/L_0 = \tanh kh$ is only 0.30 giving $h/L_0 = 0.015$ (see Table 3.5.1) so that the shallow water limit in terms of h/L_0 is significantly different from the limit in terms of h/L.

The wave steepness H/L

Fig. 3.5.6 also shows the variation of the wave steepness H/L relative to the deep water steepness H_0/L_0. Writing $H/L = H/H_0 \cdot H_0/L_0 \cdot L_0/L$, the relative steepness can obviously be written

$$\frac{H/L}{H_0/L_0} = \frac{K_s}{\tanh kh} \tag{3.5.22}$$

As could be expected from the fact that as h decreases H increases (beyond the minimum) while L decreases, we find that H/L increases quite rapidly in shoaling water.

This is relevant because one of the basic assumptions leading to the linear wave theory was that H/L was small. Hence, we can expect that as the waves move toward the shore (h/L decreasing) the linear theory rapidly becomes a rather poor approximation for the wave motion. This is one of the reasons for the importance of the nonlinear wave theories described in chapters 7 – 9.

It will be shown in those sections that the determining parameter U is formed by a combination of wave height H, wave length L and water depth h. It is called the **Ursell parameter** after F. Ursell who in (1952) defined the parameter as

$$U = \frac{HL^2}{h^3} \tag{3.5.23}$$

In fact Stokes (1847) already found that this combination of the three geometrical dimensions of the wave motion were important and therefore U is also seen termed the **Stokes parameter**.

3.5.2.2 *Determination of the refraction pattern*

We now turn to the most difficult part of determining the wave height variation on a gently sloping beach. As we have seen, it is necessary to determine the propagation pattern for the waves. This can be interpreted as determining the pattern of the wave orthogonals due to the refraction, which essentially is what we need to obtain the variation of b_1/b_2 in (3.5.11).

We first briefly derive a differential relation valid locally which governs the change $d\alpha_w$ in the direciton of the wave orthogonal at a point, and we show that for a long straight beach with parallel bottom contours, this relation can be integrated to give the global variation of the direction of the wave orthogonals. We also go through a computer oriented version of the theory. We then realize that, although physically sound, the idea of tracing the pattern of wave orthogonals directly results in numerical inconveniences.

The local refraction relation

To derive the local refraction relation, we consider the situation shown in Fig. 3.5.7. The dashed line is a depth contour and the figure shows two successive positions of a wave front, one at t and the other at $t + dt$. The orthogonals have been drawn through the two points where the two wave fronts cross the contour curve.

The angle of incidence α is defined as the angle betwen the contour curve and the wave front (or between contour normal and orthogonal). On

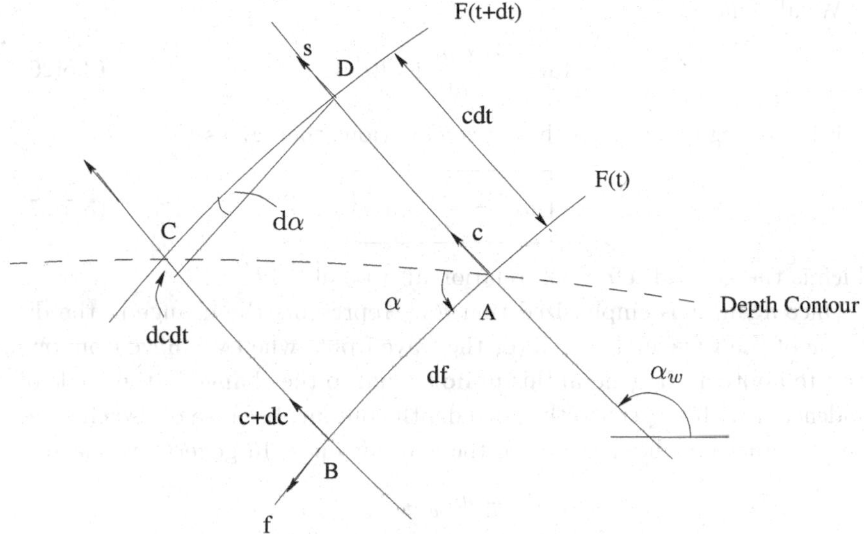

Fig. 3.5.7 The local refraction pattern at an arbitrary point.

the other hand the direction of the wave orthogonal is measured relative to a fixed (arbitrary) direction by the angle α_w. Note that α and α_w are two different quantities and changes in one are not readilly related to changes in the other.

As indicated in Fig. 3.5.7, we count both α and α_w positive counter-clockwise. The (s, f) coordinate system has s along the orthogonal, and f in the direction of the wave fronts, respectively, with directions positive as shown.

The derivation of the refraction relation is based on the observation that if we move from A to B to C, the total change in phase velocity is zero.

Between A and B, we have the change dc_f in the phase velocity. Hence, the difference in distance betwen AD and BC is

$$dc\,dt = \frac{\partial c}{\partial f} df\,dt \tag{3.5.24}$$

This corresponds to a change $d\alpha_w$ in the direction of the wave front which is

$$d\alpha_w = \frac{dc\,dt}{df} \tag{3.5.25}$$

We also have

$$\tan\alpha = \frac{cdt}{df} + O(dc) \tag{3.5.26}$$

Eliminating df between those two equations then gives

$$\boxed{d\alpha_w = \frac{dc}{c}\tan\alpha} \tag{3.5.27}$$

which is the general refraction relation at a point.

Once again it is emphasized that $d\alpha_w$ represents the change in the direction of the wave orthogonal (or the wave front) when we move from one point to another. In general this in **not** equal to the change in the angle of incidence α with respect to the local depth contour, because α also changes due to changes in the direction of the contour lines. In general, we have

$$d\alpha = d\alpha_w - dn \tag{3.5.28}$$

where dn is the change in the direction of the contour lines from point to point. Therefore, (3.5.27) cannot be integrated without including a representation of the bottom topography.

Exercise 3.5-3

Show that (3.5.27) can also be written

$$\frac{\partial\alpha_w}{\partial s} = \frac{1}{c}\frac{\partial c}{\partial f} \tag{3.5.29}$$

which links the rate of change along a wave ray of the wave angle α_w in the fixed (x, y) frame to the rate of change of c along the equivalent wave front.

Parallel bottom coutours, Snell's law

In the special case of straight, parallel depth contours, which occurs only on a cylindrical coast, the bottom topography is represented by $dn = 0$ everywhere. (3.5.28) then shows that

$$d\alpha_w = d\alpha \tag{3.5.30}$$

Substituted into (3.5.27), this means that

$$\frac{d\alpha_w}{\tan\alpha_w} = \frac{dc}{c} \tag{3.5.31}$$

which can be integrated directly to yield

$$\boxed{\frac{c}{\sin \alpha_w} = C} \tag{3.5.32}$$

where C is a constant that is given as $\frac{c}{\sin \alpha_w}$ at a reference point for the wave motion. (3.5.32) is called **Snell's law**, and it is similar to the Snell's law in optics. (3.5.32) is a powerful prediction because it expresses that if we know α_w at a point with depth h, then we know α_w, and hence, by (3.5.30), α everywhere. Thereby, the refraction pattern for the waves is known everywhere. Fig. 3.5.19 shows the pattern of fronts and orthogonals.

Exercise 3.5-4

Derive (3.5.32) from (3.5.31).

Once again, it is emphasized that Snell's law only applies globally for straight parallel depth contours. In the general case, the refraction is governed by (3.5.27).

Exercise 3.5-5

Show by simple geometrical arguments that on a cylindrical coast the ratio b_1/b_2 for the distance between two adjacent orthogonals at depth contours h_1 and h_2 will be the same for all orthogonals at their crossing of h_1 and h_2 and that

$$\frac{b_1}{b_2} = \frac{\cos \alpha_1}{\cos \alpha_2} \tag{3.5.33}$$

so that

$$\frac{H_2}{H_1} = \left\{ \frac{\cos \alpha_1}{\cos \alpha_2} \frac{(1 + G_1) \tanh k_1 h_1}{(1 + G_2) \tanh k_2 h_2} \right\}^{1/2} \tag{3.5.34}$$

where the cosines of course are given by Snell's law.

For $h_1 = \infty$, (3.5.34) can be written (neglecting the index 2)

$$\frac{H}{H_0} = \left\{ \frac{\cos \alpha}{\cos \alpha_0} (1 + G) \tanh kh \right\}^{-1/2} \tag{3.5.35}$$

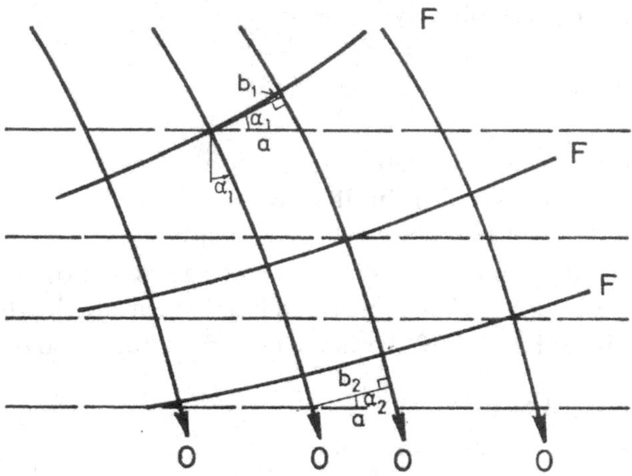

Fig. 3.5.8 The refraction pattern on a cylindrical coast (SJ76).

In parallel with the shoaling coefficient K_s, we may then introduce a **refraction coefficient** K_r

$$\frac{H}{H_0} = K_r \cdot K_s \tag{3.5.36}$$

This equation is general but in the special case of parallel bottom contours, we get

$$K_r = \left\{ \frac{\cos \alpha_0}{\cos \alpha} \right\}^{1/2} \tag{3.5.37}$$

3.5.3 *Refraction by ray tracing*

Wave rays are defined as the propagation paths for the wave energy. For waves with no underlying currents orthogonals are the same as rays. The method of determining the refraction pattern by solving (3.5.27) along wave orthogonals or wave rays is usually called **ray tracing**. Before the age of computers, this was often done semi-graphically on large scale topographical maps of the area in question and by means of special templates that helpled plot the changes in direction of the orthogonals. For a detailed description of this method see e.g., *CERC Shore Protection Manual* (1984) or Svendsen and Jonsson (1976, 1981). This approach, however, is very time consuming and requires great care to obtain reliable results.

A mathematically more refined version of the ray tracing approach was developed by Munk and Arthur (1952) and adopted for computers by e.g., Wilson (1966), Noda (1974), Skovgaard and Berthelsen (1974) and Skovgaard et al. (1975, 1976).

The basic assumptions remain the same as in the previous description: A bottom slope gentle enough to allow the use of the locally constant depth solution for linear waves. The following is a very brief overview of the method. For a more detailed description reference is made to Svendsen and Jonsson (1976, 1981) or Dean and Dalrymple (1984).

The computations use the same s, f coordinate system as used above and essentially follows the wave as it travels in time along the othogonals/rays using s as the distance. Thus,

$$\frac{ds}{dt} = c \tag{3.5.38}$$

This is related to the fixed direction x, y-coordinates by the relations

$$\frac{dx}{dt} = c \cos \alpha_w \quad ; \quad \frac{dy}{dt} = c \sin \alpha_w \tag{3.5.39}$$

and it is found that for the wave moving with speed c, the rate of change of α_w is given by

$$\frac{d\alpha_w}{dt} = \frac{\partial c}{\partial x} \sin \alpha_w - \frac{\partial c}{\partial y} \cos \alpha_w \tag{3.5.40}$$

With appropriate boundary and initial conditions (3.5.39) and (3.5.40) can then be integrated numerically which provides the refraction pattern in terms of the orthogonal patterns in the x,y-plane.

The wave height is then determined by deriving a differential equation which directly determines the variation of the distance $b(x, y)$ between the orthogonals. This is measured by the **orthogonal separation factor** β which essentially is defined as

$$\beta = \frac{b}{b_r} \tag{3.5.41}$$

where b_r is an initial reference spacing for the particular orthogonal. In most cases, the reference positions are in deep water in order to provide easy information about the starting wave condition. By combining this information, the equation for β can then, after some algebra, be written

$$\frac{d^2\beta}{dx^2} + p\frac{d\beta}{ds} + q\beta = 0 \tag{3.5.42}$$

where

$$p = -\frac{1}{c}\left(\frac{\partial c}{\partial x}\cos\alpha_w + \frac{\partial c}{\partial y}\sin\alpha_w\right) \tag{3.5.43}$$

and

$$q = \frac{1}{c}\left(\frac{\partial^2 c}{\partial x^2}\sin 2\alpha_w - \frac{\partial^2 c}{\partial x \partial y}2\sin\alpha_w\cos\alpha_w + \frac{\partial^2 c}{\partial y^2}\cos 2\alpha_w\right) \tag{3.5.44}$$

For details of this formulation see Dean and Dalrymple (1984).

The wave height is then determined by an expression analogous to (3.5.13)

$$\frac{H}{H_r} = \left(\frac{c_{gr}}{c_g}\cdot\frac{1}{\beta}\right)^{1/2} \tag{3.5.45}$$

where H_r is the wave height, c_{gr} the group velocity at the reference position. H, c_g are the values along the wave ray. We see that (3.5.45) is the equation for conservation of energy flux (3.5.17).

3.5.4 *The geometrical optics approximation*

It is instructive to realize that the somewhat intuitive results obtained above for refraction can also be obtained by a more rigorous expansion procedure. This leads to the so-called **geometrical optics approximation** which is the refraction-only parallel to the refraction-diffraction approach described in Chapter 3.7.

The core of this is the WKB method,[12] which formally expresses that the amplitude A of the wave motion (here in terms of its velocity potential) varies much more slowly than the phase θ.

Here we largely follow the approach used by Dingemans (1997), which only looks at the simplified case of a steady wave field with sinusoidal time variation so that the velocity potential Φ_t is given by

$$\Phi_t(x, y, z, t) = Re(\Phi(x, y, z)\, e^{-i\omega t}) \tag{3.5.46}$$

[12] After Wentzel, Kramer, Brillouin. It is also sometimes also called the WKBJ method, where J stands for Jeffreys.

The case of a wave motion which also varies slowly with time is sketched by Mei (1983), p 60ff.

The basic equations (3.5.1)-(3.5.3) are expressed in terms of a set of coordinates X, Y, Z given by

$$X, Y = \frac{x, y}{\Lambda} \quad ; \quad Z = \frac{z}{h_0} \quad ; \quad h' = \frac{h}{h_0} \tag{3.5.47}$$

Here x, y, z are the actual physical coordinates. The ratio between a characteristic local scale (the water depth h_0) and the slow length scale Λ (the scale of the topography variations) represents a small parameter μ given by

$$\mu = \frac{h_0}{\Lambda} \tag{3.5.48}$$

Hence, the smallness of μ is equivalent to assuming gently varying depth ($S \ll 1$).

In terms of these scaled variables, the equations become (with $\overline{\nabla_h} = \mu \left(\frac{\partial}{\partial X}, \frac{\partial}{\partial Y}\right)$)

$$\mu^2 \overline{\nabla_h}^2 \Phi + \Phi_{ZZ} = 0 \qquad -h(x, y) \leq z \leq 0 \tag{3.5.49}$$

$$-\omega^2 \Phi + \frac{g}{h_0} \Phi_Z = 0 \qquad z = 0 \tag{3.5.50}$$

$$\Phi_Z = -\mu^2 \overline{\nabla_h} \Phi \cdot \overline{\nabla_h} h' \qquad z = -h' \tag{3.5.51}$$

where we have also utilized that for sinusoidal time variation

$$\Phi_{tt} = -\omega^2 \Phi \tag{3.5.52}$$

We then formally write Φ as

$$\Phi(x, y, z, t) = Re\left\{\phi(X, Y, Z)\, e^{is}\right\} \tag{3.5.53}$$

where

$$s = s(X, Y, Z) \tag{3.5.54}$$

represents the space varying part of the phase angle θ.

To substitute this into (3.5.49), we get for the derivatives

$$\overline{\nabla_h}\Phi = \left(\overline{\nabla_h}\phi + i\phi\overline{\nabla_h}s\right)e^{is} \tag{3.5.55}$$

$$\overline{\nabla_h}^2\Phi = \overline{\nabla_h}\cdot\left(\overline{\nabla_h}\phi\right) = \left[\overline{\nabla_h}^2\phi - \phi(\overline{\nabla_h}\phi)^2 + i(\overline{\nabla_h}\phi\overline{\nabla_h}s + \overline{\nabla_h}(\phi\overline{\nabla_h}s)\right]e^{is} \tag{3.5.56}$$

$$\frac{\partial\Phi}{\partial z} = \left(\frac{\partial\phi}{\partial Z} + i\phi\frac{\partial s}{\partial Z}\right)e^{is} \tag{3.5.57}$$

$$\frac{\partial^2\Phi}{\partial Z^2} = \left[\frac{\partial^2\phi}{\partial Z^2} - \phi\left(\frac{\partial s}{\partial Z}\right)^2 + i\left(2\frac{\partial\phi}{\partial Z}\frac{\partial s}{\partial Z} + \phi\frac{\partial^2 s}{\partial Z^2}\right)\right]e^{is} \tag{3.5.58}$$

For (3.5.49) we then get a complex equation. In order for this equation to be satisfied the real and the imaginary parts of the equation must be zero separately. The real component can be written

$$\mu^2\left(\overline{\nabla_h}^2\phi - \phi(\overline{\nabla_h}^2 s)\right) + \frac{\partial^2\phi}{\partial Z^2} - \phi\left(\frac{\partial s}{\partial Z}\right)^2 = 0 \tag{3.5.59}$$

and if the imaginary component is multiplied by ϕ then it can be written

$$\mu^2\nabla_h(\phi^2\nabla_h s) + \frac{\partial}{\partial Z}\left(\phi^2\frac{\partial s}{\partial Z}\right) = 0 \tag{3.5.60}$$

The same approach is used for the boundary condition. Thus, the free surface condition becomes

$$Re: \qquad \frac{\partial\phi}{\partial Z} - \frac{\omega^2}{g}\phi = 0 \tag{3.5.61}$$

$$Z = 0$$

$$Im: \qquad \frac{\partial s}{\partial Z} = 0 \tag{3.5.62}$$

and the botton condition similarly becomes

$$Re: \qquad \frac{\partial\phi}{\partial Z} + \mu^2\overline{\nabla_h}\phi\overline{\nabla_h}h' = 0 \tag{3.5.63}$$

$$Z = -h'$$

$$Im: \qquad \frac{\partial s}{\partial Z} + \mu^2\overline{\nabla_h}\theta\overline{\nabla_h}h' = 0 \tag{3.5.64}$$

We then express that the effects of the slowly varying topography is expected to be small, by expanding the amplitude ϕ and the phase s in terms of μ. Since μ only occurs in the equations as μ^2, we use

$$\phi = \phi_0(X,Y,Z) + \mu^2\phi_1(X,Y,Z) + \cdots \tag{3.5.65}$$

For s, we further include in the expansion that the phase s is expected to vary much faster than the slowly varying amplitude.[13] Thus, in terms of the slow variables X, Y, Z, we have

$$s = \frac{1}{\mu}\left(s_0(X,Y,Z) + \mu^2 s_1(X,Y,Z) + \cdots\right) \tag{3.5.66}$$

Substitution into the Re-part of Laplace's equation (3.5.59) then gives the terms

$$\mu^{-2}\phi_0\left(\frac{\partial s_0}{\partial Z}\right)^2 + \mu^0\left(\frac{\partial^2\phi_0}{\partial Z^2} - \phi_0(\nabla_h s_0)^2\right) + O(\mu^2) = 0 \tag{3.5.67}$$

For this to be satisfied for arbitrary μ, we must require that the coefficient to each power of μ vanish. Thus, the lowest order ($O(\mu^{-2})$) gives, for $\phi_0 \neq 0$

$$\mu^{-2}: \qquad \left(\frac{\partial s_0}{\partial Z}\right)^2 = 0 \tag{3.5.68}$$

or s_0 constant over Z. For μ^0, we get

$$\mu^0: \qquad \frac{1}{\phi_0}\frac{\partial^2\phi_0}{\partial Z^2} = (\nabla_h s_0)^2 \tag{3.5.69}$$

In this equation, the consequence of (3.5.68) is that the RHS is independent of Z. Therefore, the LSH must be that too, i.e., be equal to a constant, κ^2 say. Thus, (3.5.69) separates into two equations

$$\frac{\partial^2\phi_0}{\partial Z^2} - \kappa^2\phi_0 = 0 \tag{3.5.70}$$

$$\left(\overline{\nabla_h s_0}\right)^2 = \kappa^2 \tag{3.5.71}$$

Eq. (3.5.70) can be solved directly. The bottom boundary condition for ϕ_0 is found by substituting (3.5.66) into (3.5.63) which gives

$$\frac{\partial\phi_0}{\partial Z} = 0 \qquad ; \qquad Z = -h' \tag{3.5.72}$$

[13] Dingemans shows by rational reasoning that the factor representing this in the expansion must be μ^{-1} and the factor on s_1 must be μ^2.

In dimensional form (3.5.70) can be written

$$\frac{\partial^2 \phi_0}{\partial z^2} + \frac{\kappa^2}{{h_0}^2}\phi_0 = 0 \tag{3.5.73}$$

Thus if we define $\kappa = kh_0$, we have

$$\frac{\partial^2 \phi_0}{\partial z^2} + k^2 \phi_0 = 0 \tag{3.5.74}$$

which has the solution

$$\phi_0 = a_\phi(x, y) \cdot f(z) \quad ; \quad f(z) = \frac{\cosh k(z+h)}{\cosh kh} \tag{3.5.75}$$

Similarly, using the free surface condition with (3.5.66) inserted shows that k must satisfy

$$\boxed{\omega^2 = gk \tanh kh} \tag{3.5.76}$$

Thus, we have found that the first approximation ϕ_0 has the same vertical variation as the constant depth sinusoidal wave motion analysed earlier, and that $(\omega,\ k)$ satisfy the equivalent constant depth dispersion relation.

Exercise 3.5-6

Derive the results (3.5.75) and (3.5.76).

This result verifies the fundamental assumption used in this entire section of wave propagation: **For sufficiently gently varying slope, $(S, \mu \ll 1)$, the wave motion is to the first approximation described by the constant depth solution found in Section 3.2. In other words, we have formally confirmed that the assumption of "locally constant depth", which is used in virtually all linear wave propagation models, is valid as a first approximation when** $S, \mu \ll 1$. This allows us to use the constant depth results for the waves derived in section 3.2 for description of all the wave properties.

It may be in place here, however, to issue a warning. An inspection of the derivation will show that the result only applies to the propagation patterns and amplitude variations. Because the phase motion is locked to the sinusoidal shape, the fast space and time variables such as particle velocities or pressures are not necessarily accurately modelled.

The geometrical optics equations

The second equation (3.5.71) essentially describes the variation of the phase function θ_0 in the horizontal plane. This equation is also called the **eikonal equation** and it is one of the basic equations in the geometrical optics theory. Since a wave front corresponds to θ_0 = constant and wave orthogonals are orthogonal to the wave fronts, the solution of (3.5.71) corresponds to determining the refraction pattern for the wave propagation problem by e.g. a ray tracing method.

As discussed earlier the second half of the refraction problem is then to determine the amplitude variation associated with this refraction pattern. This variation is controlled by the imaginary part of the Laplace equation (3.5.60) with the boundary conditions (3.5.62) and (3.5.64).

Exercise 3.5-7

Show by the same procedure used for the real part, that when the expansions of ϕ and s are substituted into (3.5.60) we get at the order μ

$$\overline{\nabla_h} \cdot (\phi_0 \overline{\nabla_h} s_0) + \frac{\partial}{\partial Z}\left(\phi_0 \frac{\partial s_1}{\partial Z}\right) = 0 \qquad (3.5.77)$$

and show that the boundary conditions for s_1 become

$$\frac{\partial s_1}{\partial Z} = 0 \qquad Z = 0 \qquad (3.5.78)$$

$$\frac{\partial s_1}{\partial Z} = 0 \qquad Z = -h' \qquad (3.5.79)$$

Since neither ϕ_0 nor s_0 depend on Z, we can integrate (3.5.77) over depth. Using the boundary conditions for s_1, we get

$$\int_{-h}^{0} \overline{\nabla_h} \cdot (\phi_0 \overline{\nabla_h} s_0)\, dx = 0 \qquad (3.5.80)$$

By virtue of (3.5.72) and substituting (3.5.75) for ϕ_0, this can also be written

$$\overline{\nabla_h} \cdot \left({a_\phi}^2 \overline{\nabla_h} s_0 \int_{-h}^{0} f(z)^2 dz \right) = 0 \tag{3.5.81}$$

Exercise 3.5-8

Show that

$$\int_{-h}^{0} f(z)^2 dz = \frac{c c_g}{g} \tag{3.5.82}$$

where c and c_g are the phase velocity and the group velocity, respectively, for the waves.

Returning to dimensional coordinates (with $\overline{\nabla_h} = h_0 \left(\frac{\partial}{\partial x}, \frac{\partial}{\partial y} \right) = h_0 \nabla_h$) (3.5.81) can then be written

$$\boxed{\nabla_h \cdot \left({a_\phi}^2 c\, c_g \nabla_h s_0 \right) = 0} \tag{3.5.83}$$

This equation is known as the **transport equation**. Solution of this equation will then determine the variation of the amplitude a_ϕ for ϕ_0.

In conclusion, the equations describing the geometrical optics approximation for the refraction over the (x, y) domain are

- the eikonal equation (3.5.71), which in dimensional form becomes

$$\boxed{\left(\nabla_h s_0 \right)^2 = (\mathbf{k})^2} \tag{3.5.84}$$

where $\mathbf{k}$ is the wave number vector.
- the transport equation (3.5.83)
- the dispersion relation (3.5.76)

with ϕ_0 given by the solution (3.5.75)

$$\phi_0 = a_\phi(x, y) \frac{\cosh k(z + h)}{\cosh kh} \tag{3.5.85}$$

Comparison with earlier results

It is useful to briefly show how these results compare to the earlier results for refraction.

First, it is realized that (3.5.83) is essentially the equation for conservation of energy (3.5.17) (for short termed the "energy equation") for the wave motion. We have that $c = \omega/k$ and using $\nabla_h s_0 = \mathbf{k}$, we can then write the transport equation (3.5.83) as

$$\nabla_h \cdot \left({a_\phi}^2 c\, c_g \nabla_h s_0\right) = \nabla_h \cdot \left({a_\phi}^2 c_g \frac{\mathbf{k}}{k}\omega\right) = 0 \qquad (3.5.86)$$

where

$$c_g \mathbf{k}/k = \mathbf{c_g} \qquad (3.5.87)$$

In Exercise 3.2-6, we found that the amplitude a_ϕ in (3.5.84) is given by

$$a_\phi = -\frac{g}{\omega} A \qquad (3.5.88)$$

where A is the surface amplitude. Thus, (3.5.86) can be written

$$\nabla_h \cdot \left(\mathbf{c_g} A^2\right) = 0 \qquad (3.5.89)$$

or, using $\mathbf{E_f} = \rho g \mathbf{c_g} H^2$

$$\boxed{\nabla_h \cdot \mathbf{E_f} = 0} \qquad (3.5.90)$$

which is the same as (3.5.17). Thus, we see that the transport equation is equivalent to the energy equation energy dissipation.

It may also be shown that the eikonal equation is equivalent to the refraction law (3.5.27) and in particular, for long straight bottom contours, corresponds to Snell's law (3.5.32) .

Thus the geometrical optics approximation corresponds to a mathematically more stringent formulation of the same principles we outlined earlier for the refraction with gently sloping bottom conditions. In particular this implies that the locally constant depth assumption is valid as a first approximation.

The solution of the geometrical optics equations is briefly outlined by Dingemans p 64-66.

3.5.5 *Kinematic wave theory*

From a numerical point of view, the ray tracing method in all its forms (be it the simple graphical or the β-separation method described above) has the drawback that information about the wave heights is provided only at points along the ray. Since the rays are part of the solution and may spread or approach each other in the computational domain in ways that cannot be determined until the solution is available, it is difficult to predict where information may become available for a given set of start conditions. A method providing information about the wave motion in a regular pre-chosen grid is much more convenient.

As mentioned ray tracing and the geometric optics approximation are also intimately linked to the assumption of steady waves with harmonic (i.e. sinusoidal) fast time and space variation (see (3.5.52) and (3.5.70)) but time independent amplitudes. It is possible, however, to establish a simpler wave theory, based on the same assumptions of locally constant depth and motion described by a slowly varying amplitude times a phase function, the so-called **Kinematic wave theory** (see e.g., Phillips, 1966 and later editions for a steady amplitude version).

For varying depth the wave number k is a function of the position. In the phase function θ, kx is therefore replaced by $\int k dx$. Further, in a 2DH situation, k becomes a (horizontal) vector $\mathbf{k}(x, y)$, called the **wave number vector**, which is pointed in the (local) direction of wave propagation (i.e., $\mathbf{k}$ is a tangent to the wave orthogonal). It also is possible that the frequency ω varies (slowly) with time. Therefore, the appropriate form for θ is

$$\theta = \int \omega_a dt - \int \mathbf{k} \cdot d\mathbf{x} \tag{3.5.91}$$

where $\mathbf{x}$ is the (x, y)-position vector. It is noticed that ω_a is the **absolute** frequency frequency measured at a fixed point. For waves on a current $\mathbf{V}$, ω_a is linked to the $k = |\mathbf{k}|$ and $\mathbf{V}$ by the dispersion relation

$$\boxed{\omega_a = \omega_r(kh, kH) + \mathbf{k} \cdot \mathbf{V}} \tag{3.5.92}$$

where ω_r is the relative frequency satisying (3.2.35). For details see Section 3.6 on waves and currents. Notice it has not been excluded that ω_r and hence the wave phase speed depend on the wave height in addition to the variation with depth we know from linear wave theory. In any case the method assumes that the function for ω_r is known.

We then assume that the surface variation $\eta(x, y, t)$ can be described by

$$\eta(x, y, t) = A(x, y, t) f(\theta) \tag{3.5.93}$$

where the amplitude A is a slowly varying function of position and time while the phase motion described by θ varies rapidly with θ.

This form for the wave motion has several important consequences. First, we notice that, we have

$$\mathbf{k} = -\nabla_h \theta \tag{3.5.94}$$

where $\nabla_h = \partial/\partial x,\ \partial/\partial y$ is the gradient operator θ. And since the curl of a gradient is zero $\mathbf{k}$ is irrotational implying that

$$\nabla_h \times \mathbf{k} = 0 \tag{3.5.95}$$

Similarly, we have from (3.5.91)

$$\frac{\partial \theta}{\partial t} = \omega_a \tag{3.5.96}$$

By cross-differentiating (3.5.94) and (3.5.96) (differentiate the first by t, the second by ∇_h), we can eliminate θ to get

$$\boxed{\frac{\partial \mathbf{k}}{\partial t} + \nabla_h \omega_a = 0} \tag{3.5.97}$$

(3.5.92) and (3.5.97) represents a system of equations from which $\mathbf{k}$ and ω_a can be determined. The precise form of the phase function $f(\theta)$ has not been specified. As long as the locally constant depth assumption is satisfied any constant depth wave theory can be used to provide the wave properties needed for the method. This is one of the advantages of this approach.

It is interesting that if we take the curl $\nabla_h \times$ of (3.5.97) we get

$$\nabla_h \times \left(\frac{\partial \mathbf{k}}{\partial t} + \nabla_h \omega_a\right) = 0 \tag{3.5.98}$$

and since $\nabla_h \times \nabla_h \omega_a \equiv 0$ this implies that

$$\frac{\partial}{\partial t}(\nabla_h \times \mathbf{k}) = 0 \tag{3.5.99}$$

This means that if $\nabla_h \times \mathbf{k} = 0$ initially then it remains zero at all times. Thus the condition (3.5.95) only acts as a (necessary) initial condition for the model system.

In the simple case, we have considered so far, $\mathbf{k}$ is constant in time, so that

$$\nabla_h \omega_a = 0 \tag{3.5.100}$$

or

$$\omega_a = C(t) \tag{3.5.101}$$

where again, with monochromatic waves, this means $C(t) = C$ which confirms that ω is conserved in the wave field (= the wave fronts are conserved, as shown in Section 3.5.1).

Exercise 3.5-9

Write $\mathbf{k} = k\ cos\alpha_w, k\ sin\alpha_w$ where $k = |\mathbf{k}|$ and get (3.5.93) on the form

$$\frac{\partial k \sin \alpha_w}{\partial x} - \frac{\partial k \cos \alpha_w}{\partial y} = 0 \qquad (3.5.102)$$

Show then that for a cylindrical coast with no variations in the y direction, this can be solved to recover Snell's law

$$\frac{\sin \alpha_w}{c} = \text{constant} \qquad (3.5.103)$$

In general cases solution of of the equations (3.5.92) and (3.5.97) will provide the wave pattern. This corresponds to solving the eikonal equation in traditional geometrical optics theory. The wave heights are then determined by solving the energy equation which is discussed in more detail in Chapter 4.

This method forms part of many ocean wave models such as the WAM-model (see WAMDI-group, 1988) and the SWAN-model, both discussed in Chapter 6. In those models it is used to determine the wave propagation patterns and is combined with the special version of the energy equation called the wave-action equation introduced in Chapter 4.

In the nearshore the method has so far mostly been used extensively in the simple cases where Snell's law is valid, though time varying waves have been used, and recently in an application for waves with space and time varying currents, which implies solving the general equations (3.5.92) and (3.5.97). These applications are discussed in Chapter 6.

3.6 Wave modification by currents

3.6.1 *Introduction*

In the nearshore region, the most important currents are horizontal and they have horizontal extensions (length scales) that are of the order of a wave length or much larger. Due to turbulence from various sources and the mechanisms responsible for generating the currents, they may show vertical variations in the horizontal velocity, typically in magnitude, but often the direction of the velocity is also varying over the vertical.

The presence of currents changes the waves. However, as with depth refraction, the lowest order effect turns out to be in the dispersion relation through changes in the speed of wave propagation. Because those kinematic properties do not depend on the wave amplitude, we are, in parallel with the depth refraction, able to analyze the propagation patterns in linear waves on currents separately from the analysis of the dynamics which provides the information about the amplitude variations.

A common assumption is also that we analyze the effect the current has on the waves but assume the currents remain unchanged by the waves. This of course is a simplification that is discussed extensively in the chapters about wave induced currents and nearshore circulation.

In the following, we first analyze the local effects of the simplest case, a wave motion on a steady current, uniform both over depth and in the horizontal plane. Then give a brief overview of the local effects of depth varying currents, and finally briefly discuss extension of the geometric optics approximation to the case of combined wave-current motion in the horizontal plane.

3.6.2 *Waves on a steady, locally uniform current*

Thus, we consider a current with the steady velocity **U**. The uniformity of the current means an infinitely large length scale. The primary effect is that the wave will propagate relative to the moving water.

Therefore, a wave with wave number **k** will satisfy the same dispersion relation found earlier, that is

$$\omega_r = gk \tanh kh \tag{3.6.1}$$

where ω_r, the **relative wave frequency**, represents the frequency measured by an observer moving with the current velocity and $k = \mid \mathbf{k} \mid$. In the general case, the current direction will differ from the wave direction. The wave speed, however, is measured in the direction **perpendicular** to the wave front. Therefore, the current component **parallel** to the wave front will not contribute to the speed of the wave. From a coordinate system **fixed** in space, we will therefore see the wave moving at an absolute speed c_a which is in the direction of **k** and given by

$$c_a = c_r + \mathbf{k} \cdot \mathbf{U} \tag{3.6.2}$$

or

$$c_a = c_r + U \cos \mu \tag{3.6.3}$$

where $U = \mid \mathbf{U} \mid$ and μ is the angle between $\mathbf{k}$ and $\mathbf{U}$. The situation is illustrated in Fig. 3.6.1.

However, since the wave is merely translated with the current, the wave length will look the same from the fixed and moving coordinate system, i.e., k is the same. Hence, we will have, since $\omega = ck$

$$\omega_r = c_r k \quad ; \quad \omega_a = c_a k \tag{3.6.4}$$

which substituted in (3.6.3) gives the relation

$$\boxed{\omega_a = \omega_r + kU \cos \mu} \tag{3.6.5}$$

This is also called the **Doppler relation**. It relates the absolute frequency ω_a observed from a fixed coordinate system to the relative frequency ω_r which is the frequency that satisfies the dispersion relation (3.6.1) (and which is observed from a coordinate system moving with the currents.

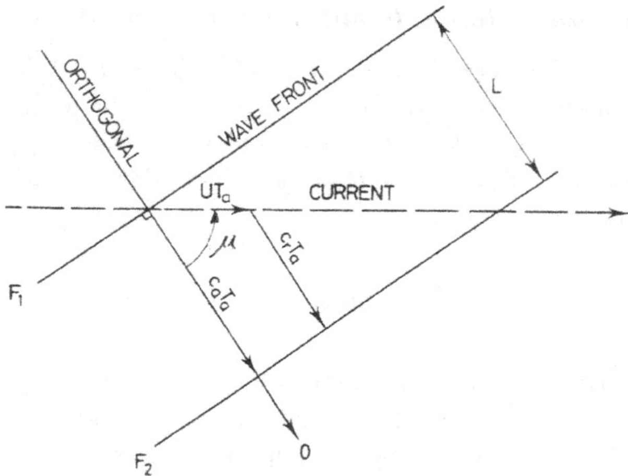

Fig. 3.6.1 The relation between absolute and relative phase velocity for a wave in a steady uniform current (from Jonsson, 1989).

Notice that for a monochromatic wave motion wave fronts will be conserved, so that in any region the number of waves moving into the region will be the same as the number of waves moving out of the region. This means that the **absolute** wave period T_a and hence the absolute frequency ω_a are conserved. That is also the frequency that in practice is easy to measure through measurements of the absolute wave period $T_a = 2\pi/\omega_a$, i.e. the period observed from a fixed point in space. On the other hand, to evaluate other features of the wave motion such as velocities, pressures, etc., we need to determine k through ω_r. Therefore, the solution of (3.6.5) is important.

Substituting (3.6.1) into (3.6.5) we get

$$\omega_a - kU\cos\mu = \pm\sqrt{gk\tanh kh} \tag{3.6.6}$$

which shows that the solution kh for given ω_a depends on $U\cos\mu$ which therefore is the parameter for the problem.

While a numerical approach is required for specific cases, we see that solving (3.6.6) essentially corresponds to seeking crossing points between the two curves,

$$\begin{aligned} y_1 &= \omega_a - kU\cos\mu \\ y_2 &= \pm\sqrt{gk\tanh kh} \end{aligned} \tag{3.6.7}$$

This is illustrated in Fig. 3.6.2.

In the figure, $U = 0$ corresponds to $\omega_a = \omega_r$ (point E in the figure), a wave moving in a region with no current. However, for $U\cos\mu \neq 0$ both the $+$ and the $-$ case in (3.6.6) potentially offer two solutions for kh.

Opposing currents

In the case of $U\cos\mu < 0$ the currents oppose the waves. There can be two, one or zero solutions depending on the magnitude of $U\cos\mu$. In the following discussion we assume that the current is changing slowly enough that locally it can be considered uniform. If a wave motion moves from a region with no current (point E) into a region with increasingly stronger opposing current the line for $\omega_a - U\cos\mu$ gets steeper and point A moves upwards on the upper curve in the figure. Clearly this means that the stronger the opposing currents the larger the wave number k becomes, that is, for a given ω_a **opposing currents reduce the wave length**.

In this general case of a weaker current, A corresponds to a situation where both the phase speed c and the group velocity (that is energy propagation velocity) $c_g = \partial\omega/\partial k$ are $>|\, U\cos\mu\,|$. Thus, both the wave train

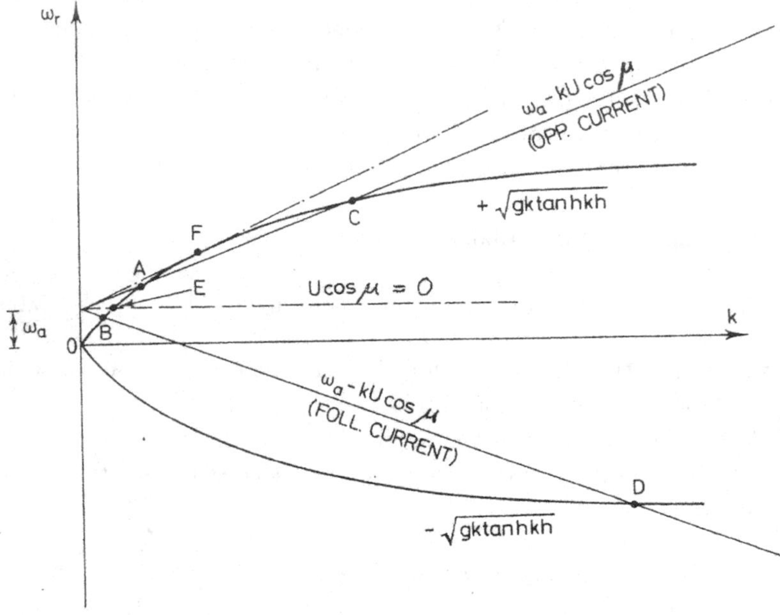

Fig. 3.6.2 Solutions for the wave number in terms of kh for given values of ω_a, h and $U \cos \mu$. Notice that $U \cos \mu$ can be both positive and negative. The case of following currents corresponds to $U \cos \mu > 0$ (upper branch of $\omega_a - kU \cos \mu$), while opposing currents correspond to $U \cos \mu < 0$ (from Jonsson, 1989).

with this kh and its associated energy are able to move against the current, i.e., a group of waves will propagate upstream.

When the current gets strong enough ($U \cos \mu$ numerically large, the straight line in the figure gets steeper) the waves are blocked by the currents from propagating further. In the figure this means A reaches point F (one solution) which is the blocking point or caustic point. Point F represents waves just able to maintain their position against the current because at F the absolute group velocity $c_{ga} = 0$ (Jonsson et al., 1970). Since the group velocity is defined as $d\omega/dk$ it is obtained by differentiating (3.6.5) with respect to k. This gives

$$c_{ga} = c_{gr} + U \cos \mu \tag{3.6.8}$$

Thus blocking of the waves occur when $c_{gr} = -U \cos \mu$.

Figure 3.6.2 is essentially based on ray theory, and if an analysis of the wave height variation approaching the blocking point is carried out according to that theory it results in infinite values (the wave height is singular). However, Peregrine (1976) conducted a nonlinear analysis in the neighborhood of the blocking point and showed that in reality the waves have large but finite steepness. It also turns out that (if no breaking occurs) the waves are reflected at the blocking point but with a change in wave number. The wave solution at point C, the second solution, corresponds to the waves reflected from the blocking point. Since the early contributions quoted above this problem has been discussed by a number of authors and lately by Madsen and Schäffer (1999), and by Chawla and Kirby (2000).

In nature blocking frequently occurs outside inlets and in river mouths with strong (often tidally dominated) currents. Where the current is strong enough to arrest the waves, the result is a region with very high/steep waves that often break before they reach the actual blocking point. Such regions represent navigational hazards particularly to smaller craft.

If the opposing current gets even stronger (the y_1-line entirely above the y_2-curve) there are no solutions, i.e., no waves are able to propagate up against the current.

Following currents

The case of $U \cos \mu > 0$ corresponds to currents following the wave motion. It is represented by the lower branch of $\omega_a - kU \cos \mu$ in Fig. 3.6.2. Here, solution B represents the simple case of a wave motion on a following current. It is seen that B corresponds to a kh-value smaller than for case E (wave with the same period on no current) indicating the effect of the following current is to increase c_a and hence L.

Wave orthogonals and wave rays

In waves with currents the wave orthogonals and the wave rays generally have different directions.

The wave orthogonals are defined as curves normal to the wave fronts. As Fig. 3.6.1 shows in a (locally) uniform current the absolute phase velocity c_a given by (3.6.1) is in the direction of the wave orthogonal.

A wave ray, on the other hand, is defined as the curve along which wave energy is moving. For a fixed system this will be in the direction of the absolute group velocity $\mathbf{c_{ga}}$. The definition of $\mathbf{c_{ga}}$ (and thereby the direction of the ray) is given by

$$\mathbf{c_{ga}} = \mathbf{c_{gr}} + \mathbf{U} \tag{3.6.9}$$

which differs from the definition (3.6.3) for c_a and has a direction different from the orthogonal. When the wave height varies along the wave front then the current component along the front matters because in addition to moving energy in the direction of the orthogonal it creates a net energy flux along the front.

3.6.3 *Vertically varying currents*

The solution of the Doppler version (3.6.5) of the dispersion relation represents a solution of the local problem of determining the parameters required to evaluate the formulas for velocity, pressure, etc. for the wave motion at a point.

Similarly, in this section we discuss the problem of determining the modification of the wave motion at a point in various cases with currents that vary over depth. This problem has been discussed in the literature by various authors. For an extensive overview of the dispersion relations for various cases see Peregrine (1976). An account of the solution for some simple current profiles is given by Dingemans (1997), while an approximate solution for more general current profiles is found in Kirby and Chen (1989). The following is a brief overview, which combines the last two versions. It also also provides some information about the changes in the wave particle velocities. For simplicity the discussion is limited to 2DV motion.

The analytical solutions available all assume that the current profile $U(z)$ is known (e.g., from measurements). The horizontal variations of the current field are assumed so slow that contribution to the equations from horizontal variations of the current can be neglected in the local relations. On the other hand, the current velocity U is assumed relatively strong in comparison to the wave particle velocity u_{wi}, w_w (u_{wi} the horizontal, w_w the vertical component, respectively, of the wave motion.)

Thus, we consider a surface elevation ζ and total velocities u_i, w that can be written

$$\eta = \eta_o(x,t) + \zeta_w(x,t) \tag{3.6.10}$$

$$u = U(x,z) + u_w(x,z,t) \tag{3.6.11}$$

$$w = W(x,z) + w_w(x,z,t) \tag{3.6.12}$$

Here it is assumed that the wave amplitude is small enough that $u_w, w_w = O(\epsilon)$, that $U = O(1), W = 0$, and that $\frac{\partial U}{\partial z} = O(1), \frac{\partial U}{\partial x} = O(\beta)$ where ϵ, β are small parameters measuring the magnitude of the wave velocities and the gradient of the current, respectively.

Substituted into the inviscid fully nonlinear (that is Euler) equations of motion and the associated (nonlinear) boundary conditions, this results in the following lowest order of the equations for inviscid motion.

$$\frac{\partial u_w}{\partial x} + \frac{\partial w_w}{\partial z} = 0 \tag{3.6.13}$$

$$\frac{\partial u_w}{\partial t} + U\frac{\partial u_w}{\partial x} + w_w \frac{\partial U}{\partial z} = -\frac{1}{\rho}\frac{\partial p_D}{\partial x} \tag{3.6.14}$$

$$\frac{\partial w_w}{\partial t} + U\frac{\partial w_w}{\partial x} = -\frac{1}{\rho}\frac{\partial p_D}{\partial z} \tag{3.6.15}$$

where p_D is the dynamic pressure created by the wave motion. We see that these equations are linearized because we assume $U \gg u_w$. The boundary conditions are

$$w_w = \frac{\partial \zeta_w}{\partial t} + U\frac{\partial \zeta_w}{\partial x} \qquad z = 0 \tag{3.6.16}$$

$$w_w = 0 \qquad z = -h \tag{3.6.17}$$

$$p_D = \rho g \zeta_w \qquad z = 0 \tag{3.6.18}$$

which are linearized in a similar way as the Euler equations.[14]

We assume the wave part of the motion is described by a stream function

$$\psi = \Re\left\{a_\psi f(z)\, e^{i\theta}\right\} \qquad \theta = \omega_a t - kx \tag{3.6.19}$$

which implies that

$$u_w = \frac{\partial \psi}{\partial y} \quad ; \quad w_w = -\frac{\partial \psi}{\partial x} \tag{3.6.20}$$

and where a_ψ and $f(z)$ are also slow functions of x_i and t. Substitution of this into the equations for the wave motion then shows that f must satisfy

[14]Since we seek only information about the effect the current has on the waves, but assume the currents unchanged, these equations assume inviscid motion also do not contain the (viscid or turbulent) terms required to actually maintain the assumed current profiles over some horizontal distance.

the equation.

$$\frac{\partial^2 f}{\partial x^2} - \left(k^2 - \frac{kU''}{\omega_a - kU}\right) f = 0 \qquad (3.6.21)$$

with the boundary conditions

$$(U - c_a)\frac{\partial f}{\partial z} - \left(\frac{g}{U - c_a} + U'\right) f = 0 \qquad z = 0 \qquad (3.6.22)$$

$$f = 0 \qquad z = -h \qquad (3.6.23)$$

where $U' = \partial U/\partial z$ at $z = 0$, $U'' = \partial^2 U/\partial z^2$

Eq. (3.6.21) is the inviscid form of the so-called Orr-Sommerfeld equation. It is also named the Rayleigh equation.[15]

Exercise 3.6-1

Derive (3.6.21), (3.6.22) and (3.6.23) from the basic equations.

Linear current profiles

Analytical solutions to (3.6.21) are only possible for a few special cases. The simplest case is the **linear current profile** for which $U''(z) = 0$. It is interesting for perspective to pursue this case. The linear profile can be written

$$U(z) = U_s(1 + \alpha\frac{z}{h}) = U_s(1 + \frac{U'}{U_s}z) \qquad (3.6.24)$$

where U_s is the current velocity at the MWL $z = 0$. Then (3.6.21) reduces to

$$\frac{\partial^2 f}{\partial z^2} - k^2 f = 0 \qquad (3.6.25)$$

which has the solution

$$f = a_\psi \frac{\sinh k(z+h)}{\cosh kh} \qquad (3.6.26)$$

[15]Equations of this type frequently emerge in hydrodynamic stability problems. For detailed discussion, reference is made to Drazin and Reid (1981).

and the surface condition (3.6.22) gives the dispersion relation (with $c = \omega/k$)

$$(U_s - c_a)^2 - \frac{c_r^2}{g} U'(U_s - c_a) - c_r^2 = 0 \qquad z = 0 \tag{3.6.27}$$

where c_r is the relative phase velocity given by ω_r/k with ω_r from (3.6.1) i.e.

$$c_r^2 = \frac{g}{k} \tanh kh \tag{3.6.28}$$

(3.6.27) is essentially the modified dispersion relation for the case of a linear current profile.

Exercise 3.6-2 Derive (3.6.27) from (3.6.22). Then show that the solution to (3.6.27) can be written

$$U_s - c_a = c_r \left[c_r \frac{U'}{2g} - \sqrt{\left(c_r \frac{U'}{2g} \right)^2 + 1} \right] \tag{3.6.29}$$

or

$$kU_s - \omega_a = \omega_r \left[c_r \frac{U'}{2g} - \sqrt{\left(c_r \frac{kU'}{2g} \right)^2 + 1} \right] \tag{3.6.30}$$

It is noticed that in the case of a linear current, it is the derivative U' of the current profile $U(z)$ that gives the modification of the dispersion relation.

For the vertical velocity distribution, it is found that the amplitude a_w is given by

$$a_w = i \, \frac{kU - \omega_a}{\tanh kh} \, a_\zeta \tag{3.6.31}$$

where a_ζ is the surface amplitude given by

$$\zeta = \Re(a_\zeta e^{i\theta}) \tag{3.6.32}$$

Thus, w_w becomes

$$w_w = \Re \left\{ i(kU - \omega_a) a_\zeta \, \frac{\sinh k(z+h)}{\sinh kh} e^{i\theta} \right\} \tag{3.6.33}$$

and u_w can be determined from the continuity equation.

In conclusion, we see that in a depth varying current the amplitudes of the wave velocities as well as the dispersion relation are modified by the current.

As a check we see that if we assume U constant over depth with $U = U(0)$, then we recover the dispersion relation (3.6.6). To complete the picture, we also realize that for $U'(0) = 0$, we have $\omega_a - kU = \omega_r$ by (3.6.30) so that (3.6.33) simplifies to

$$w_w = \Re\left\{ i\omega_r a_\varsigma \frac{\sinh k(z+h)}{\sinh kh} e^{i\theta} \right\} \tag{3.6.34}$$

which at the surface $z = 0$ gives $w_w = \frac{\partial \zeta_w}{\partial t}$ as in ordinary linear waves. This confirms the intuitively obvious result that in a depth uniform current, the wave is just convected with the current without other changes, and puts the earlier results for depth varying currents into further perspective.

General current profiles

For general current profiles, approximate solutions are available if the current is assumed weak relative to the wave speed c. This was pursued by Kirby and Chen (1989). Expanding both f and c in power series Kirby and Chen showed that in the first approximation, the dispersion relation can be written in a form similar to (3.6.5).

$$\omega_a = \omega_r + k\tilde{U} \tag{3.6.35}$$

where here $\tilde{U}$ is a weighted average over depth of the actual current profile $U(z)$. $\tilde{U}$ is given by

$$\tilde{U} = \frac{2k}{\sinh 2kh} \int_{-h}^{0} U(z) \cosh 2k(z+h) dz \tag{3.6.36}$$

For a linear current profile they confirm the exact solution (3.6.28) and find that in the first approximation $\tilde{U}$ is

$$\tilde{U} = U_s - \frac{U'h}{2} + O(kh)^2 \tag{3.6.37}$$

and

$$f(z) = \sinh\ k(z+h) \tag{3.6.38}$$

Their results imply that the parameter

$$a = \frac{U'c_r}{2g} = O(\epsilon) \tag{3.6.39}$$

is the parameter that expresses that $U \ll c$. Accordingly the second order approximation for c_a becomes (in our notation)

$$c_a = \tilde{U} + c_r \left[1 + \frac{1}{2} \left(\frac{U' c_r}{2g} \right)^2 \right] \tag{3.6.40}$$

which corresponds to a Taylor expansion of the exact solution (3.6.28) as expected.

For further details, reference is made to Kirby and Chen (1989). They also provide information about the particle motion and other aspects of the flow.

3.6.4 *The kinematics and dynamics of wave propagation on current fields*

So far we have only considered the local problem of identifying the effect of a current on the wave motion, when the horizontal variations of the current is assumed slow enough to let the waves adjust to the local current conditions "locally constant current".

However, in nature waves propagate through current fields that change in space and time and the waves change the currents. The question arises: what are the laws governing this process.

This much more complicated problem has been discussed in the literature within the framework of the geometrical approximtion, and is briefly referenced here. For the cases discussed the waves are assumed not to be amplitude dispersive (which means the phase velocity does not depend on the wave amplitude). It is then possible to separate the kinematics of the wave propagation pattern from the dynamics of how the wave amplitudes change, which of course greatly simplifies the solution. The solutions assume the currents to be so slowly varying that the "locally constant current" results developed in the previous section apply, in particular, the Doppler version of the dispersion relation (3.6.5).

This implies that we are essentially looking at an extension of the geometrical optics method to a form that includes the effects of currents as well as depth, that is a **current-depth refraction** approach. Because of the inherent limitations of the geometrical optics approach, only the principal aspects are covered. The review papers by Peregrine (1976), Peregrine and Jonsson (1983), Jonsson (1989), to mention a few, give a more detailed overview of the variety of problems and properties of the solutions. In particular Jonsson and co-workers developed a modelling system for the solution of the combined wave current motion within the framework described above. Their solution includes the effect of changes in the MWL caused by the changes in the wave motion due to varying depth and also the effect of (small) energy dissipation. This is incorporated by the introduction of a mean energy level similar to the energy line used in classical hydraulics (Christoffersen and Jonsson, 1981). See also Jonsson et al. (1970), Jonsson (1978b), Jonsson and Wang (1980).

The method of ray tracing is used for the wave patterns. The energy equation, which is used to determine the wave height variation over the domain includes the wave-current interaction as developed by Phillips (1966, 1977). The energy equation is used in the form of a wave action equation. The derivation of the wave action equation with energy dissipation and the principles of the approach is further described in Jonsson (1978a,), Christoffersen and Jonsson (1980), and in Christoffersen (1982). The details of the the energy equation in the general form is discussed further in Chapter 4.

The method, however is limited to irrotational wave motion and it only includes wave refration. Hence it is difficult to extend to conditions where diffraction is important and to the surf zone where the most active nearshore processes take place. The methods presented in the later chapters in this text do not have these limitations. For space reasons it is therefore chosen to refer readers interested in this method to the quoted references.

3.7 Combined refraction-diffraction

3.7.1 *Introduction*

The limitation of the simple ray tracing is first of all that it disregards possible variations in amplitude along the wave fronts, and disregards the effects of curvature of the wave fronts. Such variations become important, e.g., in the neighborhood of caustics, and around structures such as breakwaters that partly obstruct the wave propagation in the horizontal plane. These diffraction effects imply that wave energy propagates across wave number directions, which was specifically assumed not to happen in the simple refraction considered above. It also disregards reflections caused by variations in the borrom topography.

The classical theory for linear constant depth diffraction is based on the same basic assumption of long crested waves as the refraction methods described above. However it allows for relatively rapid variations of the wave field also in the direction perpendicular to the wave fronts. Under such conditions, the wave motion can be described by the so-called wave equation.

For the sinusoidal ("harmonic") motion this equation, which describes the wave motion in time and space, reduces to a Helmholtz equation which governs the space variation of the amplitude. It is worth already here to emphasize that the wave equation describes the propagation of the wave motion in space and time and therefore is a hyperbolic equation. Conversely, the Helmholtz equation describes the (time constant) distribution in space of the amplitude and phase of a sinusoidal wave motion, and hence is an elliptic equation.

Few analytical solutions are available for this equation, but one is the classical cases of diffraction at constant depth around the tip of a breakwater by Sommerfeld (1896) and further developed by Penney and Price (1952). For further details and discussion of these pioneering works see, e.g., *Shore Protection Manual* (1984). Also the solution for diffraction around a (large) circular cylinder by MacCamy and Fuchs (1954) should be mentioned.

For many years refraction and diffraction were by necessity treated separately because no description was available that covered the combination, although it is obvious that in most practical problems they occur in conjunction. The goal was to incorporate the effect of the vertical variation of the wave motion into an equation that describes the propagation of the wave in time over the horizontal x, y-domain. To overcome this, focus was turned to the classical (constant depth) wave equation. An extension of this equation was derived, again using the assumptions of a depth varying so slowly (or gently) that the wave motion locally can be described by the constant depth results in Chapter 3.2. The term "mild slope" has come to cover this assumption which (in more descriptive terms) implies "locally constant depth."

This resulted in the **Mild Slope Equation** (MSE) which was first derived for one horizontal dimension by Svendsen (1967), and for the general 2DH case by Berkhoff (1972).

This chapter presents and discusses the derivation of the Mild Slope Equation. Since this general equation is elliptic in the x, y-plane, it is difficult to solve numerically which has lead to development of the **parabolic approximation** for progressive waves (Radder, 1979). This and related methods are the basis for important wave models, and it is described in the last section of this chaper and in Chapter 6.

3.7.2 The wave equation for linear long waves

Like the geometric optics approximation the key feature of the various versions of the wave equation is that they provide information about the propagation pattern and variation of the wave amplitude in the $2D$ horizontal x, y-plane in which the effect of the depth variation of particle velocities and pressures is automatically incorporated.

It is instructive first to derive the classical wave equation for long waves. This is particularly simple because for long waves there is no depth variation of particle velocities and dynamic pressure.

As starting point, we take the depth integrated continuity equation

$$\frac{\partial \zeta}{\partial t} + \nabla_h(\mathbf{u}h) = 0 \tag{3.7.1}$$

where ∇_h stands for $(\frac{\partial}{\partial x}, \frac{\partial}{\partial y})$. As we see, this equation links the surface elevation $\zeta(x, y, t)$ to the depth uniform velocity vector $\mathbf{u}$.

The momentum equation is, for the inviscid problem considered here, the linearized Euler equation which reads

$$\frac{\partial \mathbf{u}}{\partial t} = -\frac{1}{\rho}\nabla_h p_D \tag{3.7.2}$$

where p_D is the dynamic pressure. Integrating this equation over the depth $(0, h)$, we get (utilizing $\mathbf{u}$ and p_D are constant over depth)

$$\frac{\partial}{\partial t}\mathbf{u}h = -\frac{1}{\rho}h\nabla_h p_D \tag{3.7.3}$$

For long waves, we have found that the pressure is hydrostatic, so that

$$-\frac{1}{\rho}h\nabla_h p_D = -gh\nabla_h \zeta \tag{3.7.4}$$

and (3.7.2) then becomes

$$\frac{\partial \mathbf{u}h}{\partial t} + gh\nabla_h \zeta = 0 \tag{3.7.5}$$

We then eliminate either $\mathbf{u}h$ or ζ from (3.7.1) and (3.7.5). Here we choose to eliminate $\mathbf{u}h$. Cross-differentiation (differentiating (3.7.1) with respect to t, (3.7.5) with respect to ∇_h) and subtracting one from the other then gives

$$\frac{\partial^2 \zeta}{\partial t^2} - \nabla_h(gh\nabla_h \zeta) = 0 \tag{3.7.6}$$

which, since $gh = c^2$ for long waves, can be written

$$\boxed{\frac{\partial^2 \zeta}{\partial t^2} - \nabla_h (c^2 \nabla_h \zeta) = 0} \tag{3.7.7}$$

We see that this is the classical wave equation for the surface elevation $\zeta(x, y, t)$ for long waves on a varying depth.

It is clearly a **hyperbolic** equation, describes the evolution in time and space of the wave motion $\zeta(x, y, t)$ as it propagates.

However, we notice that we have already found that the basic wave components of the problem have sinusoidal phase variation for which

$$\frac{\partial^2 \zeta}{\partial t^2} = -\omega^2 \zeta \tag{3.7.8}$$

Hence for constant depth, (3.7.7) is, without additional assumptions, equivalent to (using $c = \omega/k$)

$$\boxed{\nabla_h^2 \zeta + k^2 \zeta = 0} \tag{3.7.9}$$

which is a Helmholtz equation. We can write the surface variation ζ in complex form as

$$\zeta = \Re \left\{ a(x, y) e^{i\omega t} \right\} \tag{3.7.10}$$

where $a(x, y)$ is the complex amplitude of the wave motion. Thus (3.7.9) reduces to

$$\boxed{\nabla_h^2 a + k^2 a = 0} \tag{3.7.11}$$

We notice that (3.7.11) (and (3.7.9)) are **elliptic** equations which essentially describe the distribution in space of the complex wave amplitude a, (real amplitude and phase).

The general solution to the long wave equation

The shallow water wave equation is unusual in the sense that for constant depth it is possible to obtain a general solution to the equation. This is particularly interesting because it is noticed that in the derivation of (3.7.7) we did not make any assumptions about the phase motion of the wave and we even did not assume that the motion is irrotational. Hence in principle (3.7.7) could describe also waves in the surfzone which typically would be rotational.

To solve (3.7.7) we introduce two new variables instead of x and t:

$$\xi_- = x - ct \qquad \text{and} \qquad \xi_+ = x + ct \tag{3.7.12}$$

where

$$c = \sqrt{gh} \tag{3.7.13}$$

According to the chain rule we then get

$$\frac{\partial}{\partial x} = \frac{\partial}{\partial \xi_-}\frac{\partial \xi_-}{\partial x} + \frac{\partial}{\partial \xi_+}\frac{\partial \xi_+}{\partial x} = \frac{\partial}{\partial \xi_-} + \frac{\partial}{\partial \xi_+} \tag{3.7.14}$$

$$\frac{\partial}{\partial t} = \frac{\partial}{\partial \xi_-}\frac{\partial \xi_-}{\partial t} + \frac{\partial}{\partial \xi_+}\frac{\partial \xi_+}{\partial t} = -c\frac{\partial}{\partial \xi_-} + c\frac{\partial}{\partial \xi_+} \tag{3.7.15}$$

and similarly

$$\frac{\partial^2}{\partial x^2} = \frac{\partial^2}{\partial \xi_-^2} + 2\frac{\partial}{\partial \xi_- \partial \xi_+} + \frac{\partial^2}{\partial \xi_+^2} \tag{3.7.16}$$

$$\frac{\partial^2}{\partial t^2} = c^2\frac{\partial^2}{\partial \xi_-^2} - 2c^2\frac{\partial^2}{\partial \xi_- \partial \xi_+} + c^2\frac{\partial^2}{\partial \xi_+^2} \tag{3.7.17}$$

For constant depth (3.7.7) reads

$$\frac{\partial^2 \zeta}{\partial t^2} - c^2\frac{\partial^2 \zeta}{\partial x^2} = 0 \tag{3.7.18}$$

and substituting (3.7.16) and (3.7.17) yields

$$4c^2\frac{\partial^2 \zeta}{\partial \xi_- \partial \xi_+} = 0 \tag{3.7.19}$$

where $\zeta = \zeta(\xi_-, \xi_+)$. This equation can be integrated directly. Integration with respect to ξ_+ yields

$$\frac{\partial \zeta}{\partial \xi_-} = f_1(\xi_-) \tag{3.7.20}$$

Integration with respect to ξ_- then gives

$$\zeta = f(\xi_-) + g(\xi_+) \tag{3.7.21}$$

where $f(\xi_-) = \int f_1(\xi_-)d\xi_-$. Resubstituting for x and t this can also be written

$$\boxed{\zeta = f(x - ct) + g(x + ct)} \tag{3.7.22}$$

This is the general solution to (3.7.18). It is often termed the d'Alembert-solution referencing the name of the French mathematician *d'Alembert* (1717-1783) who along with other contemporary mathematicians studied the wave equation extensively. The important thing to notice here is that f and g are arbitrary functions representing surface elevations propagating in the positive (f) or negative x-direction (g). In other words: No matter which surface profile we specify for our linear long wave, it will satisfy the basic equation (3.7.18) and hence the original equations (3.7.1) and (3.7.2). The solution consists of two components which propagate with constant form, f in the positive x-direction, g in the negative x-direction. so we have again confirmed that liner wave theory does not place any constraint on the shape of the waves.

It is that property we utilize in some of the simpler applications for surfzone waves when we assume locally constant depth and use linear wave properties for waves with non-sinusoidal x and t variations.

For completeness is mentioned that f, g can be determined by the initial conditions. If these are given in the form of a wave shape $\zeta_0(x) = \zeta(x, 0)$ and the initial velocity field $u_0(x) = u(x, 0)$ at $t = 0$ then the particular solution satisfying those initial conditions can be written

$$\zeta = \frac{1}{2}(\zeta_0(x - ct) + \zeta_0(x + ct)) + \frac{1}{2c}\int_{x-ct}^{x+xt} u_0(x')dx' \qquad (3.7.23)$$

For more details see e.g Greenberg (1988, 1998).

3.7.3 *The mild slope equation*

Derivation of the mild slope equation requires a good deal more care, because we now consider an arbitrary depth where the velocity, dynamic pressure, etc. vary over depth.

It turns out here to be more convenient to use a description of the motion based on the velocity potential Φ which satisfies the Laplace equation

$$\nabla_h^2 \Phi + \frac{\partial^2 \Phi}{\partial z^2} = 0 \qquad (3.7.24)$$

and the linearized boundary conditions (see Sect. 3.2).

$$\frac{\partial \Phi}{\partial z} + \frac{1}{g}\frac{\partial^2 \Phi}{\partial t^2} = 0 \quad ; \quad \text{at} \ ; \ z = 0 \qquad (3.7.25)$$

and

$$\frac{\partial \Phi}{\partial z} + \nabla_h h \cdot \nabla_h \Phi = 0 \quad ; \quad \text{at} \quad ; \; z = h \tag{3.7.26}$$

At this point it often is assumed that the motion is harmonic in time. However, it is useful to show that this assumption is not directly needed (though (3.7.28) below implies that at least the fast time variation is harmonic). If omitted we get an equation similar to the hyperbolic time varying shallow water wave equation (3.7.7).

Instead, we assume solutions of the form

$$\Phi = \phi_s(x, y, 0, t) \cdot f(z) \tag{3.7.27}$$

where the z-variation has been factored out and ϕ_s is the velocity potential at the MWL. Note that ϕ_s can be complex and (3.7.27) includes the effect of reflected waves.

In accordance with the locally constant depth (or mild slope) assumption we found in section 3.2, that we can use

$$f(z) = \frac{\cosh k(z+h)}{\cosh kh} \tag{3.7.28}$$

Then we get

$$\frac{\partial^2 \Phi}{\partial z^2} = k^2 \Phi \tag{3.7.29}$$

and hence from (3.7.24)

$$\nabla_h^2 \Phi + k^2 \Phi = 0 \tag{3.7.30}$$

We also see that

$$\left(\frac{\partial f}{\partial z}\right)_{-h} = 0 \quad ; \quad f(0) = 1 \tag{3.7.31}$$

and

$$\left(\frac{\partial f}{\partial z}\right)_0 = k \tanh kh = \frac{\omega^2}{g} \tag{3.7.32}$$

In order to incorporate the effect of the depth variation, we use the Green's 2. theorem in the form (2.2.20) with $= \phi_1 = \Phi$ and $\phi_2 = f$, $a = -h$,

$b = 0$. This gives

$$\int_{-h}^{0}\left(\frac{\partial^2\Phi}{\partial z^2}f - \Phi\frac{\partial^2 f}{\partial z^2}\right)dz = \left(\frac{\partial\Phi}{\partial z}f - \Phi\frac{\partial f}{\partial z}\right)_0 - \left(\frac{\partial\Phi}{\partial z}f - \Phi\frac{\partial f}{\partial z}\right)_{-h} \tag{3.7.33}$$

Substituting (3.7.29), (3.7.31) and (3.7.32) for f and the equations (3.7.30), (3.7.25), and (3.7.26) for Φ changes this to

$$-\int_{-h}^{0}(f\nabla_h^2\Phi + \Phi k^2 f)\,dz = -\frac{1}{g}\frac{\partial^2\phi_s}{\partial t^2}f(0) - \phi_s\frac{\omega^2}{g} - (-\nabla_h h\cdot\nabla_h\Phi f)_{-h} \tag{3.7.34}$$

From (3.7.27), we get

$$\nabla_h\Phi = f\nabla_h\phi_s + \phi_s\nabla_h f \tag{3.7.35}$$

$$\nabla_h^2\Phi = f\nabla_h^2\phi_s + 2\nabla_h\phi_s\nabla_h f + \phi_s\nabla_h^2 f \tag{3.7.36}$$

which substituted into (3.7.34) gives

$$\begin{aligned}&\int_{-h}^{0}\{f^2\nabla_h^2\phi_s + \nabla_h\phi_s\nabla_h f^2 + \phi_s f\nabla_h^2 f + \phi_s f^2 k^2\}\,dz\\ &= \frac{1}{g}\left(\frac{\partial^2\phi_s}{\partial t^2} + \phi_s\omega^2\right) - \nabla_h h\cdot(f^2\nabla_h\phi_s + \phi_s f\nabla_h f)_{-h}\end{aligned} \tag{3.7.37}$$

Here the first two terms on the LHS may be combined to $\nabla_h(f^2\nabla_h\phi_s)$. Rearranging terms, (3.7.37) may then be written

$$\begin{aligned}&\int_{-h}^{0}\nabla_h(f^2\nabla_h\phi_s)dz + [\nabla_h h f^2\nabla_h\phi_s]_{-h} + \phi_s k^2\int_{-h}^{0}f^2dz\\ &= -\int_{-h}^{0}\phi_s f\nabla_h^2 f dz - \phi_s\nabla_h h\cdot[f\nabla_h f]_{-h}\\ &\quad + \frac{1}{g}\left(\frac{\partial^2\phi_s}{\partial t^2} + \omega^2\phi_s\right)\end{aligned} \tag{3.7.38}$$

We may then apply Leibniz' rule to the first two terms on the LHS, by which the equation can be written

$$\nabla_h \cdot \left(\nabla_h \phi_s \int_{-h}^{0} f^2 dz \right) + \phi_s k^2 \int_{-h}^{0} f^2 dz$$
$$= -\phi_s \int_{-h}^{0} f \nabla_h^2 f dz - \phi_s \nabla_h h \cdot [f \nabla_h f]_{-h} + \frac{1}{g} \left(\frac{\partial^2 \phi_s}{\partial t^2} + \omega^2 \phi_s \right) \quad (3.7.39)$$

As shown in exercise 3.5-8 using (3.7.28) for $f(z)$ gives

$$\int_{-h}^{0} f^2(z) dz = \frac{1}{2} \frac{c^2}{g} \left(1 + \frac{2kh}{\sinh 2kh} \right) \quad (3.7.40)$$

or

$$\int_{-h}^{0} f^2(z) dz = \frac{c c_g}{g} \quad (3.7.41)$$

Substituting this into (3.7.39), we get the mild slope equation on the form

$$\frac{\partial^2 \phi_s}{\partial t^2} - \nabla_h \cdot (c c_g \nabla_h \phi_s) + (\omega^2 - k^2 c c_g) \phi_s$$
$$= \left\{ \int_{-h}^{0} f \nabla_h^2 f dz + \nabla_h h \cdot [f \nabla_h f]_{-h} \right\} g \phi_s \quad (3.7.42)$$

The terms on the RHS

$$R(h) = \left\{ \int_{-h}^{2} f \nabla_h^2 f dz + \nabla_h h \cdot [f \nabla_h f]_{-h} \right\} g \phi_s \quad (3.7.43)$$

can be shown to be $O((\nabla_h h)^2,\ \nabla_h^2 h)$ (see Exercise 3.7-5). Therefore, if we introduce the mild -slope assumption that $\nabla_h h \ll kh$ corresponding to $S = \nabla_h h L/h \ll 1$ it can be argued that the $(\nabla_h h)^2$-terms $\ll$ LHS. Similarly, letting $\nabla_h^2 h \ll \nabla_h h$ is also a natural additional assumption because $\nabla_h^2 h = O(\nabla_h h)$ can only occur over short distances without changing $O(\nabla_h h)$. This is the classical approach to the mild-slope equation.[16] This means the RHS $\ll$ LHS, and in the leading approximation in $\nabla_h h$, we therefore get

$$\boxed{\frac{\partial^2 \phi_s}{\partial t^2} - \nabla_h \cdot (c c_g \nabla_h \phi_s) + (\omega^2 - k^2 c c_g) \phi_s = 0} \quad (3.7.44)$$

[16] However, as will be discussed in section 3.7.4, these assumptions cause certain limitations to the result.

which is the **time varying (or hyperbolic) version of the Mild Slope Equation**.

Exercise 3.7-1 Show that if we as a consequence of (3.7.28) introduce time harmonic motion given by

$$\phi_s = a_\phi(x,y)e^{i\omega t} \tag{3.7.45}$$

the time varying MSE (3.7.44) reduces to

$$\boxed{\nabla_h(cc_g\nabla_h a_\phi) + k^2 cc_g a_\phi = 0} \tag{3.7.46}$$

This is the equation which is normally referred to as the **mild slope equation** (MSE for short).

Exercise 3.7-2 Show by differentiating (3.7.44) with respect to t and using the dynamic free surface boundary condition

$$\zeta = -\frac{1}{g}\frac{\partial\phi_s}{\partial t} \tag{3.7.47}$$

that we can also write (3.7.44) on the form

$$\boxed{\zeta_{tt} - \nabla_h(cc_g\nabla_h\zeta) + (k^2 cc_g - \omega^2)\zeta = 0} \tag{3.7.48}$$

and hence if it is again introduced that

$$\zeta(x,y,t) = a(x,y)e^{i\omega t} \tag{3.7.49}$$

we get the MSE as

$$\boxed{\nabla_h(cc_g\nabla_h a) + k^2 cc_g\nabla_h a = 0} \tag{3.7.50}$$

This is another form of the MSE, expressing the wave motion in terms of the surface variation $a(x,y)$. It has the same limitations as (3.7.46).

Discussion of the MSE

First it is noted that in the derivation of (3.7.42) described above we have actually only utilized the mild slope assumption at one point, namely in (3.7.28) where we assume that the locally constant depth approximation applies to the vertical varation of the phase motion (and as a consequence that the dispersion relation is given by (3.7.32)). Consequently the $R(h)$-terms that account for higher order terms in the bottom slope, appear without further approximations about the bottom slope. In other words, given the locally constant depth assumption the rest of the derivation of (3.7.42) is not limited to mild slope at all. The consequences of this is discussed further in Chapter 3.7.4 where the effects of including the $R(h)$-temrs are analyzed further.

It is also emphasized again, that the locally constant depth assumption which is behind (3.7.28) implies, as shown in section 3.2, that the local time variation is given by (3.7.49). Thus the MSE equation represents a **wave motion with harmonic (i.e. "sinusoidal") time variation.**

It may be seen, however, from the derivation that $a(x, y)$ is the surface elevation and a_ϕ the value of the ϕ_s both **at a fixed time**. This means that a and a_ϕ shows the instantaneous position of the water surface and value of the velocity potential at the surface. Thus a and a_ϕ include the fast variation with (x, y) of ζ and ϕ_s, respectively, not just the amplitude. This is sometimes called a **phase resolving** model.

Exercise 3.7-3 Show that for shallow water waves (3.7.48) reduces to the shallow water wave equation (3.7.7).

Exercise 3.7-4

The MSE (3.7.46) may be changed to a simpler Helmholtz equation by changing the surface elevation a to a scaled elevation ξ by the substitution

$$\xi = \sqrt{cc_g}\, a \tag{3.7.51}$$

Show that this simplifies the MSE to the form

$$\nabla_h^2\, \xi + k_c^2\, \xi = 0 \tag{3.7.52}$$

which is a Helmholtz equation with k_c is defined by

$$k_c^2 = k^2 - \frac{\nabla_h^2 \sqrt{cc_g}}{\sqrt{cc_g}} \tag{3.7.53}$$

Relation to the geometrical optics approximation

The geometrical optics approximation is essentially based on the same assumption as the MSE of locally constant depth but with a slowly varying amplitude. The waves also have long, continuous crests though they are curved, the amplitudes vary along the crests, and they are monochromatic. Therefore it is interesting to compare the two theories.

While the MSE describes the variation of the total surface variation at a fixed time, the geometrical optics theory operates with the amplitude A of the wave motion. This means that the relation between a and A is given by

$$a(x, y) = A(x, y)\, e^{iS(x,y)} \tag{3.7.54}$$

Here A and S are real functions. Substituting this into (3.7.46) we then get

$$\begin{aligned}&\nabla_h \cdot (cc_g\, \nabla_h A) - (cc_g\, A\nabla_h S) \cdot \nabla_h S + k^2 cc_g\, A \\ &+ i\, [(cc_g \nabla_h A) \cdot \nabla_h S + \nabla_h \cdot (cc_g\, A\nabla_h S)] = 0\end{aligned} \tag{3.7.55}$$

In this equation the real and imaginary parts must balance separately. The real part becomes directly

$$-(cc_g\, A\nabla_h S) \cdot \nabla_h S + k^2 cc_g\, A = -\nabla_h \cdot (cc_g\, \nabla_h A) \tag{3.7.56}$$

which can also be written, after division by $cc_g A$

$$(\nabla_h S)^2 - k^2 = \frac{\nabla_h^2 A}{A} + \frac{\nabla_h (cc_g) \cdot \nabla_h A}{cc_g\, A} \tag{3.7.57}$$

We see that the eikonal equation of the geometrical optics approximation corresponds to neglecting the terms on the RHS of (3.7.57). Those are the terms representing the diffraction effects.

For the imaginary part we get, after multiplying by A

$$cc_g\, A\nabla_h A \cdot \nabla_h S + A\, \nabla_h \cdot (cc_g\, A\, \nabla_h S) = 0 \tag{3.7.58}$$

or

$$2\, cc_g A\nabla_h A \cdot \nabla_h S + A^2\, \nabla_h (cc_g \nabla_h S) = 0 \tag{3.7.59}$$

which combines to

$$\nabla_h \cdot (A^2\, cc_g \nabla_h S) = 0 \qquad (3.7.60)$$

This looks very similar to the transport equation of the geometrical optics theory. However, there are some subtle but important differences.

The wave number vector in diffracted waves

In the simple refraction described by the geometrical optics theory we defined the group velocity vector $\mathbf{c_g}$ by

$$\mathbf{c_g} = c_g\, \frac{\mathbf{k}}{k} \qquad (3.7.61)$$

(see (3.5.87)), where c_g and k are defined by the usual constant depth solution. This was possible because in the approximation considered the eikonal equation showed that $\nabla_h S_0 = \mathbf{k}$, where $|\mathbf{k}|$ obeys the dispersion relation.

However, as seen from (3.7.57), in diffracted waves $\nabla_h S \neq \mathbf{k}$. It also follows from (3.7.54) that $S = const$ still represents wave fronts along which the surface elevation is constant except for the slow variation with A. Hence we can still define wave rays which are perpendicular to the wave fronts but not in the direction of $\mathbf{k}$ and which are defined from (3.7.57)

$$(\nabla_h S)^2 = k^2 + \frac{\nabla_h^2 A}{A} + \frac{\nabla_h (cc_g) \cdot \nabla_h A}{cc_g\, A} \qquad (3.7.62)$$

Similarly we can define a group velocity $\mathbf{c'_g}$ by

$$\mathbf{c'_g} = c_g\, \frac{(\nabla_h S)}{k} \qquad (3.7.63)$$

which substituted into the transport equation (3.7.60) gives

$$\nabla_h \cdot (A^2 \mathbf{c'_g}) = 0 \qquad (3.7.64)$$

Here we have used that $c = \omega/k$ and divided by ω. We see that the $A^2\mathbf{c'_g}$ corresponds to the energy flux of the refracted-diffracted waves. This confirms that the wave energy moves in the direction of the group velocity $\mathbf{c'_g}$ and the wave rays. However, as (3.7.62) shows $\nabla_h S$ deviates from $\mathbf{k} = (k_x, k_y)$. It is the $k = |\mathbf{k}|$ that is used to define the wave othogonals. We therefore also conclude that in diffracted waves rays and orthogonals have different directions and wave energy does not move along the orthogonals. As shown in section 3.6 this is analogous to situations with waves on currents.

It is emphasized that it is also k that is used to calculate all the wave properties in the underlying linear wave motion, not $(\nabla_h S)^2$.

Exercise 3.7-5

The MSE is based on the same equations as the linear wave solution developed in Chapter 3.2 and which were used to derive the equation for conservation of wave energy (flux). Show that this implies that the energy flux in the diffracted waves must be in the direction of the normal to the wave fronts as indicated by (3.7.63).

Hint: use that in linear waves the energy flux is given by

$$\mathbf{E}_f = \int_{-h}^{0} p\mathbf{u}\, dz \tag{3.7.65}$$

Finally we see that a measure of the error ϵ from using the refraction theory in some form (including the Kinematic Wave theory) is represented by the last two terms in (3.7.62)

$$\epsilon = \frac{\nabla_h S}{k^2} = \frac{\nabla_h^2 A}{k^2 A} + \frac{\nabla_h (cc_g) \cdot \nabla_h A}{k^2 cc_g\, A} \tag{3.7.66}$$

If ϵ is $\ll 1$ everywhere in the domain the refraction theory gives a good approximation to the wave propagation pattern. Here the first term is recognized as the approximate measure of the error defined by Battjes (1968) for this purpose.

3.7.4 *Further developments of the MSE*

In the years since the first derivation of the MSE for water waves extensive research has been conducted into application and further extension of this important equation. One line of work lead to the parabolic approximation described in the next section. Within the frame of the MSE itself, however, two major directions have been pursued.

One is aimed at easing the numerical solution of the equation. Because the equation is elliptic the computations become very expensive for increasing size of the model domain and for even moderate size practical problems they eventually become impossible to handle. Mostly this has been done by using the hyperbolic time domain version of the equation which allows for a marching solution in time.

A second major line of research is aimed at relaxing the original constraint of a gently sloping bottom.

Both these lines are briefly described below in addition to some further extensions of the MSE to deal with energy dissipation and with a combination of waves and currents. For space reasons the review only describes the main features of the development and the reader is refered to the quoted literature for a full cover of the details.

Numerical solution of the time domain mild slope equation

As shown the MSE is generically a hyperbolic equation (3.7.44) which represents waves propagating in time across the region considered in the horizontal (x, y)-plane, although it is more often seen used in the case of steady monochromatic waves in which case it takes the elliptic form (3.7.46).

Mathematically elliptic equations describe equilibrium situations. In the case of (3.7.46) the equilibrium solution is the distribution at a fixed time of the wave surface elevation a (or surface velocity potentioal ϕ_s), which results from an incident wave motion with sinusoidal phase variation and constant amplitude in time.

Real waves, however, have time (and space) varying amplitudes. And as such waves enter the computational comain, the changes in amplitude propagate with the propagation of the wave energy. This process would be described (within the limitations of linear wave theory) by the hyperbolic form (3.7.44) of the MSE, if solved directly. Another advantage is that the hyperbolic form is much easier to solve numerically.

So the research into this topic takes the starting point in the hyperbolic MSE which we repeat here for convenience

$$\frac{\partial^2 \zeta}{\partial t^2} - \nabla \cdot (cc_g \nabla \zeta) + (\omega^2 - k^2 cc_g)\zeta = 0 \tag{3.7.67}$$

In one approach to solving this equation Copeland (1985) modified it further by reintroducing the monochromatic (i.e. time harmonic) property of ζ so that

$$-\frac{1}{\omega^2}\frac{\partial^2 \zeta}{\partial t^2} = \zeta \tag{3.7.68}$$

(whereby the capability of dealing with timevarying irregular waves is lost). Using this to replace ζ in the last term in (3.7.67) simplifies the equation to the form

$$\frac{\partial^2 \zeta}{\partial t^2} - \frac{c}{c_g} \nabla \cdot (cc_g \nabla \zeta) = 0 \tag{3.7.69}$$

Copeland further introduces the auxiliary vaiable Q_g defined by

$$\frac{\partial \zeta}{\partial t} + \frac{c}{c_g} \nabla Q_g = 0 \tag{3.7.70}$$

Substituted into (3.7.69) that equation becomes

$$\frac{\partial Q_g}{\partial t} + cc_g \nabla \zeta = 0 \tag{3.7.71}$$

Thus the original equation has been split into two simpler, coupled first order evolution equations for ζ and Q_g, (3.7.70) and (3.7.71).

These equations are much easier to solve than the elliptic form of the MSE. Copeland solved the equations using a low order finite difference scheme. Linear forms similar to (3.7.70) and (3.7.71) were also derived by Madsen and Larsen (1987) and solved using an ADI method.

It is emphasized, however, that because of the assumption of time harmonic motion Copeland's equations for the MSE and the methods derived from it only correspond to the steady state MSE (3.7.46) and do not have the capability to describe the time variations of the full hyperbolic form (3.7.44).

The approach has also been generalized by Suh et al. (1997) and Lee et al. (1998) without assuming monochromatic waves to also incoporating the small terms (see next section)

Validity of the mild-slope-assumption

Throughout the discussion of combined refraction/diffraction problems we have used the assumption of a gently sloping bottom, which in our formulation corresponds to the slope parameter $S = \frac{h_x L}{h} \ll 1$. However, in many practical problems S is not so small and the question therefore arises how large values of S can be accepted without the results becoming too inaccurate to be useful.

This question was investigated by Booij (1983) who compared finite element solutions of the MSE with similar solutions to the exact 3D problem solving the Laplace equation.

The results are very encouraging. The (numerical) tests were performed for a plane slope that forms a transition from one water depth h to another depth $h/3$. The waves were incident from the deeper section toward the slope, and a sensitive measure of the accuracy is the amplitudes of the wave reflected from the slope.

Fig. 3.7.1 shows results for the reflection coefficient versus tan $\alpha = h_x = \frac{\Delta h}{w_s}$ where Δh was 0.4 m and w_s, the length of the slope, was varied for the case of normal incidence. In the figure the $+'s$ are the 3D ("exact") solutions, the full curve the MSE solution. We see that in this comparison acceptable accuracy is lost when h_x is larger than approximately 1/3. Comparisons for the variation of the free surface show similar high accuracy over the slope of $h_x = 1/3$.[17]

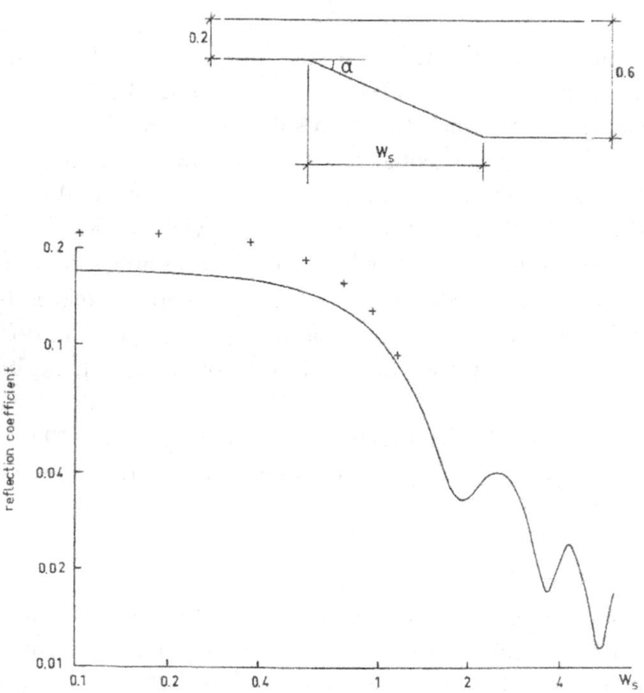

Fig. 3.7.1 Booij's computations (from Booij 1983).

The incident waves in the tests had $k_0h = 0.6$ at the large depth where k_0 is the deep water value. In terms of the slope parameter S this corresponds

[17] Fig. 3.7.2 shows later computations by Suh et al. (1997) which indicate that there are inaccuracies in the finite element results in Fig. 3.7.1.

to

$$S = \frac{h_x L}{h} = 3.47 \tag{3.7.72}$$

As the waves move toward the upper part of the slope the value of S increases even further to a value of 10.4.[18]

Thus we must conclude when it pertains to the accuracy of the MSE that our notion of a gentle slope must extend to values well beyond $S = O(1)$.

Similar comparisons for waves propagating along the slope parallel to the bottom contours show high accuracy even for a slope as steep as $h_x = 0.8$. For details see Booij (1983).

This could tempt one to conclude that the small terms in the equation (3.7.43) will for all practical purposes be negligible. However, it turns out that for certain rapidly varying bottom topographies such as a series of ripples or undular bars with wave lengths of the order the surface wave length the standard MSE is unable to predict the effect of the bottom topography.

The MSE extended to rapid depth variations

The inability of the MSE (3.7.44) or (3.7.46) to predict the reflection from periodic variations of the bottom such as a row of sand bars is caused by the following mechanism. When the spacing l between the bars equals half the wave length L of the surface waves i.e. when $2l/L = 1$, the (weak) reflected waves from each of the bars are in phase as in a resonant oscillation. Thus each reflected wave amplifies the total reflection from the bar system. The mechanism is known as resonant Bragg reflection. This phenomenon known from physics was first analysed for water waves by Davies and Heathershaw(1984) and Mei (1985).

It appears that inclusion of the effect of the small terms on the RHS of the full MSE (3.7.43) improves the accuracy of the predictions and since the first investigations mentioned above numerous studies about this effect have been published.

As may be seen from Booij's analysis in Fig. 3.7.1 the MSE also fails to accurately represent the reflection from the plane slope when h_x exceeds $0.3 - 0.4$. Therefore these studies have largely focused on the same

[18]In most publications the related parameter $S_k = h_x/kh$ is used. In the two cases mentioned above the values of S_k correspond to 1.67 and 0.56, respectively. However, from an intuitive/physical point of view S is much more transparent and makes it easier to judge what is a proper value.

benchmark tests of the reflection from the slope analysed by Booij and the Bragg scattering from ripple beds.

It may be noticed that the small terms $R(h)$ in (3.7.43) were not in the original derivation of the MSE by Berkhoff (1972), but they were included in the derivation outlined by Smith and Sprinks (1975). They analysed the terms and showed that they are $O((\nabla_h\, h)^2, \nabla_h^2 h)$.

Exercise 3.7-6 Show that the small terms $R(h)$ can be expanded to a the form

$$R(h) = R_1\, (\nabla_h h)^2 + R_2 \nabla_h^2 h \tag{3.7.73}$$

where

$$\begin{aligned} R_1 &= \int_{-h}^{0} f \frac{\partial^2 f}{\partial h^2}\, dz + \left(f \frac{\partial f}{\partial h} \right)_{-h} \\ R_2 &= \int_{-h}^{0} f \frac{\partial f}{\partial h}\, dz \end{aligned} \tag{3.7.74}$$

which confirms that the small terms are $O((\nabla_h\, h)^2, \nabla_h^2 h)$.

This is the form of the small terms derived by Smith and Sprinks (1975).

The proper form of the MSE with the small terms included remains an active research topic and the flow of publications continues, even in 2003 at the time of writing this.

As a first step toward improving the prediction of the Bragg scattering Kirby (1986) developed what has later been termed the **Extended MSE (EMSE)**. He considered the 1D problem and lets the total water depth $h'(x)$ be given as a mean (slowly varying) depth $h(x)$ with the periodic undulations as small (rapid) modulations $\delta(x)$. The assumption can be written

$$h'(x) = h(x) - \delta(x) \tag{3.7.75}$$

The modulations $\delta(x)$ can have any shape as long as they are small.

By inserting this into the MSE including the $(\nabla_h\, h)^2, \nabla_h^2 h$-terms and keeping contributions only to $O(\delta)$ Kirby arrives at the EMSE

$$\phi_{s,tt} - \nabla_h \cdot (cc_g \nabla_h \phi_s) + (\omega^2 - k^2 cc_g)\, \phi_s + \frac{g}{\cosh^2 kh} \nabla_h \cdot (\delta \nabla_h \phi_s) = 0 \tag{3.7.76}$$

Hence in (3.7.76) the effect of the small terms has been simplified to the last term.

The simulations using this equation were compared to the laboratory measurements for sinusoidal bottom undulations by Davies and Heathershaw (1984). For the resonant frequencies the results are close to the measured values. However, as is common even with carefully performed laboratory experiments unwanted disturbances create a substantial scatter in the measured data. But the trend is clear, for frequencies higher and lower than the dominant frequency the agreement is less good.

Further investigations have also shown that while the EMSE describes the resonant Bragg scattering from regular bottom undulations with a single wave length reasonably well it is much less accurate for bedforms that consist of a combination of wave lengths (O'Hare and Davies (1993)).

Recently this has lead to contributions that retain all the $(\nabla_h\, h)^2, \nabla_h^2 h)$-terms in Smith and Sprinks' original version. Starting with this version Chamberlain and Porter (1995) (CP95 in the following) essentially write the small terms on the form (3.7.73) and (3.7.74). They then consider the original mild slope equation (3.7.42) written for steady harmonic waves as

$$\nabla_h \cdot\ (cc_g \nabla_h \phi_s) + (\ k^2 cc_g +\ gR(h))\ \phi_s = 0 \tag{3.7.77}$$

CP95 terms this equation the **Modified Mild-Slope Equation or MMSE** and give explicit expressions for R_1 and R_2 derived from linear theory. They show comparisons of this and other approximations not outlined here, including the EMSE with both Booij's slope computations and with the measurements by Davies and Heathershaw (1984). Similar analysis was also conducted by Suh et al. (1997) using the MMSE separated into two evolution equations for ϕ_s and ζ, with ζ given by the linear dynamic free surface boundary condition

$$\phi_{s,t} = -g\zeta \tag{3.7.78}$$

and the equivalent equation for ζ_t. This is obtained by substituting (3.7.78) into the time dependent version of (3.7.77) which gives

$$\zeta_t + \nabla_h \cdot\ (cc_g \nabla_h \phi_s) + (\ k^2 cc_g +\ gR(h))\ \phi_s = 0 \tag{3.7.79}$$

Essentially this system is equivalent to Copeland's model extended to non-monochromatic waves and rapidly varying depth. The two equations (3.7.78) and (3.7.79) are then solved numerically and compared to new numerical solutions of the full Laplace equation for the normally incident waves on Booij's slope. Fig. 3.7.2 shows the results which clearly document that the MMSE, or equivalent equation like the two evolution equations above, give accurate predictions of the reflection coefficient even on versions of Booij's slope as steep as $\tan\alpha = 4, (\alpha = 76°)$, which corresponds to $b = 0.1$ in the figure). Similar comparisons were made by Lee et al. (1998). A truely surprising result considering the MSE and MMSE were originally derived under the assumption of a locally constant depth, but showing the effect of the $R(h)$-terms.

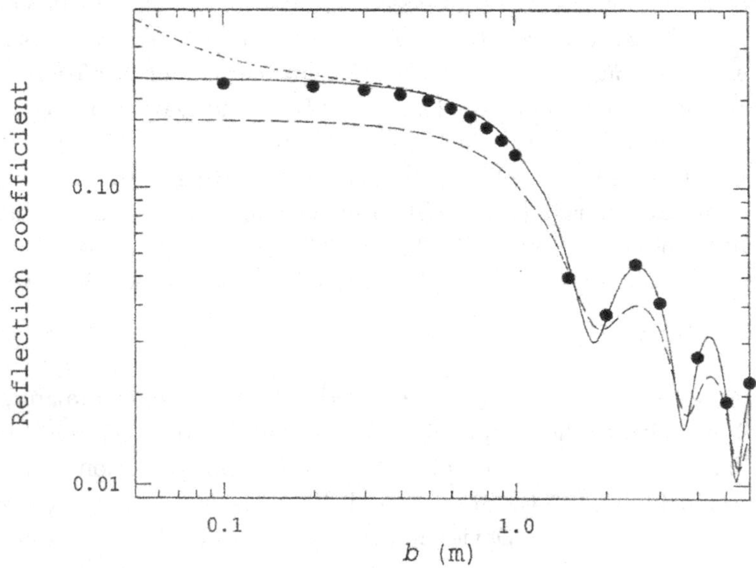

Fig. 3.7.2 Suh et al.'s (1997) computations. – represents the above equations, - - - is the MSE, - . - . - the above equations with only the curvature part of $R(h)$ included (from Suh et al., 1997).

Fig. 3.7.2 also shows the result of including only the curvature part of $R(h)$. The good agreement for this case up to very high values of the slope (small b) suggest that the curvature-term is more important than the $(\nabla_h h)^2$-term. However, on the Booij slope "curvature" only appears because in the numerical solution the discontinuity in the slope at the toe and top of the slope is represented by a smooth (but fast) transition in ∇_h. Since this is artificial it also suggests that the precise position of the dash-dot curve may depend on numerical details that are arbitrary.

Lee et al. (1998) also compare the MMSE solution to new results for the Davies and Heathershaw experiment. The new experiments show less scatter than the original data and hence are better suited than the original experiments for judging the accuracy of the model, which generally appears to be quite good also for this test.

The locally constant depth assumption revisited

In the previous sections we have discussed the importance of the (small) $R(h)$-terms in the MSE. However, considering the extreme conditions under which it has been tested it is natural also to ask how well the assumption of locally constant depth is satisfied. This was the other side of the mild slope assumption that lead to the adoption of the vertical variation of the motion given by f in (3.7.28).

Here it is important for the perspective to recall that the MSE only provides information about the wave amplitudes. It seems obvious that if we tried to used the f-variation in (3.7.28) to derive e.g. velocity distributions for the waves on the steep slope case of the computation with Booij's slope then they could be highly misleading. The surprising thing is that this does not seem to seriously impede the accuracy of the amplitude results. But it is worth issuing a warning against trying that.

What happens when the depth changes too fast is that the wave motion will not be able to adjust to the local conditions. This was mentioned by Smith and Sprinks (1975) and by CP95 and it has been analysed by Massel (1993) who finds that additional so-called evanescent terms are needed to describe the distortion of the motion. Those are non-propagating terms that acccount for the local modifications of the flow field to adjust for the inabilility of the motion to adjust to the (too rapidly varying) local bottom conditions.

Another approach in which such effects were included was the initial derivation of the 1D MSE by Svendsen (1967). This was based on a solution for the wave motion on a sloping bottom by Biesel (1952). This solution includes terms in the velocity potential proportional to $\nabla_h h$ which corresponds to including the first approximation to the deviation from the locally constant depth assumption. Jonsson and Brink-Kjær (1973) showed that for sinusoidal waves this version is equivalent to the MSE. However, the derivation does not include the $R(h)$-terms. See also Dingemans (1997) p257 for a discussion.

The MSE for waves with currents

The derivation and result for the MSE discussed so far is for a purely oscillatory wave motion. However, The MSE can also be extended to cover situations with waves on currents. For reference see Kirby (1984).

The situation considered is for depth uniform currents $\mathbf{U}(x, y)$. As described in section 3.6 the relative and the absolute wave frequencies, ω_r and ω_a respectively, are then related by

$$\omega_a = \omega_r + \mathbf{k} \cdot \mathbf{U} \tag{3.7.80}$$

Here the relative frequency ω_r satisfies the usual dispersion relation

$$\omega_r^2 = gk \tanh kh \tag{3.7.81}$$

and in MSE c and c_g are the usual (relative) phase and group velocities defined by

$$c = \frac{\omega_r}{k} \quad ; \quad c_g = \frac{\partial \omega_r}{\partial k} \tag{3.7.82}$$

The MSE for waves on currents then becomes

$$\frac{D^2\phi_s}{Dt^2} + \nabla_h \cdot \mathbf{U}\frac{D\phi_s}{Dt} - \nabla \cdot (cc_g \nabla_h \phi_s) + (\omega^2 - k^2 cc_g)\phi_s = 0 \tag{3.7.83}$$

where

$$\frac{D}{Dt} = \frac{\partial}{\partial t} + \mathbf{U} \cdot \nabla_h \tag{3.7.84}$$

The MSE with energy dissipation*

Booij (1981) showed that including a term of the form $i\omega wa$ in (3.7.50) will cause the waves to loose energy while propagating. Here w is an unspecified complex damping factor. Thus (3.7.50) becomes

$$\nabla_h(cc_g\nabla_h a) + (k^2cc_g + i\omega w)a = 0 \tag{3.7.85}$$

Different damping models for w representing w-values that simulate sources of energy loss to the waves such as a laminar bottom boundary layer, a porous bottom and other energy absorbing conditions were discussed by Dalymple et al. (1984).

3.7.5 *The parabolic approximation*

Introduction

As mentioned in section 3.7.3, the MSE equation for purely harmonic waves is an elliptic equation. This implies that a solution requires that it is solved in a closed domain with boundary conditions specified along the entire boundary.[19]

A second disadvantage of the mild slope equation is that elliptic equations are numerically more time-consuming to solve.

There is therefore a natural incentive for developing an approximation to the MSE which bypasses these problems and the so-called parabolic approximation serves that objective. It is valid for progressive waves with a main direction close to the x-direction and considers waves that vary more strongly in that direction than in the derection perpendicular to it. It transforms the elliptic equation into an equation which is of first order in the main direction of wave propagation (x) while of second order perpendicular to that (the y-direction).

Derivation from the mild slope equation

Many different approaches have been developed in the literature for deriving the parabolic approximation to the MSE, and some of those will be discussed later.

The simplest derivation of the parabolic wave theory is to derive it directly from the MSE equation. This also has the advantage of showing what the approximation is.

We therefore consider the MSE (3.7.46) (which implies considering an x, y-region with all the constraints of the MSE). Using the formulation based on the (surface) velocity potential we write it in the expanded form

$$cc_g\phi_{s,xx} + (cc_g)_x\phi_{s,x} + (cc_g\phi_{s,y})_y + k^2cc_g\phi_s = 0 \tag{3.7.86}$$

The key to the derivation is (as usual) an appropriate mathematical formulation of the assumptions underlying the theory. Thus, in a usual x, y coordinate system, we consider

- The wave motion propagates in a direction close to the x-axis. As the waves refract and diffract, they change direction and the accuracy of

[19] Alternatively, the domain can stretch to infinity in which case only part of the domain is modelled and at the boundary simulating the distant far a radiation condition is specified. For a discussion see Chapter 11.9.

the approximation decreases with increasing angle between wave and x-axis. We use a WKB-approximation in the form

$$\phi_s(x,y,t) = -\frac{ig}{\omega} A'(x,y) e^{i(\int k dx - \omega t)} + c.c. \tag{3.7.87}$$

where $c.c.$ stands for the complex conjugate component and $A'(x,y)$ is the (slowly varying) amplitude of $\zeta(x,y,t)$.

- As indicated by the fast varying exp-function in (3.7.87) the wave does not at any point move exactly in the x-direction. Therefore the deviation from this becomes absorbed in the amplitude $A(x,y)$, which is complex. The imaginary part of $A(x,y)$ represents the phase differences between the actual wave motion (moving at a small angle to the x-axis) and the motion described by the factor $\exp(i(\int k dx - \omega t))$, which represents the **primary wave**. Furthermore $A(x,y)$ will include the (slow) amplitude variations along the wave front.
- The variation of the amplitude A is assumed to have different length scales in the x and y directions. Denoting these scales by L_X and L_Y, respectively, we assume that

$$L_X = O(\epsilon^2 L_x) \tag{3.7.88}$$

where L_x is of the order of the wave length and ϵ is a small parameter. In the y-direction, the length scale is assumed to be

$$L_Y = O(\epsilon L_x) \tag{3.7.89}$$

Thus we assume that in the y-direction, the variation of the amplitude is an order of magnitude slower than in the x-direction. This can be expressed formally by introducing the slow variables X, Y given by

$$X = \epsilon^2 x = \frac{L_X}{L_x} x \tag{3.7.90}$$

$$Y = \epsilon y = \frac{L_Y}{L_x} y \tag{3.7.91}$$

This implies that A and k are $A(X,Y)$ and $k(X,Y)$, respectively. The fast variation given by the exp-function remains unchanged. Then (3.7.87) becomes

$$\phi_s = -\frac{ig}{\omega} A'(X,Y) e^{i(\int k(X,Y) dx - \omega t)} \tag{3.7.92}$$

so that in the MSE, we have

$$\frac{\partial A}{\partial x}=\frac{\partial A}{\partial X}\frac{\partial X}{\partial x}=\epsilon^2\frac{\partial A}{\partial X}\;;\;\frac{\partial^2 A}{\partial x^2}=\epsilon^4\frac{\partial^2 A}{\partial X^2} \tag{3.7.93}$$

$$\frac{\partial A}{\partial y}=\epsilon\frac{\partial A}{\partial Y}\;;\;\frac{\partial^2 A}{\partial y^2}=\epsilon^2\frac{\partial^2 A}{\partial Y^2} \tag{3.7.94}$$

- On the other hand, the depth variation is assumed of the same magnitude in the x and y directions with $\nabla_h h = O(\epsilon^2)$ so that

$$\frac{\partial k}{\partial x},\;\frac{\partial k}{\partial y}=O(\epsilon^2 k) \tag{3.7.95}$$

and similar for cc_g which is a function of k.

This approach is usually called a **multiple scale method.** Substituting (3.7.92), we then get for the differentiations

$$\phi_{s,x}=-\frac{ig}{2\omega}\left(\epsilon^2 A'_X+ikA'\right)e^{i\theta'} \tag{3.7.96}$$

$$\phi_{s,xx}=-\frac{ig}{2\omega}\left[\epsilon^4 A'_{XX}+\epsilon^2 2ikA'_X+\epsilon^2 ik_X A'-k^2A'\right]e^{i\theta'} \tag{3.7.97}$$

where

$$\theta'=\int k dx-\omega t \tag{3.7.98}$$

The MSE equation then becomes after division by $-\frac{ig}{2\omega}$

$$\epsilon^2(cc_g)_X\left(\epsilon^2 A'_X+ikA'\right)+cc_g\left(\epsilon^4 A'_{XX}+\epsilon^2 2ikA'_X+\epsilon^2 ik_X A'-k^2A'\right)$$
$$+\epsilon^2\left(cc_gA'_Y\right)_Y+k^2cc_gA'=0 \tag{3.7.99}$$

Here the $O(1)$-terms cancel, and if we neglect the $O(\epsilon^4)$-terms and consider only $O(\epsilon^2)$-term, we get the relation

$$2icc_gA'_X+i\left(kcc_g\right)_X A'+\left(cc_gA'_Y\right)_Y=0 \tag{3.7.100}$$

We see that this is a parabolic equation in X as sought. In this equation, the (complex) amplitude A' is defined to include the phase differences relative to the phase angle θ' which is based on the **local** wave number

$k(X,Y)$. It is more convenient, however, to let the phase differences incorporated in the amplitude refer to a phase angle θ based on a (constant) reference wavenumber k_0. This is done by redefining the amplitude as

$$A = A' e^{i(k_0 x - \int k dx)} \tag{3.7.101}$$

Substituting (3.7.101) into (3.7.100), we then get the parabolic equation on the final form

$$\boxed{2ikcc_g A_X + 2k(k-k_0)cc_g A + i(kcc_g)_X A + (cc_g A_Y)_Y = 0} \tag{3.7.102}$$

It is useful to repeat here that the parabolic approximation assumes a primary wave direction in the x-direction. The accuracy of the solution will depend on how much the wave direction deviates from x-axis over the domain considered. This is analysed more closely in the section below on wide angle parabolic approximations. However, in practical applications this means that in addition to how much the actual wave directions vary over the computational domain the accuracy may depend on how wisely the direction of the x-axis is chosen.

Extended with the featues of energy dissipation, wave-current interaction, nonlinear effects, etc. described briefly below the parabolic approximation form the principal basis for the REF/DIF wave model designed by Kirby and Dalrymple (1994).

History of the linear parabolic equation

Eq. (3.7.102) is the parabolic approximation to the MSE. It is one of the forms found in the literature and it is similar to the linearized version of the equation derived by Kirby and Dalrymple (1983a), who also used the same nomenclature.

However, different authors often use different symbols for the same variables, which complicates immediate comparisons of the many forms published in the literature. In addition the parabolic equation is often presented in terms of different variables.

The idea of a parabolic approximation for waves was first introduced by Leontovich and Fock (1944) for propagation of radio waves in the troposphere, and the applications have since been extended to many other areas in physics such as seismic wave propagation, nonlinear optics, plasma physics, and underwater accustics to mention a few. For a review see also Mei and Liu (1993).

The first derivation for surface water waves was given by Biésel (1972) and then by Radder (1979) who presented the equation in terms of the variable Φ^+ related to A by

$$\Phi^+ = -\frac{ig}{\omega} A\, e^{ik_0 x} \tag{3.7.103}$$

(see KD83). The result becomes

$$\Phi^+{}_X - ik\Phi^+ + \frac{1}{2kcc_g}(kcc_g)_X\Phi^+ - \frac{i}{2kcc_g}(cc_g\Phi^+{}_Y)_Y = 0 \tag{3.7.104}$$

where X, Y are the slow variables.

Radder used a derivation starting from the Helmholtz version of the MSE, (3.7.50) in which he assumed that the total wave field Φ can be divided into a transmitted wave field Φ^+ (essentially the "incident" wave) and a (much weaker) reflected field Φ^- so that

$$\Phi = \Phi^+ + \Phi^- \tag{3.7.105}$$

Writing the Helmholtz equation in the form

$$\frac{\partial^2\Phi}{\partial x^2} = -\left(k^2 + \frac{\partial^2}{\partial y^2}\right)\Phi \tag{3.7.106}$$

he then uses an operator splitting technique (Corones, 1975, McDaniel, 1975) for the operator on the RHS to split the equation into two coupled equations, one for Φ^+ and one for Φ^-. By neglecting the contributions from the Φ^- field the parabolic equation (3.7.104) is obtained.

An advantage of this approach is that it emphasizes explicitly that the parabolic approximation corresponds to neglecting the reflected part of the wave. In the derivation of the parabolic equation shown above that was implicitly behind the assumption of considering only waves of the form (3.7.87), which is a "progressive" wave. Radders derivation also shows the full version of the coupled equations which really is just a transformation of the MSE (see Radder, 1979, Eqs. (16a,b)).

Another approach was used by Losano and Liu (1980) who used a multiple scale expansion to obtain the parabolic equation, and Liu and Tsay (1983) developed an iterative technique for solving equations similar to the two coupled equations of Radder and thereby obtaining also the reflected ("backscattered") part of the wave field.

Sancho (1991) tested several different explicit and implicit finite difference schemes and boundary conditions. Since then numerous contributions have been published exploring the parabolic equation method and its applications.

Wide angle parabolic equations

One of the assumptions behind the parabolic equation is that the waves move at a small angle to the x-direction. Two questions arise. One is how to assess the effect of that assumption for wider angles and the second is to extend the parabolic equation to forms that allow wider angles.

For simplicity we consider only constant depth in the following. Under those conditions the MSE (3.7.46), which we have as a reference, reduces to the Helmholtz equation

$$\nabla_h^2 a + k^2 a = 0 \tag{3.7.107}$$

Similarly, for constant depth, we have $k - k_0 = 0$, and the parabolic equation (3.7.102) simplifies to

$$2ikA_x + A_{yy} = 0 \tag{3.7.108}$$

The relation between the amplitude $a(x, y)$ in (3.7.107) and $A(x, y)$ is determined by combining (3.7.49), (3.7.92) (which shows A' is the amplitude of the surface variation) and (3.7.101). The latter gives with $k = k_0$ that $A = A'$. Hence we have

$$Ae^{i(kx - \omega t)} = a\, e^{-\omega t} \tag{3.7.109}$$

or

$$a = A\, e^{ikx} \tag{3.7.110}$$

Substituting this into the Helmholtz equation (3.7.107) changes this to

$$A_{xx} + 2ikA_x + A_{yy} = 0 \tag{3.7.111}$$

which is the form of the MSE in the same variables as the parabolic equation (3.7.108).

One effect of wider angles is then seen in the accuracy relative to (3.7.111) which the parabolic equation provides when modelling the wave number vector $\mathbf{k} = (k_x, k_y)$ for a plane wave propagating at an angle to the $x-$axis, and given by

$$\zeta = B(x, y)\, e^{i(k_x x + k_y y - \omega t)} \tag{3.7.112}$$

In the parabolic equation the amplitude A refers to a wave defined by

$$\zeta = A(x, y)e^{i(kx-\omega t)} \tag{3.7.113}$$

The plane wave (3.7.112) therefore corresponds to an A variation obtained by equating (3.7.112) and (3.7.113) or

$$A(x, y) = B(x, y)\, e^{i[(k_x-k)x+k_y y]} \tag{3.7.114}$$

Substitution of this into (3.7.111) gives the relation

$$k_x^2 + k_y^2 = k^2 \tag{3.7.115}$$

or

$$\frac{k_x}{k} = \left[1 - \left(\frac{k_y}{k}\right)^2\right]^{1/2} \tag{3.7.116}$$

which is the exact relation between k_x/k and k_y/k that we seek to model. As (3.7.116) shows this corresponds to a unit circle in a $\frac{k_x}{k}$, $\frac{k_y}{k}$-coordinate system as shown in Fig. 3.7.3.

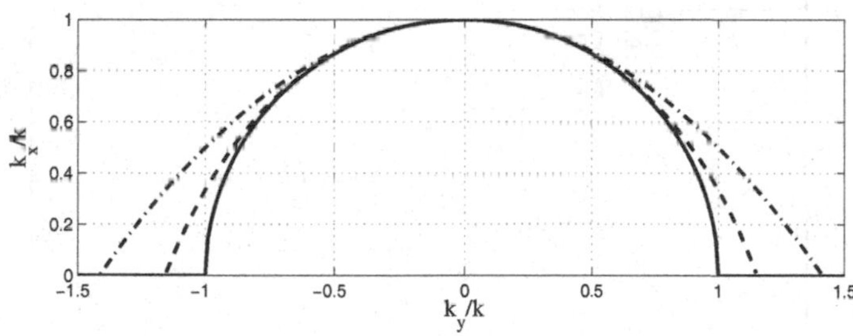

Fig. 3.7.3 Comparison between the MSE and the Parabolic approximation of the wave number vector. The solid curve corresponds to (3.7.116), the dash-dotted curve to (3.7.118), and the dashed curve to (3.7.124).

If we also substitute (3.7.114) into the parabolic relation (3.7.108) we get

$$2k(k_x - k) + k_y^2 = 0 \tag{3.7.117}$$

or

$$\frac{k_x}{k} = 1 - \frac{1}{2}\left(\frac{k_y}{k}\right)^2 \tag{3.7.118}$$

which plotted in the $\frac{k_x}{k}$, $\frac{k_y}{k}$-coordinates represents a parabola as also shown in Fig. 3.7.3. This is also called the binomial expansion (Kirby, 1986)

This result may also be viewed in terms of the wave angle α_w between the wave orthogonal and the $x-$axis. We have the exact relation

$$k_x = k \cos \alpha_w \quad ; \quad k_y = k \sin \alpha_w \tag{3.7.119}$$

and the parabolic equation gives by (3.7.118)

$$\frac{k_x}{k} = 1 - \frac{1}{2} \sin^2 \alpha_w \tag{3.7.120}$$

that is an error e of

$$e = \frac{k_x}{k} - \cos \alpha_w = 1 - \frac{1}{2} \sin^2 \alpha_w - \cos \alpha_w \tag{3.7.121}$$

Fig. 3.7.4 (Kirby 1986) shows the variation of e versus $\sin \alpha_w$.

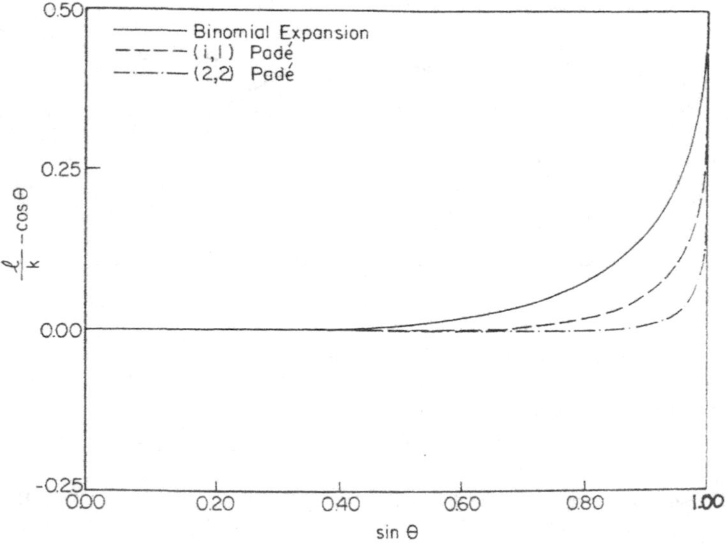

Fig. 3.7.4 The error e versus $\sin \alpha_w$ (θ in the figure) by different parabolic approximations, (from Kirby 1986). Here script l correponds to k_x. The solid curve marked "binomial expansion" corresponds to (3.7.118), the dashed curve marked (1,1) Padé is (3.7.124). The (2,2) Padé approximation shown as the dash-dot curve is not given in the text (see Kirby, 1986)

The figure also shows the errors generated by various improvements of the approximation (3.7.118), which can be considerd a first order Taylor expansion of (3.7.116).

Higher order Taylor expansions for (3.7.116) can readily be constructed. Parabolic eqations representing such higher order approximations may also be developed by a rigorous multiple scale method. For details see Kirby (1986). The next approximation (still for constant depth) in the mathematical expansion in the small parameter ϵ introduced earlier results in an equation which can be reduced to

$$2ikA_x + A_{yy} + \frac{i}{2k}A_{xyy} = 0 \tag{3.7.122}$$

This form of the parabolic approximation to the MSE was first derived by Booij (1981). Substitution into this equation of (3.7.115) for A results in the relation

$$2k(k_x - k) + k_y^2 - \frac{1}{2}i(k_x - k)k_y^2 = 0 \tag{3.7.123}$$

which solved with respect ot k_x/k gives

$$\frac{k_x}{k} = \frac{1 - \frac{3}{4}(\frac{k_y}{k})^2}{1 - \frac{1}{4}(\frac{k_y}{k})^2} \tag{3.7.124}$$

Inspection of this expression will show that it represents a (1,1) Padé approximation to (3.7.116) (Kirby 1986). The error associated with (3.7.124) is also shown in the Fig. 3.7.4, and we see it is much smaller than the error using (3.7.108).[20]

Thus improved accuracy leading to only third order derivatives in x, y can be obtained by (3.7.122). While further such extensions are possible they will also involve higher order derivatives in the parabolic equation, which makes it somewhat more difficult to solve numerically.

Instead Kirby (1986) suggested approximating (3.7.116) by an expression of the form

$$\frac{k_x}{k} = \frac{a_0 + a_1(\frac{k_y}{k})^2}{1 + +b_1(\frac{ky}{k})^2} \tag{3.7.125}$$

[20]For a brief review of the concepts of Padé approximations see the appendix to Chapter 9.

and determining the coefficients a_0, a_1, b_1 by minimizing the largest error

$$e = \frac{k_x}{k} - cos\alpha_w \tag{3.7.126}$$

that occurs in a chosen wide angle interval α_m fo α_w. This is a so-called minimax approximation.

The resulting parabolic equation can be retrieved by writing (3.7.125) on the form similar to (3.7.123) and identifying each term as a result of a differentiation (termed the method of operator correspondance). The result can be written

$$2ikA_x + 2k^2(a_0 - 1)A + 2(b_1 - a_1)A_{yy} - \frac{2ib_1}{k}A_{xyy} = 0 \tag{3.7.127}$$

For further details, including values of the constants a_0, a_1, b_1 for different wide angle intervals α_w, reference is made to Kirby (1986). The conclusion is that the smallest overall errors are obtained for $\alpha_m = 60°$. Within this interval the error in the predicted α_w using (3.7.127) is less than 0.1°, and in k no more than 0.1%.

The wide angle results significantly increase the practical value of the parabolic equation.

The nonlinear parabolic equation

An extension to include the first approximation to nonlinear effects in the parabolic equation was developed by Kirby and Dalrymple (1983a,b). Because the MSE is inherently linear, nonlinear effects can only be analysed by going back to the original equations of motion, here the Laplace equation and its (nonlinear) boundary conditions. Kirby and Dalrymple (1983a) used a multiple scale method to derive a lowest order nonlinear parabolic equation which is given as

$$2ikcc_g\, A_X + 2k(k - k_0)cc_g A + i(kcc_g)_X\, A + (cc_g\, A_Y)_Y - kcc_g\, K'|A|^2\, A = 0 \tag{3.7.128}$$

where K' is defined as

$$K' = k^3 \frac{c}{c_g} D \tag{3.7.129}$$

with

$$D = \frac{\cosh\, 4kh + 8 - 2\, \tanh^2 kh}{8\, \sinh^4 kh} \tag{3.7.130}$$

It may be noticed that the first four terms in this euation are identical with the terms in the linear version (3.7.102). Thus the only nonlinear contribution is represented by the last term.

The parabolic equation for combined wave-current motion

Similarly the parabolic approximation has been extended to situations with combined wave-currents motions. Assuming the currents (U, V) that are weak in comparison to the wave speed (as in most practical cases) Kirby and Dalrymple (1983b) found the equation takes the form

$$\begin{aligned}&(c_g + U)\, A_X + V\, A_Y + i\,(k_0 - k)(c_g + U)\, A\\&+\frac{\omega}{2}\left[\left(\frac{c_g + U}{\omega}\right)_X + \left(\frac{V}{\omega}\right)_Y\right] A - \frac{i}{2\omega}[(cc_g - V^2)\, A_Y]_Y\\&-\omega\frac{k^2}{2}\, D|A^2|\, A = 0\end{aligned} \tag{3.7.131}$$

where D is again given by (3.7.130). See also Kirby and Dalrymple (1994).

The parabolic equation with energy dissipation

The parabolic equation with energy dissipation is equivalent to the MSE with energy dissipation (see there). It was also discussed by Dalrymple et al. (1984). Results for w corresponding to different situations can also be found in Kirby and Dalrymple (1994).

3.8 References - Chapter 3

Airy, G. B. (1845) Tides and waves. *In Encyclopaedia metropolitana*, vol. 5, pp. 241-396. London

Barber, N. F. (1963). The directional resolving power of an array of wave detectors. Proc. Conf. Ocean Wave Spectra, Eaton Maryland, 1961, Nat. Acad. Science. 137–150.

Battjes, J. A. (1968). Refraction of water waves. ASCE J. Waterw., Harbors, Coast. Engrg., **94**, 437–451.

Battjes, J. A. (1978). Probabilistic aspects of ocean waves. Communications in Hydraulics, Lab Fluid Mechs., Dept. Civil Engrg. Delft Univ. Tech. Rep 77-2.

Berkhoff, J. C. W. (1972). Computation of combined refraction-diffraction. ASCE Proc 13th Int. Conf. Coasal Engrg., Vancouver. 471–490.

Biésel, F. (1972). Refraction de la houle avec diffraction modereé. ASCE Proc. 13th Int. Conf. Coast. Engrg., **1** 491–501.

Biesel, F. (1950). "Etude théorique de la houle en eau courante." La Houille Blanche. Special A, Mai 1950, pp. 3-9.

Biesel, F. (1952). Study of wave propagation in water of gradually varying depth. In "Gravity waves", Natl. Bureau of Standards, Circular 521, 243 – 253.

Booij, N. (1981). Gravity waves on water with non-uniform depth and current. Report no 81-1, Delft Univ. Tech., 130pp

Booij, N. (1983) A note on the accuracy of the mild slope equation. Coastal Engineering, **7**, 191–203.

Cartwright, D. E. and M. S. Longuet-Higgins (1956). The statistical distibution of the maxima of a random function. Proc. Roy. Soc. **A, 237**, 212–232.

Chamberlain, P. G. and D. Porter (1995). The modified mild slope equation. J. Fluid Mech. **291**, 393–407.

Chawla, A. and J. T. Kirby (2000). An experimental study on the dynamics of wave blocking and breaking on opposing currents. Univ. of Delaware, Center for Applied Coastal Res., Rep. No CACR-00-02, 157pp.

Christoffersen, J. (1982). Current depth refraction of dissipative waves. Tech. Univ. Denmark, Inst. Hydrodyn. and Hydraulic Eng., Series Paper No 30.

Christoffersen, J. and I. G. Jonsson (1980). A note on wave action conservation in a dissipative current wave motion. Appl. Ocean Res., **2**, 179–182.

Christoffersen, J. B. and I. G. Jonsson (1981). An energy reference line for dissipative water waves on a current. J. Hydr. Res., 19, No. 1.

Copeland, G. J. M. (1985). A practical alternative to the "Mild Slope" wave equation. Coastal Engineering **9**, 125–149.

Corones, J. (1975). Bremmer series that correct parabolic approximations. J. Math. Anal. Applic. **50**, 361–372.

Dalrymple, R. A., J. T. Kirby, and P. A. Hwang (1984). Wave diffraction due to areas of energy dissipation. J. Waterways, Port, Coastal and Ocean Engineering, **110**, 1, 67–79

Davies, A. G. and A. D. Heathershaw (1984). A practical alternative to the mild-slope wave equation. J. Fluid Mech. **144**, 419–443.

Dean, R. G. and R. A. Dalrymple (1984). Water wave mechanics for engineers and scientists. Prentice Hall, 353pp. (later editions: World Scientific, Singapore)

Dingemans, M. W. (1997). Water wave propagation over uneven bottoms. Vol 1 & 2. World Scientific, Singapore. pp 967

Drazin, P.G. and W.H. Reid (1981). Hydrodynamic Stability. Cambridge Univ. Press, 527 pp.

Green, L. (1837). On the motion of waves in a variable canal of small depth and width. Trans. Phil. Soc. London, **6**.

Greenberg, M. D. (1988, 1998). Advanced Engineering Mathematics, Prentice Hall.

Hasselmann, K. et al. (1973). Measurements of wind-wave growth and swell decay during the Joint North Sea Wave Project (JONSWAP), Deutsche Hydrographische Zeitschrift, Hamburg, Reihe A (8°), No 12, 95pp.

Jenkins, G. M. and D. G. Watts (1968). Spectral analysis and its applications. Holden-Day, San Francisco, 525pp.

Jonsson, I. G. (1978a). Energy flux and wave action in gravity waves propagating on a current. J. Hydrulic Res. **16**. 223–234.

Jonsson, I. G. (1978b). Combination of waves and currents. In P. Bruun, (ed.) Stability of tidal inlets. 162–203. Elsevier, Devel. Geotech. Eng **23**, Amsterdam.

Jonsson, I. G. (1989) Wave current interactions. Report S49, Danish Center for Applied Mathematics and Mechanics (DCAMM), Tech. University Denmark, 86pp.

Jonsson, I. G., O. Skovgaard, and J. D. Wang (1970). Interaction between waves and currents. ASCE Proc 12th Int. Conf. Coastal Engrg., Washington DC. vol I, 489-507.

Jonsson, I. G. and O. Brink-Kjær (1973). A comparison between two reduced wave equations for gradually varying depth. ISVA Technical University of Denmark. Progr. Rep. **31**, 13-18.

Jonsson, I. G. and J. D. Wang (1980). Current-depth refraction of water waves. Ocean Eng., **7**, 153–171.

Kirby, J. T. (1984). A note on linear surface wave-current interaction over slowly varying topography. J. Geophys. Res., **89**, 745–747.

Kirby, J. T. (1986). Rational approximations in the parabolic equation method. Coast. Engrg., **10**, 355-378.

Kirby, J. T. (1986). Rational approximations in the parabolic equation method for water waves. Coastal Engrg., **10**, 355–378.

Kirby, J. T. (2005). Analysis of regular and random ocean waves. Text in preparation.

Kirby, J. T. and R. A. Dalrymple (1983a). A parabolic equation for the combined refraction-diffraction of Stokes waves by mildly varying topography. J. Fluid Mech. **136**, 453–466.

Kirby, J. T. and R. A. Dalrymple (1983b). The propagation of weakly nonlinear waves in the presence of varying depth and currents. Proc XXth IAHR congr., vol VII, p 168, IAHR, Moscou.

Kirby, J. T. and T.-M. Chen (1989). "Surface waves on vertically sheared flows: Approximate dispersion relations." *J. Geophys. Res.*, **94**, C1, 1013-1027.

Kirby, J. T. and R. A. Dalrymple (1994). Combined Refraction/Diffraction model REF/DIF 1, version 2.5. Research Report no CACR-94-22. Center for Applied Coastal Research, University of Delaware, 171pp.

Komen, G. L., L. Cavaleri, M. Donelan, K. Hasselmann, S. Hasselmann and P. A. E. M. Janssen (1994). Dynamics and modelling of ocean waves. Cambridge Univ. Press, 532pp.

Lee, C., W. S. Park, Y.-S. Cho, and K. D. Suh (1998). Hyperbolic mild-slope equations extended to account for rapidly varying topography. Coastal Engineering, **34**, 243–257.

Liu P. L.-F. and T.-K. Tsay (1983). On weak reflection of water waves. J. Fluid Mech. **131**, 59–71.

Leontovich, M. and V. A. Fock (1944). A method of solution of problems of electromagnetic wave propagation along the earth's surface. Izv. Akad. Nank., SSSR, **8**, 16–22.

Longuet-Higgins, M. S., D. E. Cartwright, and N. D. Smith (1963). Observations of the directional spectrum of sea waves using the motions of a floating buoy. Proc. Conf. Ocean Wave Spectra, Eaton Maryland, 1961, Nat. Acad. Science. 111–132.

Losano, C. and P. L-F. Liu (1980). Refraction-diffraction model for linear surface waves. J. Fluid Mech. **101**, 705–720.

MacCamy, R. C. and R. A. Fuchs (1954). Wave forces on piles: a diffraction theory. US Army Corps of Engineers. Tech Memo 69.

Madsen, P. A. and J. Larsen (1987). An efficient finite-difference approach to the mild slope equation. Coast. Engrg. **11**, 329–351.

Madsen, P. A. and H. A. Schäffer (1999). A review of Boussinesq-type equations for surface gravity waves. Advances in Coastal Engineering, **5**, 1–93.

Massel, S. R. (1993). Extended refraction-diffraction equation for surface waves. Coastal Engineering, **19**, 97–126.

McDaniel, S. T. (1975). Parabolic approximations for underwater sound propagation. J. Accoust. Soc. Am. **58**, 1178–1185.

Mei, C. C. (1983). The applied dynamics of ocean surface waves. John Wiley & Sons, Inc., 740pp.

Mei, C. C. (1985). Resonant relection of surface water waves by periodic sand bars. J. Fluid Mech., **152**, 315–335.

Mei, C. C. and P. L-F. Liu (1993). Surface waves and coastal dynamics. Ann. Rev. Fluid Mech. **25**, 215–240.

Mitsuyasu, H., F. Tasai, T. Suhara, S. Mizuno, M. Ohkusu, T. Honda, and K. Rikiisi (1975). Observations of the directional spectrum of ocean waves using a cloverleaf buoy. J. Phys. Oceanography, **5**, 750–760.

Munk, W. H. and R. S. Arthur (1952). Wave intensity along a refracted ray in gravity waves. Natl. Bur. Stand. Wash. DC. Circ 521.

Noda, E. K. (1974). Wave-induced nearshore circulation. Jour. Geophys. Res., **75**, 27.

Ochi, M. K. (1982). Stochastic analysis and probabilistic prediction of random seas. Advances in Hydroscience, **13**, Academic Press, 218–375.

O'Hare, T. J. and A. G. Davies (1993). A comparison of two models for surface-wave propagation over rapid varying topography. Appl. Ocean Res., **15**, 1–11.

Penney, W. G. and A. T. Price (1952). The diffraction theory of sea waves and the shelter afforded by breakwaters. Phil. Trans. Roy. Soc. A**244**, 236–253.

Pierson, W. J. and L. Moskowitz (1964). A proposed spectral form for fully developed wind seas based on the similarity theory of S. A. Kitaigoroskii. J. Geophys. Res, **69**, 24, 5101–5190.

Peregrine, D. H. (1976). Interaction of waves and currents. Adv. Appl. Mech., **16**, 9-117.

Peregrine, D. H. and I. J. Jonsson (1983). "Interaction of waves and currents." U.S. Army Corps of Engineers Res. Center, *Misc. Rep. No. 83-6.*

Phillips, O. M. (1966, 1977). Dynamics of the upper ocean. Cambridge University Press.

Radder, A. C. (1979). On the parabolic equation method for water wave propagation. J. Fluid Mech. **95**, 159–176.

Sancho, F. E. P. (1991). A finite difference wave refraction-diffraction model. Proceedings of the 2nd International Conference in Computational Methods in Ocean Eng., Barcelona, pp. 333-341.

Sand, S. E. (1979). Three dimensional deterministic structure of ocean waves. Inst. Hydrodyn. and Hydraulic Engrg. (ISVA), Tech. Univ. Denmark, Lyngby. 177pp.

Seymour, R. J. and A. L. Higgins (1977). A slope array for measuring wave direction. Proc Workshop on Coastal Processes Instrumentation, La Jolla, Univ. Calif, San Diego, Sea Grant Publ No 62, IMR Ref No. 78-102, 133–142.

Shore Protection Manual (1984). Waterways Experiment Station, Corps of Engineers, Coast. Eng. Res. Center, Dept of the Army.

Skougaard O. and J. A. Bertelsen (1974). Refraction computation for practical applications. ASCE Int. Symp. Ocean Wave Measurements and Analysis, New Orleans, **I**, 761–773.

Skougaard O., I. G. Jonsson, and J. A. Bertelsen (1975). Computation of wave heights due to refraction and friction. Proc. ASCE, Waterways, Harbors and Coastal Engr., **101**, WW1, 15–32.

Sommerfeld, A. (1896). Matematische Theorie der Refraction, Math. Ann, **47**, 317–374.

Smith, R. and T. Sprinks (1975). Scattering surface waves by a conical island, J. Fluid Mech. **72**, 373–384.

Stokes, G. G. (1847). On the theory of oscillatory waves. Trans. Cambridge Phil. Soc., **8**, 441–473.

Svendsen, I. A. (1967). The wave equation for gravity waves in water of gradually varying depth. ISVA, Tech. Univ. of Denmark, Basic Research Progress Report **15**, 2–7.

Svendsen, I. A. and I. G. Jonsson (1976, 1980). Nearshore Hydrodynamics. Den Danske Ingeniør Fond, Copenhagen. 285pp.

Suh, K. D., C. Lee, Y-H. Park (1997). TIme deoendent equations for wave propagation on rapidly varying topography. Coast. Eng., **32**, 91–117.

Suh, K. D., C. Lee, Y-H. Park, T. H. Lee (2001). Experimental verification of horizontal two-dimensional modified mild-slope equation model, Coast. Eng., **44**, 1–12.

Thomas, G. P. (1981). "Wave current interactions: An experimental and numerical study. Part 1. Linear waves." *J. Fluid Mech.*, **110**, 457–474.

Ursell, F. (1952). The long wave paradox. Proc. Cambr. Phil. Soc., **49**, 685–694.

WAMDI Group (1988): The WAM model: A third generation ocean wave prediction model. *J. of Physical Oceanography* **18**, 1775–1810.

Wilson, W. S. (1966). A method for calculating and plotting surface wave rays, Tech. Memo 17, USArmy Coast. Eng. Res. Center.

Young, I. (1999). Wind generated ocean waves, Elsevier, 288pp.

Chapter 4

Energy Balance in the Nearshore Region

4.1 Introduction

In this chapter we analyse the energy balance for non-breaking and breaking waves. In Section 4.2 we derive a crude version of the energy equation for waves and currents in a situation with energy dissipation, and in Section 4.3 this equation is written in terms of the wave height, and it is shown that under certain conditions a closed form solution can be found to the equation. This solution is a generalization of the simple equation for conservation of energy used for 1DH wave shoaling with no dissipation. It illustrates how, for breaking waves inside a surfzone in the simple 1-D cross-shore case, two mechanisms - the shoaling and the energy dissipation - counteract one another in deciding the variation of the wave height. Section 4.4 gives the general form of the energy equation in the 2D horizontal plane for combined wave-current motion. This equation is valid for any slowly varying wave motion on a depth uniform current. In Section 4.5 it is briefly outlined how for linear waves with currents the energy equation can be expressed in particularly simple form in terms of the socalled wave action.

4.2 The energy equation

We first derive the energy equation on a heuristic basis for combined waves and currents with energy dissipation. This will help illustrate the basic interactions between different types of energy such as organized mechanical energy (waves, currents), random turbulent energy and energy "loss" to heat. In this derivation we seek to determine the change in organized mechanical energy under the influence of waves and currents. It provides a generic form of the energy equation which can be used for simple purposes.

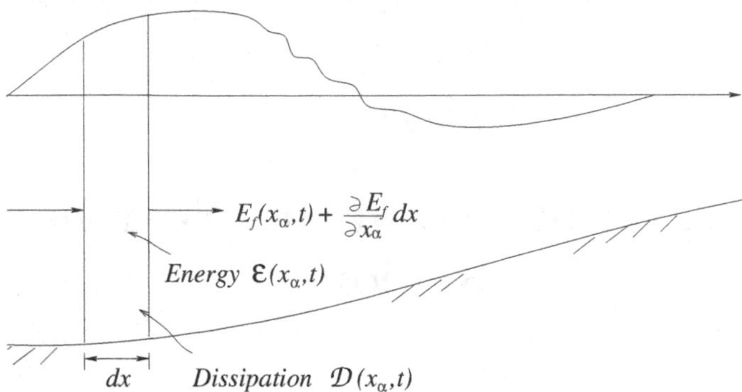

Fig. 4.2.1 The variables in the energy balance for a water column - shown for 1D horizontal only.

In chapter 3 we found that for waves only with no energy dissipation we have

$$\frac{\partial E_{fw\alpha}}{\partial x_\alpha} = 0 \tag{4.2.1}$$

where $E_{fw\alpha}$ is the energy flux which for steady waves only is given by (2.6.8). In vector notation this can be written

$$\nabla_h \cdot \mathbf{E}_{fw} = 0 \tag{4.2.2}$$

In a combined wave-current motion with a time varying current and energy dissipation, however, the picture is a little more complex. The following analysis is a (somewhat intuitive) derivation that tries to avoid going into details that are unimportant for the final result.

We consider the water column in Fig. 4.2.1. Inside this column we have a certain amount of mechanical energy $\mathcal{E}(x_\alpha, t)$, consisting of both kinetic and potential energy (the latter relative to a horizontal reference level). The kinetic energy can be both in the form of organized energy (waves and currents) and in the form of turbulent kinetic energy. If we use the still water level (SWL) ($= x-$axis equivalent to $z = 0$) as reference for potential energy, then we have the total mechanical energy $\mathcal{E}$ (per unit area) is given by

$$\mathcal{E}(x_\alpha, t) = \int_{-h_0}^{\zeta} \left(\rho g z + \frac{1}{2}\rho \left(u^2 + v^2 + w^2\right)\right) dz \tag{4.2.3}$$

where u, v, w are the total particle velocities including both waves, currents, and turbulent components.

The energy flux through the vertical walls of the column is still given by the same (exact) expression, that is

$$E_{f\alpha}(x_\alpha, t) = \int_{-h_0}^{\eta} \left(p + \rho g z + \frac{1}{2}\left(u^2 + v^2 + w^2\right)\right) u_\alpha dz \qquad \text{(exact)} \quad (4.2.4)$$

where so far $E_{f\alpha}$ is the flux of total mechanical energy, i.e., both wave, current and turbulent energy because p and u, v, w represent the total values of those quantities.

One of the basic principles of physics is that the total energy is conserved. The total energy is the sum of the mechanical and the thermal energy. Here, however, we are only interested in the mechanical energy. Therefore we introduce the concept of *energy dissipation* $\mathcal{D}_h$, which strictly speaking means: "the change in the type of energy we have not accounted for otherwise".

If we consider the energy conservation for the column in Fig. 4.2.1, we can express this by realizing that the dissipation of energy $\mathcal{D}_h$ is the amount of energy transformed from mechanical energy to heat.

Therefore, the rate of change of total *mechanical* energy $\mathcal{E}$ inside the column is

$$\frac{\partial \mathcal{E}(x_\alpha, t)}{\partial t} = -\frac{\partial E_{f\alpha}(x_\alpha, t)}{\partial x_\alpha} + \mathcal{D}_h(x_\alpha, t) \qquad (4.2.5)$$

which means: $\frac{\partial \mathcal{E}(x_\alpha, t)}{\partial t}$ results from a net inflow of mechanical energy $-\partial E_f(x_\alpha, t)/dx_\alpha$ (minus because the net inflow is positive when $\partial E_f(x_\alpha, t)/\partial x_\alpha < 0$), minus the energy transformed into heat (i.e. $\mathcal{D}_h < 0$ for loss to heat).

It is also clear from the definition of $E_{f\alpha}$ that it can be divided into a turbulent and a wave-current part by a Reynolds decomposition (see section 2.5), so we can write

$$E_{f\alpha}(x_\alpha, t) = \overbrace{E_{f\alpha}(x_\alpha, t)} + E'_{f\alpha}(x_\alpha, t) \qquad (4.2.6)$$

where $\overbrace{E_{f\alpha}(x_\alpha, t)}$ is the wave-current (also termed the "organized mechanical") part of the energy flux, $E'_{f\alpha}(x_\alpha, t)$ the turbulent part.

The total dissipation $\mathcal{D}_h(x_\alpha, t)$ can also be divided into two parts: a small amount of wave and current energy, $\overbrace{\mathcal{D}_h(x_\alpha, t)}$ *is* turned directly into heat by dissipative forces, whereas the major part $\mathcal{D}'_h(x_\alpha, t)$ of $\mathcal{D}_h(x_\alpha, t)$ is the transformation of (small scale) turbulent kinetic energy to heat. Thus we may write

$$\mathcal{D}_h(x_\alpha, t) = \widehat{\mathcal{D}_h(x_\alpha, t)} + \mathcal{D}'_h(x_\alpha, t) \tag{4.2.7}$$

Finally the energy density $\mathcal{E}$ may be separated into an organized part $\widehat{\mathcal{E}}$ and a turbulent part $\mathcal{E}'$ by

$$\mathcal{E}(x_\alpha, t) = \widehat{\mathcal{E}}(x_\alpha, t) + \mathcal{E}'(x_\alpha, t) \tag{4.2.8}$$

Therefore (4.2.5) can be written

$$\frac{\partial \widehat{\mathcal{E}}}{\partial t} + \frac{\partial \mathcal{E}'}{\partial t} = -\frac{\partial \widehat{E_f}}{\partial x_\alpha} - \frac{\partial E'_f}{\partial x_\alpha} + \widehat{\mathcal{D}_h} + \mathcal{D}'_h \tag{4.2.9}$$

However, for turbulent energy $\mathcal{E}'$ we have: Turbulent kinetic energy is added ("produced") from the wave-current energy and extracted from $\mathcal{E}'$ due to dissipation. Thus the energy balance for the turbulent part $\mathcal{E}'$ can be written

$$\frac{\partial \mathcal{E}'}{\partial t} = -\frac{\partial E'_f}{\partial x_\alpha} + \text{Prod} + \mathcal{D}'_h \tag{4.2.10}$$

where "Prod" means the production of turbulent kinetic energy in the column per unit time from ordered wave-current energy.

Substituting (4.2.10) into (4.2.9) we get

$$\frac{\partial \widehat{\mathcal{E}(x_\alpha, t)}}{\partial t} = -\frac{\partial \widehat{E_f(x_\alpha, t)}}{\partial x_\alpha} + \widehat{\mathcal{D}_h(x_\alpha, t)} - \text{Prod}(x_\alpha, t) \tag{4.2.11}$$

Eq. (4.2.11) is an instantaneous balance of organized mechanical energy in the column. We now consider the wave (or time) average of (4.2.11) given by the usual $\overline{}$. We then get

$$\frac{\partial \overline{\mathcal{E}}}{\partial t} = -\frac{\partial \overline{E_f}}{\partial x_\alpha} + \overline{\mathcal{D}}_h - \overline{\text{Prod}} \tag{4.2.12}$$

It is emphasied that the minus in front of Prod is because Prod is considered positive when wave energy is transformed into turbulent energy.

If the wave motion we consider is steady even over a time scale much longer than the wave period then

$$\frac{\partial \overline{\mathcal{E}}}{\partial t} \sim 0 \tag{4.2.13}$$

and consequently we have

$$-\frac{\partial \overline{E_f}}{\partial x_\alpha} + \overline{\mathcal{D}}_h - \overline{\text{Prod}} = 0 \tag{4.2.14}$$

which is the equation for simple regular waves in a steady current environment.

Discussion

We see equation (4.2.11) shows the balance of wave-current energy which is similar in structure to the equation (4.2.10) for turbulent energy except the Prod-term occurs with the opposite sign. In hindsight we might have constructed (4.2.11) directly because it says: The rate of change of wave-current energy inside the column originates from three contributions:

i) The net inflow $-\dfrac{\partial \widehat{E_f(x_\alpha, t)}}{\partial x_\alpha}$ of wave-current energy.
ii) The loss "-Prod" due to production of turbulent energy from wave-current energy inside the column.
iii) The direct dissipation $\widehat{\mathcal{D}_h}$ of wave-current energy into heat.

A difference between (4.2.10) and (4.2.11) or (4.2.14) is the sign in front of the Prod-term, the production of turbulent energy from organized wave-current energy. The production of turbulent energy of course emerge as a loss in the equation (4.2.11) for wave energy and as a gain in (4.2.10) the balance of turbulent energy, but otherwise the two equations are similar.

For simplicity we drop the $\widehat{\ }$ in the following so that from now on $E_{f\alpha}(x_\alpha, t)$, say, means flux of organized energy (rather than of total energy).

The balances and processes are illustrated graphically in Fig. 4.2.2. In the figure, the fluxes of $E_f(x_\alpha, t)$ and $E'_f(x_\alpha, t)$ are shown as going out because $\partial/\partial x$ of those quantities positive corresponds to a net outflow.

Several observations can be made here.

First we realize that in this heuristic derivation we have not yet distinguished between wave and current energy. Both $\mathcal{E}$ and $E_{f\alpha}$ represent the total mechanical energy. To get a description separating the wave and the current energy we have to use the analytical method described in

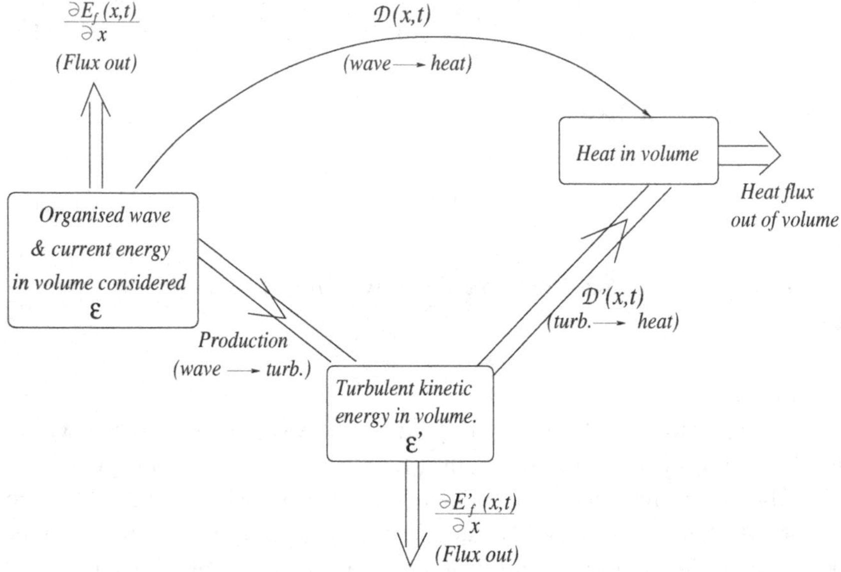

Fig. 4.2.2 A qualitative sketch of the energy transfer processes in a turbulent wave-current flow.

section 4.4. This will also reveal the mechanisms for the exchange of energy between the waves and the currents.

Second it is clear that at this level of description we would not be able to tell which part of the energy loss goes directly to heat ($\mathcal{D}_h$) and which part is used in producing turbulence (the Prod-term). It is conceivable, however, that in all cases of breaking waves $\mathcal{D}_w \ll$ Prod. To be able to separate $\mathcal{D}_h$ and Prod in (4.2.12), we will have to go into much more detail than we have done so far and at least develop the conservation equation for turbulent energy (4.2.11) sufficiently to be able to evaluate "Prod". This is beyond the scope of the present analysis, and in fact this is far from trivial even with today's increasingly advanced knowledge of wave breaking.

Another intuitive estimate is that when wave breaking is involved the production of turbulent energy due to the currents is negligible relative to the turbulence production in the wave breaking.

For all practical applications for waves and currents, however, it suffices to combine the two contributions on the RHS of (4.2.12) into one and term it "the loss (or "dissipation") of wave (and current) energy" $\mathcal{D}$ and hence to write (4.2.12) in the simple form (dropping also the $\overline{}$ signs of wave averaging)

$$\frac{\partial \mathcal{E}}{\partial t} = -\frac{\partial E_f}{\partial x_\alpha} + \mathcal{D}$$

where (4.2.15)

$$\mathcal{D} \equiv \mathcal{D}_h - \text{Prod}$$

and $\mathcal{E}$ represents the energy density of the combined wave current motion.

It is then straightforward to realize that if the wave motion is a steady wave train whose characteristics do not change (slowly) with time, then the energy equation simply becomes

$$\frac{\partial E_f}{\partial x_\alpha} = \mathcal{D} \tag{4.2.16}$$

4.3 The energy balance for periodic waves

4.3.1 *Introduction of dimensionless parameters for 1-D wave motion*

In the following we assume that the motion consists only of 1-D regular periodic waves. We can then introduce the dimensionless parameter B already used for the wave part E_{fw} of E_f

$$E_{fw} = \rho g c H^2 B \tag{4.3.1}$$

and add a similar definition for the dimensionless energy dissipation D

$$\mathcal{D} = \frac{\rho g H^3}{4hT} D \tag{4.3.2}$$

The reason for the choice of this form will become apparent later (see Chapter 5).

Substituted into (4.2.15) this becomes

$$\frac{\partial}{\partial x}\left(cH^2 B\right) = \frac{H^3}{4hT} D \tag{4.3.3}$$

By expanding the differentiation on the LHS and rearranging terms, this equation may be transformed into the following differential form:

$$\boxed{H_x + \left(\frac{c_x}{c} + \frac{B_x}{B}\right)\frac{H}{2} = \frac{D}{8hLB} H^2} \tag{4.3.4}$$

where index $_x$ means $\partial/\partial x$ and

$$L \equiv cT \tag{4.3.5}$$

is a wave length defined the usual way on the basis of the local phase speed c. This local wave length L is well defined because c and T are, but this L is fictitious in the sense that it cannot be measured between any two points in the wave motion.

If $c(x)$, $B(x)$ and $D(x)$ are known as functions of h, H and T, then (4.3.4) is a differential equation for H.

Alternatively, we can write (4.3.4) as an equation for H/h. Simple rearrangement of terms yields

$$\boxed{\left(\frac{H}{h}\right)_x = -\left(\frac{h_x}{h} + \frac{c_x}{2c} + \frac{B_x}{2B}\right)\frac{H}{h} + \frac{D}{8LB}\left(\frac{H}{h}\right)^2} \tag{4.3.6}$$

Exercise 4.1

Derive equations (4.3.4) and (4.3.6).

It is important to emphasize that the equations (4.3.4) and (4.3.6) assume very little about the nature of the wave motion. In fact, all that is needed to derive (4.3.4) or (4.3.6) is the assumption that the waves are periodic and progressive on a slowly varying depth and with no currents. It has not even been assumed that the waves are breaking and as mentioned earlier: $D > 0$ could correspond to the *generation* of waves by wind if we include in D the energy added to the column due to the wind action on the surface.

(4.3.4) and (4.3.6) are quite convenient forms of the energy equation for waves only.

Discussion

If we consider (4.3.6), it can be seen that in a nearshore region with decreasing depth the two terms on the RHS represent two counteracting mechanisms:

The H/h-term represents the shoaling effect of changing (decreasing or increasing) water depth. If the depth decreases then $h_x < 0$ and $c_x < 0$ so the term normally becomes positive: the shoaling tends to increase H/h.

Similarly, bottom friction or breaking corresponds to $D < 0$ and the second term consequently is negative if those mechanisms are in effect. Then the last term will tend to *decrease* H/h.

Closer analysis will show that for breaking waves the last term will dominate immediately after breaking where H/h is large, provided D is sufficiently large and h_x sufficiently small. However, as H/h decreases in consequence of this, $(H/h)^2$ will decrease faster than H/h and hence there is a fair chance that the two terms become equal. Hence if the coefficients of H/h and $(H/h)^2$ were constant, that would mean that a breaking wave on a steadily decreasing depth would tend to a situation where $(H/h)_x = 0$ or H/h = constant as is often assumed.

In general, however, the coefficients are not constant. Both c and L are decreasing with h but not as fast as h_x/h (as e.g. Green's law indicates for linear waves). Therefore, as the wave propagates under continued breaking into smaller and smaller depth, we find the first term becomes the largest so that near the shore H/h starts increasing again. Consequently H/h often has a **minimum** at some point in the surfzone.

4.3.2 *A closed form solution of the energy equation*

As the previous discussion indicates the energy equation for short waves can be used to determine the wave height variation in the nearshore. It turns out that for some quite plausible approximations we are able to give a closed form solution for the 1-D cross-shore energy equation (4.3.6) (or (4.3.4)), and for sufficiently simple bottom contours the integrals of the solution may even be solved in terms of known functions. This solution can be useful as illustration of how the wave height depends on the parameters involved.

There are two ways this equation can be solved.

1st solution method

In both methods it is necessary in (4.3.6) to assume that h, c, B are (known) functions of h and T but not of H (as in linear waves). We also assume that the dimensionless dissipation D is independent of H. If we introduce the definition

$$y \equiv H/h \tag{4.3.7}$$

then (4.3.6) can be written

$$\frac{dy}{dx} + a_1(x)y = a_2(x)y^2$$

where

$$a_1(x) = \frac{h_x}{h} + \frac{c_x}{2c} + \frac{B_x}{2B}$$
$$a_2(x) = \frac{D}{8LB}$$

This is a first order ordinary differential equation of the **Bernoulli type**. The general form is

$$\frac{dy}{dx} + a_1(x)y = a_2(x)y^n \tag{4.3.8}$$

which can be solved by the substitution $v = y^{1-n}$. Hence, in (4.3.7) where $n = 2$ we introduce

$$\left.\begin{array}{l} v = y^{-1} \\ y = v^{-1} \end{array}\right\} \; ; \quad \frac{dy}{dx} = -\frac{1}{v^2}\frac{dv}{dx} \tag{4.3.9}$$

This transforms the equation into

$$\frac{dv}{dx} + f_1(x)v = f_2(x) \qquad \begin{array}{l} f_1(x) = -a_1(x) \\ f_2(x) = -a_2(x) \end{array} \tag{4.3.10}$$

which is a linear inhomogeneous first order differential equation. The solution to (4.3.10) is given by (see e.g. Greenberg, 1998).

$$v = e^{-\int f_1(x)dx} \left\{ \int f_2(x) e^{\int f_1(x)dx} dx + C \right\} \tag{4.3.11}$$

(where C is the integration constant) or

$$\frac{H}{h} = v^{-1} = e^{\int f_1(x)dx} \left\{ \int f_2(x) e^{\int f_1(x)dx} dx + C \right\}^{-1} \tag{4.3.12}$$

C is to be determined from H/h at some reference depth h_r where we choose to specify the wave initially. This is of course equivalent to specifying the boundary condition for (4.3.7).

The solution (4.3.12) is called a **closed form solution** because it is expressed in terms of integrals of known functions.

Hence, a solution of the energy equation is possible provided we can disregard the influence of H/h on the two functions $f_1(x)$ and $f_2(x)$. Clearly, this means that we cannot have a phase velocity c, dimensionless energy flux B or dimensionless dissipation D that depends on H/h.

We will later find expressions for B and D in the surf zone (see Section 5.6) that correspond to a moderate variation only of those quantities with respect to H/h. However, in particular, D turns out to be almost constant for a reasonable range of H/h. B depends slightly more on H/h but we notice that $B_x/2B$ is only one of three terms in $f_1(x)$ and the h_x and c_x terms turn out to be much larger than the B_x-term. Hence, provided reasonable values of the dimensionless parameters are used, the solution we can determine from (4.3.12) will clearly give some quite relevant guidelines as to the variation of H/h in a surf zone.

2nd solution method

An alternative solution method utilizes the physical balance between shoaling and dissipation discussed in Sect. 4.2. This method is useful for illustrating that actually the problem only has one parameter. The two methods of course lead to the same final result.

To develop the second approach, we define the shoaling coefficient $K_s(h)$, and a dissipation coefficient $K_d(h)$.

Consider the reference depth $h = h_r$ where we intend to specify the wave by giving the wave height H_r:

$$\text{At} \qquad h = h_r \qquad \text{we have} \qquad H = H_r \tag{4.3.13}$$

and hence

$$E_f(h_r) = \rho g H_r^2 c_r B_r \tag{4.3.14}$$

We now define $K_s(h)$ so that, if there were no dissipation, then at depth h we would have the wave height H' given by

$$H'(h) = K_s(h) H_r \tag{4.3.15}$$

Thus, without dissipation $E_f(h)$ can be written

$$E_f(h) = \rho g c (K_s H_r)^2 B \tag{4.3.16}$$

With no dissipation, however, we have

$$\frac{\partial E_f}{\partial x_\alpha} = 0 \tag{4.3.17}$$

or since H_r is a constant, we get substituting (4.3.16) into (4.3.17)

$$\frac{\partial}{\partial x_\alpha}\left(cK_s^2 B\right) = 0 \tag{4.3.18}$$

In particular, we can of course write

$$cK_s^2(h)B = c_r B_r K_s^2(h_r) \qquad \text{where} \qquad K_s(h_r) \equiv 1 \tag{4.3.19}$$

which gives

$$K_s = \left(\frac{c_r}{c}\frac{B_r}{B}\right)^{1/2} \tag{4.3.20}$$

It is important to emphasize that (4.3.18) is a consequence of how K_s is defined. Hence, (4.3.18) applies also when we consider situations *with* dissipation.

With dissipation, the wave height at h may be then written as

$$H = K_d(h)K_s(h)H_r \tag{4.3.21}$$

where K_d is a dissipation coefficient that accounts for the reduction in wave height relative to the value $H' = K_s H_r$ we would have with no dissipation.

Hence, E_f now becomes

$$\frac{E_f}{\rho g} = cK_d^2 K_s^2 H_r^2 B \tag{4.3.22}$$

The x-derivative of this can then be written

$$\frac{1}{\rho g}\frac{\partial E_f}{\partial x} = \frac{\partial}{\partial x}\left(cH^2 B\right) = \frac{\partial}{\partial x}\left(cK_s^2 K_d^2 H_r^2 B\right) \tag{4.3.23}$$

which using (4.3.18) becomes

$$\begin{aligned}\frac{1}{\rho g}\frac{dE_f}{dx} &= 2cBK_s^2 H_r^2 K_d \frac{dK_d}{dx} \\ &= 2\frac{E_f}{\rho g}\frac{K_d'}{K_d}\end{aligned} \tag{4.3.24}$$

where

$$K_d' \equiv \frac{dK_d}{dx} \tag{4.3.25}$$

The energy equation (4.2.15) then reads

$$2cH_r^2 K_s^2 K_d^2 B \frac{K_d'}{K_d} = \frac{K_d^3 K_s^3 H_r^3}{4hT} D$$

or

$$\frac{K_d'}{K_d^2} = -\left(\frac{1}{K_d}\right)' = \frac{K_s H_r D}{8chTB} \tag{4.3.26}$$

The latter of those equations can be formally integrated to give

$$\frac{1}{K_d} = \frac{1}{K_{dr}} - \int_{x_r}^{x} \frac{K_s H_r D}{8chTB} dx \tag{4.3.27}$$

where we for convenience have named the integration constant $\frac{1}{K_{dr}}$. And since $K_{dr} \equiv 1$ by definition, we get

$$K_d = \left[1 - \int_{x_r}^{x} \frac{K_s H_r D}{8chTB} dx\right]^{-1} \tag{4.3.28}$$

Substituting this and K_s from (4.3.20) into (4.3.21) then yields

$$\frac{H}{H_r} = K_s \left[1 - \frac{H_r}{8c_r B_r T} \int_{x_r}^{x} \frac{DK_s^3}{h} dx\right]^{-1} \tag{4.3.29}$$

Note that we can define the parameter

$$K = \frac{H_r}{8c_r B_r T} = \frac{1}{8B_r} \frac{H_r}{h_r} \frac{h_r}{L_r} \tag{4.3.30}$$

where $L_r = c_r T$, so that H/H_r can be written

$$\boxed{\frac{H}{H_r} = K_s \left[1 - K \int_{x_r}^{x} \frac{DK_s^3}{h} dx\right]^{-1}} \tag{4.3.31}$$

The function under the integration sign

$$f(h) = \frac{K_s^3(h)}{hh_x(h)} D(h) \tag{4.3.32}$$

represents the way in which the variation of h influences the solution.

So far, no assumptions have been used with respect to waves except: the waves are progressive and periodic (progressive because K_s is only defined for progressive waves), and that c, D, B **are independent of** H

$$c,\ D,\ B = f(h, T) \tag{4.3.33}$$

This implies that the expression for H/H_r is on the form

$$\frac{H}{H_r} = g(h, h_r, T) \tag{4.3.34}$$

where the function g is given by (4.3.31).

We see that the only parameter in (4.3.29) is $K = \frac{H_r}{8c_r B_r T}$. Hence, we conclude that for a wave model (which describes how D and K_s depend on h) **all waves, that at the reference depth h_r have the same K value will have the same H/H_r-variation throughout the region where the solution applies**, no matter which combination of H_r, h_r, B_r, etc. is used to compose the K value in question.

In the special case where h *is monotonous in* x we can express this more explicitly because then we can change the ingration variable to h. We get

$$h = h(x) \Leftrightarrow x = x(h) \tag{4.3.35}$$

$$\Downarrow$$

$$dh = h_x dx \Rightarrow dx = \frac{dh}{h_x} \tag{4.3.36}$$

$$\Downarrow$$

$$\frac{H}{H_r} = K_s \left[1 - \frac{H_r}{8c_r B_r T} \int_{h_r}^{h} \frac{DK_s^3}{hh_x} dh \right]^{-1} \tag{4.3.37}$$

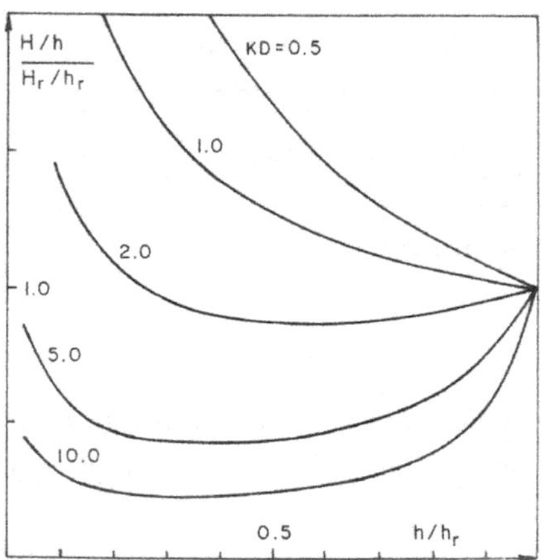

Fig. 4.3.1 The variation of H/h in a wave motion with energy dissipation such as in a surfzone. It is assumed the dissiation starts at $h = h_r$ and that the dissipation rate D = conastant. The value of the factor KD represents the combination of the incoming wave characteristics (K), defined by (4.3.30), and the dissipation rate (D) (from Svendsen, 1984).

Using K, the result for H/H_r can be written.

$$\frac{H}{H_r} = K_s \left[1 - K \int_{h_r}^{h} \frac{D(h)K_s^3(h)}{hh_x(h)} dh\right]^{-1} \tag{4.3.38}$$

Eq. (4.3.31) (or (4.3.38)) is again a closed form solution which may be evaluated analytically for some of the simpler variations of the integrand. For any integrand, however, it can easily be evaluated numerically.

Fig. 4.3.1 shows an example of the variation of H/h for a wave motion with dissipation such as in the surfzone. The variation is determined from (4.3.38) using the parameters for a plane beach and D = constant. Then the factor KD, where K is defined by (4.3.30), becomes the only parameter for the problem. It is seen that, as decribed earlier, right after breaking the wave height to water depth ratio decreases rapidly but as the depth decreases the value of H/h reaches a minimum and then starts increasing toward the shoreline. If KD is too small the shoaling will dominate over the dissipation even from the beginning and the value of H/h never decreases. A similar variation was found experimentally by Horikawa and Kuo (1966). For further details see Svendsen (1984).

4.3.3 *The energy equation for steady irregular waves*

The concept of a "steady wave train" can be extended to mean a train of random waves with a constant spectrum and hence a time independent H_{rms}. If we assume the time averaging underlying (4.2.16) is done over sufficiently long time (many wave periods) for the steady wave train then the definition of E_f and $\mathcal{D}$ simply means the longterm average of those quantities. Thus for the energy flux we can write

$$E_f = \overline{\rho g c H^2 B} \tag{4.3.39}$$

where B is the dimensionless shape factor for the energy flux defined by (3.3.61). Often this averaging is performed while assuming the spectral variation of the phase velocity c and the shape factor B are small enough to be neglected. Then the longterm averaged expression for E_f in a random wave train simply becomes

$$E_f = \rho g c H_{rms}^2 B \tag{4.3.40}$$

where c and B are evaluated for the frequency of the spectrum corresponding to H_{rms}. Notice that strictly speaking this averaging does not require sinusoidal waves for c or B though this is often assumed. Similarly for $\mathcal{D}$ averaging is possible over many waves, but the result depends on which expression is used for $\mathcal{D}$ (see Section 5.7.1).

4.4 The general energy equation: Unsteady wave-current motion

In this section we return to the derivation of the energy equation and outline a more formal and complete derivation. This approach will also provide more details for the generic terms in (4.2.11), including the separation of energies between the waves and the currents and the exchange of energy between those two components of the motion. The derivation will only be outlined because it is algebraically complex. For a more detailed (though still short) derivation reference is made to Phillips (1966, 1977).

To derive the general equation it is necessary to go back to the Reynolds equations and multiply scalarly those equations by u_α. This means we create an equation with terms of the form

$$\overrightarrow{\text{Force}} \cdot \overrightarrow{\text{velocity}} = \text{Force}_\alpha \cdot u_\alpha \tag{4.4.1}$$

which is the *effect* (or work-done-per-unit-time) of the forces in those equations. We know that the equivalent acceleration terms then can be written

$$u_\alpha \frac{du_\alpha}{dt} = \frac{1}{2}\frac{du_\alpha^2}{dt} \tag{4.4.2}$$

The result represents the well known principle rate of increase in kinetic energy equals the work done per unit time of all external forces.

Thus we have created an equation for the instantaneous, local variation of energy caused by the external forces.

Those equations then need to be integrated over depth and averaged over T. The details of this rather complicated operation is beyond the scope of this text, and we will only discuss the outcome.

First we realize that the equation that can be derived this way will describe the transformation of the *total* energy, that is the energy associated with both wave and current motion. We do want also to have equations for the wave motion only (including the effect the currents have on that motion) and for the current motion only (analogously with terms describing the effect of the waves on the current energy).

Hence there will in principle be *three* energy equations:

- One for the wave (or fluctuating) part of the energy,
- One for the (slowly varying) current part of the energy,
- One for the total energy of the combined waves and currents, which essentially is the sum of the other two.

We first introduce some convenient abbreviations of terminology. Thus we define the total volume flux Q_α through a vertical section

$$Q_\alpha(t) = \int_{-h_o}^{\zeta} u_\alpha dz \tag{4.4.3}$$

This is used to define the shorthand terms

$$Q_\alpha = \overline{Q_\alpha(t)} \qquad\qquad Q^2 = Q_\alpha Q_\alpha \tag{4.4.4}$$

Similarly the volume flux in the waves $Q_{w\alpha}$ (see Section 3.3) is defined by

$$Q_{w\alpha} = \overline{\int_{-h_0}^{\zeta} u_{w\alpha} dz} = \left(\overline{\int_{\bar{\zeta}}^{\zeta} u_{w\alpha} dz}\right) \tag{4.4.5}$$

where $u_{w\alpha}$ is the wave part of the total horizontal velocity u_α which is defined so that $\overline{u_w} = 0$ below wave trough level. This is used to define

$$(Q_w)^2 = Q_{w\alpha} Q_{w\alpha} \tag{4.4.6}$$

Both Q_α and $Q_{w\alpha}$ are allowed to be slowly varying functions of t.

The current velocity is defined by splitting the total velocity u_α using the definition

$$u_\alpha = V_\alpha + u_{w\alpha} + u'_\alpha \tag{4.4.7}$$

where $\overline{u'_\alpha} = 0$ and V_α is the current. We assume the current velocity is depth uniform and given by

$$V_\alpha = \frac{Q_\alpha - Q_{w\alpha}}{h} \tag{4.4.8}$$

which means we extract the wave volume flux from the total volume flux when assessing what is currents and what is waves.

Further we notice that, since we write the equations in the general form for 2D horizontal motion, the energy flux E_f becomes the vectorial quantity $E_{f\alpha}$ given by (4.2.4).

All these definitions will be discussed in more details later, in chapter 11.

We then get for the wave averaged energy flux due to the wave motion only

$$E_{fw\alpha} = \overline{\int_{-h_0}^{\zeta} \left[p + \rho g z + \frac{1}{2}\rho \left(u_{w\alpha} u_{w\alpha} + w_w^2 \right) \right] u_{w\alpha} dz} \tag{4.4.9}$$

Similarly the energy density E for the wave part of the motion is given by

$$\begin{aligned} E &= \overline{\int_{-h_0}^{\zeta} \rho g z + \frac{1}{2}\rho \left(u_{w\alpha} u_{w\alpha} + w_w^2 \right) dz} - \int_{-h_0}^{\bar{\zeta}} \rho g z dz \\ &= \frac{1}{2}\rho g \overline{\eta^2} + \overline{\int_{-h_0}^{\zeta} \frac{1}{2} \rho u_{wi} u_{wi} dz} \qquad i = 1,\ 2,\ 3 \end{aligned} \tag{4.4.10}$$

After lengthy derivations (for reference see O. M. Phillips, (1966, 1977) "The dynamics of the upper Ocean," where a brief account is given) we get the following equation for the *total* energy (waves and currents).

$$\begin{aligned} &\frac{\partial}{\partial t} \left\{ E - \frac{1}{2}\rho \frac{Q_w^2}{h} + \frac{1}{2}\rho g \left(\bar{\zeta}^2 - h_0^2 \right) + \frac{1}{2}\rho V_\alpha Q_\alpha + \frac{1}{2}\rho \frac{Q_\alpha Q_{w\alpha}}{h} \right\} \\ &+ \frac{\partial}{\partial x_\alpha} \left\{ E_{fw\alpha} + V_\alpha E + \rho Q_\alpha \left(\frac{1}{2} \left(\frac{Q_\alpha}{h} \right)^2 + g\bar{\zeta} \right) \right. \\ &\left. - \frac{1}{2}\rho Q_\alpha \left(\frac{Q_w}{h} \right)^2 + V_\beta S'_{\alpha\beta} \right\} = \mathcal{D} \end{aligned} \tag{4.4.11}$$

where, as before, $\mathcal{D}$ represents the total dissipation (4.2.15) of organized energy.

Since the motion is considered unsteady, there are of course some $\partial/\partial t$ terms, some of which represent the wave motion, some the current. Under the $\partial/\partial x_\alpha$ operator we find numerous terms representing the current or the interaction between waves and the currents. Before considering the different versions of the energy equation separately, notice that this total energy equation is on the form

$$\frac{\partial A}{\partial t} + \frac{\partial B_\alpha}{\partial x_\alpha} = \mathcal{D} \tag{4.4.12}$$

Had $\mathcal{D}$ been equal to zero this form would have been similar to that of the continuity equation which expresses that mass (or volume for incompressible flow) is conserved. $\mathcal{D} = 0$ means mechanical energy is conserved, and, in general, equations on the form

$$\frac{\partial a}{\partial t} + \frac{\partial b_\alpha}{\partial x_\alpha} = 0 \tag{4.4.13}$$

express that something is conserved, whence such equations are said to be on *conservation form.* Extensive mathematical/numerical results and tools are available for such equations or equation systems.

The reason why (4.4.11) is not quite on conservation form, in spite of the fact that we do assume that energy is conserved, is - as described earlier - that in (4.4.11) we only consider the total *mechanical* energy, not the total thermodynamical energy (see Section 4.2 for discussion).

The energy equation for the fluctuating part of the motion

The equation for the fluctuating part of the total wave-current motion (the waves) can be obtained by subtracting V_α times the momentum equation from the total balance (4.4.11). Again, the details are omitted here, but the result for the wave related energy becomes:

$$\frac{\partial}{\partial t}\left\{E - \frac{1}{2}\rho\frac{Q_w^2}{h}\right\} + \frac{\partial}{\partial x_\alpha}\left\{E_{fw\alpha} + V_\alpha E - \frac{1}{2}\rho Q_\alpha\left(\frac{Q_w}{h}\right)^2\right\} + S_{\alpha\beta}\frac{\partial V_\beta}{\partial x_\alpha}$$
$$= \mathcal{D} - V_\alpha \overline{\tau_{b\alpha}} \tag{4.4.14}$$

For our purposes, this is the important equation because given the currents this would be an equation which we could use to compute the wave height variation in combined wave and current motion.

For a start, we notice that this equation is not on conservation form. In addition to the RHS of (4.4.12) we now have terms which represent the interaction between waves and currents.

We also notice that if we extract the equation for time varying wave motion only by putting $V_\alpha = 0$ in (4.4.14) then we get

$$\frac{\partial}{\partial t}\left\{E - \frac{1}{2}\rho\frac{Q_w^2}{h}\right\} + \frac{\partial E_{f\alpha}}{\partial x_\alpha} = \mathcal{D} \tag{4.4.15}$$

which is not the trivial extension of (4.2.15) we would have expected from the heuristic derivation of the steady case in Section 4.2.

Comparison of (4.4.15) with (4.4.14) clearly shows the additional terms that describe the interaction between the waves and the currents. Essentially, there are three groups:

a)
$$\frac{\partial}{\partial x_\alpha}\left(V_\alpha E - \frac{1}{2}\rho Q_\alpha \left(\frac{Q_w}{h}\right)^2\right) \tag{4.4.16}$$

which represents the enhanced (or decreased depending on direction between waves and currents) energy flux in the waves: The current moves wave energy but not quite the simple amount $V_\alpha E$ we would have expected.

b)
$$S_{\alpha\beta}\frac{\partial V_\alpha}{\partial x_\beta} \tag{4.4.17}$$

is a term which represents the work that the radiation stress does on the current. This is equivalent to the forcing term in the time averaged momentum equation (see Chapter 11) which accounts for how the waves generate the currents.

c)
$$V_\alpha \overline{\tau_{b\alpha}} \tag{4.4.18}$$

is the energy dissipation of the current due to the mean bottom shear stress $\overline{\tau_{b\alpha}}$. This term is included because we have defined $\mathcal{D}$ as the *total* energy dissipation in the wave-current motion. Hence $\mathcal{D} - V_\alpha \overline{\tau_{\beta\alpha}}$ is the energy disipation in the waves only.

The energy equation for the current motion

For completeness, we also show the energy equation for the current (or mean motion). It can be found directly by subtracting (4.4.14) from (4.4.11) which gives

$$\frac{\partial}{\partial t}\left\{\frac{1}{2}\rho\frac{Q_\alpha^2}{h} + \frac{1}{2}\rho g\left(\bar{\zeta}^2 - h^2\right)\right\} + \frac{\partial}{\partial x_\alpha}\left\{\rho\frac{Q_\alpha}{h}\left(\frac{1}{2}\left(\frac{Q_\alpha}{h}\right)^2 + g\bar{\zeta}\right)\right\} + V_\beta\frac{\partial S_{\alpha\beta}}{\partial x_\alpha} = V_\alpha\overline{\tau_{b\alpha}} \tag{4.4.19}$$

The terms in this equation represent the following:

$\frac{1}{2}\rho\frac{Q_\alpha^2}{h}$: Kinetic energy in the current

$\frac{1}{2}\rho g\left(\overline{\zeta}^2 - h_0^2\right)$: Potential energy for the current motion (< 0 because the x-axis = the MWL is used as reference level, and most of the water is below that level)

$\frac{\partial}{\partial x_\alpha}\frac{1}{2}\rho\frac{Q_\alpha^3}{h^2}$: Divergence of the flux of kinetic current energy

$\frac{\partial}{\partial x_\alpha}\left(\rho g Q_\alpha \frac{\overline{\zeta}}{h}\right)$: Divergence of the flux of potential energy

$V_\beta\frac{\partial S_{\alpha\beta}}{\partial x_\alpha}$: Work done by the current on the waves

$V_\alpha\overline{\tau_{b\alpha}}$: Energy dissipation due to bottom friction.

4.5 The wave action equation

All the energy equations discussed above are expressed in terms of quantities such as E, $E_{f\alpha}$, Q, Q_w, etc., that can be evaluated for any linear or nonlinear wave motion, with any depth uniform current. However, it was shown by Bretherton and Garrett (1968) that for linear ("sinusoidal") waves and in a non-dissipative environment (D and $\overline{\tau_{b\alpha}} = 0$) the wave part of the energy equation can be written in the following simple form

$$\frac{\partial}{\partial t}\left(\frac{E}{\omega_r}\right) + \frac{\partial}{\partial x_\alpha}\left[(U_\alpha + c_{gr\alpha})\frac{E}{\omega_r}\right] = 0 \tag{4.5.1}$$

where $c_{gr\alpha}$ is the relative group velocity.

Thus, when there are no energy losses in the motion, the quantity E/ω_r is conserved. E/ω_r is called the **wave action**. For details of the derivation see Mei (1983). Christoffersen and Jonsson (1980) (for the steady case) and Christoffersen (1982) (for the general unsteady case) showed that for waves and currents **with** energy dissipation (4.5.1) is extended to

$$\frac{\partial}{\partial t}\left(\frac{E}{\omega_r}\right) + \frac{\partial}{\partial x_\alpha}\left[(U_\alpha + c_{gr\alpha})\frac{E}{\omega_r}\right] = \frac{\mathcal{D} + V_\alpha\overline{\tau_{b\alpha}}}{\omega_r} \tag{4.5.2}$$

where $\mathcal{D}$ and $\overline{\tau_{b\alpha}}$ have the same meaning as above.

We see that in this Equation the energy dissipation terms act as (negative) source terms that modify the wave action as it propagates.

The strength of this equation is of course that the effect of the current and its interaction with the wave motion, which in the general energy equation for the wave motion (4.4.14) are explicitly accounted for by the many terms in the equation, are here simplified into the terms on the LHS.

The wave action form of the energy equation for waves with currents is used in several wave models, usually enhanced with various other source terms (see Chapter 6).

4.6 References - Chapter 4

Bretherton, F. P. and C. J. R. Garrett (1968). Wavetrains in inhomogenious moving media. Proc. Roy. Soc. London, **A, 302**, 529–554.

Christoffersen, J. B. (1982). Current-depth refraction of dissipative water waves. Tech. Univ. Denmark, Inst. Hydrodyn. and Hydraulic Engrg., Series Paper 30.

Christoffersen, J. B. and I. G. Jonsson (1980). A note on wave action conservation in a dissipative current-wave action. Appl. Ocean Res. **2**, 179–182.

Greenberg, M. D. (1998). Advanced engineering mathematics. Prentice Hall.

Horikawa, K. and C. Kuo (1966). Wave transformation after a breaking point. ASCE Proc 10th Int. Conf Coast. Engrg., Chap 15, 217–233.

Mei, C. C. (1983). The applied dynamics of ocean surface waves. World Scientific, Singapore.

Phillips, O. M. (1966, 1977). Dynamics of the upper ocean. Cambridge University Press.

Svendsen, I. A. (1984). Wave heights and setup in the surfzone. Coastal Engrg. **8**, 4, 303–330.

Chapter 5

Properties of Breaking Waves

5.1 Introduction

The process of wave breaking on a beach is both one of the most dramatic visually and one of the most important physically for the wave motion and for the development of the nearshore currents.

Breaking of a wave "dissipates" energy (a term defined more closely in Section 4.2) and hence causes the height of the wave to decrease. Once a wave starts breaking this process has a tendency to continue. However, the sustained breaking found in the surfzone on a beach requires a reasonably large wave height relative to the water depth to maintain the breaking. Therefore the sustained breaking of the type we see in a nearshore surfzone will always be associated with a decreasing water depth. In fact if the depth is not decreasing fast enough for the (decreasing) wave height to remain large enough relative to the water depth, the breaking will stop. In particular if a breaking wave passes over the shoreward edge of a longshore bar and into the (even slightly) deeper water in the trough behind the bar, breaking often ceases almost immediately because the wave height to water depth ratio becomes too small.

Our knowledge about the processes involved in the initiation of and the sustained wave breaking is still far from complete. Therefore experimental results play an important role in the clarification of these processes. In this chapter we will outline the present state of knowledge and seek to systematize the insight we can gain using a combination of experimental results and simple theoretical approximations.

In section 5.2 we first briefly discuss the concept of the highest possible wave on a given (constant) water depth. Though it will be shown that this concept necessarily has nothing to do with wave breaking it is of academic interest. It has also played a prominent role, particularly in older literature, and it is still used for (inappropriate) assessments of breaking wave heigths. Section 5.3 gives a qualitative discription of the breaking process and the classification of breaker types and section 5.4 provides an analysis of empirical data for the height of waves at the initiation of breaking. In a way this can be said to describe the highest possible **breaking** waves. In section 5.5 we then look at the characteristics of waves in the surfzone on a beach, and Section 5.6 seek to throw further light on the properties of surfzone waves through the use of simple theoretical methods.

5.2 The highest possible wave on constant depth

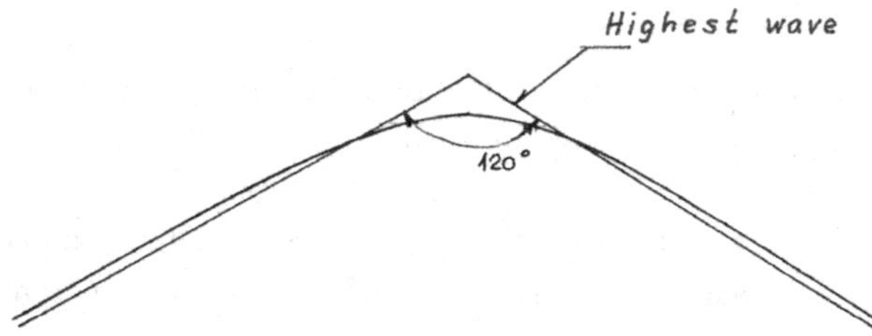

Fig. 5.2.1 Surface elevation for Stokes corner flow for the highest possible wave. Shown is also a the surface profile of slightly smaller wave.

The question of what is the highest possible wave of constant form that can propagate on a constant depth has been intensively discussed in the literature since Stokes' time. Here we only give a very brief account of the basic ideas and results.

Since we consider waves of constant (or "permanent") form, the flow underneath the surface in such a wave can be made steady by observing the wave from a coordinate system following the wave at its speed c of propagation. Hence the motion in a wave propagating to the right will be seen as a steady flow moving towards the left. In such a flow the highest point - the wave crest - must have the velocity zero (corresponding to all kinetic energy being converted to potential enegy). Thus, seen from a fixed coordinate system trough which the wave moves with the phase velocity c, the particle velocity at the wave crest equals c.

The first important contribution to the discussion was made by Stokes (1880), who showed that for irrotational waves in the neighbourhood of the crest the surface has the shape of a corner with opening angle 120° (see

Fig. 5.2.1). However, this solution is only valid locally at the crest and says nothing about the shape of the the rest of the wave, so the question of the maximum possible height requires further analysis.

On limited depth h the problem has two dimensionless parameters, h/L and H/h, where L is the wave length. However, for many years the attention was largely focused on waves of infinite length, i.e. $L/h \to \infty$ (so called **solitary waves**) because that reduces the parameters to one, $(H/h)_{max}$. Table 5.2.1 shows a list some of the results obtained over a period of almost 100 years. The last value is considered the most accurate.

Author	year	$(H/h)_{max}$
Boussinesq	(1871)	0.73
McCowan	(1894)	0.78
Gwyther	(1900)	0.83
Davies	(1952)	0.83
Packham	(1952)	1.03
Fenton	(1972)	0.85
Longuet-Higgins and Fenton	(1974)	0.8261

Of particular interest here is the result by McCowan (1894) which was based on the correct criterion of a particle velocity $u = c$ at the wave crest. His value of $(H/h)_{max} = 0.78$ is still (unjustifiably) used even today as the breaking index for periodic waves on a slope.

In **deep** water the results for the highest wave have been more consistent. In this case the only parameter is of course the wave steepness H/L, and the values in Table 5.2.2 have been obtained by various authors where again the last result is considered correct.

Author	year	$(H/L)_{max}$
Mitchell	(1893)	0.142
Havelock	(1918)	0.1418
Longuet-Higgins	(1975)	0.1412

It is worth to mention that these two cases, infinitely large and infinitely small values of h/L represent the extremes. The question of the highest possible wave for arbitrary h/L were not propery analysed until Yamada and Shiotani (1968) and Cokelet (1977). However, a heuristic interpolation formula for arbitrary h/L by Miche (1944) is frequently quoted in the literature. It gives the highest wave steepness as

$$\left(\frac{H}{L}\right)_{max} = 0.142 \tanh kh \tag{5.2.1}$$

where $k = 2\pi/L$ as usual.

Exercise 5.2-1

In the deep water limit $kh \to \infty$ this formula clearly gives $(H/L)_{max} = 0.142$ which is almost the exact value. Show that for a solitary wave the formula predicts $(H/L)_{max} = 0.89$, compared to the exact value of 0.8261.

For further details of the full variation of $(H/h)_{max}$ and $(H/L)_{max}$ on arbitrary depth reference is made to Yamada and Shiotani (1968) and Cokelet (1977).

It is also worth to mention that although c generally increases with H the largest phase speed actually occurs for a wave height that is slightly smaller than the highest possible wave.

For the present context, however, it is important to emphasize that all waves of constant form on a constant depth are strictly symmetrical with respect to vertical lines through the wave crests and the wave troughs. Thus the fronts and the rear of the waves have exactly the same slope. Therefore such waves are **not** at the brink of breaking. They bear no physical similarity to a wave breaking on a beach which always becomes skew with the front side of the wave crest becoming increasingly steeper than the rear side as the wave approaches breaking until the crest tumbles forward.

5.3 Qualitative description of wave breaking

Wave breaking is usually classified according to a visual impression of the process. This description was systematized by Galvin (1968) partly based on his own experiments on a plane beach partly on the experiments by Iversen (1952). The classification is inspired by wave breaking phenomena in coastal regions where the widest range of breaking forms occur. Even though such a description is insufficient for modelling purposes the terminology is useful as a reference frame.

The classifiaction operates with the following three breaker types

(1) Spilling breakers
(2) Plunging breakers
(3) Surging breakers.

Figs. 5.3.1 - 5.3.4 show the three types.

These three types of breakers are of course canonical examples, and on a coast we will find a continuous spectrum of transitions between the canonical cases.

On a plane slope it is primarily the bottom slope h_x in combination with the wave period, i.e. the relative bottom slope $S = h_x L/h$, that determines the type of breaking. Here index$B$ refers to values at the breaking point. In addition the initial wave steepness, usually refered to as the equivalent deep water steepness H_0/L_0 plays a role (Iversen (1952), Galvin (1968)). Galvin presents his results in terms of the parameter B_G defined as

$$B_G = \frac{H_0/L_0}{h_x^2} \tag{5.3.1}$$

Utilizing the relation (5.4.3) below for the connection between H_0/L_0 and the value of L_B/h_B at the breaking point we get

$$S_B = \frac{2.30}{\sqrt{B_G}} \tag{5.3.2}$$

where S_B is defined as $h_x L_B/h_B$.

Transforming the values of B_G given by Galvin into values of S_B gives the limits for the three breaker types as

- Spilling breaking occurs for $S_B < 1.05$
- Plunging breaking occurs for $1.05 < S_B < 7.6$
- Surging breaking occurs for $S_B > 7.6$

Battjes (1974) points out that B_G is also connected to a deep water version $\xi_0 = h_x/\sqrt{H_0/L_0}$ of the so-called surf similarity parameter ξ. Utilizing again the relation (5.4.3) we get

$$\xi_0 = \frac{1}{2.30} S_B \tag{5.3.3}$$

which shows that the frequently used surf similarity parameter is also connected to the slope parameter S_B. Notice the parameter S_B is a **local** parameter while the other two parameters include the deep water wave steepness.

However, it is also pointed out by Kjeldsen (1968) that the limits between the types suggested by Iversen and by Galvin differ somewhat.

Spilling breakers

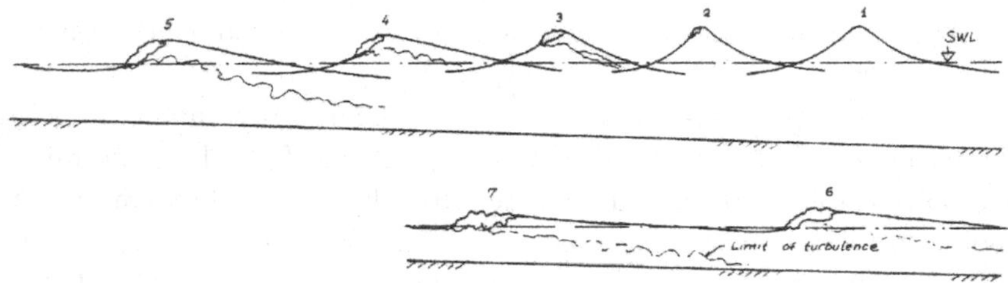

Fig. 5.3.1 Development of a spilling breaker. The waves are propagating towards the left.

In the ideal spilling breaker the turbulence starts in a very small scale at the front side of the wave crest. It has been hypothesized that this is actually just a very small scale plunging breaker, but Longuet-Higgins (1992) has shown, and Duncan and Dimasc (1996) have confirmed experimentally, that at least for some cases the initiation of the spilling breaker occurs as a very fast development of a series of small scale undulations just in the front of the crest which grow rapidly and turn into turbulence.

As the wave propagates the extent of the turbulent region expands down the slope to cover the entire front. The water closest to the surface is tumbling down the front while the flow is constantly fed from beneath. Thus a (turbulent and air-entrained) amount of water (called the **roller**) is riding on the front of the wave much as in a moving hydraulic jump or **bore**. Experiments with a special illustration technique by Peregrine and Svendsen (1978) show, however, that the turbulence is actually spreading beneath the roller as illustrated in Fig. 5.3.2.

If the water depth continues to decrease as on a plane beach the front gradually develops similarity with a bore, and the term a **periodic bore** can properly be used.

Plunging breakers

The plunging breaker occurs for steeper slopes (i.e. larger values of S).

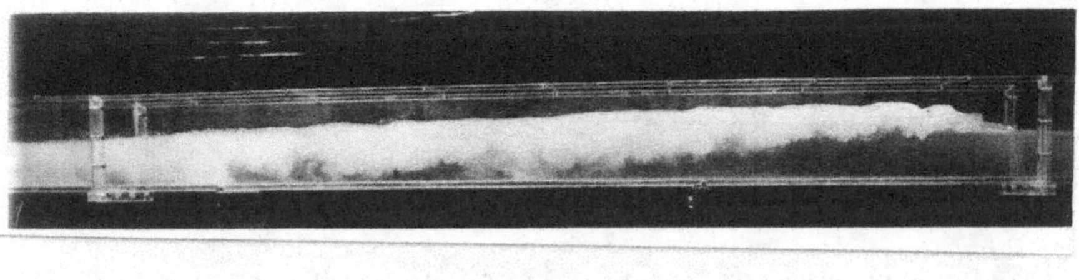

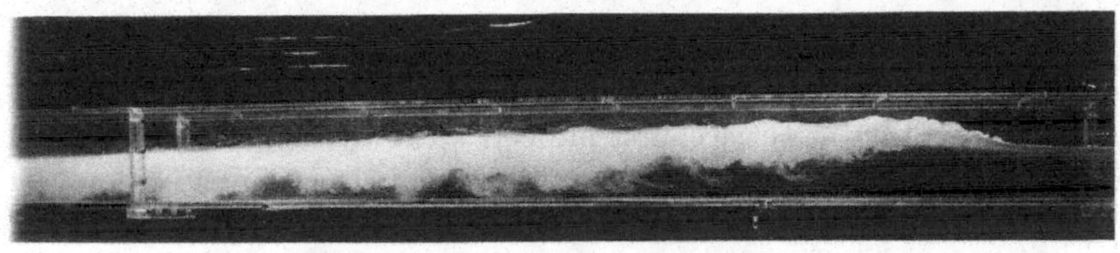

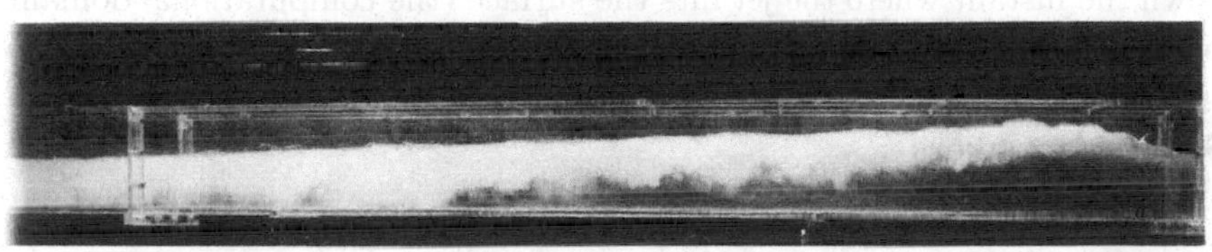

Fig. 5.3.2 Experiments showing turbulence spreading underneath a spilling breaker. The roller clearly does not represent the limit of the turbulent region, and dissipation due to the breaking must take place in the entire turbulent region. The white area emerges when tiny bubbles positioned on the free surface before the breaking wave arrives are entrained by the turbulence (Peregrine and Svendsen, 1978).

Also, waves with a smaller deep water steepness H_0/L_0 will travel to smaller values of h/L before they become large enough to break. Hence at breaking the value of S will be larger. Towards breaking the wave becomes very skew and the front eventually passes a vertical position shortly before the crest shoots forward as a jet that plunges down in a free fall, hitting the trough of the wave in front of the crest (Fig. 5.3.3).

Further shoreward the wave motion attains forms that very much resemble the periodic bore of the spilling breaker. Ting and Kirby (1995, 1996) have found experimentally, however, that there are differences between the two type of breakers in the values of some of the wave parameters.

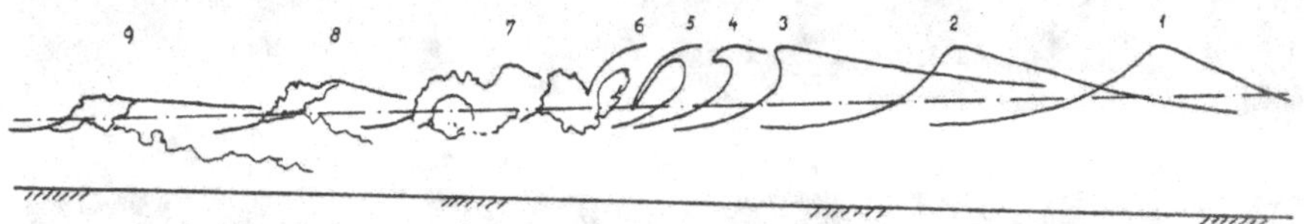

Fig. 5.3.3 The development of a plunging breaker.

Use of the Boundary Element Method (BEM) has made it possible computationally to follow the initial overturning of the front till the point where the jet hits the surface, first in deep water by Longuet-Higgins and Cokelet (1976), and later in arbitrary depth in numerous investigations, by Dold and Peregrine (1986), Svendsen and Grilli (1990), Otta et al. (1992), Grilli and Subramanya (1994), Grilli et al. (1997) to mention just a few.

Unfortunately the BEM, which is based on the Gauss theorem, breaks down the instant where the jet hits the surface (the computational domain ceases to be a simply connected region).

Surging breakers

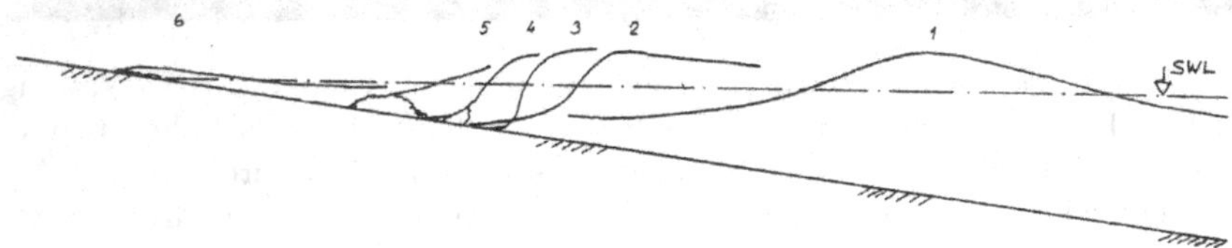

Fig. 5.3.4 Surging breakers.

If the relative bottom slope S is increased further we reach a stage where the water depth in front of the breaking wave becomes very small. The breaking virtually occurs **on** the beach itself. Fig. 5.3.4 shows the process. It will then be seen, that at the time when the front almost reaches the vertical position, the toe of the wave mountain all of a sudden shoots forward and runs up the beach.

In the first two types of breaking (spilling and plunging) essentially all the incident wave energy is turned into turbulence and dissipated. In a surging breaker, however, the turbulence generation is often limited. Therefore only part of the incoming energy is dissipated. The remaining energy is reflected and carried back seaward as a (partially) reflected wave. Thus this breaking form provides a continuous transition to the full reflection/no breaking that occurs when the relative slope S becomes large enough.

Swash

The surging breaking also is a situation that today is known as **swash**, and the description above corresponds to the swash motion we encounter on plane beaches.

Typically, however, real beaches are not plane, although that may in some cases be a good approximation to the beach profile in the region away from the shore. In the immediate neighborhood of the shoreline the beach profile often gets rather steep and the beach front just above the mean water line has slopes of typically 1 : 10 – 1 : 3. If the tidal variations are small enough there also often is a step in the bottom almost at the mean waterline. The situation is shown in Fig. 5.3.5, which also shows the breaking and runup that occurs on such beach profiles.

On the relatively gently sloping bottom before the shoreline the waves are often showing spilling breaking or, if the beach has an offshore bar, may not be breaking at all in the trough shoreward of the bar and in front of the shoreline. Close to the shoreline, however, the wave rapidly steepens and a violently plunging breaker develops with only the little water in front of it that comes from the downrush on the beach front from the previous breaker. Within fractions of a second the large pressure gradients represented by the steep surface slope in the breaker accelerate the water and send it up the slope at high velocity. In this phase the motion is almost entirely under the control of gravity and the relatively weak bottom friction and thus resembles a free fall on a slope. The downrush is usually met by and merges with the next surge.

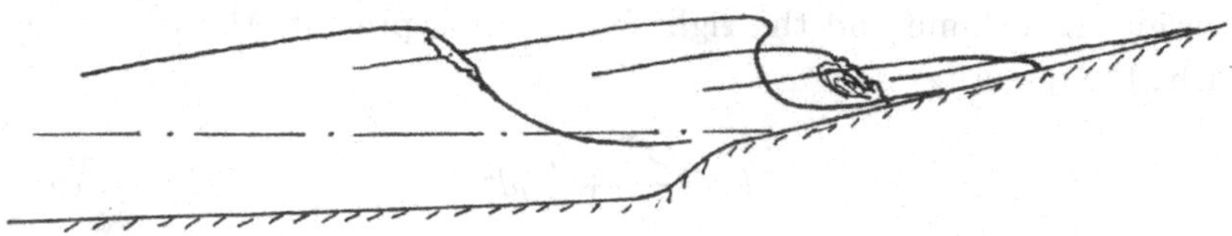

Fig. 5.3.5 Swash on a beach with a bottom step at the shoreline.

5.3.1 *Analysis of the momentum variation in a transition (Or: Why do the waves break?)*

It is interesting to consider why waves break. In this section we analyse the mechanisms that on one side make non-breaking waves unstable so they steepen and turn over as breakers, and on the other side seem to stabilize the broken waves in a form that only changes very slowly once the breaking has been established as in the inner region of the surfzone.

If we consider the transition in water level in a non-breaking bore or hydraulic jump seen from a coordinate system "following" the transition and assume all velocities are uniformly distributed over depth, then we can analyze the motion using the depth integrated momentum equation. Since we are essentially only looking at a large scale nonuniform current, there are no wave or turbulence components and the x-momentum equation hence simply becomes (omitting the index α since we are only dealing with one-dimensional variation in the x-direction, see Fig. 5.3.7)

$$\rho\frac{\partial Q}{\partial t}+\rho\frac{\partial}{\partial x}\left(\frac{Q^2}{h}\right)=-\rho g d\frac{\partial d}{\partial x} \tag{5.3.4}$$

In this form of the equation, we have also utilized that since we consider a horizontal bottom we have that

$$\frac{\partial \zeta}{\partial x}=\frac{\partial d}{\partial x} \tag{5.3.5}$$

We will use d for the total local depth (i.e. $d = h_o + \zeta$).[1]

The equation (5.3.4) describes the momentum balance for a column as shown in Fig. 5.3.6. Write (5.3.4) on the form

$$\rho\frac{\partial Q}{\partial t}=-\rho\frac{\partial}{\partial x}\left(\frac{Q^2}{d}+\frac{1}{2}gd^2\right) \tag{5.3.6}$$

Then we see that the left hand side is the rate of change of momentum ρQ inside the column and the right hand side represents the total forces. Clearly, the quantity

$$M=\frac{Q^2}{d}+\frac{1}{2}gd^2 \tag{5.3.7}$$

[1]The derivation of these equations can be found in books on classical open channel flow. See also Whitham (1974), or Lighthill (1978). However, they can also be derived as simplified versions of the nearshore circulation equations for depth uniform currents derived in Chapter 11 by assuming constant depth, no wave forcing, i.e. $Q_{w\alpha}, S_{\alpha\beta}=0$, and neglecting the turbulent stresses.

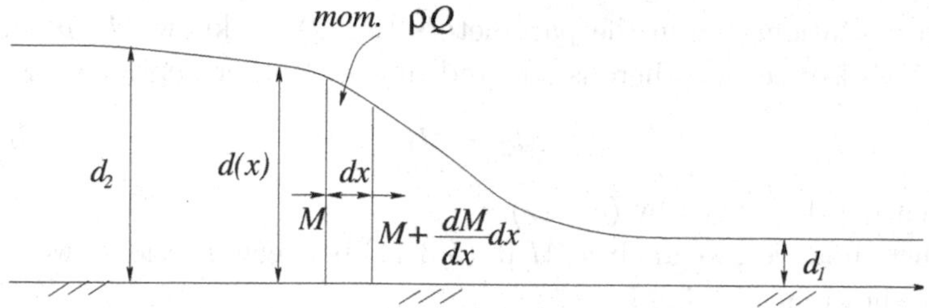

Fig. 5.3.6 Analysis of momentum flux in a bore.

is the force on (or "momentum flux through") any vertical section (as can be ascertained directly using the assumptions of depth uniform current velocities and hydrostatic pressure).

From (5.3.6) we may conclude that if the transition in water level is going to propagate **without change in form**, then $\partial Q/\partial t$ must be zero everywhere; i.e., M must be constant.

We therefore analyze the variation from d_1 through d_2 of the momentum flux, M, and show that with depth uniform velocity, hydrostatic pressure and no turbulence, M cannot be constant.

Let

$$\mathcal{F}_1^2 = \frac{U_1^2}{gd_1} = \frac{Q_1^2 d_1}{g} \tag{5.3.8}$$

Then we can rewrite M as ($d = d(x)$)

$$\begin{aligned} M &= \frac{U_1^2 d_1^2}{d} + \frac{1}{2} g d^2 \quad \text{assuming } Q = U_1 d_1, \text{ everywhere} \\ &= \mathcal{F}_1^2 \cdot g d_1^2 \cdot \frac{d_1}{d} + \frac{1}{2} g d_1^2 \frac{d^2}{d_1^2} \end{aligned} \tag{5.3.9}$$

or

$$M = g d_1^2 \left(\frac{\mathcal{F}_1^2}{\xi} + \frac{1}{2} \xi^2 \right) \tag{5.3.10}$$

where $$\xi = \xi(x) = d/d_1 \tag{5.3.11}$$

The variation of M versus ξ through the bore is shown in Fig. 5.3.7. We see that the maximum value of ξ is at d_2 where $\xi = \xi_2$ is given by

$$\xi_2 = \frac{d_2}{d_1} = \frac{1}{2} \left(\sqrt{1 + 8\mathcal{F}_1^2} - 1 \right) \tag{5.3.12}$$

(Note: ξ_2 is the quantity called ξ in (5.6.63)).

Interpretation: Given the parameters (d_1, Q) we know $M_1/\rho = \frac{Q^2}{d_1} + \frac{1}{2}gd_1^2$. We also see that there is one and only one other depth d_2 for which

$$M_2 = M_1 \tag{5.3.13}$$

and that depth is given by (5.3.12).

When, however, we analyze M in (5.3.11) between d_1 and d_2 we get the picture shown in Fig. 5.3.7.

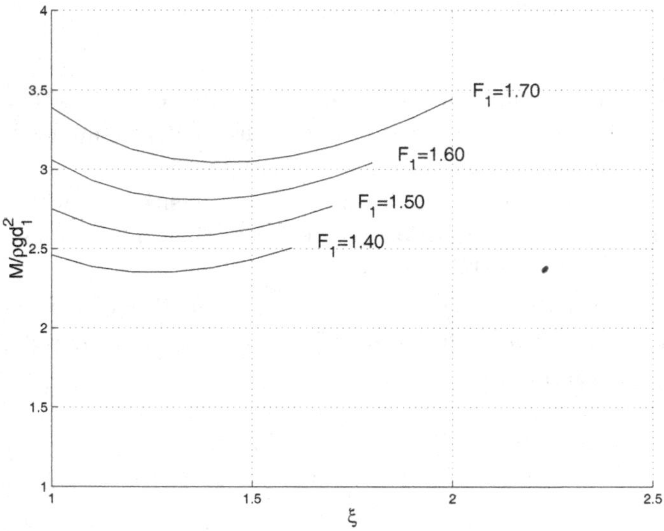

Fig. 5.3.7 Variation of the momentum flux M versus ξ defined by (5.3.11) through a bore if we assume the velocity is depth uniform and the pressure hydrostatic from the toe to the top as given by (5.3.11).

This shows that between the two corresponding depths d_1 and d_2 there is a deficit in M relative to what is required to keep $M = \text{const}$. Consequently, at any given point $d_1 < d < d_2$, we will according to (5.3.5) observe a $\partial Q/\partial t$, the sign of which depends on $\partial M/\partial x$.

Using the depth integrated continuity equation

$$\frac{\partial \zeta}{\partial t} + \frac{\partial Q}{\partial x} = 0 \tag{5.3.14}$$

together with the $\partial Q/\partial t$ determined form (5.3.5), we can realize qualitatively that the changes we should expect from the above analysis is a **steepening** of the front as is actually observed.

And, of course, the steepening will eventually lead to breaking in the real world (which would here be represented by a vertical tangent on the surface profile).

Let us assume the bore is initially smooth and non-breaking as shown above. Then, we can understand that in a bore (or a wave with the same conditions at trough and crest), the following mechanism is then responsible for the turbulent breaking:

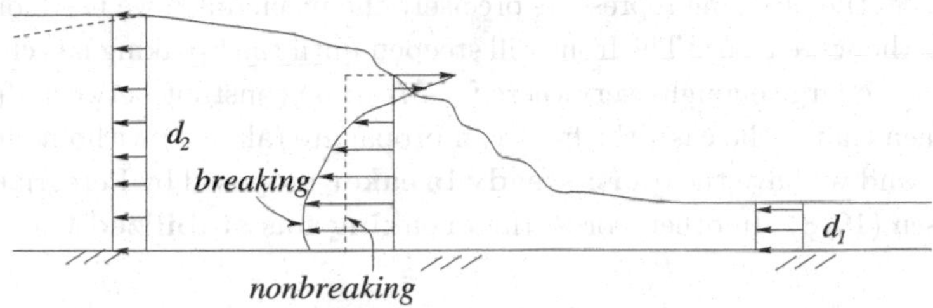

Fig. 5.3.8 Similarity between the velocity profiles in a (non-breaking) bore and a breaking bore seen from a coordinate system moving at the speed of the bore/wave so that the motion appears steady. The shape of the equivalent wave has been shown as a dashed line.

Because M is too small between d_1 and d_2 to satisfy the requirement of steady momentum balance, the wave will steepen. How fast depends on its relative height.

When the wave becomes steep enough it will start breaking and the well known turbulent front is formed.

This front, however, represents water particles tumbling down the slope near the surface. Hence, we must have a velocity distribution as shown in Fig. 5.3.8.

The nonuniform velocity profile under the front means that the momentum flux changes from

$$M = \rho \left(\frac{Q^2}{d} + \frac{1}{2} g d^2 \right) \tag{5.3.15}$$

to

$$M_t = \rho \left(\alpha_v \frac{Q^2}{d} + \frac{1}{2} g d^2 \right) \tag{5.3.16}$$

where

$$\alpha_v \equiv \int_0^h \left(\frac{U(z)}{Q/d}\right)^2 dz \tag{5.3.17}$$

is > 1; i.e., $M_t > M$.

The steeper the front, the larger the velocity of the particles tumbling down the front; i.e., the larger the velocity variation under the front, which means α_v becomes larger.

Hence, the breaking represents precisely the mechanism we need for stabilizing the wave front: The front will steepen until the breaking has created an α_v-value large enough everywhere for M_t to be constant between d_1 and d_2. When that is the case, the front can propagate (almost) without change of form and we have the **quasi-steady breaker** proposed by Peregrine and Svendsen (1978). In other words, the **breaking has stabilized the wave form**.

5.4 Wave characteristics at the breakpoint

Ideally the start of breaking should be defined as the point where the energy dissipation begins. That point, however, is very difficult to identify experimentally. So is the point where the water surface becomes vertical as in a plunging breaker. Instead one may consider either the point of maximum wave height H_{max} or the point of maximum value of the breaking index $(H/h)_{max}$, which are both easy to measure using a wave gage. Here we use $(H/h)_{max}$ as the definition of the start of breaking (Svendsen and Veeramony (2000) found that the two definitions were virtually indistinguishable).

There are two important questions that need to be answered about the characteristics at the break point:

- At which point (depth) do the waves start to break?
- What is the value of the breaker index H/h?

We first notice that computationally these two questions cannot be answered by using simple theories. A simple linear wave theory will of course provide no information in itself about when the wave is breaking (unless an empirical breaking criterion is introduced), because linear waves imply a fixed wave shape. Perhaps more surprisingly, however, even the more advanced nonlinear Boussinesq theories are unable to predict when breaking starts, because in Boussinesq waves the stabilizing dispersive effects grow very large when the surface curvature becomes large, which prevents anything that looks like breaking from developing. The wave height on a slope can just continue to grow. Also, none of these theories, which describe the surface position as the vertical distance above a certain level, would be able to describe the overturning of a plunging breaker. The only theoretical approach which can predict the initiation of breaking is the BEM, which is too computer intensive for most practical applications.

Therefore in practical models the initiation of breaking is identified based on empirical input to the model.

The breaker index

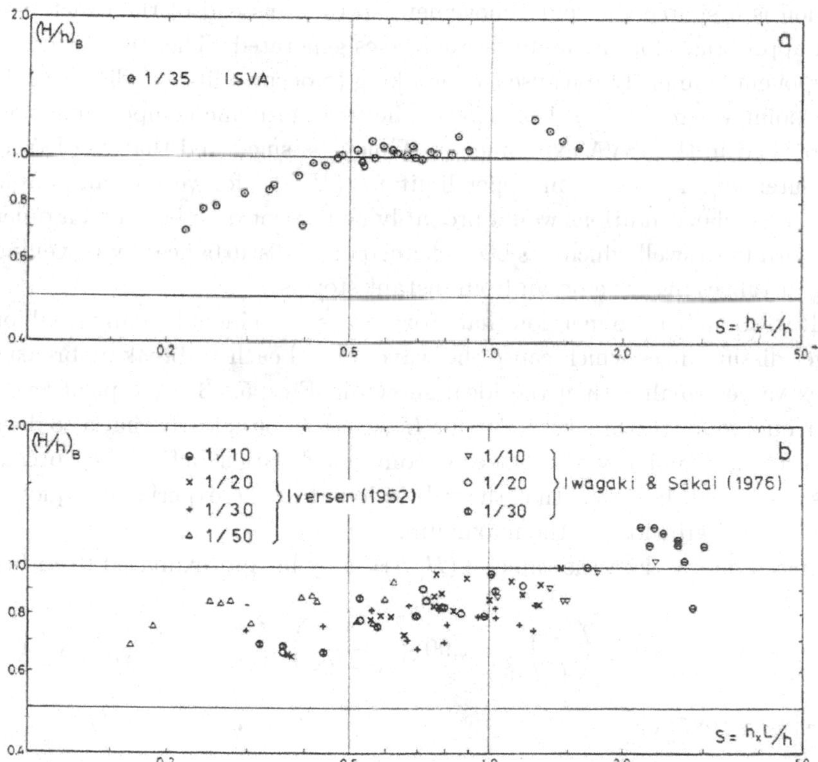

Fig. 5.4.1 The breaker index versus the relative bottom slope $S = h_x(L/h)_B$ (from Svendsen and Hansen, 1976).

Fig. 5.4.1 shows a plot of experimental values of $(H/h)_B$ at the breaking point versus the relative bottom slope S at breaking. After numerous attempts based on a wide selection of laboratory data Svendsen and Hansen (1976) (SH76 in the following) found that this way of plotting the data gave the most consistent results. The data used here are from Iversen (1952), Iwagaki and Sakai (1976) and data from SH76 (marked "ISVA"). It is noticed that in the ISVA-data, which are measurements on a 1/35 plane slope, the values of $(H/h)_B$ show very little scatter, while there is more scatter in the data from Iversen's (1952) and Iwagaki and Sakai's (1976) measurements, which also show somewhat smaller values for $(H/h)_B$. It is suspected that the smaller values are caused by small disturbances on the intended stable wave form generated in the wave tank. In particular so-called free harmonic wave componenets are generated when the wavemaker motion is a simple sinusoidal movement in time instead of the exact motion appropriate for the finite height waves generated. The free harmonic components are likely to cause the breaking to occur a little earlier, that is at a point where $(H/h)_B$ is smaller. The free harmonic components were suppressed in the ISVA-experiments. Thus it is suggested that the ISVA-measurements represent an upper limit for $(H/h)_B$ for very clean, stable waves. Similar conditions would probably be present on a beach on a quiet day with pure swell which has been cleaned of all disturbances by travelling long distances over the ocean from distant storms.

It is worth here to mention that storm waves also include many small or larger disturbances which cause the waves on a beach to break at breaker index values smaller than the ideal shown in Fig. 5.4.3. A typical value often quoted for the breaker index for H_{rms} is 0.6, but clearly that also does not exclude that individual waves become much larger before they break. It is suspected, however, that the values found in the experiments quoted in Fig. 5.4.3 are close to the maximum.

Based on the data the value of $(H/h)_B$ may be approximated by either

$$\left(\frac{H}{h}\right)_B = 1.90 \left(\frac{S}{1+2S}\right)^{1/2} \tag{5.4.1}$$

(Svendsen 1987) or

$$\left(\frac{H}{h}\right)_B = S^{0.25} \tag{5.4.2}$$

the latter valid for $0.25 < S \leq 1$ (Hansen, 1990).

We also see that $(H/h)_B$ increases for increasing S-values. This is consistent with the perception that the breaking process takes some time to develop: the steeper the slope the further the wave will manage to travel before the breaking actually starts and the larger the breaker index. This is in particular the case for the surging breaker. Notice also that $(H/h)_B$ may be both smaller and larger than the value of 0.78 which is so frequently quoted as the value of the breaker index.

Breaking position

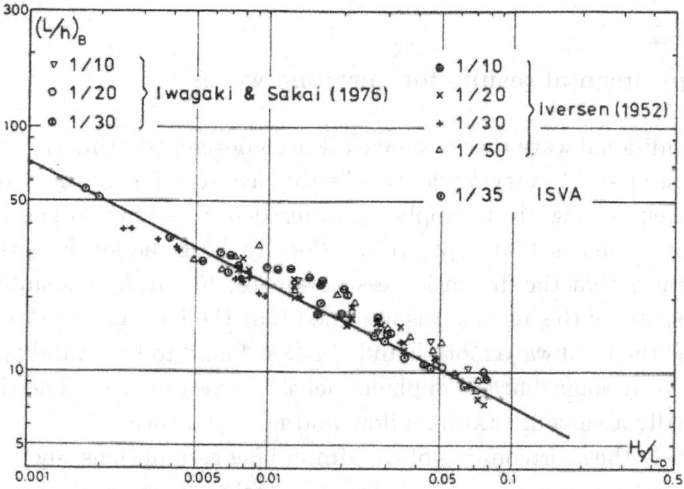

Fig. 5.4.2 The value of L/h at breaking versus the deep water wave steepness H_0/L_0. From Svendsen and Hansen (1976).

The diagram Fig. 5.4.1 is only useful for predicting $(H/h)_B$ if the value of $(h/L)_B$ is known at the break point. Since none of the commonly used wave theories are able to provide this information empirical input is required for this too. Analysis of the same breaker data as above (using cnoidal theory to determine the wave length L because $h/L < 0.10$ for all the data) gives the result shown in Fig. 5.4.2 for the relation between the deep water steepness H_0/L_0 and the values of $(L/h)_B$ at the breaking point. The straight line is given by the empirical relation

$$(L/h)_B = 2.30\,(H_0/L_0)^{-1/2} \tag{5.4.3}$$

The scatter in this plot is generally small enough to allow the relationship (5.4.3) to act as a guideline. An exception is perhaps the results for the steepest slopes 1/10 both from Iversen and from Iwagaki and Sakai's measurements, which gives slightly larger values for $(L/h)_B$ than (5.4.3). That implies the waves break at a smaller depth than indicated by (5.4.3). Again this is consistent with the perception that wave breaking takes a while to develop, which particularly shows up on the steeper slopes.

As can be seen from the analysis above, however, the prediction of the wave characteristics at the break point, even from quality laboratory data for the simplest case of regular waves on a plane slope, is subject to substantial uncertainty.

5.5 Experimental results for surfzone waves

The traditional wave theories have all been developed primarily for constant depth or slowly varying depth. As we have seen for linear waves this can be extended slightly to apply to water depth varying so gently that we can apply the constant depth results locally using the local depth. The requirement is that the dimensionless parameter $S = h_x L/h$ is sufficiently small. Essentially this means it is assumed that the local conditions suffice to describe the local wave motion and this was found to be a valid approximation for e.g. some diffraction phenomena (see Section 3.7). The theories also normally assume irrotational flow and no turbulence.

In spite of the shortcomings of the simplifying assumptions, such models may be useful because sine wave theory does include many of the important mechanisms that characterise also surfzone waves. Hence simplified models may give a **qualitatively** correct picture also of the surfzone processes.

However, for a closer study of waves before and after breaking this is not a good approximation. Measurements show that real breaking waves are quantitatively very different from the simplified picture provided by the simple theoretical assumptions, i.e. the qualitatively correct picture may be **quantitatively** quite inaccurate. For the reasons described earlier, experimental results are important for examining the characteristics of waves in the surf zone and for deciding what are acceptable assumptions.

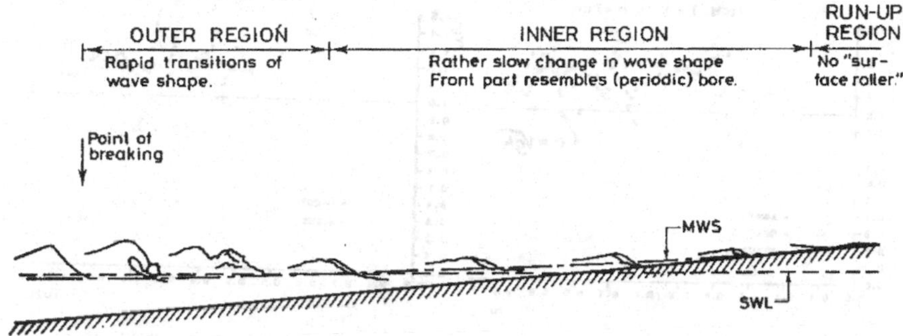

Fig. 5.5.1 Schematic representation of the surf zone on a gently sloping beach with no bars (from Svendsen et al., 1978).

5.5.1 *Qualitative surfzone characteristics*

Fig. 5.5.1 shows a schematic of a typical surf zone on a gently sloping beach with no bars. The waves will initially break at some depth and often continue to break until they reach the shore. The term **surfzone** refers to this region.

Immediately after breaking the waves will usually change quite rapidly in the so called outer region which generally has a width of a few to about 5 - 10 times the breaker depth. Further shoreward of the outer region the waves begin increasingly to resemble periodic bores. This has been termed the inner or the bore region. In this chapter we review experimental data for various wave properties in the surf zone and discuss approximations for some of the most important parameters for the wave motion.

Detailed measurements of surf zone wave properties meet with difficulties much more serious than in non-breaking waves. Major reasons are the bubble entrainment usually associated with breaking, and the highly turbulent flow conditions in the breaking wave front where it is difficult even to define where the free surface is.

Most of the following measurements are ensemble averaged from results for many waves in experiments with carefully controlled periodic waves.

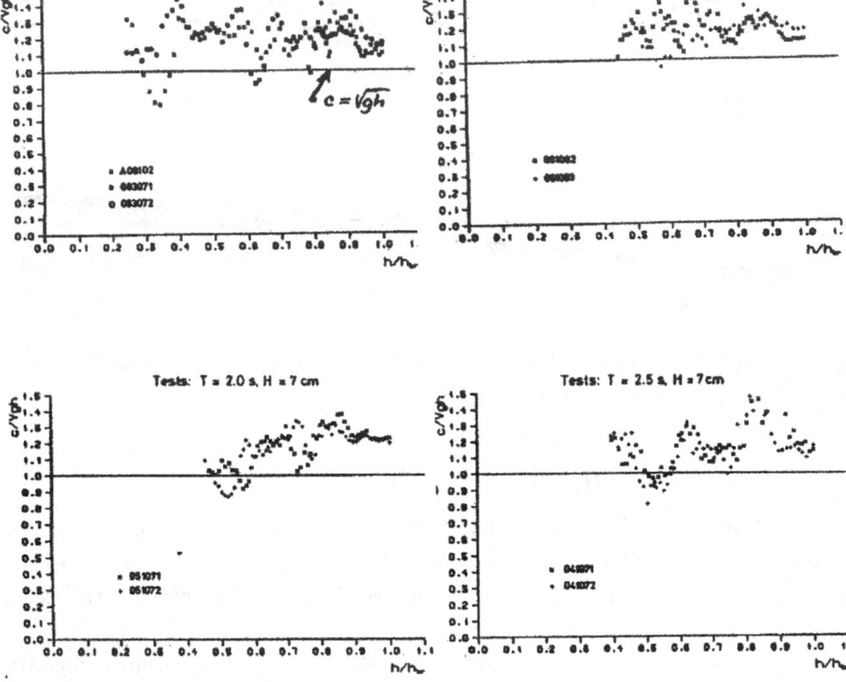

Fig. 5.5.2 Measured phase velocities c_a^m in the surf zone. It is seen that the actual phase velocities are somewhat larger than $\sqrt{gh}$ (Svendsen, 1986).

5.5.2 *The phase velocity c*

The phase velocities measured in experiments are absolute velocities c_a. In a laboratory wave flume there is a weak return current, U. averaged, over the depth it has the value of

$$U = -\frac{Q_w}{h} \tag{5.5.1}$$

where Q_w is the volume flux caused by the waves. This current compensates for the forward volume flux Q_w created by the waves, since there can be no net volume flux in a closed laboratory flume.

We would then expect the relative phase velocities c_r (that is the phase velocity relative to the water) to satisfy

$$c_a^m = c_r + \quad \text{current} \quad = c_r - Q_w/h \tag{5.5.2}$$

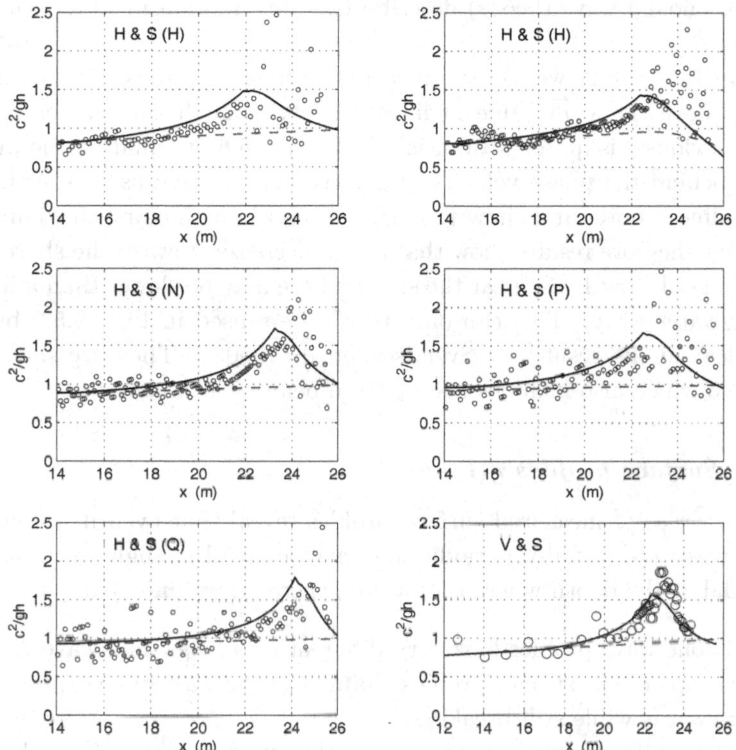

Fig. 5.5.3 Comparisons of c^2/gh between the experimental data (o) (from Hansen and Svendsen (1979)), and Svendsen and Veeramony (2001) and the cnoidal-bore model (—), and a model based on sinusoidal phase motion (–·) (from Svendsen et al., 2003).

or

$$c_r = c_a^m + \frac{Q_w}{h} \tag{5.5.3}$$

where index a stands for "absolute", i.e., relative to a fixed observer, and m stands for measured. Notice again that c_r is the velocity we would determine from a theory for waves in water without currents such as the the linear wave theory.[2] In an example in Section 5.6, we will see that Q_w/h could typically be 5–8% of c which means the actual c_r is $\sim$ 1.05–1.08 times the measured c_a^m. Notice also that most wave theories (such as

[2] c_r is also sometimes called the "intrinsic" celerity, a term we will avoid here.

linear or cnoidal wave theory) describe the wave motion in water without a current.[3]

From Fig. 5.5.2, we see the measured phase velocities c_a^m are significantly above the $\sqrt{gh}$ value of linear theory, though the scatter in the results obviously is quite substantial. In Section 5.6, we analyze the mechanisms behind the phase velocity of a wave more closely using the theory of a nonlinear bore, and show that this results in a similar behaviour. In particular the bore results show that c^2/gh decreases toward the shore and can even get below 1, i.e. near the shore a bore may be slower than a linear shallow water wave. This can only barely be sensed in Fig. 5.5.2 but is quite clear in the results by Svendsen et al. (2003). They are shown in Fig. 5.5.3 which also includes results from before breaking.

5.5.3 *Surface profiles* $\eta(t)$

Time series of measured surface profiles reveal that even if the generated waves were initially periodic and maybe small enough to be nearly sinusoidal at the depth where they were generated we find that

- surf zone wave profiles look very different from, e.g., sine waves.
- the shape of the surface profiles changes as the waves move/propagate shorewards while still breaking.
- the variability from wave to wave, although they were all equal when generated, is quite significant, in particular in the inner surfzone.

In Fig. 5.5.4 is shown η/H versus t/T taken from **time-profiles** (i.e. temporal variations at a fixed point). It exposes more clearly the change in shape towards the shore. Note that time increases to the right so that these time profiles have the shape of waves propagating to the **left**, and the turbulent front of the wave is on the left hand side of the "crest" of the time profile.

One of the features that can be observed in this figure is that for all profiles plotted against t/T, the front seems equally steep in time. This means that in space the length λ of the front is a constant fraction of the local wave length L. To the first approximation we have

$$L = T\sqrt{gh} \propto \sqrt{h}$$
$$H \propto h \tag{5.5.4}$$

[3]The phase velocity for waves with the return current is also analysed in Chapter 8 under the heading "Stokes' two definitions of c".

so that the wave steepness H/L changes as

$$\frac{H}{L} \propto \sqrt{h} \tag{5.5.5}$$

which means the steepness of the breaking wave is **decreasing** as it approaches the shore. Since the length λ of the front is $\propto L$, this means the front steepness will change as

$$\frac{H}{\lambda} \propto \sqrt{h} \tag{5.5.6}$$

i.e., the steepness of turbulent front of the breaking waves in Fig. 5.5.4 also decreases towards the shore.

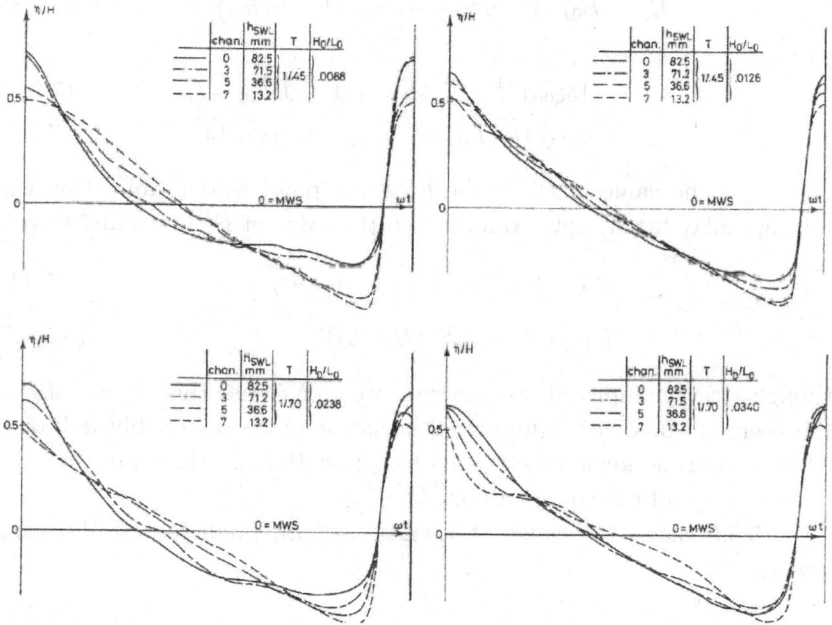

Fig. 5.5.4 Development of the surface wave profiles in the surf zone (from Svendsen et al., 1978). The figure shows that as the waves propagate toward the shore, the rear side of the waves becomes more straight so that the waves eventually approach a sawtooth shape.

The rear side of the wave time profile is changing too: from a relatively hollow shape, the rear side becomes more straight so that the waves eventually approach a **sawtooth shape**.

5.5.4 *The surface shape parameter B_0*

The surface shape parameter B_0 defined as

$$B_0 = \frac{\overline{\eta^2}}{H^2} = \overline{(\eta/H)^2} \tag{5.5.7}$$

is fairly easy to measure. It will appear later in approximate expressions for several of the wave averaged quantities in surfzone waves such as radiation stress and energy flux. Fig. 5.5.5 shows the variation of B_0 for a large number of published laboratory experiments for surfzone waves. The figure is from Hansen (1990) and it also shows (as solid curves) the values of B_0 according to the empirical formulas developed by Hansen:

$$B_0 = B_{0B}\left[1 - a\left(b - h/h_B\right)\left(1 - h/h_B\right)\right] \tag{5.5.8}$$

where

$$a = \left(15\xi_{00}\right)^{-1} \quad ; \quad b = 1.3 - 10\left(\xi_0 - \xi_{00}\right) \tag{5.5.9}$$

$$\xi_o = h_x/\sqrt{H_0/L_0} \quad ; \quad \xi_{00} = h_x/\sqrt{0.142} \tag{5.5.10}$$

Here B_{0B} is the value of B_0 at the breaking point as determined by the following (curve fitted) approximation to the value of B_0 in cnoidal waves.

$$B_{0b} = 0.125 \tanh\left(11.40/\sqrt{U_B}\right) \tag{5.5.11}$$

$$U_B = 10.1\, h_x^{0.20}\left(H_0/L_0\right)^{-1} \tag{5.5.12}$$

Although strictly empirical, we see from Fig. 5.5.5 that this system of formulas seems to predict the measured variation of B_0 reasonably well over the entire surfzone for a wide range of h_x and H_0/L_0, which are the only real parameters of the empirical formulas.

Fig. 5.5.6 shows the range of variation of B_0 predicted by Hansen's formulas.

Discussion of B_0

The value of B_0 reflects the shape of the wave surface profile. Thus a sinusoidal wave has $B_0 = 1/8 = 0.125$, whereas a very long wave with a short, peaky crest may have any B_0 down to 0 (solitary wave), but often has $B_0 \sim 0.04$–0.05. This development is qualitatively predicted by the cnoidal theory for waves approaching breaking (Svendsen et al. (2003)). As Fig. 5.5.5 shows at the breaking point, B_0 is often as low as 0.05, i.e. much smaller than for a sinusoidal wave. However, as Fig. 5.5.5 (or Eqs. (5.5.9)) also show, depending on the deep water wave steepness H_0/L_0 and bottom

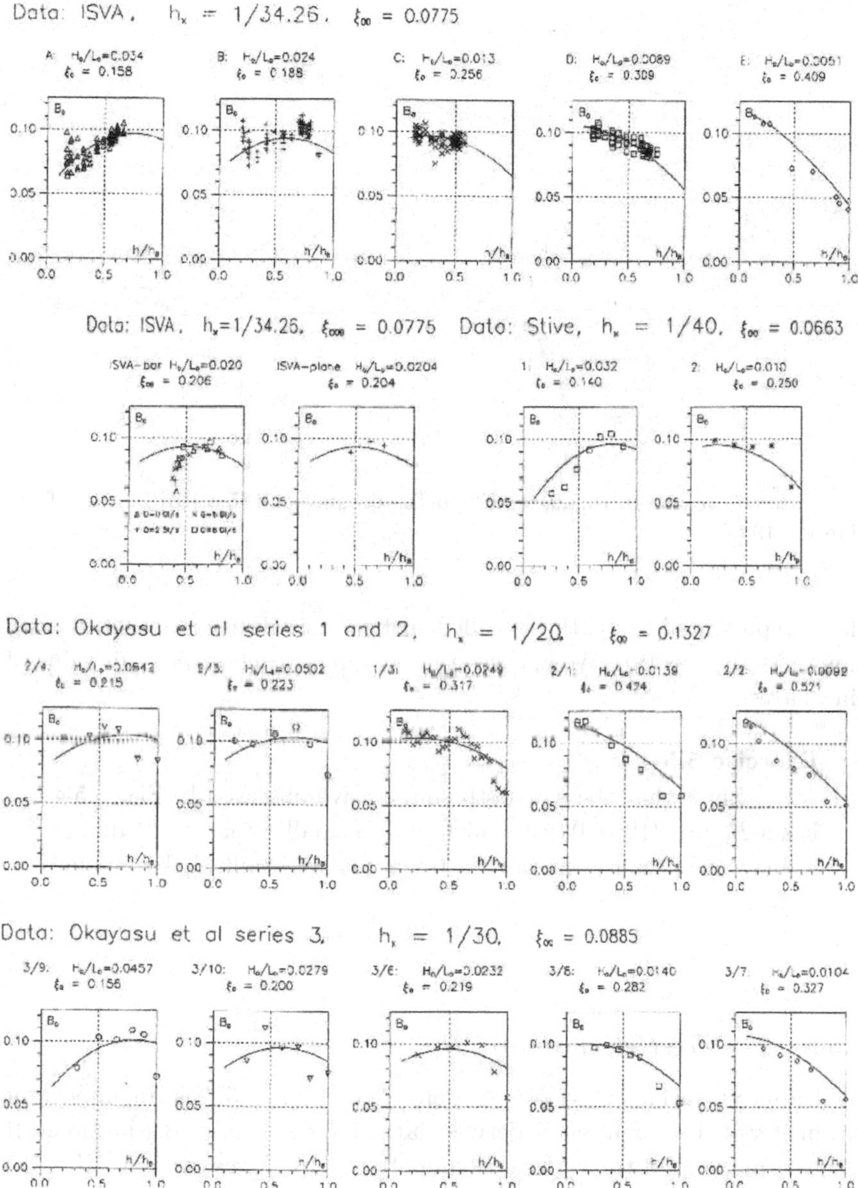

Fig. 5.5.5 The variation of B_0 for laboratory experiments in the literature (from Hansen, 1990).

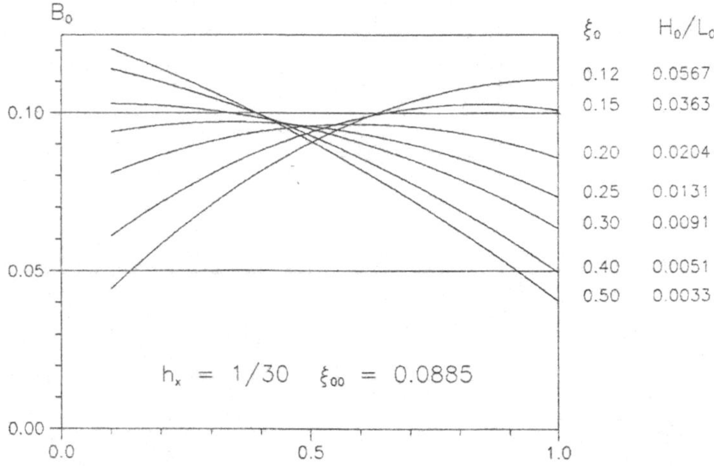

Fig. 5.5.6 B_0 from Eq. (5.5.9) and associated equations (5.5.10)-(5.5.10) (from Hansen, 1990).

slope (represented by ξ_0, the so called surf zone similarity parameter), B_{0B} can be as large as 0.1. We also see that it rarely reaches the high value of sine waves.

Exercise 5.5-1

Show that the sawtooth shape wave observed in Fig. 5.5.4 has a $B_0 = 1/12 = 0.083$, which is also smaller than the value of a sine wave because the sawtooth wave is less "bulky" than a sine wave.

5.5.5 *The crest elevation η_c/H*

The crest surface elevation η_c defined in Fig. 5.5.7 is an important parameter in the expression derived later for the energy dissipation. It also is a measure of the vertical asymmetry of the waves, with $\eta_c/H = 0.5$ representing no asymmetry (as in sine waves or sawtooth waves).[4]

[4]This vertical assymmetry is often in the analysis of wind waves termed the "skewness" of the wave, a term we here (more illustratively it seems) use for the difference in front and rear slope of the wave profile.

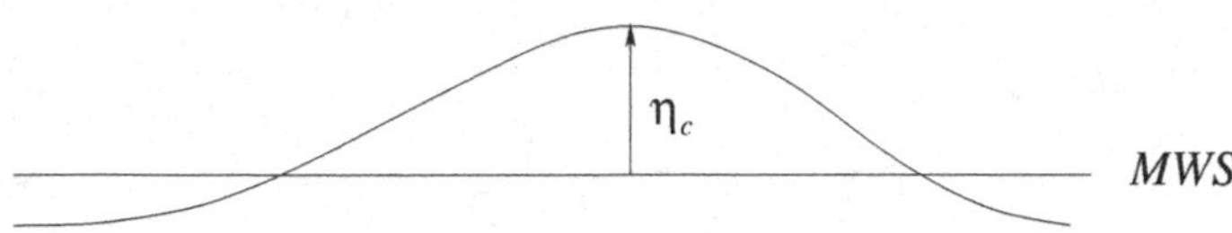

Fig. 5.5.7 Definition of the crest elevation η_c.

Fig. 5.5.8 shows the η_c/H in surf zone waves from the same selection of experiments used in Fig. 5.5.5. Clearly, at breaking the waves have fairly large η_c/H ratios (0.7–0.8) and the trend is toward decreasing vertical asymmetry shoreward from the breaking point. Closest to the shore, the waves are almost vertically symmetrical ($\eta_c/H \rightarrow 0.5$) as the waves approach a sawtooth shape (see also Fig. 5.5.5).

The continuous curves are given by the empirical expressions (Hansen, 1990):

$$\frac{\eta_c}{H} = 0.5 + \left[\left(\frac{\eta_c}{H}\right)_B - 0.5\right](h/h_B)^2 \qquad (5.5.13)$$

where

$$\left(\frac{\eta_c}{H}\right)_B = 1 - 0.5\tanh\left(4.85/\sqrt{U_B}\right) \qquad (5.5.14)$$

with U_B given by (5.5.12). We see again that these empirical formulas cover the data really well and that the η_c/H variation is virtually independent of the bottom slope once the value at the breaking point is determined.

5.5.6 *The roller area*

As mentioned a continuously breaking wave has a turbulent front in which water is tumbling down towards the trough as the wave propagates forward, and this water essentially is carried with the wave and actually has particle velocities at the surface slightly larger than the phase speed c of the wave. This volume of water is called the **roller**. Fig. 5.5.9 shows schematically the velocity profile in and under the roller in a surfzone wave. It is also pointed out that the roller must end at the wave crest, because surface particles in the roller by definition flow down the wave front ($u_w \geq c$). This is not possible mechanically for a particle on the rear side of the crest.

The roller is a well defined region of the wave. This can be understood by viewing the flow inside the wave from a coordinate system that follows

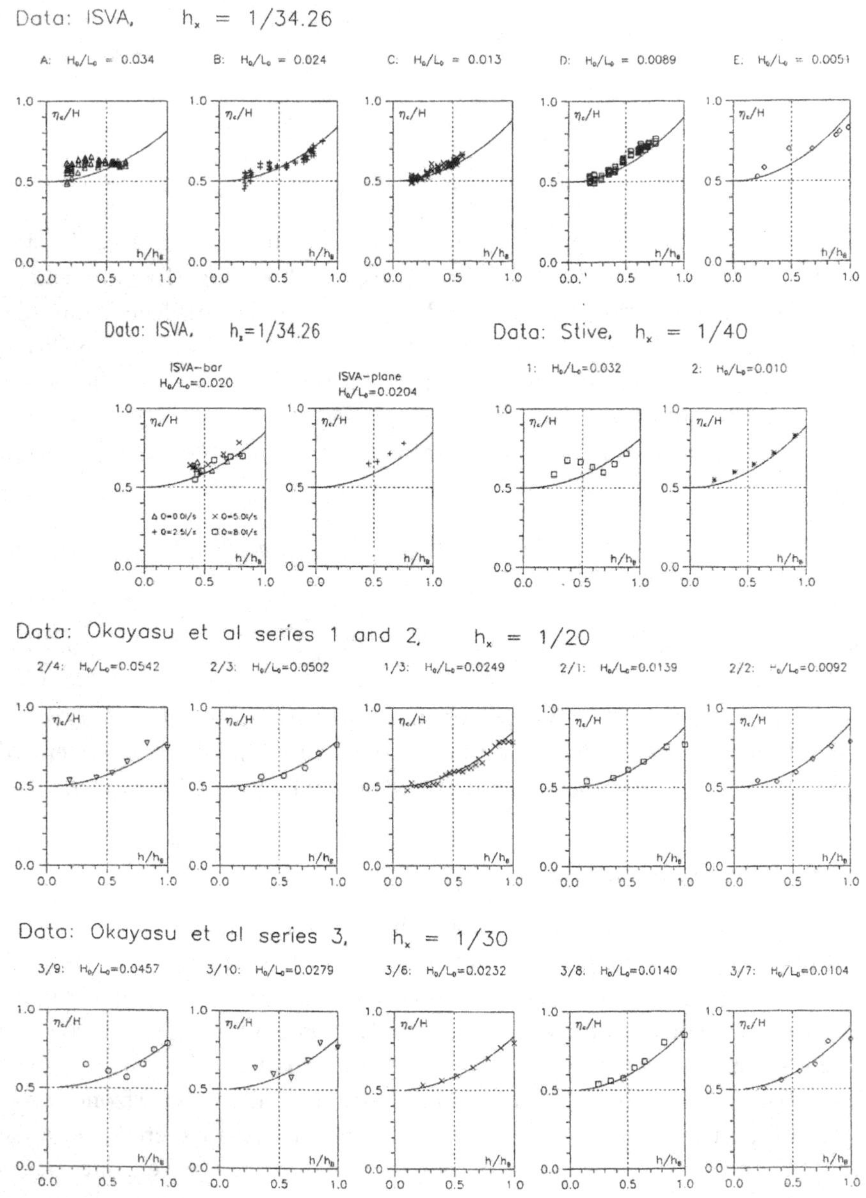

Fig. 5.5.8 Experimental values of η_c/H. — is the prediction by (5.5.14) (from Hansen, 1990).

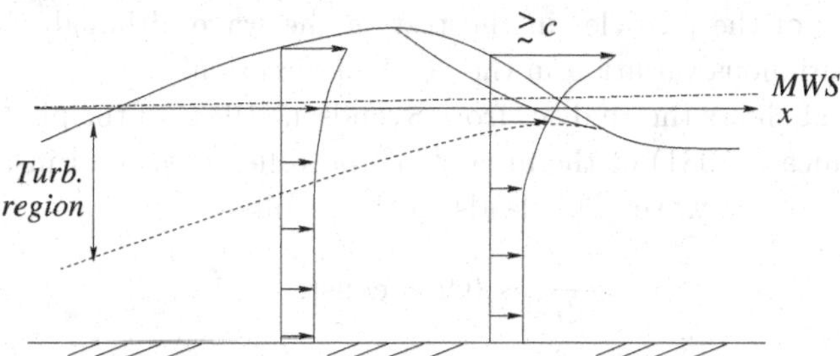

Fig. 5.5.9 Illustration of the characteristics of the velocity field under a wave in the surf zone.

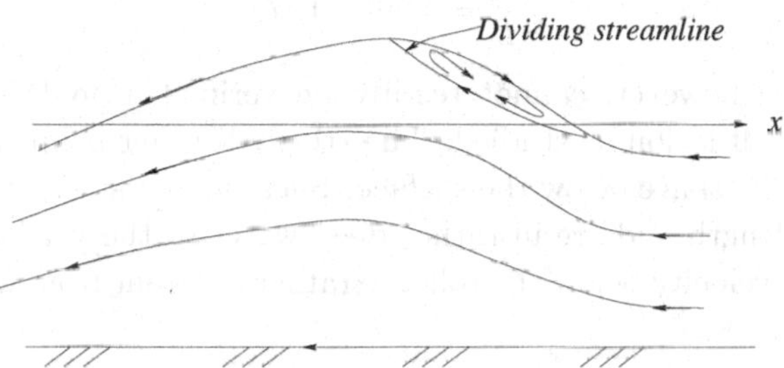

Fig. 5.5.10 Velocity field under a surf zone wave viewed from a coordinate system following the wave.

the wave with velocity c. We get the velocity field observed from such a system by subtracting c from the velocities observed from the fixed system. It is emphasized that this of course does not change the flow pattern as such - only the way we view it. Fig. 5.5.10 shows the result. In this system the flow is nearly steady and going against the direction of wave propagation except in the upper part of the roller where the actual velocity is larger than c. Hence the free surface is a streamline. The roller now becomes a region of recirculation where the water above the lower limit of the roller stays in the roller, the water below passes through the picture from the right to the left. The two regions are separated by the **dividing streamline** shown in the figure. This streamline forms the lower limit of the roller.

The consequence of this is that inside the roller the velocity is larger than c in the upper part, smaller than c in the lower part of the roller and in average close to c. This also means that the surface roller has velocities far in excess of the particles in the rest of the wave although there is of course a continuous variation in the vertical direction.

Fig. 5.5.11 shows the analysis from Svendsen (1984) of the photographic data by Duncan (1981) of the area A of the roller measured in a vertical cross section of the wave. This leads to the result

$$\frac{A}{H^2} \sim 0.9 = \text{const.} \tag{5.5.15}$$

Later investigations (Okayasu, 1989) have suggested that an alternative approximation is to use A/HL constant (instead of A/H^2 constant) over the surfzone where $L = cT$ is the local wave length. These measurements suggest that we may use

$$\frac{A}{HL} = 0.06 - 0.07 \tag{5.5.16}$$

This result, however, cannot readily be verified from Duncan's data because of the situation he studied: The steady breaker behind a hydrofoil towed at some distance below the surface. Such breakers do not have a well defined wave length and are in rather deep water so the vertical variation of the particle velocity below the roller is rather different from what we find in surf zone waves.

5.5.7 *Measurements of particle velocities*

So far, most measurements of particle velocities has only covered the region up to or slightly above trough level. The reason is that the instruments used (laser-doppler anemometers, accoustic-doppler anemometers) cannot easily record between wave trough and crest where there is intermittently water and air.

However, Iwagaki and Sakai (1976) used photographic images obtained with stroboscobic lighting in an early Particle Image Velocimetry (PIV) method to obtain velocity measurements under the crest of waves at the breaking point. This is exactly the instant when the largest velocities can be expected. The extensive measurements for two different bottom slopes (1 : 20 and 1 : 30) and a range of deep water wave steepnesses from 0.008 to 0.051 are shown in Fig. 5.5.12. The measurements of u/c are plotted as function of the dimensionless vertical coordinate $\xi = (z + h)/(\eta_c + h)$ (η_c the surface elevation at the wave crest) which varies from zero at the bottom to 1 at the wave crest. The figure also shows the empirical formula suggested by Van Dorn (1978) for the velocity profiles under the crest of

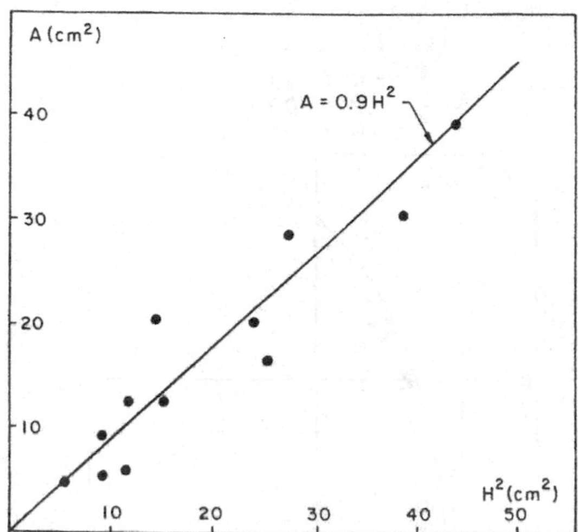

Fig. 5.5.11 Cross section area A for a roller. Measurements by Duncan (1981) (from Svendsen, 1984).

breaking waves. It is given by

$$\frac{u}{c} = 0.2 + \frac{0.1\xi}{1.125 - \xi} \tag{5.5.17}$$

As can be seen this formula assumes that the particle velociy at the crest of the breaking wave is exactly equal to the phase velocity c. We also see that the measurements give some support to this conjecture which is in keeping with the physical ideas of what happens when a wave starts breaking. Van Dorn's (1978) measurements (not shown here) were done for slopes of $h_x = 1:45, 1:25$, and $1:12$ but do not distinguish between different wave steepnesses. They showed similar good agreement with (5.5.17) for the first two bottom slopes, but less so for $h_x = 1:12$. Though there is significant scatter in the figure in particular near the crest the empirical expression (5.5.17) at least gives guidance.

Figure 5.5.13 shows measurements by Cox et al. (1995) of the entire velocity field shortly after breaking in a spilling breaker. Not surprisingly the dominating velocity direction is horizontal. More surprisingly perhaps is that apart from the region right under the crest the horizontal velocity appears to be nearly constant over depth. A linear approximation to the

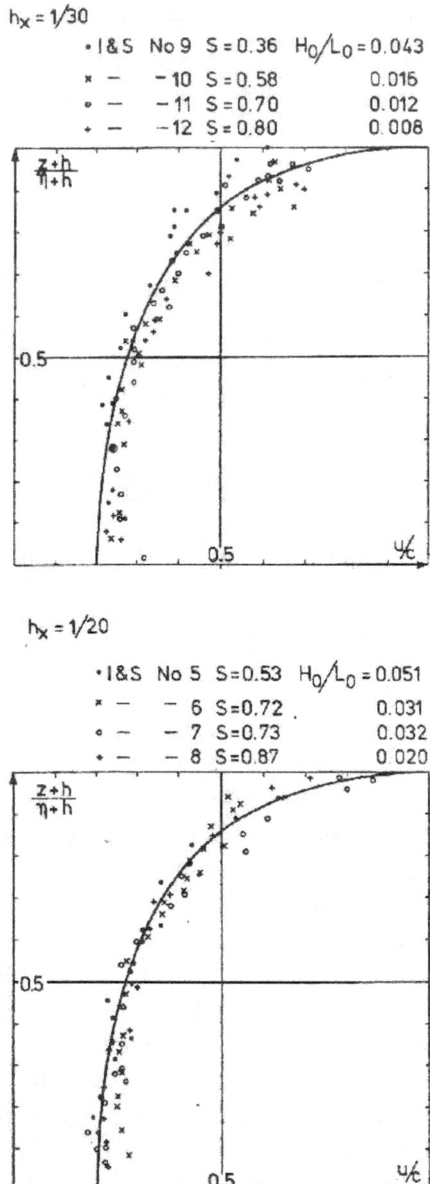

Fig. 5.5.12 Measured velocity profiles under the crests of breaking waves. The measurements are by Iwagaki and Sakai (1976), the curve corresponds to (5.5.17).

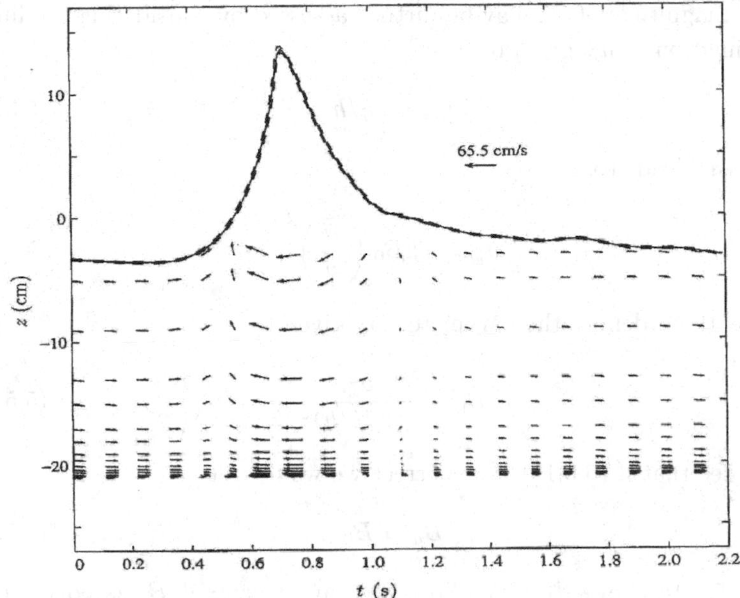

Fig. 5.5.13 Measured velocity field in a surfzone wave. Notice the front is to the left of the crest (from Cox et al., 1994).

equation for the wave component u_w would be

$$\frac{\partial u_w}{\partial t} = -\frac{1}{\rho}\frac{\partial p}{\partial x} + \frac{\partial \tau_{zx}}{\partial z} \tag{5.5.18}$$

where τ_{zx} represent the horizontal turbulent stresses. Thus from the nearly constant values of u_w over the vertical we may deduce that the RHS of (5.5.18) is nearly constant over depth. This means the turbulent stresses do not strongly influence the horizontal velocity u_w below trough level, but also the pressure gradient has to be nearly depth uniform in that region, suggesting the pressure may be nearly hydrostatic.

This is also confirmed by measurements of $\overline{u_w^2}$ which is of particular interest because it occurs for example in the exact expression for the radiation stress and the energy flux. Figure 5.5.14 shows that $\overline{u_w^2}$ is also nearly uniform over depth below trough level.

We also see that the combination of variables $\frac{\overline{u_w^2}}{gh(H/h)^2}$ plotted in Fig. 5.5.14 and discussed below only varies a little over the entire surfzone. This is confirmed by the experiments by Hansen and Svendsen (1986).

The magnitude of $\overline{u_w^2}$ may be further assessed by considering the linear approximation to u_w given by

$$u_w = c\eta/h \tag{5.5.19}$$

For $\overline{u_w^2}$ this leads to

$$\overline{u_w^2} = ghB_0 \left(\frac{H}{h}\right)^2 \tag{5.5.20}$$

Hansen (1990) defines the parameter B_u given by

$$B_u = \frac{\overline{u_w^2}}{gh(H/h)^2} \tag{5.5.21}$$

and we see that if (5.5.19) were correct we would get

$$B_u \simeq B_0 \tag{5.5.22}$$

Fig. 5.5.15 shows B_u/B_0. We see that in general B_u is equal to or slightly smaller than B_0 which means that (5.5.19) gives a little too large values of u_w. For the present, however, it may be taken as a reasonable justification for using (5.5.19) for the estimation of surfzone properties in Section 5.6.

5.5.8 *Turbulence intensities*

Measured results for the *rms*-value, $\sqrt{\overline{u'^2}}$, of the horizontal turbulent fluctuations (the horizontal turbulent normal stress) are shown in Fig. 5.5.16. Similar to the particle velocities u_w the turbulent fluctuations are nearly uniformly distributed over depth below trough level. The values vary between 2 and 8% of $\sqrt{gh}$. Similar results were obtained by Svendsen (1987) from analysing published experimental data and, with smaller values of $\sqrt{\overline{u'^2}/gh}$ ($\sim 0.02 - 0.03\sqrt{gh}$, by Nadaoka and Kondoh (1982) and by Hattori and Aono (1985)). When compared with the results above for $\overline{u_w^2}$ it appears that the turbulent normal stress is less than 10% of the $\overline{u_w^2}$ value and hence within the uncertainty of the calculations in the surfzone. This was already observed by Stive and Wind (1982). Therefore the normal Reynolds stresses are normally neglected in nearshore computations. Other measurements of velocities and turbulent energy are found in Ting and Kirby (1994, 1995, 1996).

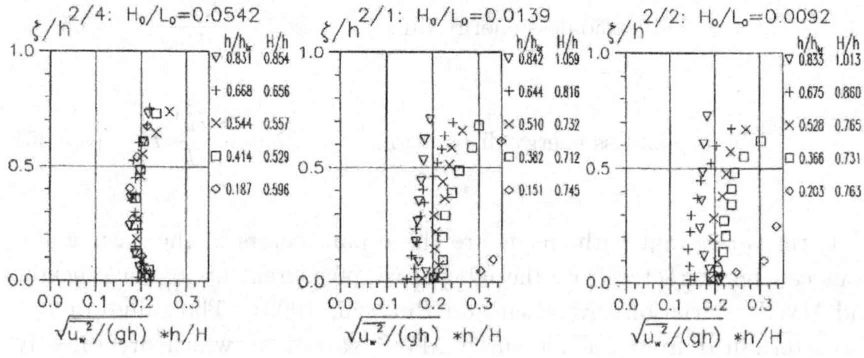

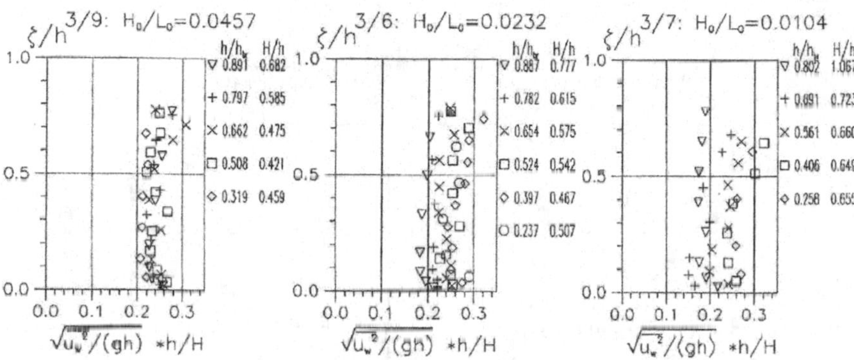

Fig. 5.5.14 The values of $\overline{u_w^2}$ in experiments by Okayasu (from Hansen, 1990). The experiments numbered 2/ are on a 1/20 slope, experiments 3/ are on a 1/30 slope.

It may also be noted that that $\sqrt{\overline{u'^2}/gh}$ decreases relative to $\sqrt{gh}$ toward the shoreline while $\overline{u_w^2/gh(H/h)^2}$ largely stays constant.

5.5.9 *The values of P_{xx}, B_x and D*

The shape factors P_{xx} and B_x defined in Section 3.3 and D defined in Chapter 4, represent the non-dimensional radiation stress, energy flux and energy dissipation, respectively. The definitions were

Dimensionless radiation stress $$P_{xx} = \frac{S_{xx}}{\rho g H^2} \tag{5.5.23}$$

Dimensionless energy flux $$B_x = \frac{E_{fx}}{\rho g H^2 c} \tag{5.5.24}$$

Dimensionless energy dissipation $$\mathcal{D} = \frac{\rho g H^3}{4hT} D \tag{5.5.25}$$

It turns out that with some care these parameters in the basic equations can be extracted from the laboratory measurements of wave height and MWS - variation (Svendsen and Putrevu, 1993). The values of P_{xx} are determined from the measured MWS variations which are directly linked to S_{xx}, and the values of B_x are then deduced from the wave height

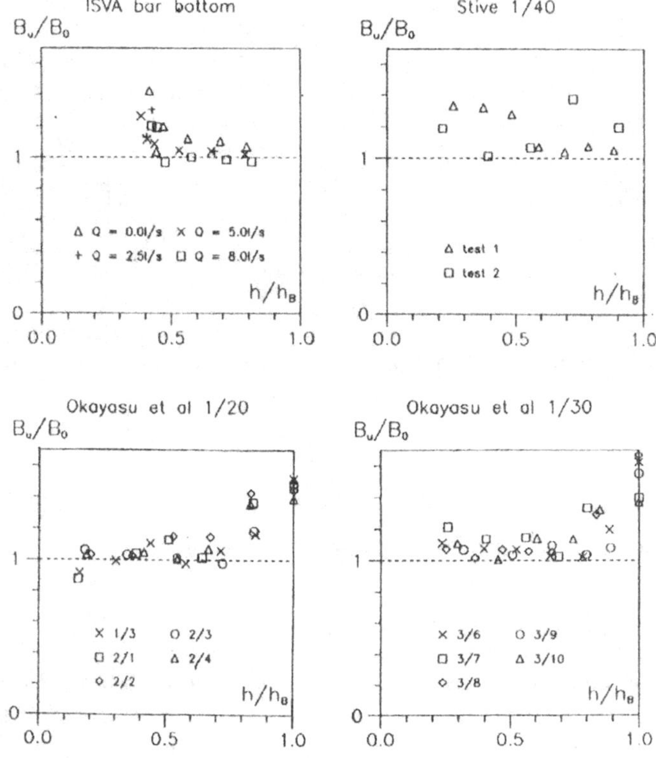

Fig. 5.5.15 B_u/B_o (from Hansen, 1990).

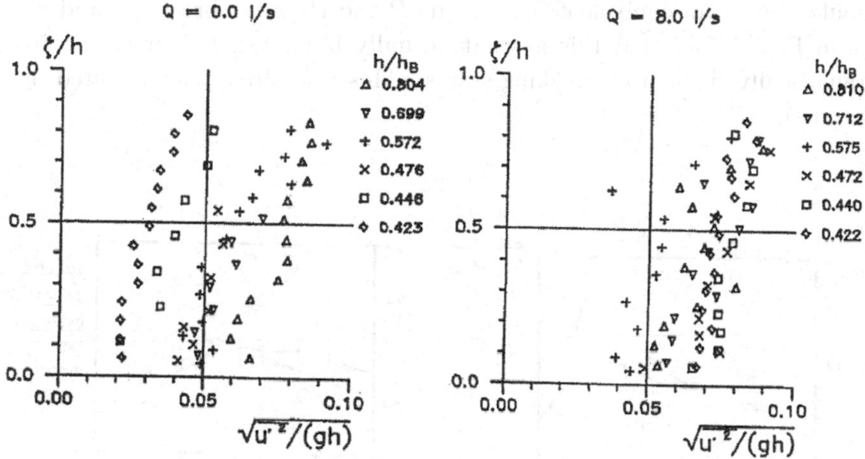

Fig. 5.5.16 The figure shows the wave averaged turbulent kinetic energy component in the x-direction, $\overline{u'^2}$. The parameter Q represents a net current flux moving with the waves in the experiments (from Hansen and Svendsen, 1986).

measurements. Some additional assumptions about the roller etc. are made to obtain the results and reference for the details are made to the quoted paper. Fig. 5.5.17 shows the value of P_{xx} (here P for short) derived that way. The experimental results have been divided into four regions of the parameter S at the breaking point

$$S_b = \frac{h_x L_B}{h_B} \tag{5.5.26}$$

For comparison the value of P for shallow water sinusoidal waves is

$$P = 3/16 = 0.1875 \tag{5.5.27}$$

In all intervals of S_b, the value of P at the break point is lower than for sinusoidal waves. This is in accordance with the observations made earlier that at the start of breaking the waves have short, peaky crests and long shallow troughs. The breaking, however, essentially makes the wave collapse. This transforms some of the potential energy stored in the crest into forward oriented momentum (kinetic energy). Measurements of the MWS in this region indicate little change, however, which means the radiation stress $S_{xx} = \rho g H^2 P$ must be almost constant, in spite of the rapidly decreasing wave height after the start of breaking. Therefore we should expect an equivalent increase in P shortly after breaking, and we see in Fig. 5.5.17 that this is what actually happens: P increases quite dramatically right after breaking starts. This was already conjectured by Svendsen (1984).

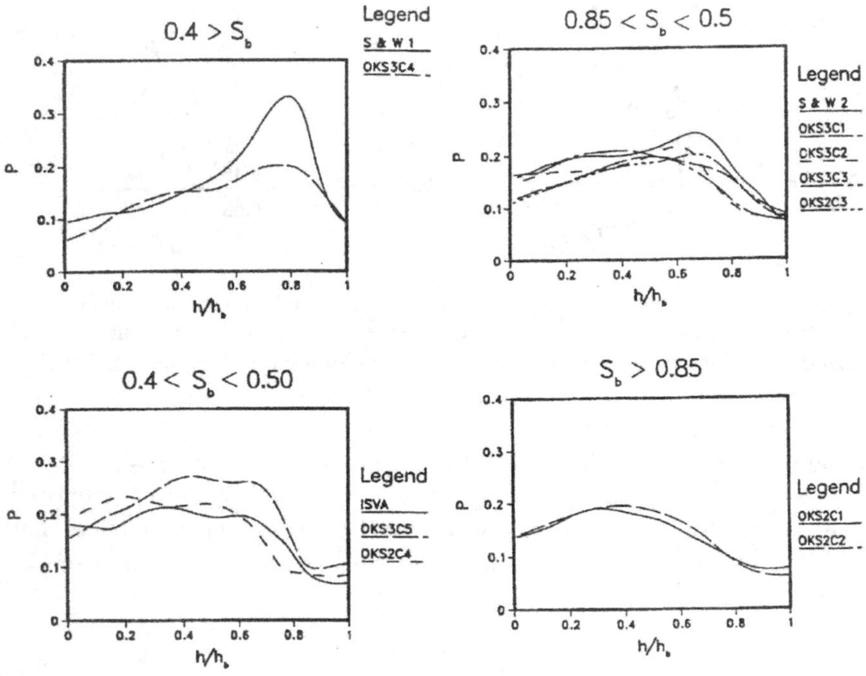

Fig. 5.5.17 The variation of $P = S_{xx}/\rho g H^2$ in laboratory waves. S & W refers to the experiments by Stive and Wind (1982), OK is based on data from Okayasu et al. (1986), while "ISVA" represents data from Hansen and Svendsen (1986) (modified from Svendsen and Putrevu, 1993).

Further shorewards the value of P decreases again, in some cases to well below 0.1.

Comparing with the linear long wave result of $P = 3/16$, we see how different surf zone waves are. It will be shown in Chapter 11 that the forcing of currents and set-up is given by $\partial S_{\alpha\beta}/\partial x_\alpha$ and the figure shows that the variations of P (i.e., the change in wave shape) contributes to the variation of this forcing.

Similarly Fig. 5.5.18 shows results for the dimensionless energy flux B (short for B_x). Here the relevant sinewave comparison is $B = 0.125$ and again it is the derivative of the results that appears in the equation of energy. The sine wave theory may be off by as much as a factor of 2 for the energy flux because it assumes $B =$ constant and at a wrong value.

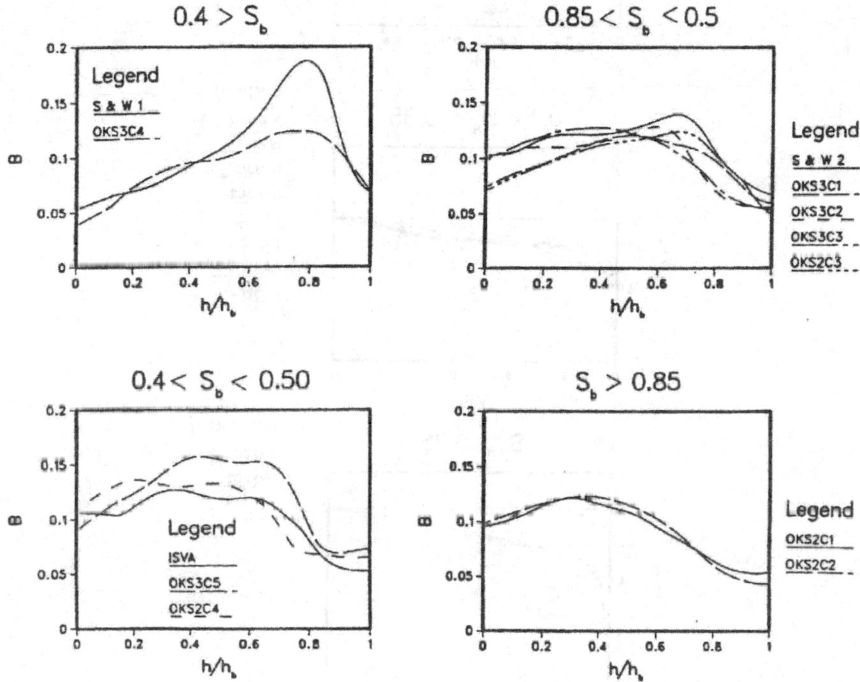

Fig. 5.5.18 The variation of $B = E_{fw}/\rho g c H^2$ in laboratory waves (modified from Svendsen and Putrevu, 1993).

Finally, we may look at the value D of the dimensionless energy dissipation. In Fig. 5.5.19, D is compared to the value D_b in a bore or moving hydraulic jump. Thus $D/D_b = -1$ as shown in the figure implies the energy dissipation in the surf zone wave is the same as it would be in a bore of the same height.

We see that for a sufficiently gentle slope ($S_b < 0.4$), there is a wide part of the surf zone where D/D_b is constant and not much larger than

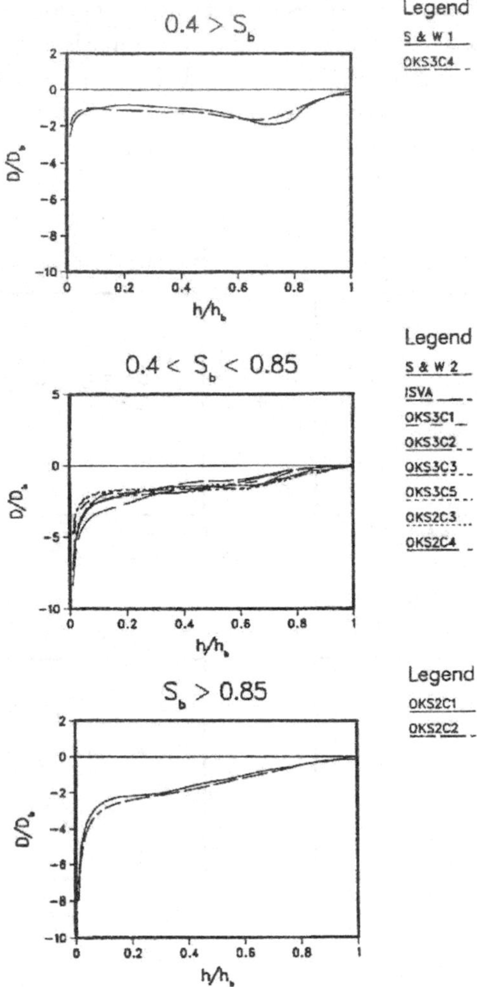

Fig. 5.5.19 Energy dissipation D relative to the dissipation D_b in a bore of the same height (modified from Svendsen and Putrevu, 1993).

1. This probably indicates the wave is nearly in local equilibrium (locally constant depth conditions are valid).

For steeper slopes, relative to h/L $(S_b > 0.4)$ the D/D_b becomes increasingly larger in absolute value, and the same happens for $S_b < 0.4$ as the waves approach the shoreline. As described in Chapter 4 this leads to increasing H/h-values as the waves appraoch the shoreline. In terms of breaking, this suggests that the dissipation is larger than in a bore of the same height: the wave is breaking more violently and most likely is unable to keep up with the shoaling effect (see discussion of equation (4.3.6)). This is actually happening in many experimental results as shown by Horikawa and Kuo (1966).

Exercise 5.5-2 Show that if the waves are modelled by the simple energy wave model described in Section 4.3.2, and have a sawtooth shape, then as they approach the shoreline we get $H \to 0$, but with $H/h \to 2$ at the shoreline (see Svendsen and Hansen (1986)).

5.5.10 *The wave generated shear stress* $\overline{u_w w_w}$

In symmetrical waves such as sinusoidal waves and other steady uniform waves on a constant depth and with no friction we have $\overline{u_w w_w} = 0$. In a bottom boundary layer, however, it is found that $\overline{u_w w_w} \neq 0$ (Chapter 10).

The $\tau_{w,xz} = -\rho\overline{u_w w_w}$ represents a horizontal shear stress that comes from deformation of the wave motion relative to the steady motion on a constant depth. In the local momentum balance for the vertical velocity profiles in cross-shore circulation - and also in general 3D models - the vertical gradient $\partial\overline{u_w w_w}/\partial z$ represents a contribution to the horizontal momentum balance in line with the turbulent shear stresses. It is therefore of interest to seek information about this term.

It turns out that in many of the measurements available u_w and w_w cannot be combined to calculate the value of $\overline{u_w w_w}$ because the time series for the two components have not been collected simultaneously. Thus slightly different waves would provide the data for each of the two components and this turns out to disturb the accuracy of the time averaging sufficiently to make the results difficult to interpret. However Figs. 5.5.20, 5.5.21 and 5.5.22 (Ting, 2003) show results of $\overline{u_w w_w}$ obtained from simultaneous measurement of u_w and w_w for a wave breaking on a plane beach with slope $1:50$. $z = 0$ is at the MWS. These figures represent measurements in a spilling breaker. The first at $h/h_b = 0.874$ is relatively close to the breaking point whereas the second and third are in the center of the surfzone at $h/h_b = 0.590$ and $h/h_b = 0.538$, respectively.

In the measurements shown in Figs. 5.5.20 and 5.5.22 it is obvious that the values of $\overline{u_w w_w}$ are much smaller than the turbulent shear stress $\overline{u'w'}$ which is the relevant magnitude to compare with.

For the measurements shown in Fig. 5.5.21 there is quite a bit of scatter so the situation is not so clear. However, it turns out that the vertical column can be divided into three regions with quite distinct differences in the value of $\overline{u_w w_w}$. There is an upper region of approximately 10 mm, ($z = -10$ to -20 mm, approximatly 10% of the depth of 98 mm) where the $\overline{u_w w_w}$ is in average $35 - 40\%$ of the turbulent shear stress $\overline{u'w'}$. This is probably a region strongly dominated by the turbulence from the breaking at the surface. In the central region ($z = -20$ to -80 mm), however, the average value of $\overline{u_w w_w}$ is less than 1% of $\overline{u'w'}$, which is at least an order of magnitude smaller than $\overline{u'w'}$. In fact in this region it is not possible within the accuracy of the measurements to distinguish $\overline{u_w w_w}$ from zero. The same consequently applies also to the vertical gradient $\partial\overline{u_w w_w}/\partial z$. Finally in the lowest approximately $16 - 18$ mm we cannot distinguish between the two types of shear stresses. At the other two locations $\overline{u_w w_w}$ is actually larger than $\overline{u'w'}$ in the region closest to the bottom. This is likely to be mainly in the bottom boundary layer where we would expect the value of $\overline{u_w w_w}$ to be quite significant anyway due to the boundary layer flow (see Chapter 10).

Attempts have also been made to calculate the magnitude of $\overline{u_w w_w}$ by assuming the wave motion is a combination of two sine waves phase shifted 90° from each other (Deigaard and Fredsøe, 1989, Stive and DeVriend, 1994). This approach, however, results in values of $\overline{u_w w_w}$ that are of the same order of magnitude as the turbulent shear stress $\overline{u'w'}$, a result which is not supported by the mesurements shown above. The work is based on the sensible assumption that since wave energy is dissipated primarily near the surface causing the entire wave motion to decay, energy must be transported vertically toward the surface. In the wave decay due to the dissipation in a bottom boundary layer (see Chapter 10) it is found that this generates a weak distortion of the wave motion above the boundary layer that provides the necessary vertical energy transport in the potential part of the flow above the boundary layer. For shallow water waves this transport turns out to be propotional to $\overline{u_w w_w}$.

In a surfzone wave, however, the turbulence extends over the entire depth and the assumption that the wave motion can be approximated by two sinusoidal components phase shifted 90° does not seem to catch the main mechanisms of the flow. The interpretation of the measurements is that it is primarily the turbulent shear stresses that provide the necessary vertical energy transport, but this remains to be proved.

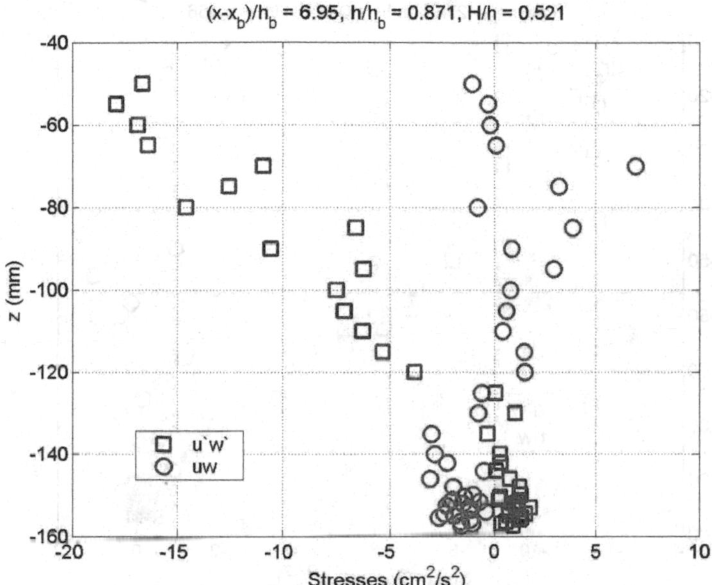

Fig. 5.5.20 Measured values of the shear stress $\tau_{w,zx}/\rho = \overline{u_w w_w}$ generated by the deformation of the wave breaking in the surfzone and the turbulent Reynolds stress $\overline{u'w'}$. Squares represent the $\overline{u_w w_w}$, o's the values of $\overline{u'w'}$. z is the vertical distance below SWL. The wave is a spilling breaker, and the meaurements were taken in the outer surf zone at $h/h_b = 0.874$ (courtesy F. Ting, 2003).

Thus at present there is no satisfactory understanding of what generates the $\overline{u_w w_w}$ stress. The few measurements available suggest that this wave shear stress is small in comparison to the turbulent stresses., and this will be the assumption used in the analysis of the cross-shore circulation currents called undertow (see Section 12.2) and in general 3D currents (Chapter 13).

5.6 Surfzone wave modelling

5.6.1 *Surfzone assumptions*

In the following we review some simple theoretical approximations for the wave averaged properties such as volume, momentum and energy flux in surf zone waves, much as we have already done for linear waves outside

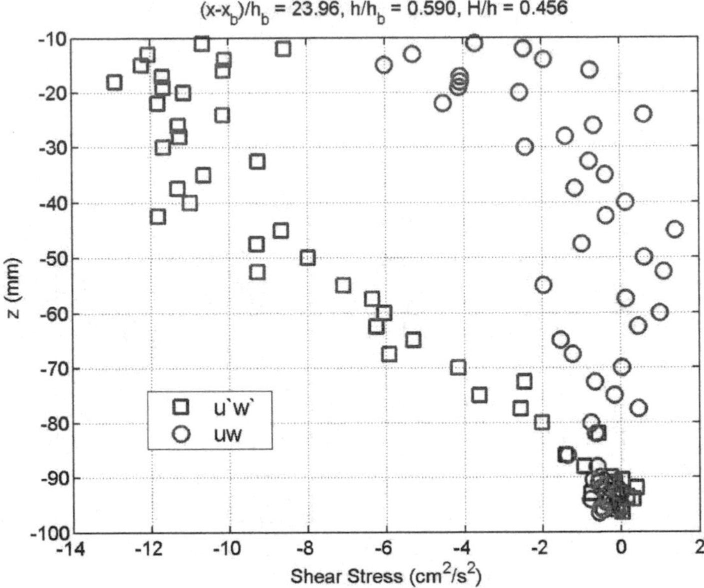

Fig. 5.5.21 Measured values of the shear stress $\tau_{w,zx}/\rho = \overline{u_w w_w}$ generated by the deformation of the wave breaking in the surfzone and the turbulent Reynolds stress $\overline{u'w'}$. Symbols as in Fig. 5.5.20. The wave is a spilling breaker, and the meaurements were taken in the central surf zone at $h/h_b = 0.590$ including setup (courtesy F. Ting, 2003).

the surf zone. We also examine the similarities between a surfzone wave and a periodic bore and use this to determine approximations for the phase velocity and energy dissipation in surfzone waves.

The basic ideas in these estimates are:

(1) The approximations of linear shallow water theory can be applied. By superposition of waves linear theory basically allows **any** wave form $\eta(x, t)$ (sine waves is just one, see Section 3.4.3). However, linear long waves have hydrostatic pressure

$$p_D = \rho g \eta \tag{5.6.1}$$

and a depth uniform velocity given by

$$u_w = c\,\frac{\eta}{h} \quad ; \quad \eta = \zeta - \bar{\zeta} \tag{5.6.2}$$

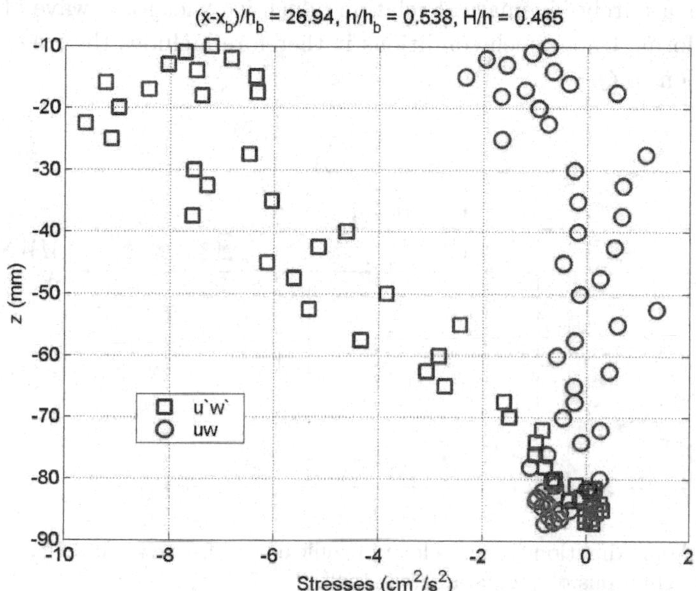

Fig. 5.5.22 Measured values of the shear stress $\tau_{w,zx}/\rho = \overline{u_w w_w}$ generated by the deformation of the wave breaking in the surfzone and the turbulent Reynolds stress $\overline{u'w'}$. Symbols as in Fig. 5.5.20. The wave is a spilling breaker, and the meaurements were taken in the central surf zone at $h/h_b = 0.538$ including setup (courtesy F. Ting, 2003).

where c is the phase velocity. This also implies that

$$w \ll u_w \tag{5.6.3}$$

(2) The actual particle velocities under a broken wave are indicated by the velocity profiles shown in Fig. 5.5.9. The major feature distinguishing the velocity field in a surfzone wave is the "roller".

(3) We simplify this velocity distribution as shown in Fig. 5.6.1. In the roller the average velocity is constant and equal to c. The cross-sectional area of the roller in an x, z-plot is A, which is a geometrically well defined quantity that must (and can in principle) be measured or estimated. Below the roller the velocity is also depth uniform and equal to the value required to satisfy that the instantaneous volume flux Q in the wave motion is given by

$$Q = c\,\eta \tag{5.6.4}$$

This is a purely kinematical relation which is exact for a wave of constant form. Under such conditions it therefore includes the roller contribution to Q.

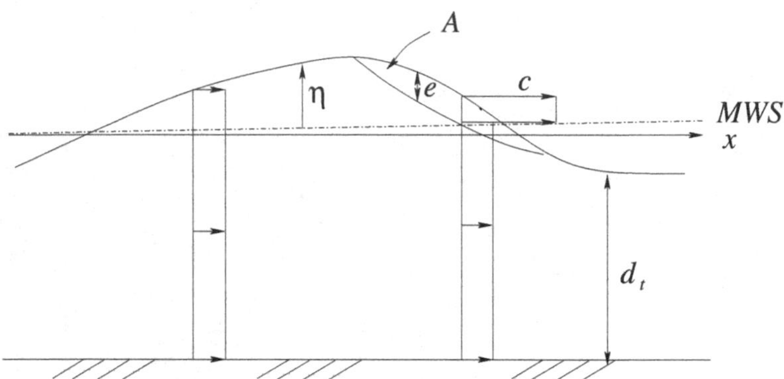

Fig. 5.6.1 Approximation for the velocity profile in a surf zone wave with a surface roller, and definitions of the parameters used.

It turns out that the roller represents an important contribution from the breaking process to the wave averaged quantities such as the volume flux, radiation stress and energy flux for surf zone waves.

The linear long wave theory leads to a phase velocity of

$$c = \sqrt{gh} \tag{5.6.5}$$

However, the experimental results shown in Sect. 5.5 showed that the actual phase velocity can be $10 - 20\%$ higher than the linear value and this will also be confirmed by the theoretical results below. Hence, in the following derivations we will simply assume that

$$c = a\sqrt{gh} \tag{5.6.6}$$

5.6.2 *Energy flux E_f for surfzone waves*

We first consider the energy flux in the form

$$E_{fx} = \rho g c H^2 B \tag{5.6.7}$$

where for plane waves ($v = 0$)

$$B = \frac{1}{\rho g c H^2} \overline{\int_{-h_0}^{\zeta} \left[p_D + \frac{1}{2} \rho \left(u_w^2 + w_w^2 \right) \right] u dz} \tag{5.6.8}$$

For the first term we get (with $e(x,t)$ equal to the local height or thickness of the roller as defined in Fig. 5.6.1)

$$\overline{\int_{-h_0}^{\zeta} p_D u dz} = \overline{\int_{-h_0}^{\zeta - e} \rho g \eta \cdot c \frac{\eta}{h} dz} + \overline{\int_{\zeta - e}^{\zeta} \rho g \eta \cdot c dz} \tag{5.6.9}$$

$$= \rho g c \overline{\frac{\eta^2}{h} (h + \eta - e)} + \rho g c \overline{\eta e} \quad \text{(since } h_0 + \zeta - e = h + \eta - e\text{)}$$

$$= \rho g c H^2 \left(\frac{\overline{\eta^2}}{H^2} + \frac{\overline{\eta^3}}{H^2 h} - \frac{\overline{e \eta^2}}{H^2 h} \right) + \rho g c H^2 \frac{\overline{\eta e}}{H^2}$$

$$= \rho g c \frac{\overline{\eta^2}}{H^2} + 0 \overline{\left(\frac{\eta}{H} \right)^3} \tag{5.6.10}$$

because

$$\frac{\overline{\eta^3}}{H^2 h}, \frac{\overline{e \eta^2}}{H^2 h}, \frac{\overline{\eta e}}{H^2} \ll \frac{\overline{\eta^2}}{H^2} \tag{5.6.11}$$

Hence we can write

$$\overline{\int_{-h_0}^{\zeta} p_D u dz} \sim \rho g c H^2 B_0 \tag{5.6.12}$$

where

$$B_0 \equiv \frac{1}{T} \int_0^T \left(\frac{\eta}{H} \right)^2 dt \tag{5.6.13}$$

To evaluate the second term in (5.6.8), we use the approximation for the velocity distribution shown in Fig. 5.6.1. We then get, using (5.6.2)

$$\overline{\int_{-h_0}^{\zeta} \frac{1}{2} \rho u_w^3 dz} = \frac{1}{2} \rho \overline{\int_{-h_0}^{\zeta - e} u_0^3 dz} + \frac{1}{2} \rho \overline{\int_{\zeta - e}^{\zeta} c^3 dz} \tag{5.6.14}$$

For the first term in (5.6.14) we have

$$\frac{1}{2} \rho \overline{\int_{-h_0}^{\zeta - e} u_0^3 dz} = \frac{1}{2} \rho c^3 \overline{\frac{\eta^3}{h^3} (h + \eta - e)}$$

$$\simeq \frac{1}{2} \rho g a^2 c H^2 \frac{H}{h} \overline{\left(\frac{\eta}{H} \right)^3 \left[1 + \frac{\eta}{h} - \frac{e}{h} \right]} \tag{5.6.15}$$

where a is given by (5.6.6).

Hence to the first approximation this term becomes

$$\frac{1}{2}\rho\overline{\int_{-h_0}^{\zeta-e} u_0^3 dz} = \frac{1}{2}\rho g a^2 c H^2 \frac{H}{h}\overline{\left(\frac{\eta}{H}\right)^3} \tag{5.6.16}$$

Because η^3 is both positive and negative over a wave period whereas η^2 is positive only and $\eta/H < 1$ we see that we must have

$$\overline{\left(\frac{\eta}{H}\right)^3} \ll \overline{\left(\frac{\eta}{H}\right)^2} \tag{5.6.17}$$

We therefore conclude that the first term in (5.6.14) is $\ll$ the first term in (5.6.8).

For the second term in (5.6.14) we get

$$\frac{1}{2}\rho\overline{\int_{\zeta-e}^{\zeta} c^3 dz} = \frac{1}{2}\rho c^3 \bar{e} \tag{5.6.18}$$

where for $\bar{e}$ we have

$$\begin{aligned} \bar{e} &= \frac{1}{T}\int_0^T e(t)dt \\ &\simeq \frac{1}{cT}\int_0^L e dx = \frac{1}{cT}\int_0^\lambda e dx = \frac{A}{cT} = \frac{A}{L} \end{aligned} \tag{5.6.19}$$

since $dx = cdt$ if the waves are of permanent form.

Thus the final result for (5.6.14) becomes

$$\overline{\int_{-h_0}^{\zeta} \frac{1}{2}\rho u_w^3 dz} = 0 + \frac{1}{2}\rho c^3 \frac{A}{L} = \rho g c H^2 \frac{1}{2}\frac{c}{gT}\frac{A}{H^2} \tag{5.6.20}$$

For completeness it is mentioned that, as could be expected, the last term in (5.6.8) is even smaller than the second because $w_w \ll u_w$.

Combining (5.6.12) and (5.6.20) then yields

$$\boxed{E_{f,w} = \rho g c H^2 \left(B_0 + \frac{1}{2}\frac{A}{H^2}\frac{c}{gT}\right)} \tag{5.6.21}$$

Here c/gT can also be written as

$$\frac{c}{gT} = \frac{ch}{gh\,T} = a^2\,\frac{h}{L} \tag{5.6.22}$$

$E_{f,w}$ can also be written

$$\boxed{E_{f,w} = \rho g c H^2 \left(B_0 + \frac{1}{2} a^2 \frac{A}{HL} \frac{h}{H} \right)} \tag{5.6.23}$$

Therefore B can be used in two different forms

$$B = B_0 + \begin{cases} \frac{1}{2}\, a^2 \frac{A}{H^2}\, \frac{h}{L} \\ \frac{1}{2} a^2 \frac{A}{HL}\, \frac{h}{H} \end{cases} \tag{5.6.24}$$

Notice that A/H^2 = const. and A/HL = const represent different variations of B with h:

$$\frac{A}{H^2} \text{ const } \Rightarrow \frac{A}{H^2}\, \frac{h}{L} \sim \sqrt{h} \quad \text{over the surfzone} \tag{5.6.25}$$

$$\frac{A}{HL} \text{ const } \Rightarrow \frac{A}{HL}\, \frac{h}{H} \sim \text{ const over the surfzone if } H/h \text{ is constant} \tag{5.6.26}$$

Therefore it is not quite without consequence which expression is used for the area of the roller.

Exercise 5.6-1

Show that non-breaking sine waves have

$$A = 0 \quad ; \quad B = \frac{1}{16}(1 + G) \qquad G = \frac{2kh}{\sinh 2kh} \tag{5.6.27}$$

so that

$$E_{fw} = \frac{1}{16} \rho g c H^2 \left(1 + \frac{2kh}{\sinh 2kh} \right) \tag{5.6.28}$$

as found in Section 3.3.

Show that for long waves (already assumed) this means

$$E_{fw} = \frac{1}{8} \rho g c H^2 \qquad \text{corresponding to} \qquad B = B_0 = 0.125 \tag{5.6.29}$$

Exercise 5.6-2

Consider breaking waves with

$$\frac{A}{H^2} \sim 0.9 \quad ; \quad c = \sqrt{gh} \ \text{(i.e. } a = 1) \tag{5.6.30}$$

Show that if we have $h/L = 0.05$ we get

$$\frac{1}{2}\frac{A}{H^2}\cdot\frac{h}{L} \sim 0.0225 \tag{5.6.31}$$

or

$$E_{fw} = \rho g c H^2 \left(B_0 + 0.0225\right) \tag{5.6.32}$$

Here B_0 must be estimated from measurements or using Hansen's (1990) formulas (see Section 5.4). If $B_0 = 0.125$ (sinusoidal waves) we see that the roller contributes close to 20% to the energy flux, but much more when $B_0 \sim 0.06 - 0.08$ However, by using $A = const \cdot H^2$ the value of E_{fw} decreases toward the shorelline.

Exercise 5.6-3

Some numerical results (Okayasu 1989) suggest

$$\frac{A}{HL} \sim 0.06 - 0.07 \quad \text{is constant.} \tag{5.6.33}$$

Assuming $H/h = 0.6$ show that we get

$$\frac{1}{2}\frac{A}{HL}\cdot\frac{h}{H} \sim 0.05 - 0.06 \tag{5.6.34}$$

and hence

$$E_{fw} = \rho g c H^2 \left(B_0 + 0.05\right) \tag{5.6.35}$$

Thus if $B_0 = 0.125$ the roller in this approximation contributes approximately $40 - 45\%$ to the energy flux.

Measurements (see Fig. 5.5.5) also show that typically

$$B_0 \sim 0.065 - 0.10 \tag{5.6.36}$$

In conclusion the likely range for B—obtained by heuristically combining the smallest values of B_0 and A and the largest values of B_0 and A—becomes

$$B \sim \left\{ \begin{array}{ll} 0.065 + 0.0225 = 0.0875 & \text{(possible minimum)} \\ 0.10 + 0.050 = 0.150 & \text{(possible maximum)} \end{array} \right\} \tag{5.6.37}$$

When this is compared with the results in Fig. 5.5.18 of analyzing measurements directly we see that, except for a short interval in one of the experiments, the value of B generally stays below 0.15 and generally also is above 0.065. Thus (5.6.37) is within the range of Fig. 5.5.18.

5.6.3 *Radiation stress in surfzone waves*

We next look at the two S_{xx} components:

$$S_m = \overline{\int_{-h_0}^{\zeta} \rho u_w^2 dz} \tag{5.6.38}$$

$$S_p = \int_{-h_0}^{\zeta} \left(-\rho w^2\right) dz + \frac{1}{2}\rho g \overline{\eta^2} \tag{5.6.39}$$

For the first we get

$$\begin{aligned} S_m &= \overline{\int_{-h_0}^{\zeta - e} \rho c^2 \frac{\eta^2}{h^2} dz} + \overline{\int_{\zeta - e}^{\zeta} \rho c^2 dz} \qquad \text{where : } \zeta = \eta + \bar{\zeta} \\ &= \rho c^2 \overline{\frac{\eta^2}{h^2} \cdot (h - e)} + \rho c^2 \cdot \bar{e} \\ &\cong \rho c^2 \overline{\frac{\eta^2}{h^2}} \cdot h + \rho c^2 \frac{A}{L} \end{aligned} \tag{5.6.40}$$

because $e/h \ll 1$.

Hence we get for S_m the approximation

$$S_m = \rho g H^2 \frac{c^2}{gh} \left[\overline{\frac{\eta^2}{H^2}} + \frac{A}{H^2} \frac{h}{L} \right] \tag{5.6.41}$$

which means

$$S_m = \rho g H^2 \frac{c^2}{gh} \left[B_0 + \frac{A}{H^2} \frac{h}{L} \right] \tag{5.6.42}$$

or

$$S_m = \rho g H^2 a^2 \left[B_0 + \frac{A}{LH} \cdot \frac{h}{H} \right] \tag{5.6.43}$$

Similarly we get for S_p, neglecting the w^2-term, which represents the deviation from hydrostatic pressure

$$S_p = 0 + \frac{1}{2}\rho g H^2 \left(\frac{\overline{\eta}}{H}\right)^2 = \frac{1}{2}\rho g H^2 B_0 \tag{5.6.44}$$

which results in

$$\boxed{S_{xx} = \rho g H^2 \left(\left(a^2 + \frac{1}{2}\right) B_0 + a^2 \frac{A}{H^2} \frac{h}{L}\right)} \tag{5.6.45}$$

or

$$\boxed{S_{xx} = \rho g H^2 \left(\left(a^2 + \frac{1}{2}\right) B_0 + a^2 \frac{A}{HL} \frac{h}{H}\right)} \tag{5.6.46}$$

Some examples will further illustrate this.

Exercise 5.6-4 Long, non-breaking sine waves have

$$A = 0 \quad , \quad a = 1 \quad , \quad B_0 = \frac{1}{8} \tag{5.6.47}$$

Confirm using the above expressions that this gives

$$S_{xx} = \frac{3}{16}\rho g H^2 = 0.183 \rho g H^2 \tag{5.6.48}$$

as we know from earlier.

Exercise 5.6-5

Consider surf zone waves, and assume the waves are close to the sawtooth shape found in measurements of waves far into the surf zone (Section 5.5), which means that

$$B_0 = 0.083 \tag{5.6.49}$$

We again assume that $a = 1$ and

$$A/HL \sim 0.06 \quad ; \quad H/h = 0.6 \tag{5.6.50}$$

Show that we then get

$$S_{xx} \sim 0.22 \rho g H^2 \tag{5.6.51}$$

or

$$P_{xx} = 0.22 \tag{5.6.52}$$

That means, even though the sawtooth shaped waves have smaller B_0, S_{xx} is larger than for a sine wave of the same height because of the roller, which here contributes almost half of the total S_{xx}.

The values found in these examples clearly are in the same range as the empirical results shown for B and P in Section 5.4, Fig. 5.5.17 and Fig. 5.5.18. However, these figures also suggest that the real picture is more complicated than the nearly constant values for B and P across the surfzone determine above.

5.6.4 *Volume flux in surfzone waves*

The volume flux in the waves is defined as

$$Q_w = \overline{\int_{\zeta_t}^{\zeta} u_w dz} \tag{5.6.53}$$

Q_w can also be calculated for the simplified breaker used above. As before we assume that

$$u_w = \begin{cases} c\eta/h & -h_0 < z < \zeta - e \\ c & \zeta - e < z < \zeta \end{cases} \tag{5.6.54}$$

and get from the second equality in (5.6.53)

$$\begin{aligned} Q_w &= \frac{c}{h}\overline{(\eta - e - \eta_t)\,\eta} + c\bar{e} \\ &= \frac{c}{h}\overline{\eta^2} + \overline{e\eta} + c\bar{e} \end{aligned} \tag{5.6.55}$$

since $\bar{\eta} = 0 \Rightarrow \overline{\eta\eta_t} = 0$. Furthermore, $e(\geq 0)$ is non-zero only in an interval of t where η is both positive and negative. Therefore, we have

$$\overline{\eta e} \ll \overline{\eta^2} \tag{5.6.56}$$

Hence we get, using (5.6.19) for $\bar{e}$

$$\boxed{Q_w = \frac{gH^2}{c}\, a^2 \left[B_0 + \frac{A}{H^2}\frac{h}{L}\right]} \tag{5.6.57}$$

or

$$\boxed{Q_w = \frac{gH^2}{c}\, a^2 \left[B_0 + \frac{A}{HL}\,\frac{h}{H}\right]} \tag{5.6.58}$$

For non-breaking waves this result is of course identical with the result found in Section 3.3.

Exercise 5.6-6

If we consider a sawtooth shaped wave we again have

$$B_0 = 0.083 \tag{5.6.59}$$

and we also again take

$$\frac{A}{HL} \sim 0.06 \tag{5.6.60}$$

and consider a case where $H/h = 0.6$. Show that this gives

$$Q_w \sim ch \cdot 0.066 \tag{5.6.61}$$

In this example the roller contributes the most to Q_w The result means that in a wave tank (where $Q(= \overline{\widehat{Q}}) = 0$ and hence the return current $U = -Q_w/h$) we have

$$U \simeq -0.066c \tag{5.6.62}$$

i.e., a depth averaged return current in the surf zone which equals 6.6% of c! This result is very close to the actually measured values (for further details see Section 12.2).

5.6.5 *The phase velocities for quasi-steady breaking waves*

A bore is essentially a moving hydraulic jump. In the following the quasi steady breakers in the surfzone are assumed to have uniform velocities over depth and hydrostatic pressure. Those assumptions are usually also made in classical hydraulics when dealing with bores and hydraulic jumps. Figure 5.6.2 shows equivalent features of the two types of flow.

In a fixed frame of reference a bore and a quasi steady breaker propagate with a certain speed c_b and c_w, respectively. We know that the bore speed

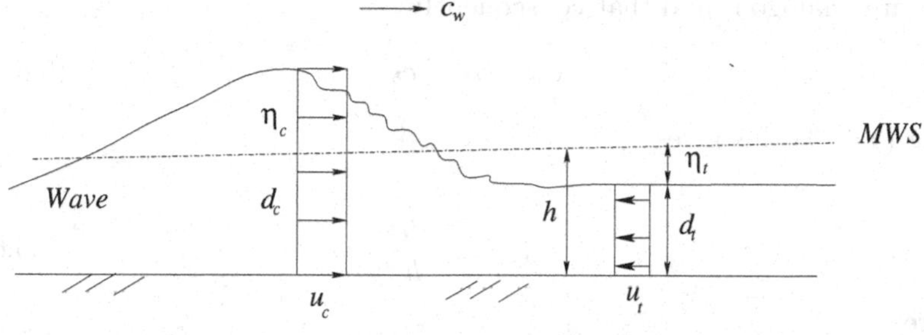

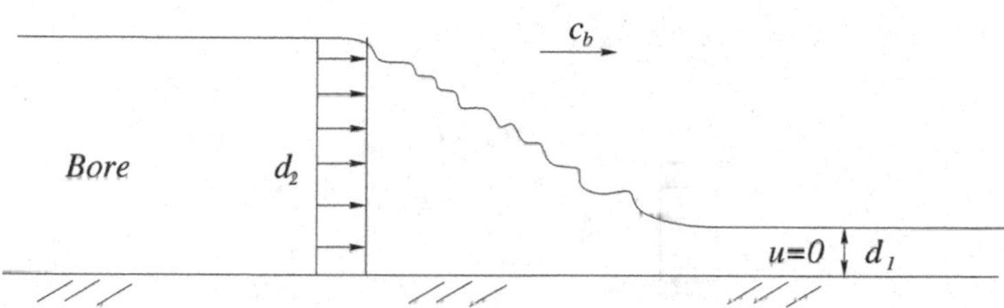

Fig. 5.6.2 Similarities between a breaking wave and hydraulic bore.

c_b is given by

$$c_b^2 = gd_1 \cdot \frac{1}{2}\xi(\xi + 1) \quad ; \quad \xi = d_2/d_1 \tag{5.6.63}$$

and seek to determine the equivalent value of c_w. Fig. 5.6.3 shows the definitions of the variables used.

One obvious difference is: the bore propagates into quiescent water, whereas in a fixed frame of reference the breaking wave has an opposing velocity in the wave trough in front of it which is (exactly if the wave has constant form and no net volume flux):

$$u_t = c_w \frac{\eta_t}{h + \eta_t} \tag{5.6.64}$$

where η_t is the surface elevation in the trough.

However, if we observe the wave from a coordinate system moving with the particle speed $u_t(< 0)$ in the trough (figure 5.6.3), we see that the two flows are analogous and that consequently

$$c_w - u_t = c_b \tag{5.6.65}$$

Using (5.6.64) this gives

$$c_w = c_b \cdot \frac{d_t}{h} \tag{5.6.66}$$

where

$$d_t = h + \eta_t \tag{5.6.67}$$

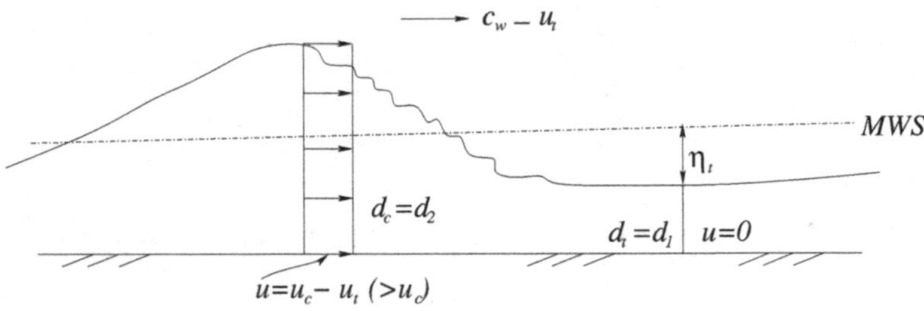

Fig. 5.6.3 Definitions of variables used in the comparison of bores and breaking waves. The wave is here viewed from a coordinate system moving with the velocity $c_w - u_t$ so that the velocity in the trough is zero. This corresponds to the flow in a bore moving into quiescent water.

Since $d_1 = d_t$ we therefore have

$$\boxed{c_w^2 = \frac{1}{2} gh \cdot \left(\frac{d_t}{h}\right)^3 \xi\,(\xi + 1)} \tag{5.6.68}$$

Notice that thought d_t/h may be close to 1, the third power of this factor makes it worth keeping it in expression (5.6.68).

Hence the bore-equivalent of the wave speed is **not** $\sqrt{gh}$ as i the linear theory. The reason is that (5.6.68) is an exact solution that contains non-linear terms. If we linearize (5.6.68) by letting $\xi = d_2/d_1 \sim 1$ and $d_t \sim h$ then obviously we get

$$c_w^2 \sim gh \tag{5.6.69}$$

In terms of the parameter a defined by (5.6.6), we get

$$a^2 = \frac{c^2}{gh} = \frac{1}{2}\left(\frac{d_t}{h}\right)^3 \xi\,(\xi+1) \tag{5.6.70}$$

As mentioned above, these results emerge if we require volume and momentum conserved in a frame of reference following the wave on a horizontal bottom as in hydraulics. (5.6.70) is also the expression used in the surfzone in the comparisons with measurements in Fig. 5.5.3 and we see that the results from this formula generally fit the measured values in the surfzone very well.

Exercise 5.6-7

Show that if we introduce the crest elevation η_c/H the expression (5.6.68) can (without further approximations) be written

$$\frac{c^2}{gh} = 1 + \left(-\frac{3}{2} + 3\frac{\eta_c}{H}\right)\frac{H}{h} + \left(\frac{1}{2} - 3\frac{\eta_c}{H} + 3\frac{\eta_c^2}{H^2}\right)\left(\frac{H}{h}\right)^2$$
$$+ \left(\frac{1}{2}\frac{\eta_c}{H} - \frac{3}{2}\frac{\eta_c^2}{H^2} + \frac{\eta_c^3}{H^3}\right)\left(\frac{H}{h}\right)^3 \tag{5.6.71}$$

which shows explicitly how the phase velocity depends on the two parameters H/h and η_c/H.

5.6.6 *The energy dissipation in quasi steady surfzone waves*

The total energy dissipation in a hydraulic jump/bore can be written as

$$\Delta E = -\rho g Q d_1 \frac{(\xi-1)^3}{4\xi} \tag{5.6.72}$$

where Q is the volume flux in the jump. As before we have defined $\Delta E < 0$ for energy loss.

If we observe the wave from a coordinate system moving with speed c_w so that the wave is stationary, then the volume flux, Q, in the flow is constant and given by

$$Q = c_w h \tag{5.6.73}$$

where h is the local depth, and the wave height H is given by

$$H = d_c - d_t = (\xi - 1) d_t \tag{5.6.74}$$

Inserting (5.6.73) and (5.6.74) into (5.6.72) this can then be written

$$\Delta E = -\rho g c_w h \frac{H^3}{4 d_t d_c} \tag{5.6.75}$$

Since the energy dissipation is invariant to a Galilei transformation ΔE is also the energy dissipation in a wave observed from a fixed reference frame. This energy dissipation ΔE is per unit time and it happens once every time a wave passes. We seek the dissipation $\mathcal{D}$ in average over a wave length. Therefore

$$\boxed{\mathcal{D} = \frac{\Delta E}{L} = -\rho g h \frac{H^3}{4 d_t d_c T}} \tag{5.6.76}$$

where also $\mathcal{D} < 0$ for loss as usual. Thus the dimensionless energy dissipation D defined as

$$D = \mathcal{D} \frac{4hT}{\rho g H^3} \tag{5.6.77}$$

can be written as

$$\boxed{D = -\frac{h^2}{d_t d_c}} \tag{5.6.78}$$

Exercise 5.6-8

Show that (5.6.78) can be written

$$D = -\frac{1}{(1 + \frac{\eta_c}{H}\frac{H}{h})(1 + \frac{H}{h}(\frac{\eta_c}{H} - 1))} \tag{5.6.79}$$

The bore similarity was first utilized by LeMehauté (1962). The derivation above was given by Svendsen et al. (1978) and also used by Svendsen (1984a). While no assumptions have been made about the surface shape of the wave, this means that the relative crest elevation η_c/H is an important parameter for the energy dissipation in a breaking wave.

Discussion

The energy dissipation we have found is approximate due to several factors:

(1) The energy dissipation calculated is the amount of wave (or organized mechanical) energy which is turned into turbulent energy between the

two sections under the trough and under the crest. Hence, it neglects the production of turbulent energy after the section under the crest, that is under the rear part of the wave. This is expected to be a small error, but it means that the actual energy dissipation in a wave can be expected to be (a little) larger than given by (5.6.76). See also discussion for D in Section 5.5.

(2) The assumption of depth uniform velocity and hydrostatic pressure may be realistic for the verticals used in the calculation of energy dissipation in a bore, but this is not necessarily the case for the trough and crest sections in a wave.

(3) In the calculation of c_w we neglect the momentum flux due to turbulence, both in the trough and the crest motion. Again a (small) error that would further increase c_w and hence ΔE. For a given value of c_w, however, this does not influence the calculation of ΔE.

5.7 Further analysis of the energy dissipation

Along with the flux of mass, momentum and energy, the wave energy dissipation is one of the most important elements in the surfzone processes. Several other methods have been developed for determining the dissipation in surfzone waves, and the most commonly used are briefly described here.

5.7.1 *Energy dissipation for random waves*

The bore expression (5.6.76) essentially states the dissipation for each breaking wave event (averaged over the wave). In an irregular sea waves of different height breaking starts at different places. On a uniform slope we can assume that the waves will continue to break once they have started. Thus at a given point a certain percentage Q_b% of the (originally highest) waves will be breaking. By averaging over many wave periods we can then determine the mean dissipation rate as a function of Q_b. Thereby we can express the effect of the wave randomness as a dissipation which is a certain fraction of the bore dissipation we would have if the wave height had been constant and equal to, say, the H_{rms}.

Among the contributions in the literature pursuing This idea was first developed by Battjes and Janssen (1978). Thornton and Guza (1983) have given a slightly different version.

The Battjes-Janssen approach

Battjes and Janssen (BJ78 for short) assume that the heights of the unbroken waves follow the Rayleigh distribution given by

$$P(H \leq H_0) = 1 - e^{-\frac{H_0^2}{2\hat{H}^2}} \tag{5.7.1}$$

where $\hat{H}$ is a parameter of the distribution. It is then assumed that as the waves move shoreward they start breaking when their height is H_m which depends on the local depth h. Once the waves have started breaking it is assumed their height remains equal to the value of H_m at each depth. Thus the percentage Q_b of waves breaking is then given by

$$Q_b = P(H > H_m) = e^{-\frac{H_m^2}{2\hat{H}^2}} \tag{5.7.2}$$

The probability that the wave height is smaller than a certain height H_0 then changes to

$$\begin{aligned} P(H \leq H_0) &= 1 - e^{-\frac{H_0^2}{2\hat{H}^2}} \quad \text{for } H_0 \leq H_m \\ &= 1 \qquad \text{for} \qquad H_0 > H_m \end{aligned} \tag{5.7.3}$$

This is used to determine the H_{rms} which, as usual (see Section 3.4.3), is defined by

$$H_{rms}^2 = \int_0^\infty H^2 p(H)\, dH \tag{5.7.4}$$

where $p(H) = dP/dH$ is the probability density function. For $p(H)$ we have

$$\begin{aligned} p(H) &= e^{\frac{H^2}{2\hat{H}^2}} \cdot \frac{H}{\hat{H}^2} && \text{for } H \leq H_m - \epsilon \\ &= \delta(H_m)(1 - P(H_m)) && \text{for } H_m - \epsilon \leq H \leq H_m + \epsilon \\ &= 0 && \text{for } H_m + \epsilon \leq H \leq \infty \end{aligned} \tag{5.7.5}$$

If (5.7.5) is substituted into (5.7.4) we get $H_{rms} = \hat{H}$. However, the probability distribution (5.7.3) gives, using (5.7.2) to express the exponential function in terms of Q_b after some algebra

$$H_{rms}^2 = 2(1 - Q_b)\hat{H}^2 \tag{5.7.6}$$

Exercise 5.7-1

Derive equation (5.7.6)

It is then possible to eliminate $\hat{H}$ from (5.7.2) and (5.7.6). This gives an equation for the breaking probability Q_b in terms of H_m, H_{rms}. This is a transcendental equation in Q_b

$$\frac{1 - Q_b}{\ln Q_b} = -\left(\frac{H_{rms}}{H_m}\right)^2 \tag{5.7.7}$$

Thus it is possible to determine Q_b at each point once the two parameters H_m, H_{rms} are known. Near the shore H_m is often approximated as a fraction γ of the depth h. BJ78 uses a version of Miche's interpolation formula (5.2.1) for arbitrary depth, modified to give $H_m = \gamma h$ in shallow water.

The H_{rms} is then the only parameter in the energy equation (4.2.16) for the random waves which is solved numerically but requires specification of the dissipation rate $\mathcal{D}$.

Equation (5.6.76) is used as a starting point for the assessment of the energy dissipation $\mathcal{D}$ in the breaking waves. However, two approximations are made. The first is to set

$$D = \frac{h^2}{d_c d_t} = 1 \tag{5.7.8}$$

which changes $\mathcal{D}$ to

$$\mathcal{D} = -\frac{1}{4}\rho g \frac{H_m^3}{hT} \tag{5.7.9}$$

The second is to assume $H_m/h \sim 1$ which means the final result for $\mathcal{D}$ is

$$\mathcal{D} = -\frac{\alpha}{4}\rho g \frac{H_m^2}{T} \tag{5.7.10}$$

Notice that this reduces the dependence of $\mathcal{D}$ on H from H^3 to H^2. α is a factor that is introduced to account for inaccuracies and approximations in the form used for $\mathcal{D}$. α is determined by fitting results to measurements.

When averaging over many waves, only Q_b% of which are breaking, the average dissipation at each point becomes

$$\overline{\mathcal{D}} = -\frac{\alpha}{4}\rho g \frac{H_m^2}{T_m} Q_b \tag{5.7.11}$$

where T_m is the mean spectral period ($= 1/f_m$, f being the spectral frequency).

It turns out that for waves passing over a shallow bar Q_b decreases causing a rapid decline of the calcualted dissipation rate, which is physically realistic.

The Thornton-Guza approach

Thornton and Guza (1983) (TG83 for short) also use the bore expression for the energy dissipation but in the more accurate form (5.7.9). However, they also introduce a "breaking coefficient" B to give leeway to adjustment of the results by comparing to measurements. Their dissipation then reads

$$\mathcal{D} = \frac{\overline{f}}{4}\rho g \frac{(BH)^3}{h} \tag{5.7.12}$$

where $\overline{f} = 1/\overline{T}$ represents the mean frequency in the spectrum.

The major difference from BJ78 is in the determination of the probability that the waves are breaking. The choice of a fixed breaking height H_m used by BJ78 is judged not to be realistic. Instead TG83 write the probality function $p_b(H)$ for the breaking waves as

$$p_b(H) = W(H)\, p(H) \tag{5.7.13}$$

where $p(H)$ again is the probability density distribution for the unbroken waves, and $W(H)$ is the probability that a wave is breaking. A Rayleigh distribution for $p(H)$ is in fact found to give remarkably good estimates.

Using extensive analyses of, and comparison with, field data two forms for $W(H)$ are suggested

$$W(H) = \left(\frac{H_{rms}}{\gamma h}\right)^n \tag{5.7.14}$$

and

$$W(H) = \left(\frac{H_{rms}}{\gamma h}\right)^n \left[1 - \exp\left(-\left(\frac{H}{\gamma h}\right)^2\right)\right] \tag{5.7.15}$$

In contrast to BJ78 these expressions imply a more complicated variation of the height of the breaking waves. Therefore the longterm averaged energy dissipation is calculated as

$$\overline{\mathcal{D}} = \int_0^\infty \mathcal{D}(H)\, p_b(H)\, dH = \frac{B^3}{4} \rho g \frac{\overline{f}}{h} \int_0^\infty H^3\, p_b(H)\, dH \tag{5.7.16}$$

The result for $\mathcal{D}$ for the first weighting function (5.7.14) is given by

$$\overline{\mathcal{D}} = \frac{3\sqrt{\pi}}{16} \rho g \frac{B^3 \overline{f}}{\gamma^4 h^5} H_{rms}^7 \tag{5.7.17}$$

Similarly the result for (5.7.15) becomes

$$\overline{\mathcal{D}} = \frac{3\sqrt{\pi}}{16} \rho g \frac{B^3 \overline{f}}{\gamma^2 h^3} H_{rms}^5 \left[1 - \frac{1}{(1 + (H_{rms}/\gamma h)^2)^{5/2}} \right] \tag{5.7.18}$$

For a discussion of these results see TG83.

Again these expressions are used in the energy equation (4.2.16) which - it turns out - for the first case can be solved analytically. TG83 also include bottom friction and find as expected that the loss from bottom friction is negligible. An exception is, though, the region closest to the shore where the general scale of the wave motion becomes very small due to the vanishing depth.

Thus it is possible with relatively simple measures to establish wave models that include the most important effect of irregularitiy of the waves, namely that the breaking point changes with the wave height. The effect this has on the nearshore wave climate is analyzed further in Chapter 14, where it is shown how short wave energy is transformed into long wave energy through the breaking process, so called Infra Gravity (IG) waves.

5.7.2 *Energy dissipation with a threshold*

Dally et al. (1985) observed that if H/h decreased below a certain level (roughly between 0.35 and 0.40), the waves stop breaking. They used this to assume that below that level the waves will become stable even if they were originally breaking. Thus—using linear wave results—they define the empirical concept: the stable energy flux $(Ec_g)_s$ which is defined as the energy flux at $H/h = 0.35$–0.40.

They then assume that the energy dissipation at any point is given by

$$\mathcal{D} \propto \frac{-\kappa}{h} \left(Ec_g - (Ec_g)_s \right) \tag{5.7.19}$$

Here Ec_g is essentially the energy flux E_f in the wave motion (sine waves implicitly assumed) and κ is a dimensionless decay coefficient adjusted to fit experimental data.

The value of $(Ec_g)_s$ is determined so that $Ec_g = (Ec_g)_s$ for $H/h = 0.35 - 0.40$, which causes the dissipation to vanish for H/h below that limit.

The dissipation in (5.7.19) differs from the bore dissipation (5.6.76) by assuming $\mathcal{D} \propto E_f/h$. It is not clear how this is justified physically. It implies that the dissipation varies as H^2/h versus H^3/h for the bore. However, the comparisons with measurements show that with a proper selection of the empirical constants the difference from the coefficient-free bore expression is not so significant and the method has been extensively used e.g. in the REF/DIF wave model.

The merit of the model, however, lies in the recognition of the threshold in the wave height to water depth ratio beneath which waves usually stop breaking. This particular feature is realistic for example when waves pass over a bar into the deeper water behind in which case breaking usually stops quite abruptly. It may be noticed that the idea can readily be incorporated in many other dissipation expressions.

5.7.3 *A model for roller energy decay*

From the description in Section 5.6.6 it is clear that the roller concept is not needed to determine the total dissipation as e.g. in a bore. However, it is also clear that the roller contributes significantly to the time averaged properties such as volume, momentum and energy flux. Therefore it is necessary to develop a tool that keeps track of how the roller area develops through the surfzone. The two enmpirical relations (5.5.15) and (5.5.16) represent such a tool.

An alternative approach was initiated by Nairn et al. (1990) and further developed by Reniers and Battjes (1997). It establishes a separate equation for the solution of the roller energy E_r using an empirical formula for the energy dissipation of the roller energy given by Duncan (1981).

The evolution of the total wave energy is still controlled by the energy equation for the wave motion (4.4.14) developed by Phillips (1966, 1977), and often the dissipation rate in this equation is approximated by the bore dissipation (5.6.76).[5]

[5]The concept of the roller as a mass of fluid carried with the wave implies that there must be a shear stress along the boundary between the roller and the flow underneath

The evolution of the roller energy then provides information about the area of the roller and thereby the roller contribution to time averaged wave parameters, in particular the radiation stress.

5.7.4 *Advanced computational methods for surzone waves*

In recent years powerful computational fluid dynamics (CFD) methods have been brought to bearing on the problem of modelling the broken waves in the surfzone. As mentioned in Section 5.3 the very successful Boundary Element (or rather Boundary Integral) method breaks down for fundamental mathematical reasons when the jet hits the surface in front of the wave crest. Entirely different approaches are needed to deal with the highly turbulent flow in the surfzone, where even the definition of where the surface is becomes an important problem.

The methods applied all deal with two major problems. One is to describe in some way the turbulence and its effect on the flow. In the surfzone we have high intensity turbulence (relative to the mean flow) that also has large scale fluctuations in both time and space. Even defining the mean flow is not trivial (see also the discussion in Section 2.5).

The second problem is developing a method for identifying a highly irregular and rapidly fluctuating free surface (or more generally a fluid interface) in a computatinal grid and tracking its motion in time and space. Since both these problems are present in the surfzone any efficient model must deal with both.

In addition we are facing the numerical problem of solving the Navier-Stokes (NS) or Reynolds equations in more or less unabridged form. Though numerical techniques are now available for this task they are not simple.

The following is just a very brief list of examples of various types of modelling efforts for breaking waves. A review of the application of the methods is given by Bradford (2000). A more thorough introduction to the general methods of turbulence modelling is found in Wilcox (1998).

that keeps the roller from sliding down the front. It was suggested by Deigaard and Fredsøe (1989) to calculate the energy dissipation as the work done by the shear stress just along that boundary as if those were the only shear stresses in the flow. This approach has been used in a number of applications. The model, however, is not in accordance with the actual fluid dynamic processes where energy dissipation takes place over the entire turbulent region, both inside and outside the roller.

Surface tracking methods

A common method for dealing with the free surface is the Volume Of Fluid (VOF) method, introduced by Hirt and Nichols (1981) using an algorithm for solving the NS equations called SOLAVOF. The VOF method is designed for fluid interfaces in general and is here used for the air-water interface. It represents the position of the interface in a grid cell by the fraction of the cell occupied by water. The VOF method has been improved through many developments that seek higher accuracy. Some of those are included in the review of the applications to breaking waves given below. In one such development leading to a SL-VOF by Guignard et al. (2001) the NS-equations are solved directly using a pseudo-compressibility method in curvilinear coordinates.

Traditional turbulence approach (RANS methods)

The traditional turbulence approach is to determine turbulence parameters - particularly the time and space varying eddy viscosity - by two equation or even higher order turbulence models. Petit et al. (1994) used a constant eddy viscosity while Lemos (1992) used a standard $k - \epsilon$ model. Another reference using a standard $k-\epsilon$ model is Christensen et al. (2000).

More recently Lin and Liu (1998a,b) solved the Reynolds Averaged NS (RANS) equations using a $k - \epsilon$ model along with a VOF method. Instead of the usual linear relationship for the Reynolds stresses, they introduced a nonlinear, but still algebraic, relationship which helped predict the initiation of the breaking.

A characteristic for the results of RANS-models seem to be that while many of the surface features are quite well represented the initiation of breaking is not and the undertow is underestimated. This is an indication that the mass flux in the surfzone waves is underestimated, an inaccuracy which originates in the turbulent breaking region at the surface.

LES methods

Large Eddy Simulation (LES) methods do not use Reynolds averaging but use a filtering technique to develop the turbulence equations. In essence this method implies that only the small scale turbulence is modelled by empirical relations such as the Reynolds stress model by means of an eddy viscosity. For this is often used a model suggested by Smagorinsky (1963). The larger eddies are represented directly in the numerical simulations. Examples are Zhao and Tanimoto (1998), Christensen and Deigaard (2001), Zhao et al. (2003).

5.8 Swash

The wave breaking that occurs on the shoreline was described qualitatively in Section 5.3. The phenomenon of swash is the runup on the beach which occurs as a result and it requires a whole chapter in itself in the theoretical analysis of nearshore wave motion. The reason is that the two basic assumptions underlying many of the wave theories available fail completely to be satisfied in the swashzone even to the first approximation. Therefore the simple sine wave, or even traditional Boussinesq wave theories cannot be expected to give even a first approximation without close scrutiny of the conditions. The two conditions that are so radically different are

- The wave height is not small in comparison to the water depth. In fact in the runup process the wave height can be many times the water depth, which is different from the rest of the surfzone where we can get at least a first approximation by recognizing that the wave height is not negligible.
- The change in depth is not small over a characteristic horizontal length for the motion. It is even dificult to define the traditional parameters such as depth and wave length. Therefore the depth is not "slowly varying" and the locally constant depth assumption cannot be justified. The change in depth is so rapid, seen from the moving wave, that no local equilibrium has a chance to get established.

These conditions for the swash motion essentially require that we start from the basics.

It turns out that there is one version of the "processed" equations for nearshore wave motion that comes close to satisfying the conditions in the swash zone. That is the special version of the long wave equations called the Nonlinear Shallow Water (NSW) equations described in Section 9.12. These equations, which can be viewed as a limiting case of the Boussinesq equations, assume $H/h = O(1)$. They assume the characteristic length scale λ for the wave motion is very long in comparison to the depth (which is **not** in keeping with the conditions near the toe of the swash). However, this leads to hydrostatic pressure in the vertical direction and a velocity variation which is constant over depth and, in a modified form discussed below, that is in reasonable agreement with the conditions observed in the swash region. The only difference is that the hydrostatic pressure may be in the direction normal to the slope rather than in the vertical direction (which is not a quite negligible difference on a relatively steep slope).

The use of the NSW equations for analysis of swash has been pursued quite extensively for many years and the literature is rich with publications.

1DH swash on a frictionless slope

On the relatively steep slopes found in the swashzone on natural beaches the hydrostatic pressure occurs in the direction normal to the slope rather than in the vertical direction assumed in the usual NSW-equations. Thus the proper equations are a modified version of the normal NSW-equations.

To derive the equations we use a coordinate system with the x-axis along, the z-axis normal to the slope as shown in Fig. 5.8.1. This figure also shows the definition of the variable used.

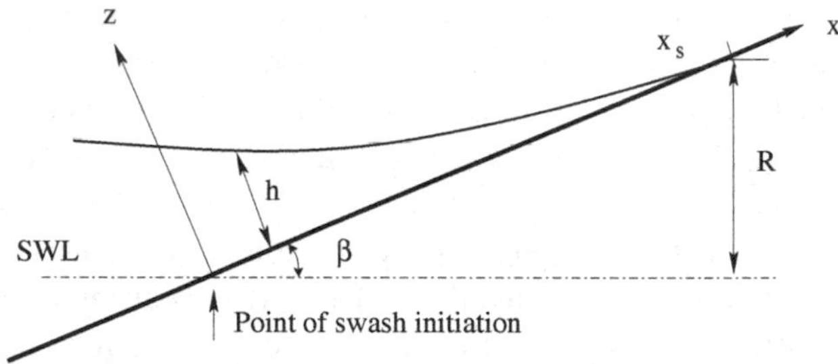

Fig. 5.8.1 The coordinate system and definition of variables used in the swash solution. x_s represents the maximum runup.

Along the axes gravity has the components

$$g_x, g_z = g \ \sin\beta, g \ \cos\beta \tag{5.8.1}$$

The approaching wave is either breaking or close to. Often it is considered a bore (Shen and Meyer, 1963, and others). However, as described in Section 5.3, at the moment the water depth in front of the wave becomes zero (or nearly so) the wave collapses and shoots up the beach forming the swash motion. In the solution this is supposed to happen at the undisturbed shorline $x = 0$ and the velocity in the ensuing swash motion is assumed to be uniform over depth.

The conservation of volume at a point of this motion is best represented by the depth integrated version of the continuity equation which reads

$$\boxed{\frac{\partial h}{\partial t} + \frac{\partial}{\partial x}(uh) = 0} \tag{5.8.2}$$

Since the motion is mostly parallel to the slope we also have that the velocity w in the slope-normal direction is so small that the normal acceleration dw/dt can be neglected. Neglecting also the friction the motion is governed by the Euler equations. The z-component then becomes

$$0 = -\frac{1}{\rho}\frac{\partial p}{\partial z} - g\cos\beta \tag{5.8.3}$$

with the boundary condition $p(h) = 0$ at the free surface. Integrated (5.8.3) then yields

$$p = \rho g(h - z)\cos\beta \tag{5.8.4}$$

which represents the alledged hydrostatic pressure in the slope-normal direction with the reduced gravity $g_x = g\cos\beta$.

In the slope-parallel x-direction we then get

$$\frac{du}{dt} = -\frac{1}{\rho}\frac{\partial p}{\partial x} - g\sin\beta \tag{5.8.5}$$

Using (5.8.4) and expanding the acceleration therm this can be written

$$\boxed{\frac{\partial u}{\partial t} + u\frac{\partial u}{\partial x} + g\cos\beta\frac{\partial h}{\partial x} = -g\sin\beta} \tag{5.8.6}$$

Equations (5.8.2) and (5.8.6) are the version of the Nonlinear shallow water equations (NSW) that governs the frictionless motion of the swash.

Those are essentially the equations solved by Shen and Meyer (1963) with direct reference to swash. However, there are numerous other references to the discussion for the wave motion on steep slopes. In particular is mentioned the analytical solution of similar equations for non-breaking, fully reflected waves by Carrier and Greenspan (1958).

The solution by Shen and Meyer uses the method of characteristics. A summary of results for the entire runup flow are given by Peregrine and Williams (2001). To derive this the equations are written on characteristic form. The procedure for that is outlined in Section 11.9. However, the details of the method are beyond the scope of the of the present text.[6]

The initial and boundary conditions for the Shen and Meyer solution, which are an idealization of the collapse of the bore similar to a moving dam break, cause the equations to be singular at the origin $x, t = 0, 0$ of

[6] A detailed description of the method of characteristics for flows with a horizontal bottom is given by Abbott and Basco (1989). The method is also described briefly in Johnson (1997).

the motion. The method of characteristics allows this, but this idealization means that for small x, t the solution is not an accurate representation of the real swash flow (Peregrine and Wiliams, 2001).

However, it was pointed out already by Shen and Meyer that surprisingly the entire swash flow in their solution depends very little on the boundary conditions after the initial collapse of the bore. To quote Meyer and Taylor (1972), the solution "forgets its boundary conditions after a short finite time". This particularly applies to the region near the tip of the runup. See also Meyer and Taylor (1972) and many references quoted therein for an extensive discussion of this issue. This makes the ballistic solution described below for the movement of the tip of the swash more relevant.

The NSW equations were solved numerically by Hibberd and Peregrine (1979) for the swash of a solitary wave using a dissipative Lax-Wendroff scheme, which considers the breaking waves as bores and represent them in the solution as discontinuities. Such schemes also allow the formation of new bores inside the solution. Hibberd and Peregrine found that frequently a new bore is formed during the downrush phase of the flow.

Similar numerical approaches have since been further explored extensively in the literature, see e.g. Kobayashi et al. (1989), Kobayashi and Wurjanto (1992). More advanced methods for "bore capture" such as the "Weighted Average Flux" method (WAF) has been used by Brocchini et al. (2001), Hubbard and Dodd (2002) and Brocchini and Bellotti (2002) to mention a few.

The ballistic model for the tip of the runup

Another feature of the Shen and Meyer solution, is that near the tip of the runup the swash layer becomes very thin and actually the depth h approaches zero tangentially at the tip. This means that near the tip the motion is similar to a frictionless ballistic motion of a particle shooting up the slope with a velocity U_0.

This has lead to an approximate formulation of the motion of the runup tip which also allows analysis of cases that include friction. The weak variation with x near the tip makes it possible to neglect the nonlinear term in (5.8.6) and also the term proportional to $\partial h/\partial x$ originating from the pressure variation along the tip. It may also be assumed that the waves are incident on the beach at an angle α_w relative to the shore normal so that the velocity in the swash has the components u, v. For a long straight beach and adding further a friction term (5.8.6) can then be written

$$\frac{\partial u}{\partial t} = -g \sin\beta - \frac{1}{2}\frac{f}{h}u|U| \tag{5.8.7}$$

together with the longshore component

$$\frac{\partial v}{\partial t} = -\frac{1}{2}\frac{f}{h}v|U| \tag{5.8.8}$$

Here $U = (u^2 + v^2)^{1/2}$ is the total velocity of the flow. Because of the similarity with a heavy particle moving under the only influence of gravity and friction it is often called the **ballistic solution** but is also often just accredited to Shen and Meyer.

This is again a nonlineear equation. However, a linearization is natural if we assume that the friction only changes the speed of the particle slightly over a cycle. Then U can be considered constant (equal for example to the starting value U_0) and the equations become

$$\begin{aligned} \frac{\partial u}{\partial t} &= -g \sin\beta - \frac{1}{2}\frac{f}{h}u|U_0| \\ \frac{\partial v}{\partial t} &= -\frac{1}{2}\frac{f}{h}v|U_0| \end{aligned} \tag{5.8.9}$$

These two linear equations are uncorrelated and hence can be solved independently using standard methods for ordinary linear differential equations. The solution can be written

$$\begin{aligned} u &= -\frac{g \sin\beta}{C_f} + \left(U_0 \cos\alpha_w + \frac{g \sin\beta}{C_f}\right) e^{-C_f t} \\ v &= U_0 \sin\alpha_w \, e^{-C_f t} \end{aligned} \tag{5.8.10}$$

where $C_f = fU_0/2h$. For the coordinate x_s of the tip of the runup we get by integrating (5.8.10)

$$x_s = \left(\frac{U_0}{C_f}\cos\alpha_w + \frac{g_x}{C_f^2}\right)(1 - e^{-C_f t}) - \frac{g_x t}{C_f} \tag{5.8.11}$$

Fig. 5.8.2 shows the variation with time of the 1DH rupup for some values of the friction factor. The water depth in the expression for C_f has (somewhat arbitrarily) been set to $0.2U_0^2/2g$ and U_0 has been chosen to 2 m/s. It is seen that the friction reduces not just the magnitude of the runup, which might be expected, but also the duration of each swash event. Surprisingly the larger the bottom friction the shorter the total swash duration T_s, counted as the time between start of runup and time where the tip is back at the SWL. For $x_s = 0$ (5.8.11) gives

$$0 = \left(\frac{U_0}{C_f}\cos\alpha_w + \frac{g_x}{C_f^2}\right)(1 - e^{-C_f T_s}) - \frac{1}{2} g_x T_s^2 \tag{5.8.12}$$

which is an transcendental equation for T_s. For small values of the parameter C_f we can expand the exponential function and we need to keep terms to order $(C_f T_s)^2$. For $C_f = 0$ we then get, assuming also shorenormal waves ($\cos\alpha_w = 1$)

$$T_s = \frac{2U_0}{g_x} = T_{s0} \tag{5.8.13}$$

For small values of C_f we get

$$T_s = T_{s0} \frac{1}{1 + \frac{U_0 C_f}{g_x}} \tag{5.8.14}$$

which confirms, since $C_f \propto f$, that the friction *decreases* the swash duration.

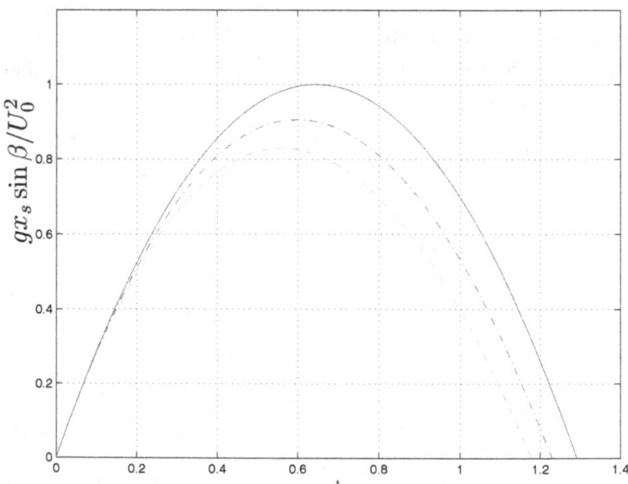

Fig. 5.8.2 The position of the swash tip versus time for three different value of the friction factor f for the solution given by (5.8.11). In the calculation of C_f has been used a depth of $h = 0.2\ U_0^2/2g$ where U_0 has been set to 2 m. The three curves correspond to $f = 0$ (–), $f = 0.01$ (-.-.-), and $f = 0.02$ (- - -).

Connection to the incident waves

The maximum runup predicted by the ballistic solution is obtained from (5.8.11). With no friction and shore-normal waves we get along the slope

$$x_{s,max} = \frac{U_0^2}{2g_x} \tag{5.8.15}$$

or, for the vertical runup $R = x_{s,max}\ \sin\beta$ we get

$$R = \frac{U_0^2}{2g} \tag{5.8.16}$$

The question of how to link the initial speed U_0 at the shoreline to the incident wave motion remains open. U_0 clearly depends on the nature of the bore collaps. However we have to remember that the notion of the actual wave motion represented by a bore collapse is an idealization, which can cover situations from waves not or barely breaking (as in the Carrier and Greenspan solution) to the type of swash described in Fig. 5.3.5. In waves with an infragravity component included the runup of the individual short waves also changes over the IG wave period. In order to cover this level of variability it is convenient to introduce a dimensionless empirical factor that can be adjusted according to the situation, and Baldock and Holmes (1999) suggested a coefficient C defined as

$$C = \frac{U_0}{\sqrt{gH_B}} \tag{5.8.17}$$

where H_B is the perceived height of the incident waves at the shoreline. Introducing this into (5.8.16) gives

$$R = \frac{C^2 H_B}{2} \tag{5.8.18}$$

Simultaneous measurements of R and H_B then essentially provides values of C. Baldock and Holmes find that $C \sim 1.8 - 2$ fit measurements in both regular and irregular waves. To interpret the meaning of C we should realize that this coefficient accounts for a number of effects.

However, even after we have determined C the task remains of determining H_B (or U_0) for the incident wave motion and it is not obvious how that is done. For example the expression $\sqrt{gH_B}$ does not represent the particle or phase velocity according to any known wave theory. In a linear wave on constant depth the phase velocity would be $\sqrt{gh_B}$ where h_B is the depth, which is not helpful for guidance because the water depth at the shore is 0. The value of $\sqrt{gH_B}$ is also not related to the bore velocity c_b given by expression (5.6.63) or the equivalent wave phase speed (5.6.68) which both tend to zero too for $h \to 0$. Finally it is recalled that all mild-slope wave models except the NSW equations predict that the wave height goes to zero as the depth goes to zero at the shoreline.

Though there are many analyses of swash in the literature a simple procedure for connecting H_B (or U_0) to the characteristics of the incoming waves which is applicable to engineering problems is still to come. Potentially a Boussinesq model for breaking waves could yield meaningfull results because the Boussinesq equations actually are valid for quite steep bottom slopes, but this has not been tried yet. In the meantime numerical simulations with the NSW equations using the appropriate software or more advanced CFD methods such as the VOF method seem the most reliable approach. That will also provide more detailed information about the entire flow field in the swash motion, not just the tip of the runup.

Applications and comparisons with measurements

In particular the ballistic (or Shen and Meyer) solution has been investigated quite extensively in recent years in attempts to test the theory.

The field measurement by for example Raubenheimer and Guza (1996), Raubenheimer (2002), Holland and Puleo (2001), Hughes and Baldock (2004) are important sources of information. So are laboratory experiments that have been reported. Reference is made to e.g. Mase (1988), Baldock and Holmes (1999).

A central issue given much attention is the possible value of the friction factor f and its effect on the ballistic solution. Typical values found for turbulent boundary layers are reported in Chapter 10. However, in the swash motion boundary layers do not have time to develop fully which influences the value of f in a friction factor approximation, and so does beach porosity (Packwood, 1983). In addition, as the solution (5.8.10) or (5.8.11) shows, to evaluate the results the parameter C_f must be known and this requires information about the depth h in the swash motion. The ballistic solution does not provided by this information. Recent references where these problems are discussed include Hughes (1995), Puleo and Holland (2001) and Raubenheimer (2002).

As can be seen the swashzone mechanics is a highly active research area with several unsolved questions and new publications appearing every year.

Therefore the reader interested in this area is particularly advised to make a fresh literature search for the latest information.

5.9 References - Chapter 5

Abbott, M. B. and D. R. Basco (1989). Computational fluid dynamics. An introduction for engineers. Longman Scientific and Technical/John Wiley and Sons. pp 425.

Baldock, T. E. and P. Holmes (1999). Simulation and prediction of swash oscillations on a steep beach. Coastal Engineering, **36**, 219-242.

Battjes, J. A. (1974). Computation of set-up, longshore currents, run-up and overtopping due to wind-generated waves. Delft Univ. Tech. PhD-dissertation.

Battjes, J. A. and J. P. F. M. Janssen (1978). Energy loss and set-up due to breaking of random waves. ASCE Proc. 16th Int. Conf. Coast. Engr., Chap 32, 569–582.

Boussinesq, J. (1871). Theorie de l'intumscence liquid appeleé onde solitaire ou de translation. Comptes Rendus Ac. Sci., **72**, 755.

Bradford, S. F. (2000). Numerical simulation of surfzone dynamics. ASCE J. Waterw., Port, Coast. and Ocean Engrg., **126**, 1, 1–13.

Brocchini M., R. Bernetti, A. Mancinelli and G. Albertini (2001). An efficient solver for nearshore flows based on the WAF method. Coastal Engrg., **43**, 105-129.

Brocchini, M. and G. Bellotti (2002). Integral flow properties of swash zone and averaging. Part 2. Shoreline boundary conditions for wave averaged models. J. Fluid Mech., **458**, 269-281.

Carrier, G. F. and H. P. Greenspan (1958). Water waves of finite amplitude on a sloping beach. J. Fluid Mech., **4**, 97–109.

Christensen, E. D., J. H. Jensen and S. Mayer (2000). Sediment transport under breaking waves. ASCE Proc. 27th Int. Conf. Coast. Engrg., Australia, 2767–2780.

Christensen, E. D. and R. Deigaard (2001). Large eddy simulation of breaking waves. Coastal engineering, **42**, 53–86.

Cokelet, E. D. (1977). Steep gravity waves on arbitrary uniform depth. Phil. Trans. Roy. Soc. Lond., **286**, 183–230.

Cox, D. T., N. Kobayashi, and A. Okayasu (1995). Experimental and numerical modeling of surfzone hydrodynamics. Res. Rep. No. CACR-95-07, University of Delaware. 293 pp

Dally, W. R., R. G. Dean and R. A. Dalrymple, (1985). Wave height variation across beaches of arbitrary profile. J. Geophys. Res. **90**, p.11917,

Davies, T. V. (1952). Symmetrical, finite amplitude gravity waves. In: Gravity Waves, Nat. Bur. Stan., Wash DC. Circ no 521, 55–60.

Deigaard, R. and J. Fredsøe (1989). Shear stress distributions in dissipative water waves. *Coastal Engrg.* **13**, 357–378.

Dold, J. and D. H. Peregrine (1986). An efficient boundary integral method for steep unsteady water waves. Numerical methods for fluid dynamics II (K. W. Morton and M. J. Baines, eds.), Clarendon Press, Oxford, 671–679.

Duncan, J. H. (1981). An experimental investigation of a wave breaking produced by a towed hydrofoil. Proc. Roy. Soc. Lond. A **377**, 331-348.

Duncan, J. H. and A. A. Dimas (1996). Surface ripples due to steady breaking waves. J. Fluid Mech., **329**, 309–339.

Fenton, J. D. (1972). Ninth order solution for solitary wave. J. Fluid Mech., **53**, 2, 257–271.

Galvin, C. J. (1968). Breaker type classification on three laboratory beaches. J. Geophys. Res., **73**, 12.

Grilli, S. T. and R. Subramanya (1994). Numerical modeling of wave breaking induced by fixed and moving boundaries. Comp. Mech. **17**, 6, 374–391.

Grilli, S. T., I. A. Svendsen and R. Subramanya (1997). Breaking criterion and characteristics for solitary waves on slopes. ASCE J. Waterway, Port, Coastal and Ocean Engineering, **123**, 3, 102–112.

Guignard, S., R. Marcer, V. Rey, C. Kharif, P. Fraunié (2001). Solitary wave breaking on sloping beaches: 2-D two phase flow numeriacl simulation by SL-VOF method. Europ. J. Mech., B - Fluids, **20**, 57–74.

Gwyther, R. F. (1900). The classes of long progressive waves. Phil. Mag., **50**, 5, 213.

Hansen, J. B. (1990). Periodic waves in the surf zone: Analysis of experimental data. *Coast. Eng.*, **14**, 14–41.

Hansen, J. B. and I. A. Svendsen (1979). Regular waves in shoaling water, experimental data. Inst. Hydrodyn. Hydraul. Eng. (ISVA), Tech. Univ. Denmark, Lyngby, Ser. Pap., 21.

Hansen, J. B. and I. A. Svendsen (1986). Experimental investigation of the wave and current motion over a longshore bar. ASCE Proc. 20th Int. Conf. Coast. Engrg., Ch. 86, 1166–1179.

Hattori, M. and T. Aono (1985). Experimental study on turbulence structures under spilling breakers. In The ocean surface (Toba and Mitsuyasu eds.) Reidel Publ Comp., Dordrecht.

Havelock, E. T. (1918). Periodic irrotational waves of finite height. Proc. Roy. Soc. Lond., A **95**, 38–51.

Hirt, C. W. and B. D. Nichols (1981). Volume Of Fluid method for the dynamics of free boundaries. J. Comp. Phys., **39**, 323–345.

Holland, K. T. and J. A. Puleo (2001). Variable swash motions associated with foreshore profile change. J. Geophys. Res., **106**, 4613–4623.

Horikawa, K. and C. Kuo (1966). Wave transformationn after breaking point. ASCE Proc. 10th Int. Conf. Coastal Engrg., Tokyo, Chap 15, 217–233.

Hubbard, M. E. and N. Dodd (2002). A 2D numerical model of wave run-up and overtopping. Coastal Engineering, **47**, 1, 1-26.

Hughes, M. G. (1992). Application of a non-linear shallow water wave theory to swash following bore collapse on a sandy beach. J. Coast. Res., **8**, 562–578.

Hughes, M. G. (1995). Friction factors for wave uprush. J. Coast. Res., **11**, 4, 1089–1098.

Hughes, M. G. and T. E. Baldock (2004). Eulerian flow velocities in the swash zone: Field data and model predictions. J. Geophys. Res., **109**, C080009, doi:10, 10291/2003JC002213

Iversen, H. W. (1952). Laboratory study of breakers. In: Gravity waves, Nat. Bur. Stan., Wash DC., Circ no 521, 9–32.

Iwagaki, Y. and T. Sakai (1976). Representation of particle velocity of breaking waves on beaches by Dean's stream function. Memoirs of The Faculty of Engineering, Kyoto Univ. **38**, 1.

Johnson, R. S. (1997). A Modern Introduction to the Mathematical Theory of Water Waves. Cambridge Univ. Press.

Kjeldsen, S. (1968). Wave breaking. Physical description. Dissertation in Danish. Coastal Engineering Laboratory, Tech. Univ. Denmark.

Kobayashi, N., G. S. DeSilva, and A. Wurjanto (1989). Wave transformation and swash oscillation in gentle and steep slopes. J. Geophys. Res., **94**, 951–966.

Kobayashi, N. and A. Wurjanto (1992). Irregular wave setup and runup on beaches. ASCE J. Waterways, Port, Coast. and Ocean Engrg., **118**, 368–386.

LeMehauté, B. (1962). On the non-saturated abreakers and wave run-up, ASCE Proc. 8th Int. Conf. Coast. Engrg. 77–92.

Lemos, C. M. (1992). A simple numerical technique for turbulent flows with free surfaces. Int. Jour. Num. Meth. Fluids, **15**, 127–146.

Lighthill, J. (1978). Waves in fluids. Cambridge Univ. Press.

Lin, P. and P. L.-F. Liu (1998a). A numerical study of breaking waves in the surfzone. J. Fluid Mech., **359**, 239–264.

Lin, P. and P. L.-F. Liu (1998b). Turbulence transport, vorticity dynamics, and solute mixing under plunging breaking waves in the surfzone. J. Geophys. Res., **103**, C8, 15677–15694.

Mase, H. (1988). Spectral characteristics of random wave runup. Coast. Engrg., **12**, 175–189.

Longuet-Higgins, M. S. (1992). Capillary rollers and bores. J. Fluid Mech., **240**, 659–679.

Longuet-Higgins, M. S. and J. D. Fenton (1974). On mass, momentum, energy and circulation of a solitary wave, II. Proc. Roy. Soc. Lond. A **340**, 471–493.

Mase, H. (1988). Spectral characteristics of random wave-runup. Coastal Engrg., **12**, 175–189.

McCowan, J. (1894). On the highest wave of permanent type. Phil. Mag. **38**, 351–357.

Meyer, R. E. and A. D. Taylor (1972). Run-up on beaches. In Waves on Beaches (ed. R. E. Meyer), 357-411.

Mitchell, J. H. (1893). On the highest wave in water. Phil. Mag. Series 5, **36**, No 222, 430–437.

Miche, M. (1944). Mouvement ondulatoires de la mer. Annales des ponts et chaussee, **114**, 25–78, 131–164, 270–292, 369–406.

Nadaoka, K. and T. Kondo (1982). Laboratory measurements of velocity field structure in the surfzone by LDV. Coastal Engrg. Japan, **25**, 125–145.

Nairn, R. B., J. A. Roelvink, and H. N. Southgate (1990). Transition zone width and implications for modelling surfzone hydrodynamics. ASCE Proc. 22nd Int. Conf. Coast. Engrg., 68–81.

Okayasu, A. (1989) Characteristics of turbulence structures and undertow in the surf zone. PhD Thesis, Yokohama Nat. Univ.

Okayasu, A. T., Shibayama, and N. Nimura (1986). Velocity field under plunging waves. ASCE Proc. 20th Int. Conf. Coast. Engrg., Ch. 116, 1580–1594.

Otta, A., I. A. Svendsen and S. T. Grilli (1992). The breaking and run-up of solitary waves. ASCE Proc. 23th Int. Conf. Coast. Engrg., 660–674.

Packham, B. A. (1952). The theory of symmetrical gravity waves on finite amplitude: II, the solitary wave. Proc. Roy. Soc. Lond., A **213**, 238.

Packwood (1983). The influence of beach porosity on wave uprush and backwash. Coast. Engrg., **7**, 29-40.

Peregrine, D. H. and I. A. Svendsen (1978). Spilling breakers, bores and hydraulic jumps. ASCE Proc. 16th Int. Conf. Coast. Engrg., Chap 30, 540–550.

Peregrine, D. H. and S. M. Williams (2001). Swash overtopping a truncated plane beach. J. Fluid Mech., **440**, 391–399

Petit, H. A., P. Tonjes, M. R. A. Van Gent, and P. Van Den Bosch (1994). Numerical simulation and validation of plunging breaker using a 2D Navier Stokes model. ASCE Proc. 24th Int. Conf. Coast. Engrg., 522–524.

Phillips, O. M. (1966, 1977). Dynamics of the upper ocean. Cambridge University Press.

Puleo, J. A. and K. T. Holland (2001). Estimating swash zone friction coefficients on a sandy beach. Coast. Engrg., **43**, 25–40.

Raubenheimer, B. (2002). Observations and predictions of fluid velocities in the surf and swash zones. J. Geophys. Res., **107**, C10, 3190.

Raubenheimer, B. and R. T. Guza (1996). Observations and predictions of runup. J. Geophys. Rs., **101**, 25575–25587.

Reniers, A. J. H. M. and J. A. Battjes (1997). A laboratory study of longshore currents over barred and non-barred beaches. Coast. Engrg., **30**, 1–22.

Shen, M. C. and R. E. Meyer (1963). Climb of a bore on a beach, part 3. Run-up. J. Fluid Mech., **16**, 113–125.

Smagorinsky, J. (1963). General circulation expriments with primitive equations. I. The basic experiment. Mon. Weather Rev., **91**, 99–164.

Stive, M. J. F. and H. J. deVriend (1994). Shear stresses and mean flow in shoaling and breaking waves. Proceedings of the 24th International Conference on Coastal Engineering, 594-608.

Stive, M. J. F. and H. Wind (1982). A study of radiation stress and set-up in the nearshore region. *Coast. Eng.* **6**, 1, 1–26.

Stive, M. J. F. (1984). Energy dissipation in waves breaking on gentle slopes. Coastal Engineering, **8**, 99–127.

Stokes, G. G. (1880). On the theory of oscillatory waves, mathematical and physical papers **1**, 225–228. Cambridge Univ. Press.

Svendsen, I. A. (1984). Wave height and set-up in the surf zone. *Coast. Eng.*, **8**, 4, 303–329.

Svendsen, I. A. (1986). Mass flux and undertow in a surf zone, reply to discussion by W. R. Dally and R. G. Dean. Coast. Emgrg., **10**, 299-307.

Svendsen, I. A. (1987). Analysis of surfzone turbulence. J. Geophys. Res. **92**, C5, 5115–5124.

Svendsen, I. A. and J. B. Hansen (1976). Deformation up to breaking of periodic waves on a beach. ASCE Proc. 15th Int. Conf. Coast. Engrg., Chap 27, 477–496.

Svendsen, I. A., P. A. Madsen and J. B. Hansen (1978). Wave characteristics in the surf zone. *Proc. 16 ICCE*, Hamburg, Chap. 29, p. 529–539.

Svendsen, I. A. and J. B. Hansen (1986). The interaction of waves and currents over a longshore bar. ASCE Proc. 20th Int. Conf. Coast. Engrg., Ch. 116, 1580–1594.

Svendsen, I. A. and S. T. Grilli (1990). Nonlinear waves on steep slopes. J. Coast. Res., S17, 185–202.

Svendsen, I. A. and U. Putruvu, (1993). Surf zone wave parameters from experimental data. Coastal Engineering, **19**, 283 - 310.

Svendsen, I. A. and J. Veeramony (2001). Wave breaking in wave groups. J. of Wtrwy., Port, Coast., and Oc. Engrg., ASCE, **127**(4), 200-212.

Svendsen, I. A., W. Qin, and B. A. Ebersole (2003). Modelling waves and currents at the LSTF and other laboratory facilities. Coastal Engrg. **50**, 19–45.

Thornton, E. B. and R. T. Guza (1983). Transformation of wave height distribution. J. Geophys. Res., **88**, C10, 5925–5938.

Ting, F. C. K. and J. T. Kirby (1994). Observation of undertow and turbulence in a laboratory surf zone. Coastal Engineering, **24**, 51–80.

Ting, F. C. K. and J. T. Kirby (1995). Dynamics of surf-zone turbulence in a strong plunging breaker. Coastal Engineering, **24**, 177–204.

Ting, F. C. K. and J. T. Kirby (1996). Dynamics of surf-zone turbulence in a spilling breaker. Coastal Engineering, **27**, 131–160.

Ting, F. C. K. (2003). Measurements of stresses in breaking waves. Private communication.

Van Dongeren, A. R., A. Reniers, J. Battjes and I. A. Svendsen (2003). Numerical modelling of infragravity wave response during DELILAH. J. Geophys. Res., **108**, C9, 3288.

Van Dorn, W. G. (1978). Breaking invariants in shoaling waves. J. Geophys. Res., **83**, 2981–2988.

Wilcox, D. C. (1998). Turbulence modeling for CFD. DCW Industries, Inc., La Cañada, Calif 91011.

Whitham, G. B. (1974). Linear and nonlinear waves. Wiley, New York.

Yamada, H. and T. Shiotani (1968). On the highest water waves of permanent type. Bull Dissas. Pev. Inst., **18**, 135, 1-22.

Zhao, Q. and K. Tanimoto (1998). Numerical simulation of breaking waves by large eddy simulation and VOF method. ASCE Proc. 26th Int. Conf. Coast. Engrg., 892–905.

Zhao, Q., S. Armfield and K. Tanimoto (2003). Numerical simulation of breaking waves by a multi-scale turbulence model. Coast. Engrg., **51**, 52–80.

Chapter 6

Wave Models Based on Linear Wave Theory

6.1 Introduction

This chapter briefly reviews the methods available for modelling wave motion in nearshore regions based on linear wave theory. While most of the theoretical background material has been presented in the previous chapters the objective is to discuss some of the practical aspects in terms of strengths and limitations of using the methods in question.

Useful models based on linear wave theory were derived by averaging over a wave period (socalled **phase averaging**) and assuming locally constant depth. This lead to the classical refraction theory, ray tracing methods and the geometrical optics theory. Simply assuming locally constant depth and writing the wave motion as a product of a slowly varying amplitude and a phase function we obtained the kinematic wave model. These models all provide information about the propagation pattern of the wave motion (wave ray or wave number vector variation) and they all must be combined with an equation expressing the conservation of energy (minus various dissipative factors such as bottom friction and wave breaking) to obtain information about the wave amplitude variation. Though the theoretical basis for those methods may seem similar in that they only describe the depth refraction process they nevertheless have different properties.

Another branch of models were based on, again, the locally constant depth assumption but combined with the assumption that the wave motion can be written as a product of a slowly varying wave amplitude $H(x, y)$ and a phase function which is given by $e^{i\omega t}$. This leads to the mild slope equation MSE and the parabolic approximation to the MSE. Those models include the effect of combined refraction-diffraction.

Therefore phase averaged models essentially consist of a component that determine the propagation pattern for the waves and a component that solves the energy equation to obtain the amplitudes. In the case of the MSE and the parabolic approximation to the MSE the two components have been combined into one by using a complex amplitude.

The concept of **phase resolution** is often associated with the MSE and parabolic wave models because the solution of those equations provides both the amplitude and the spacial phase variation for the wave motion, which is slowly varying in space with locally sinusoidal time variation. However, the term also applies to the growing selection of nonlinear wave models, such as the Boussinesq models described in Chapter 9, Boundary Element models, etc., that are time domain models, and spectral Boussinesq models that operate in the wave frequency space. In addition to the phase variation, those models also determine the actual shape of the waves.

It is characteristic for many of the model equations that they represent propagation patterns and amplitude variations that correspond to steady situations. As an example of the consequences of this a change in the incident wave conditions is not followed in time as it propagates through the model domain. A change in the incident wave characteristics (amplitude and wave number vector) will be represented by an equivalent instantaneous change in the amplitude and wave number vector in the entire domain. This also applies to the steady state MSE and to the parabolic approximation to the MSE. Exceptions to this are the kinematic wave models, and models based on the time varying version of the mild slope equation.

As mentioned all the models we can construct using the linear theory imply the locally constant depth assumption. While this has turned out to be less of a constraint in e.g. the case of the MSE and its extensions (see Section 3.7), it has consequences for the model behavior at the shoreline. Basically all those models predict that the wave heights are going to zero when the water depth goes to zero. That is: these models cannot predict for example swash.

A review of shallow water wave modelling was given by Battjes (1994).

Application as wave drivers for circulation models

Because these models provide wave amplitude information at all points in the computational domain they can readily be used to determine the radiation stresses and volume fluxes in a nearshore region. Those two quantities essentially represent the driving mechanisms for nearshore circulation phenomena such as wave generated currents and their shear instabilities, very long waves called **infragravity waves** generated by the variation of short waves, etc. These phenomena play an important role in the coastal wave climate and longterm morphological development due to sediment transport and they are described in Chapters 11 - 14.2. The models discussed in this chapter are therefore, with all their limitations, well suited and often used as **wave drivers** for the models describing nearshore circulation problems.

6.2 1DH shoaling-breaking model

In this, the simplest possible nearshore wave model, the direction of the wave number vector **k** is given by the shorenormal direction and its value is determined from the dispersion relation. Therefore the only component to this model is the energy equation

$$\frac{dE_f}{dx} = \mathcal{D} \tag{6.2.1}$$

where, as before, $\mathcal{D} < 0$ for energy loss. To obtain an equation in a useful variable, usually the wave height, this requires that we express the energy flux E_f and the dissipation $\mathcal{D}$ as functions of that variable.

This model was discussed at length in Chapter 4. Examples of the use of the model are many analyses of the situations in 1D wave tanks. As described in Section 4.3.3, with suitable choice of parameters the model includes random wave motion.

6.3 2DH refraction models

This group of models include all traditional refraction models and thus have two components: one that describes the wave propagation pattern and one that provides solution to an energy equation.

6.3.1 *The wave propagation pattern*

Determination of the wave propagation pattern is based on the principles described in Section 3.5.2. The changes in wave propagation are caused by changes in the phase velocity for the wave motion, primarily due to depth variation but also due to currents as described in Section 3.6. If the phase speed is independent of the wave height, as in linear waves, the refraction pattern can be determined independently of the wave amplitude variation. We therefore first analyze this part of the problem.

Long straight coasts

The simple case of longshore uniformity as on a long straight beach is often assumed, which implies Snell's law is valid. This of course greatly simplifies the refraction pattern which is the same along all cross-shore transects and given analytically by

$$\frac{c}{\sin\,\alpha_w} = C_0 \tag{6.3.1}$$

where C_0 is determined as $c/\sin\,\alpha_w$ at a reference point, usually offshore.

For obvious reasons the use of Snell's law to describe the refraction pattern is very popular and the literature contains numerous examples of such applications. A frequent case is related to analysis of laboratory experiments with longshore currents in wave basins such as Visser (1984), Reniers and Battjes (1997) to mention just two. However, Snell's law is also often assumed for the wave pattern on real coasts as e.g. in Thornton and Guza (1983), Van Dongeren et al. (2003) and in numerous analyses of field data from the large field experiments on (nearly) straight beaches.

As mentioned in Section 3.5.2 a warning is appropriate against uncritical use of Snell's law in cases where sufficient knowledge about the topography and general wave conditions is not available. In cases like the analysis of intensive field experiments, or testing of models against measurements from such experiments, there is usually sufficient additional information available about the wave pattern and wave height variation to verify that the assumption of a long straight coast is justified. In particular for the experiments at the Field Research Facility at Duck in North Carolina the incident waves are measured right offshore of the test site so that the risk of unnoticed deviations from the assumption of longshore uniformity is limited.

However, in general one has to remember that small depth variations at some distance offshore can have profound effects on the refraction pattern. Shoals or pits from offshore sand mining for beach nourishment are typical examples. This can result in considerable longshore variations of the wave heights in the nearshore region of interest. Therefore the warning is against uncritically assuming longshore uniformity as an expedient for analysis of the nearshore wave height distributions on a coast that may look relatively straight. It can give very misleading results.

General coastal topography

The general case of refraction on an arbitrary coastal topography requires a fully fledged refraction analysis. In the past this was usually done using the ray tracing approach described in Section 3.5.2. This approach is difficult to implement into a regular computational grid. An alternative could be to solve the eikonal equation of the geometrical optics approximation. However, the refraction approach also has the limitation, that on a general topography there is a risk that locally wave rays may intersect or come very close to do so. Under such conditions the refraction method breaks down as it predicts infinite wave heights locally. The reason is that at such locations the diffraction effects become important which will prevent the rays from intersecting. In numerical models based on fixed grids numerical dissipation usually prevents this from actually occuring in the model, which is useful although it is an artificial effect. In many cases of a general topography it is preferred today to use the Mild Slope Equation (MSE) or the parabolic approximation to the MSE to analyse the wave pattern.

The kinematic wave model

The kinematic wave model is an alternative method for determining the wave refraction pattern. From a physical point of view this model is based on the same fundamental ideas of slowly varying depth and locally constant depth as other refraction methods, but it expresses those ideas in a different mathematical form. The model was described in Section 3.5.2, and it utilizes that the wave number vector $\mathbf{k}$ is irrotational

$$\nabla_h \times \mathbf{k} = 0 \tag{6.3.2}$$

which is an initial condition for the system. The evolution equation for $\mathbf{k}$ is given by

$$\frac{\partial \mathbf{k}}{\partial t} + \nabla_h \omega_a = 0 \tag{6.3.3}$$

where ω_a is the absolute frequency, and the wave number is linked to ω_a by the dispersion relation

$$\omega_a^2 = f(kh, kH) + \mathbf{k} \cdot \mathbf{V} \tag{6.3.4}$$

where $\mathbf{V}$ is the velocity of the ambient current. f could typically be the linear dispersion relation but does not have to be so.

Though this method was developed a long time ago (see e.g. Phillips 1966, 1977) it is only recently that it has caught the attention of nearshore researchers.

The kinematic wave model has the advantage over the traditional ray tracing approach that the basic equation (6.3.3) is an evolution equation that gives the development of the wave number vector (i.e. the refraction pattern) in space and time. Thus it is possible to analyse time varying wave and current fields.

This has been utilized by Özkan-Haller and Li (2003) who studied shear wave development (see Section 14.2). Though the incident wave motion is steady in time the currents fluctuate because shear instabilities of the longshore currents develop. The influence on the wave motion from the time and space varying currents are studied using the kinematic wave model to describe the waves. The paper also give some simple examples of how the kinematic wave model works.

Both the ray tracing and the kinematic wave models have the advantage over e.g. the geometrical optics method and the MSE that they allow use of non-sinusoidal phase motion. An example of this would be the use of cnoidal wave theory (see Section 9.5) which represents a better approximation to the wave surface profile in the nearshore region. This means a dispersion relation different from the usual linear dispersion relation applies which influences the refraction pattern. Similarly different dimensionless time averaged wave parameters such as the coefficients for mass, momentum (radiation stress) and energy flux, which influencies the energy equation, the forcing of nearshore currents, etc. Another eample could be using a stream function method described in Section 8.4 for representation of the waves.

The wave number variation is only directly influenced by the phase velocity. Cnoidal waves and stream function waves, however, are both frequency and amplitude dispersive which means the phase velocity depends (weakly) on the wave height. This means the refraction pattern and the amplitude variation need to be determined simultaneously, even on a long straight coast with Snell's law valid. Due to the weak amplitude influence, though, a properly designed iterative procedure will generally work fast. An example with cnoidal waves was shown in Svendsen et al. (2003).

6.3.2 *Determination of the wave amplitude variation*

The wave amplitude is determined by solving the energy equation which was discussed in Chapter 4. As shown for an irregular wave field with steady spectral parameters and no currents the energy equation simply reads

$$\nabla_h \cdot E_{f,rms} = \mathcal{D} \tag{6.3.5}$$

where $E_{f,rms}$ is the rms-value of the energy flux. The dissipation $\mathcal{D}(H)$ (due to breaking, bottom friction etc.) for such a wave field is given as the mean-value of the dissipation in the individual waves

$$\mathcal{D} = \int_0^\infty \mathcal{D}(H) p(\mathcal{D}(H)) dH \tag{6.3.6}$$

where $p(\mathcal{D}(H))$ is the probability density for $\mathcal{D}(H)$. An example was given in Section 5.6 for the case where $\mathcal{D}(H) \propto H^3$.

In the general case of waves propagating in a (varying) current field the general energy equation for the wave motion is given by (4.4.14). In wave models of the type discussed here the Q_w-terms in this equation are normally regarded as small and therefore omitted. The equation then reads (with E short for E_{rms} and similar for $E_{f\alpha}$)

$$\frac{\partial E}{\partial t} + \frac{\partial}{\partial x_\alpha}(E_{f\alpha} + V_\alpha E) = -S_{\alpha\beta}\frac{\partial V_\beta}{\partial x_\alpha} + \mathcal{D} - V_\alpha \tau_{b\alpha} \tag{6.3.7}$$

or written in terms of the individual coordinate contributions

$$\frac{\partial E}{\partial t} + \frac{\partial}{\partial x}(E_f \cos\alpha_w + V_x E) + \frac{\partial}{\partial y}(E_f \sin\alpha_w + V_y E) =$$

$$-S_{xx}\frac{\partial V_x}{\partial x} - S_{xy}(\frac{\partial V_x}{\partial y} + \frac{\partial V_y}{\partial x}) - S_{yy}\frac{\partial V_y}{\partial y} + \mathcal{D} - V_x\tau_{bx} - V_y\tau_{by} \tag{6.3.8}$$

where $E_f = |E_{f\alpha}|$ is the energy flux in the direction of the wave motion, and α_w the angle between the wave number vector and the x-axis.

In order to solve this equation we need again to express the variables E, E_f, $S_{\alpha\beta}$, $\mathcal{D}$ and $\tau_{b\alpha}$ in terms of a single parameter, usually the wave height H_{rms}. This is conveniently done by introducing the dimensionless form of these quantities defined in Section 3.3 for E_f, E and $S_{\alpha\beta}$ and in

Section 4.2 for $\mathcal{D}$. We have

$$\begin{aligned} E &= \rho g H^2 B_E \\ E_f &= \rho g c H^2 B \\ S_{\alpha\beta} &= \rho g H^2 P_{\alpha\beta} \\ \mathcal{D} &= \rho g \frac{H^3}{4hT_m} D \end{aligned} \tag{6.3.9}$$

The dimensionless shape factors B_E, B, and $P_{\alpha\beta}$ are then determined from the prescribed phase variations of the chosen wave model using the (exact) definitions of these parameters shown in Section 3.3.

6.4 Wave action models

Another type of wave model based on linear wave theory is the wave acion equation described in Section 4.5. The (homogeneous) wave action equation (7.5.1)

$$\frac{\partial}{\partial t}\left(\frac{E}{\omega_r}\right) + \frac{\partial}{\partial x_\alpha}\left[(U_\alpha + c_{gr\alpha})\frac{E}{\omega_r}\right] = 0 \tag{6.4.1}$$

is a special version of the energy equation for waves on currents. It expresses the evolution in time and space of the wave action E/ω_r in terms of the energy density E of a single monochromatic wave with relative frequency ω_r and group velocity $c_{gr\alpha}$. The equation says that if we move with the group velocity of the wave then the wave action is conserved.

Wave action models generally deal with irregular waves which then are represented by their wave spectra. The wave action spectrum $N(f, \alpha_w)$ is then defined on the basis of the directional frequency spectrum $F(f, \alpha_w)$ (see Section 3.4.3) as

$$N(f, \alpha_w) = \frac{F(f, \alpha_w)}{\omega_r(f, \alpha_w)} \tag{6.4.2}$$

In addition the wave action equation is modified by adding source terms that represent the change in wave action from various energy sources, positive and negative. The equation (4.5.2) derived by Christoffersen and Jonsson (1982) is an example of such an equation based on rigorous derivation. It takes into account the drain of wave action caused be the energy dissipation due to wave breaking (S_{br}) and to bottom friction (S_{bf}).

In operational wave action models the total amount of source contributions S_t is written as the sum of each of the contributions. In deep water they include, in addition to the sources mentioned above:

- S_w = energy input from wind stresses which contributes to the growth of the wind waves.
- S_{wc} = energy dissipation due to whitecap breaking.
- S_{nl} = energy transfer due to nonlinear wave-wave interaction between spectral components. This contributes to the modification of the spectrum during the wind generation or decay of the wave motion.

The total source term can then be written

$$S_t = S_{br} + S_{bf} + S_w + S_{wc} + S_{nl} \tag{6.4.3}$$

and the wave action equation for each spectral component takes the form

$$\frac{\partial N(f, \alpha_w)}{\partial t} + \frac{\partial}{\partial x_\alpha}((U_\alpha + c_{gr\alpha})N(f, \alpha_w)) = S_t \tag{6.4.4}$$

This is solved in combination with the kinematic evolution equation for **k** which is also considered for each spectral component.

The basis for this type of wave model was developed for mainly deep water waves by Hasselmann et al. (1973). An application especially for nearshore wave and current conditions has been developed from the WAM deep water model (see WAMDI Group (1988), and Komen et al. (1994)). It is described by Booij et al. (1999) and Ris et al. (1999) and named the SWAN-model. The model does not account for diffraction effects since the wave action equation is based on refraction principles only.

6.5 Models based on the mild slope equation and the parabolic approximation

Wave models based on the MSE and its parabolic approximation were discussed extensively in Section 3.7. As emphasized there they are based on the same assumptions of slowly varying depth and locally constant depth as the refraction models. In the MSE and parabolic wave models, however, the wave propagation pattern and the amplitude variations are determined simultaneously through the use of a complex amplitude. Then the phase information for the wave motion (and hence the propagation pattern) is imbedded in the imaginary part of the amplitude.

Reference to MSE and parabolic wave models described in the literature were given in Section 3.7 including the widely used parabolic REF/DIF model. The REF/DIF model is also available in a spectral version called REF/DIF-S. In this model the wave patterns and wave heights are determined for a large number of wave situations (angles of incidence, frequency and incident wave height) determined from the values in a incident directional spectrum. The result corresponds to the spectral values of height and direction for the given incident spectrum at each point. The rms wave direction and wave height at each point can then be determined by (linear) superposition using the spectral weights of all the wave components found at that point.

6.6 References - Chapter 6

Battjes, J. A. (1994). Shallow water wave modelling. Proc IAHR Int. Symp. Waves - Phys. and Num. Modelling, Univ. Brit. Col., Canada. 1-23.

Booij, N., R. C. Ris, and L. H. Holthuisen (1999). A third generation wave model for coastal regions. 1. Model description and validation. J. Geophys. Res., **104**, C4, 7649–7666.

Hasselmann, K. et al. (1973). Measurement of wind-wave growth and swell decay during the Joint North Sea Wave Project (JONSWAP). Deutsche Hydrogr. Zeitschrift, Suppl., **12**, A8, 1-95.

Komen, G. L., L. Cavaleri, M. Donelan, K. Hasselmann, S. Hasselmann and P. A. E. M. Janssen (1994). Dynamics and modelling of ocean waves. Cambridge Univ. Press, 532 pp.

Özkan-Haller, H. T. and Y. Li (2003). Effects of wave-current intraction on the shear insatabilities of longshore currents. J. Geophys. Res., **108**, C5, 3139.

Phillips, O. M. (1966, 1977). Dynamics of the upper ocean. Cambridge University Press.

Qin, W., I. A. Svendsen, B. A. Ebersole and E. R. Smith (2002). Modeling sediment transport in the LSTF at DHL. Proc. 28th Int. Conf. Coastal Engrg., Cardiff, July 2002.

Reniers, A. J. H. M. and J. A. Battjes (1997). A laboratory study of longshore currents over barred and non-barred beaches. Coast. Engrg., **30**, 1–22.

Ris, R. C., L. H. Holthuisen and N. Booij (1999). A third generation wave model for coastal regions. 2. Verification. J. Geophys. Res., **104**, C4, 7667–7681.

Svendsen, I. A., W. Qin and B. A. Ebersole (2003). Modelling waves and currents in the LSTF and other laoboratory facilities. Coast. Engrg. **50**, 19–45.

Thornton, E. B. and R. T. Guza (1983). Transformation of wave height distribution. J. Geophys. Res., **88**, C10, 5925–5938.

Van Dongeren, A. R., A. Reniers, J. Battjes and I. A. Svendsen (2003). Numerical modeliing of infragravity wave response during DELILAH. J. Geophys. Res., **108**, C9, 3288.

Visser, P. J. (1984). A mathematical model of uniform longshore currents and comparison with laboratory data. *Communications on Hydraulics. Report 84-2*, Department of Civil Engineering, Delft University of Technology, 151 pp.

WAMDI Group (1988): The WAM model: A third generation ocean wave prediction model. *J. of Physical Oceanography* **18**, 1775-1810.

Chapter 7

Nonlinear Waves: Analysis of Parameters

7.1 Introduction

As described in Chapter 3 the linear water wave theory — also called "sinusoidal wave theory" due to the shape of the waves — assumes that the wave height to wave length ratio H/L (the wave steepness) is so small that the nonlinear terms in the free surface boundary conditions can be neglected.

Although we found that the linear theory in many respects yields meaningful results, there are also situations where it is insufficient. This, of course, applies to cases where one wants to study wave properties that originate from the neglected nonlinearity of the waves. Some of these phenomena will be discussed in the following. It also applies to situations where the wave steepness is so large that linear wave theory becomes too inaccurate. And third, it is relevant to realize that the confidence in linear wave results, where they **are** known to suffice, essentially comes from comparing with more accurate higher order wave theories.

The more accurate wave theories all include the effects of the nonlinear terms in some form. Hence, all such wave theories are nonlinear and therefore also more complicated than the linear wave theory.

The nonlinear wave theory most widely used today was developed by Stokes (1847) (who by the way at the same time invented the **perturbation method** which still is the most powerful mathematical tool for solving (weakly) nonlinear problems in all areas of physics), and for that reason, the theory is know as **Stokes' wave theory** (see also Stokes (1880) for a more direct version). This theory utilizes that the wave steepness H/L is small but not infinitely small as for sinusoidal waves. This assumption is reasonable since in actual waves the steepness never exceeds 0.10–0.15.

The result is a wave description for which the linear wave theory appears as a first approximation.

Stokes' wave theory, however, assumes that $kh = O(1)$ ($k = 2\pi/L$ being the wave number and h the water depth) which is not always the case. In coastal regions and for long waves in moderate depths on the continental shelf, the h/L ratio may become too small for Stokes' theory to work properly. For such cases, another wave theory was developed by Boussinesq (1872) and by Korteweg and De Vries (1895), in which h/L was assumed small. The constant form wave solution for shallow water waves resulting from this approach is often termed **cnoidal waves** (an analogy to sinusoidal waves) because the wave profile is described by Jacobi's elliptic cn-function. Such waves are also termed **long waves since** L/h is large.

A particularly simple, limiting case of cnoidal waves obtained for $L/h \to \infty$ is the **solitary wave** which physically looks like one single wave crest on an otherwise undisturbed water surface.

The approach, however, has far wider application than the constant form cnoidal and solitary solutions, and its general version, which does not assume constant form for the waves is the basis for some of the most advanced and successful computational wave models available today. In this form it is usually known as **Boussinesq wave theory**.

These theories are all based on the same fundamental hydrodynamical equations and their boundary conditions. In this chapter we will analyze more precisely how the different assumptions behind these classical wave theories lead to different forms of the equations to be solved. In particular the treatment of the nonlinear boundary conditions at the free surface is important for distinction between the different theories.

The perturbation method used in all the wave theories described above in essence generates solutions by successive approximations in the equations. Stokes himself only carried the solution to third order. De (1955) and Chappelear (1961) gave various 5th order solutions for the Stokes waves. Later on, numerical results have been obtained on a computer corresponding to 110th order (Cokelet, 1977). These results will be discussed further in Chapter 8.

In Chapter 8, we derive and discuss important aspects of the Stokes waves theory, and in Chapter 9 the nonlinear theories for long waves are described.

In all the above described classical wave theories, a powerful element in the solution is the assumption that the waves are of constant or permanent form. This assumption is also an essential part of the so-called **stream**

function theory by Dean (1965) and has since been improved by several authors including Rienecker and Fenton (1981). This is essentially a computer solution based on a truncated Fourier expansion of the exact non-linear equations. The description of this method given in Section 8.4 is based on Rienecker and Fenton's version.

7.2 The equations for the classical nonlinear wave theories

For many years there was no precise account of how the classical wave theories were related and what the differences were. One consequence of this was the so-called "long wave paradox" (Ursell, 1953). The content of the paradox was that seemingly the same assumptions could lead to different versions of the nonlinear long wave theory depending on the method of derivation. The reason turned out to be that during the derivation different **additional** assumptions were implicitly invoked by different authors without being noticed.

Here we give a more general derivation which includes both Stokes' wave theory, which applies to intermediate and deep water waves, and to the different long wave theories. Thus it is shown how the following theories and equations are related:

- Stokes' waves
- Long waves – small amplitude
- Long waves – moderate amplitude (leading to Boussinesq theory)
- Long waves – large amplitude
  (leading to the socalled non-linear shallow water equations)

So, one objective is to answer the question: Which assumptions lead to which of these theories? Or, phrased in a different way: Under which conditions does which theory apply? The second question is: How do the different assumptions influence the form of the basic equations from which the theories are derived?

We answer these questions by analyzing the relative order of magnitude of each term in the inviscid hydrodynamical equations. The information thus gained turns out also to be important for understanding the different procedures for the solutions leading to the different wave theories.

For all cases, we consider the following simple situations:

(1) Constant depth h and period T.

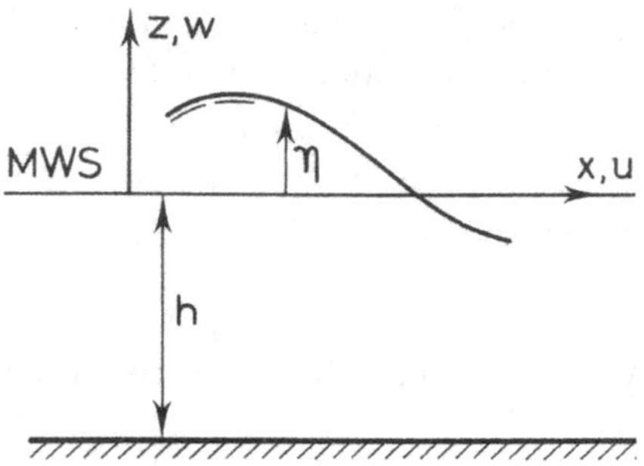

Fig. 7.2.1 Definition of nomenclature.

(2) Long-crested waves (two-dimensional vertical motion - 2DV motion) of permanent form.
(3) Incompressible flow.
(4) Effects of viscosity and turbulence are neglected.

The symbols used are shown in Fig. 7.2.1.
We use a velocity potential ϕ defined so that[1]:

$$(u, w) = + \ \nabla\phi = + \ \text{grad} \ \phi \tag{7.2.1}$$

Therefore, the **basic equations** are:

$$\text{Laplace}: \ \nabla^2\phi = \frac{\partial^2\phi}{\partial x^2} + \frac{\partial^2\phi}{\partial z^2} = \phi_{xx} + \phi_Z = 0 \tag{7.2.2}$$

with the boundary conditions:

$$\phi_z = 0 \, ; \qquad z = -h \tag{7.2.3}$$

$$\eta_t + \phi_x\eta_x - \phi_z = 0 \, ; \qquad z = \eta \tag{7.2.4}$$

$$\eta + \frac{1}{2g}\left(\phi_x^2 + \phi_z^2\right) + \frac{1}{g}\,\phi_t = \frac{C(t)}{g} \, ; \qquad z = \eta \tag{7.2.5}$$

where $C(t)$ is the arbitrary function in the generalized Bernoulli equation. In the linear wave theory, it was chosen to include $C(t)$ in ϕ_t. In the order of magnitude considerations considered in this chapter we can freely choose to put $C(t) = 0$.

[1]The + in front of the gradient operator is included to emphasize the difference from older textbooks that sometimes use − in the definition of ϕ.

In addition, we assume that the waves are periodic in space which is expressed in the periodicity condition

$$\phi_x(0, z, t) - \phi_x(L, z, t) = 0 \tag{7.2.6}$$

For pure geometrical reasons, the specification of a wave of permanent form apparently involves **three length** scales:

h : water depth
H : scale of wave height (or amplitude)
L : scale of wave length

Two independent dimensionless parameters can be formed from these three scales. We choose:

$$\mu = \frac{h}{L} = \text{the wave length parameter} \tag{7.2.7}$$

$$\delta = \frac{H}{h} = \text{the amplitude parameter} \tag{7.2.8}$$

and we must **expect** that specification of μ and δ specifies the problem uniquely except for the absolute scale.

To arrive at this "expectation", let μ and δ be given. This is not enough to specify the **scale** of the wave but we can choose (freely) $h = h_0$ and realize that μ and δ then allow a sketch of the surface profile wave to be drawn. The precise shape of the wave is yet unknown, but all important positions are fixed relative to each other (crest, trough, bottom, etc.). The shape is what the equations are going to give us when we solve them. The solution to the equations will also tell us at what speed this wave moves – that is the dispersion relation.

We first notice that through the free surface boundary conditions, the equations are nonlinear.

Experience shows that for most situations we can assume that the **linear solution at least gives the order of magnitude of all relevant quantities**. The only exceptions are situations governed by the so-called Nonlinear Shallow Water (NSW) equations, which are strongly nonlinear so that no linear first approximation exists.

7.3 The system of dimensionless variables used

As mentioned in the introduction the objective is to determine the relative order of magnitude of each term in the equations. This is essentially done in two steps. We first introduce a system of dimensionless variables based on characteristic scales for the motion. The scales used are

- h_0 a characteristic depth
- λ a characteristic horizontal scale (not necessarily the wave length)
- A a characteristic amplitude
- c_0 a characteristic velocity (typically the phase velocity for the wave)

Notice that thereby the characteristic timescale is implicitly fixed as λ / c_0.

When applied to the independent variables this gives the following dimensionless variables (denoted by $'$)

$$(x, y) = \lambda(x', y'); \qquad (z, h) = h_0(z', h') \tag{7.3.1}$$

$$t = \frac{\lambda}{c_0} t'; \qquad \eta = A\eta' \tag{7.3.2}$$

This implies, following the chain rule, that differentiation with respect to these variables can be written

$$\frac{\partial}{\partial x} = \frac{\partial}{\partial x'} \frac{\partial x'}{\partial x} = \frac{1}{\lambda} \frac{\partial}{\partial x'} \tag{7.3.3}$$

$$\frac{\partial}{\partial z} = \frac{1}{h_0} \frac{\partial}{\partial z'} \tag{7.3.4}$$

$$\frac{\partial}{\partial t} = \frac{c_0}{\lambda} \frac{\partial}{\partial t'} \tag{7.3.5}$$

We then use linear theory to express the important basic variables η and ϕ in a similar way as the product of a dimensionless quantity and a dimensional factor. In linear theory we have

$$\eta = A \cos \theta \; ; \qquad \theta = \omega t - kx \tag{7.3.6}$$

Hence we define η' by

$$\eta = A \, \eta' \tag{7.3.7}$$

Similarly we have for ϕ

$$\phi = -Ac \, \frac{\cosh k(z+h)}{\sinh kh} \sin \theta \tag{7.3.8}$$

This expression requires some modification to identify the proper separation between the dimensionless ϕ' and the dimensional factor. We see that (7.3.8) can be written

$$\phi = -\frac{Ac}{\tanh kh} \frac{\cosh\ k(z+h)}{\cosh\ kh} \sin\theta \tag{7.3.9}$$

and in order to unify all cases in the same formula it turns out to be convenient in the order of magnitude estimates to choose the following definition for ϕ'

$$\phi = \frac{Ac_0}{\tanh\ kh}\ \phi' \tag{7.3.10}$$

which implies

$$\phi' = \frac{\cosh\ k(z+h)}{\cosh\ kh} \sin\theta \tag{7.3.11}$$

It may be emphasized here that this is as far as linear theory can give guidance. The actual η' and ϕ' will essentially be the unknown functions in the study. Hence for the nonlinear wave motions we are going to study we can expect that

$$\eta' \neq \cos\ \theta \tag{7.3.12}$$

$$\phi' \neq \frac{\cosh\ k(z+h)}{\cosh\ kh} \sin\ \theta \tag{7.3.13}$$

The dimensionless version of the basic equations

We can then determine the dimensionless version of each quantity occurring in the basic equations and thereby the form of each term in the equations. We get

$$\phi_x = \frac{Ac_0}{\tanh\ kh_0} \frac{1}{\lambda}\ \phi'_{x'} \tag{7.3.14}$$

$$\phi_z = \frac{Ac_0}{\tanh\ kh_0} \frac{1}{h_0} \phi'_{z'} \tag{7.3.15}$$

$$\phi_t = \frac{Ac_0}{\tanh\ kh_0} \frac{c}{\lambda} \phi'_{t'} \tag{7.3.16}$$

$$\eta_t = \frac{Ac_0}{\lambda}\ \eta'_{t'} \tag{7.3.17}$$

Similarly we get

$$\phi_{xx} = \frac{Ac_0}{\tanh kh_0} \frac{1}{\lambda^2} \phi'_{x'x'} \tag{7.3.18}$$

$$\phi_z = \frac{Ac_0}{\tanh kh_0} \frac{1}{h_0{}^2} \phi'_{z'} \tag{7.3.19}$$

Substituting this into the basic equations (7.2.2) - (7.2.5) we then get the dimensionless version of the equations. For the Laplace equation we get

$$\boxed{\phi'_{x'x'} + \frac{1}{\mu^2} \phi'_{z'z'} = 0} \tag{7.3.20}$$

and for the bottom boundary condition we we get (constant depth)

$$\boxed{\frac{1}{\mu^2} \phi'_{z'} = 0; \qquad z' = -h'} \tag{7.3.21}$$

For the kinematic free surface boundary condition (KFSBC) we have after some reorganization

$$\boxed{\mu \tanh kh_0 \, \eta'_{t'} - \phi'_{z'} + \delta\mu^2 \eta'_{x'} \phi'_{x'} = 0; \qquad z' = \delta\eta'} \tag{7.3.22}$$

and finally the dynamic free surface boundary condition (DFSBC) yields

$$\boxed{\phi'_{t'} + \frac{gh_0}{c_0{}^2} \frac{\tanh kh_0}{\mu} \eta' + \frac{1}{2} \frac{\delta\mu}{\tanh kh_0} \left(\phi_{x'}{}^2 + \frac{1}{\mu^2} \phi_{z'}{}^2 \right) = 0; \qquad z' = \delta\eta'} \tag{7.3.23}$$

both written here in a form convenient for the following. We notice that in dimensionless form the surface $z = \eta$ changes to $z' = \delta\eta'$.

Exercise 7.3-1

Derive equations (7.3.20) - (7.3.23).

It is worth noting that although use of linear theory and introduction of characteristic scales give relevant expressions for the dimensionless variables we have no guarantee that all the derivatives shown above are $O(1)$ for all the cases to be studied. In fact we will find that for long waves a few of them are not.

The equations (7.3.20)–(7.3.23) are now the basic equations on a form where all terms are expressed as a product of a combination of the external parameters δ, μ, and tanh kh_0, and some dimensionless functions. The δ, μ-combinations give the order of magnitudes we are looking for, the dimensionless functions the more precise variation in space and time, which it is the purpose of an actual solution to describe.

7.4 Stokes waves

Stokes wave theory emerges from the equations (7.3.20)–(7.3.23) if we assume intermediate to deep water relative to the characteristic horizontal scale λ. This mens assuming that

$$\mu = \frac{h_0}{\lambda} = O(1) \tag{7.4.1}$$

When this is the case we see that

$$\tanh kh_0 = O(1) \tag{7.4.2}$$

and using the linear dispersion relation for the waves we also have ${c_0}^2 \sim g\lambda$ so that

$$\frac{gh_0}{{c_0}^2} \sim \frac{gh_0}{g\lambda} = \mu = O(1) \tag{7.4.3}$$

Substituting this the four basic equations become
Laplace:

$$\phi'_{xx} + \phi'_{zz} = 0 \tag{7.4.4}$$

with the boundary conditions

$$\phi'_z = 0 \;; \qquad z' = -h' \tag{7.4.5}$$

$$\eta'_t + \delta\mu\phi'_x\,\eta'_x - \phi'_z = 0 \;; \qquad z' = \delta\eta' \tag{7.4.6}$$

$$\eta' + \frac{1}{2}\,\delta\mu\,\left(\phi_x'^2 + \phi_z'^2\right) + \phi'_t = 0 \;; \qquad z' = \delta\eta' \tag{7.4.7}$$

Exercise 7.4-1

Derive equations (7.4.4) - (7.4.7).

This is the case corresponding to **Stokes waves.** As the equations show, there is only one small parameter $\delta\mu = A/\lambda \sim H/L$, and it occurs as a factor on all nonlinear terms.

Hence in Stokes' waves, all nonlinear terms in the basic equations will be of the same order of magnitude, and we can keep nonlinear terms small by requiring

$$H/L \ll 1 \tag{7.4.8}$$

H/L is, of course then, the proper expansion parameter in the perturbation expansion used in Stokes wave theory.

In this case all the dimensionless variables in the equations turn out to be $O(1)$. As an example we determine $\phi'_{x'}$ and $\phi'_{z'}$, the horizontal and vertical velocity components, by comparing with the values of linear theory.

Differentiating the dimensional ϕ (7.3.8) with respect to x and using (7.3.14) we get

$$\phi_x = u = \frac{2\pi Ac}{L} \frac{\cosh\ k(z+h)}{\sinh\ kh} \cos\ \theta \sim \frac{Ac}{\lambda\ \tanh\ kh} \phi'_{x'} \tag{7.4.9}$$

Solving with respect to $\phi'_{x'}$ and letting $\lambda/L = 1$ then gives

$$\phi'_{x'} \sim 2\pi \frac{\cosh\ k(z+h)}{\cosh\ kh} \cos\ \theta \tag{7.4.10}$$

which is $O(1)$. Similarly we get for $\phi'_{z'}$ differentiating (7.3.8) with respect to z and using (7.3.15)

$$\phi_z = w = \frac{2\pi Ac}{L} \frac{\sinh\ k(z+h)}{\sinh\ kh} \cos\ \theta \sim \frac{Ac_0}{h_0\ \tanh\ kh_0} \phi'_{z'} \tag{7.4.11}$$

which means

$$\phi'_{z'} \sim 2\pi \frac{h_0\ \tanh\ kh_0}{L} \frac{\sinh\ k(z+h)}{\sinh\ kh} \sin\ \theta \tag{7.4.12}$$

And since $h_0\ \tanh kh_0/L = O(1)$ we get

$$\phi'_{z'} \sim 2\pi\ \frac{\sinh\ k(z+h)}{\sinh\ kh} \sin\ \theta \tag{7.4.13}$$

which again is $O(1)$

7.5 Long waves

In the region near the shore the water depth becomes so small that the incident, windgenerated waves have wave lengths L much longer than the water depth h. This is the "shallow water region". These waves are therefore also termed "long waves". In terms of μ it means that long waves are characterized by

$$\boxed{\mu = \frac{h}{L} \ll 1}$$

At the same time the wave height is often comparable to the water depth h.

Under these circumstances the linear theory predicts that

$$\frac{{c_0}^2}{gh_0} = O(1) \tag{7.5.1}$$

$$\tanh\, kh_0 = kh_0 = O(\mu) \tag{7.5.2}$$

Here we have assumed $kh_0 = O(h_0/L)$ which underlines that O only stands for "order of magnitude". For long waves, however, it turns out that the non-dimensional functions in the equations are not all $O(1)$. To reveal this we use the same approach as for the Stokes wave case, and again we use $\phi'_{x'}$ and $\phi'_{z'}$ as examples. For the first of those we get

$$\phi'_{x'} = 2\pi \frac{\cosh\, k(z+h)}{\cosh\, kh} \cos\, \theta = O(1) \tag{7.5.3}$$

However, for $\phi'_{z'}$ we now get

$$\phi'_{z'} = 2\pi \frac{h}{L} \tanh\, kh \frac{\sinh\, k(z+h)}{\sinh\, kh} \sin\, \theta \sim 2\pi\mu^2\, \frac{\sinh\, k(z+h)}{\sinh\, kh} \sin\, \theta \tag{7.5.4}$$

From this we conclude that

$$\phi'_{z'} = O(\mu^2) \tag{7.5.5}$$

which implies it is the product $\phi'_{z'}/\mu^2$ which is $O(1)$.

Exercise 7.5-1 Show that for long waves all other dimensionless functions in the equations are $O(1)$ except $\phi'_{z'z'}$ which is also $O(\mu^2)$.

Hence in the Laplace equation (7.3.20) the $\phi'_{z'z'}/\mu^2$ term is of the same order as the $\phi'_{x'x'}$ (which of course is consistent with the fact that the equation only has two terms!), and the equation remains the same as (7.3.20)

$$\phi'_{x'x'} + \frac{1}{\mu^2}\phi'_{z'z'} = 0 \tag{7.5.6}$$

Similarly the other linear equation, the bottom boundary condition, remains the same

$$\frac{1}{\mu^2}\phi'_{z'} = 0 \;; \quad z = -h \tag{7.5.7}$$

In the two nonlinear surface conditions, however, the assumptions about the magnitude of δ and μ change the form of the equations. We get For (7.3.22) and (7.3.23)

$$\eta'_t + \delta\, \phi'_{x'}\, \eta'_{x'} - \frac{1}{\mu^2}\phi'_{z'} = 0 \qquad ; \quad z' = \delta\eta' \tag{7.5.8}$$

$$\eta' + \frac{1}{2}\delta\left(\phi'^2_{x'} + \frac{1}{\mu^2}\phi'^2_{z'}\right) + \phi'_{t'} = 0\,; \quad z' = \delta\eta' \tag{7.5.9}$$

where the terms $\delta\, \phi'_{x'}\, \eta'_{x'}$ and $\delta(\phi'^2_{x'} + \frac{1}{\mu^2}\phi'^2_{z'})$ are now the small terms.

This clearly is much more complicated case than for Stokes waves, because the two surface conditions contain two (small) parameters: δ and μ. The proper approximation depends on the relative magnitude of δ and μ. There are essentially 3 cases to consider:

$$\left.\begin{array}{l} 1.\ \delta \ll O(\mu^2) \\ 2.\ \delta = O(\mu^2) \\ 3.\ \delta \gg O(\mu^2) \end{array}\right\} \tag{7.5.10}$$

This can also be written

$$\left.\begin{array}{l} 1.\ \frac{\delta}{\mu^2} = \frac{A\lambda^2}{h^3} \ll O(1) \\ 2.\ \frac{\delta}{\mu^2} = \frac{A\lambda^2}{h^3} = O(1) \\ 3.\ \frac{\delta}{\mu^2} = \frac{A\lambda^2}{h^3} \gg O(1) \end{array}\right\} \tag{7.5.11}$$

In the following we consider each of these cases separately starting with the second condition which turns out to include the other two conditions as special cases.

7.5.1 *The Stokes or Ursell parameter*

The significance of the parameter $\frac{A\lambda^2}{h^3}$ was already pointed out by Stokes (1847). He showed the value of this parameter is a measure of the shallow water limit for the validity of his second order theory which will be discussed further in chapter 8. For this reason the parameter is sometimes called the **Stokes parameter**. However, the parameter is also often called the **Ursell parameter** because Ursell (1953) showed the result indicated above that this parameter is important in distinguishing between the three different long wave cases.

In practical applications the wave amplitude A is often replaced by the wave height H and the characteristic horizontal scale λ is often set equal to the wavelength L. We see that then $\frac{A\lambda^2}{h^3}$ changes to

$$\frac{A\,\lambda^2}{h^3} \sim U = \frac{H\,L^2}{h^3} \tag{7.5.12}$$

so that the proper approximation for long waves can also be said to depend on whether the parameter

$$U \equiv \frac{H\,L^2}{h^3} \begin{cases} \ll O(1) \text{ linear shallow water waves} \\ = O(1) \text{ cnoidal and solitary waves} \\ \gg O(1) \text{ nonlinear shallow water waves} \end{cases} \tag{7.5.13}$$

However, depending on the estimate of λ, the differences in numerical values of $\frac{A\,\lambda^2}{h^3}$ and $\frac{H\,L^2}{h^3}$ can be quite substantial.

7.5.2 *Long waves of moderate amplitude*

The assumption here is

$$\boxed{\delta = O(\mu^2) \ll 1 \quad \textit{or} \quad \frac{\delta}{\mu^2} = O(1)} \tag{7.5.14}$$

Inspection of the equations (7.5.6) - (7.5.9) shows that this means all the nonlinear terms must be retained. The relevant equations therefore are simply (7.5.6) - (7.5.9):

$$\phi'_{x'x'} + \frac{1}{\mu^2}\phi'_{z'z'} = 0 \tag{7.5.15}$$

$$\frac{1}{\mu^2}\phi'_{z'} = 0 \; ; \qquad z' = -h' \tag{7.5.16}$$

$$\eta'_{t'} + \delta\phi'_{x'}\eta'_{x'} - \frac{1}{\mu^2}\phi'_{z'} = 0 \; ; \qquad z' = \delta\eta' \tag{7.5.17}$$

$$\eta' + \frac{1}{2}\delta\left(\phi'^2_{x'} + \frac{1}{\mu^2}\phi'^2_{z'}\right) + \phi'_{t'} = 0 \; ; \qquad z' = \delta\eta' \tag{7.5.18}$$

The linear terms are clearly $O(1)$ as is ${\phi'_{x'}}^2$. However, the nonlinear terms have factors δ which, as mentioned, indicates that these terms are proportional to the amplitude of the waves. For $\delta \ll 1$ all the nonlinear terms are small, and the waves are therefore often termed weakly nonlinear. It is also recalled that ${\phi'_{z'}}^2$ is $O(\mu^2)$ so the term $\frac{1}{\mu^2}{\phi'_{z'}}^2$ is $O(\mu^2)$.

It will be shown later that these equations lead to the so-called Boussinesq waves. solutions for the constant form waves called **cnoidal** and **solitary waves**. However, substantial modifications, shown in Chapter 9, are needed to bring the equations on a form suitable for closer analysis. The analysis of this case forms the major part of Chapter 9.

Because these equations contain all the terms of the original equations we can formally derive the other two cases from these equations by assuming δ an order of magnitude smaller than μ^2 (giving the case of very small values of the Ursell parameter) or δ an order of magnitude larger than μ^2 (which gives the case of very large values of the Ursell parameter.

7.5.3 *Long waves of small amplitude*

$$\boxed{\frac{\delta}{\mu^2} = \frac{A\lambda^2}{h^3} \ll O(1)} \tag{7.5.19}$$

This implies that all nonlinear terms are very small. The equations simplify to :

$$\phi'_{xx} + \frac{1}{\mu^2}\phi'_{zz} = 0 \tag{7.5.20}$$

$$\phi'_z = 0 \quad ; \quad z' = -h' \tag{7.5.21}$$

$$\eta'_t - \frac{1}{\mu^2}\phi'_z = 0 \quad ; \quad z = \eta \tag{7.5.22}$$

$$\eta' + \phi'_t = 0 \quad ; \quad z = \eta \tag{7.5.23}$$

where the 0 on the RHS actually stands for $O(\mu^4)$.

Closer inspection and comparison with Chapter 3 shows that these equations are identical to the equations solved to get the linear wave theory. Their relevance for the Boussinesq waves will be discussed in Chapter 9.

7.5.4 *Long waves of large amplitude*

$$\boxed{\delta = O(1) \gg O(\mu^2)} \tag{7.5.24}$$

When $\delta = O(1)$ the waves have heights of the same magnitude as the water depth, and some of the nonlinear terms are large, (i.e., of the same order of magnitude as the linear terms).

However, we still have $\phi'_{z'} = O(\mu^2)$ according to (7.5.5). As for waves of moderate amplitude this implies that the linear term $\frac{1}{\mu^2}\phi'_{z'} = O(1)$ whereas the term $\frac{1}{\mu^2}(\phi'_{z'})^2 = O(\mu^2)$.

To the O(μ^2) the equations then reduce to:

$$\phi'_{xx} + \phi'_{zz} = 0 \tag{7.5.25}$$

$$\phi'_z = 0\,; \quad z' = -h' \tag{7.5.26}$$

$$\eta'_t + \phi'_x\, \eta'_x - \phi'_z = 0\,; \quad z' = \lambda\eta' \tag{7.5.27}$$

$$\eta' + \frac{1}{2}\,\phi'^2_x + \phi'_t = 0\,; \quad z' = \lambda\eta' \tag{7.5.28}$$

We see that all terms in these equations are of the same order of magnitude because the only small nonlinear term (the ϕ'^2_z term representing the effect of the vertical velocities) has been omitted.

For these equations all nonlinear terms are essential and of the same order of magnitude as the linear terms. This implies that the results from linear wave theory cannot be used as a first approximation. Also no analytical or even approximate solution is possible. The case is of significant interest as it represents many situations occurring in practice, and the analysis of the properties of the equations will be discussed further in Section 9.12. As will be shown there, this case leads to the so-called **nonlinear shallow water equations** (NSW) or **finite amplitude shallow water equations**

7.6 Conclusion

In conclusion, the analysis in this chapter shows how the magnitude of the two parameters δ and μ leads to the four different sets of equations and thereby four different wave theories known from the classical literature. Derivation of the actual equations for those four cases, solutions for constant form waves where they exist, and discussion of many of the other results and methods associated with classical nonlinear wave theory, are discussed in the following chapters.

7.7 References - Chapter 7

Boussinesq, J. (1872). Theorie des onde et des resous qui se propagent le long d'un canal rectangulaire horizontal, en communiquant au liquide contenu dans ce canal des vitesses sensiblement pareilles de la surface au fond. Journal de Math. Pures et Appl., Deuxieme Serie, **17**, 55–108.

Chappelear, J. E. (1961). Direct numerical calculation of wave properties. J. Geophys. Res., **66**, 501–508.

Cokelet, E. D. (1977). Steep gravity waves in water of arbitrary uniform depth. Phil. Trans. Roy. Soc. London, **386**, 179–206.

De, S. C. (1955). Contributions to the theory of Stokes waves. Proc. Cambr. Phil. Soc., **51**, 713–736.

Dean, R. G. (1965). Stream function representation of nonlinear water waves. J. Geophys. Res., **70**, 18, 4561–4572.

Korteweg, D. J. and G. DeVries (1895). On the change of form of long waves advancing in a canal, and on a new type of long stationary waves. Phil. Mag., Ser. 5, **39**, 422–443.

Rienecker, M. and J. D. Fenton (1981). Numerical solution of the exact equations of water waves. J. Fluid Mech. **104**, 119–137.

Stokes, G. G. (1847). On the theory of oscillatory waves. Trans. Cambridge Phil. Soc., **8**, 441–473.

Stokes, G. G. (1880). Mathematical and Physical Papers, Vol 1. Cambridge University Press.

Ursell, F. (1953). The long wave paradox. Proc. Cambr. Phil. Soc., **49**, 685–694.

Chapter 8

Stokes Wave Theory

8.1 Introduction

The Stokes wave theory is both the oldest and most well-studied of the nonlinear theories. The reason for studying it in some detail is, however, that in addition to being the simplest it also exhibits most of the effects associated with nonlinear waves.

The theory was first developed by Stokes (1947), whence its name, and reprinted with some additions in the collected papers Stokes (1880).

One of those characteristics is illustrated in Fig. 8.1.1 which shows the difference between the surface profile of a sinusoidal wave in intermediate depth of water and that of a "real" wave. While the sinusoidal wave has equally high and equally long crests and troughs the real wave has shorter and higher crests and longer and shallower troughs.

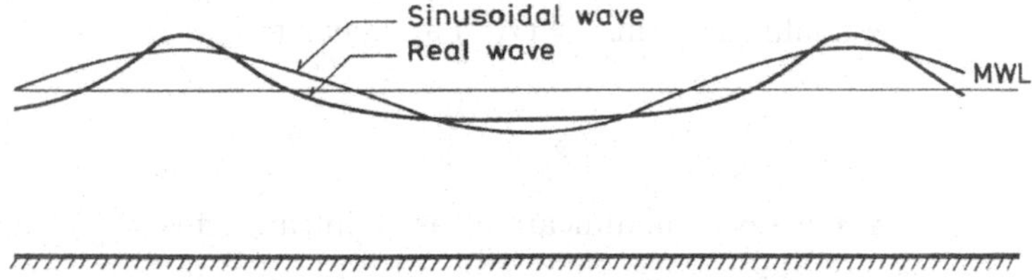

Fig. 8.1.1 Comparison of wave profiles.

In consequence of the results of Chapter 7, we again consider the basic equations. For convenience we repeat the equations in dimensional form here. They are:

$$\text{Laplace}: \quad \nabla^2\phi = \frac{\partial^2\phi}{\partial x^2} + \frac{\partial^2\phi}{\partial x^2} = \phi_{xx} + \phi_{zz} = 0 \tag{8.1.1}$$

with the boundary conditions:

$$\phi_z = 0 \quad ; \quad z = -h \tag{8.1.2}$$

$$\eta_t + \phi_x\eta_x - \phi_z = 0 \quad ; \quad z = \eta \tag{8.1.3}$$

$$\eta + \frac{1}{2g}\left(\phi_x^2 + \phi_z^2\right) + \frac{1}{g}\,\phi_t = \frac{C(t)}{g} \quad ; \quad z = \eta \tag{8.1.4}$$

where $C(t)$ is the arbitrary function in the generalized Bernoulli equation. The variables used are defined in Fig. 7.2.1 which also shows the coordinate system is place with the x-axis (i.e. $z = 0$) on the Mean water surface MWS.MWS

In addition we assume the waves are periodic in x which we express as

$$\phi_x(0, z, t) = \phi_x(L, z, t) \tag{8.1.5}$$

Thus we are returning to dimensional form of the equations and seek solutions based on the assumption, that if we define the parameter γ as

$$\gamma = \delta\epsilon = \frac{H}{h}\frac{h}{L} = \frac{H}{L} \tag{8.1.6}$$

then we have

$$\text{nonlinear terms} \;= O(\gamma) \;\cdot\; \text{linear terms} \tag{8.1.7}$$

or

$$\gamma \ll 1 <=> \text{nonlinear terms} \ll \text{linear terms} \tag{8.1.8}$$

8.2 Second order Stokes waves

8.2.1 *Development of the perturbation expansion*

So γ is the wave steepness which we know is the proper expansion parameter for Stokes waves.

We now **formally** expand all quantities in a power series in γ:

$$\phi = \gamma\phi_1' + \gamma^2\phi_2' + \cdots \tag{8.2.1}$$

$$\eta = \gamma\eta_1' + \gamma^2\eta_2' + \cdots \tag{8.2.2}$$

$$p = -\rho g z + \gamma p_1' + \gamma^2 p_2' + \cdots \tag{8.2.3}$$

$$\omega = \omega_0' + \gamma\omega_1' + \gamma^2\omega_2' + \cdots \tag{8.2.4}$$

$$C(t) = \gamma C_1' + \gamma^2 C_2' + \cdots \tag{8.2.5}$$

where ω_0 is included because even waves of infinitesimal steepness $\delta = 0$ of course have a finite wave frequency.

In the following, we will leave ω unexpanded when met as an argument to the trigonometric functions (like $\cos(\omega t - kx)$) and only substitute (8.2.5) when ω occurs as a factor outside the functions. Notice that expanding the frequency instead of the wave number k is convenient. It implies that as we go to higher order approximations, we consider the changes in ω for given k. In the numerical applications, however, we often solve the resulting equations in respect to k for given ω. (This is discussed later.)

When we substitute into the equations it of course makes no difference whether we write $\gamma\phi_1'$, $\gamma^2\phi_2', \ldots$ or just $\phi_1(=\gamma\phi_1')$, $\phi_2(=\gamma^2\phi_2')$, ... So for simplicity, we use the expansions:

$$\phi = \phi_1 + \phi_2 + \cdots \tag{8.2.6}$$

$$\eta = \eta_1 + \eta_2 + \cdots \tag{8.2.7}$$

$$p = -\rho g z + p_1 + p_2 + \cdots \tag{8.2.8}$$

$$\omega = \omega_0 + \omega_1 + \omega_2 + \cdots \tag{8.2.9}$$

$$C(t) = C_1 + C_2 + \cdots \tag{8.2.10}$$

and recall that quantities with index_1 are $O(\gamma)$, quantities with index_2 are $O(\gamma^2)$, etc. (Notice that this change is not necessary. It just reduces the number of symbols in some of the equations.)

Substitution into the equations yields:

Laplace:

$$(\phi_{1xx} + \phi_{1zz}) + (\phi_{2xx} + \phi_{2zz}) + \cdots = 0 \tag{8.2.11}$$

Bottom boundary condition:

$$\phi_{1z} + \phi_{2z} + \cdots = 0 \quad ; \qquad z = -h \tag{8.2.12}$$

Periodicity condition:

$$(\phi_{1x}(0,z,t) - \phi_{1x}(L,z,t,)) + (\phi_{2x}(0,z,t,) - \phi_{2x}(L,z,t)) + \cdots \quad (8.2.13)$$

Kinematic free surface conditions:

$$(\eta_{1t} - \phi_{1z}) + (\eta_{2t} - \phi_{2z} + \phi_{1x}\eta_{1x}) + \cdots = 0 \;; \qquad z = \eta \quad (8.2.14)$$

Dynamic free surface condition:

$$\left(\eta_1 + \frac{1}{g}\phi_{1t} - \frac{C_1(t)}{g}\right) + \left(\eta_2 + \frac{1}{g}\,\phi_{2t} + \frac{1}{2g}\left(\phi_{1x}^2 + \phi_{1z}^2\right) - \frac{C_2(t)}{g}\right) + \ldots$$
$$= 0\;; \qquad z = \eta \quad (8.2.15)$$

Here we have collected terms proportional to the same power of γ and utilized that since both ϕ_{1x} and η_{1x} are $O(\gamma)$, we get $\phi_{1x}\eta_{1x} = O(\gamma^2)$, etc.

This procedure is followed everywhere in the following. As in linear wave theory, we now realize that we cannot handle the free surface boundary conditions at the unknown and varying $z = \eta$-level.

So, using $z = 0$ as expansion point, we introduce a Taylor expansion for all ϕ-derivatives. This implies writting:

$$f(\eta) = f(0) + \eta\left(\frac{\partial f}{\partial z}\right)_{z=0} + \frac{1}{2}\,\eta^2\left(\frac{\partial^2 f}{\partial z^2}\right)_{z=0} + \cdots \quad (8.2.16)$$

for all z-dependent variables in (8.2.14) and (8.2.15). We get (e.g.)

$$\begin{aligned}\phi_{1x}(x,\eta,t) &= \phi_{1x}(x,0,t) + (\eta_1 + \eta_2 + \cdots)\phi_{1xz}(x,0,t) + \cdots \\ &= \phi_{1x} + \eta_1\phi_{1xz} + \left(\frac{1}{2}\eta_1^2\phi_{1xzz} + \eta_2\phi_{1x}\right) + \cdots \end{aligned} \quad (8.2.17)$$

and similar expressions for other variables.

Substitution of all these expansions into (8.2.14) and (8.2.15) yields:

$$(\eta_{1t} - \phi_{1z}) + (\eta_{2t} - \phi_{2z} + \phi_{1x}\eta_{1x} - \eta_{1x}\phi_{1zz}) + \ldots = 0 \quad ; \quad z = 0 \quad (8.2.18)$$

and

$$(g\eta_1 + \phi_{1t}) + \left(g\eta_2 + \phi_{2t} + \frac{1}{2}\left(\phi_{1x}^2 + \phi_{1z}^2\right)\right) + \eta_1\phi_{1tz} + \ldots = 0 \;;\quad z = 0 \quad (8.2.19)$$

Exercise 8.2-1

Derive (8.2.18) and (8.2.19) by following the procedure outlined above.

The fundamental set of arguments which makes the perturbation technique a method at all may now be described in the following way:

If we consider the equations we have derived, viz. (8.2.11), 8.2.12), (8.2.13), (8.2.18) and (8.2.19), we realize that in each parenthesis all terms are proportional to γ^n (where n is the number of the parenthesis). We can then write each equation on the form:

$$\gamma f_1 + \gamma^2 f_2 + \gamma^3 f_3 + \cdots = 0 \tag{8.2.20}$$

Eq. (8.2.20) is a polynomium in γ. Since we want the results of our theory (i.e., the results for f_1, f_2, etc.) to apply for all values of $\gamma = H/L$, this means the **polynomium must be zero for arbitrarily many values of the variable** γ. And clearly this can only happen if **all** coefficients f_1, f_2, ... are identically zero.

Thus we must conclude that to satisfy (8.2.20) for arbitrary values of $\gamma = H/L$, we **must** require:

$$f_1 = 0 \tag{8.2.21}$$

$$f_2 = 0 \qquad \text{etc.} \tag{8.2.22}$$

which means that each of the above mentioned equations split into one equation for the expression in each of the parentheses.

8.2.2 *First order approximation*

This is the idea that opens for the solution of the problem. If we now consider the first parentheses of all equations, we get the following system for the first order approximation:

$$\phi_{1xx} + \phi_{1zz} = 0 \tag{8.2.23}$$

$$\phi_{1z} = 0 \quad ; \quad z = -h \tag{8.2.24}$$

$$\phi_{1x}(O, z, t) - \phi_{1x}(L, z, t) = 0 \tag{8.2.25}$$

$$\eta_{1t} - \phi_{1z} = 0 \quad ; \quad z = 0 \tag{8.2.26}$$

$$g\eta_1 + \phi_{1t} - C_1(t) = 0 \quad ; \quad z = 0 \tag{8.2.27}$$

which is exactly the system of equations we have solved previously to find the **linear wave theory**.

Hence, we have by this formal procedure shown that the linear waves do indeed constitute the first approximation of what turns out (when we go on) to be a wave theory that can be developed to arbitrarily high order (and (hopefully) accuracy) provided we are willing to do the work.

Recall that the results for linear waves are:

$$\phi_1 = -\frac{Hc}{2} \frac{\cosh k(z+h)}{\sinh kh} \sin\theta \quad ; \quad \theta = \omega t - kx \tag{8.2.28}$$

$$\eta_1 = \frac{H}{2} \cos\theta \tag{8.2.29}$$

$$p_1 = \rho g \frac{H}{2} \frac{\cosh k(z+h)}{\cosh kh} \cos\theta \tag{8.2.30}$$

$$\omega_0^2 = gk \tanh kh \qquad \text{or} \qquad c_0^2 = \frac{g}{k} \tanh kh \tag{8.2.31}$$

and $C_1(t) = 0$. So, until here everything is well known and we can go on to the next approximation.

8.2.3 *Second order approximation*

Obviously the equations determining the second approximation also come from (8.2.11), (8.2.12, (8.2.13), (8.2.18) and (8.2.19). It turns out that it is convenient to separate $C_2(t)$ into a mean part $C_2 = \overline{C_2(t)}$ and a time varying part $C_2'(t)$ by writing

$$C_2(t) = C_2 + C_2'(t) \tag{8.2.32}$$

and to include $C_2'(t)$ in ϕ_2. For the second approximation we then get:

$$\phi_{2xx} + \phi_{2zz} = 0 \tag{8.2.33}$$

$$\phi_{2z} = 0 \quad ; \quad z = -h \tag{8.2.34}$$

$$\phi_{2x}(0,z,t) - \phi_{2x}(L,z,t) = 0 \tag{8.2.35}$$

$$\eta_{2t} - \phi_{2z} = \eta_1 \phi_{1zz} - \eta_{1x}\phi_{1x} \quad ; \quad z = 0 \tag{8.2.36}$$

$$g\eta_2 + \phi_{2t} - C_2 = -\eta_1 \phi_{1tz} - \frac{1}{2}\left(\phi_{1x}^2 + \phi_{1z}^2\right) \quad ; \quad z = 0 \tag{8.2.37}$$

The solution to these equations will give us η_2 and ϕ_2. As in all potential flow problems, the pressure is then afterwards determined from the Bernoulli equation. For this purpose, we introduce the dynamic pressure p_D, again defined as the pressure over and above the hydrostatic pressure from the MWL.

$$p = -\rho g z + p_D \tag{8.2.38}$$

which in combination with (8.2.8) shows that

$$p_D = p_1 + p_2 + \cdots \tag{8.2.39}$$

The Bernoulli equation for p_2 then becomes:

$$p_2 = \rho\,\phi_{2t} + \rho C_2 - \frac{1}{2}\rho\left(\phi_{1x}^2 + \phi_{1z}^2\right) \tag{8.2.40}$$

Exercise 8.2-2

Derive (8.2.40) from the Bernoulli equation.

The next step towards a solution for the second order approximation is the same as for linear waves: η_2 only occurs in the two surface conditions, so we eliminate η_2. Differentiation of (8.2.37) with respect to t and subtraction from (8.2.36) multiplied by g yields (using that C_2 is a constant):

$$\phi_{2tt} + g\phi_{2z} = -\left(\eta_1\phi_{1tz}\right)_t - \frac{1}{2}\left(\phi_{1x}^2 + \phi_{1z}^2\right)_t - g\eta_1\phi_{1zz} + g\eta_{1x}\phi_{1x}\ ; \qquad z = 0 \tag{8.2.41}$$

This may also be written

$$\phi_{2tt} + g\phi_{2z} = -\eta_1\left(gw_{1z} + w_{1tt}\right) - \left(u_1^2 + w_1^2\right)_t \tag{8.2.42}$$

(This transcription is not necessary, though, (8.2.41) suffices).

Substitution of the results for η_1, u_1 and w_1 from first order theory then yields:

$$\phi_{2tt} + g\phi_{2z} = \frac{3}{4}gc(kH)^2\frac{\sin 2(\omega t - kx)}{\sinh 2kh}\ ; \qquad z = 0 \tag{8.2.43}$$

which is the combined kinematic and dynamic free surface boundary condition.

Exercise 8.2-3

Derive (8.2.43) from (8.2.41) by utilizing that u_1, w_1 satisfy the equations for linear waves.

Discussion

There are two important observations to be made here.

First: We notice that the initial equations (8.1.1), (8.1.2) and (8.1.5) which are linear produce the same equations (8.2.33), (8.2.34), and (8.2.35) for second (and higher) order as the equations (8.2.23), (8.2.24), and (8.2.25) for first order.

Second: In the second order approximation the nonlinear boundary condition (8.2.42) or (8.2.43) differs from first order version, but in a very particular way. On the LHS ϕ_2 and η_2 occur in exactly the same way as ϕ_1 and η_1 do in the first order equations. The difference is that instead of 0 on the RHS (as in (8.2.26)), we now have a combination of terms all depending on the **first** order solution only. And since that solution is by now **known**, the RHSs of (8.2.36-8.2.37) represent known mathematical expressions.

In mathematical terms, the first order expressions (8.2.23–8.2.27) are **homogeneous**, the second order (8.2.33-8.2.37) are **inhomogeneous**.

If we think of a simple linear mechanical oscillation, a **homogeneous** equation represents free oscillations (with the natural frequency). Such oscillations may have any amplitude we want. In our wave case, this property shows up in the first order solution: we can choose any wave height we want. (The analogy with a simple one-dimensional oscillator fails at another point, however: we have infinitely many degrees of freedom and hence not **one** natural frequency. We can choose any wave frequency.)

Similarly, in the one dimensional oscillator, an inhomogeneous equation represents a **forced** oscillation with an amplitude and frequency determined entirely by the external forcing term (i.e., the inhomogeneity). In our second order approximation, the inhomogeneities in the equations are formed by combinations of first order terms. We therefore arrive at the following physical interpretation:

> The second order approximation for the wave motion is a forced oscillation, generated ("forced") by the first order wave. Hence, the second order solution will have the same frequency as the forcing term and its amplitude will also be fixed by the magnitude of that term.

If we go to higher approximations we will, of course, find that at each approximation the wave component of that order is forced by combinations of all the lower order solutions already determined.

The reason why a second order contribution is generated is apparently that the first order solution does not satisfy the free surface boundary conditions completely: we left out non-linear terms and let the conditions be satisfied at $z = 0$ instead of $z = \eta$. The second order terms that appear are the first approximation to correcting this. This second order improvement, however, will neglect terms $O(H/L)^3$ and hence a correction of that order of magnitude is required in the third order approximation. And so forth.

8.2.4 *The solution for ϕ_2*

As mentioned, the second order solution must have the same frequency as the forcing term. Hence, ϕ_2 must vary as $2(\omega t - kx)$ and we therefore seek solutions of the form:

$$\phi_2 = \phi_2' + Kx = F(2\theta) \cdot Z(z) + Kx \tag{8.2.44}$$

$$\theta = \omega t - kx \tag{8.2.45}$$

where we continue to consider ω the full unexpanded value. The significance of the last term will become apparent soon. The important thing is that it turns out that we can allow such a term to be included.

Substitution into Laplace's equation (8.2.33) yields:

$$F_{xx}Z + FZ'' = 0 \tag{8.2.46}$$

By the usual arguments for the separation method (see Chap 3) this splits into two equations

$$4k^2\frac{F''}{F} = -\frac{Z''}{Z} = -\lambda^2 \tag{8.2.47}$$

where λ is the separation constant. The complete solutions to (8.2.47) are

$$F = A_1 \sin\frac{\lambda}{2k}\,2\theta + A_2 \cos\frac{\lambda}{2k}\,2\theta \tag{8.2.48}$$

$$Z = B_1 \sinh \lambda z + B_2 \cosh \lambda z \tag{8.2.49}$$

The bottom boundary condition yields:

$$B_1 = B_2 \tanh \lambda h \tag{8.2.50}$$

$$Z = B_2\frac{\cosh\lambda(z+h)}{\cosh\lambda h} \tag{8.2.51}$$

The free surface condition (8.2.43) then gives:

$$g\left[A_1 \sin\frac{\lambda}{2k}2\theta + A_2\cos\frac{\lambda}{2k}2\theta\right] B_2\lambda\tanh\lambda h$$
$$-\omega^2\frac{\lambda^2}{k^2}\left[A_1 \sin\frac{\lambda}{2k}2\theta + A_2\cos\frac{\lambda}{2k}2\theta\right] B_2$$
$$= \frac{3}{4}gc(kH)^2\frac{\sin 2\theta}{\sinh 2kh} \tag{8.2.52}$$

To satisfy this equation for all θ, we must require that the cos 2θ and the sin 2θ terms vanish separately which gives the requirements:

$$\text{a)}\quad A_2 = 0 \tag{8.2.53}$$
$$\text{b)}\quad \lambda = 2k \tag{8.2.54}$$

and

$$\text{c)}\ 2gA_1B_2k\ \tanh\ 2kh - 4\omega^2A_1B_2 = \frac{3}{4}gc(kH)^2\frac{1}{\sinh 2kh} \tag{8.2.55}$$

At this point we substitute (8.2.9) for ω. However, since (8.2.55) is already $O(kH)^2$ contributions from ω_1 would be a term $O(kH)^3$, etc. Therefore to the second order approximation we still have

$$\omega^2 = \omega_0^2 = gk\ \tanh\ kh \tag{8.2.56}$$

which means that the dispersion relation for the second order approximation is the same as for linear waves. Using this, plus the trigonometric identities, we then find:

$$A_1B_2 = -\frac{3}{32}\frac{c^2}{k}(kH)^2\frac{\cosh 2kh}{\sinh^4 kh} \tag{8.2.57}$$

where c is given by (8.2.56) in combination with $c = \omega/k$.

Hence, we get for ϕ_2:

$$\boxed{\phi_2 = -\frac{3}{32}\,ckH^2\,\frac{\cosh 2k(z+h)}{\sinh^4 kh}\,\sin 2\theta + Kx} \tag{8.2.58}$$

Notice that as anticipated, the amplitude A_1B_2 of ϕ_2 cannot be chosen freely. It **is** a "forced oscillation".

Exercise 8.2-4

Show that in fact the equations allow a solution with $A_2 \neq 0$. Show that this solution corresponds to a wave with the frequency $\omega_2 = 2\omega$ that satisfies a dispersion relation similar to (8.2.56).

Deduce that this means that in addition to the forced solution (8.2.58), a **free second harmonic wave** is a solution to the second oder equations. This solution turns out to appear as an unwanted disturbance when waves are generated by a wave maker in a wave flume.

Exercise 8.2-5 Show that in deep water ($kh \to \infty$) we get:

$$\phi = \phi_1 + \phi_2 + \ldots = \phi_1 + 0(kH)^3 \tag{8.2.59}$$

that is, in deep water ϕ_2 vanishes.

Show also that in shallow water ($kh \to 0$) we have:

$$\phi_2 \gg \phi_1$$

and hence deduce that there must be a limit to how small a water depth the theory can be applied at.

8.2.5 *The surface elevation η*

The η_2 component of the surface elevation was eliminated from the two free surface conditions (8.2.36-8.2.37). We now go back and use (8.2.37) to determine η_2 (exactly as done for linear waves) inserting the expressions for ϕ_2, η_1 and ϕ_1.

We get:

$$\eta_2 = \frac{1}{8}kH^2\left[\frac{3 + 2\sinh^2 kh}{2\sinh^2 kh}\,\frac{\cos 2\theta}{\tanh kh} - \frac{1}{\sinh 2kh}\right] + \frac{C_2}{g} \tag{8.2.60}$$

Exercise 8.2-6 Derive (8.2.60).

We see that the expression for η_2 corresponds to a mean value $\overline{\eta}$ of η which is non-zero and can be written

$$\overline{\eta} = \overline{\eta_2} = -\frac{1}{8}\,\frac{kH^2}{\sinh 2kh} + \frac{C_2}{g} \tag{8.2.61}$$

However, we have defined η as the surface elevation above the MWS which means we must have $\overline{\eta} = 0$. Therefore we must require $\overline{\eta_2} = 0$ which means we must put

$$C_2 = g \, \frac{1}{8} \, \frac{kH^2}{\sinh 2kh} \tag{8.2.62}$$

Exercise 8.2-7

Show that this value of C_2 can also be written

$$C_2 = -\frac{1}{16} \, \frac{H^2}{h} \, G \tag{8.2.63}$$

where G as usual is defined as

$$G = \frac{2kh}{\sinh 2kh} \tag{8.2.64}$$

We then get the following result for η_2, the second order approximation to the surface elevation measured from the MWS

$$\eta_2 = \frac{1}{8} kH^2 \left[\frac{3 + 2\sinh^2 kh}{2\sinh^2 kh} \, \frac{\cos 2\theta}{\tanh kh} \right] \tag{8.2.65}$$

Exercise 8.2-8

Show that (8.2.65) may also be written:

$$\boxed{\eta_2 = \frac{1}{16} kH^2 (3\coth^2 kh - \coth kh) \cos 2\theta} \tag{8.2.66}$$

For completeness, the total result for η in second order Stokes waves can then be written

$$\boxed{\eta = \eta_1 + \eta_2 = \frac{H}{2} \cos\theta + \frac{1}{16} kH^2 \left(3\coth^3 kh - \coth kh\right) \cos 2\theta} \tag{8.2.67}$$

Fig. 8.2.1 shows surface profiles in terms of η for second order Stokes' waves for two different wave steepnesses at $kh = 1$. We see that the steeper the wave the higher and shorter the crest and the shallower and longer the trough, which is exactly what is observed in nature.

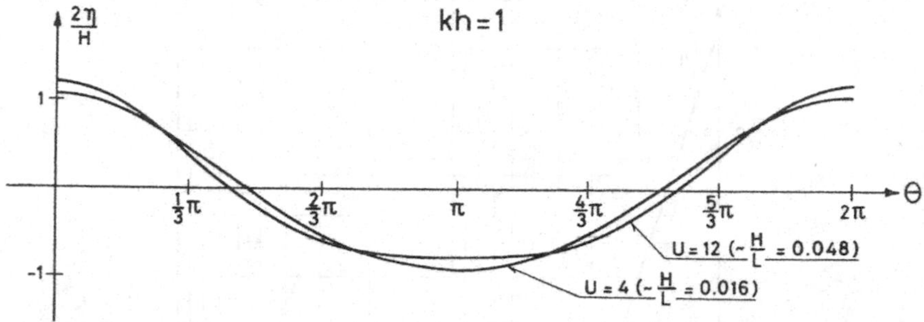

Fig. 8.2.1 Surface profiles for second order Stokes' waves for two different wave steepnesses at $kh = 1$ (SJ76).

Following Svendsen and Jonsson (1976) (denoted SJ76 in the following) the result for η and η_2 can be written in a more illustrative form if we introduce the definition

$$a_1 = \frac{H}{2} \tag{8.2.68}$$

We can then write η_2 as

$$\eta_2 = a_2 \cos 2\theta = a_1 \frac{H}{L} f_\eta(kh) \cos 2\theta \tag{8.2.69}$$

where f_η is defined as

$$f_\eta = \frac{\pi}{4} (3 \coth^3 kh - \coth kh) \tag{8.2.70}$$

We see that this confirms that the second order apprximation is O(H/L) times the first order solution as postulated in Section 3.2. Fig. 8.2.2 shows the variation of f_η with kh. As indicated earlier it is clear that f_η grows dramatically as $kh \to 0$, signifying the breakdown of the theory because we started out assuming $\eta_2 \ll \eta_1$.

Exercise 8.2-9

Show that, in the limit $kh \to \infty$, we get

$$f_\eta \to \frac{\pi}{2} \tag{8.2.71}$$

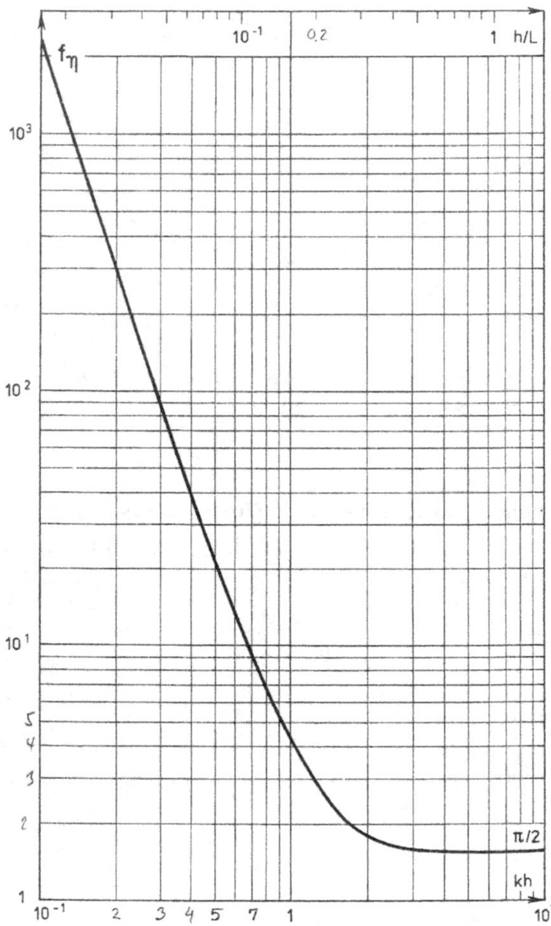

Fig. 8.2.2 Variation of f_η with kh (from SJ76).

as shown in the Fig. 8.1.7. Or

$$\eta_2 = \frac{1}{8}\, kH^2 \cos 2\theta \tag{8.2.72}$$

Thus we have the paradoxical result that for $kh \to \infty$ the second order contribution ϕ_2 to the velocity potential vanishes (see exercise 8.2-5) but the second order contribution η_2 to the surface elevation does not.

We also find that in the limit $kh \to \infty$ we get

$$\overline{\eta_2} \to 0 \tag{8.2.73}$$

Exercise 8.2-10

Show that for $kh \to 0$ we get

$$\frac{a_2}{a_1} \to \frac{3}{32\pi^2}\frac{H\,L^2}{h^3} \tag{8.2.74}$$

so that for shallow water we have

$$\eta = \eta_1 + \eta_2 = a_1\left(\cos\,\theta + \frac{3}{32\pi^2}\frac{H\,L^2}{h^3}\cos\,2\theta\right) \tag{8.2.75}$$

which shows that in shallow water the Ursell parameter HL^2/h^3 is the parameter that determines the magnitude of the second order term relative to the first order solution. This result was already found by Stokes (1880), which is why the parameter is also called the Stokes parameter.

We also notice that the factor $\frac{3}{32\pi^2} = 0.0095 = O(10^{-2})$ is actually quite small. Therefore even when the value of $\frac{H\,L^2}{h^3} \sim 25$ (as e.g. for $H/h = 0.5$ and $L/h = 7$) the second order term in (8.2.75) is only 25% of the first order term. For further discussion of this see Section 8.2.10.

8.2.6 *The pressure p*

As usual in potential theory the pressure is obtained from the Bernoulli equation. For p_2 (8.2.40) yields:

$$p_2 = \frac{1}{8}\rho g\,kH^2\left[\left(3\frac{\cosh 2k(z+h)}{\sinh^2 kh} - 1\right)\frac{\cos 2\theta}{\sinh 2kh} - \frac{\cosh 2k(z+h)}{\sinh 2kh}\right] + \rho C_2 \tag{8.2.76}$$

and if we substitute C_2 from (8.2.62), we get

$$p_2 = \frac{1}{8}\frac{\rho g k H^2}{\sinh 2kh}\left[\left(3\frac{\cosh 2k(z+h)}{\sinh^2 kh} - 1\right)\cos 2\theta + 1 - \cosh 2k(z+h)\right] \tag{8.2.77}$$

The total pressure p is then given by

$$p = -\rho\, g\, z + p_1 + p_2 = -\rho\, g\, z p_1 + p_2 \tag{8.2.78}$$

The expression for p_{D2} consists of a timevarying part changing with x and t as cos 2θ and some terms that do not depend on x and t.

It is instructive here to check the value of the mean pressure at the bottom. For $\overline{p(-h)}$ we have

$$\overline{p(-h)} = \rho\, g\, h + \overline{p_1} + \overline{p_2} = \rho\, g\, h \tag{8.2.79}$$

This is exactly the result that can be derived from the vertical momentum balance: at the bottom the mean pressure equals the weight of the water in the column above, or: the mean pressure at the bottom is the same with and without progressive waves. Hence the value assigned to C_2 above also ensures that this requirement is satisfied.

Often it is convenient to introduce the dynamic pressure as the pressure over and above the hydrostatic pressure defined as the pressure relative to the local MWS. That is we define the dynamic pressure p_D by

$$p = p_D - \rho\, g\, z \tag{8.2.80}$$

Going back to the definition of (8.2.8) for the perturbation expansion for p we see that this means writing p_D as

$$p_D = p_1 + p_2 \tag{8.2.81}$$

or,

$$p_{D2} = p_2 \tag{8.2.82}$$

Exercise 8.2-11

Show that for $kh \to \infty$ the amplitude of the oscillating part of the p_2 goes to

$$p_{2,max} \to -\frac{1}{8}\rho\, g\, kH^2\, \frac{\cosh\, 2k(z+h)}{\sinh\, 2kh} \tag{8.2.83}$$

and consequently $p_2 \to 0$ at the bottom.

Interpret this result.

Discussion of the physical implications of the results for η, p and C_2 *

Varying depth

The value of C_2 given by (8.2.62) clearly depends on the depth h. Therefore, if we consider the results presented above in a case where the depth changes (slowly) from one point to another then obviously the value of C_2 will change with x which is in conflict with the knowledge that C_2 is independent of x. The solution (see Chapter 11) is to place the x-axis (which **has** to be horizontal) at a level that differs from the MWS, and put $C_2 = 0$. Then the surface elevation above the x-axis will be measured by a new variable ζ for which we have $\overline{\zeta}(\neq 0)$ is the elevation of the MWS above the x-axis and

$$\eta = \zeta - \overline{\zeta} \tag{8.2.84}$$

where η remains the surface elevation measured from the MWS and given by the expressions derived above for Stokes waves.

Notice that the result (8.2.73) means that **in deep water** $\overline{\zeta} = 0$. It is also mentioned that the expression (8.2.63) will later (see Chapter 11.7) be shown to be the vertical distance between the MWS for a wave in deep water and the MWS for the same wave when it reaches depth h, that is $\overline{\zeta} = -C_2$.

The mean pressure

It is noted that for points above the bottom, $\overline{p} \neq \rho g z$. The reason for this is discussed in Chapter 11. However, we also see from the Bernoulli equation (8.2.40) that at arbitrary z (since $\overline{\phi_{2t}} = 0$), we have the mean second order pressure

$$\overline{p_2} = -\frac{1}{2}\rho\left(\overline{\phi_{1x}^2} + \overline{\phi_{1z}^2}\right) + \rho C_2 \tag{8.2.85}$$

on account of $\overline{p_1} = 0$. At the bottom (where $\phi_{1z} = 0$, and $\overline{p_2} = 0$ when the MWS is at $z = 0$), this expression implies that C_2 can be written

$$C_2 \equiv -\frac{1}{2g}\overline{(\phi_{1x}^2)_b} \tag{8.2.86}$$

(b = value at the bottom).

Exercise 8.2-12

Show that (8.2.86) can be generalized to the **exact** result for the Bernoulli constant C

$$C = -\frac{1}{2g}\,\overline{\phi_{x,b}^2} \tag{8.2.87}$$

for $z = 0$ at the MWS (i.e. the level defined by $\overline{\eta} = 0$). This simple result is useful for application in high order theories with $\overline{\eta} = 0$.

8.2.7 *The volume flux and determination of K*

In (8.2.44), we added, somewhat arbitrarily, a term Kx to the form of ϕ_2. In solving for ϕ_2 to find (8.2.58), we met no restriction on K. Thus the question arises: what determines K? Since the answer to that question turns out to be quite important for both higher order approximations and for treating waves on a current, the following discussion is kept more general than a second order theory requires.

First, we notice that the K-term in (8.2.58) represents a steady uniform flow with velocity

$$K = \frac{1}{T}\int_0^T \frac{\partial \phi}{\partial x} dt \tag{8.2.88}$$

Since this K did not influence the oscillatory term in the solution for ϕ_2, we conclude that we **can** freely add to the wave motion a steady uniform current with velocity K. Yet recall: K occurs in the second order equations so K can **at most** be $O(H/L)^2$

Within this limit, however, K can in principle be any current, both positive and negative. Before discussing how to determine K for a specific motion we first analyse the concept of mass flux in waves.

Volume flux in waves

One of the important nonlinear properties for surface waves is the mean volume flux $\overline{Q}$ for the waves defined by

$$\overline{Q} = \overline{\int_{-h}^{\eta} u dz} = \frac{1}{T}\int_0^T dt \int_{-h}^{\eta} u\, dz \tag{8.2.89}$$

where the overbar again means average over T.

In Section 3.3 we derived $\overline{Q}$ for linear waves and found that such waves have a volume flux Q_w given by (3.3.17). Here we extend this calculation to more general conditions.

In calculating this integral, it is useful first to define the time averaged velocity U under the wave trough level as:

$$U = \frac{1}{T}\int_0^T u dt = \overline{u} \qquad \text{for} \qquad z < \eta_{tr} \tag{8.2.90}$$

where η_{tr} is η at the wave trough. Thus, U is the mean velocity in the region of the flow where there is water all the time. We then split the velocity u into a purely oscillatory part u_w which has:

$$\overline{u_w} = 0 \tag{8.2.91}$$

below trough level and a steady part U so that:

$$u = u_w + U \tag{8.2.92}$$

Here we have assumed that U is independent of z. The integral in (8.2.89) is now determined in the following way:

$$\begin{aligned}\overline{Q} &= \overline{\int_{-h}^{\eta_{tr}} u_w \, dz} + \overline{\int_{\eta_{tr}}^{\eta} u_w \, dz} + \overline{\int_{-h}^{\eta} U \, dz} \\ &= 0 + \overline{\int_{\eta_{tr}}^{\eta} u_w \, dz} + U\overline{(\eta + h)}\end{aligned} \tag{8.2.93}$$

where the first integral is 0 on account of (8.2.91) and the fact that η_{tr} is independent of t. In the third term we have also assumed the value of U is extended above $z = \eta_{tr}$ to $z = \overline{\eta}$.

From (8.2.93) we can see that it is convenient in general to define the quantity:

$$Q_w = \overline{\int_{\eta_{tr}}^{\eta} u_w \, dz} = \overline{\int_{-h}^{\eta} (u - U) dz} \tag{8.2.94}$$

so that:

$$\boxed{\overline{Q} = Q_w + U(h + \overline{\eta})} \qquad \text{exact for } U \text{ independent of } z \tag{8.2.95}$$

Q_w is seen to be the volume flux of the purely oscillatory part of the wave motion. As found in Section 3.3 and seen from (8.2.102) below, this is a nonzero quantity.[1]

Second order approximation

In the lowest (second) order approximation considered above, we have:

$$u_w = \phi_{1x} + \phi'_{2x} \quad ; \quad U = K \tag{8.2.96}$$

where ϕ'_{2x} corresponds to the oscillatory part of ϕ_{2x}. (8.2.95) yields (choosing $\bar{\eta} = 0$)

$$\overline{Q} = \overline{\int_{\eta_{tr}}^{\eta} u_w \, dz} + Uh \qquad \text{exact} \tag{8.2.97}$$

$$= \overline{\int_{(\eta_1+\eta_2)_{tr}}^{\eta_1+\eta_2} (\phi_{1x} + \phi'_{2x}) \, dz} + Kh \qquad \text{correct to second order} \tag{8.2.98}$$

For the integrals in (8.2.98), we have:

$$\overline{\int_{(\eta_1+\eta_2)_{tr}}^{\eta_1+\eta_2} \phi_{1x} \, dz} = \overline{\int_{\eta_{1tr}}^{\eta_1} \phi_{1x} \, dz} + 0(H^3) \tag{8.2.99}$$

$$\overline{\int_{(\eta_1+\eta_2)_{tr}}^{\eta_1+\eta_2} \phi'_{2x} ez} = 0(H^3) \tag{8.2.100}$$

Thus the only contribution of $O(H^2)$ in the integral of (8.2.98) is the integral of (8.2.99) and so we get:

$$\bar{Q} = \overline{\int_{\eta_{1tr}}^{\eta_1} \phi_{1x} \, dz} + Kh \qquad \text{Correct to second order}$$

$$= \overline{(\eta_1 - \eta_{1tr}) \left(\phi_{1x}\right)_{z=0}} + Kh \qquad \text{Correct to second order} \tag{8.2.101}$$

or

$$\bar{Q} = \overline{\eta_1 \left(\phi_{1x}\right)_{z=0}} + Kh \qquad \text{Correct to second order} \tag{8.2.102}$$

[1] Q_w is also called the **Stokes drift**. Unfortunalely this name can be ambiguous because it is sometimes also used for other mean flows in waves. Thus Mei (1983) uses the term for the mean flow induced in a wave boundary layer. We therefore avoid the term here.

As found already in Section 3.3 the terms $\overline{(\eta_1 \phi_{1x})_{z=0}}$ is the lowest order volume flux and as (8.2.102) shows this is a $O(H^2)$ contribution which can be determined **knowing only the first order solution**. It is not until we want to determine the volume flux to $O(H^3)$, $O(H^4)$ etc. – accuracy that we need higher order approximations.

The value of K

The reason why K above seems to be a free parameter is that the problem we have considered so far is actually incomplete. Obviously, K must be determined by the boundary conditions in the x-direction and/or the initial conditions for the wave motion. This question is hardly ever discussed in the literature, but it is instructive to mention at least two canonical cases.

The **first** is a steady situation in a wave flume where the wave maker has been operating for a long time. This case was already mentioned in Section 3.3. If no current is added or extracted at the ends of the flume, the boundary conditions in the x-direction require that:

$$\overline{Q} = \overline{\int_{-h}^{\eta} u dz} = 0 \tag{8.2.103}$$

Obviously, to obtain this we must to the purely oscillatory motion add a net (or return) flow U so that:

$$\overline{Q} = Q_w + Uh = 0 \tag{8.2.104}$$

In the special case of zero net volume flux, the wave motion corresponds to a purely oscillatory part (giving a volume flux Q_w) and a ("return") current U in the opposite direction that compensates for Q_w. Thus we have:

$$U = -\frac{Q_w}{h} \qquad \text{exact} \tag{8.2.105}$$

$$= K \qquad \text{to second order} \tag{8.2.106}$$

and:

$$K = -\frac{1}{8}\, g\, \frac{H^2}{c} \qquad \text{to second order} \tag{8.2.107}$$

It is stressed that this value of K implies that at any point below the wave trough we have

$$\overline{u} = \overline{u_1} + \overline{u_2} = 0 + K \tag{8.2.108}$$

The **second** canonical case we want to consider corresponds to $U = 0$ (or to second order $K = 0$). This obviously means:

$$\bar{Q} = Q_w \qquad \text{exact} \tag{8.2.109}$$

$$= \frac{1}{8}\, g\, \frac{H^2}{c} \qquad \text{to second order} \tag{8.2.110}$$

i.e., **the waves have a volume flux**. In fact, any other value of U than (8.2.105) (i.e., K given by (8.2.107)) represents waves with a volume flux.

We notice that $K = 0$ means including only oscillatory terms in ϕ. This is the way results for Stokes waves are usually presented. For extensions of the theory to higher order of approximation, this turns out to give a far simpler description than the form with $\bar{Q} = 0$. Furthermore, in the ocean, far from horizontal boundaries, this seems a natural form if other effects, such as wind stresses, are disregarded. It is emphasized again, however, that strictly speaking the value of K is determined by the boundary conditions along the horizontal boundaries, and - if the flow is not steady - by the initial conditions.

For completeness it may be noted that both η, η and p are independent of the value of K.

8.2.8 *Stokes' two definitions of the phase velocity*

We are now ready to discuss the so-called "Stokes two definitions of the phase velocity c". Notice that unless otherwise stated, the following remarks apply to approximations of arbitrary high order.

In his 1880 papers, pp. 202–203, Stokes suggests that two definitions for the phase velocity c are possible. This has since been discussed widely in the literature. Some of the following points of view, however, do not seem to be represented.

The first definition

The "first definition" c originates from including only the oscillatory terms in ϕ, which — we have seen — for second order waves, means $K = 0$. Stokes states this clearly and calls this "definition" "the most convenient." Thus c corresponds to:

$$\bar{u} = \frac{1}{T} \int_0^T u dt = 0 \tag{8.2.111}$$

at any point **below the wave trough**, and the waves have a volume flux which is given by (8.2.109).

This also means that since the water below the trough-level has no mean velocity, a frame of reference from which the wave is seen to be stationary must move with the velocity c.

Or, in other words: c **is the speed of the wave relative to the water**. Hence, from a physical point of view, this **is** c for the wave, and also the c **determined from the dispersion relation.**

The second definition The "second definition" $c*$ corresponds to assuming (8.2.103). Since the wave still propagates with the velocity c relative to the water, we therefore have:

$$c* = c - \frac{Q_w}{h} \qquad \text{(exact)} \qquad (8.2.112)$$

$$= c - \frac{1}{8}\, g\, \frac{H^2}{c} \qquad \text{(to second order)} \qquad (8.2.113)$$

Thus $c*$ is actually the speed of the wave with a current $-Q_w/h$. (Or, more precisely, a wave seen from a coordinate system relative to which there is a current $-Q_w/h$.)

Stokes terms this "definition" "the most natural." Notice that by virtue of (8.2.112) c and $c*$ differ only by an amount $O(H^2)$.

Conclusion about c

To clarify, the conclusion of the above analysis is:

a) There is for physical reasons of course only one value of the phase velocity. A wave of given period (and height) on a given depth of water propagates with one, well defined speed relative to the water: c. And this is the c (also called the relative phase velocity c_r in Section 3.6) which comes out of the dispersion relation for the case $\bar{Q} = Q_w$.
b) Stokes' second "definition" $c*$ is a special case which actually corresponds to a wave on a particular current.

To second order we have seen that this zero-mass-flux-description gives no problems: The extra "return-flow" K required to balance the Stokes drift can freely be added.

For higher order approximations, however, the zero-volume-flux-description turns out to require a much more elaborate derivation because K and its higher order equivalents interact with the oscillatory terms. Such a derivation has never been carried out to higher than third order (see e.g., SJ76). Also for this reason it is, therefore, more convenient to use the description with only oscillatory terms and accept that this corresponds to a net volume flux $\bar{Q} = Q_w$.

8.2.9 *The particle motion*

The particle velocities

The particle velocities u, w are readiliy obtain by differentiating the expression

$$\phi = \phi_1 + \phi_2 \tag{8.2.114}$$

with (8.2.28) for ϕ_1 and (8.2.58) for ϕ_2.

Exercise 8.2-13 Show that the result is for the second order contributions u_2, w_2 are

$$u_2 = \frac{3}{16}\, c(kH)^2 \,\frac{\cosh\, 2k(z+h)}{\sinh^4 kh}\, \cos\, 2\theta \tag{8.2.115}$$

$$w_2 = -\frac{3}{16}\, \frac{\sinh\, 2k(z+h)}{\sinh^4 kh}\, \sin\, 2\theta \tag{8.2.116}$$

It is pointed out that in this Eulerian description of the velocity field the the average of the velocities over a wave period are zero.

Finally it is mentioned that if we go back to the derivation of the solution for ϕ_2 we will find that the boundary conditions at "the free surface" strictly speaking are only evaluated at the level of the first order solution $z = \eta_1$, not at the actual second order position of the surface (because the Taylor expansions from $z = 0$ were only carried to first order in z). Therefore the velocity field we have found is strictly speaking only valid below $z = \eta_1$. This limitation is often disregarded in practical applications of higher order wave theories, as is the equivalent limitation on velocities in the linear wave theory (which is only valid up to $z = 0$).

The particle paths

For the instantaneous particle position (x, z) we get to second order

$$x = \xi + x_1 + x_2 \tag{8.2.117}$$

$$z = \eta + z_1 + z_2 \tag{8.2.118}$$

where (ξ, η) denote the (imaginary) particle positions at rest. In parallel with the linear wave particle motion (Section 3.2) this represents a Lagrangian description of the wave motion.

Here the first order contributions (x_1, z_1) are the contributions already determined for the linear waves. To obtain the second order terms in (8.2.117) and (8.2.118) we need to integrate the particle velocities.

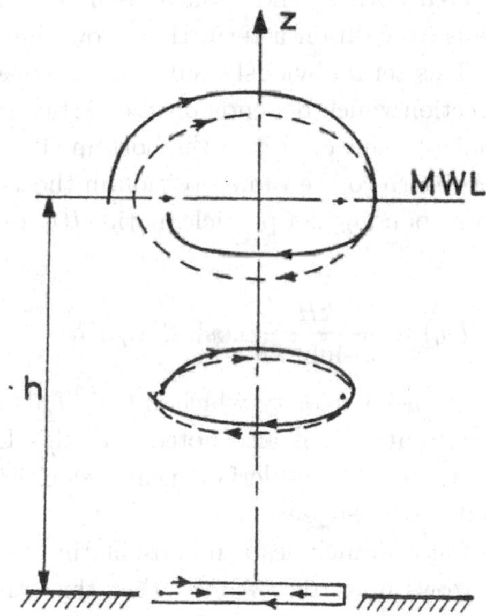

Fig. 8.2.3 The particle paths. The full curves are the paths in a second order Stokes wave motion, the dashed lines the equivalent linear wave solution. The equilibrium positions (ξ, η) of the particles are in the center of each first order particle path. In the second order approximation the particle paths are not closed curcuits but represent a net movement (SJ76).

Excercise 8.2-14

Show that the result of the integration is

$$x_2 = \frac{k\,H^2}{32\,\sinh^4 kh}\left(3\cosh\,2k(\eta+h) - 2\,\sinh^2 kh\right)\,\sin\,2\theta$$
$$+\frac{k\,H^2}{8\,\sinh^2 kh}\,(\cosh\,2k(\eta+h))\,\omega t \tag{8.2.119}$$

$$z_2 = \frac{k\,H^2}{32\,\sinh^4 kh}\left(3\,\cos\,2\theta + 2\,\sinh^2 kh\right)\,\sinh\,2k(\eta+h) \tag{8.2.120}$$

As could be expected both x_2 and z_2 have oscillatory terms depending on 2θ. For x_2 there is in addition a term that grows linearly with t, and independently of x. This term obviously represents a constant net motion in the positive x-direction which depends on the vertical position η of the particle with the smallest value closest to the bottom. Because of this term the particle does not return to the same position in the x-direction after a wave period. The corresponding net particle motion $l(\eta)$ over a wave period is given by

$$l(\eta) = \frac{\pi k H^2}{4\,\sinh^2 kh}\cosh\,2k(\eta+h) \tag{8.2.121}$$

which corresponds to a mean velocity which is l/T. This is also called the **Lagrangian drift velocity**. It is also noted that this Lagrangian mean velocity occurs even though the Eulerian mean velocities of the particle motion, as mentioned above, are zero.

The vertical postion also includes a non-oscillating term which is zero at the bottom and grows upwards. We see that this implies that in the vertical direction the oscillating motion of the particle takes place relative to a position which is elevated above the equilibrium position $z=\eta$. This can also be sensed in Fig. 8.2.4.

8.2.10 *Convergence and accuracy*

The presumption in perturbation theory is that the more terms we include in the expansion (i.e., the higher order the approximation) the more accurate is the approximation to the exact (albeit unknown) solution.

For convergence, this **at least** requires that beyond some value of n the terms A_n in the perturbation expansion are decreasing for increasing order. That is:

$$\frac{A_{n+1}}{A_n} < 1 \tag{8.2.122}$$

And to expect a good accuracy for a limited number of terms, we must rather require that for all n, even $n = 1$:

$$\frac{A_{n+1}}{A_n} \ll 1 \quad \text{as indication that} \quad \sum_{n+1}^{\infty} A_i \ll \sum_{1}^{n} A_i \tag{8.2.123}$$

so that the terms omitted represent insignificant corrections to the result.[2]

Since we have only derived two terms in the expansion there is not much we can do in terms of assessing the accuracy, except requiring that:

$$\frac{A_2}{A_1} < 1 \tag{8.2.124}$$

For ϕ this yields for $z = 0$ and $\sin\theta = 1$ the requirement:

$$\left[\frac{\phi_2}{\phi_1}\right] = \frac{3}{16} \frac{kH \cosh 2kh}{\cosh kh \cdot \sinh^3 kh} < 1 \tag{8.2.125}$$

Since in deep water $\phi_2 \to 0$, this means no restriction at all from ϕ for deep water waves. This is, of course, not realistic and may be changed by considering η instead.

For shallow water, (8.2.125) becomes:

$$U = \frac{HL^2}{h^3} < \frac{64\pi^2}{3} \simeq 210 \tag{8.2.126}$$

Notice that the Stokes-Ursell parameter HL^2/h^3 shows up here. (That was exactly where Stokes found it.)

$U = 210$, however, will give very unrealistic waves. The reason is that when the second order term becomes sufficiently large, it causes a secondary maximum to occur in the trough of the main wave both in η and (in a generalized sense) in other variables too. And this is not a real physical phenomenon for waves of constant form.

[2] As is known from mathematics even the first part of (8.2.123) is not a sufficient condition for ensuring the second part.

Hence, we may set up another and more realistic limit for the use of the second order Stokes theory: **the steepness must not be so large that a secondary maximum occurs in the wave trough for any of the variables.** It turns out that this criterion gives the sharpest constraint when applied to the surface elevation.

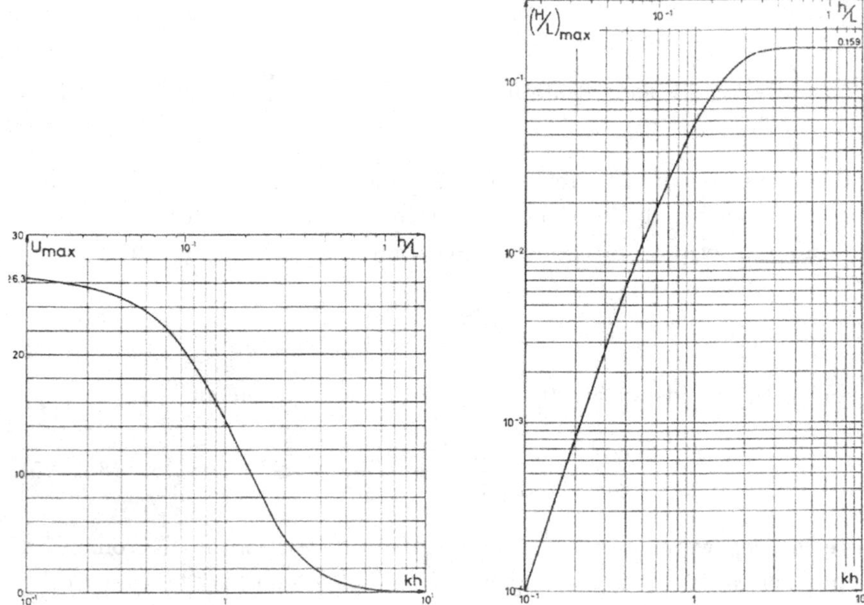

Fig. 8.2.4 Maximum Ursell parameter U and wave steepness H/L for the application of Stokes' second order theory based on the requirement of no secondary crest in the trough of the surface profile.

Exercise 8.2-15 Show that for η the requirement of no secondary crest in the trough of the surface profile means that:

$$a_2 < \frac{1}{4} a_1 \tag{8.2.127}$$

where a_1, a_2 are the amplitudes of η_1, η_2. In terms of U this means a maximum value of

$$U_{max} = \frac{1}{\pi} \left(\frac{L}{h}\right)^3 (3 \coth^3 kh - \coth kh)^{-1} \tag{8.2.128}$$

or, in terms of H/L:

$$\left(\frac{H}{L}\right)_{max} = \frac{1}{\pi}\left(3\coth^3 kh - \coth kh\right)^{-1} \tag{8.2.129}$$

which is far more restrictive than (8.2.126). Fig. 8.2.4 shows the variation of (8.2.128) and (8.2.129) with kh.

This is the criterion normally used today, and it clearly emphasizes the original finding that Stokes wave theory is a theory for intermediate and deep water. The restrictions on H/L become increasingly severe as h/L decreases and essentially prevents practical applications for $h/L <$ $0.07 - 0.10$.

Exercise 8.2-16

Work out similar curves for H/h.

Finally, it is mentioned that although extension to very high order improves the accuracy of the theory, the above mentioned conclusion cannot be changed: Stokes theory only works well for $h/L > 0.05$.

8.3 Higher order Stokes waves

8.3.1 *Introduction*

In the previous section, the second order approximation for Stokes waves was derived. The presentation went through the derivation and analysis of the results in quite some detail in order to illustrate the complications and the new features entering the theory when extending from linear waves to just the second approximation in the wave amplitude. In this section, we briefly describe how higher order approximations further modify the wave description, some of which may not be readily anticipated.

Of particular notice in the second order approximation was the fact that we could freely define a wave with zero mass flux simply by adding a net contribution to the phase velocity.

8.3.2 *Stokes third order theory*

Extension of the theory to third order will add two important features to the wave characteristics. These new features carry over to all higher order theories as well.

(1) The wave height H is not just $2a_1$ where a_1 is the amplitude of the first order term in the expression for the surface elevation.
If we consider the surface elevation in the third order approximation, this can be written

$$\eta = a_1 \cos\theta + a_2 \cos 2\theta + a_3 \cos 3\theta \tag{8.3.1}$$

We see that for this expression the wave height defined as usual as the vertical distance between wave trough and crest is given by

$$H = 2(a_1 + a_3) \tag{8.3.2}$$

and the analog expression for higher order approximations including all uneven terms. Thus, the wave height becomes a parameter separate from the first order amplitude.
If we want to express the surface elevation η in terms of the wave height, we therefore get in the third approximation (in dimensionless form)

$$k\eta = \frac{kH}{2}\cos\theta + \left(\frac{kH}{2}\right)^2 b_2 \cos 2\theta + \left(\frac{kH}{2}\right)^2 b_3(\cos 3\theta - \cos\theta) + O(kH)^5 \tag{8.3.3}$$

so that the third order term will contain contributions varying both as $\cos 3\theta$ and as $\cos\theta$, etc.

(2) The second new feature is that in third and higher order approximation, the phase velocity c_r relative to the water depends not just on the water depth h and the wave period T but also on the wave height, that is $c_r = c_r(kh, kH)$. As discussed below, this has profound implications for all situations of waves with a current. Thus we have

$$c = c_0 \left[1 + \frac{c_2}{c_0}(kH)^2 + \frac{c_4}{c_0}(kH)^4 + \dots\right] \tag{8.3.4}$$

where c_0 is the already known first order result

$$c_0 = \sqrt{\frac{g}{k} \tanh kh} \tag{8.3.5}$$

To third order, the expression becomes

$$c = \sqrt{\frac{g}{k} \tanh kh} \left[1 + \frac{(kH)^2}{32 \sinh 4kh} (9 + 8 \cosh 4kh - 8 \cosh 2kh) \right] + O(kH)^4 \tag{8.3.6}$$

which means that third order Stokes waves are both frequency and amplitude dispersive.

8.3.3 *Waves with currents*

In general situations, waves will usually occur in combination with a net current, where we continue to define the current velocity U as the time averaged part of the total velocity u below the wave trough, that is

$$U = \overline{u} \tag{8.3.7}$$

If a current U was added, this changes the absolute wave speed to c_a by the Doppler shift given by

$$c_a = c_r + U \tag{8.3.8}$$

In the first approximation, the dispersion relationship gave the connection between the relative wave frequency $\omega_r = \omega_r(kh)$ and the wave number k. For waves with a current, the equivalent dispersion relationship is

$$\omega_a = \omega_r(kh) + kU \tag{8.3.9}$$

When the wave is specified in terms of H, h and the absolute wave period T_a – as is usually the case – this equation can for first and second order Stokes waves be solved independently of the wave height. For details see Section 3.6 on the interaction between waves and currents.

However, since now we have the dispersion relation in the general form

$$\omega_r = (gk)^{1/2} f(kh, kH) \tag{8.3.10}$$

(8.3.9) takes the form

$$\omega_a = (gk)^{1/2} f(kh, kH) + kU \tag{8.3.11}$$

which means the wave height will influence the solution for k. Therefore, this dispersion relationship cannot be solved in advance. It needs to be solved along with all the other equations for the motion to determine k when the wave is specified in terms of T_a.

8.3.4 *Stokes fifth order theory*

The fifth order approximation for the Stokes wave theory has for many years been a standard reference wave theory used in offshore applications. A number of versions that differ at minor points are available in the literature. Examples are De (1955), Skjelbreia and Hendrickson (1960), Chappelear (1961). The very brief introduction given here is meant as an introduction to the version by Fenton (1985). See also Fenton (1990). The solution is derived in a coordinate system that follows the wave so that the flow is stationary. A stream function is used which leads to the same formulation with the same boundary conditions as for the Stream Function Method described in Section 8.4. The solution, is presented however, in terms of a velocity potential. The expansion used for the velocity potential ϕ has the form (still in a coordinate system following the wave that has y as the vertical coordinate and $y = 0$ at the bottom)

$$\phi = -c_r x + c_0 \left(\frac{g}{k3}\right)^{1/2} \sum_{i=1}^{5} \epsilon^i \sum_{j=1}^{i} A_{ij} \cosh jky \sin jkx_1 + O(\epsilon^6) \tag{8.3.12}$$

where $\epsilon = kH/2$ is the dimensionless first order wave amplitude proportional to the wave steepness H/L. This means the solution for ϕ is given in terms of a power series in ϵ with coefficients each of which are a series in $\sin\, jkx_1$ with a set of double indexed coefficients A_{ij}.

As before c_r is the relative phase velocity which has the expansion

$$c_r \left(\frac{k}{g}\right)^{1/2} = c_0 + \epsilon^2 c_2 + \epsilon^4 c_4 + O(\epsilon^6) \tag{8.3.13}$$

Similarly the solution operates with an expansion for the surface elevation η (measured all the way from the bottom rather than from the MWL) in the form

$$k\eta(x) = kh + \epsilon \cos kx_1 + \epsilon^2 B_{22} \cos 2kx_1 + \epsilon^3 B_{31} (\cos kx_1 - \cos 3kx_1) \\ + \epsilon^4 (B_{42} \cos 2kx_1 + B_{44} \cos 4kx_1) + \epsilon^5 (-(B_{53} + B_{55}) \cos kx_1 \\ + B_{53} \cos 3kx_1 + B_{55} \cos 5kx_1) + O(\epsilon^6) \quad (8.3.14)$$

Here x_1 is the horizontal coordinate in the moving system which is linked to a fixed x-system by

$$x_1 = x - c_a t \quad (8.3.15)$$

Furthermore the solution requires expansion of the volume flux Q in the steady flow and of the Bernoulli constant C still in powers of ϵ. In practical applications the moving wave is obtained by substituting (8.3.15) for x_1 in the results.

In Fenton (1985) the final results are given as a list of the analytical expressions for all the coefficients in the expansions described above. Substituted into the expansions this makes it possible to give both analytical and numerical evaluations of the fifth order approximation.

The above indicates how complicated the solutions become at higher order. For further details the reader is refered to Fenton (1985).

8.3.5 *Very high order Stokes waves*

The Stokes expansion has been carried to, in principle, arbitrary high order by Cokelet (1977). In practice this means about 110th order. The results are obtained by a combination of clever choice for the expansion variable, which is not just the wave height, combined with extrapolations using Padé approximations. In this way results have been obtained for properties of the highest possible waves. The results show several surprising features including the fact that the phase velocity and other parameters reach their maximum value at wave heights slightly smaller than the highest waves. For further details reference is made to Cokelet (1977).

8.4 The stream function method

8.4.1 *Introduction*

The Stream Function Method is a numerical solution to the exact governing equations and their boundary conditions. It is based on expanding the unknown stream function and surface elevation into fourier series. The solution is for a constant form wave on constant depth and therefore leads to wave solutions that are horizontally symmetrical with respect to verticals through crest and trough.

The Stream Function Method was first developed by Dean (1965). Later Chaplin (1980) found the solution technique used by Dean does not apply to the highest waves. Chaplin suggested an alternative procedure which, however, is rather complicated to implement and use. The approach described here is due to Rienecker and Fenton (1981) and it applies even to the highest possible waves (see also Svendsen and Justesen, 1984). Thus if sufficiently many terms are included in the expansions it can supplement the results for the highest waves described earlier in Chapter 5.

8.4.2 *Description of the stream function method*

The wave motion considered is assumed incompressible and irrotational. In a fixed (x, z)-coordinate system the waves are propagating in the positive x-direction with an absolute phase velocity c_a, including the effect of currents. In the fixed coordinate system the particle velocities are (u, w).

Because the waves are of constant form observing them from a coordinate system moving with the wave will show a steady flow. Thus in a (x_1, z) coordinate system moving with speed c_a we have

$$x_1 = x - c_a t \tag{8.4.1}$$

and the velocities (u_1, w)

$$u_1 = u - c_a \quad ; \qquad w = w \tag{8.4.2}$$

The stream function ψ is defined so that in the moving coordinate system we have

$$u_1 = \frac{\partial \psi}{\partial z} \quad ; \qquad w = -\frac{\partial \psi}{\partial x} \tag{8.4.3}$$

and ψ satisfies the Laplace equation

$$\frac{\partial^2 \psi}{\partial x_1^2} + \frac{\partial^2 \psi}{\partial z^2} = 0 \qquad (8.4.4)$$

The boundary conditions express that in the moving coordinate system the sea bottom and the free surface are streamlines along which ψ is constant with the values chosen as

$$\psi(x_1, -h) = 0 \quad ; \quad \psi(x_1, \eta(x_1)) = -Q \qquad (8.4.5)$$

The discharge Q is the total flow between bottom and surface and it may be written

$$-Q = \psi(x_1, \eta(x_1)) - \psi(x_1, -h) = \int_{-h}^{\eta(x_1)} u_1(x_1, z)dz \qquad (8.4.6)$$

In addition, the solution must satisfy the dynamic surface condition of constant pressure at the free surface. Thus at $z = \eta(x_1)$ we have

$$\frac{1}{2}\left(\left(\frac{\partial \psi}{\partial x_1}\right)^2 + \left(\frac{\partial \psi}{\partial z}\right)^2\right) + \eta(x_1) = C \qquad (8.4.7)$$

where C is the Bernoulli constant for the steady flow.

Following Rienecker and Fenton (1981) function can be expressed as the series

$$\psi(x_1, z) = B_0\ (z + h) + \sum_{j=1}^{N} B_j\ \frac{\sinh\ jk(z+h)}{\cosh\ jkD}\ \cos\ jkx_1 \qquad (8.4.8)$$

In this expression the denominator in the terms in the summation is a scaling factor that helps keeping the values of the the B-coefficients limited for large depths. Hence the value of D can be chosen freely, but a value of $D = h$ is appropriate.

In (8.4.8) each of the terms in the sum satisfy the Laplace equation (8.4.4) and the bottom boundary condition (8.4.5). The purpose is therefore to determine the coefficients $B_0, ..., B_N$ so that the surface boundary conditions are satisfied, and at the same time determine η, k, R and Q.

Substitution of (8.4.8) into (8.4.5) gives

$$B_0\ (\eta + h) + \sum_{j=1}^{N} B_j\ \frac{\sinh\ jk(\eta+h)}{\cosh\ kD}\ \cos\ jkx_1 = 0 \qquad (8.4.9)$$

and (8.4.7) gives

$$(\eta + h) + \frac{1}{2}\left(B_0 + k\sum_{j=1}^{N} jB_j\,\frac{\cosh\, jk(\eta+h)}{\cosh\, jkD}\,\cos jkx_1\right)^2$$

$$+\frac{1}{2}\left(k\sum_{j=1}^{N} jB_j\,\frac{\sinh\, jk(\eta+h)}{\cosh\, jkD}\,\sin jkx_1\right)^2 = C \quad (8.4.10)$$

Here (8.4.9) and (8.4.10) represent equations with the B_j's and the values of the η's as well as k, Q, and C are unknowns.

The solution is then obtained by requiring that (8.4.9) and (8.4.10) are satisfied at $N + 1$ points in x_1 where point 0 is at the wave crest, point N at the wave trough. This yields $2N + 2$ equations with $N + 1$ B_j's and η_j's plus k, Q and C as unknowns, that is a total of $2N + 5$ unknowns. The additional three equations are then obtained, the first by requiring that the wave height is H so that

$$\eta_0 - \eta_N - H = 0 \quad (8.4.11)$$

The second requirement is that the wave length L is linked to the phase speed c_a by $L = c_a T_a$ so that

$$kc_aT_a - 2\pi = 0 \quad (8.4.12)$$

and finally we require that the mean surface elevation is zero, which gives

$$\overline{\eta} = \frac{2}{L}\int_0^{L/2} \eta dx_1 = 0 \quad (8.4.13)$$

In these three equations c_a is also unknown. However, c_a is the speed of the wave in a fixed coordinate system so we have

$$c_a = c_r + U \quad (8.4.14)$$

where c_r is the speed of the wave relative to the water, and U the iniform current velocity. Further, since in the fixed coordinate system $\overline{u} = U$, (8.4.2) implies that $\overline{u_1} = U - c_a = -c_r = -B_0$ so that

$$B_0 = U - c_a \quad (8.4.15)$$

Therefore equations we have $2N + 6$ (transcendental) equations with as many unknowns. These equations can be solved on a given depth h by specifying the wave motion by its height H, period T and the current velocity U. It is mentioned that through the B_j-coefficients the solution obviously provides both the stream function and the velocity field determined by differentiating (8.4.8). However, part of the solution is also numerical values of the surface elevation given at as many points as there are terms in the expansions used. Hence with e.g. 64 terms in the expansion we get information about η at 64 points over half the wave length.

The solution clearly requires a generalized iterative solution scheme such as a Newton-Raphson method. For details see Rienecker and Fenton (1981).

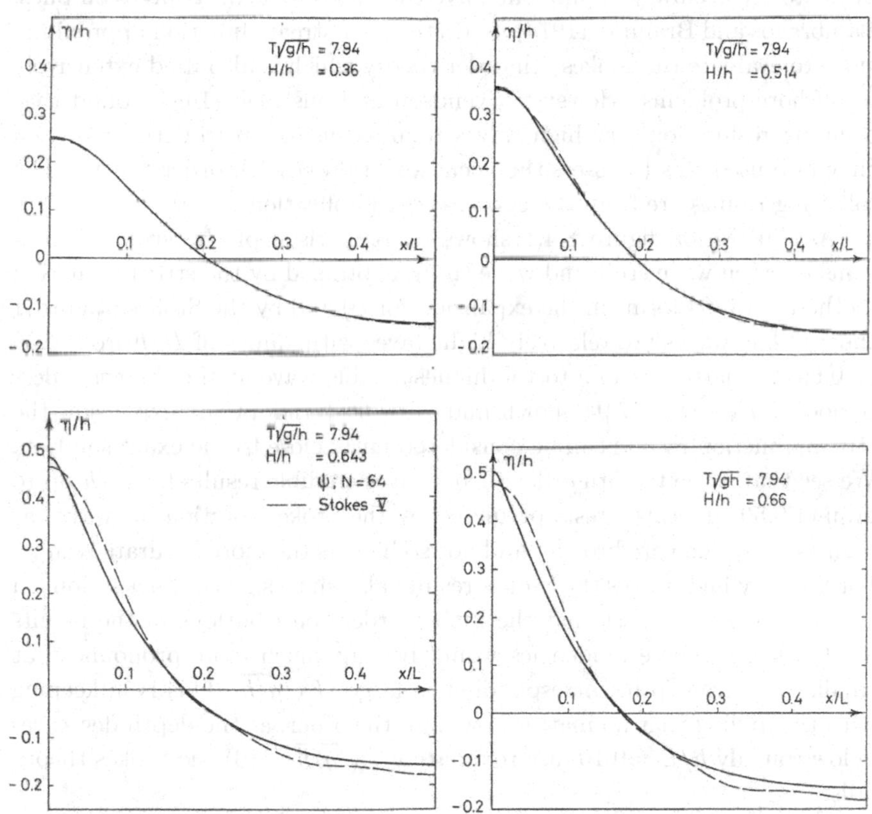

Fig. 8.4.1 Surface profiles for waves with different height H/h but all with $T\sqrt{g/h} = 7.94$. The dased curves represent Stokes 5th order theory, the full curve the stream function result with 64 terms in the expansion for ψ. From Svendsen and Justesten (1984).

8.4.3 *Comparison of stream function results with a Stokes 5th order solution*

The Stream Function method has been applied in particular in the offshore industry to assess the effect of very high waves on offshore structures. Dalrymple (1974) extended the method to cases with currents varying linearly over depth.

Since the stream function results can be made as accurate as wanted by including enough terms in the expansion it can be used to provide reference results for analytical wave theories. Thus Dean (1970) used the method to evaluate the accuracy of different wave theories for the drag forces on piles. Lambrakos and Brannon (1974) used 7th order stream function approximations to evaluate the Stokes 5th order theory which is also used extensively in offshore problems. However, Svendsen and Justesen (1984) found that accurate results for very high waves required using up to 64th order and they also used this to assess the accuracy of Stokes 5th order theory. The following results are from the last of those publications.

As illustration figure 8.4.1 shows a comparison of the surface variations between wave crest and wave trough obtained by the stream function method (with 64 terms in the expansion for ψ) and by the Stokes 5th order theory. The waves are relatively high waves with values of H/h from 0.36 to 0.66 The latter is close to the highest stable wave at the dimensionless period of $T\sqrt{g/h} = 7.94$ shown and with 64 terms in the expansion the stream function results can be considered fairly close to the exact solution. We see that while the 5th order theory gives credible results for H/h up to around 0.50 the wave crests predicted by the Stokes solution for waves as high as $0.64 - 0.66$ are broader and not so high as the more accurate results. For the very high waves the Stokes results also show signs of fluctuations in the trough that suggest that the higher order contributions in the results are too large. These tendencies would become much more pronounced at smaller relative depth, corresponding to larger $T\sqrt{g/h}$. This is in keeping with the finding that no matter how high the order, as the depth decreases below roughly $h/L = 0.10$ (approximately $T\sqrt{g/h} > 10$) the Stokes theory fails.

Similarly Fig. 8.4.2 shows results for the vertical profiles of the horizontal velocities under the wave crests. In addition to the results from the stream function method and from the 5th order Stokes theory the figure also shows for comparison the profiles predicted by the linear wave theory. As can be expected the largest deviations occur for the highest waves. It is particularly important from an application point of view that the velocities near the crest of the waves are the most inaccurately predicted by the linear wave theory and even the 5th order theory shows too small values near the top of the wave.

For a further and much more detailed discussion about these issues reference is made to the literature cited.

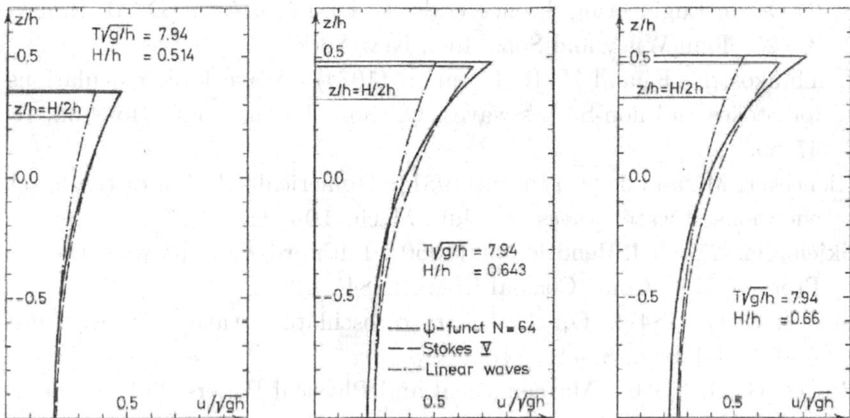

Fig. 8.4.2 Velocity profiles under the wave crests for three of the same waves shown in Fig. 8.4.1. In addition to the stream function results (full curve) and the Stokes 5th order theory (_ _ _) the profiles for linear waves are also shown (_ . _). The two nonlinear velocity profiles are extended to the z/h-value corresponding to the crest height in the respective wave theory, the linear profile is continued beyond that. From Svendsen and Justesten (1984).

8.5 References - Chapter 8

Chappelear, J. E. (1961). Direct numerical calculation of wave properties. J. Geophys. Re., **66**, 501 –508.

Chaplin, J. R. (1980). Developments of stream function theory. Coastal Engineering **3**, 3, 179 – 206.

Cokelet, E. D. (1977). Steep gravity waves in water of arbitrary uniform depth. Phil. Trans. Roy. Soc. London, **386**, 179 – 206.

Dalrymple, R. A. (1974). A finite amplitude wave on a linear shear current. J. Geophys. Res., **79**, 4498 – 4505.

De, S. C. (1955). Contributions to the theory of Stokes waves. Proc. Cambr. Phil. Soc., **51**, 713 – 736.

Dean, R. G. (1965). Stream function representation of nonlinear water waves. J. Geophys. Res., **70**, 18, 4561 – 4572.

Dean, R. G. (1970). Relative validities of water wave theories. Proc. ASCE, **90**, WW1, 105 – 119.

Fenton, J. D. (1985). A fifth-order Stokes theory for steady waves. Proc. ASCE J. WPCOE, **111**, 2.

Fenton, J. D. (1990). Nonlinear wave theories. Chapter 1 of *The Sea*, **9**: Ocean Engineering Science (Eds. B. LeMehauté and D. M. Hanes) 3 – 25. John Wiley and Sons, Inc., New York.

Lambrakos, K. F. and H. R. Brannon (1974). Wave force calculations for Stokes and non-Stokes waves. Offshore Techn. Conf., Houston. II, 47–60.

Rienecker, M. and J. D. Fenton (1981). Numerical solution of the exact equations of water waves. J. Fluid Mech. **104**, 119 –137.

Skjelbreia, L. and J. Hendrickson (1960). Fifth order gravity wave theory. Proc. 7th Int. Conf. Coastal Engrg., 184 – 196.

Stokes, G. G. (1847). On the theory of oscillatory waves. Trans. Cambridge Phil. Soc., **8**, 441 – 473.

Stokes, G. G. (1880). Mathematical and Physical Papers, Vol. 1. Cambridge University Press.

Svendsen, I. A. and I. G. Jonsson (1976, 1980) (SJ76). Hydrodynamics of coastal regions. Den Private Ingeiørfond, Technical University of Denmark. 285 pp.

Svendsen, I. A. and P. Justesen (1984). Forces on slender cylinders from very high waves and spilling breakers. Proc. Symp. Description and modelling of directional seas. Danish Hydraulic Institute, Hørsholm, Denmark. Paper D-7-1. 16 pp.

Chapter 9

Long Wave Theory

9.1 Introduction

In the present chapter we analyze closer the important case where the characteristic depth h_0 is much smaller than the characteristic horizontal length λ in the wave motion i.e. $\mu = \lambda/h_0 \ll 1$). Though λ is not always equal to the wave length these waves are usually called "long waves". At first we will particularly examine the case where the dimensionless wave amplitude $\delta = A/h_0 = O(\mu^2)$) so that the Ursell parameter HL^2/h^3 is $O(1)$. This is also meaningful because we found in Chapter 7 that the equations for the other two cases ($\delta \ll \mu^2$) and ($\delta \gg \mu^2$) can actually be deduced from this general case simply by omitting the proper terms in the basic equations.

For clarity, we look at the simplest case of waves moving in one horizontal direction only (chosen as the x-axis) and on a constant depth h. This is defendable since all the important mechanisms will still be present. We will also continue to consider the equations in dimensionless form. The advantage of working with the dimensionless version is that the magnitudes of terms extracted in Chapter 7 remain explicit which greatly helps guiding the derivations. However, to simplify the equations we omit the $'$ that indicated dimensionless variables in Chapter 7. In order to keep contact with the physical content essential results will be transformed back into physical variables in exercises.

For completeness it is recalled that for long waves the dimensionless system introduced in Chapter 7 reduces to

$$\begin{aligned} x &= \lambda x' \\ (z,h) &= h_0(z',h') \\ \eta &= A\,\eta' \\ t &= \frac{\lambda}{c_0}t' \\ \phi &= \frac{A\,c_0}{\mu}\phi' \end{aligned} \tag{9.1.1}$$

which implies the rules for differentiation are given by

$$\frac{\partial}{\partial x} = \frac{1}{\lambda}\frac{\partial}{\partial x'} \quad ; \quad \frac{\partial}{\partial z} = \frac{1}{h_0}\frac{\partial}{\partial z'} \quad ; \quad \frac{\partial}{\partial t} = \frac{c_0}{\lambda}\frac{\partial}{\partial t'} \tag{9.1.2}$$

The velocity scales thus follow from ϕ as δc_0. It is also recalled from Chapter 7 that while most dimensionless variables in this system are $O(1)$ we get $w = \phi_z = O(\mu^2)$.

Exercise 9.1-1

Derive the following useful relations:
From the difinition $c_0 = \sqrt{gh}$ it follows that

$$c_0' = \sqrt{h'} \tag{9.1.3}$$

Also show that since $u = \phi_x$ we get

$$u = \delta c_0 u' \tag{9.1.4}$$

where $\delta = A/h_0$.

On a domain with constant depth h it might seem natural to choose that depth as the characteristic depth h_0. That is frequently seen in the literature. However, doing so implies $h' = h/h_0 = 1$ which means the variation with h disappears from the equations, which is impractical for the physical understanding of the results. Hence, in the following, we use $h_0 \neq h$ and hence $h'(= \text{“}h\text{”}) \neq 1$.

Thus, the equations (4.2-37a-d) can be written as the Laplace equation

$$\phi_{xx} + \frac{1}{\mu^2}\phi_{zz} = 0 \tag{9.1.5}$$

with the bottom boundary condition

$$\phi_z = 0; \qquad z = -h \tag{9.1.6}$$

and the two free surface conditions

$$\eta_t + \delta\phi_x\eta_x - \frac{1}{\mu^2}\phi_z = 0 \quad ; \quad z = \delta\eta \tag{9.1.7}$$

and

$$\eta + \phi_t + \frac{1}{2}\delta\left[(\phi_x)^2 + \frac{1}{\mu^2}(\phi_z)^2\right] = 0 \quad ; \quad z = \delta\eta \tag{9.1.8}$$

9.2 Solution for the Laplace equation

The first step toward obtaining a solution to the problem of long waves is to develop a preliminary (or partial) solution for the Laplace equation (9.1.5) using only the bottom boundary condition (9.1.6). This will provide information about the vertical variation of ϕ.

For constant depth we assume ϕ can be written as an expansion from the bottom in the form[1]

$$\begin{aligned}\phi(x,z,t) &= \sum_{n=0}^{\infty}(z+h)^n\ \phi_n(x,t)\\ &= \phi_0(x,t) + (z+h)\phi_1(x,t) + (z+h)^2\phi_2(x,t) + \cdots\\ &\quad +(z+h)^n\phi_n(x,t) + \cdots\end{aligned} \tag{9.2.1}$$

This implies that (omitting the (x,t))

$$\phi_{xx} = \phi_{0xx} + (z+h)\phi_{1xx} + (z+h)^2\phi_{2xx} + (z+h)^n\phi_{nxx} + \cdots \tag{9.2.2}$$

and

$$\begin{aligned}\phi_{zz} &= 2\phi_2 + 2\cdot 3(z+h)\phi_3 + 3\cdot 4(z+h)^2\phi_4 + \cdots\\ &\quad +(n+1)(n+2)\phi_n + \cdots\end{aligned} \tag{9.2.3}$$

[1]For completenes it is mentioned that in the general case of varying depth it is more convenient to introduce an expansion from the undisturbed still water level, SWL.

The Laplace equation then becomes

$$\phi_{0xx} + (z+h)\phi_{1xx} + \cdots + (z+h)^n \phi_{nxx} + \cdots + \frac{2}{\mu^2}\phi_2 + \frac{6}{\mu^2}(z+h)\phi_3 + \cdots + \frac{(n+1)(n+2)}{\mu^2}(z+h)^n \phi_n + \cdots = 0 \tag{9.2.4}$$

or in compact form

$$\sum_{n=0}^{\infty} \left[\mu^2 \phi_{nxx} + (n+1)(n+2)\phi_{n+2}\right](z+h)^n = 0 \tag{9.2.5}$$

The philosophy is now the following: (9.2.4) or (9.2.5) represents a polynomium in $(z+h)$ which is required to be zero for all (i.e., infinitely many) values of $z+h$ in the interval $0 \le z+h \le h+\eta$. This is only possible if all the coefficients in the polynomium are zero separately. This leads to the following set of equations for $n = 0, 1, 2, \cdots$.

$$n = 0: \phi_{0xx} + \frac{2}{\mu^2}\phi_2 = 0 \tag{9.2.6}$$

$$n = 1: \phi_{1xx} + \frac{2\cdot 3}{\mu^2}\phi_3 = 0 \tag{9.2.7}$$

$$n = 2: \phi_{2xx} + \frac{3\cdot 4}{\mu^2}\phi_4 = 0 \tag{9.2.8}$$

$$\vdots$$

$$n: \phi_{nxx} + \frac{(n+1)(n+2)}{\mu^2}\phi_{n+2} = 0 \tag{9.2.9}$$

The latter of those corresponds to the general recursion formula

$$\boxed{\phi_{n+2} = \frac{-\mu^2}{(n+1)(n+2)}\phi_{nxx}} \tag{9.2.10}$$

which links ϕ_{n+2} to ϕ_n.

Using (9.2.10) repeatedly for the terms in (9.2.6–9.2.9), we then get (to $O(\mu^4)$ included)

$$\phi = \phi_0 + (z+h)\phi_1 - \frac{\mu^2}{1\cdot 2}(z+h)^2\phi_{0xx} - \frac{\mu^2}{2\cdot 3}(z+h)^3\phi_{1xx} + \frac{\mu^4}{2\cdot 3\cdot 4}(z+h)^4\phi_{0xxxx} + \frac{\mu^4}{2\cdot 3\cdot 4\cdot 5}(z+h)^5\phi_{1xxxx} + O(\mu^6) \tag{9.2.11}$$

In this equation, we have been able to express all the terms for n even by means of ϕ_0 and all uneven terms by means of ϕ_1.

We next substitute (9.2.1) into the bottom boundary condition (9.1.6) and get

$$\sum_{n=0}^{\infty}(n+1)\phi_{n+1}(z+h)^n = 0; \qquad z = -h \tag{9.2.12}$$

We see that the terms in this sum are all 0 at $z = -h$ except for $n = 0$ which gives

$$\phi_1 = 0 \tag{9.2.13}$$

Hence, combining (9.2.10) and (9.2.13), the first terms in the solution to the Laplace equation become

$$\boxed{\phi = \phi_0 - \frac{\mu^2}{2}(z+h)^2\phi_{0xx} + \frac{\mu^4}{24}(z+h)^4\phi_{0xxxx} + O(\mu^6)} \tag{9.2.14}$$

Notice that ϕ is an infinite series. Therefore, when we truncate the series we must indicate the magnitude of the first omitted term, here $O(\mu^6)$, which will be the magnitude of the sum of all omitted terms if the expansion (9.2.1) converges uniformly for $\mu \to 0$.

Exercise 9.2-1

Show that the complete solution for ϕ can also be written

$$\boxed{\phi = \sum_{n=0}^{\infty}(-1)^n\frac{\mu^{2n}}{2n!}(z+h)^{2n}\phi_{0(2nx)}} \tag{9.2.15}$$

where index $(2nx)$ means $2n$ differentiations with respect to x.

Discussion of the results

In conclusion, we have found that by solving the Laplace equation and utilizing only the bottom boundary condition, we have determined an expression for the vertical variation of the velocity potential ϕ expressed in terms of ϕ_0, and its horizontal derivatives where ϕ_0 is the value of ϕ at the bottom $z = -h$.

It is an important feature of the solution that, while the vertical variation of ϕ is represented by an infinite polynomial in $z + h$ with coefficients that include horizontal derivatives of ϕ_0, we have so far not met any restrictions on the surface elevation η (or δ).

It is also noticed that this implies that the solution found for ϕ is the same for all the three domains for the Ursell parameter discussed in Chapter 7.

Hence, ϕ_0 is from now on one of only two remaining unknowns, ϕ_0 and η. The second is the surface elevation η. For determining these two variables, we have left the two free surface boundary conditions (9.1.7) and (9.1.8). The analysis of this problem will be described in Section 9.3.

Exercise 9.2-2

For future reference, it is useful to show that

$$\phi_t = \phi_{0t} - \frac{\mu^2}{2}(z+h)^2\phi_{0xxt} + O(\mu^4) \tag{9.2.16}$$

$$\phi_z = -\mu^2(z+h)\phi_{0xx} + \frac{\mu^4}{6}(z+h)^3\phi_{0xxxx} + O(\mu^6) \tag{9.2.17}$$

$$\phi_x = \phi_{0x} - \frac{\mu^2}{2}(z+h)^2\phi_{0xxx} + O(\mu^4) \tag{9.2.18}$$

It will soon become clear why it is relevant to extend the expansion for ϕ_z to include $O(\mu^4)$ terms, while the expansions for ϕ_t and ϕ_x are only shown to $O(\mu^2)$.

Exercise 9.2-3

In fact, it is not necessary to assume the polynomial variation (9.2.1) of ϕ with respect to (z^n). Show for the 1-DH case with h = const that assuming

$$\phi = \sum_{j=0}^{\infty} A_j(z)\phi_j(x,t) \tag{9.2.19}$$

and solving the Laplace equation with the bottom boundary condition leads to the same result

$$\phi = \sum_{j=0}^{\infty}(-1)^j \frac{\mu^{2j}}{2j!}(z+h)^j\ \phi_{0(2jx)} \tag{9.2.20}$$

Exercise 9.2-4

Show that the dimensional form of ϕ is

$$\phi = \phi_0 - \frac{1}{2}(z+h)^2\phi_{xx} + \frac{1}{24}(z+h)^4\phi_{xxxx} + O(\mu^6) \qquad (9.2.21)$$

and for the velocity $u = \phi_x$

$$u = u_0 - \frac{1}{2}(z+h)^2 u_{xx} + \frac{1}{24}(z+h)^4 u_{xxxx} + O(\mu^6) \qquad (9.2.22)$$

9.3 The Boussinesq equations

The next step is to examine the two free surface boundary conditions using the expression (9.2.14) for ϕ. It turns out there are many ways this can be done and each way leads to a different form of the Boussinesq equations. These forms usually differ in ways which turn out to be essential for their suitability for computational purposes. This will be discussed in detail in Section 9.7 along with the transformation of the equations between the different forms.

For reference, it is recalled that in 1DH the two free surface boundary conditions are given by (9.1.7) and (9.1.8)

$$\eta_t + \delta\phi_x\eta_x - \frac{1}{\mu^2}\phi_z = 0 \qquad z = \delta\eta \qquad (9.3.1)$$

$$\eta + \phi_t + \frac{1}{2}\delta\left[(\phi_x)^2 + \frac{1}{\mu^2}(\phi_z)^2\right] = 0 \qquad z = \delta\eta \qquad (9.3.2)$$

The most direct approach is to substitute the expression (9.2.14) for ϕ into (9.3.1) and (9.3.2). For simplicity this is the procedure first followed here, although the resulting equations will later be found to have less desirable computational properties.

For convenience we define the local depth d as

$$d = h + \delta\eta \qquad (9.3.3)$$

When substituting for ϕ, the kinematic condition (9.3.1), we have from (9.2.17) that at the surface $z = \delta\eta$ the ϕ_z-term can be written

$$\frac{1}{\mu^2}\phi_z(z=\delta\eta) = -d\phi_{0xx} + \frac{\mu^2}{6}d^3\phi_{0xxxx} + O(\mu^4) \qquad (9.3.4)$$

We see that this expression includes terms with

$$d = (\delta\eta + h) \tag{9.3.5}$$
$$d^3 = h^3 + 3\delta h^2\eta + 3\delta^2 h\eta^2 + \delta^3\eta^3 \tag{9.3.6}$$

or, in other words, terms $O(1, \delta,\ \delta^2,\ \delta^3)$ which in the case of $U = O(1)$ (i.e., $\delta = O(\mu^2)$) are $O(\mu^2,\ \mu^4,\ \mu^6)$. In the final equation, we will neglect terms $O(\mu^4)$ and smaller. Hence, we will neglect terms $O(\mu^4,\ \delta\mu^2,\ \delta^2, \delta^3)$. Consequently, the version of (9.3.4) used in the kinematic condition can be written

$$\frac{1}{\mu^2}\phi_z = -h\phi_{0xx} - \frac{\mu^2}{6}h^3\phi_{0xxxx} + O(\mu^4,\ \delta\mu^2,\ \delta^2) \tag{9.3.7}$$

Notice that at this point it becomes clear why we needed in (9.2.17) to include $O(\mu^4)$ terms in the expression for ϕ_z: The ϕ_z-term is multiplied by $1/\mu^2$ which makes the $O(\mu^4)$ term in the expression for ϕ_z appear as a μ^2-term in the equation.

Similarly, the term $\delta\eta_x\phi_x$ in the kinematic condition (9.3.1) is truncated to

$$\begin{aligned}\delta\eta_x\phi_x &= \delta\eta_x\left(\phi_{0x} - \frac{\mu^2}{2}d^2\phi_{0xxx} + O(\mu^4)\right)\\ &= \delta\eta_x\phi_{0x} + O(\mu^4,\ \delta\mu^2,\ \delta^2)\end{aligned} \tag{9.3.8}$$

Consequently, the total kinematic condition becomes, to the order of accuracy chosen here

$$\eta_t + [(h + \delta\eta)\phi_{0x}]_x - \frac{\mu^2}{6}h^3\phi_{0xxxx} = O(\mu^4,\ \delta\mu^2,\ \delta^2) \tag{9.3.9}$$

By similar reasoning, we find the dynamic free surface condition can be written

$$\phi_{0t} + \eta + \frac{1}{2}\delta(\phi_{0x})^2 - \frac{\mu^2}{2}h^2\phi_{0txx} = O(\mu^4,\ \delta\mu^2,\ \delta^2) \tag{9.3.10}$$

The equations in terms of the bottom velocity

The two equations (9.3.9) and (9.3.10) can also be written in terms of the bottom velocity u_0. We get, after differentiation of (9.3.10) with respect to x

$$\eta_t + [(h + \delta\eta)u_0]_x - \frac{\mu^2}{6} h^3 u_{0xxx} = O(\mu^4,\, \delta\mu^2,\, \delta^2) \qquad (9.3.11)$$

$$u_{0t} + \eta_x + \frac{1}{2}\delta \left(u_0^2\right)_x - \frac{\mu^2}{2} h^2 u_{0txx} = O(\mu^4,\, \delta\mu^2,\, \delta^2) \qquad (9.3.12)$$

Both (9.3.9 and 9.3.10) and (9.3.11 and 9.3.12) are possible forms of the **Boussinesq equations.** As can be seen these equations are in the form of evolution equations for η and u_0 They will be analyzed and discussed in detail later. We notice that the nonlinear terms in these equations are all $O(\delta)$. Since $\delta = O(\mu^2)$ is small, the equations are often termed **"weakly nonlinear"**.

Exercise 9.3-1

Derive (9.3.10) in detail.

Exercise 9.3-2

Show that in dimensional form the Boussinesq equations (9.3.11 and 9.3.12) become

$$\eta_t + ((h + \eta)u_0)_x - \frac{1}{6}\, h^3\, u_{0xxx} = O(\mu^4,\, \delta\mu^2,\, \delta^2) \qquad (9.3.13)$$

$$u_{0t} + g\eta_x + \frac{1}{2}\left(u_0^2\right)_x - \frac{1}{2}\, h^2\, u_{0xxt} = O(\mu^4,\, \delta\mu^2,\, \delta) \qquad (9.3.14)$$

9.4 Boussinesq equations in one variable

9.4.1 *The fourth order Boussinesq equation*

To the order considered in the derivation of the Boussinesq equations either η or ϕ_0 can in fact be eliminated to generate equations in the other variable only. It appears that and equation in η only requires the extra constraint that the wave travel in one direction only, while an equation in ϕ_0 can be obtained without that constraint.

Here we first derive an equation in η only and do this by eliminating u_0 from the large terms in equations (9.3.11) and (9.3.12). Differentiating (9.3.11) with respect to t and (9.3.12) with respect to x and multiplying by h gives (omitting the order indications on the RHS)

$$\eta_{tt} + hu_{0xt} + \delta(\eta u_0)_x - \frac{\mu^2}{6} h^3 u_{0xxx} = 0 \tag{9.4.1}$$

$$h\eta_{xx} + hu_{0tx} - \frac{1}{2}\delta h(u_0^2)_{xx} + \frac{\mu^2}{2} h^3 u_{0xxxt} = 0 \tag{9.4.2}$$

Subtracting these two equations yields

$$\eta_{tt} - h\eta_{xx} + \delta(\eta u_0)_{xt} - \frac{1}{2}\delta h(u_0)^2_{xt} + \frac{\mu^2}{3} h^3 u_{0xxxt} = 0 \tag{9.4.3}$$

Lowest order approximation for $u_0(\eta)$

We see that to the lowest order of approximation (9.3.11) and (9.3.12) can be written

$$\eta_{tt} - h\eta_{xx} = O(\delta, \mu^2) \tag{9.4.4}$$

As found in Section 3.7 this is the linear shallow water wave equation, which has solutions that are the sum of constant form waves propagating in the positive and negative x-directions, respectively, that is (in dimensional form)

$$\eta = f(x - c_0 t) + g(x + c_0 t) \tag{9.4.5}$$

In order to eliminate u_0 from the small terms in (9.4.3) we must assume here that the waves are moving in one direction only, say the positive x-direction. Then we get (in dimensional form)

$$\frac{\partial}{\partial t} = -c_0 \frac{\partial}{\partial x} + O(\delta, \mu^2) \tag{9.4.6}$$

which in dimensionless form (using the definitions (9.1.2)) becomes

$$\frac{\partial}{\partial t} = -\frac{\partial}{\partial x} + O(\delta, \mu^2) \tag{9.4.7}$$

The wave equation also implies that to lowest order we have (dimensionally)

$$u_0 = c_0 \frac{\eta}{h} + O(\delta, \mu^2) \tag{9.4.8}$$

which in dimensionless form becomes

$$u_0 = \frac{\eta}{h} + O(\delta, \mu^2) \tag{9.4.9}$$

In the small terms in (9.4.3) these relations can then be used to eliminate u_0 and changing $\partial/\partial t$ to $\partial/\partial x$. Substituting we then get

$$\eta_{tt} - h\eta_{xx} - \frac{\delta}{h}(\eta^2)_{xx} - \frac{1}{2}\frac{\delta}{h}(\eta)^2_{xt} + \frac{\mu^2}{3}h^2\eta_{xxxx} = 0 \tag{9.4.10}$$

or, rearranging the terms

$$\boxed{\eta_{tt} - h\eta_{xx} - \left(\frac{3}{2}\frac{\delta}{h}\eta^2 + \frac{\mu^2}{3}h^2\eta_{xx}\right)_{xx} = 0} \tag{9.4.11}$$

This equation was first obtained by Boussinesq (1872).[2]

Exercise 9.4-1

The Boussinesq equation can be written in several different forms depending on the way (9.4.5) and (9.4.7) are used. It is also possible to derive a version in ϕ_0 in which η is eliminated. Show uisng a similar procedure as above that this equation reads

$$\begin{aligned}\phi_{0tt} - h\phi_{0xx} - \frac{\mu^2}{3}\, h^2(\phi_{0xx})_{tt} + \delta\left[({\phi^2}_{0x})_t + \frac{1}{2h}({\phi^2}_{0t})\right]_t \\ = O(\mu^4,\ \delta\mu^2,\ \delta^2)\end{aligned} \tag{9.4.12}$$

and that the derivation of this equation does not require the constraint of waves in one direction only. This is essentially the form given by Mei (1983).

Exercise 9.4-2 Show that in dimensional form (9.4.11) reads

$$\eta_{tt} - gh\eta_{xx} - c_0^2\left(\frac{3}{2h}\eta^2 + \frac{h^2}{3}\eta_{xx}\right)_{xx} = 0 \tag{9.4.13}$$

---

[2]Therefore, the equation was originally called "the Boussinesq equation" – not to be confused with the term "the Boussinesq equations" which is now normally used for (9.3.11) and (9.3.12).

It is also useful to realize that omitting the $O(\mu^2, \delta)$ terms in (9.4.11) or (9.4.12) gives the linear wave equation. Obviously the two fourth order equations are extensions of that equation that include (the first approximation in terms of $O(\delta)$ to) nonlinear processes in the wave and (the first approximation in terms of $O(\mu^2)$ to) effects of non-uniform vertical velocity profiles and non-hydrostatic pressures over the vertical. This will be discussed extensively in section 9.7.5.

9.4.2 *The third order Korteweg-deVries (KdV) equation*

In the case of uni-directional waves (waves moving in one direction only), it turns out that it is possible to derive a third order equation in one variable only. This was first done by Korteweg and deVries (1895) and the equation is therefore called the Korteweg-deVries (or just KdV) equation. This equation essentially represents a one time integrated version of the 4th order Boussinesq equation derived above. We again let $h = h_0$ and hence write the equation as

$$\eta_{tt} - \eta_{xx} - \left(\frac{3}{2}\delta\eta^2 + \frac{\mu^2}{3}\eta_{xx}\right)_{xx} = 0 \tag{9.4.14}$$

The derivation requires a trick, however. To describe this, we change to a coordinate system moving with the velocity $c_0 = \sqrt{gh}$. In dimensionaless variables this means

$$\xi = x - t \tag{9.4.15}$$

This system does not completely follow the waves because the actual phase velocity is $c = c_0 + O(\delta)$. We will therefore in the ξ-system observe slow changes of the motion as the waves pass and also as the waves change shape (non-constant form). To account for this we also need to include a slow variable τ given by

$$\tau = \delta t \tag{9.4.16}$$

Changing to the variable ξ, τ we have

$$\frac{\partial}{\partial x} = \frac{\partial}{\partial \xi}\frac{\partial \xi}{\partial x} = \frac{\partial}{\partial \xi} \tag{9.4.17}$$

$$\frac{\partial}{\partial t} = \frac{\partial}{\partial \xi}\frac{\partial \xi}{\partial t} + \frac{\partial}{\partial \tau}\frac{\partial \tau}{\partial t} = -\frac{\partial}{\partial \xi} + \delta\frac{\partial}{\partial \tau} \tag{9.4.18}$$

This is traditionally used to derive the KdV-equation from the 4th order equation (9.4.13) in ϕ (see e.g., Mei (1983)). Since, however, we are assuming unidirectional waves, we here choose (9.4.14), the 4th order equation in η.

Using (9.4.17) and (9.4.18) we get

$$\eta_{tt} = \eta_{\xi\xi} - 2\delta + \delta^2 \eta_{\tau\tau} \tag{9.4.19}$$

where the last term can be neglected, and

$$\eta_{xx} = \eta_{\xi\xi} \tag{9.4.20}$$

$$(\eta^2)_{xx} = (\eta^2)_{\xi\xi} = 2(\eta\eta_\xi)_\xi \tag{9.4.21}$$

$$\eta_{xxxx} = \eta_{\xi\xi\xi\xi} \tag{9.4.22}$$

Inserted into (9.4.12) that gives

$$-2\delta\eta_{\xi\tau} + \eta_{\xi\xi} - \eta_{\xi\xi} - 3\delta(\eta^2)_{\xi\xi} - \frac{\mu^2}{3}\eta_{\xi\xi\xi\xi} = O(\mu^4,\, \delta\mu^2,\, \delta^2) \tag{9.4.23}$$

or

$$\delta\eta_{\xi\tau} + \frac{3}{2}\delta(\eta\eta_\xi)_\xi + \frac{\mu^2}{6}\eta_{\xi\xi\xi\xi} = O(\mu^4,\, \delta\mu^2,\, \delta^2) \tag{9.4.24}$$

Notice that his equation only contains terms $O(\delta, \mu^2)$. It can be integrated once with respect to ξ which gives

$$\delta\eta_\tau + \frac{3}{2}\delta\eta\eta_\xi + \frac{\mu^2}{6}\eta_{\xi\xi\xi} + C(\tau) = O(\mu^4,\, \delta\mu^2,\, \delta^2) \tag{9.4.25}$$

Here $C(\tau)$ is an arbitrary function, which represents a ξ- (i.e. space-) independent contributions to η_τ. However, averaged over ξ such changes must for continuity reasons be zero, so that $C(\tau) = 0$, and the resulting equation becomes

$$\delta\eta_\tau + \frac{3}{2}\delta\eta\eta_\xi + \frac{\mu^2}{6}\eta_{\xi\xi\xi} = O(\mu^4,\, \delta\mu^2,\, \delta^2) \tag{9.4.26}$$

Eq. (9.4.26) is the KdV-equation in a coordinate system moving with the speed c_0.

Eq. (9.4.26) is transformed back to fixed (x, t) coordinates by inverting the equations (9.4.15) and (9.4.16) for $(\xi,\ \tau)$ to give

$$x = \xi + \frac{\tau}{\delta} \qquad t = \frac{\tau}{\delta} \tag{9.4.27}$$

This implies substituting into (9.4.26) that

$$\frac{\partial}{\partial \xi} = \frac{\partial}{\partial x} \quad ; \quad \frac{\partial}{\partial \tau} = \frac{1}{\delta}\frac{\partial}{\partial x} + \frac{1}{\delta}\frac{\partial}{\partial t} \tag{9.4.28}$$

Used in (9.4.26) this results in the KdV-equation for fixed coordinates

$$\boxed{\eta_t + \eta_x + \frac{3}{2}\delta\eta\eta_x + \frac{\mu^2}{6}\eta_{xxx} = 0} \tag{9.4.29}$$

It is noted that the η_τ, which was $O(\delta)$ in the coordinate system moving with velocity c_0, becomes $\eta_t + \eta_x$ in the fixed coordinate system. While this sum of course remains $O(\delta)$, η_t and η_x separately are $O(1)$.

Exercise 9.4-3

Show that in dimensional form, this equation becomes

$$\eta_t + c_0\eta_x + \frac{3c_0}{2h}\eta\eta_x + \frac{c_0 h^2}{6}\eta_{xxx} = 0 \tag{9.4.30}$$

Discussion of the KdV-equation

It is of interest to notice here that (9.4.29) is not the only possible form of the KdV-equation. And neither is (9.4.11) the only possible form of the 4th order Boussinesq equation. This follows from the fact that (9.4.6) applies and can be used to interchange the derivatives with respect to x and t, in the small terms of (9.4.29). This is of particular interest for the third order term, because changing any of the x-differentiations of the highest order term changes the nature of the equation.

This ambiguity was first pointed out by Benjamin et al. (1972) who suggested that using (9.4.6) to change one of the x-differentiations in (9.4.29) results in the alternative form

$$\eta_t + c_0\eta_x + \frac{3}{2}\frac{c_0}{h}\,\eta\eta_x - \frac{\mu^2}{6}h^2\eta_{xxt} = 0 \tag{9.4.31}$$

Used in its full consequences, this means that to the order considered (9.4.29) has several other equivalent forms. Following Mei (1983), the following forms are all equivalent to the order of approximation for (9.4.30).

$$\eta_t + c_0\eta_x + \frac{3c_0}{2}\begin{Bmatrix}\eta_x \\ -\frac{1}{c_0}\eta_t\end{Bmatrix} + \frac{\mu^2}{6}c_0 h^2 \begin{Bmatrix}\eta_{xxx} \\ -\frac{1}{c_0}\eta_{xxt} \\ \frac{1}{c_0^2}\eta_{xtt} \\ -\frac{1}{c_0^3}\eta_{ttt}\end{Bmatrix} = 0 \tag{9.4.32}$$

Benjamin et al. (1972) already mentioned the better numerical behaviour of (9.4.31). The reasons behind this will be analysed and discussed extensively in Section 9.7 for Boussinesq equations in general.

As mentioned, this equation was first derived in 1895 by Korteweg and deVries, and used by them to also derive the analytical solution for cnoidal waves described in Section 9.5. However, it was not until the advent of digital computers in the 1950's and 1960's that the equation itself became the focal point of extensive studies that have helped provide valuable insight into the mechanisms of nonlinear water waves. The interest in the equation was further enhanced by the fact that similar equations were found to apply to the propagations of nonlinear waves in many other physical contexts including nonlinear optics, nonlinear accoustics, nonlinear elastic materials, etc. FInally immense attention was generated by the development of the so-called **inverse scattering method** which proved the existense of and ways to determine exact time-varying solutions to the KdV equation for a wide range of initial conditions.

Briefly described, the lowest order of the equation, i.e. the first two terms in the equation

$$\eta_t + c_0\eta_x = 0 \tag{9.4.33}$$

describes the motion of a linear wave that moves in the $+x$ direction with velocity c_0 without change of shape.

However, the two small terms influence both the propagation velocity c and the shape of the wave. As can be seen from (9.4.29), the first of those terms is nonlinear and represents (the first approximation to) the effect of the finite amplitude of the waves. The effect that the amplitude of the wave influences the the phase speed is termed **amplitude dispersion**. See also Chapter 9. This effect will have a tendency to destabilize the wave by making the crest move faster than the trough (see Section 9.12).

The second term, which is linear, represents (as can be seen by going back to the origin of this term in (9.2.14)) the effect of the depth variation of the horizontal velocity. Further inspection would reveal that this depth variation is also associated with deviation of the pressure from simple hydrostatic pressure relative to the local surface level in the wave motion. This feature has a tendency to stabilize the wave by preventing curvatures on the surface from becoming too large. It will be shown in Section 9.7 that this also affects the speed c of the wave. Since this effect depends on the depth to wave length ratio (as indicated by the $-\mu^2$-factor in (9.4.29)) this means the waves are also **frequency dispersive**.

In Boussinesq and KdV-waves it is assumed that $U = \delta/\mu^2 = O(1)$ which means that the two small terms are of the same order of magnitude. This is sometimes described as frequency and amplitude dispersion tend to balance each other.

As we can see from the discussion in Chapter 7, the case where the Ursell parameter $U = \frac{\delta}{\mu^2} \ll 1$ will cause the μ^2-term in (9.4.26) to dominate over the nonlinear δ-term, making (9.4.26) linear. Similarly, if $U \gg 1$, we have $\delta \gg \mu^2$ and the linear term becomes negligible.

Further aspects of all this are discussed in succeeding chapters.

9.5 Cnoidal waves – Solitary waves

Within the assumption of $U = O(1)$, 1DH and constant depth, the Boussinesq equations (9.3.11) and (9.3.12) do not place any constraint on the type of waves considered. In fact, when solved numerically, the equations can describe the development in x, and t of random waves. Under such general conditions, it is not surprising that no analytical solution is possible.

However, if the very simple restriction is introduced that the waves considered propagate in one direction only — the direction toward positive x, say — then it turns out that it is possible to solve the equations analytically for **waves of constant form**.

The solution was termed **cnoidal waves** (in analogy to sinusoidal waves) by Korteweg and DeVries (1895) who first discovered this solution. The reason is that the surface profile is described by the elliptic cn-function. An advantage of cnoidal wave theory is that it gives the analytical solution for constant form weakly nonlinear Boussinesq waves. Hence it represents the equivalent to sinusloidal wave theory for linear waves. This makes it possible to use cnoidal wave theory to check some of the principal properties of weakly nonlinear Boussinesq waves.

Cnoidal wave theory is applicable in sufficently shallow water where it is more accurate than linear wave theory. For a discsussion of the limitations of the theory see Section 9.6.4. It also avoids the limitations of 2nd and higher order Stokes theory for small depths discussed in Chapter 8. A characteristic feature is that the dispersion relation includes the wave height so that the propagation speed depends on the wave height. Another important characteristic we shall see is that the shape of the surface variation for a wave on a given depth and with a given period depends on the wave height. in fact it will be shown, more precisely, that the wave shape and most other cnoidal wave properties are functions of the Ursell parameter U only.

To solve (9.4.11) we assume waves of constant or "permanent" form. In Chapter 3 it was shown that assuming constant form implies that the wave motion depends on x and t only in the combination

$$\theta = \frac{t}{T} - \frac{x}{L} \tag{9.5.1}$$

so that $\eta(x,t) = \eta(\theta)$. By the chain rule this implies that

$$\eta_t = \frac{1}{T}\,\eta_\theta \qquad ; \qquad \eta_x = -\frac{1}{L}\,\eta_\theta \tag{9.5.2}$$

By eliminating η_θ between these two equations, we see that this also means that

$$\eta_t = -c\eta_x \tag{9.5.3}$$

where as usual

$$c = \frac{L}{T} \tag{9.5.4}$$

is the phase velocity for the wave.

Since we are considering uni-directional waves the Bossinesq equation (9.4.11) can be used for the analysis. We see that introducing θ into the Boussinesq equation (9.4.11) brings this equation on the form (henceforth omitting the order information $O(\mu^4,\ \delta\mu^2,\ \delta^2)$)

$$(c^2 - h)\eta_{\theta\theta} - \frac{3}{2}\delta(\eta^2)_{\theta\theta} - \mu^2\frac{h^3}{3}\theta_x^2\eta_{\theta\theta\theta\theta} = 0 \tag{9.5.5}$$

For convenience, we introduce the following shorthand

$$h\theta_x = h^* = -\frac{h}{L} \tag{9.5.6}$$

where we remember that both h and L are dimensionless quantities (based on the characteristic length λ). We also introduce the definition

$$b \equiv c^2 - h \tag{9.5.7}$$

Hence we get for (9.5.5)

$$b\eta_{\theta\theta} - \frac{3}{2}\delta(\eta^2)_{\theta\theta} - \mu^2 \frac{h^{*2}h}{3}\,\eta_{\theta\theta\theta\theta} = 0 \tag{9.5.8}$$

This equation can be integrated directly twice, to give

$$b\eta - \frac{3}{2}\delta\eta^2 - \mu^2\frac{h^{*2}h}{3}\,\eta_{\theta\theta} + C_0\theta + \frac{1}{2}\delta C_1 = 0 \tag{9.5.9}$$

However, if we want periodic solutions with some period θ_0, we must require that $\eta(\theta) = \eta(\theta + \theta_0)$, etc. so that $C_0 = 0$. We then get for (9.5.9)

$$b\eta - \frac{3}{2}\delta\eta^2 - \mu^2\frac{h^{*2}h}{3}\eta_{\theta\theta} + \frac{1}{2}\delta C_1 = 0 \tag{9.5.10}$$

We next multiply (9.5.10) by $6\eta_\theta$ which gives

$$(6b\eta - 9\delta\eta^2 + 3\delta C_1)\eta_\theta - 2\mu^2 h^{*2} h \eta_{\theta\theta}\eta_\theta = 0 \tag{9.5.11}$$

Here we have that

$$\left.\begin{aligned} \eta\eta_\theta &= \tfrac{1}{2}(\eta^2)_\theta \\ \eta^2\eta_\theta &= \tfrac{1}{3}(\eta^3)_\theta \\ \eta_{\theta\theta}\eta_\theta &= \tfrac{1}{2}(\eta_\theta^2)_\theta \end{aligned}\right\} \tag{9.5.12}$$

so that (9.5.11) can be written

$$3b(\eta^2)_\theta - 3\delta(\eta^3)_\theta + 3\delta C_1\eta_\theta - \mu^2 h^{*2}(\eta_\theta^2)_\theta = 0 \tag{9.5.13}$$

We see that this equation can be integrated once more to yield

$$3\delta C_1\eta + 3b\eta^2 - 3\delta\eta^3 - \mu^2 h^{*2} h\eta_\theta^2 + 3\delta C_2 = 0 \tag{9.5.14}$$

where $3\delta C_2$ is the integration constant. Thus we have

$$\eta_\theta^2 = \frac{3\delta}{\mu^2 h^{*2} h}\left\{C_2 + C_1\eta + \frac{b\eta^2}{\delta} - \eta^3\right\} \tag{9.5.15}$$

C_1 and C_2 remain unknown for a while and so does b. (9.5.15) is an ordinary first order differential equation, the solution of which can be obtained just by separating variables. We get

$$\int \frac{d\eta}{\sqrt{-\eta^3 + \frac{b}{\delta}\eta^2 + C_1\eta + C_2}} = \frac{1}{\mu h*}\sqrt{\frac{3\delta}{h}} \int d\theta + C \tag{9.5.16}$$

9.5.1 *The periodic case: Cnoidal waves*

This is the general case and, as we shall see, it turns out that in this case $C_1, C_2 \neq 0$. This implies that the integral on the left hand side is an elliptic integral, the behavior of which depends on the cubic under the square root. The simpler case with $C_1, C_2 = 0$ is treated in section 9.5.3. Directions for practical applications and how to evaluate cnoidal wave formulas are given in Appendix 9C.

The detailed derivation given below is standard for what is necessary to resolve elliptic integrals involving the $\sqrt{\ }$ of a cubic, but is a little more complicated here by the fact that the cubic contains three unknown constants, namely b, C_1, and C_2.[3]

It is convenient to define

$$M^2 = \frac{3\delta}{\mu^2 h^{*2} h} \tag{9.5.17}$$

and to write the cubic which represents η_θ^2 as

$$-P(\eta) = -\eta^3 + \frac{b\eta^2}{\delta} + C_1\eta + C_2 \tag{9.5.18}$$

so that η_θ can be written

$$\eta_\theta^2 = M^2(-P(\eta)) \tag{9.5.19}$$

If we now choose $\theta = 0$ in the wave crest where $\eta = \eta_1$ is assumed, we can write (9.5.16) as

$$\int_\eta^{\eta_1} \frac{d\eta}{\sqrt{-P}} = M \int_\theta^0 d\theta = -M\theta \tag{9.5.20}$$

[3] A minor simplification may be obtained by shifting the zero-level for η to the wave trough where also η_θ is zero. This was done by Korteweg and de Vries in their presentation (1895) and yields that one of the roots in the cubic (corresponding to η_2 below) becomes zero. The gain is small, however, and for reasons of later practical applications we choose the general approach here.

Here the LHS is in one of the standard forms for elliptic integrals listed in Abramowitz and Stegun (1964) (A & S in the following). In order to completely identify the type of solution to the integral we first show that P has three real roots. For this we factorize $-P$ as

$$P(\eta) = (\eta - \eta_1)(\eta - \eta_2)(\eta - \eta_3) \tag{9.5.21}$$

where η_1, η_2 and η_3 are the roots. By identification we find that

$$\frac{b}{\delta} = \eta_1 + \eta_2 + \eta_3 \tag{9.5.22}$$

$$C_1 = -(\eta_1\eta_2 + \eta_1\eta_3 + \eta_2\eta_3) \tag{9.5.23}$$

$$C_2 = \eta_1\eta_2\eta_3 \tag{9.5.24}$$

and we may without loss of generality assume that

$$|\eta_1| \geqq |\eta_2| \geqq |\eta_3| \tag{9.5.25}$$

In order for a wave with a real η_θ to exist we must have an interval where η_θ^2 is positive. We see that $-P$ and hence $\eta_\theta^2 \to -\infty$ for $\eta \to \infty$ and $\eta_\theta^2 \to \infty$ for $\eta \to -\infty$. Hence the area of interest will be between η_2 and η_1 where $-P > 0$ i.e.

$$\eta_2 \leqq \eta \leqq \eta_1 \tag{9.5.26}$$

Therefore η_1 will represent the largest η possible (the wave crest), η_2 the smallest (the wave trough), and we can conclude that the wave height H is given by

$$H = \eta_1 - \eta_2 \tag{9.5.27}$$

If we assume that the mean value of η over a wave length is zero (which just implies a choice of the vertical position of the x-axis) then we must have $\eta_2, \eta_3 < 0$, $\eta_1 > 0$. It also implies, however, that η_1 and η_2 are real and hence η_3 must be real too, so that $-P$ has three real roots.

Fig. 9.5.1 shows a sketch of $-P$ satisfying these conditions.

For a $-P$ with three real roots the integral on the LHS of (9.5.20) corresponds directly to the case described in A & S equation (17.4.69), which reads

$$\int_\eta^{\eta_1} \frac{d\eta}{\sqrt{-P}} = \frac{1}{\lambda} F(\phi, m) \tag{9.5.28}$$

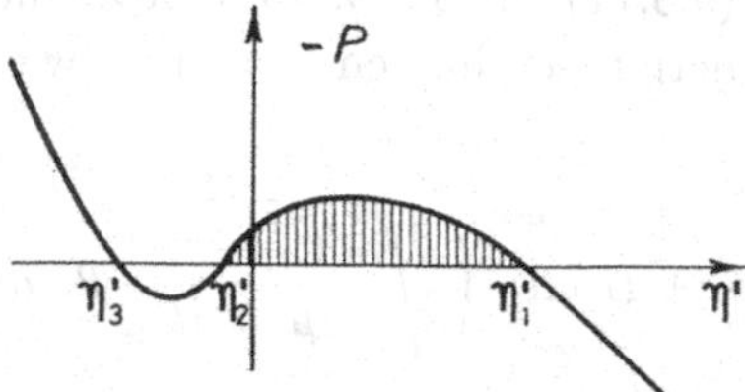

Fig. 9.5.1 Variation of $-P$.

where $F(\phi/m)$ is the **incomplete elliptic integral of the first kind**, with amplitude ϕ and parameter m. For the case considered here Abramowitz and Stegun gives ϕ by the relation (17.4.69)

$$\cos^2\phi = \frac{\eta - \eta_2}{\eta_1 - \eta_2} \tag{9.5.29}$$

and m is given by (17.4.61) as

$$m = \frac{\eta_1 - \eta_2}{\eta_1 - \eta_3} \tag{9.5.30}$$

while the constant λ is (again from (17.4.61)) given by

$$\lambda = \frac{1}{2}\sqrt{\eta_1 - \eta_3} \tag{9.5.31}$$

Substituting (9.5.28) into (9.5.20) we then get

$$\frac{1}{\lambda}F(\phi, m) = -M\theta \tag{9.5.32}$$

The Jacobian elliptic function *cn* is defined so that it links ϕ to $F(\phi, m)$ by the formula

$$\cos^2\phi \equiv \mathrm{cn}^2(F(\phi, m)) \tag{9.5.33}$$

Into this we sustitute (9.5.32) for $F(\phi, m)$ and (9.5.29) for $cos^2\phi$ which gives

$$\frac{\eta - \eta_2}{\eta_1 - \eta_2} = \mathrm{cn}^2\,(-\lambda\; M\; \theta) \tag{9.5.34}$$

which can be solved with respect to η. That gives

$$\eta = \eta_2 + (\eta_1 - \eta_2)\mathrm{cn}^2\,(-\lambda\; M\; \theta) \tag{9.5.35}$$

We then substitute (9.5.17) for M, (9.5.31) for λ, and (9.5.27) for $\eta_1 - \eta_2$. Since cn is an even function so that cn$-$ = cn$+$, we then finally end up with the result

$$\eta = \eta_2 + H \,\mathrm{cn}^2 \left(\sqrt{\frac{3\delta(\eta_1 - \eta_3)}{4\mu^2 h^{*2} h}}\, \theta,\ m \right) \tag{9.5.36}$$

where the parameter m has been added in the bracket for the cn-function.[4]

In (9.5.36), the three roots η_1, η_2 and η_3 of the cubic in the elliptic integral still remain undetermined. However, with η_1, η_2 and η_3 known m can be obtained from (9.5.30).

Physically, this means that although the variation of η is determined by (9.5.36) the level of this variation relative to our $\eta = 0$ level (represented by η_2) is unknown and the two parameters, the wavelength L and the wave propagation speed c are also undetermined. On the other hand, to specify the wave we only have to choose L and c or T in addition to H, because L, c and T are already related by (9.5.4)

Thus, we only have to produce two more equations by imposing conditions on our solution.

9.5.2 *Final cnoidal wave expressions*

The first additional constraint is that the mean value $\overline{\eta}$ of η taken over the period of 1 for η is 0, i.e. that

$$\overline{\eta} = \int_0^1 \eta d\theta = 0 \tag{9.5.37}$$

This esentially states that we want the x-axis (i.e. $z = 0$) placed in the mean water surface MWS. The detailed evaluation of this expression is shown in Appendix 9B. It results in the relation

$$\eta_3 K(m) = -(\eta_1 - \eta_3) E(m) \tag{9.5.38}$$

where $K(m)$ and $E(m)$ are complete elliptic integrals of the first and second kind.

[4]As can be seen in this approach for the resolution of the elliptic integral we do not need to transform the elliptic integral on the LHS of (9.5.28) to the standard form (9.A.1), which is most often associated with the notion of elliptic integrals. It utillizes directly results already established in the literature for the particular case we have. However, for an ilustration of the transformations that change the LHS of (9.5.28) to (9.A.1) see Mei (1983), sect 11.5.3.

The second condition consists of establishing a relation between the variation of x (or t) and the variation of the cn^2-function in (9.5.36).

The variation of the cn^2-function is sketched in Fig. 9.5.2. It depends somewhat on the value of m, but for all m the maximum value is +1 and the minimum is 0, and it has the period $2K$. Thus, in (9.5.36) there will be wave crests at $\theta = 0$ and at $\sqrt{\frac{3(\eta_1-\eta_3)}{4kh^{*2}h}}\,\theta = 2K$ (whereas a wave trough will be at $\sqrt{\frac{3(\eta_1-\eta_3)}{4kh^{*2}h}}\,\theta = K$).

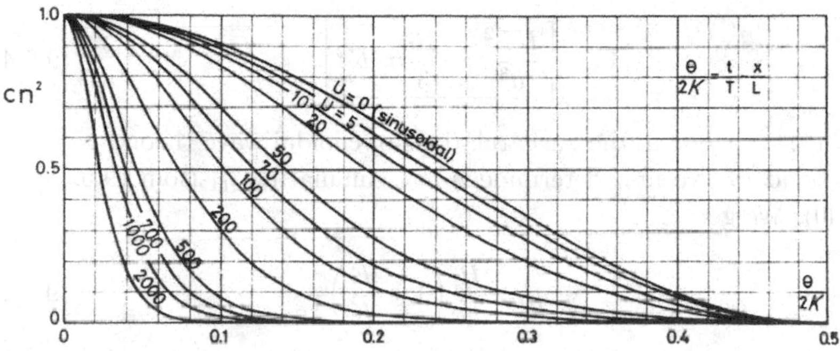

Fig. 9.5.2 Variation of cn^2 in the expression for the surface profile of a cnoidal wave for different values of the Ursell parameter U. For $U = 0$ the profile equals a sinusoidal wave, for $U \to \infty$ the profile becomes that of a solitary wave.

As we want the distance between two wave crests, which correspond to a θ-interval of 1, to be the wave length L, we obtain the other relation as

$$\sqrt{\frac{3(\eta_1-\eta_3)}{4h^{*2}h}}\cdot 1 = 2K \tag{9.5.39}$$

The rest is mathematical manipulations. We first eliminate $\eta_1 - \eta_2$ from the expression (9.5.30) for m using (9.5.27) and get

$$\eta_1 - \eta_3 = \frac{H}{m} \tag{9.5.40}$$

which we can use in both the above derived equations. We substitute (9.5.40) into (9.5.39) together with (9.5.6) for $h*$ and get

$$L\sqrt{\frac{3\delta H}{4m\mu^2\, h^3}} = 2K \tag{9.5.41}$$

which can be rearranged into

$$\frac{\delta H L^2}{\mu^2 h^3} = \frac{16}{3} m\, K^2 \tag{9.5.42}$$

As the formula in its nature is dimensionless and as only lengths normalized by the same λ are involved, it does not change its form when the dimensional quantities H, L and h are reintroduced. Substituting the definitions for δ, μ and the other dimensionless variables the dimensional version of (9.5.42) becomes

$$\boxed{\frac{H L^2}{h^3} = \frac{16}{3} m\, K^2} \tag{9.5.43}$$

Eq. (9.5.43) is one of the basic relations of cnoidal wave theory.

To find η_2, we first determine η_1 by eliminating η_3 from (9.5.38) and (9.5.40). We get

$$\eta_1 = \frac{H}{m}\left(1 - \frac{E}{K}\right) \tag{9.5.44}$$

and

$$\eta_3 = -\frac{H}{m}\frac{E}{K} \tag{9.5.45}$$

and from (9.5.27) then

$$\boxed{\eta_2 = H\left[\frac{1}{m}\left(1 - \frac{E}{K}\right) - 1\right]} \tag{9.5.46}$$

which is the same as the value of η in the wave trough. Hence (9.5.46) represents the vertical distance from the MWS to the wave trough. The complete expression for η may now be written as

$$\boxed{\eta = H\left[\frac{1}{m}\left(1 - \frac{E}{K}\right) - 1 + \mathrm{cn}^2\left(2K\left(\frac{t}{T} - \frac{x}{L}\right), m\right)\right]} \tag{9.5.47}$$

which is obtained by substituting (9.5.46) and (9.5.39) into (9.5.36).

The last of the integral-parameters of the wave is the phase velocty c. This is obtained from (9.5.4), (9.5.7) and (9.5.22) as

$$c^2 - h = \eta_1 + \eta_2 + \eta_3 \tag{9.5.48}$$

Substituting (9.5.44), (9.5.45) and (9.5.46) and rearranging terms yields

$$c^2 = h + \frac{H}{m}\left(2 - m - 3\frac{E}{K}\right) \tag{9.5.49}$$

or, by reintroduction of the dimensional c, we get

$$\boxed{\frac{c^2}{gh} = 1 + \frac{H}{mh}\left(2 - m - 3\frac{E}{K}\right)} \tag{9.5.50}$$

Notice, as a check, that the deviation of c^2/gh from 1 is a term of $O(\delta)$ since $H/h = O(\delta)$.

By comparison with sinusoidal theory, we see that 2π in the argument of $\sin\left(2\pi\left(\frac{t}{T} - \frac{x}{L}\right)\right)$ has been replaced by $2K$.

In the limit $m \to 0$ we have

$$E, K \to \pi/2 \tag{9.5.51}$$

but

$$\frac{E}{K} \to 1 - \frac{m}{2} \tag{9.5.52}$$

so we get

$$\eta_2 \to -\frac{H}{2} \tag{9.5.53}$$

and therefore

$$\mathrm{cn}^2 2K\theta \to \frac{1}{2}(1 + \cos 2\pi\theta) \tag{9.5.54}$$

so that

$$\eta \to \frac{H}{2}\cos 2\pi\theta \tag{9.5.55}$$

This means that in the limit $m \to 0$ (i.e. $U \to 0$) the cnoidal waves are equal to sinusoidal waves. This is consistent with the basic assumptions of the theory and shows that the cnoidal wave theory is smoothly linked to the sinusoidal wave theory.

In case we had chosen to base our derivation on the Korteweg-de Vries equation, we would have got for c the expression

$$\frac{c}{\sqrt{gh}} = 1 + \frac{H}{2mh}\left(2 - m - 3\frac{E}{K}\right) \tag{9.5.56}$$

which is equal to (9.5.50) to the degree of accuracy of that equation, i.e., to terms $O(\mu^4)$. Squaring (9.5.56) directly gives (9.5.50).

In some textbooks (e.g., Wiegel 1964) and in the tables of functions by Masch and Wiegel (1961), the expression for c is based on the depth h_t below wave trough instead of below MWL. The expression by Masch and Wiegel can be obtained by substituting

$$h \equiv h_t - \eta_2 \tag{9.5.57}$$

(h_t being the depth under the trough and $\eta_2 < 0$) into e.g., (9.5.56). Doing so we get

$$\frac{c}{\sqrt{gh_t}} = \sqrt{1 - \frac{\eta_2}{h_t}} \left(1 + \frac{H}{2mh}\left(2 - m - 3\frac{E}{K}\right)\right) \tag{9.5.58}$$

which be means of (9.5.46) for η_2 and by neglecting small terms becomes

$$\frac{c}{\sqrt{gh_t}} = 1 + \frac{H}{h_t m}\left(\frac{1}{2} - \frac{E}{K}\right) \tag{9.5.59}$$

which is the formula (2.160 a) in Wiegel 1964.[5]

However, the formula (9.5.59) is not as convenient for practical applications as (9.5.50) or (9.5.56) because h_t depends on the wave data and hence is not suited as a reference parameter.

9.5.3 *Infinitely long waves: Solitary waves*

The results derived above for the periodic cnoidal waves are valid for any wave length larger than approximately 10 times the water depth h. For analysis of Boussinesq waves shorter than $10h$ (i.e., waves in deeper water), reference is made to Section 9.9.

Closer inspection of periodic wave results shows, however, that in the opposite end of the spectrum, the case of infinitely long waves, the cnoidal results reduce to a simpler form. Such waves are know as **Solitary waves** and they essentially consist of one wave crest only (because the next crest is infinitely far away).

Mathematically, we find that for $L/h \to \infty$, the parameter $U = HL^2/h^3 \to \infty$

[5] In the tables of functions (Masch and Wiegel 1961), an unfortunate misprint has emerged in (2) pag. 4: d should be replaced by h_t at least on the left hand side corresponding to (9.5.59). This wrong formula has later been used (Masch 1964) to investigate the shoaling of cnoidal waves.

From (9.5.43), we see that this means

$$mK^2 \to \infty \tag{9.5.60}$$

which occurs when $m \to 1$, $K \to \infty$. In this limit, we also have $E \to 1$. Hence, in (9.5.37), we get

$$\frac{\eta_3}{\eta_1 - \eta_3} = -\frac{E}{K} \to 0 \tag{9.5.61}$$

which means $\eta_3 \to 0$. (9.5.40) then gives

$$\frac{\eta_2}{H} \to 0 \tag{9.5.62}$$

Therefore, in this case η_2 and η_3 are both zero so the polynomial $-P$ has a double root at $\eta = 0$.

The result (9.5.62) also means, however, that the distance between the wave trough and the MWS is zero, or, in other words, the surface is positioned entirely above the MWS. At first this may seem in conflict with the requirement expressed in (9.5.37) that $\overline{\eta} = 0$. However, if we transform (9.5.37) to dimensional form, it reads

$$\overline{\eta} = \frac{1}{L}\int_0^L \eta \, dx = 0 \tag{9.5.63}$$

which clearly yields $\overline{\eta} = 0$ for $L \to \infty$.

As in periodic waves $\eta = \eta_1$ represents the wave crest. Hence, continuing the results above, we have everywhere

$$0 \leq \eta \leq \eta_1 \tag{9.5.64}$$

and we see from (9.5.22) that

$$\frac{b}{\delta} = \frac{c^2 - h}{\delta} = \eta_1 \tag{9.5.65}$$

Since η_1 represents the wave crest which according to (9.5.64) is also the amplitude, we conclude that (the dimensionless) $\eta_1 = \frac{\eta_{\max}}{\text{amplitude}} = 1$ or from (9.5.65)

$$\frac{c^2 - h}{\delta} = 1 \tag{9.5.66}$$

or

$$c^2 = h + \delta \tag{9.5.67}$$

We also see from (9.5.23) and (9.5.24) that in this case

$$C_1 = C_2 = 0 \tag{9.5.68}$$

Substituting $\eta_2 = \eta_3 = 0$ into (9.5.22) shows the expression for P reduces to

$$P(\eta) = (\eta - \eta_1)\eta^2 \tag{9.5.69}$$

and hence

$$\eta_\theta^2 = -A^2\, \eta^2(\eta - \eta_1) \tag{9.5.70}$$

and with $\eta_1 = 1$ this simplifies to

$$\eta_\theta^2 = A^2\, \eta^2(1 - \eta) \tag{9.5.71}$$

It turns out that in this case the integral on the LHS of (9.5.20) is no longer elliptic but can be solved directly and expressed in terms of simpler, ordinary functions.

Thus, (9.5.71) can be written

$$\frac{d\eta}{d\theta} = A\eta\sqrt{1-\eta} \tag{9.5.72}$$

which gives

$$\int \frac{d\eta}{\eta\sqrt{1-\eta}} = A \int d\theta \tag{9.5.73}$$

The integral on the LHS can be solved by the substitution $y = \sqrt{1-\eta}$ which gives

$$-2\int \frac{dy}{1-y^2} = A(\theta - \theta_0) \tag{9.5.74}$$

This gives

$$2\tanh^{-1}\sqrt{1-\eta} = A(\theta_0 - \theta) \tag{9.5.75}$$

or

$$\sqrt{1-\eta} = \tanh\frac{A}{2}(\theta_0 - \theta) \tag{9.5.76}$$

Solving with respect to η then gives

$$\eta = 1 - \tanh^2\frac{A}{2}(\theta_0 - \theta) \tag{9.5.77}$$

and since $\tanh^2$ is an even function, we can replace $\theta_0 - \theta$ with $\theta - \theta_0$ and write

$$\eta = 1 - \tanh^2 \frac{a}{2}(\theta - \theta_0) \tag{9.5.78}$$

This expression is usually transformed using that $\tanh = \sinh / \cosh$ and $\cosh^2 - \sinh^2 = 1$ which brings (9.5.78) on the form

$$\eta = \text{sech}^2 \frac{A}{2}(\theta - \theta_0) \tag{9.5.79}$$

where $\text{sech} \equiv 1/\cosh$.

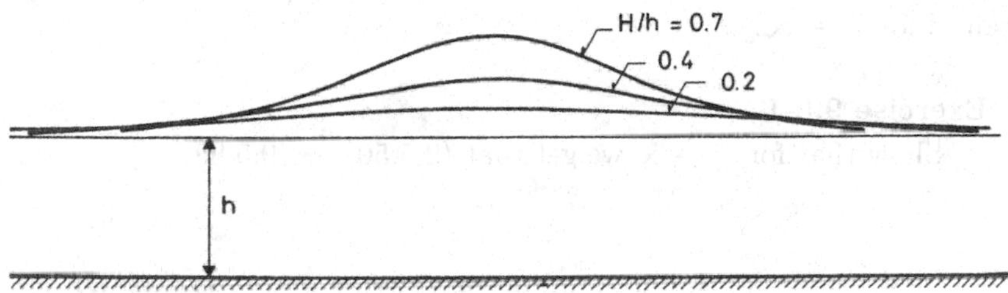

Fig. 9.5.3 Solitary wave surface profiles for three different values of H/h.

We recall that in dimensional form, we have

$$\begin{aligned} \delta &= \frac{a}{h_0} = \frac{H}{h_0} \quad ; \quad \mu = \frac{h_0}{\lambda} \\ h &= \frac{h}{h_0}; \eta = \frac{\eta}{a} \quad ; \quad L = \frac{L}{\lambda} \end{aligned} \tag{9.5.80}$$

Hence, we get (with $\theta = \frac{t}{T} - \frac{x}{L} = \frac{1}{L}(ct - x)$)

$$\frac{A}{2}(\theta - \theta_0) = \sqrt{\frac{3H}{4h^3}}\,(ct - x - \theta_0) \tag{9.5.81}$$

As (9.5.79) shows, θ_0 represents the position x_0 of the crest at $t = 0$. Hence, we get in dimensional form

$$\eta = H \text{sech}^2 \sqrt{\frac{3H}{4h^3}}(ct - (x - x_0)) \tag{9.5.82}$$

which is the solitary wave solution. We see from (9.5.82) that the variation with x (or t) is scaled by $\sqrt{H}$; the larger H, the stronger variation of the argument of the sech^2 function. That is, the larger wave height, the shorter crest of the waves, as shown in Fig. 9.5.3 that gives the shape of the surface for three different values of H/h.

From (9.5.67) we also get in dimensional form

$$\frac{c^2}{gh} = 1 + \frac{H}{h} \tag{9.5.83}$$

Inspection of the cnoidal wave results will show that this is the largest value of c obtained for $L \to \infty$. Hence, the cnoidal/solitary waves show the same property as found for linear waves: the largest phase speed is obtained for $L \to \infty$.

Exercise 9.5-1

Show that for $L \to \infty$ we get that (9.5.50) $\to$ (9.5.83)

9.6 Analysis of cnoidal waves for practical applications

(In this section, we use all formulas on dimensional form.)

The practical application of cnoidal waves has been very limited. This is probably partly due to the difficulties in evaluating the results and the lack of information about the results for important parameters such as velocities, pressure, wave averaged quantities as energy flux, E_f, radiation stress, S_{xx}, volume flux, Q_w, etc. The aim of this section is to help alleviate that. In this, we draw heavily on Svendsen (1974) and Svendsen and Jonsson (1976, 1980), which is used in the following without further quoting.

In order to obtain numerical information about the cnoidal wave parameters, we need to define the wave from practical parameters such as H, h, and L or T and then establish practical methods for calculating the elliptic functions and integrals.

In principle, numbers could be obtained from standard tables for elliptic integrals such as, e.g., Abramowitz and Stegun (1964) or Milne-Thomson (1968). However, the values of m which emerge in practical applications are so close to unity that standard tables are of no real use. Hence, for practical purposes, evaluation of cnoidal waves will either require special tables or special computer methods. Such tables were developed by Masch and Wiegel (1961) and by Skovgaard et al. (1974) (see also Svendsen, 1974).

In the following, we discuss these problems. Simple tables for a first estimate are prescribed along with simple computer methods for solution of the basic cnoidal equations.

9.6.1 *Specification of the wave motion*

We first notice that similarly to the problem discussed in Section 3.2.2, the wave motion can either be specified by H, L and h or by H, T and h. Basically, these variables can be reduced to two independent dimensionless parameters which can be either H/h and L/h (if H, L, h are specified) or H/h and $T\sqrt{g/h}$ (if H, T, h are specified). The tables by Masch and Wiegel are almost entirely two parameter tables. However, it turns out that most of the cnoidal wave parameters can be determined as a function of one parameter (e.g., U) only.

We discuss the two cases separately.

Case 1: H/h and L/h specified

In this case, the value of the Ursell parameter U can be calculated directly as

$$U = \frac{HL^2}{h^3} \tag{9.6.1}$$

However, to obtain the second parameter, m, we need to solve the transcendental equation (9.5.43)

$$U = \frac{16}{3} m\, K^2 \tag{9.6.2}$$

where $K = K(m)$. This equation establishes m as a monotonic function of U or vise versa. A simple and robust computer method for doing this is outlined in Appendix. State-of-the-are routines for computing E and K from m are given by Press et al. (1986).

We notice then that once m has been computed, all other elliptic quantities such as the cn-function, the elliptic integrals, $E(m)$, and $K(m)$, etc. can in principle be determined too. Table 9.6.1 shows results for both m, $m_1 = 1 - m$, $E(m)$ and $K(m)$ for a range of U-values encountered in most practical applications.

We also see that as U increases, the value of m gets so close to unity that an increasing number of digits are needed to separate m from 1 with

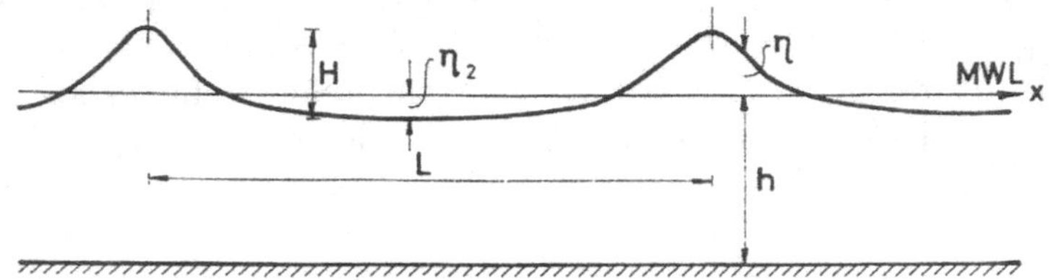

Fig. 9.6.1 Definition sketch for cnoidal waves.

the loss of accuracy as a consequence. Therefore, $m_1 = 1 - m$ is generally a more convenient parameter for controlling the computations.

To calculate the phase velocity we write c on the form

$$\frac{c^2}{gh} = 1 + \frac{H}{h} A \tag{9.6.3}$$

where

$$A = \frac{2}{m} - 1 - 3\frac{E}{mK} \tag{9.6.4}$$

We have $A = A(m)$ and through (9.6.2) this means $A = A(U)$, i.e., A is only a function of the two nondimensional parameters H/h and L/h in the combination given as U.

In other words, A can be determined as a function of one variable, U, and although c is a function of both H/h and A, the variation with H/h is a simple algebraic dependence. The variation of A versus U is also given in Table 9.6.1.

When c is determined, the wave period T can then be found from $T = L/c$.

Likewise, η_2 defined in Fig. 9.6.1 and by (9.5.46) is a function of U only and given too in Table 9.6.1. The variation of $A = A(U)$ with U is shown in Fig. 9.6.2.

It may be noticed that for $U < 47$, the value of A becomes negative. As (9.6.3) shows, this means that we get $c < \sqrt{gh}$. U small is most often associated with L/h small, i.e., deeper water. This feature mirrors a similar property of sinusoidal waves. However, there are negative aspects to this which are discussed for Boussinesq waves in general in Section 9.7.5.

We also notice that $A \to 1$ for $U \to \infty$ which confirms the result that in a solitary wave $c^2/gh = 1 + H/h$. The convergence is slow, though, and even for $U = 100$, A is only 0.31.

Table 9.6.1 The variation of cnoidal wave parameters versus U (Skovgaard et al. 1974).

U	m	m_1	K	E	$\frac{\eta_{min}}{H}$	A	B
1	(-2)7.317	(-1)9.268	1.601	1.542	-0.495	-13.152	0.1250
2	(-1)1.410	(-1)8.590	1.631	1.514	-0.491	-6.565	0.1250
3	(-1)2.038	(-1)7.962	1.662	1.487	-0.486	-4.365	0.1249
4	(-1)2.619	(-1)7.381	1.692	1.462	-0.481	-3.261	0.1249
5	(-1)3.157	(-1)6.843	1.723	1.438	-0.476	-2.596	0.1249
6	(-1)3.655	(-1)6.345	1.754	1.416	-0.472	-2.151	0.1248
7	(-1)4.116	(-1)5.884	1.786	1.394	-0.467	-1.830	0.1247
8	(-1)4.543	(-1)5.457	1.817	1.373	-0.462	-1.588	0.1246
9	(-1)4.937	(-1)5.063	1.849	1.354	-0.458	-1.399	0.1245
10	(-1)5.302	(-1)4.698	1.881	1.335	-0.453	-1.245	0.1244
11	(-1)5.639	(-1)4.361	1.912	1.318	-0.449	-1.119	0.1243
12	(-1)5.952	(-1)4.048	1.944	1.301	-0.444	-1.012	0.1242
13	(-1)6.240	(-1)3.760	1.976	1.285	-0.440	-0.921	0.1241
14	(-1)6.507	(-1)3.493	2.008	1.270	-0.435	-0.842	0.1239
15	(-1)6.754	(-1)3.246	2.041	1.256	-0.431	-0.773	0.1238
16	(-1)6.982	(-1)3.018	2.073	1.243	-0.426	-0.712	0.1236
17	(-1)7.194	(-1)2.806	2.105	1.230	-0.422	-0.657	0.1234
18	(-1)7.389	(-1)2.611	2.137	1.218	-0.418	-0.607	0.1233
19	(-1)7.570	(-1)2.430	2.169	1.207	-0.414	-0.562	0.1231
20	(-1)7.738	(-1)2.262	2.201	1.196	-0.410	-0.521	0.1229
22	(-1)8.036	(-1)1.964	2.266	1.176	-0.402	-0.449	0.1225
24	(-1)8.293	(-1)1.707	2.329	1.158	-0.394	-0.387	0.1220
26	(-1)8.513	(-1)1.487	2.393	1.142	-0.386	-0.333	0.1216
28	(-1)8.702	(-1)1.298	2.456	1.128	-0.379	-0.285	0.1211
30	(-1)8.866	(-1)1.134	2.519	1.116	-0.372	-0.243	0.1206
32	(-1)9.006	(-2)9.937	2.581	1.104	-0.365	-0.204	0.1201
34	(-1)9.128	(-2)8.720	2.643	1.094	-0.358	-0.170	0.1195
36	(-1)9.233	(-2)7.666	2.704	1.085	-0.352	-0.138	0.1190
38	(-1)9.325	(-2)6.768	2.764	1.077	-0.345	-0.109	0.1184
40	(-1)9.404	(-2)5.959	2.824	1.070	-0.339	-0.082	0.1179
42	(-1)9.473	(-2)5.268	2.883	1.063	-0.334	-0.056	0.1173
44	(-1)9.533	(-2)4.665	2.942	1.057	-0.328	-0.033	0.1167
46	(-1)9.586	(-2)4.139	3.000	1.052	-0.323	-0.011	0.1161
48	(-1)9.632	(-2)3.678	3.057	1.047	-0.317	0.009	0.1155
50	(-1)9.673	(-2)3.274	3.113	1.043	-0.312	0.029	0.1149

Integers in parentheses indicate powers of 10 by which the following numbers are to be multiplied

U	m	m_1	K	E	$\frac{\eta_{min}}{H}$	A	B
50	(-1)9.673	(-2)3.274	3.113	1.043	-0.312	0.029	0.1149
55	(-1)9.753	(-2)2.466	3.252	1.034	-0.301	0.072	0.1134
60	(-1)9.813	(-2)1.874	3.386	1.027	-0.290	0.111	0.1119
65	(-1)9.856	(-2)1.438	3.516	1.022	-0.280	0.145	0.1104
70	(-1)9.889	(-2)1.112	3.643	1.017	-0.271	0.175	0.1090
75	(-1)9.913	(-3)8.668	3.766	1.014	-0.263	0.203	0.1075
80	(-1)9.932	(-3)6.805	3.886	1.012	-0.255	0.227	0.1061
85	(-1)9.946	(-3)5.379	4.003	1.009	-0.248	0.250	0.1048
90	(-1)9.957	(-3)4.278	4.117	1.008	-0.242	0.271	0.1034
95	(-1)9.966	(-3)3.423	4.228	1.006	-0.235	0.290	0.1021
100	(-1)9.972	(-3)2.753	4.336	1.005	-0.230	0.308	0.1009
150	(-1)9.996	(-4)3.955	5.304	1.001	-0.188	0.434	0.0902
200	(-1)9.999	(-5)7.674	6.124	1.000	-0.163	0.510	0.0822
250	1.000	(-5)1.808	6.847	1.000	-0.146	0.562	0.0760
300	1.000	(-6)4.894	7.500	1.000	-0.133	0.600	0.0711
350	1.000	(-6)1.471	8.101	1.000	-0.123	0.630	0.0671
400	1.000	(-7)4.807	8.660	1.000	-0.115	0.654	0.0636
450	1.000	(-7)1.681	9.186	1.000	-0.109	0.673	0.0607
500	1.000	(-8)6.224	9.682	1.000	-0.103	0.690	0.0582
550	1.000	(-8)2.419	10.155	1.000	-0.098	0.705	0.0560
600	1.000	(-9)9.803	10.607	1.000	-0.094	0.717	0.0540
650	1.000	(-9)4.122	11.040	1.000	-0.091	0.728	0.0522
700	1.000	(-9)1.791	11.456	1.000	-0.087	0.738	0.0506
750	1.000	(-10)8.015	11.859	1.000	-0.084	0.747	0.0491
800	1.000	(-10)3.682	12.247	1.000	-0.082	0.755	0.0478
850	1.000	(-10)1.733	12.624	1.000	-0.079	0.762	0.0465
900	1.000	(-11)8.333	12.990	1.000	-0.077	0.769	0.0454
950	1.000	(-11)4.089	13.346	1.000	-0.075	0.775	0.0443
1000	1.000	(-11)2.044	13.693	1.000	-0.073	0.781	0.0434
2000	1.000	(-16)2.421	19.365	1.000	-0.052	0.845	0.0318
3000	1.000	(-20)4.015	23.717	1.000	-0.042	0.874	0.0263
4000	1.000	(-23)2.611	27.386	1.000	-0.037	0.890	0.0230
5000	1.000	(-26)4.066	30.619	1.000	-0.033	0.902	0.0207
6000	1.000	(-28)1.177	33.541	1.000	-0.030	0.911	0.0190
7000	1.000	(-31)5.451	36.228	1.000	-0.028	0.917	0.0176
8000	1.000	(-33)3.663	38.730	1.000	-0.026	0.923	0.0165
9000	1.000	(-35)3.336	41.070	1.000	-0.024	0.927	0.0156
10000	1.000	(-37)3.918	43.301	1.000	-0.023	0.931	0.0149
∞	1.000	0.000	∞	1.000	-0.000	1.000	0.0000

The surface profile η can, in terms of η_2, be written

$$\eta = \eta_2 + H\mathrm{cn}^2\left(2K\left(\frac{t}{T} - \frac{x}{L}\right), m\right) \tag{9.6.5}$$

Fig. 9.5.2 shows the variation of the cn^2 function versus θ for different values of $U(m)$.

For computational purposes, an algorithm is needed for the cn function. For state-of-the-art routines, reference is gain made to Press et al. (1986).

Note that it is the variation of cn^2 with m that makes the cnoidal waves change shape as the main parameters H/h and L/h change.

Case 2. H/h **and** $T\sqrt{g/h}$ **specified**

In the second case where T and hence $T\sqrt{g/h}$ is specified instead of L/h, it is not possible to determine U directly. This case is actually the most common because it is usually much easier to determine the period T of a wave than the length L.

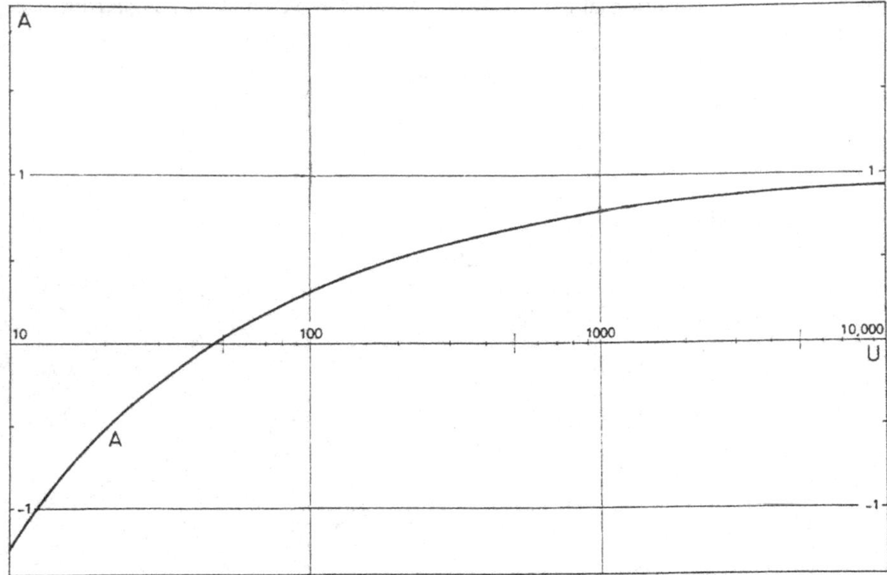

Fig. 9.6.2 The variation of $A = A(U)$ with U.

The problem then becomes to determine the wave length L. We see that $L = cT$ can be written

$$L = T\sqrt{gh\left(1 + \frac{H}{h}A\right)} \tag{9.6.6}$$

or

$$\frac{L}{h} = T\sqrt{g/h}\,\sqrt{1 + \frac{H}{h}A} \tag{9.6.7}$$

which also leads to

$$U = \left(T\sqrt{g/h}\right)^2 \frac{H}{h}\left(1 + \frac{H}{h}A\right) \tag{9.6.8}$$

which shows that L (or U) are solutions to transcendental equations with two independent variables H/h and $T\sqrt{g/h}$. Again an efficient computational procedure for solving this problem is described in Appendix.

Once L/h has been determined, U can be calculated and the problem is reduced to Case 1.

Discussion of results: The values of U versus the assumption of $\delta/\mu^2 = O(1)$.

First it is pointed out that L/h can also be written

$$\frac{L}{h} = \frac{c}{\sqrt{gh}}\, T\,\sqrt{g/h} \tag{9.6.9}$$

Thus values of $L/h < T\sqrt{g/h}$ indicate conditions where $c/\sqrt{gh} < 1$ that is the speed of the cnoidal wave is smaller than that of a linear wave. We see from figure 9.6.2 that this occurs for small values of U. This is what happens in deeper water, and it is discused further in Section 9.7.

A simple example will verify that values of U for typical practical wave conditions are actually $\gg 1$. Thus if we assume $L/h = 10$ (which is close to the lower limit for long wave approximations) and $H/h = 0.5$ then $U = 50$, and in many practical cases U is much larger. This is in seeming contrast to the basic assumptions for Boussinesq waves that $\delta/\mu^2 = O(1)$. Part of the explanation for this paradox lies in the fact that, as we have seen, the terms that occur in the dimensionless equations are actually of order $\mu = h_0/\kappa^2$ where $\kappa = 2\pi\mu$, and $(2\pi)^2 \sim 40$ is not really small. Further $H \sim 2a$. So when $U = 50$ we have $\delta/\kappa^2 = al^2/h^3 \simeq 50/2 \cdot 40 = 0.62$ which is $O(1)$

However, in many cases U becomes even 500 or more. For waves with so large U-values, the characteristic length λ in $\mu = h_0/\lambda$ is $\ll L$, the wave length. λ is assumed characteristic in the sense that "significant changes" in wave properties such as η or u occur over the distance λ. As Fig. 9.5.2 shows, when U is large, the wave crest is short and steep and most of the changes in η occur within a horizontal length which more typically could be set at

$$\lambda = O(H/\eta_{x,\max})$$

Considering the first approximation $u \sim c\eta/h$, the same applies to the horizontal particle velocity. Hence, values of $U = O(500)$ do not invalidate the basic assumption that $\delta/\mu^2 = O(1)$.

9.6.2 *Velocities and pressures*

The horizontal velocity u

The variation over depth of the horizontal velocity u is essentially determined by (9.2.18) which with $u_0 = \phi_{0x}$ inserted reads (in dimensional form)

$$u = u_0 - \frac{1}{2}(z+h)^2 u_{0xx} + O(\mu^4) \tag{9.6.10}$$

This requires we determine u_0, the velocity at the bottom, which can be done in several ways.

One approach is to use either of the equations (9.3.13) or (9.3.14), from which we initially eliminated u_0 to get the 4th order Boussinesq equation in η only. Since the cnoidal waves have constant form, we have (see 9.4.7)

$$u_0 = c\frac{\eta}{h} + O(\delta\ \mu^2) \tag{9.6.11}$$

and also exactly

$$\frac{\partial}{\partial t} = -c\frac{\partial}{\partial x} \tag{9.6.12}$$

Substituting (9.6.11) into the small terms of (9.3.13) and using (9.6.12) on η_t, we get to the order considered here

$$-c\eta_x + hu_{0x} + \frac{c}{h}(\eta^2)_x - \frac{1}{6}ch^2\eta_{xxx} = 0 \tag{9.6.13}$$

or

$$u_0 = c\frac{\eta}{h} - c\frac{\eta^2}{h^2} + \frac{1}{6}ch\ \eta_{xx}c_1(t) \tag{9.6.14}$$

where $C_1(t)$ is an arbitrary function of t. However, since the wave is assumed periodic $C_1(t) = C_1$.

Substituting (9.6.14) into (9.6.10), then gives

$$u = c\frac{\eta}{h} - c\frac{\eta^2}{h^2} + \frac{1}{2}ch\left(\frac{1}{3} - \frac{(z+h)^2}{h^2}\right)\eta_{xx} + C_1 \tag{9.6.15}$$

which is the expression for u as a function of z and of x, t expressed in terms of $\eta(x,t)$.

Since $\overline{\eta} = 0$, we see the constant C_1 is a measure of the net flow in the wave. In parallel with both sine waves and also definition of the wave motion in combined waves and currents we choose to determine C_1 so that the average $\overline{u}$ of u over a wave period is zero. Hence, we get

$$0 = -c\frac{\overline{\eta^2}}{h^2} + C_1 \tag{9.6.16}$$

which gives

$$u = c\frac{\eta}{h} - c\left(\frac{\eta^2}{h^2} + \frac{\overline{\eta^2}}{h^2}\right) + \frac{1}{2}ch\left(\frac{1}{3} - \frac{(z+h)^2}{h^2}\right)\eta_{xx} \tag{9.6.17}$$

Discussion of the result for u

We see that (9.6.17) determines the velocity profiles from the surface profiles of the cnoidal wave.

To the first approximation, the velocity equals the depth uniform velocity given by (9.6.11).

The total depth averaged velocity $\tilde{u}$ is given by

$$\begin{aligned}\tilde{u} &= \frac{1}{h+\eta}\int_{-h}^{\eta} u dz \\ &= \frac{1}{h+\eta}\left[\left(c\frac{\eta}{h} - c\frac{\eta^2 - \overline{\eta^2}}{h^2}\right)(h+\eta) + \frac{1}{2}ch\eta_{xx}\int_{-h}^{\eta}\left(\frac{1}{3} - \frac{(z+h)^2}{h^2}\right)dz\right] \\ &= c\frac{\eta}{h} - c\frac{\eta^2 - \overline{\eta^2}}{h^2} + \frac{1}{2}ch\,\eta_{xx}\left[\frac{1}{h}\int_{-h}^{0}\left(\frac{1}{3} - \frac{(z+h)^2}{h^2}\right)dz\right] + O(\mu^4)\end{aligned} \tag{9.6.18}$$

where the last term has been approximated by the leading contribution. We see that to the $O(\mu^4)$ the last term is zero, so that $\tilde{u}$ become

$$\tilde{u} = c\frac{\eta}{h} - c\frac{\eta^2 - \overline{\eta^2}}{h^2} + O(\mu^4) \tag{9.6.19}$$

which can also be written

$$\tilde{u}_1 = c\frac{\eta}{h+\eta} + c\frac{\overline{\eta^2}}{h^2} + O(\mu^4) \tag{9.6.20}$$

We recognize (see Section 11.2) that the first term in this expression represents the (exact) depth averaged velocity $\tilde{u}_0$ in a wave with a net volume flux $\overline{Q} = 0$.

$$\tilde{u}_0 = c\frac{\eta}{h+\eta} \tag{9.6.21}$$

However, we also found in Section 3.3 that a wave with zero net volume flux $\overline{Q} = 0$ includes a return current $-Q_w/h$ where Q_w is the volume flux in a wave with $\overline{u} = 0$ at all points (as specified here). This means that

$$\tilde{u}_0 = \tilde{u} - \frac{Q_w}{h} \tag{9.6.22}$$

so that

$$\tilde{u} = \tilde{u}_0 + \frac{Q_w}{h} = c\frac{\eta}{h+\eta} + \frac{Q_w}{h} \tag{9.6.23}$$

Hence, we conclude by comparison of (9.6.20) and (9.6.23) that in cnoidal waves

$$Q_w = c\frac{\overline{\eta^2}}{h} \tag{9.6.24}$$

This result for Q_w is confirmed in Section 3.3.

We also notice that the variation of the velocity over depth is proportional to η_{xx}, the curvature of the surface profile.

The vertical velocity w

Exercise 9.6-1 Show that the vertical velocity w can be written

$$w = -c(z+h)\left[\frac{\eta_x}{h}\left(1 - \frac{2\eta}{h}\right) + \frac{1}{6}h\left(1 - \frac{(z+h)^2}{h^2}\right)\eta_{xxx}\right] \tag{9.6.25}$$

However, since $w = O(\mu^2 u)$ and w is less important in practical applications, it would often suffice to use the first approximation which is

$$w = -c(z+h)\frac{\eta_x}{h} \tag{9.6.26}$$

which varies linearly with the vertical coordinate.

η_x can be determined by differentiating (9.5.42) with respect to x. Svendsen and Jonsson (1976, 1980) find (with cn θ short for cn(θ, m)

$$\eta_x = \frac{4KH}{L}\text{cn}\,\theta\sqrt{1-\text{cn}^2\theta}\,\sqrt{1-m+m\text{cn}^2\theta} \tag{9.6.27}$$

It turns out that this has its maximum for

$$\text{cn}^2\theta = \frac{1}{3}(2m - 1 + \sqrt{1-m+m^2})) \tag{9.6.28}$$

We see that for $m \to 1$ (U large), we get $\text{cn}^2\theta \to 2/3$ in this expression. The corresponding maximum value $\eta_{x,m}$ for η_x is given by

$$\eta_{x,m} \simeq \pm\frac{8\sqrt{3}}{9} K \frac{H}{L} \quad ; \quad m \sim 1 \tag{9.6.29}$$

which from (9.6.27) provides the numerically largest vertical velocity as

$$w_m \simeq \pm\frac{8\sqrt{3}}{9} K \frac{H}{L} c \frac{z+h}{h} \quad ; \quad m \sim 1 \tag{9.6.30}$$

Exercise 9.6-2

Show that for $m \sim 1$ this also corresponds to

$$w_m \sim \pm\frac{2}{3}\left(\frac{H}{h}\right)^{3/2} c\,\frac{z+h}{h} \quad ; \quad m \sim 1 \tag{9.6.31}$$

Even for U as small as 30, the actual value of $\text{cn}^2\theta$ from (9.6.28) is $\text{cn}^2\theta = 0.647 \sim 2/3$. Hence even for relatively small values of U the correct expression for w_m will differ only by a few % from the result (9.6.30).

The pressure variation

The presure in the cnoidal waves is determined from the Bernoulli equation which reads

$$gz - \frac{p}{\rho} + \frac{1}{2}(u^2 + w^2) + \phi_t = 0 \tag{9.6.32}$$

Here we as usual introduce the dynamic pressure p_D (relative to the mean water level) defined so that

$$p = \rho g z + p_D \tag{9.6.33}$$

Then (9.6.32) reduces to

$$p_D = -\rho\phi_t - \frac{1}{2}\rho(u^2 + w^2) \tag{9.6.34}$$

or, to the order of accuracy used here

$$p_D = -\rho\phi_t - \frac{1}{2}\rho u^2 + O(\mu^4) \tag{9.6.35}$$

While u^2 in this equation can be calculated from (9.6.11) (because u^2 is a small term) we determine ϕ_t from (9.3.10) for ϕ_{0t} and (9.2.16) for ϕ_t.

The result for ϕ_{0t} is (in dimensional form)

$$\phi_{0t} = -g\eta - \frac{1}{2}u_0^2 + \frac{1}{2}h^2 u_{0xt} \tag{9.6.36}$$

and (9.2.16) then gives

$$\phi_t = -g\eta - \frac{1}{2}u_0^2 + \frac{1}{2}h^2 \left(1 - \frac{(z+h)^2}{h^2} u_{0xt}\right) \tag{9.6.37}$$

Here we can use (9.6.11) in small terms and since $\partial/\partial t = -c\partial/\partial x$ we get

$$\phi_t = -g\eta - \frac{1}{2}c^2\frac{\eta^2}{h^2} - \frac{1}{2}c^2 h \left(1 - \frac{(z+h)^2}{h^2}\right)\eta_{xx} \tag{9.6.38}$$

where $c^2 = c_0{}^2 = gh + O(\mu^2)$. Inserting ϕ_t into (9.6.35) then gives

$$\boxed{p_D = \rho g \left[\eta + \frac{1}{2}h^2 \left(1 - \frac{(z+h)^2}{h^2}\right)\eta_{xx}\right]} \tag{9.6.39}$$

9.6.3 *Wave averaged properties of cnoidal waves*

As in sinusoidal waves, the wave averaged properties of the motion such as the volume flux Q_w, the radiation stress S_{xx} and the energy flux E_f are all proportional to H^2. This means that even the leading order contribution is $O(\delta^2) = O(\mu^4)$. It is therefore possible to use low order approximations for the parameters involved to get a first approximation to these quantities.

The volume flux Q_w

The volume flux is as usual defined as

$$Q_w = \overline{\int_{-h}^{\eta} u \, dz} \tag{9.6.40}$$

where $\bar{\cdot}$ again represents time averaging over a wave period. The integral is divided into two parts

$$Q_w = \overline{\int_{-h}^{0} udz} + \overline{\int_{0}^{\eta} udz} \tag{9.6.41}$$

where the first integral vanishes because $\overline{u} = 0$. In the second integral we get the leading conribution by using (9.6.3) for u. This yields

$$\boxed{Q_w = c\frac{\overline{\eta^2}}{h} + O(\mu^6)} \tag{9.6.42}$$

which is the same as (9.6.24).

The radiation stress

As shown in Section 3.3.3, a convenient form to write the radiation stress is

$$S_{xx} = \overline{\int_{-h}^{\eta} \rho(u^2 - w^2)dz} + \frac{1}{2}\rho g\overline{\eta^2} \tag{9.6.43}$$

(which is exact).

In cnoidal waves, we found the w^2-term is an order of magnitude smaller than the other two terms. Again, we only look for the leading order of approximation and therefore can use (9.6.3) for u. We then get

$$S_{xx} = \rho h\, c^2\frac{\overline{\eta^2}}{h^2} + \frac{1}{2}\rho g\overline{\eta^2} + O(\mu^6) \tag{9.6.44}$$

whre $c^2 = gh + O(\mu^2)$. We, therefore, get

$$\boxed{S_{xx} = \frac{3}{2}\rho g\overline{\eta^2} + O(\mu^2)} \tag{9.6.45}$$

The energy flux E_f

The exact expression for the wave averaged flux of energy through a vertical section is

$$E_f = \overline{\int_{-h}^{\eta} \left\{p_D + \frac{1}{2}\rho(u^2 + w^2)\right\} udx} \tag{9.6.46}$$

The leading term is $p_D u$, so we get

$$E_f = \overline{\int_{-h}^{\eta} p_D\, u\, dz} + O(\mu^6) \tag{9.6.47}$$

We also see from (9.6.39) that the leading term for p_D is $\rho g\eta$. Hence, we get (replacing also η with 0 in the upper limit of the integral because the difference is a small term and using (9.6.39) for u)

$$\boxed{E_f = \rho g c\overline{\eta^2} + O(\mu^6)} \tag{9.6.48}$$

Discussion of the results for Q_w, S_{xx} and E_f

Thus, we find that all wave averaged quantities are determined by $\overline{\eta^2}$. Substituting (9.5.42) for η, we get

$$\eta^2 = H^2 \left[\eta_2^2 + 2\eta_2 cu^2(u,m) + cu^4(u,m)\right] \tag{9.6.49}$$

In the averaging over a wave period we use (9.B.3) (with $a = 2K$) to get

$$\int_0^{2K} cu^2\ u\ du = \frac{2}{m}(E - (1-m)K) \tag{9.6.50}$$

Similarly, it can be shown that (see Gradshteyn and Ryzhik (1965) p. 629)

$$\int_0^{2K} cu^4 udu = \frac{2K}{3m^2}\left[3m^2 - 5m + 2 + (4m-2)\frac{E}{K}\right] \tag{9.6.51}$$

Thus, we find that

$$\frac{\overline{\eta^2}}{H^2} = B(m) = \frac{1}{m^2}\left[\frac{1}{3}\left(3m^2 - 5m + 2 + (4m-2)\frac{E}{K}\right) - \left(1 - m - \frac{E}{K}\right)^2\right] \tag{9.6.52}$$

Thus, defining the quantity B in (9.6.52), we see that we can write the wave averaged properties for cnoidal waves as

$$Q_w = c\frac{H^2}{h}B(m) \tag{9.6.53}$$

$$S_{xx} = \frac{3}{2}\rho g H^2 B(m) \tag{9.6.54}$$

$$E_f = \rho g c H^2 B(m) \tag{9.6.55}$$

The variation of $B(m) = B(U)$ with U is shown in Fig. 9.6.3.

It is interesting to compare $B(U)$ with the value of 0.125 found for sinusoidal shallow water waves. We see that the results for all three wave averaged quantities essentially are equivalent to those found in linear shallow water theory. The only difference lies in the value of $\overline{\eta^2}$ or B. In linear waves, B has one value only: 0.125. In cnoidal waves, B depends on the wave shape and since cnoidal waves for all cases have steeper and shorter wave crests and have shallower, longer wave troughs, B is always smaller than 0.125 and $B \to 0$ for $U \to \infty$. In brief, it can be said that such waves

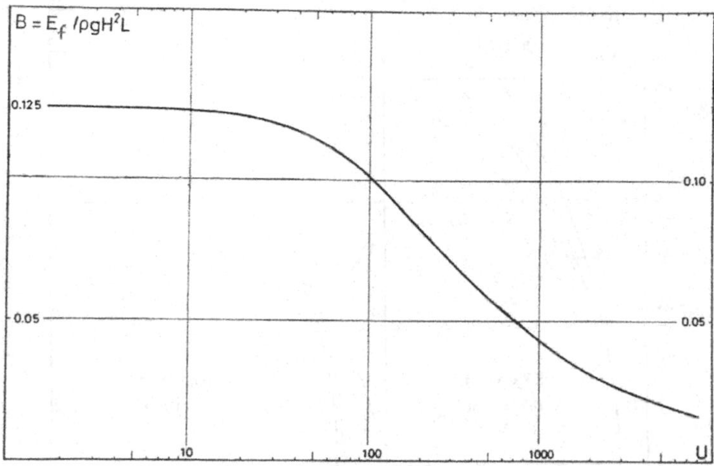

Fig. 9.6.3 The variation of $B = B(U)$ with U.

are less bulky than sine waves and therefore, for a given wave height, they produce less Q_w, S_{xx} and E_f.

This is a general feature for non-sinusoidal waves.

9.6.4 *Limitations for cnoidal waves*

When we consider only he lowest order cnoidal waves derived above the most important limitation is that even for waves small enough to be considered linear we will fail to find solutions for the h/L for a given period if

$$h/L_0 > 3/8\pi \sim 0.1193 \tag{9.6.56}$$

where L_0 is the equivalent deep water wave length $L_0 = gT^2/2\pi$ (Svendsen, 1974). As explained further in Section 9.7 the reason for this is that the linear dispersion relation underlying the cnoidal wave equations has no solution when the value of h/L or kh becomes too large. For the details of this see Exercise 9.7-7. Thus there is a deep water limit given by (9.6.56) even for very small values of H/h .

The weakly nonlinear theory presented here has no upper limit for H/h (just as linear wave theory does not). It is found, however, that the second order polynomial expression for the horizontal velocity profiles given by

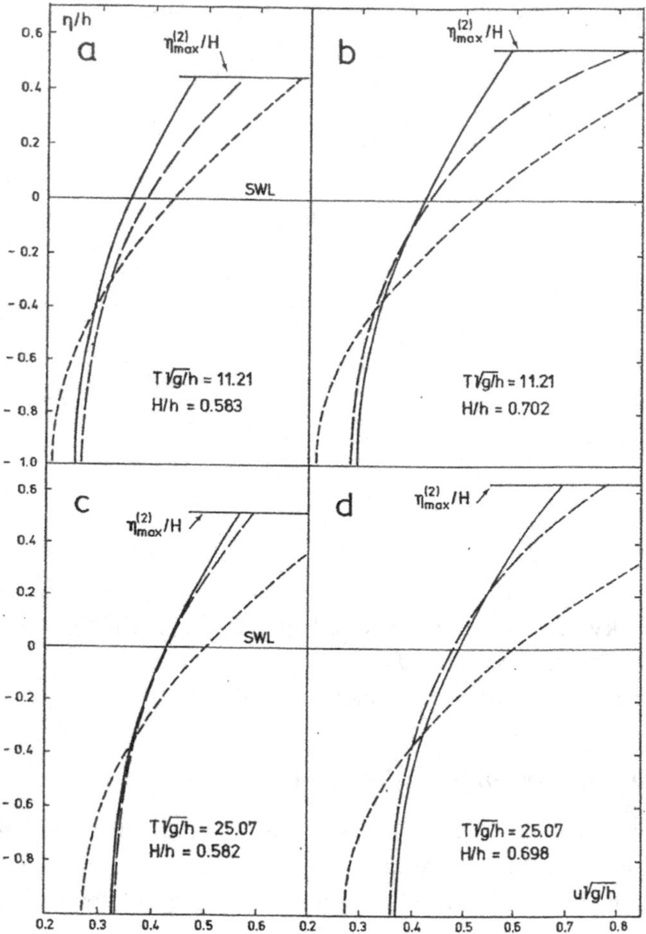

Fig. 9.6.4 Comparison of the horizontal velocity profiles for different wave theories. The full line curves represent the stream function results, - - - are the results using (9.6.17) and – – – represent (9.6.57). From Svendsen and Staub (1981).

(9.6.17) becomes quite inaccurate for higher waves. Svendsen and Staub (1981) (SS81 in the following) showed that a simple expression can be derived under the crest and trough of the wave (where $\eta_x = 0$) for the second order polynomial approximation in which all terms in δ are. This is the type of approximation later termed **fully nonlinear** even though it only includes terms p to $O(\mu^2)$. The result for the modified velocity profiles under the wave crest/wave trough becomes

$$u = c\frac{\eta}{h+\eta} + \frac{1}{2}ch\left(\frac{1}{3} - \frac{(z+h)^2}{h^2}\right)\left(1 - \frac{1}{3}(\eta+h)\eta_{xx}\right)^{-1}\eta_{xx} \qquad (9.6.57)$$

In Fig. 9.6.4 the numerical results using (9.6.17) and (9.6.57) have been compared with velocity profiles obtained by using accurate calcualtion with the stream function method (Chaplin's 1978 version) which for the present purpose can be considered exact. The waves are very steep waves with $H/h = 0.58 - 0.70$ and values of $T\sqrt{g/h} = 11.21$ and 25.07 respectively. The lower of those values is close to the deep water limit and hence it is not surprising that both the velocity profiles show inaccuracies, though the modified profile is clearly more accurate. For the case with the longer waves the modified profiles are clearly the more accurate. The results indicate the level of accuracy one can expect with a second order polynomial approximation for the horizontal velocities. The paper SS81 also shows extensive comparisons with measured velocity profiles.

If the cnoidal theory is consistently developed to higher order in δ (in the sense that all terms consistent with $U = O(1)$ are retained but non smaller) the velocity profiles will start showing unrealistic behaviour for sufficiently large values of H/h. Fenton (1979) developed the theory to ninth order and found the ninth order results for the velocities under the wave crest were "wildly divergent from the fifth order results". It was shown by SS81 that this anomaly exists already in the third order approximation. However, their results also indicated that the anomaly is essentially eliminated if the coefficients of the polynomial velocity profiles are derived so that all nonlinear terms are retained even though they formally are smaller than corresponding to $U = O(1)$. The advantage of keeping all nonlinear terms is discussed further in Section 9.10.

9.7 Alternative forms of the Boussinesq equations - The linear dispersion relation

The choice of the bottom velocity u_0 as the reference velocity in the Boussinesq equations is arbitrary. At the same time, it turns out that this choice has profound consequences for the behavior of the equations, in particular for cases where the water depth is not quite small in comparison to the characteristic length λ.

The effect of choosing different reference velocities for the Boussinesq equations is discussed further in this section. To facilitate the discussion we first derive the Boussinesq equations for different reference velocities.

9.7.1 *Equations in terms of the velocity* u_0 *at the bottom*

We first recall that the horizontal velocity is given by

$$u = \phi_{0x} - \frac{\mu^2}{2}(z+h)^2 \phi_{0xxx} + \frac{\mu^4}{24} \phi_{0xxxxx} + O(\mu^6) \tag{9.7.1}$$

which can also be written

$$u = u_0 - \frac{\mu^2}{2}(z+h)^2 u_{0xx} + \frac{\mu^2}{24}(z+h)^4 u_{0xxxx} + O(\mu^6) \tag{9.7.2}$$

Here we have for illustration of the following included $O(\mu^4)$ terms.

We found that in terms of the bottom velocity u_0, the equations were given by (9.3.7) and (9.3.8) which for reference read

$$\eta_t + ((h+\delta\eta)u_0)_x - \frac{\mu^2}{6} h^3 u_{0xxx} = 0 \tag{9.7.3}$$

$$u_{0t} + \eta_x + \delta u_0 u_{0x} - \frac{\mu^2}{2} h^2 u_{0txx} = 0 \tag{9.7.4}$$

9.7.2 *Equations in terms of the velocity* u_s *at the MWS*

To replace the bottom velocity u_0 by the velocity u_s at $z = 0$ we consider u_s given by (9.7.1)

$$u_s = u_0 - \frac{\mu^2}{2} h^2 u_{0xx} + \frac{\mu^4}{24} h^4 u_{0xxx} + O(\mu^6) \tag{9.7.5}$$

This equation can be reversed to express u_0 in terms of u_s by successive approximations. We first realize that to first order

$$u_s = u_0 + O(\mu^2) \tag{9.7.6}$$

or

$$u_0 = u_s + O(\mu^2) \tag{9.7.7}$$

This approximation we can insert into the second term in (9.7.5) which gives

$$u_s = u_0 - \frac{\mu^2}{2} h^2 u_{sxx} - O(\mu^4) \tag{9.7.8}$$

Solving with respect to u_0 yields

$$u_0 = u_s + \frac{\mu^2}{2} h^2 u_{sxx} + O(\mu^4) \tag{9.7.9}$$

Similarly we can obtain the expression to $O(\mu^6)$ for $u_0 = u_0(u_s)$ by inserting (9.7.9) into the second term of (9.7.5), and (9.7.7) into the last term and solve with respect to u_0. We get

$$u_0 = u_s + \frac{\mu^2}{2} h^2 \left(u_s + \frac{\mu^2}{2} h^2 u_{sxx} \right)_{xx} - \frac{\mu^4}{24} h^4 u_{sxxxx} + O(\mu^6) \tag{9.7.10}$$

or

$$u_0 = u_s + \frac{\mu^2}{2} h^2 u_{sxx} + \frac{5}{24} \mu^4 h^4 u_{sxxxx} + O(\mu^6) \tag{9.7.11}$$

This expression can now be used in (9.7.3) and (9.7.4) to replace u_0 with u_s. The result is

$$\eta_t + ((h + \delta\eta) u_s)_x + \frac{\mu^2}{3} h^3 u_{sxxx} = 0 \tag{9.7.12}$$

$$u_{st} + \eta_x + \delta u_s u_{sx} = 0 \tag{9.7.13}$$

which are the Boussinesq equations in terms of the surface velocity u_s. We see that changing from u_0 to u_s changes the μ^2-terms but leaves the rest of the terms in the equations unchanged (as long as we only consider equations correct to $O(\mu^4)$).

9.7.3 *The equations in terms of the depth averaged velocity* $\tilde{u}$

The equations (9.3.11) and (9.3.12) can also be expressed in terms of the depth averaged velocity $\tilde{u}$ which is defined as

$$\tilde{u} = \frac{1}{h + \delta\eta} \int_{-h}^{\delta\eta} u(z) dz \tag{9.7.14}$$

where $u(z) = \phi_x(z)$ is determined by (9.2.16). Notice that the depth averaging is over the **local** time varying depth $d = h + \delta\eta$.

Exercise 9.7-1

To calculate $\tilde{u}$, it is convenient to change variable in (9.7.14) to

$$\xi = z + h \tag{9.7.15}$$

and define

$$d = h + \delta\eta \qquad (9.7.16)$$

so that (9.7.1) becomes

$$\tilde{u} = \frac{1}{d}\int_0^d u d\xi \qquad (9.7.17)$$

Show that this yields

$$\tilde{u} = u_0 - \frac{\mu^2}{6} d^2 u_{0xx} + O(\mu^4) \qquad (9.7.18)$$

This series expansion for $\tilde{u} = \tilde{u}(u_0)$ can be inverted to express $u_0 = u_0(\tilde{u})$ by the same technique as used for u_s. The result becomes

$$u_0 = \tilde{u} + \frac{\mu^2}{6} d^2 \tilde{u}_{xx} + O(\mu^4) \qquad (9.7.19)$$

Exercise 9.7-2

Show that if this technique is continued to the next order of approximation, we get

$$u_0 = \tilde{u} + \frac{\mu^2}{6} d^2 \tilde{u}_{xx} - \frac{7}{360} u^4 d^4 \tilde{u}_{xxxx} + O(\mu 6) \qquad (9.7.20)$$

Substituting (9.7.19) into (9.7.1) and (9.7.2), we then get the Boussinesq equations in terms of the depth averaged velocity $\tilde{u}$. The result is

$$\eta_t + ((h + \delta\eta)\tilde{u})_x = 0 \qquad (9.7.21)$$

$$\tilde{u}_t + \eta_x + \delta\tilde{u}\tilde{u}_x - \frac{\mu^2}{3} h^2 \tilde{u}_{xxt} = 0 \qquad (9.7.22)$$

where we have reinserted that

$$d = h + \delta\eta \qquad (9.7.23)$$

It is of particular interest to notice that not only is the kinematic boundary condition (9.7.21) in a very simple form in this case, but it is in fact exact to any order in μ and δ.

Exercise 9.7-3 Show that this is not surprising by integrating the continuity equation

$$\frac{\partial u}{\partial x} + \frac{\partial w}{\partial z} = 0 \tag{9.7.24}$$

over depth. Show that without any approximation, this yields

$$\frac{\partial \eta}{\partial t} + \frac{\partial}{\partial x} \int_{-h}^{\eta} u dz = 0 \tag{9.7.25}$$

Inserting here the definition (9.7.14) of $\tilde{u}$, show this simply is identical with (9.7.21).

Hence, the depth integrated continuity equation can replace the kinematic free surface boundary condition.

9.7.4 *The equations in terms of Q*

Finally, it is useful to consider the equations in terms of the total, instantaneous volume flux Q defined by

$$Q = \int_{-h}^{\eta} u dz = (h + \delta\eta)\tilde{u} \tag{9.7.26}$$

We see directly from (9.7.25) that the continuity equation integrated over depth can be written as

$$\frac{\partial \eta}{\partial t} + \frac{\partial Q}{\partial x} = 0 \tag{9.7.27}$$

which is exact and represents the equivalent of the KFSBC. The dynamic equation is then most directly obtained by expressing Q in terms of $\tilde{u}$. We have (again exactly)

$$Q = \tilde{u}d = \tilde{u}h + \delta\tilde{u}\eta \tag{9.7.28}$$

or

$$\tilde{u} = \frac{Q}{d} \tag{9.7.29}$$

Inserting this into (9.7.22) then gives

$$\left(\frac{Q}{d}\right)_t + \eta_x + \delta\frac{Q}{d}\left(\frac{Q}{d}\right)_x - \frac{\mu^2}{3}h^2\left(\frac{Q}{d}\right)_{xxt} = 0 \tag{9.7.30}$$

Here we can use the continuity equation (9.7.27) to eliminate d_t. We first realize that $d_t = (h + \delta\eta)_t$ which means

$$d_t = -Q_x \tag{9.7.31}$$

so that after multiplication by d (9.7.30) can also be written

$$Q_t + \eta_x + \delta 2\,\frac{Q}{d}\left(\frac{Q}{d}\right)_x - \frac{\mu^2}{3}h^3\left(\frac{Q}{d}\right)_{xxt} = 0 \tag{9.7.32}$$

This form clearly contains $O(\delta^2, \delta\mu^2)$ terms (through d in the denominator of the Q terms). Omitting the $O(\delta\mu^2,\ \delta^2)$-terms gives the dynamic equation in Q as

$$Q_t + \eta_x + \delta 2\,\frac{Q}{h}\left(\frac{Q}{h}\right)_x - \frac{\mu^2}{3}h^2 Q_{xxt} = 0 \tag{9.7.33}$$

9.7.5 *The linear dispersion relation*

As direct computer solutions of the Boussinesq equations in time and space progressed along with the development of the capacity of computers, it became evident that the various alternative forms of the equations developed in Sections 6.3 and 6.7 were not equally well behaved. In particular, it became clear that some were more prone to instability or unrealistic break down due to high frequency noise, i.e., due to explosive development of initially small amplitude disturbances with large values of kh.

At the same time, the increasing success of numerically simulating more and more complex wave phenomena - in particular by using the even more advanced models described later in this chapter on varying depth, 2D-horizontal domains and allowing even larger waves through accepting more nonlinear effects - made the restriction of the Boussinesq methods to shallow water (i.e. small kh) even more severe. In consequence, the recent 10-15 years have seen first of all extensive and very successful work aimed at understanding the nature of the limitation in kh, including why the various forms derived above are not equally useful, and also modifications of the Boussinesq equations that allow application to larger and larger kh.

It turns out the clue is found in analysis of how small amplitude waves behave when kh is not really small. This understanding will account both for the numerical growth of small disturbances with wave length much shorter than the main wave, and for the behavior of the main waves in areas further offshore where the wave height may be smaller but kh larger due to larger depth.

In the following, we examine these ideas for the Boussinesq models developed above.

Hence we turn to the linear part of the Boussinesq equations. We first examine the behavior of small sinusoidal waves in the original form of the equations based on the bottom velocity u_0 developed in Section 6.3. The linear version of the equation is obtained by neglecting all δ-terms.

$$\eta_t + hu_{0x} - \frac{\mu^2}{6} h^3 u_{0xxx} = 0 \tag{9.7.34}$$

$$u_{0t} + \eta_x - \frac{\mu^2}{2} h^2 u_{0txx} = 0 \tag{9.7.35}$$

The small sinusoidal waves are assumed in the form

$$\left.\begin{aligned} \eta &= \eta_0 e^{i\theta} \\ u_0 &= U_0 e^{i\theta} \end{aligned}\right\} \quad \theta = \omega t - kx \tag{9.7.36}$$

Substituting this into (9.7.34) and (9.7.35) yields

$$\omega\eta_0 - kh\left(1 - \frac{\mu^2}{6}(kh)^2\right) U_0 = 0 \tag{9.7.37}$$

$$-k\eta_0 + \omega\left(1 + \frac{\mu^2}{2}(kh)^2\right) U_0 = 0 \tag{9.7.38}$$

This is a set of homogeneous linear equations in $(\eta_0,\ U_0)$ which can be written as

$$\overline{\overline{A}}\,\overline{X} = 0 \qquad\qquad \overline{X} = \begin{Bmatrix} \eta_0 \\ U_0 \end{Bmatrix} \tag{9.7.39}$$

where

$$\overline{\overline{A}} = \begin{Bmatrix} \omega & -kh(1 + \frac{\mu^2}{6}(kh)^2 \\ -k & \omega(1 + \frac{\mu^2}{2}(kh)^2 \end{Bmatrix} \tag{9.7.40}$$

This homogeneous system only has solutions provided the determinant of $\overline{\overline{A}}$ is zero. That is, provided

$$\omega^2\left(1 + \frac{\mu^2}{2}(kh)^2\right) - k^2 h\left(1 + \frac{\mu^2}{6}(kh)^2\right) = 0 \tag{9.7.41}$$

or by rearranging, provided

$$\frac{\omega^2}{k^2} = c^2 = h\frac{1 + \frac{\mu^2}{6}(kh)^2}{1 + \frac{\mu^2}{2}(kh)^2} \tag{9.7.42}$$

In dimensional form this is

$$\frac{c^2}{gh} = \frac{1 + \frac{1}{6}(kh)^2}{1 + \frac{1}{2}(kh)^2} \tag{9.7.43}$$

Thus, we find that in the η, u_0 version of the equation, the small sinusoidal waves will propagate with a phase speed given by (9.7.43).

Equations in terms of the surface velocity u_s

Exercise 9.7-4

Show by a similar analysis for the equations (9.7.12) and (9.7.13) that those equations will propagate small sinusoidal waves with the velocity

$$\frac{c^2}{gh} = 1 - \frac{1}{3}(kh)^2 \tag{9.7.44}$$

Equations in terms of the depth mean velocity $\tilde{u}$

Exercise 9.7-5

Show that for the equations in terms of η, $\tilde{u}$, we get

$$\frac{c^2}{gh} = \frac{1}{1 + \frac{1}{3}(kh)^2} \tag{9.7.45}$$

Discussion of results

The important feature here is that for kh sufficiently large, the linear dispersion relation for the Boussinesq models analyzed above obviously behave differently. Since these are small amplitude waves, one would ideally

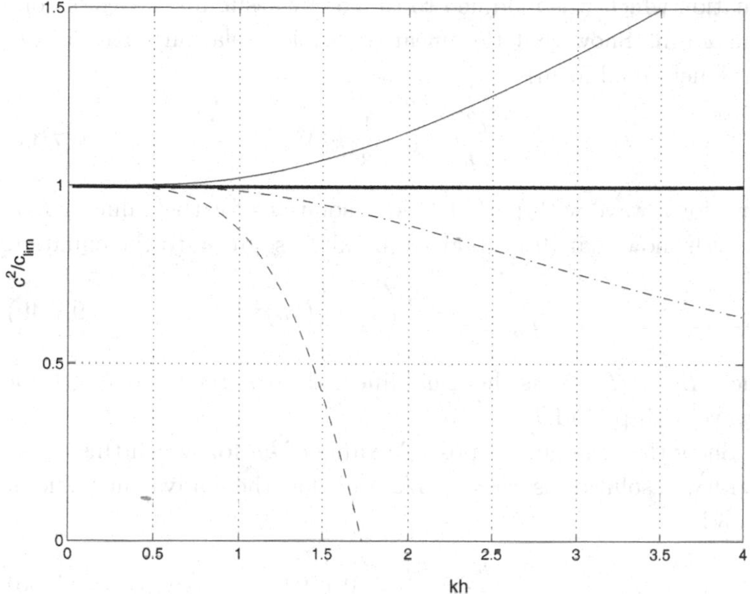

Fig. 9.7.1 The variation of c^2/c_{lin} for linear dispersion relations of the equations based on u_0, u_s and $\tilde{u}$. - - - corresponds to (9.7.44), - · - · - to (9.7.45), and —— to (9.7.43). The linear dispersion relation is the line at $c/c_{lin} = 1$.

expect them all to emulate the linear wave dispersion relation for which we have

$$\frac{c_{lin}^2}{gh} = \frac{\tanh kh}{kh} \tag{9.7.46}$$

We see that the Taylor expansion for (9.7.45) can be written

$$\frac{c_{lin}^2}{gh} = 1 - \frac{1}{3}(kh)^2 + O(kh)^4 \tag{9.7.47}$$

Exercise 9.7-6

Show that to terms $O(kh)^2$ the dispersion relations for the equations in u_0, u_s and $\tilde{u}$, the linear dispersion relations all resemble (9.7.47) (which in itself is only an approximation to (9.7.46).

Exercise 9.7-7

Cnoidal waves were derived from the fourth order Boussinesq

equation which was modified to (9.5.5) by assuming waves of constant form. Show that the linear dispersion relation for (9.5.5) is (in dimensional form)

$$\frac{c^2}{gh} = 1 - \frac{1}{3}(kh)^2 \tag{9.7.48}$$

If we for a wave with period T seek solutions for the value of L/h (or kh), show that (9.7.48) gives the kh as solution to the equation

$$\frac{2\pi h}{L_0} = (kh)^2 \left(1 - \frac{1}{3}(kh)^2\right) \tag{9.7.49}$$

where $L_0 = gT^2/2\pi$ is the usual linear deep water wave length for a wave with period T.

Show that the largest possible value of kh for which this equation has a solution is $kh = \sqrt{3/2}$ and that the equivalent value of h/L_0 is

$$\frac{h}{L_0} = \frac{3}{8\pi} = 0.1193 \tag{9.7.50}$$

(Madsen et al. 1991) as mentioned in Section 9.6.4.

Thus to the first approximation in (kh) (that is for very small kh), all the linear dispersion relations are the same. However, Fig. 9.7.1 shows the variation of the linear dispersion relations for the u_0, u_s and $\tilde{u}$ models for a wider range of kh-values and we see that even for the moderately large kh values of 2 (i.e., $\mu = h/L \gtrsim 0.30$) the deviation from (9.7.46) is substantial in particular for the versions based on u_s and u_0.

A particularly bad behavior is observed for the u_s model for which c^2 becomes negative for $kh > \sqrt{3} = 1.73$. Hence, for values of kh larger than that, c will be imaginary. We can write θ as $\theta = k(ct - x)$. We see that if an imaginary value is substituted for c this means the disturbances given by (9.7.36) will grow exponentially in time.

In conclusion, since disturbances with larger kh can not be avoided in numerical computations, it is obvious that the behavior of the linear dispersion relation is of crucial importance for the performance of Boussinesq models and the various forms of the equations discussed so far all show significant inaccuracies for kh growing. Possible solutions to this problem will be discussed further in the following chapter.

9.8 Equations for 2DH and varying depth

Before continuing it is useful to extend the equations to two dimensions in the horizontal plane (2DH) and to varying depth. In principle that implies going back to the starting point and solve the Laplace equation for ϕ_0 with the full set of boundary conditions under the more general assumption that the waves are now propagating in the horizontal x, y-plane and the depth h changes with position so that $h = h(x, y)$. Though there are no principal difficulties in doing this the computations become a good deal more complicated.

The extension to two horizontal dimensions is in principle only a question of changing the $\partial/\partial x$ operator to

$$\nabla_h = \frac{\partial}{\partial x}, \frac{\partial}{\partial y} \tag{9.8.1}$$

though it needs to be remembered that ∇_h represents a vector operator.

The extension to varying depth requires a little more consideration because the bottom slope is usually assumed small. Thus for varying depth we are dealing with a third small parameter $\nabla_h h$ in addition to μ, δ. For the weakly nonlinear Boussinesq waves discussed so far it is normal to assume that $\nabla_h h = O(\mu^2)$ so that in an equation correct to $O(\mu^4)$ terms with $(\nabla_h h)^2, \nabla_h^2 h$ can be neglected. Then the expression (9.2.14) for the velocity potential up to terms $O(\mu^4)$ becomes

$$\phi = \phi_0 - \mu^2(z+h)\nabla_h h \nabla_h \phi_0 - \frac{\mu^2}{2}(z+h)^2 \, \nabla_h^2 \phi_0 + O(\mu^4) \tag{9.8.2}$$

which shows that in the first approximation the sloping bottom gives raise to an extra term that varies linearly over the depth.

Exercise 9.8-1 Derive equation (9.8.2)

As in the case with constant depth the further derivations require that we decide which reference velocity potential or velocity we use to describe the equations. In principle the derivations, as mentioned above, follow the same lines as for constant depth. However, they get rather involved and it is beyond the scope of the present text to go into details. For reference we give the equivalent of the Boussinesq equations (9.3.11) and (9.3.12) (here in dimensional form) in terms of the bottom velocity u_0.

$$\eta_t + \nabla_h(\mathbf{u_0}(h+\eta)) - \nabla_h\left(\frac{1}{2}h^2\,\nabla_h(\nabla_h\cdot(h\mathbf{u_0})) - \frac{1}{3}h^3\,\nabla_h(\nabla_h\cdot\mathbf{u_0})\right) = O(\mu^4) \tag{9.8.3}$$

and

$$\mathbf{u_0}_t + (\mathbf{u_0}\cdot\nabla_h)\nabla_h\mathbf{u_0} + g\,\nabla_h\eta - \frac{\partial}{\partial t}\left(h\,\nabla_h(\nabla_h\cdot(h\mathbf{u_0})) - \frac{1}{2}h^2\,\nabla_h(\nabla_h\cdot\mathbf{u_0})\right) = O(\mu^4) \tag{9.8.4}$$

where the actual magnitude of the RHS of $O(\mu^4, \delta^2, \mu^2\delta, (\nabla_h h)^2, \nabla_h^2 h)$ has been abbreviated to $O(\mu^4)$ because of our assumptions of the relative magnitude of the small parameters.

Exercise 9.8-2 It is natural here to pose the exercise to derive equations (9.8.3) and (9.8.4). This essentially requires generalizing to varying depth all the steps in the derivation for constant depth. However, the reader is warned that it is a little more involved than the constant depth derivation shown in the previous chapters.

Similarly the equations in terms of the depth averaged velocity $\tilde{\mathbf{u}}$ become

$$\eta_t + \nabla_h(\tilde{\mathbf{u}}(h+\eta)) = 0 \tag{9.8.5}$$

(which is exact) and

$$\tilde{\mathbf{u}}_t + (\tilde{\mathbf{u}}\cdot\nabla_h)\nabla_h\tilde{\mathbf{u}} + g\,\nabla_h\eta - \frac{1}{2}\frac{\partial}{\partial t}\left(h\,\nabla_h(\nabla_h\cdot(h\tilde{\mathbf{u}})) - \frac{1}{3}h^2\,\nabla_h(\nabla_h\cdot\tilde{\mathbf{u}})\right) = O(\mu^4) \tag{9.8.6}$$

Equations (9.8.5) and (9.8.6) were first derived by Peregrine (1967), and this is one of the most frequently used forms of the weakly nonlinear equations because of the relatively benign linear dispersive behaviour.

Derivations of the various forms of the 2DH and varying bottom equations are available from various sources but one of the most detailed and convenient reference is Madsen and Schäffer (1998).

9.9 Equations with enhanced deep water properties

9.9.1 *Introduction*

Peregrine (1967), Benjamin et al. (1972) and others already observed that the lowest order, linear version of the equations, which give relationships between the time and space derivatives of η and the velocity parameter in the equations (be it u_0 , u_s, or $\tilde{u}$) can be used to change differentiations in the highest order (μ^2)-terms from x to t and vice-versa without changing the formal accuracy of the equations. An example of this was demonstrated in Section 9.5 which showed four different versions of the KdV-equation that were equally valid. Benjamin et al. also pointed out that the different versions obtained that way behaved differently numerically.

As discussed in the previous section the behavior of the phase velocity, when the water depth increases, is entirely linked to the highest order linear terms in the equations.

Combining these two observations suggests that, within certain limits, the accuracy of the dispersion properties of the equations in deeper water can be improved by suitable modifications of the highest order linear terms.

In order to judge what we should expect in deeper water we recall that since the Stokes theory is considered a good approximation to waves in that region it is natural to compare the performance of the Boussinesq theory in a more broad sense with the results for Stokes waves. A closer investigation then reveals, however, that the poor predictions of the phase velocity in deeper water are not the only inaccuracies relative to Stokes waves that show up as the depth-to-wavelength ratio increases. In recent years this line of thought has been investigated and developed extensively, in particular by Madsen, Schäffer and co-workers and also along different lines by Nwogu (1993) and by Kirby and his co-workers. To illustrate the ideas pursued we briefly outline some of the results in this section.

9.9.2 *Improvement of the linear dispersion properties*

Direct change of highest order terms

The start of the development was based on Witting (1984). He essentially argued that since the coefficients of the μ^2 terms in the Boussinesq equations can to a large extent be changed freely one can choose them as unknowns. As the results in section 9.7.5 indicate this can also be interpreted as using an unknown velocity U as reference velocity for the series expansion for ϕ (instead of, say, u_0). In this general formulation the linear part of the two Boussinesq equations can be written (in a 1D-horizontal form of Madsen et al., 1991) (M91 in the following)

$$\eta_t + hU_x - \mu^2 b_1 h^3 U_{xxx} = 0 \tag{9.9.1}$$

$$U_t + \eta_x - \mu^2 a_1 h^2 U_{xxt} = 0 \tag{9.9.2}$$

where a_1 and b_1 are the, so far, unknown coefficients.

Following the procedure outlined in Section 9.7.5 we see that this leads to the linear dispersion relation

$$\frac{c^2}{gh} = \frac{1 + b_1(kh)^2}{1 + a_1(kh)^2} \tag{9.9.3}$$

The question then becomes, how do we compare this approximation with the dispersion relation (9.7.46) for the Stokes waves?

We notice that (9.9.3) is a rational fraction approximation. Determining the coefficients a_1 and b_1 by equating (9.9.3) to (9.7.46) creates a so-called Padé approximation. Priciples of such approximations are given in Appendix. Essentially it means equating the Taylor expansion of (9.9.3) with the Taylor expansion of (9.7.46).

The result is (see M91) that

$$b_1 = \frac{1}{15}\,; \qquad a_1 = b_1 + \frac{1}{3} = \frac{2}{5} \tag{9.9.4}$$

This gives for (9.9.3)

$$\frac{c^2}{gh} = \frac{1 + \frac{1}{15}(kh)^2}{1 + \frac{2}{5}(kh)^2} \tag{9.9.5}$$

Because the numerator and denominator here both consist of second order polynamials in kh this is called a [2/2] Padé approximation to the linear Stokes dispersion relation (9.7.46).

M91 noted that if b_1 in (9.9.3) is considered a free parameter B, then (9.9.5) can be written

$$\frac{c^2}{gh} = \frac{1 + B(kh)^2}{1 + (B + \frac{1}{3})(kh)^2} \tag{9.9.6}$$

In this form (9.9.5) and the other dispersion relations found in Section 9.7.5

can be recovered by the following values of B

$$B = \frac{1}{6} \quad ; \quad \text{for} \quad u_0 \tag{9.9.7}$$

$$B = -\frac{1}{3} \quad ; \quad \text{for} \quad u_s \tag{9.9.8}$$

$$B = 0 \quad ; \quad \text{for} \quad \tilde{u} \tag{9.9.9}$$

$$B = \frac{1}{15} \quad ; \quad \text{for} \quad U \tag{9.9.10}$$

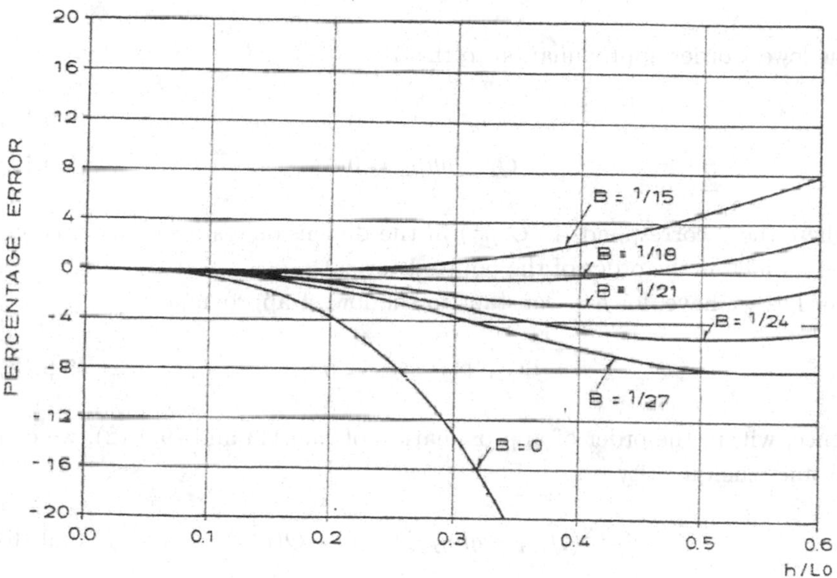

Fig. 9.9.1 The percentage error in the dispersion relation relative to the linear dispersion relation for different values of B (from Madsen and Sørensen, 1991).

Fig. 9.9.1 shows a plot of these and other cases illustrating that the value of $B = \frac{1}{15}$ is indeed a far better approximation to (9.7.46) than the other values found in Section 9.7.5.

The derivation of (9.9.5) shows that the expression agrees with the Taylor espansion for (9.7.46) up to $(kh)^4$. However, the direct Taylor expansion to $(kh)^4$ is not nearly as good an approximation to (9.7.46) as (9.7.46) is. The remarkable feature of the Padé approximation is that, while the traditional Taylor expansion has a radius of conversion of only $kh < \pi/2$

(and therefore start to diverge when kh approaches that value), the Padé approximation is well behaved in the entire domain $0 < kh < \infty$.

It is not immediately clear which assumption about the reference velocity in the Boussinesq equations will lead to the linear dispersion relation (9.9.5). To clarify this M91 considered the equations in terms of the volume flux Q derived in the previous section. In dimensional form they read

$$\eta_t + Q_x = 0 \tag{9.9.11}$$

$$Q_t + \left(\frac{Q}{d}\right)^2_x + g\, d\, \eta_x - \frac{1}{3}\, h^2\, Q_{xxt} = 0 \tag{9.9.12}$$

The lowest order approximation to this is

$$\eta_t + Q_x = 0 \tag{9.9.13}$$

$$Q_t + gh\eta_x = 0 \tag{9.9.14}$$

(where the 0 corresponds to $O(\mu^2)$ in the dimensionless form of the equations, which is the order of the omitted terms).

Thus we have, for h = constant to the lowest approximation

$$Q_{xxt} + gh\eta_{xxx} = 0 \tag{9.9.15}$$

Hence, within the order of approximation of (9.9.11) and (9.9.12), we can subtract the quantity

$$B\, h^2(Q_{xxt} + gh\eta_{xxx}) = 0(\sim O(\mu^2)) \tag{9.9.16}$$

from the momentum equation so that the equations become

$$\eta_t + Q_x = 0 \tag{9.9.17}$$

$$Q_t + \frac{Q^2}{d}_x + g\, d\, \eta_x - \left(B + \frac{1}{3}\right) h^2\, Q_{xxt} - g\, B\, h^3 \eta_{xxx} = 0 \tag{9.9.18}$$

We see that these equations have the linear dispersion relation (9.9.5) as wanted.

Since the parameter B can be chosen at any value, a value that gives a least squares fit to (9.7.46) within a given interval of kh can also be used. M91 found that $B = 1/18$ actually gives significantly better results in the interval $0 < h/L_0 < 0.6$ (corresponding to $0 < kh < 3.8$) than $B = 1/15$.

This interval is beyond the limit for deep water which is normally set as $h/L(\sim h/L_0) = 0.5, kh = 3.15$. The error for $B = 1/18$ reaches a maximum of only 3% at the largest value of $h/L_0 = 0.6$.

Madsen and Schäffer (1998) showed that the manipulations of the equations leading to introduction of the B-terms can be simulated by applying a linear dfferential operator L to the equations where L is given by

$$L = 1 + B\mu^2 h^2 \frac{\partial^2}{\partial x^2} \tag{9.9.19}$$

This "linear enhancement operator" greatly helps and simplifies introduction of this enhancement to other versions of the Boussinesq equations.

Use of a reference velocity specified at an arbitrary z-level

An alternative approach to improving the dispersion characteristics of the Boussinesq equations was followed by Nwogu (1993).

Realizing the significance of choosing the appropriate reference velocity Nwogu expressed the Boussinesq equations in terms of the velocity $\mathbf{u}_\alpha$ at an arbitrary depth z_α.

The following steps are involved in that derivation. For varying depth $h(x, y)$ the velocity profile $\mathbf{u}(x, y, z, t)$ expressed in terms of the bottom velocity $\mathbf{u}_0$ is given by

$$\mathbf{u} = \mathbf{u}_0 - \mu^2(h + z)\nabla_h[\nabla_h \cdot (h\mathbf{u}_0)] + \frac{\mu^2}{2}(h^2 - z^2)\nabla_h(\nabla_h \cdot \mathbf{u}_\alpha) + O(\mu^4) \tag{9.9.20}$$

where ∇_h as before stands for the horizontal vector gradient $(\frac{\partial}{\partial x}, \frac{\partial}{\partial y})$.

From this expression we can get the velocity $\mathbf{u}_\alpha$ at a selected depth z_α in terms of $\mathbf{u}_0$. We invert that expression to get $\mathbf{u}_0$ in terms of $\mathbf{u}_\alpha$ and by substituting that into (9.9.20) we then have the expression for the general velocity $\mathbf{u}(x, y, z, t)$ in terms of $\mathbf{u}_\alpha$

$$\mathbf{u} = \mathbf{u}_\alpha + \mu^2(z_\alpha - z)\nabla_h[\nabla_h \cdot (h\mathbf{u}_\alpha)] + \frac{\mu^2}{2}(z_\alpha^2 - z^2)\nabla_h(\nabla_h \cdot \mathbf{u}_\alpha) + O(\mu^4) \tag{9.9.21}$$

Nwogu then developed the Boussinesq equations in the general $2D_h$-form on an arbitrary varying depth $h(x, y)$ in terms of the velocity $\mathbf{u}_\alpha$. The result is

$$\eta_t + \nabla_h[(h+\delta\eta)\mathbf{u}_\alpha] + \frac{\mu^2}{2}\nabla_h \cdot \left\{\left(z_\alpha^2 - \frac{h^2}{3}\right)h\nabla_h(\nabla_h \cdot \mathbf{u}_\alpha) + 2\left(z_\alpha + \frac{h}{2}\right)h\nabla_h\left[\nabla_h \cdot (h\mathbf{u}_\alpha)\right]\right\} = 0 \tag{9.9.22}$$

$$\mathbf{u}_\alpha + \nabla_h\eta + \delta(\mathbf{u}\cdot\nabla_h)\mathbf{u}_\alpha + \frac{\mu^2}{2}\left\{{z_\alpha}^2\nabla_h(\nabla_h \cdot \mathbf{u}_{\alpha t}) + 2z_\alpha\nabla_h[\nabla_h \cdot (h\mathbf{u}_{\alpha t})]\right\} = 0 \tag{9.9.23}$$

However, as was the case with M91 his analysis of the dispersion properties are limited to the $1D_h$ version of the equations. The linearized form of the equations then become

$$\eta_t + h\, u_{\alpha x} + \left(\alpha + \frac{1}{3}\right)h^3 u_{\alpha xxx} = 0 \tag{9.9.24}$$

$$u_{\alpha t} + g\,\eta_x + \alpha h^2\, u_{\alpha xxt} = 0 \tag{9.9.25}$$

Again small amplitude waves as given by (9.7.36) lead to the dispersion relation

$$\frac{c^2}{gh} = \frac{1 - (\alpha + \frac{1}{3})\,(kh)^2}{1 - \alpha\,(kh)^2} \tag{9.9.26}$$

Nwogu (1993) found that a value of $\alpha = -0.39$ gives the best agreement with the linear dispersion relation. Fig. 9.9.2 shows the results for some α values.

We see that for $\alpha = -(B + \frac{1}{3})$ (9.9.26) is identical with (9.9.6). Based on this we notice that the value of $\alpha = -2/5$ corresponds to $B = 1/15$, and that $\alpha = -0,39$ closely corresponds to the value of $B = 1/18$ found by M91 to give the best approximation in a least square sense in the interval up to $h/L_0 = 0.6$

Thus by comparing the two results we have also established the connection between the level z_α of the reference velocity u_α in the Boussinesq equations and the value of B in the equations by M91. The table below gives an overview of the z_α values discussed here.

9.9.3 *Improvement of other properties*

Thus it is possible to vastly improve the accuracy of the the equations for their linear dispersion properties.

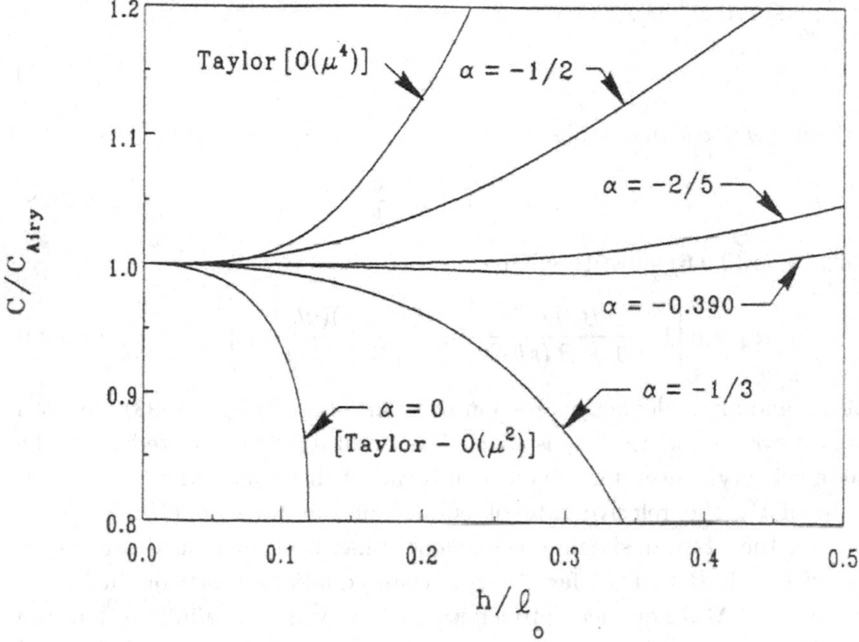

Fig. 9.9.2 The values of c/c_{Airy} for different values of α. Airy theory is linear theory (from Nwogu, 1993).

α	z_α	B
0	0	$-1/3$
$-1/3$	$-0.42\,h$	0
$-2/5$	$-0.55\,h$	$1/15$
-0.39	$-0.53\,h$	$1/17.6 \sim 1/18$
-0.5	$-h$	$1/6$

Work on improving the deep water performance has continued through the development of improved models and through the testing of other properties of the models. The question asked is: how accurate are the equations for other properties when h/L increases?

Thus the variation of the group velocity c_g was already analysed by M91 by Nwogu (1993).

The group velocity is given by

$$c_g = \frac{\partial \omega}{\partial k} \tag{9.9.27}$$

and since $\omega = c\,k$ this implies

$$c_g = c + k\,\frac{\partial c}{\partial k} \tag{9.9.28}$$

which with (9.9.6) substituted gives

$$c_g = c\left[1 + \frac{B\,(kh)^2}{1 + B\,(kh)^2} - \frac{(B+\frac{1}{3})(kh)^2}{1+(B+\frac{1}{3})\,(kh)^2}\right] \tag{9.9.29}$$

(M91), and an analogous expression in terms of α (Nwogu, 1993). In both cases it was found that c_g is much less accurately represented than the phase velocity c, even for the optimal forms of the enhanced equations.

Similarly the relative rate of change of the wave amplitude $\frac{\partial A}{\partial x}/A$ (termed the "Linear shoaling coefficient") has been examined and again a model with $B = 1/15$ fares better than equations based on (for comparison) $\tilde{u}$. Also nonlinear properties such as wave-wave interaction and energy transfer between wave components and the nonlinear (3rd-order) dependence of the celerity on the wave height are among the tests. For details about these and other aspects of the development of the deep water characteristics reference is made to Nwogu (1993), Madsen and Sørensen (1992), Schäffer and Madsen (1995), Madsen and Schäffer (1998, 1999).

9.10 Further developments of Boussinesq modelling

The advances described in the previous section were aimed primarily at improving and testing the deep water characteristics of the Boussinesq models. These are important because they influence the wave celerity which controls basic transformation processes such as wave refraction and diffraction and also wave shoaling.

Meanwhile the general accuracy and sophistication of the Boussinesq models have also undergone a wealth of other developments that dramatically improve the general performance of the models. These modifications have been dealing with a wide spectrum of important aspects of the models.

Some of these steps are discussed a little further below but it is only possible here to give a brief introduction to the type of problems considered and the concepts introduced and analysed in these developments. And as with other advanced subjects in this text the problems are so complex that this description cannot be exhaustive and neither is the literature review. However, very extensive reviews are provided by Kirby (1997), Madsen and Schäffer (1998, 1999), Kirby (2002), and Brocchini and Landrini (2004).

9.10.1 *Fully nonlinear models*

One of the most important developments is based on the observation that the derivation of the weakly nonlinear equations does not require an expansion of the solution in terms of the amplitude or "nonlinearity" parameter δ. Going back through the derivation it will be found that consistent with the basic assumption for the weakly nonlinear model of $U = O(1)$ or $\delta = O(\mu^2)$ we deliberately omitted terms $O(\delta^2, \delta\mu^2)$. However, it is perfectly possible to conduct the derivations while keeping all terms in δ. Such equations were first derived by Serre (1953) and it was shown by Wei et al. (1995) that retaining all δ-terms leads to a set of nonlinear equations that only contains terms in δ which are $O(\delta, \delta^2, \delta^3)$, while still being accurate to $O(\mu^2)$ in μ. Since these equations include all nonlinear terms to the accuracy of μ^2 they are called **fully nonlinear equations**, though they do not contain neither linear nor nonlinear terms of order higher than μ^2.

Tests of the fully nonlinear versions of the models show considerable improvements of accuracy in particular for the very high waves approaching breaking. These ideas are now broadly accepted and they can in principle be pursued also for higher order models.

9.10.2 *Extension of equations to* $O(\mu^4)$ *accuracy*

As the model versatiliy expands one of the limiting factors is in the fact that in models with $O(\mu^2)$-accuracy velocities and pressures are still only described by a second order polynomial variation over depth which is a very poor representation of the conditions in deeper water and also for very high waves near breaking. Extension to $O(\mu^4)$-accuracy improves this. As a generalization of Nwogu's approach Gobbi and Kirby (1999), and Gobbi et al. (2000) chose a model version that averages two reference velocities. However, the direct extension of the model to $O(\mu^4)$-accuracy also introduces fifth order derivatives into the equations which make the numerical schemes (somewhat) more complicated. Therefore attempts have also been made to find forms of the equations correct to $O(\mu^4)$ and retaining Padé [4,4] order dispersive accuracy but without the high order derivative terms (Madsen and Schäffer, 1998)

9.10.3 *Waves with currents*

The interaction of long waves with currents was already studied by Yoon and Liu (1989), Chen et al. (1998) used the the enhanced Boussinesq approach to develop versions that correctly represent the doppler shift of currents also in deeper water.

9.10.4 *Models of high order*

This type of model represents a new approach to Boussinesq modelling outlined by Agnon et al. (1999), Madsen et al. (2002), and Madsen et al. (2003).

The new procedure achieves the same acceuracy in nonlinear properties as in linear properties. This is done by expressing the boundary conditions at the freee surface and at the bottom in terms of both the vertical and the horizontal velocity components at the still water level SWL, while using a truncated expansion solution for the Laplace equation. This results in a coupled system of equations with a total of 6 unknowns, including the surface elevations and several velocity components. The initial version of this system (Agnon et al. (1999)) allows an accurate description of the dispersive nonlinear waves up to $kh = 6$. In the extensions by Masdsen et al. (2002, 2003) these results are further improved by expanding the Laplace solution from an arbitray z-level and minimizing the depth averaged error in the velocity profile. The resulting equations have linear and nonlinear wave characteristics that are accurate up to $kh = 40$ and reliable velocity profiles up to $kh = 12$.

In a recent paper Schäffer (2004) describes how to determine the velocity and other kinematic information from the results generated by such numerical models.

9.10.5 *Robust numerical methods*

One of the problems associated with Boussinesq equations is that they include third or even higher order derivatives. A consistent numerical scheme is therefore required to represent the lower order derivatives to an order where the truncation errors are smaller than the numerical representation of the high order terms. While several numerical schemes have been devised that satisfy this requirement it is probably fair to say that the simplest, most transparent, and most flexible is the method developed by Wei an Kirby (1995).

This method is a generalization of the Predictor-Corrector scheme introduced for ordinary differential equations by Adams-Bashford-Moulton. It is used to integrate the equations in time over a domain in the horizontal coordinates. The equations are written in the form

$$\frac{\partial \mathbf{E}}{\partial t} = \mathbf{F} \tag{9.10.1}$$

where **E** is the vector quantity given by the unknowns of the formulation in question and **F** is the vector with components given by the rest of the terms in each of the governing equations. In ordinary differential equations **F** is just a combination of the independent and dependent variables and thus can be evaluated exactly. In partial differential equations **F** also include space derivatives of the unknowns. Hence the accuracy of the representation depends on the order to which those derivatives are evaluated.

For each time step the horizontal spatial derivatives are calculated using schemes of sufficiently high order to satisfy the requirement of consistency. With third order predictor and a fourth order corrector steps often used the lowest order derivatives in space are determined with fourth order finite difference expressions.

The method has turned out to be quite versatile and has also been used for the nearshore circulation equations described later.

9.10.6 *Frequency domain methods for solving the equations*

An entirely different way of solving the governing equations for the wave motion is represented by the frequncy domain methods. This involves using Fourier transform to the equations governing the wave motion. This means assuming the wave motion (in terms of the surface elevation η) can be described by the sum

$$\eta(x, y, t) = \sum_{n=1}^{N} a_n(x, y) e^{-i\omega_n t} + c.c. \tag{9.10.2}$$

Here N is the total number of frequency components included in the computation. It is noted that this expansion assumes periodicity of the motion in time. Thus if irregular waves are considered the period of the smallest frequency ω_1 must be the length of the entire time series. Therefore to describe an irregular wave sequence a large number of frequencies will be needed.

When substituted into the governing equations the nonlinear terms give raise to products of the series. This leads to terms that represent nonlinear interaction between the individual sinusoidal wave components in the expansion. The overall computation time is a function of the number of terms kept in the expansion and of the way the nonlinear interaction terms are handled.

The method is best suited for non-breaking waves because in breaking waves a particular problem evolves from having to decide how the energy dissipation due to the breaking process is distributed to the individual wave components

For a extensive and well written review of the aspects of and the literature describing this method reference is made to Kaihatu (2003). This paper gives detailed overview of the form of the many different versions of the Boussinesq equations in the frequency domain and also includes such information for the nonlinear MSE models and a discussion of spectral breaking wave models.

9.11 Boussinesq models for breaking waves

In recent years the Boussinesq models have been extended to include the effect of wave breaking. This is done essentially in three different ways that differ in the details but in principle function the same way, which is by enhancing the momentum flux in the momentum equation.

It is first noticed that Boussinesq models and other phase resolving models give a description of the flow based on the continuity and momentum equations. We know from the fundamental principle of hydrodynamics that if solved correctly such a description provides all the necessary details of the flow including the effect of energy dissipations. However, in contrast to the wave models described in Chapters 4 and 6 based on analysing the energy balance, in the momentum equation the energy dissipation only shows up as modifications of the momentum fluxes.

This was also illustrated in the description in Section 5.3.1 of "why waves break?" It is recalled that in the turbulent front of a quasi-steady breaking wave we can expect the breaking process to enhance the momentum flux over and above the flux from the velocities and pressures in a non-breaking wave.

In a Boussinesq wave the turbulence will generate a similar enhancement of the momentum flux which represents the signature of the breaking process. The Boussinesq base model describes a potential flow which conserves energy and the enhancement will change the energy balance.

A noticeable weakness of all these methods is that they cannot predict the position where the breaking starts. In Boussinesq waves the wave front never becomes vertical because as the waves steepen the dispersive effects, which tend to stabilize the wave shape, from a certain point become so strong that they prevent the front from steepening further. Therefore breaking must always be initiated by artificial empirical mechanisms such as a limit on the front steepness or the wave height to water depth ratio.

The following is a very brief description of the principles behind the three model types. They all need empirical input or calibration in some form in order to function. In addition the use of the nonlinear shallow water (NSW) equations for describing breaking waves is briefly mentioned.

9.11.1 *Eddy viscosity models*

One way of including the signature of breaking in the Boussinesq equations is based on adding a viscosity or diffusion term, which has the form $(\nu_t u_x)_x$. Examples are the models by Zelt (1991), Karambas and Koutitas (1992), Wei et al. (1995), Kennedy et al. (2000) to mention some. The value of the eddy viscosity term is calibrated against experimental data both with respect to its magnitude and with respect to its variation with the wave phase. With proper choices for the term it is possible to obtain good agreement with data for the variation of the wave height. However, there is no physical justification for a term of the magnitude needed to simulate the breaking. The flow is also still modelled as a potential flow which means the velocity profile is unchanged from the polynomial representation in the underlying Boussinesq model (usually the second order polynomial approximation corresponding to $O(\mu^2)$-models). This approach has since been developed extensively with addition of further terms to better simulate the variation of the wave heights.

9.11.2 *Models with roller enhancement*

Roller methods add the effect of breaking in the form of terms that simulate the effect of the surface roller. Brocchini et al. (1992) used the concept introduced by Deigaard and Fredsøe (1989) where the weight of the roller causes a change in the pressure variation in the wave which enhances the pressure part of the momentum flux. On the other hand the model introduced by Schäffer et al. (1993) considers the enhancement in momentum flux from a roller riding on the front face of the wave and hence moving with speed equal to the phase speed c of the wave, as suggested in the wave averaged model by Svendsen (1984) Section 5.6. This leads to enhancement of the momentum flux both from the velocity of the roller and from the pressure variation. The empirical input to this model is in the specification of the roller thickness and extent in the horizontal direction.

This approach was developed and tested in a series of papers by Madsen et al. (1997a,b), Sørensen et al. (1998).

9.11.3 *Vorticity models*

In the last type of model the motion is divided into a potential part, which is similar to the nonbreaking solution, and a rotational part. The rotational part is a result of the vorticity generated by the breaking and this is the part that generates the enhancement of the momentum flux due to both velocity and pressure variation. Reference is made to Veeramony and Svendsen (1998, 2000).

It is worth to mention that this model approach utilizes a different way of deriving the Boussinesq equations which is based on the depth integrated equations of motion (derived later in Chapter 11). This method was first developed by Yoon and Liu (1989) in another context. The advantage in breaking waves is that the role of stresses on the free surface such as turbulent stresses at the turbulent mean surface of the wave front is more explicitly exposed.

This model also requires empirical input in the form of the boundary condition along the surface in the roller for the vorticity. The data for this boundary condition is taken from detailed measurements in a hydraulic jump by Svendsen et al. (2000).

Of the three methods described here this model gives the most detailed description of the internal flow in the breaking wave. The rotational part of the velocity profiles, which is not restricted to a polynomial, gives a much stronger variation over the vertical in particular just beneath the roller than the traditional polynomial variation in potential flow. This is in closer agreement with the measured profiles than the polynomial profiles that can be obtained using $O(\mu^2)$ or even $O(\mu^4)$ approximation, as in the eddy viscosity method.

9.11.4 *Wave breaking modelled by the nonlinear shallow water equations*

Like the Boussinesq equations the NSW-equations conserve mass and momentum. And since they contain no terms that represent dissipation of energy the exact solution to these equations will also conserve energy. However, these quations are often solved numerically by means of a Lax-Wendroff (or similar dissipative) numerical scheme in which artificial dissipation is introduced in such a way that the steepening of the front of the wave stops just before the front becomes vertical. We then have a permanent-form long wave of finite amplitude for which mss and momentum is conserved and it was shown in Section 5.6, that in such a wave energy is dissipated at the same rate as in a bore of the same height. The use of these equations to describe breaking waves is then based on the fact that the dissipative schemes of the Lax-Wendroff type freeze the front when it becomes steep enough. Therefore the solution using such schemes essentially reproduce the wave decay due to breaking corresponding to bore dissipation of the energy and with it some representation of the phase variation.

Starting with Hibberd and Peregrine (1979), this method for modelling breaking waves has been developed extensively. Other references are Packwood and Peregrine, (1980), Watson and Peregrine (1992), Watson et al. (1992) to mention a few. Also Kobayashi et al. (1989), Kobayashi and Wurjanto (1992), Cox et al. (1992, 1994) to mention some, have extended the method to analyzing swash zone motion on very steep slopes, and wave overtopping over structures.

The method has two major weaknesses. One is that the position where the front becomes steep enough for the dissipation to start (the front is frozen) is determined only as a certain distance from the offshore boundary of the computation. Hence the method cannot predict the position of the onset of breaking. The second is that the length/steepness of the frozen front is fixed to a certain number of grid points, i.e. a few times the Δx in the numerical grid, rather than to a length related to the actual physical processes.

9.12 Large amplitude long waves $U \gg 1$: The nonlinear shallow water equations (NSW)

The third case identified in Chapter 7 is for long waves with amplitudes of the same order as the water depth, i.e., $\delta = A/h_0 = O(1)$. Since $\mu = \lambda/h_0$

is still $\ll 1$, this implies that $U = \delta/\mu \gg 1$. In this case, δ is not small and the resulting equations will therefore be strongly nonlinear. Therefore, in the equations found in Chapter 7 for this case, we only need to include terms $O(1)$ to get equations with the first approximation to the nonlinear terms. This results in

$$\eta_t + \delta\phi_x\eta_x - \frac{1}{\mu^2}\phi_z = 0 \qquad z = \delta\eta \tag{9.12.1}$$

$$\eta + \phi_t + \frac{1}{2}\delta(\phi_x)^2 = 0 \qquad z - \delta\eta \tag{9.12.2}$$

Following the same procedure as before, we substitute the solution for ϕ to get for the kinematic condition. In 1DH ϕ for varying depth from (9.8.2) is given by

$$\phi = \phi_0 - \mu^2(z+h)h_x\phi_{0x} - \frac{\mu^2}{2}(z+h)^2\,\phi_{0xx} + O(\mu^4) \tag{9.12.3}$$

This implies that

$$\frac{1}{\mu^2}\,\phi_z = -h_x\phi_{0x} - (z+h)\,\phi_{0xx} \tag{9.12.4}$$

which shows that at $z = \delta\eta$ we get for the last two terms in the kinematic condition

$$\phi_{0x}\delta\,\eta_x - \frac{1}{\mu^2}\,\phi_z = ((h+\delta\eta)\phi_0 x)_x \tag{9.12.5}$$

The kinematic condition then becomes

$$\eta_t + ((\delta\eta + h)\phi_{0x})_x = O(\mu^2) \tag{9.12.6}$$

Similarly, we get for the dynamic condition

$$\eta + \phi_{0t} + \frac{1}{2}\delta(\phi_{0x})^2 = O(\mu^2) \tag{9.12.7}$$

Again, we can express these two equations in terms of the bottom velocity u_0 to get

$$\begin{aligned} \eta_t + ((\delta\eta + h)u_0)_x &= O(\mu^2) \\ \eta_x + u_{0t} + u_0u_{0x} &= O(\mu^2) \end{aligned}$$

However, since the approximation for ϕ used in deriving these equations is just $\phi(x,z,t) = \phi_0(x,t)$ the assumed velocity is constant over the depth and consequently $u_0 = \tilde{u}$.

In the general 2DH version and dimensional form the equations become

$$\eta_t + \nabla_h(\tilde{\mathbf{u}}(h+\eta)) = 0 \tag{9.12.8}$$

$$\tilde{\mathbf{u}}_t + \tilde{\mathbf{u}} \cdot \nabla_h \tilde{\mathbf{u}} + g\,\nabla_h \eta = 0 \tag{9.12.9}$$

This is the set of equations that is called the **Nonlinear Shallow Water (NSW) equations**. They have interesting features that are discussed briefly in Section 9.11. Note that the Boussinesq waves described in earlier parts of this chapter include the NSW equations as a subset.

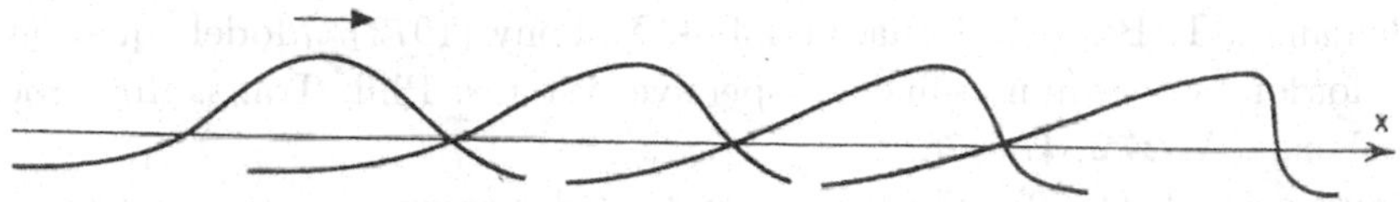

Fig. 9.12.1 The deformation toward breaking of a wave propagating according to the nonlinear shallow water equations.

It may also be mentioned that these equations are often extended by adding the effect of a bottom friction by replacing the 0 on the RHS of (9.12.9) with a term of the form $-\frac{1}{2}f\;\tilde{\mathbf{u}}|\tilde{\mathbf{u}}|/h$, where f is the friction factor (see Chapter 10).

The NSW-equations have no solutions for waves of constant form. Analysis of the above derivation shows that they correspond to hydrostatic pressue relative to the **instantaneous** (i.e. local) surface level and as mentioned to constant velocity over depth. Using the method of characteristics (see e.g. Abbott and Basco, 1989) it is found that in waves propagating according to these equations each part of the wave will propagate with a speed of $c = u + \sqrt{g(h+\eta)}$, i.e. a c corresponding to the local depth of water and including the local particle velocity. This means that higher parts of a wave will move faster than lower parts, and overtake the lower parts as shown in Fig. 9.12.1, and eventually the waves break. Of course the basic assumptions behind the equations have broken down long before this happens, because the characteristic horizontal length is no longer large in comparison to the water depth.

As with linear shallow water waves the phase velocity for these waves is independent of the wave period so they are only amplitude dispersive.

9.13 References - Chapter 9

Abbott, M. B. and D. R. Basco (1989). Computational fluid dynamics. Longman Scientific and Technical. 425 pp.

Abramowitz, M. and I. A. Stegun (1964). Handbook of mathematical functions. Dover Publications, New York.

Agnon, Y., P. A. Madsen, and H. A. Schäffer (1999). A new approach to high-order Boussinsq models. J. Fluid Mech., **399**, 319–333.

Baker, G. S. Jr. and P. Graves-Morris (1980). Padé approximants, Cambridge University Press, 745 pp.

Benjamin, T. B., J. L. Bona, and J. J. Mahony (1972). Model equations for long waves in non-linear dispersive systems. Phil. Trans. Roy. Soc. Lond., A, **272**, 47 - 78.

Boussinesq, J. (1872). Theorie des onde et des resous qui se propagent le long d'un canal rectangulaire horizontal, en communiquant au liquide contenu dans ce canal des vitesses sensiblement pareilles de la surface au fond. Journal de Math. Pures et Appl., Deuxieme Serie, **17**, 55–108.

Brocchini, M., M. Drago, and L. Iovenitti (1992). The modelling of short waves in shallow waters. Comparison of numerical models based on Boussuinesq and Serre euqations. ASCE Proc. 23rd Int. Conf. Coastal Engrg., 76–88.

Brocchini, M. and M. Landrini (2004). Water waves for engineers. Springer Verlag, Berlin (Approx 250 pp, in preparation)

Chaplin, J. D. (1978). Developments of stream function theory. Rep. MCE/2/78, Dept. Civ. Engrg., Univ. Liverpool, UK.

Chen, Q., P. A. Madsen, H. A. Schäffer, and D. R. Basco (1998). Wave-current interaction based on an enhanced Boussinesq approach. Coastal Engrg., **33**, 11 -39.

Cox, D. T., N. Kobayashi, and A. Wurjanto (1992). Irregular wave transformation processes in surf and swash zones. ASCE Proc. 23rd Int. Conf. Coastal Engrg., Chap 10, 156 - 169

Cox, D. T., N. Kobayashi, and D. L. Kriebel (1994). Numerical model verification using Supertank data in surf and swash zones. ASCE Proc. Coastal Dynamics '94, Barcelona Spain.

Deigaard, R. and J. Fredsøe (1989). Shear stress distribution in dissipative water waves. Coastal Engineering, **13**, 357–378.

Dingemans, M. W. (1997). Water wave propagation over unenven bottoms. World Scientific, Singapore, 967 pp.

Erdélyi, A. (editor) (1953). Higher transcendental functions, Vol II. McGraw-Hill, New York.

Fenton, J. D. (1979). A high order cnoidal wave theory. J. Fluid Mech., **94**, 1, 129–161.

Gobbi, M. F. and J. T. Kirby (1999). Wave evolution over submerged sills: test of a high order Boussinesq model. Coastal Engrg., **37**, 57–96.

Gobbi, M. F., J. T. Kirby, and G. Wei (2000). A fully nonlinear Boussinesq model for surface waves. Part 2. Extension to $O(kh)^4$. J. Fluid Mech., **405**, 181–210.

Gradshteyn, I. S. and I. M. Ryzhik (1965). Tables of integrals, series and products. Academic Press, New York.

Hibberd, S. and D. H. Peregrine (1979). Surf and runup on a beach: a uniform bore. J. Fluid Mech. **95**, 323–345.

Karambas, T. and C. Koutitas (1992). A breaking wave propagation model based on the Boussinesq equations. Coastal Engineering, **18**, 1–19.

Kaihatu, J. M. (2003). Frequency domain wave models in the nearshore and surfzones. In Advances in Coastal Modeling, (V. C. Lakhan ed.), Elsevier Science, 43–72.

Kennedy, A. B., Q. Chen, J. T. Kirby, and R. A. Dalrymple (2000). Boussinesq modeling of wave transformation, breaking and runup. I: 1D. J. Waterway, Port, Coastal and Ocean Engineering, **126**, 206–214

Kirby, J. T. (1997). Nonlinear dispersive long waves in water of variable depth. In: Gravity waves in water of finite depth (J. N. Hunt, ed.) Advances in Fluid Mechanics, **10**, 55-125. Comp. Mech. Publications.

Kirby, J. T. (2002). Boussinesq models and appications to nearshore wave propagation surf zonne processes and wave induced currents. In: Advances in nearhsore modelling (Lakhan ed.)

Kobayashi, N., G. S. DeSilva, and K. D. Watson (1989). Wave transformation and swash oscillation on gentle and steep slopes. J. Geophys. Res., **94**, C1, 951–966.

Kobayashi, N. and Wurjanto, A. (1992). Irregular wave setup and runup on beaches. ASCE J. Waterw., Port, Coast, and Ocean Engrg., **118**, 368–386.

Korteweg, D. J. and G. DeVries (1895). On the change of form of long waves advancing in a canal, and on a new type of long stationary waves. Phil. Mag., Ser. 5, **39**, 422 – 443.

Madsen, P. A., R. Murray, and O. R. Sørensen (1991). A new form of the Boussinesq equations with improved linear dispersion charactieristics. Coast. Engrg. **15**, 371–388.

Madsen, P. A. and O. R. Sørensen (1992). A new form of the Boussinesq equations with improved linear dispersion characteristics. Part 2. A slowly-varying bathymetry. Coast. Engrg. **18**, 183–204.

Madsen, P. A., O. R. Sørensen and H. A. Schäffer (1997a) Surf zone dynamics simulated by a Boussinesq type model.Part I. Model description and cross-shore motion of regular waves. Coast. Engrg, **32**, 255–288.

Madsen, P. A., O. R. Sørensen and H. A. Schäffer (1997b) Surf zone dynamics simulated by a Boussinesq type model.Part II. surf beat and swash oscillations for wave groups and irregular waves. Coast. Engrg, **32**, 289–319.

Madsen, P. A. and H. A. Schäffer (1998). Higher order Boussinesq equations for surface waves: derivation and analysis. Phil. Trans. R. Soc. Lond. A **356**, 3123–3184.

Madsen, P. A. and H. A. Schäffer (1999). A review of Boussinesq-type equations for surface gravity waves. Advances in Coastal Engineering, **5**, 1–93.

Madsen, P. A., H. B. Bingham, and H. Liu (2002). A new Boussinesq method for fully nonlinear waves from shallow to deep water. J. Fluid Mech., **462**, 1–30.

Madsen, P. A., H. B. Bingham, and H. A. Schäffer (2003). Boussinesq-type formulations for fully nonlinear and extremly dispersive water waves. derivation and analysis. Proc. Roy. Soc. Lond. A, **459**, 1075-1104.

Madsen, P. A. and Y. Agnon (2003). Accuracy and convergence of velocity formulations for water waves in the framework of Boussinesq theory. J. Fluid Mech. **477**, 285-319.

Masch, F. D. and R. L. Wiegel (1961). Cnoidal waves, tables of functions. Council of Wave Research, The Engineering Foundation, Univ. Calif., Richmond, Calif., 129 pp.

Milne-Thomson, L. M. (1968). Theoretical hydrodynamics, 5th ed. Macmillan, New York.

Mei, C. C. (1983). The applied dynamics of ocean surface waves. World Scientific, Singapore.

Nwogu, O. (1993). Alternative form os Boussinesq equations for nearshore wave propagation. J. Waterway, Port Coastal and Ocean Engrg., ASCE, **119**, (6) 618–638.

Packwood, A. and D. H. Peregrine (1980). The propagation of solitary waves and bores over a porous bed. Coast. Engrg., **3**, 221–242.

Peregrine, D. H. (1967). Long waves on a beach. J. Fluid Mech., **27**, 815–827.

Peregrine, D. H. (1972). Equations for water waves and the approximations behind them. In "Waves on Beaches" (Ed R. Meyer), 95–121.

Press, W. H., B. P. Flannery, S. A. Teukolsky and W. T. Wetterling (1986). Numerical recipes, Camebridge University Press.

Russell, J. S. (1844). Report on waves. Brit. Ass. Adv. Sci. Rep.

Schäffer, H. A. (2004). Accurate determination of internal kinematics from numerical wave model results. Coastal Engrg., **50**, 4, 199–212.

Schäffer, H. A., P. A. Madsen, and R. Deigaard (1993). A Boussinesq model for waves breaking in shallow water. Coastal Engrg., **29**, 185–202.

Schäffer, H. A. and P. A. Madsen (1995). Further enhancements of Boussinesq-type equations. Coast. Engrg, **26** 1 – 14.

Serre, P. F. (1953). Contribution *á* l'*é*tude des *é*coulement permanents et variables dans les canaux. La Houille Blanche, 374–488 and 830–872.

Skovgaard, O., I. A. Svendsen, I. G. Jonsson, and O. Brink-Kjær (1974). Sinusoidal and cnoidal gravity waves - Formulae and tables. Inst. Hydrodyn. and Hydraulic Engrg., Tech. Univ. Denmark, 1974, 8 pp.

Svendsen, I. A. (1974). Cnoidal waves over gently sloping bottoms. Series paper no 6, Institute of Hydrodynamics and Hydraulic engineering. Technical University of Denmark.

Svendsen, I. A. and C. Staub (1981). Horizontal velocity profiles in long waves. J. Geophys. Res., **86**, c5, pp. 4138–4148.

Svendsen, I. A. and I. G. Jonsson (1976, 1980). Hydrodynamics of coastal regions. Den Private Engineering Fund, Technical University of Denmark.

Svendsen, I. A., J. Veeramony, J. Bakunin, and J. T. Kirby (2000). Analysis of the flow in weakly turbulent hydraulic jumps. J. Fluid Mech. **418**, 25 – 57

Sørensen, O. R., H. A. Schäffer, and P. A. Madsen (1997) Surf zone dynamics simulated by a Boussinesq type model. Part III. Wave induced horizontal nearshore circulations.

Veeramony, J. and I. A. Svendsen (1998). A Boussinesq model for surf zone waves: Comparison with experiments. *Proc* 26th *Intl. Conf. Coastal Engrg.*. 258–271.

Veeramony, J. and I. A. Svendsen (2000). The flow in surf zone waves. Coastal Engrg. **39**, 93 – 122.

Watson, G. and D. H. Peregrine (1992). Low frequancy waves in the surf zone. Proc 23rd Int. Conf. Coastal Engrg. 818 – 831.

Watson, G., D. H. Peregrine and E. F. Toro (1992). Numerical solution if the shallow water equations on a beach using the weighted average flux method. Computational Fluid Dynamics, **1**, 495–502.

Wei, G. and J. T. Kirby (1995). A time dependent numerical code for extended Boussinesq equations. ASCE J. WPCOE, **120**, 251 – 261.

Wei, G., J. T. Kirby, S. T. Grilli, and R. Subramanya (1995). A fully nonlinear Boussinesq model for surface waves. Part I. Highly nonlinear unsteady waves. J. Fluid Mech., **294**, 71 – 92.

Wiegel, R. L. (1960). A presentation of cnoidal wave theory for practical application. J. Fluid Mech. **7**, 273 – 286.

Wiegel, R. L. (1964) Oceanographical engineering. Prentice Hall, New York.

Witting, G. B. (1984). A unified model for the evolution of nonlinear water waves. J. Comp. Phys. **56**, 203-236.

Yoon, S. B. and P. L.-F. Liu (1989). interaction of currents and weakly nonlinear waves in shallow water. J. Fluid Mech., **205**, 397 – 419.

Zelt, J. A. (1991). The runup of nonbreaking and breaking solitary waves. Coastal Engineering, **15**, 205 – 246

Appendix

Appendix 9A: Elliptic integrals and functions

This appendix gives a very brief review of the definitions of elliptic functions and integrals used in the derivations in Chapter 6.5.

Incomplete elliptic integrals and functions

The elliptic functions used here are of the Jacobian type. The central definition is that of $F(\phi, m)$ which is called an incomplete elliptic integral of the first kind. F is defined by

$$F(\phi, m) \equiv \int_0^\phi \left(1 - m \sin^2 \theta\right)^{-\frac{1}{2}} \, d\theta \tag{9.A.1}$$

m is called the parameter of F, ϕ the amplitude.

The inverse function of F with respect to ϕ is defined within the necessary restrictions of monotony so that

$$\phi \equiv \arccos(cn(F, m)) \tag{9.A.2}$$

or

$$\cos\phi \equiv cn(F, m) \tag{9.A.3}$$

which defines the elliptic function cn. this is written

$$\cos\phi = cn\ u \tag{9.A.4}$$

by the simplifying definition

$$u \equiv F(\phi, m) \tag{9.A.5}$$

One also encounters **incomplete elliptic integrals of the second kind** defined by

$$E(\phi, m) \equiv \int_0^{\phi} \left(1 - m\sin^2\theta\right)^{\frac{1}{2}} d\theta \tag{9.A.6}$$

and since $\phi = \phi(u, m)$, $E(\phi, m)$ is usually written $E(u, m)$.

These two forms of elliptic integrals represent the standard forms to which all other elliptic integrals can be reduced.

By various auxiliary definitions, the basic forms may be written in many different forms (see e.g., Abramowitz and Stegun (1964), Chapter 17). The reduction of arbitrary elliptic integrals to standard forms is well described by Erdélyi et al. (1953), vol. II.

Complete elliptic integrals

When $\phi = \pi/2$, the integrals are called **complete elliptic integrals.** We define

$$K(m) \equiv F(\pi/2, m) \tag{9.A.7}$$

$$E(m) \equiv E(K(m), m) \tag{9.A.8}$$

where (9.A.5) has been used.

Appendix 9B: Derivation of Eq. (9.5.38)

This appendix gives the derivation of equation (9.5.38).

Substituting (9.5.36) into (9.5.37) yields

$$\int_0^1 \eta d\theta = \int_0^1 \{\eta_2 + Hcn^2\, 2K\theta\}\, d\theta = 0 \tag{9.B.1}$$

where we have also used (9.5.39). This can also be written

$$\eta_2 \cdot 1 + \frac{H}{2K}\int_o^{2K} cn^2 u\, du = 0 \tag{9.B.2}$$

From Abramowitz and Stegun (1964) (A & S in the following) equations (16.25) and (16.26) we have

$$\int_o^a cn^2 u\, du = \frac{1}{m}E(a, m) - \frac{m_1}{m}a \tag{9.B.3}$$

where

$$m_1 = 1 - m \tag{9.B.4}$$

Hence, (9.B.2) becomes

$$\eta_2 + \frac{H}{2mK}\left(E\left(2K, m\right) - 2m_1 K\right) = 0 \tag{9.B.5}$$

A & S eq. (17.4.4) yields

$$E(2K, m) = 2E(m) \tag{9.B.6}$$

which brings (9.B.5) on the form (after multiplication by mK)

$$mK\eta_2 + H(E - m_1 K) = 0 \tag{9.B.7}$$

As m is given by (9.5.30), we have for m_1

$$m_1 = 1 - \frac{\eta_1 - \eta_2}{\eta_1 - \eta_3} = \frac{\eta_2 - \eta_3}{\eta_1 - \eta_3} \tag{9.B.8}$$

Substituting this into (9.B.7) together with (9.5.27) for H and (9.5.30) for m and rearranging terms finally yields

$$\eta_3 K(m) = -\left(\eta_1 - \eta_3\right) E(m) \tag{9.B.9}$$

which is eq. (9.5.38).

Appendix 9C: Numerical solution for cnoidal properties

This appendix outlines simple numerical algorithms for obtaining the parameter m and for solving for the wave parameters when the waves are specified by H, h, and T.

Numerical solution of $U = 16/3\ m\ K^2$

Solution of the equation

$$U = \frac{16}{3}\ m\ K^2(m) \tag{9.C.1}$$

for given values of U provides the value of m in the elliptic functions and integrals in the cnoidal wave solution. Here $0 \le m < 1$. For practical cases we normally have $U > 10$ which means the relevant values of m is close to 1, and $K(m)$ is large.

(9.C.1) is a transcendental equation for m, and a Newton-Raphson iterative solution will give m with only a few iterations. The problem is evaluation of $K(m)$. To obtain a convenient formulation of the problem we introduce the complimentary parameter $m_1 = 1 - m$ for the elliptic functions and write (9.C.1) as

$$U = \frac{16}{3}\ (1 - m_1)\ K^2(m_1) \tag{9.C.2}$$

It turns out that in the range of solutions in question $K(m_1)$ can with sufficient accuracy be approximated by (see Abramowitz and Stegun, 1964, p 592, (17.3.35))

$$K(m_1) = [\ a_0 + a_1\ m_1 + a_2\ m_1^2] - [\ b_0 + b_1\ m_1 + b_2\ m_1^2] \ln\ m_1 + \epsilon(m_1) \tag{9.C.3}$$

where

$$\begin{aligned} a_0 &= 1.3862944 \quad ; \quad b_0 = 0.5 \\ a_1 &= 0.1119723 \quad : \quad b_1 = 0.1213478 \\ a_2 &= 0.0725296 \quad ; \quad b_2 = 0.0288729 \end{aligned} \tag{9.C.4}$$

and $\epsilon < 3 \cdot 10^{-5}$.

Writing (9.C.2) as

$$f(m_1) = 1 - m_1 - \frac{3\,U}{16\,K^2} \tag{9.C.5}$$

We seek values of m_1 for which $f(m_1) = 0$. The Newton-Raphson iterative formula for m_1 becomes

$$m_{1,i+1} = m_{1,i} - \frac{f(m_{1,i})}{f'(m_{1,i})} \tag{9.C.6}$$

where the derivative f' is obtained directly by differentiation of (9.C.5)

$$f'(m_1) = \frac{3}{8}\frac{U}{K^3}\frac{dK}{dm_1} - 1 \tag{9.C.7}$$

and

$$\frac{dK}{dm_1} = a_1 + 2a_2 m_1 + [\, b_1 + 2b_2 m_1 \,] \ln m_1 - [\, b_0 + b_1 m_1 + b_2 m_1^2 \,] \frac{1}{m_1} \tag{9.C.8}$$

Initial estimate for m_1

Using a suitable initial estimate m_1 (9.C.5) will rapidly converge toward the solution for m_1. An initial estimate for m_1 can be obtained from (9.C.2) by substituting (9.C.3) for K and assuming $m_1 \ll 1$. This gives

$$U \sim \frac{16}{3} [\, a_0 - b_0 \ln m_1]^2 \tag{9.C.9}$$

or

$$m_1 \sim \exp\left(\frac{a_0 - (\frac{3}{16}\, U)^{1/2}}{b_0} \right) \tag{9.C.10}$$

For $U > 30$ (9.C.10) will even function as a quite accurate direct solution for m_1.

With m_1 known K is determined from (9.C.3).

Calculation of L/h

Since in most cases the wave is specified by H, h and T it is necessary to first to calculate L/h in order to determine U. For L/h we have

$$\frac{L}{h} = T\,\sqrt{g/h}\, \left(1 + \frac{h}{h}\, A \right)^{1/2} \tag{9.C.11}$$

where

$$A(m) = \frac{2}{m} - 1 - \frac{3\,E(m)}{m\,K(m)} \tag{9.C.12}$$

and $E(m)$ is the complete elliptic intergral of the second kind. (9.C.11) must be solved iteratively along with the solution of (9.C.1) for m. Substituting the definition for U into (9.C.1) gives

$$\frac{L}{h} = \left(\frac{16}{3} \, \frac{m\,K^2}{H/h} \right)^{1/2} \tag{9.C.13}$$

In the expression for A E can be approximated by (see Abramowitz and Stegun (1964), p591)

$$E(m_1) = [\,1 + e_1 m_1 + e_2 m_1^2\,] - [\,f_1 m_1 + f_2 m_1^2\,] \ln m_1 + \epsilon(m) \tag{9.C.14}$$

where

$$\begin{array}{lcl} e_1 = 0.4630151 & : & f_1 = 0.2452727 \\ e_2 = 0.1077812 & ; & f_2 = 0.0412496 \end{array} \tag{9.C.15}$$

and $\epsilon < 4 \cdot 10^{-5}$.

The iterative solution is started by assuming $A = 1$ as the first approximation. A direct iteration can then be performed by the following procedure

(1) Set $A = 1$
(2) Calculate L/h from (9.C.11).
(3) Calculate $U = HL^2/h^3$.
(4) Calculate m by solving (9.C.1) iteratively as described above.
(5) Calculate a new approximation for A using (9.C.12).
(6) Return to (2), and so on.

The algorithm converges to an accurate solution usually within 4-5 iterations (depending on the required axccuracy).

It may be mentioned that there is a false root which is avoided by the initial choice of $A = 1$.

Appendix 9D: Padé approximations: A brief outline

Padé approximations (or "approximants as they are often called) are approximations for arbitrary functions $f(z)$ in the form of rational fractions

$$\left[\frac{L}{M}\right] = \frac{a_0 + a_1\,z + a_2\,z^2 + \ldots + a_L\,z^L}{b_0 + b_1\,z + b_2\,z^2 + \ldots + b_M\,z^M} + O(z^{L+M+1}) \tag{9.D.1}$$

Usually the fraction is divided by b_0 so that the denominator becomes $1 + b_1\, z + b_2\, z^2 + ...$

This appendix gives a very brief outline of how the Padé coeefficients in (9.D.1) are determined.

Determination of a_n, b_n is (9.D.1) can only be done by equating the RHS of (9.D.1) to the Taylor expansion of $f(z)$, which is written

$$f(z) = \sum_{i=0}^{\infty} c_i\, z^i \tag{9.D.2}$$

Thus determining the Padé approximation is equivalent of determining a_n, b_n in the equation

$$\sum_{i=0}^{\infty} c_i\, z^i = \frac{a_0 + a_1\, z + a_2\, z^2 + ... + a_L\, z^L}{1 + b_1\, z + b_2\, z^2 + ... + b_M\, z^M} + O(z^{L+M+1}) \tag{9.D.3}$$

for known c_i.

The powers L and M in the numerator and denominator polynomiums, respectively, must be chosen in advance. Thus we have the following number of unknowns

$L + 1$ numerator coefficients
M denominator coefficients

or a total of $L + M + 1$ unknowns.

The procedure for determining the a_n, b_n coefficients is best illustrated by an example.

Example Consider the function

$$\frac{c^2}{gh} = \frac{\tanh z}{z} \tag{9.D.4}$$

where we imagine $z = kh$. It can be shown that this function has the Taylor expansion

$$\frac{\tanh z}{z} = 1 - \frac{1}{3}\, z^2 + \frac{2}{15}\, z^4 - \frac{17}{315}\, z^6 + \frac{62}{2865}\, z^8 + O(z^{10}) \tag{9.D.5}$$

We first seek the Padé approximation with $L = 0, M = 2$ (called the [0/2] Padé approximation). This means seeking values of b_1 and b_2 that satisfy

$$1 - \frac{1}{3}\, z^2 + \frac{2}{15}\, z^4 = \frac{1}{1 + b_1\, z + b_2\, z^2} \tag{9.D.6}$$

Multiplication by the denominator on the RHS gives

$$1 + b_1\, z + \left(b_2 - \frac{1}{3}\right) z^2 - \frac{1}{3} b_1\, z^3 + \ldots = 1 \qquad (9.D.7)$$

which gives the equations (by equating equal powers of z)

$$1 = 1 \;; \qquad b_1 = 0 \;; \qquad b_2 - \frac{1}{3} = 0 \qquad (9.D.8)$$

Solving the equations, the last one first, gives

$$b_2 = \frac{1}{3} \;; \qquad b_1 = 0 \qquad (9.D.9)$$

which means that the [0/2] Padé approximation is

$$\frac{\tanh z}{z} = \frac{1}{1 + \frac{1}{3}\, z^2} + O(z^3) \qquad (9.D.10)$$

Exercise D-1

Using the same procedure as in the example above show that the [2/2] Padé approximation becomes

$$\frac{\tanh z}{z} = \frac{1 + \frac{1}{15}\, z^2}{1 + \frac{2}{5}\, z^2} + O(z^6) \qquad (9.D.11)$$

Exercise D-2

Still using the same procedure as in the example show that the [0/1] and the [1/0] approximation only give the trivial answer

$$\frac{\tanh z}{z} = 1 \qquad (9.D.12)$$

which is only valid for $z = 0$

Exercise D-3

Show that the [2/4] approximation becomes

$$\frac{\tanh z}{z} = \frac{1 + \frac{2}{21}\, z^2}{1 + \frac{3}{7}\, z^2 + \frac{1}{105}\, z^4} + O(z^8) \qquad (9.D.13)$$

See Witting (1984) , who also gives the [4/4] approximation as

$$\frac{\tanh z}{z} = \frac{1 + \frac{1}{9}\, z^2 + \frac{1}{945}\, z^4}{1 + \frac{4}{9}\, z^2 + \frac{1}{63}\, z^4} + O(z^{10}) \qquad (9.D.14)$$

It is noticed that **all** the coefficients in the approximations change when the order changes.

The general version

The general equation corresponding to (9.D.7) becomes

$$\begin{aligned}(1 + b_1\, z + b_2\, z^2 + ... + b_M\, z^M)(c_0 + c_1\, z + c_2\, z^2 + ... + c_i\, z^i)& \\ -(a_0 + a_1\, z + a_2\, z^2 + ... + a_m\, z^M) = 0& \qquad (9.D.15)\end{aligned}$$

which gives the equations

$$\begin{aligned}
c_0 &= 0\,; &&\text{for } z^0 \\
c_1 + c_0 b_1 &= 0\,; &&\text{for } z^1 \\
c_2 + c_1 b_1 + c_0 b_2 &= 0\,; &&\text{for } z^2 \\
... &= ... && \qquad (9.D.16)\\
c_L + c_{L-1} b_1 + ... + c_0 b_L &= 0\,; &&\text{for } z^L \\
c_{L+1} + c_L b_1 + ... + c_0 b_{L+1} &= 0\,; &&\text{for } z^{L+1} \\
... &= ... \\
c_{L+M} + c_{L+M-1} b_1 + ... + c_L b_M &= 0\,; &&\text{for } z^{L+M}
\end{aligned}$$

Hence we have established $L + M + 1$ linear equations for the $a_0, a_1, ..., a_L, b_1, ..., b_M = L + M + 1$ unknown coefficients.

These equations are called the Padé equations. The solution to the equations obviously gives the $L + M + 1$ coefficients in the [L/M] Padé approximation.

For further information about Padé approximations ("approximants") reference is made to Baker and Graves-Morris (1980).

Chapter 10

Boundary Layers

10.1 Introduction

One of the most important discoveries in fluid mechanics was the fact that vorticity is not generated inside a normal flow. It always originates from disturbances along some boundary or discontinuity in the flow. This causes boundary layers to develop. The boundary layer is the layer with significant vorticity due to the large velocity gradients in the direction perpendicular to the boundary.

Examples of vorticity generating flow phenomena are:

Boundary layers at solid walls.

Shear layers Jets Wave breaking	Two parts of fluid with different speeds are meeting

Boundary layers can be both laminar and turbulent, but they will always start laminar.

In nearshore regions, the boundary layer that develops at the sea bottom in the oscillatory motion caused by the waves (and currents) is of particular interest as it accounts for substantial energy dissipation when active over longer distances. The bottom shear stress may be small locally in comparison to other forces acting on the fluid column, but as the waves travel the energy dissipation due to bottom friction accumulates. The bottom boundary layer also plays a central role in how the waves and currents move sediments.

Also, internal "boundary layers" such as the vortex sheets that develop when the jet from a breaking wave hits the water in front of the crest are of importance. However, so far, nobody has been able to give a rational analyzes of those details that are of any significance for our understanding of the processes and even less of practical use.

Here we only consider boundary layers along plane walls and concentrate on the oscillatory bottom boundary layer caused by waves. To some extent, the effects of currents will be included. Fig. 10.1.1 shows the concept of the bottom boundary layer and basic definitions of variables used in the following. Note that for simplicity in this chapter, we use $z = 0$ at the bottom.

Although such boundary layers are often turbulent (at least on natural beaches, though not always under the conditions used in laboratory experiments), substantial insight is gained by considering laminar boundary layer flows. In addition to being the simplest, those flows are also the only ones for which generally accepted theoretical analyses are available. In the last section in this chapter we will look at some aspects of turbulent boundary layers.

10.1.1 *The boundary layer equations. Formulation of the problem*

The boundary layer develops because the viscosity causes the fluid to adhere to the wall, which means that at the wall the velocity is zero. The assumption of a plane boundary and (essentially) laminar flow implies that we can restrict our examinations to the 2D flow in the vertical plane of wave propagation.

In oscillatory flow such as waves, this layer with high velocity shear perpendicular to the bottom is very thin. Outside the boundary layer, the flow can often (—not in breaking waves, though) be considered a potential flow, which means our well known wave theories apply outside the boundary layer.

Near a plane horizontal bottom we know from linear wave theory that the bottom parallel velocity U in the irrotational wave flow is characterized by

$$\frac{\partial U}{\partial z} = 0 + O(kz) \qquad z \geq 0 \tag{10.1.1}$$

or $U \simeq U_b + O(kz)^2$. This means that in the immediate neighborhood above the boundary layer ($z \ll h$), we can assume the potential flow velocity equals the value at $z = 0$ predicted by the potential theory.

The equations for the boundary layer are further based on the assumption that in the boundary layer the velocity variation perpendicular to the wall is much bigger than the variations over similar distances along the wall. This leads to the simplifications of the Navier-Stokes equations known as the boundary layer approximation (see Section 2.3.4).

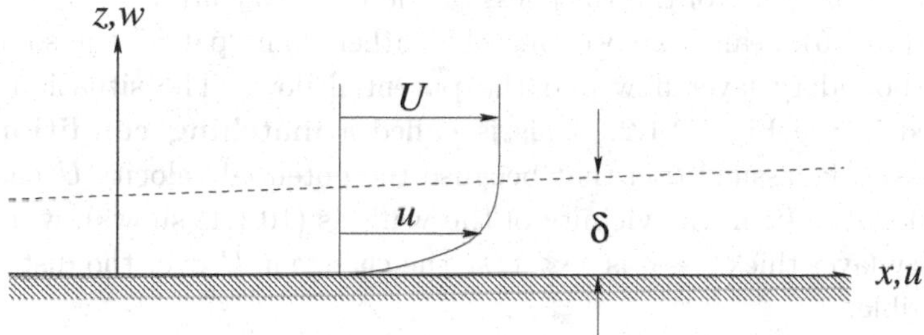

Fig. 10.1.1 Definition sketch for the boundary layer variables.

In the boundary layer approximation, the continuity equation is:

$$\frac{\partial u}{\partial x} + \frac{\partial w}{\partial z} = 0 \tag{10.1.2}$$

The momentum equations can be reduced to the so called **boundary layer equation** in the x-direction which, for the velocity u inside the boundary layer, reads

$$x: \quad \frac{\partial u}{\partial t} + u\frac{\partial u}{\partial x} + w\frac{\partial u}{\partial z} = -\frac{1}{\rho}\frac{\partial p_D}{\partial x} + \nu\frac{\partial^2 u}{\partial z^2} \tag{10.1.3}$$

where p_D is the dynamic pressure.

In the direction perpendicular to the wall, the momentum equation simplifies to

$$z: \quad \frac{\partial p_D}{\partial z} = 0 \tag{10.1.4}$$

which implies that inside the boundary layer the pressure variation perpendicular to the wall can be neglected. Thus, over the entire thickness of the boundary layer, we have the same dynamic pressure $p_D(x, t)$.

The boundary conditions for (10.1.2) and (10.1.3) are

$$\text{i)} \quad u = 0 \;, \quad w = 0 \qquad \text{at} \quad z = 0 \tag{10.1.5}$$

$$\text{ii)} \quad u \to U_b \qquad \text{at} \quad z/\delta \to \infty \tag{10.1.6}$$

where δ is a measure of the thickness of the boundary layer.

Eq. (10.1.6) means that we "match" rather than "patch" the solutions for the boundary layer flow and the potential flow. The situation is illustrated in the Fig. 10.1.2. This is called a **matching condition**. It is based on the assumption that because the potential velocity U changes proportional to kz in the vicinity of the wall (as (10.1.1) shows), while the boundary layer thickness δ is $\delta \ll 1/k$, the change in U over the distance δ is negligible.

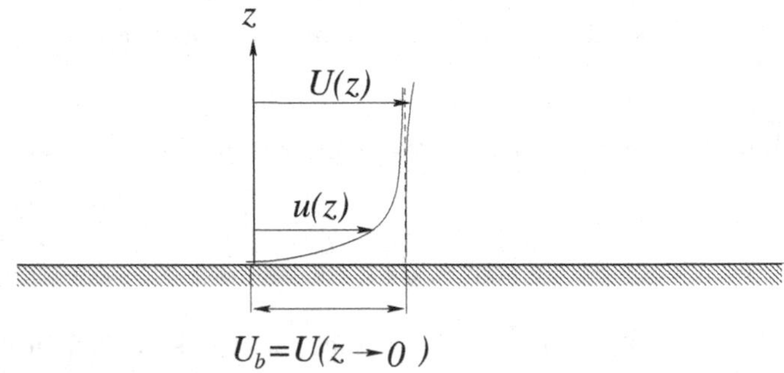

Fig. 10.1.2 The concept of a matching (boundary) condition: At the upper limit of the boundary layer we assume that $u(z)$ approaches the bottom value U_b of the potential flow velocity even though the actual value of $U(z)$ at that level may deviate (slightly) from U_b. This is consistent because the boundary layer solution assumes that the thickness of the boundary layer is small in comparison to the vertical scale for the variation of the outside flow $U(z)$.

The inviscid flow outside the boundary layer satisfies the Euler equation with $w = 0$

$$\frac{\partial U}{\partial t} + U\frac{\partial U}{\partial x} = -\frac{1}{\rho}\frac{\partial p_D}{\partial x} \qquad z \to 0 \tag{10.1.7}$$

In (10.1.7), we have already introduced the assumption (10.1.1) that the boundary layer does not reach further from the wall than what allows us to let (10.1.7) be valid.

In the solution of the equations for u, it is assumed that U is known. p_D can then be eliminated from (10.1.3) using (10.1.7) which gives

$$\frac{\partial u}{\partial t}+u\frac{\partial u}{\partial x}+w\frac{\partial u}{\partial z}=\frac{\partial U}{\partial t}+U\frac{\partial U}{\partial x}+\nu\frac{\partial^2 u}{\partial z^2} \tag{10.1.8}$$

10.1.2 *Perturbation expansion for u*

(10.1.3), (10.1.7) and (10.1.8) are nonlinear equations. Hence, we seek a perturbation solution for u, w given by

$$\begin{aligned} u &= u^{(1)}+u^{(2)}+\cdots \\ w &= w^{(1)}+w^{(2)}+\cdots \end{aligned} \tag{10.1.9}$$

and assume that U is given by a similar (known) solution

$$U=U^{(1)}+U^{(2)}+\cdots \tag{10.1.10}$$

where $U^{(1)}$ is supposed to correspond to the first order Stokes solution, $U^{(2)}$ the second order, etc., that is, we assume

$$U=\sum_{n=1}^{\infty}U_n\cos\left[n(\omega t-kx)\right] \tag{10.1.11}$$

so that (with $\theta=\omega t-kx$)

$$U^{(1)}=U_1\cos(\omega t-kx)\equiv U_1\cos\theta \tag{10.1.12}$$

$$U^{(2)}=U_2\cos 2\theta, \tag{10.1.13}$$

etc.

When (10.1.9) and (10.1.10) are substituted into the continuity equation (10.1.2) and the boundary layer equation (10.1.8) and terms are collected according to order of approximation, we get

1. order approximation

$$\frac{\partial u^{(1)}}{\partial t}-\nu\frac{\partial^2 u^{(1)}}{\partial z^2}=\frac{\partial U^{(1)}}{\partial t}$$

or just

$$u_t^{(1)} - \nu u_{zz}^{(1)} = U_t^{(1)} \tag{10.1.14}$$

and (10.1.2) gives

$$u_x^{(1)} + w_z^{(1)} = 0 \tag{10.1.15}$$

2. order approximation

$$u_t^{(2)} - \nu u_{zz}^{(2)} = U_t^{(2)} + U^{(1)}U_x^{(1)} - u^{(1)}u_x^{(1)} - w^{(1)}u_z^{(1)} \tag{10.1.16}$$

and

$$u_x^{(2)} + w_z^{(2)} = 0 \tag{10.1.17}$$

10.1.3 *The 1st order solution*

The solution to the 1 order problem is easiest to obtain if we consider the complex equivalent to (10.1.14). We first introduce the **defect velocity** v given by (see Fig. 10.1.3)

$$v = u^{(1)} - U^{(1)} \tag{10.1.18}$$

which (using $\frac{\partial U}{\partial z} = 0$) transforms (10.1.14) into

$$\frac{\partial v}{\partial t} = \nu \frac{\partial^2 v}{\partial z^2} \tag{10.1.19}$$

and the boundary conditions

$$z = 0 \implies v = -U^{(1)} \tag{10.1.20}$$

$$z/\delta \to \infty \implies v \to 0 \tag{10.1.21}$$

We further look for solutions of complex form so that we assume that the external flow is

$$U^{(1)} = U_1 e^{i\theta} \tag{10.1.22}$$

and the solutions for v have the form

$$v_c = V(z)e^{i\theta} \qquad \text{with} \qquad v = \text{Re}(v_c) \tag{10.1.23}$$

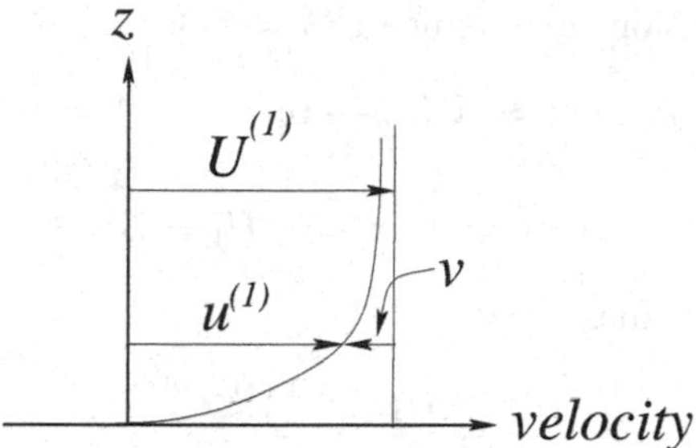

Fig. 10.1.3 Definition of the defect velocity v. Note that v has the opposite sign of u.

where subscript c stands for "complex." Substituted into (10.1.19) it yields

$$i\omega V(z)e^{i\theta} = \nu V''(z)e^{i\theta} \tag{10.1.24}$$

or

$$V''(z) - i\frac{\omega}{\nu}V(z) = 0 \tag{10.1.25}$$

The complete solution to this equaiton is of the form

$$V(z) = A\, e^{R_1 z} + B\, e^{R_2 z} \tag{10.1.26}$$

where R_1 and R_2 are solutions to the characteristic equation

$$\mathrm{R}^2 - i\frac{\omega}{\nu} = 0$$

from which we get

$$\mathrm{R} = \pm\sqrt{\frac{\omega}{\nu}}\sqrt{i} = \pm\sqrt{\frac{\omega}{2\nu}}\,(1+i) \tag{10.1.27}$$

$$\left(\text{since } i = 1\cdot e^{i\pi/2} \Longrightarrow \sqrt{i} = e^{i\pi/4} = \frac{1}{\sqrt{2}} + \frac{i}{\sqrt{2}}\right) \tag{10.1.28}$$

We define

$$\frac{1}{\delta} = \beta \equiv \sqrt{\frac{\omega}{2\nu}} \qquad \Longrightarrow \qquad \mathrm{R} = \pm\beta(1+i) \tag{10.1.29}$$

Hence, the complete solution to (10.1.25) is

$$V(z) = Ae^{\beta(1+i)z} + Be^{-\beta(1+i)z} \tag{10.1.30}$$

The boundary conditions give (since $z/\delta = \beta z$):

$$\beta z \to \infty \Rightarrow V(z) \to 0; \qquad \Rightarrow A = 0 \tag{10.1.31}$$

$$z = 0 \Rightarrow V(z) = -U_1 = B \tag{10.1.32}$$

so that (10.1.23) becomes

$$v_c = -U_1 e^{-\beta(1+i)z} e^{i\theta} \tag{10.1.33}$$

We then have from (10.1.18)

$$\begin{aligned} u_c &= U_c + v_c \\ &= U_1 \left(e^{i\theta} - e^{-\beta z} e^{i(\theta - \beta z)} \right) \end{aligned} \tag{10.1.34}$$

Taking the real part, we then get

$$u^{(1)} = \mathrm{Re}\,\{u_c\}$$

or

$$\boxed{u^{(1)} = U_1 \left[\cos\theta - e^{-\beta z} \cos(\theta - \beta z) \right]} \tag{10.1.35}$$

The parameter $\delta = 1/\beta$ is often considered a measure of the boundary layer thickness. Since there is no well defined upper limit to the boundary layer this is a definition and as we will see from the numerical examples 2δ or 3δ is actually a more realistic measure of the thickness of the boundary layer if we think in terms of the height at which it appears reasonable to say that $u^{(1)} \approx U^{(1)}$ or the shear stress is negligible. Fig. 10.1.4 shows velocity profiles for 8 different phases over half a wave period.

If we write

$$U^{(1)} = U_1 \cos\theta \left(= \mathrm{Re}\left\{U_1 e^{i\theta}\right\}\right) \tag{10.1.36}$$

and

$$u^{(1)} = u_1(z) \cos(\theta + \phi) \tag{10.1.37}$$

then, equating u_1 from (10.1.35) and (10.1.37), we get for the velocity amplitude $u_1(z)$ and the phase $\phi_1(z)$ relative to the exterior velocity U_1

$$\begin{aligned} u_1(z) &= U_1 \sqrt{1 + e^{-2\beta z} - 2e^{-\beta z} \cos\beta z} \\ \tan\phi &= \frac{e^{-\beta z} \sin\beta z}{1 - e^{-\beta z} \cos\beta z} \end{aligned} \tag{10.1.38}$$

Exercise 10.1-1

Derive the above expressions for $u_1(z)$ and $\tan\phi$.

This shows that for $\beta z < \pi$ we havesin $\beta z > 0$ and $\phi > 0$ which implies that u is <u>ahead</u> of U, at least in the lower part of the boundary layer where sin $\beta z > 0$. Fig. 10.1.5 shows the variation of ϕ with z/δ.

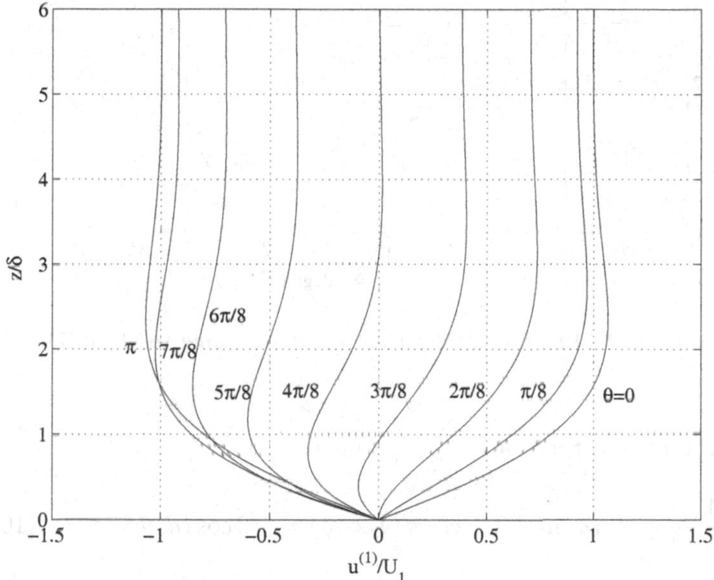

Fig. 10.1.4 The first order velocity profiles $u^{(1)}(z,\theta)$ according to (10.1.35).

We can also find the shear stress τ.

$$\tau = \rho\nu\frac{\partial u^{(1)}}{\partial z} = \rho\nu U_1\left[\beta e^{-\beta z}\cos\left(\theta - \beta z\right) - e^{-\beta z}\beta\sin\left(\theta - \beta z\right)\right]$$

$$= \rho\nu\beta U_1 e^{-\beta z}\left[\cos\left(\theta - \beta z\right) - \sin\left(\theta - \beta z\right)\right] \qquad (10.1.39)$$

In (10.1.39) we consider the [ ] an expression of the form (with $a \equiv \theta - \beta z$)

$$-1 \cdot \sin a + \cos a$$

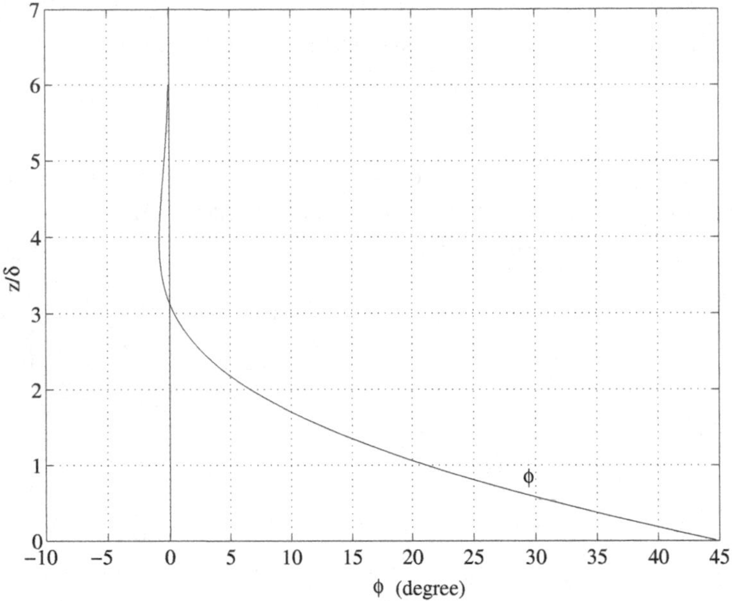

Fig. 10.1.5 Phase angle ϕ for $u^{(1)}(z, \theta)$ according to (10.1.37).

and introduce $1 = \tan \pi/4 = \frac{\sin \pi/4}{\cos \pi/4}$ to get

$$\frac{1}{\cos \pi/4} (-\sin a \sin \pi/4 + \cos \pi/4 \cos a) = \sqrt{2} \cos (a + \pi/4) \qquad (10.1.40)$$

which means

$$\tau = \rho \nu \beta U_1 e^{-\beta z} \sqrt{2} \cos (\theta - \beta z + \pi/4) \qquad (10.1.41)$$

Especially at the bottom $z = 0$ we get the bottom shear stress τ_b as

$$\tau_b = \rho \nu \beta U_1 \sqrt{2} \cos (\theta + \pi/4) \qquad (10.1.42)$$

i.e.: τ_b is ahead of the free stream velocity $U^{(1)} = U_1 \cos \theta$ by an angle $\pi/4 = 45°$. This, however, only applies near the bottom. For increasing βz, the phase difference decreases as (10.1.41) shows, and for $\beta z > \pi/4$ τ is lagging behind $U^{(1)}$. At the same time, τ decreases rapidly as we move away from the boundary due to the $e^{-\beta z}$-factor. Fig. 10.1.6 shows the phase variation of τ versus z/δ.

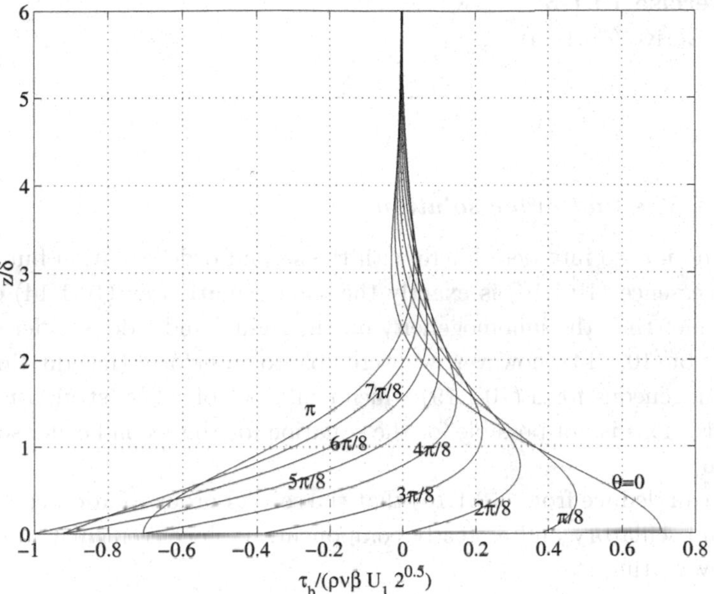

Fig. 10.1.6 The first order shear stress profiles $\tau(z, \theta)$ according to (10.1.41).

Exercise 10.1-2

Derive (10.1.41) and (10.1.42). Show also that

$$|\tau_b|_{max} = \rho U_1 \sqrt{\nu\omega} \tag{10.1.43}$$

For completeness it is mentioned that once $u^{(1)}$ has been determined, $w^{(1)}$ can be found from the continuity equation (10.1.15). Thus we get

$$w^{(1)} = \int_0^z -\frac{\partial u^{(1)}}{\partial x} dz \tag{10.1.44}$$

(where we have used as boundary condition that $w^{(1)}(0) = 0$). Differentiating (10.1.35), inserting into (10.1.44), and integrating gives the result (after some algebra)

$$w^{(1)} = -\frac{U_1 k}{\beta} \left\{ \beta z \sin\theta + \frac{1}{\sqrt{2}} e^{-\beta z} \sin\left(\theta - \beta z - \frac{\pi}{4}\right) - \frac{1}{\sqrt{2}} \sin\left(\theta - \frac{\pi}{4}\right) \right\} \tag{10.1.45}$$

Exercise 10.1-3
Derive (10.1.45).

10.1.4 *The 2nd order solution*

We do not go into detail here with the second order solution but notice that in essence (10.1.16) is exactly the same equation as (10.1.14) except for the fact that the inhomogeneity on the right hand side is different. In the case of (10.1.14), however, we were able to transform the equation into the homogeneous form (10.1.19) which could be solved by straightforward methods. This is not possible for the equation for the second order solution (10.1.16).

We can deduce from (10.1.16) that the second order solution consists of both an oscillatory and a steady component. It is convenient to separate those by writing

$$u^{(2)} = u_a + u_s \tag{10.1.46}$$

where u_a is the oscillatory part, and u_s the mean part of $u^{(2)}$.

We also realize that since both $U^{(1)}$ and $u^{(1)}$ have the form

$$\left(u^{(1)},\ U^{(1)}\right) = (u_1,\ U_1)\, e^{i\theta} \tag{10.1.47}$$

all terms on the right hand side of (10.1.16) must be proportional to $e^{2i\theta}$. Thus, we assume that the oscillatory part u_a can be written

$$u_a = \text{Re}\left(u_0^{(2)} e^{2i\theta}\right) \tag{10.1.48}$$

To recapitulate, the second order equation is (10.1.16)

$$u_t^{(2)} - \nu u_{zz}^{(2)} = U_t^{(2)} + U^{(1)} U_x^{(1)} - u^{(1)} u_x^{(1)} - w^{(1)} u_z^{(1)} \tag{10.1.49}$$

where

$$u^{(1)} = u_1 e^{i\theta} \qquad , \qquad U^{(1)} = U_1 e^{i\theta} \tag{10.1.50}$$

and

$$U^{(2)} = U_2 e^{2i\theta} \tag{10.1.51}$$

In the following, we consider the complex solution to this equation but omit for simplicity the subscript c used in the derivation of the first order solution. Hence, $u^{(2)}$, u_a etc. stands for both the complex solution and (later) also for the real value of those solutions.

Combining (10.1.46) and (10.1.48) $u^{(2)}$ is taken as:

$$u^{(2)} = u_o^{(2)} e^{2i\theta} + u_s \tag{10.1.52}$$

Substituting this into (10.1.16), we get

$$u_{a,t} + u_{s,t} - \nu \left(u_{a,zz} + u_{s,zz} \right) = U_t^{(2)} + U^{(1)} U_x^{(1)} - u^{(1)} u_x^{(1)} - w^{(1)} u_z^{(1)} \tag{10.1.53}$$

In order to separate this equation into equations for u_a and u_s, we use that by virtue of (10.1.48) the time average (over a wave period) of u_a is zero so that

$$\overline{u^{(2)}} = u_s \tag{10.1.54}$$

Hence, u_s will satisfy the equation

$$u_{s,t} - \nu u_{szz} = \overline{U^{(1)} U_1^{(1)}} - \overline{u^{(1)} u_x^{(1)}} - \overline{w^{(1)} u_z^{(1)}} \tag{10.1.55}$$

When we compute the righthand side in this equation we get (see Sect. 10.2)

$$U^{(1)} U_x^{(1)} = a e^{2i\theta} \quad \Rightarrow \quad \overline{U^{(1)} U_x^{(1)}} = 0 \tag{10.1.56}$$

$$u^{(1)} u_x^{(1)} = b e^{2i\theta} \quad \Rightarrow \quad \overline{u^{(1)} u_x^{(1)}} = 0 \tag{10.1.57}$$

$$w^{(1)} u_z^{(1)} = c e^{2i\theta} + \overline{w^{(1)} u_z^{(1)}} \text{ where } \overline{w^{(1)} u_z^{(1)}} \neq 0 \tag{10.1.58}$$

where a, b and c are the factors we get by calculating $U^{(1)} U_x^{(1)}$, etc. from the expressions for $U^{(1)}$, $u^{(1)}$ and $w^{(1)}$.

Thus, the equation for u_s becomes

$$u_{s,t} - \nu u_{szz} = -\overline{w^{(1)} u_z^{(1)}} \tag{10.1.59}$$

The equation for u_a is then obtained by subtracting (10.1.59) from (10.1.53) and using (10.1.57)–(10.1.58). We get

$$u_{at} - \nu u_{a,zz} = U_t^{(2)} + (a - b - c) \, e^{2i\theta} \tag{10.1.60}$$

Here $U_t^{(2)}$ is of course known from the outer solution. We will see later that $U^{(2)}$ gives by far the largest contribution to u_a.

The oscillatory part of the second order solution u_a

It turns out that the solution to (10.1.60) after quite some algebra can be written

$$\begin{aligned} u_a = \frac{U_1^2}{c} \Bigg\{ & \frac{3}{4\sinh^2 kh} \cos 2\theta \\ & - \left[\frac{3}{4\sinh^2 kh} + \frac{1}{2} \right] e^{-\sqrt{2}\beta z} \cos\left(2\theta - \sqrt{2}\beta z\right) \\ & + \frac{1}{2}\beta z e^{-\beta z} \cos(2\theta - \beta z) \\ & - \frac{1}{2}\beta z e^{-\beta z} \cos\left(2\theta - \beta z + \frac{\pi}{4}\right) \Bigg\} \end{aligned} \tag{10.1.61}$$

By the same procedure as for $w^{(1)}$ we then get

$$\begin{aligned} w^{(2)} = -\frac{U_1^2 k}{\beta c} \Bigg\{ & \frac{3}{2\sinh^2 kh} \beta z \sin 2\theta \\ & + \left[\frac{3}{4\sinh^2 kh} + \frac{1}{2} \right] \sin\left(2\theta - \frac{\pi}{4}\right) \\ & - \beta z e^{-\beta z} \sin(2\theta - \beta z) \\ & - \left[\frac{3}{4\sinh^2 kh} + \frac{1}{2} \right] e^{-\sqrt{2}\beta z} \sin\left(2\theta - \sqrt{2}\beta z - \frac{\pi}{4}\right) \Bigg\} \end{aligned} \tag{10.1.62}$$

Exercise 10.1-4

Derive (10.1.61) and (10.1.62).

These two solutions are listed here for reference. The steady component u_s is discussed in the next section.

The final result for the boundary layer motion is then (to second order)

$$u(x, z, t) = u(\theta, z) = u^{(1)} + u_a + u_s + O(kH)^3 \tag{10.1.63}$$

where u_s is given by (10.1.84) determined in the next section.

Numerical results

Fig. 10.1.7 shows an example of $u^{(1)}$ and u_a, both relative to U_1, the bottom velocity amplitude in the 1. order potential motion. It is the velocity amplitudes that are shown.

In some literature the second order component $U^{(2)}$ of U has been omitted in the result for u_a (that is $U^{(2)}$ is put equal to zero in (10.1.16)) (curve marked $U^{(2)} = 0$ in Fig. 10.1.7). As seen from the figure, this gives a radically different result for u_a which becomes an order of magnitude smaller. In other words, $U^{(2)}$, which is the second order component of the (outer) Stokes wave solution, is important for $u^{(2)}$.

We also see that for the wave example considered (which has a value of the Ursell parameter of $HL^2/h^3 = 18.1$, meaning Stokes expansion can be used to describe the wave) the second order contribution is quite small relative to $u^{(1)}$ (O(10%)).

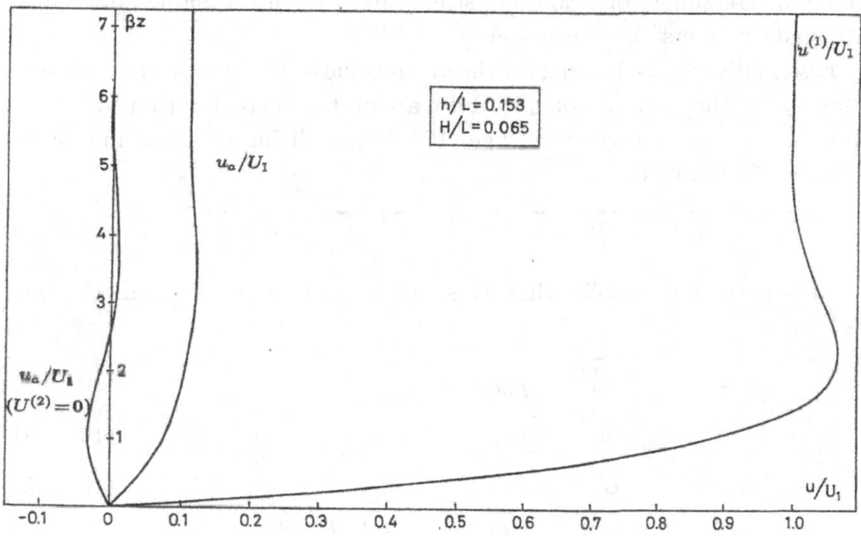

Fig. 10.1.7 Amplitudes of first and second order oscillatory velocities for a case with $h/L = 0.153$ and $H/L = 0.065$. Shown is also the second order amplitude we would obtain if we neglected the second order component of the potential motion $U^{(2)}$.

10.1.5 *The steady streaming u_s in wave boundary layers*

The steady component u_s of the second order solution to the wave boundary layer problem is actually more interesting than the oscillatory part u_a. Where u_a merely changes the details of how the time variation of u is, u_s adds a new element to the flow: a net velocity is induced by the oscillatory flow in the boundary layer. This is termed the **steady**

streaming velocity. This is of significant importance for the sediment transport.

In addition to the steady streaming, we will have a Lagrangian drift of the individual particles, which is the net **drift** of individual particles in a wave motion where we at a fixed point have $\bar{u} = 0$, $\bar{w} = 0$. As described in Chapter 8 this occurs at all points over the vertical also in the potential part of the wave motion. To avoid confusion let us make it clear already here: the steady streaming has nothing to do with the Lagrangian particle motion. The steady streaming occurs in the boundary layer and gives $\bar{u} \neq 0$ at a fixed point!

In the literature, we sometimes find the steady streaming termed "induced streaming" or "Eulerian streaming." Even the name "mass flux in boundary layers" has been seen.

Essentially, we are looking for the mean value of $u^{(2)}$ which was excluded from u_a by the assumption (10.1.48) about the form for that solution. To find $\overline{u^{(2)}} = u_s$ we time average (10.1.55), utilizing that the motion is periodic. Thus we get

$$\overline{u_t^{(2)}} - \nu \overline{u_{zz}^{(2)}} = \overline{U_t^{(2)}} + \overline{U^{(1)} U_x^{(1)}} - \overline{u^{(1)} u_x^{(1)}} - \overline{w^{(1)} u_z^{(1)}} \tag{10.1.64}$$

The periodicity implies that u_s is a constant in time. Thus in (10.1.64) we have

$$\overline{u_t^{(2)}} = u_{s,t} = 0 \tag{10.1.65}$$

$$\overline{u_{zz}^{(2)}} = u_{s,zz} \tag{10.1.66}$$

$$\overline{U_t^{(2)}} = 0 \tag{10.1.67}$$

$$\overline{U^{(1)} U_x^{(1)}} = \frac{1}{2} \frac{\partial \overline{U^{(1)2}}}{\partial x} = \frac{-1}{2c} \frac{\partial \overline{U^{(1)2}}}{\partial t} = 0 \tag{10.1.68}$$

and so also

$$\overline{u^{(1)} u_x^{(1)}} = 0 \tag{10.1.69}$$

Exercise 10.1-5

Show in detail that equations (10.1.64) - (10.1.69) hold.

Hence we are left with the following equation for u_s:

$$\nu u_{s,zz} = \overline{w^{(1)} u_z^{(1)}} \tag{10.1.70}$$

The RHS of (10.1.70) can be written

$$\overline{w^{(1)}u_z^{(1)}} = \overline{(w^{(1)}u^{(1)})_z} - \overline{w_z^{(1)}u^{(1)}} \tag{10.1.71}$$

which, using (10.1.15) and (10.1.69) above, becomes

$$\begin{aligned} &= \overline{(w^{(1)}u^{(1)})_z} + \overline{u_x^{(1)}u^{(1)}} \\ &= \overline{(w^{(1)}u^{(1)})_z} \end{aligned} \tag{10.1.72}$$

Therefore (10.1.70) is equivalent to

$$\nu u_{s,zz} = \overline{(w^{(1)}u^{(1)})_z} \tag{10.1.73}$$

which can be integrated once directly. First, however, consider the boundary conditions: Clearly, u_s must satisfy

$$u_s = 0 \qquad \text{at } z = 0 \tag{10.1.74}$$

The second boundary condition is less obvious. It is customary to specify the condition that the shear stress τ_s created by u_s vanishes as $z/\delta \to \infty$, which means

$$u_{s,z} \to 0 \qquad \text{as } z/\delta \to \infty \tag{10.1.75}$$

We will use this as the second boundary condition and the appropriateness of this assumption will be discussed after we have found the solution.

Integrating (10.1.73) yields

$$u_{s,z} = \frac{1}{\nu}\overline{w^{(1)}u^{(1)}} + C \tag{10.1.76}$$

Using (10.1.75) to determine C then gives

$$u_{s,z} = \frac{1}{\nu}\left(\overline{w^{(1)}u^{(1)}} - \overline{(w^{(1)}u^{(1)})}_\infty\right) \tag{10.1.77}$$

where the last term is the (non zero) limit for $\overline{w^{(1)}u^{(1)}}$ at $z/\delta \to \infty$.

Thus the final solution for u_s can be found by directly integrating (10.1.77):

$$u_s = u_s(0) + \frac{1}{\nu}\int_0^z \left(\overline{w^{(1)}u^{(1)}} - \overline{(w^{(1)}u^{(1)})}_\infty\right) dz \tag{10.1.78}$$

which by virtue of (10.1.74) becomes

$$\boxed{u_s = \frac{1}{\nu}\int_0^z \left(\overline{w^{(1)}u^{(1)}} - \overline{(w^{(1)}u^{(1)})}_\infty\right) dz} \tag{10.1.79}$$

This expression was first derived by Longuet-Higgins (1956)

10.1.6 *Results for u_s*

Evaluation of the RHS in (10.1.79) using $u^{(1)}$ and $w^{(1)}$ from Section 10.1 is tedious but straight forward. The results we find are the following

$$\overline{u^{(1)}w^{(1)}} = \nu\beta\frac{U_1^2}{c}\left\{e^{-\beta z}\left(\beta z\sin\beta z + \cos\beta z\right) - \frac{1}{2}e^{-2\beta z} - \frac{1}{2}\right\} \quad (10.1.80)$$

which also yields

$$\overline{\left(u^{(1)}w^{(1)}\right)}_\infty = -\frac{1}{2}\nu\beta\frac{U_1^2}{c} \quad (10.1.81)$$

where U_1 is the velocity amplitude in the first order potential wave solution at the bottom (see Section 10.1):

$$U^{(1)} = U_1\cos\theta \quad (10.1.82)$$

Thus the integrand in (10.1.79) becomes

$$\frac{1}{\nu}\left(\overline{u^{(1)}w^{(1)}} - \overline{\left(u^{(1)}w^{(1)}\right)}_\infty\right) = \beta\frac{U_1^2}{c}\left\{e^{-\beta z}\left(\beta z\sin\beta z + \cos\beta z\right) - \frac{1}{2}e^{-2\beta z}\right\} \quad (10.1.83)$$

which, we notice, **is independent of ν!** This finally results in

$$\boxed{u_s = \frac{U_1^2}{4c}\left\{3 + 3^{-2\beta z} - 2e^{-\beta z}\left[(\beta z - 1)\sin\beta z + (\beta z + 2)\cos\beta z\right]\right\}} \quad (10.1.84)$$

which is the steady streaming velocity.

Exercise 10.1-6

Derive equations (10.1.80) - (10.1.84).

Once again: it is remarkable that this result is independent of ν. Hence **even if the boundary layer flow is turbulent (but with a <u>constant</u> ν), we get the same formula (10.1.84) for the steady streaming.** The only difference lies in the value of β which with a large eddy viscosity

$\nu_t \gg \nu$ and

$$\beta = \beta_t = \sqrt{\frac{\omega}{2\nu_t}} \tag{10.1.85}$$

will be much smaller in a turbulent boundary layer than in a laminar boundary layer.

Numerical results

Figure 10.1.8 shows u_s for the same wave as used in the example in Sect. 10.1. We see that u_s is somewhat larger than u_a. By comparison with the figure in Sect. 10.1 we find that in this example u_s is approximately 1/7 of the amplitude (i.e., the maximum) of $u^{(1)}$.

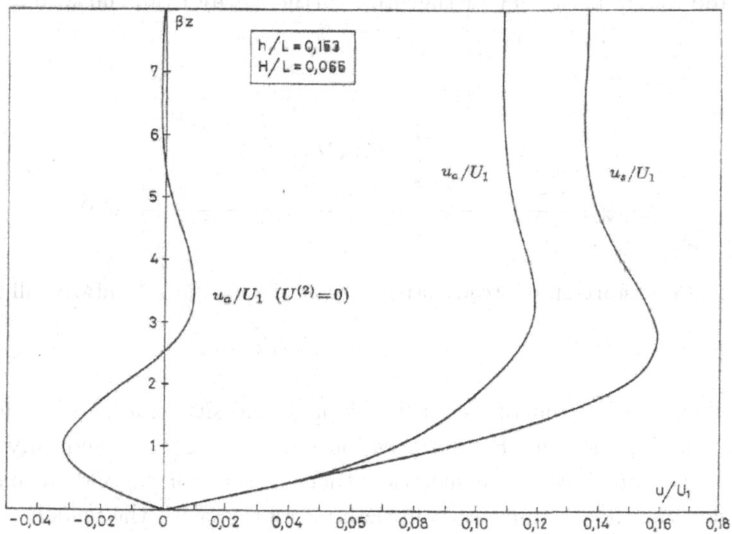

Fig. 10.1.8 Second order steady streaming velocities and oscillatory velocity amplitudes.

Discussion

If we know what to look for it is fairly easy to see that (10.1.75) cannot be true until after a very long time.

To understand this we assume that the motion is started from rest. We first realize that when the wave motion is started then the u_s low in the boundary layer will develop very rapidly, because it is driven by the $\overline{u^{(1)}w^{(1)}} - (\overline{u^{(1)}w^{(1)}})_\infty$, that is present at each point of the boundary layer and fairly large near the bottom of the boundary layer. Therefore, shortly after we have established the wave motion and the bottom boundary layer, we have for moderate z/δ a situation for steady streaming approximately as given by (10.1.84) and Fig. 10.1.9. "Approximately" because of the reasons listed below.

As we approach larger values of z/δ, however, the driving force $\overline{u^{(1)}w^{(1)}} - (\overline{u^{(1)}w^{(1)}})_\infty$ decreases as $\overline{u^{(1)}w^{(1)}}$ approaches its ∞-value. At the same time the u_s-value at those elevations above the bottom are the largest in the steady case (see Fig. 10.1.8). Hence the time it takes to establish the (large) steady state u_s with the smaller driving forces become longer and longer as we get further and further away from the wall.

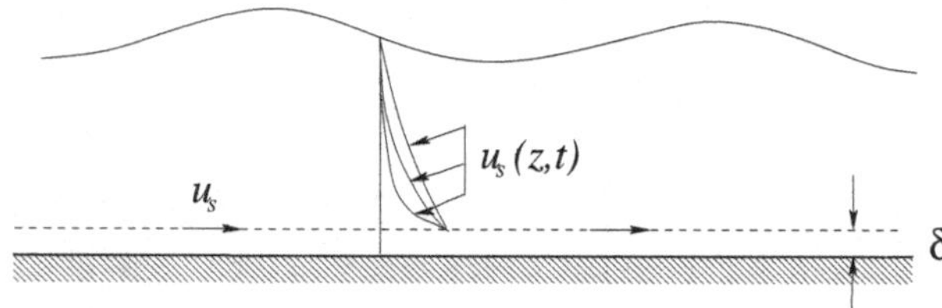

Fig. 10.1.9 The start-up of steady streaming in a wave tank (tentative illustration).

In a large scale (that of the entire depth), the situation can be viewed as shown in Fig. 10.1.9. For large values of z/δ, we are essentially outside the boundary layer. The situation there closely corresponds to having (suddenly) put the bottom into a horizontal motion with the velocity u_s at the top of the boundary layer. That would lead to an unsteady flow above the bottom boundary layer described by

$$u_t = \nu u_{zz} \qquad \begin{aligned} u &= u_s \qquad z \sim 0 \\ u &= 0 \qquad z \to \infty \end{aligned} \tag{10.1.86}$$

"Suddenly" because the time scale of starting up the u_s inside the boundary layer is much shorter than the time scale for spreading the velocity away from the wall outside the boundary layer by the diffusion mechanism described by (10.1.86).

In our case, we get a slightly modified equation. By assuming that $u_s = u_s(z,t)$ equation (10.1.80) becomes

$$\overline{u_t^{(2)}} = u_{s,t}(\neq 0) \tag{10.1.87}$$

Then (10.1.73) takes the form

$$u_{s,t} = \nu u_{s.zz} - \overline{(w^{(1)}u^{(1)})}_z \tag{10.1.88}$$

which includes the effect of the local driving force $\overline{w^{(1)}u^{(1)}}$ in addition to the diffusive effect caused by u_s growing fastest at the lower levels of z/δ and hence dragging the fluid along at higher levels much as the impulsively started wall drags the fluid with it.

The boundary conditions for (10.1.88) as $z/\delta \to \infty$ in, say, a wave tank, are complicated to specify because as the layer with significant u_s expands due to the upward diffusion of momentum it starts to interfere with other parts of the motion such as currents, mass flux at the surface, etc. In a wave tank there will also be a "return flow" (= a current) to compensate for the mass flux generated by the short waves, and there will be variations in the x-direction due to the finite length of tank. Hence, the problem becomes very messy both in x, z and t, in particular as time increases. Reference is made to Longuet-Higgins (1953) for some further discussion. The situation is sketched in Fig. 10.1.10.

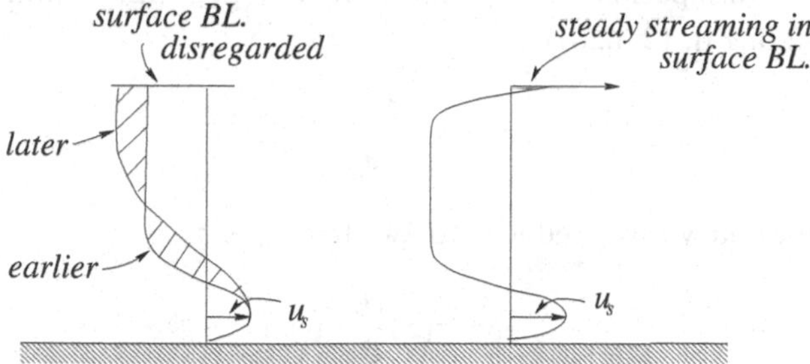

Fig. 10.1.10 The mean velocities in a long wave tank at two different times if we disregard the (weak) boundary layer generated by the no-stress condition at the free surface (the left part of the figure). The right hand part of the fugure gives a qualitative illustration of the effect of the surface boundary layer. End effects due to the finite length of the tank are ignored.

Note on turbulent flow

If we, instead of considering laminar flow with a constant ν, assume the boundary layer flow is turbulent with an eddy viscosity defined so that

$$\tau_{zx} = \rho \nu_t \frac{\partial u}{\partial z} \tag{10.1.89}$$

then we can readily account for a z-variation of ν_t in the solution for u_s. If $\nu_t = \nu_t(z)$ then (10.1.70) becomes

$$\left(\nu_t u_{s,z}\right)_z = \overline{w^{(1)} u_z^{(1)}} \tag{10.1.90}$$

which following the same procedures as above leads to

$$u_s = \int_0^z \frac{1}{\nu_t(z)} \left(\overline{w^{(1)} u^{(1)}} - \left(\overline{w^{(1)} u^{(1)}}\right)_\infty \right) dz \tag{10.1.91}$$

which may be integrated—analytically or perhaps only numerically—when $\nu_t(z)$ is specified.

Unfortunately, it is not equally straight forward to include a time variation $\nu_t = \nu_t(t,z)$. For further details about a time varying ν_t see Trowbridge and Madsen (1984)

10.2 Energy dissipation in a linear wave boundary layer

The energy dissipation can be shown to be (per unit volume of fluid) (see Section 2.3, Eq. (2.3.20))

$$\epsilon = \tau_{ij} \frac{\partial u_i}{\partial x_j} \tag{10.2.1}$$

which in a boundary layer reduces to (with $\tau(z) \equiv \tau_{zx}$)

$$\epsilon = \tau(z) \frac{\partial u}{\partial z} \tag{10.2.2}$$

Over the entire boundary layer we therefore have the instantaneous energy loss

$$\begin{aligned} \mathcal{D}(t) &= -\int_0^\infty \tau(z) \frac{\partial u}{\partial z} dz \\ &= -\tau(z) u \Big]_0^\infty + \int_0^\infty u \frac{\partial \tau}{\partial z} dz \end{aligned} \tag{10.2.3}$$

where the boundary conditions $\tau(\infty) = 0$ and $u(0) = 0$ cancel the first term. Hence,

$$\mathcal{D}(t) = \int_0^\infty u \frac{\partial \tau}{\partial z} dz \tag{10.2.4}$$

Substituting (10.1.14) in the form

$$\rho \frac{\partial (u-U)}{\partial t} = \frac{\partial \tau}{\partial z} \tag{10.2.5}$$

yields

$$\begin{aligned}
\mathcal{D}(t) &= \int_0^\infty \rho u \frac{\partial (u-U)}{\partial t} dz \\
&= \frac{1}{2}\frac{\partial}{\partial t}\int_0^\infty \rho (u-U)^2 dz + U\rho \int_0^\infty \frac{\partial (u-U)}{\partial t} dz \\
&= \frac{1}{2}\frac{\partial}{\partial t}\int_0^\infty \rho (u-U)^2 dz + U \int_0^\infty \frac{\partial \tau}{\partial z} dz \\
&= \frac{1}{2}\frac{\partial}{\partial t}\int_0^\infty \rho (u-U)^2 dz - U\tau_b
\end{aligned} \tag{10.2.6}$$

again using $\tau(\infty) = 0$ and $\tau(0) \equiv \tau_b$

Averaging over a wave period then yields

$$\mathcal{D} \equiv \frac{1}{T}\int_0^T \mathcal{D}(t) dt = -\overline{U\tau_b} \tag{10.2.7}$$

Hence, the energy dissipation in a linear wave boundary layer can be determined simply from the bottom shear stress τ_b and U, the velocity outside the boundary layer.

This result happens to be valid also for a turbulent boundary layer, but as can be seen from the use of (10.2.5) used for the derivation, it is only valid for linear wave boundary layers.

Exercise 10.2-1

Show that in a linear shallow water wave the energy dissipation can be written

$$D = -\frac{1}{4}\rho\omega\frac{U_0^2}{\beta} \tag{10.2.8}$$

where U_0 is the velocity amplitude in the wave.

Vertical energy flux in waves with dissipation

The interaction between the wave motion and the boundary layer is particularly easy to analyse in the case of shallow water waves.

Consider a shallow water wave propagating on a constant depth. As energy is being slowly dissipated the entire wave motion is decaying. The question to be analysed is how is the energy, that is taken out of the wave when it decays, moved to the boundary layer where it is dissipated?

We assume in analogy with the locally constant depth assumption of gently sloping bottoms that the energy dissipation is so slow that at each point the motion is described by the steady linear solution. Then for the shallow water wave the boundary layer solution (10.1.35) is

$$u = U_0 \left(\cos\,\theta - \; e^{-\beta\, z}\; \cos(\theta - \beta\, z)\right) \tag{10.2.9}$$

which is valid over the entire water depth, and so is the expression (10.1.45) for w. If we consider the linear kinematic boundary condition at the free surface it is

$$\frac{\partial \zeta}{\partial t} = w(\xi = h) \tag{10.2.10}$$

Sustituting the expression for w into this we get for ζ the expression

$$\zeta = \frac{H}{2}\left(\cos\,\theta + \frac{e^{-\beta h}}{2\beta h}(\cos(\theta - \beta h) + \sin(\theta - \beta h)) - \frac{1}{2\beta h}(\sin\,\theta + \cos\,\theta)\right) \tag{10.2.11}$$

For a thin boundary layer $\beta h \ll 1$ so that $e^{-\beta h} \ll 2\beta h$ which makes the second parenthesis negligible in comparison to the last. Thus the last term represents the first approximation to the disturbance of the free surface due to the boundary layer. We therefore have

$$\zeta = \frac{H}{2}\left(\cos\,\theta - \frac{1}{2\beta h}(\sin\,\theta + \cos\,\theta)\right) \tag{10.2.12}$$

We see that there is a small disturbance of the wave due to the boundary layer.

Furthermore, in shallow water waves the pressure is hydrostatic and given by

$$p_D = \rho g \zeta \tag{10.2.13}$$

The disturbance means that in average over the wave period there is a vertical energy flux e_{fz} inside the wave motion which is given by

$$e_{fz} = \overline{wp_D} \tag{10.2.14}$$

Substituting the expressions for w and p_D into this we get after some lengthy but trivial algebraic calculations that

$$e_{fz} = -\frac{1}{4}\rho\omega\frac{U_0^2}{\beta}\left(1 - \frac{\xi}{h}\right) + O(e^{-\beta\xi}) \tag{10.2.15}$$

where again U_0 is the velocity amplitude in the wave motion.

This shows there is a negative (= downward oriented) energy flux in the disturbed wave motion. The flux is zero at the surface $\xi = h$. It increases linearly from the surface to the bottom. This means that equal amounts of energy is extracted from the wave motion at each level. At the bottom we get the value

$$e_{fz} = -\frac{1}{4}\,\rho\omega\,\frac{U_0^2}{\beta} \tag{10.2.16}$$

which is equal to the energy dissipation in the wave motion as shown in (10.2.7). Hence the slight disturbance of the potential wave motion above the boundary layer wave mirrored in the surface profile is actually responsible for taking energy from each point of the vertical in the wave and moving it down to the boundary layer at exactly the rate at which it is dissipated.

10.3 Turbulent wave boundary layers

Turbulent boundary layer problems are an order of magnitude more difficult than laminar boundary layers because it is the flow itself that creates the turbulence and hence the essential contribution to the "viscosity." Since the basic hydrodynamics of this flow remain unsolved, the results available are limited to semi-empirical formulas.

In addition in many nearshore wave and circulation modelling problems it suffices to consider only the effect that the bottom boundary layer has on the flow above namely to generate a bottom shear stress and to dissipate energy from the flow above the boundary layer. In essence that means in such modelling problems that the boundary layer is regarded as a black box that generates just a bottom shear stress and an energy dissipation. This simplification is the basis for the following discussion about turbulent boundary layers where empirical results are used to determine the bottom shear stress and the energy dissipation without resolving the details of the boundary layer flow.

Needless to say this approach will generally not be sufficient when dealing with sediment transport. This, however, is beyond the scope of the present text. For further discussion reference is made e.g. to Nielsen (1993), Sleath (1984), and Fredsøe and Deigaard (1992). A review of some of the models available is also given by Soulsby et al. (1993), and detailed measurements of the hydrodynamical details in 2D-vertical the turbulent flow in waves with currents were published by Kemp and Simons (1982, 1983).

The first contribution to the problem of turbulent wave boundary layers using this approach was by Jonsson (1966) who introduced the **wave friction factor** f_w in analogy to the current friction factor f in hydraulics. These factors are dimensionless, empirical coefficients that link the bottom shear stress to the free stream velocity (or depth averaged velocity in the case of a current). Hence, like the friction factor in hydraulics, f_w cannot be measured directly but has to be inferred from other results. f_w is - in other words - what is sometimes called a "fudge factor" that allows us to treat the boundary layer as a black box so that we do not need to resolve the details of the boundary layer flow in order to calculate the shear stress from flow information outside the boundary layer.

In the case of waves, the first step is to determine the amplitude or maximum value of the time-varying shear stress. Thus if we think of the bottom shear stress as

$$\tau(z = -h_0,\ \theta) = \tau_{b,o} \cdot f(\theta) \tag{10.3.1}$$

we are actually first seeking results for $\tau_{b,o}$. Using the free stream velocity

$$U = U_o \cos \omega t \tag{10.3.2}$$

the wave friction factor f_w is **defined** so that

$$\tau_{bo} = \frac{1}{2}\rho f_w U_0^2 \tag{10.3.3}$$

The question of what would be appropriate values of f_w then becomes a question of establishing methods for analyzing measurements in turbulent wave boundary layers. Originally, it was very difficult to measure τ directly in water, since the turbulent shear stress is defined as

$$\tau = -\rho \widehat{u'w'} \tag{10.3.4}$$

where $\frown$ represents turbulent (i.e., ensemble) averaging. Measuring u' and w' simultaneously at the same point was not really possible. Therefore, in Jonsson (1966) and later in Jonsson and Carlsen (1976), τ was derived from measured profiles of turbulent averaged velocities. This essentially means solving

$$\frac{\partial v}{\partial t} = \frac{1}{\rho}\frac{\partial \tau}{\partial z} \tag{10.3.5}$$

where v is the defect velocity, numerically from measurements of v.

Today Laser Doppler Anemometers make it possible to measure u', w' and $u'w'$ directly. For such results reference is made to e.g. Cox et al. (1995). Even then, however, it is of interest to analyse which **parameters** are relevant for the variation of f_w or τ.

Assuming a sinusoidal time variation in the flow outside the boundary layer, the important variables that determine the nature of the flow near the boundary are the following:

(1) the velocity amplitude U_o in the oscillation outside the boundary layer.
(2) the particle excursion amplitude (or stroke) a_b in that oscillation.
(3) the roughness k_N of the wall. k_N is a length that characterizes the size of the irregularities on the surface of the boundary (such as the grain size d of a granular bottom, but k_N is not directly equal to d). The concept was introducd by Nikuradse for steady hydraulic flows.
(4) for low Reynolds numbers the viscosity ν of the fluid is important. ν is also important in the **laminar sublayer** very close to the wall (where the velocity is small).

Notice that for sinusoidal motion, the wave period T or frequency $\omega = 2\pi/T$ is related to U_o and a_b by

$$a_b = U_o/\omega \tag{10.3.6}$$

so that ω can replace a_b or U_o in the list above if this is convenient.

The variables U_o, a_b, k_N and ν can be combined into two independent, dimensionless parameters:

- A boundary layer Reynolds number RE defined by

$$\mathrm{RE} = \frac{U_o a_b}{\nu} \tag{10.3.7}$$

and

- an amplitude/roughness ratio

$$a_b/k_N \tag{10.3.8}$$

These two parameters turn out to be the dominating parameters for the variation of both f_w and the thickness $\delta(t)$ of the boundary layer.

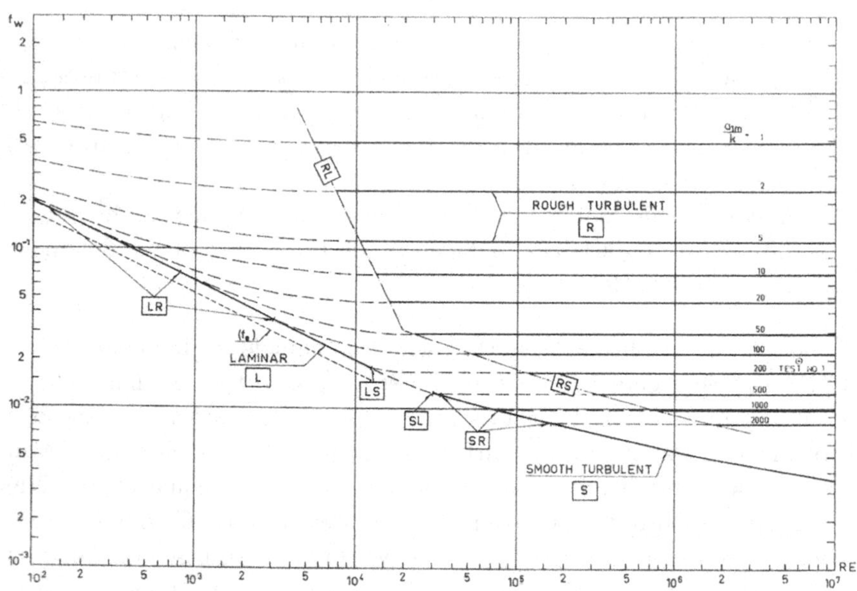

Fig. 10.3.1 The wave friction factor, f_w versus the boundary layer Reynolds number RE (Jonsson, 1966).

It also turns out that in analogy to pipe flow or open channel flow there are a number of important cases or flow regimes corresponding to different physical situations. Fig. 10.3.1 shows the variation of f_w for the most important flow situations, using RE and a_b/k_N as parameters:

- Laminar flow: f_w is independent of the roughness a_b/k_N so RE is the only parameter

$$f_w = f_w(RE) \tag{10.3.9}$$

- Smooth turbulent flow: The roughness of the wall is so small that the irregularities are buried in a layer in the immediate vicinity of the wall (0.1–0.2 mm) where the flow remains laminar (the laminar sublayer). See Fig. 10.3.2.

- A transition region: a relatively wide range of RE values where f_w essentially has the value of rough turbulent flow, though substantial variation occurs.
- Rough turbulent flow: the most frequently occurring case in nature. Here the roughnesses are so large that the flow around the individual roughness elements is turbulent (Fig. 10.3.2).

Fig. 10.3.2 Smooth turbulent flow (left). Rough turbulent flow (right).

The important characteristic of the **rough turbulent flow** is that f_w is essentially independent of RE, so that a_b/k_N is the only parameter, i.e.,

$$f_w = f_w(a_b/k_N) \qquad \text{(rough turb. flow)} \tag{10.3.10}$$

where in general we have

$$f_w = f_w(a_b/k_N,\ \text{RE}) \tag{10.3.11}$$

In the following, we concentrate on the rough turbulent flow and seek to establish an empirical relationship for $f_w(a_b/k_N)$ in (10.3.10), and also for the maximum thickness δ_b of the wave boundary layer, and for the energy dissipation $\overline{\mathcal{D}}$.

10.3.1 *Rough turbulent flow*

Although the variation of f_w with a_b/k in the rough turbulent region may be determined from Fig. 10.3.1, it is more convenient to plot it directly as a function of a_b/k. We should remember, however, that the relationship

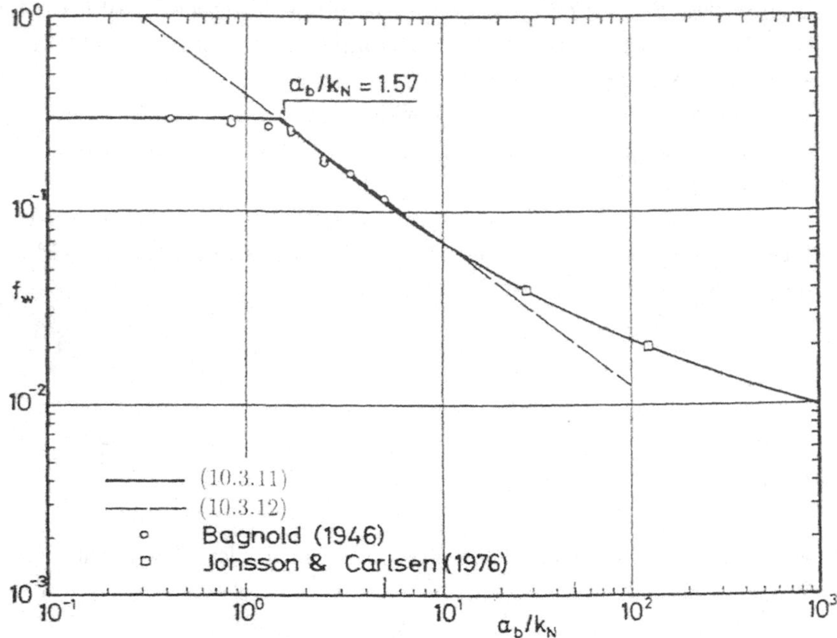

Fig. 10.3.3 Friction factor versus dimensionless particle amplitude a_b/k_N (in the figure $\alpha_b = a_b$). at the sea bed for fully turbulent flow. The full line corresponds to (10.3.12), the dashed line to (10.3.13) (not to (10.3.11) and (10.3.12) as indicated in the figure) (SJ76).

is essentially empirical and rather few experiments are available from which f_w can be determined. Hence, the curve shown in Fig. 10.3.3 is the result of a theoretical derivation which requires substantial simplification and approximations. It results in the following expression for (10.3.10) (Jonsson, 1966)

$$\frac{1}{4\sqrt{f_w}} + \log_{10} \frac{1}{4\sqrt{f_w}} = -0.08 + \log_{10} \frac{a_b}{k_N} \qquad a_b/k_N > 1.57 \qquad (10.3.12)$$

where the constant -0.08 is determined so that the curve fits the experimental results. Since (10.3.12) can only be solved iteratively for f_w, several approximations to (10.3.12) and the experimental data have been proposed.

Thus, Kamphuis (1975) suggests that f_w is approximated by

$$f_w = 0.4\,(a_b/k_N)^{-0.75} \qquad a_b/k_N < 50 \tag{10.3.13}$$

which is the straight line shown in Fig. 10.3.3, clearly only a good approximation below the a_b/k_N limit indicated.

Alternatively, Swart (1974) proposed

$$\ln f_w = -5.977 + 5.213(a_b/k_N)^{-0.194} \tag{10.3.14}$$

which actually fits (10.3.12) very well.

In the course of deriving (10.3.12), a result for the boundary layer thickness, δ_b, is also obtained. It is recalled that the boundary layer does not have a well defined thickness, and the maximum thickness, δ_b, given below represents one of the approximations introduced. Fig. 10.3.5 shows how δ_b is defined.

The quantity $\delta_b = \max\,(\delta(t))$ is given by

$$\frac{30\delta_b}{k_N}\log_{10}\frac{30\delta_b}{k_N} = 1.2\frac{a_b}{k_N} \tag{10.3.15}$$

Fig. 10.3.4 shows the solution to this equation in terms of $\frac{\delta_b}{a_b} = f\left(\frac{a_b}{k_N}\right)$ together with the alternative approximation

$$\frac{\delta_b}{a_b} = 0.072\left(\frac{a_b}{k_N}\right)^{-1/4} \tag{10.3.16}$$

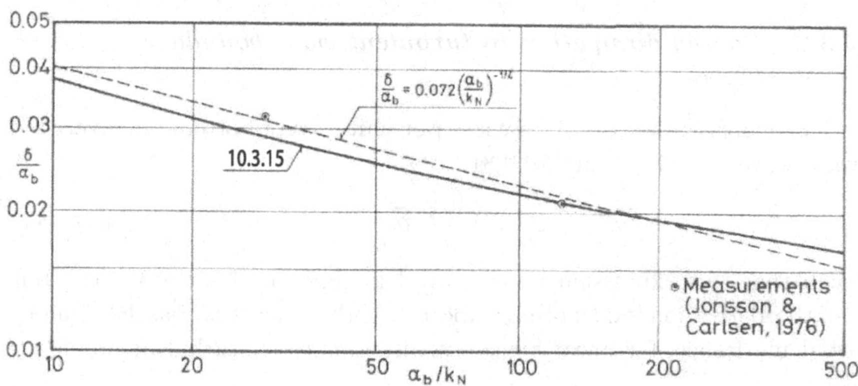

Fig. 10.3.4 Solution for the wave boundary layer thickness δ_b according to (10.3.15) and (10.3.16) (SJ76).

Variation of u and τ with z

Fig. 10.3.5 and Fig. 10.3.6 show typical variation of $u(z)$ at the time of maximum U outside the boundary layer, and the variation of the maximum of $\tau(z)$ over the maximum τ_b. We see that at a distance of $2\delta_b$ above the bottom $u(z)$ is close to U, and $\tau_{\max}$ is less than 10% of the bottom value, and at $3\delta_b$ there are essentially no shear stresses left.

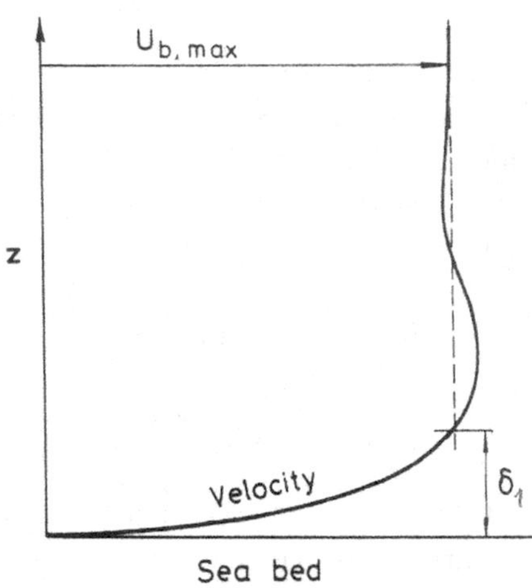

Fig. 10.3.5 Typical variation of the velocity $u(z)$ and sketch for the definition of the wave boundary layer thickness δ_b used in (10.3.15) (SJ76).

10.3.2 *Energy dissipation in turbulent wave boundary layers*

The expression for the energy loss per unit area of bottom and averaged over a wave period is from (10.2.8)

$$\mathcal{D} = \overline{U\tau_b} \tag{10.3.17}$$

Recall that this expression is only valid for linearized wave motion, but that the derivation said nothing about whether the flow was laminar or turbulent. Hence, for linear motion it also applies for turbulent boundary layers.

It does require, however, that we know not just the maximum value, τ_{bo} of τ_b (named $\tau_{b,max}$ in the figure) but the full time variation of the shear stress.

It is therefore assumed that the **instantaneous** bottom shear stress $\tau_b(t)$ is given by the generalization of (10.3.3) to arbitrary time:

$$\tau_b(t) = \frac{1}{2}\rho f_w |U(t)| U(t) \tag{10.3.18}$$

where $U(t)$ is given by (10.3.2) and f_w is assumed constant.

Substituting (10.3.2) and (10.3.18) into (10.3.17) and performing the averaging yields, the result

$$\mathcal{D} = -\frac{2}{3\pi}\rho f_w U_o^3 \tag{10.3.19}$$

It turns out that writing $\mathcal{D}$ as

$$\mathcal{D} = -m_e \rho f_w U_o^3 \tag{10.3.20}$$

and determining m_e from measurements of the real τ_b values leads to practically the same value of $m_e = 2/3\pi$ as determined using the heuristic variation (10.3.18) for $\tau_b(t)$. Thus (10.3.19) is not far from being correct (in fact it is found that $m_e \sim 0.8 \cdot 2/3\pi$) and in recognition of the uncertainties of the result $\mathcal{D}$ is often simply determined by (10.3.19) (Jonsson, 1966).

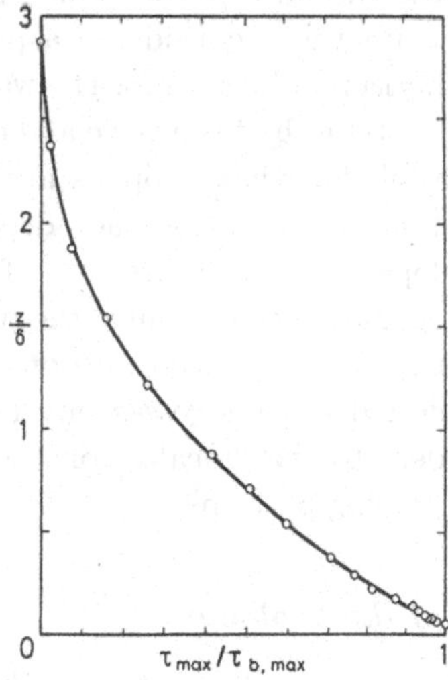

Fig. 10.3.6 Typical dimensionless shear stress amplitude in a turbulent flow (SJ76).

In this context, it may be noticed that we can always write

$$\tau_b(t) = \frac{1}{2}\rho f \mid U(t) \mid U(t) \tag{10.3.21}$$

The real approximation in (10.3.18) or (10.3.21) is to assume that f is constant over the wave period and equal to the value f_w found for $\tau_{bo} = \max(\tau_b(t))$.

10.4 Bottom shear stress in 3D wave-current boundary layers

10.4.1 *Introduction*

In the general case of waves and currents at an angle relative to each other and waves approaching the depth contours at an angle, the flow very rapidly becomes complicated. Again, we only look for the bottom shear stress. However, in order for the results to be useful for nearshore applications, we must extend the analysis to the time variation of $\vec{\tau}_b$ and also seek results for the wave averaged $\overline{\vec{\tau}_b}$.

As an introductory observation, it is mentioned that the wave-current interaction in the boundary layer is considered a purely local process in which we assume locally horizontal bottom for the wave motion. Therefore at each point only the local strength of the wave and current, and the angle between them are responsible for what happens in the boundary layer at that point. The bottom shear stress does not depend on the direction in which the bottom is sloping (i.e., the direction of the bottom contours relative to the wave orthogonals does not enter the problem).

Hence, in the following, we will concentrate on examining the nature of the wave-current generated shear stresses, given waves and currents. Essentially this is equivalent to the "locally constant depth assumption" used for waves on gently sloping bottoms.

10.4.2 *Formulation of the problem*

Thus we consider the situation illustrated in Fig. 10.4.1. The angle between the wave direction and the current velocity direction at the bottom is μ, and the wave direction relative to the (arbitrary but fixed) (x, y) coordinate system is ϕ. It is assumed that the flow is turbulent and that the instantaneous bottom shear stress $\vec{\tau}_b(t)$ can be expressed by

$$\vec{\tau}_b = \frac{1}{2}\rho f_{cw}\vec{u}_b(t) \mid \vec{u}_b(t) \mid \tag{10.4.1}$$

where $u_b(t)$ is the instantaneous total wave-current velocity at the bottom without considering the presense of the boundary layer (the "slip-velocity")

$$\vec{u}_b = \vec{u}_w + \vec{V}_b \tag{10.4.2}$$

Here u_w is the wave particle velocity just outside the boundary layer. The waves are propagating in the direction of the wave number vector $\vec{k}$ and we define the unit vector $\vec{e}_w$ in that direction by

$$\vec{e}_w \equiv \frac{\vec{k}}{k} \quad \text{where} \quad k \equiv \mid \vec{k} \mid = \left(k_x^2 + k_y^2\right)^{1/2} \tag{10.4.3}$$

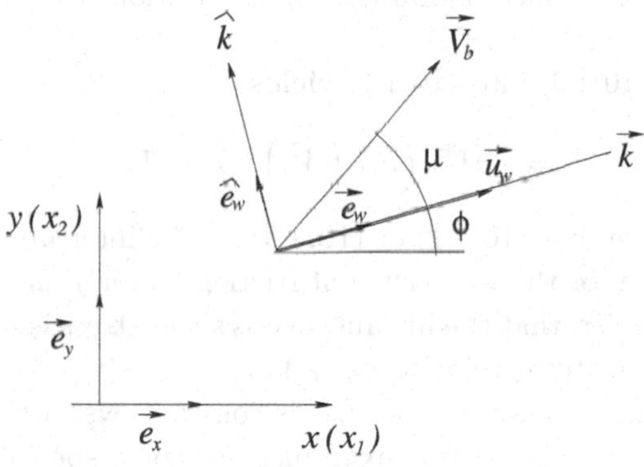

Fig. 10.4.1 Definition sketch for nomenclature used.

We let the wave induced particel velocity be given by

$$\vec{u}_w = \vec{u}_o \cos\theta \ ; \quad \text{and} \ u_o \equiv \mid u_o \mid \tag{10.4.4}$$

On varying depth we have that

$$\theta = \omega t - \int \vec{k} \cdot d\vec{x} \quad \left(\text{or} \ \omega t - \int k_\alpha dx_\alpha\right) \tag{10.4.5}$$

is the phase angle including the effect of depth variation along the wave orthogonal which makes $k_\alpha = k_\alpha(x_\alpha)$.

Similarly, we have a current with the local bottom velocity $\vec{V_b}$ and we define

$$V_b = \mid \vec{V_b} \mid = \left(V_{bx}^2 + V_{by}^2\right)^{1/2} \tag{10.4.6}$$

We also want to use the vector $\hat{e}_w$ which equals $\vec{e}_w$ rotated 90° in the counter clockwise direction. Thus with

$$\vec{e}_w \equiv \left(\frac{k_x}{k}, \frac{k_y}{k}\right) \quad \left(= \frac{\vec{k}}{k} = \frac{k_i}{k}\right) \tag{10.4.7}$$

we have

$$\hat{e}_w = \left(-\frac{k_y}{k}, \frac{k_x}{k}\right) \tag{10.4.8}$$

At any time the total velocity is the vector sum of the steady current velocity and the time varying wave particle velocity. The situation is illustrated in Fig. 10.4.2 which shows the time variation of the total velocity vector.

Substituting (10.4.2) into (10.4.1) yields

$$\vec{\tau_b} = \frac{1}{2}\rho f_{cw}\left(\vec{u}_w + \vec{V_b}\right) \mid \vec{u}_w + \vec{V_b} \mid \tag{10.4.9}$$

We also observe that (10.4.1) or (10.4.9) can be interpreted as simply a definition equation for the wave-current friction factor f_{cw}. A consequence of (10.4.1) is, however, that the instantaneous shear stress is in the direction of the instantaneous total velocity $\vec{u}_w + \vec{V_b}$.

In the following we assume that f_{cw} is constant over a wave period.

The variation of (10.4.9) was examined for some special cases by Liu and Dalrymple (1978). Here we follow the more general approach described by Svendsen and Putrevu (1990), Putrevu and Svendsen (1991).

10.4.3 *The mean shear stress*

For the wave-averaged nearshore circulation equations (see Chapter 11), we are particularly interested in the wave averaged value of $\vec{\tau_b}$. Averaging (10.4.9), we get using (10.4.4)

$$\overline{\vec{\tau_b}} = \frac{1}{2}\rho f_{cw}\overline{\left(\vec{u}_o \cos\theta + \vec{V_b}\right) \mid \vec{u}_o \cos\theta + \vec{V_b} \mid} \tag{10.4.10}$$

To calculate the time average, it is convenient first to separate $\vec{u}_w + \vec{V}$ into their components with respect to $\vec{e}_w$ and $\hat{e}_w$. We see that (10.4.2) may

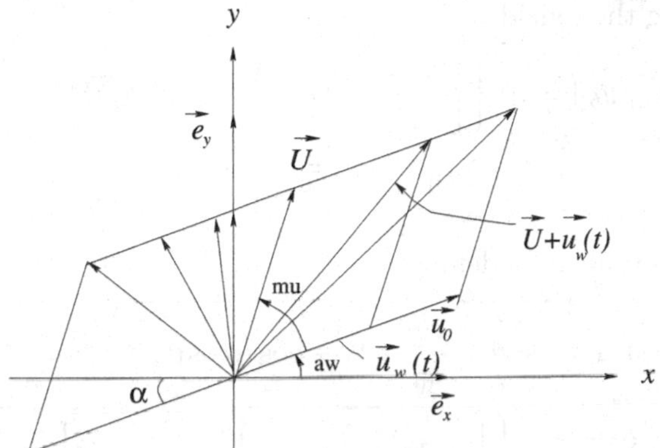

Fig. 10.4.2 The temporal variation of the total velocity vector in a situation with waves and currents at angle to each other. ψ is the angle between the current and the x-axis, α the angle between the wave orthogonal and the x-axis, and ϕ is the angle of the total current relative to the x-axis. Clearly ϕ is varying with time while the other two angles are constant.

be written as

$$\vec{u}_b = (u_o \cos\theta + V_b \cos\mu)\,\vec{e}_w + V_b \sin\mu \hat{e}_w \tag{10.4.11}$$

so that

$$\begin{aligned} |\,\vec{u}_b\,| &= \left[u_0^2\cos^2\theta + V_b^2\cos^2\mu + 2u_0V_b\cos\theta\cos\mu + V_b^2\sin^2\mu\right]^{1/2} \\ &= u_0\left[\cos^2\theta + \frac{V_b^2}{u_0^2} + 2\frac{V_b}{u_0}\cos\mu\cos\theta\right]^{1/2} \\ &\equiv u_0\ \beta_1(t) \end{aligned} \tag{10.4.12}$$

where $\beta_1(t)$ represents the bracket. Thus we have

$$\vec{u}_b\,|\,\vec{u}_b\,| = \left(\vec{u}_0\cos\theta + \vec{V}_b\right)\,|\,\vec{u}_0\cos\theta + \vec{V}_b\,| \tag{10.4.13}$$

$$\begin{aligned} &= \left[(u_o\cos\theta + V_b\cos\mu)\,\vec{e}_w + V_b\sin\mu\hat{e}_w\right]\cdot u_o\beta_1(t) \\ &= \left[u_o^2\left(\cos\theta\ \beta_1(t) + \frac{V_b}{u_o}\cos\mu\ \beta_1(t)\right)\right]\vec{e}_w + u_oV_b\sin\mu\ \beta_1(t)\hat{e}_w \end{aligned} \tag{10.4.14}$$

Time averaging then yields

$$\overline{\vec{u}_b \mid \vec{u}_b \mid} = u_0^2 \left\{ \left[\overline{\cos\theta\, \beta_1(t)} + \frac{V_b}{u_0} \cos\mu\, \overline{\beta_1(t)} \right] \vec{e}_w + \frac{V_b}{u_0} \sin\mu\; \overline{\beta_1(t)} \hat{e}_w \right\} \tag{10.4.15}$$

We see it is convenient to define

$$\beta_1 \equiv \overline{\beta_1(t)} \equiv \overline{\left[\cos^2\theta + \frac{V_b^2}{u_0^2} + 2\frac{V_b}{u_0} \cos\mu \cos\theta \right]^{1/2}} \tag{10.4.16}$$

$$\beta_2 \equiv \overline{\beta_1(t) \cos\theta} \equiv \overline{\left\{ \left[\cos^2\theta + \frac{V_b^2}{u_0^2} + 2\frac{V_b}{u_0} \cos\mu \cos\theta \right]^{1/2} \cos\theta \right\}} \tag{10.4.17}$$

and we see that $(\beta_1,\ \beta_2) = f(\mu,\ V_b/u_0)$.

Using those definitions, we can write (10.4.15)

$$\overline{\vec{u}_b \mid \vec{u}_b \mid} = u_0^2 \left\{ \left(\beta_2 + \frac{V_b}{u_0} \beta_1 \cos\mu \right) \vec{e}_w + \frac{V_b}{u_0} \beta_1 \sin\mu\; \hat{e}_w \right\} \tag{10.4.18}$$

Notice that, as mentioned earlier, this confirms that at a given point the bottom shear stress only depends on the angle μ between the waves and their currents (in addition to V_b/u_o), not on the angle of incidence α_w for the waves relative to the bottom contours.

With β_1 and β_2 defined, we can now change reference coordinates from $(\vec{e}_w, \hat{e}_w)$ to $(\vec{e}_x, \vec{e}_y)$. In terms of ϕ defined in Fig. 10.4.1, we have

$$\begin{aligned} \vec{e}_w &= \vec{e}_x \cos\phi + \vec{e}_y \sin\phi \\ \hat{e}_w &= \vec{e}_y \cos\phi - \vec{e}_x \sin\phi \end{aligned} \tag{10.4.19}$$

Substituting into (10.4.18) this yields after some algebra

$$\overline{\vec{u}_b \mid \vec{u}_b \mid} = u_0^2 \left\{ \left[\frac{V_b}{u_o} \beta_1 \cos(\mu + \phi) + \beta_2 \cos\phi \right] \vec{e}_x + \left[\frac{V_b}{u_0} \beta_1 \sin(\mu + \phi) + \beta_2 \sin\phi \right] \vec{e}_y \right\} \tag{10.4.20}$$

Here we introduce the x, y-components of V_b and u_0 given by

$$\left. \begin{aligned} V_{bx} &= V_b \cos(\mu + \phi) \\ V_{by} &= V_b \sin(\mu + \phi) \end{aligned} \right\} \tag{10.4.21}$$

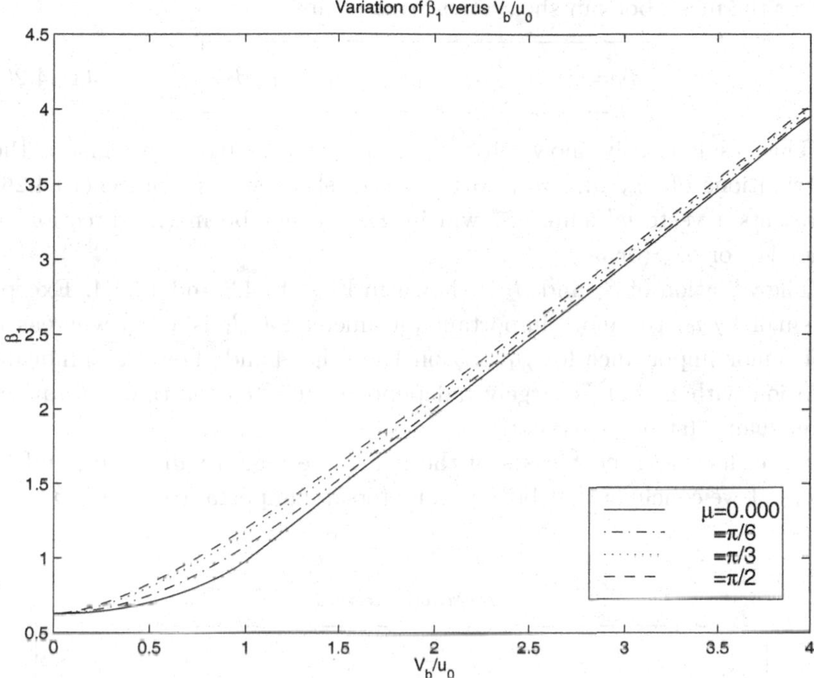

Fig. 10.4.3 The variation of β_1 with sinusoidal bottom velocity variation versus V_b/u_0 for different values of μ.

$$\left.\begin{aligned} u_{0x} &= u_0 \cos\phi \\ u_{0y} &= u_0 \sin\phi \end{aligned}\right\} \tag{10.4.22}$$

by which we can write (10.4.20) as

$$\overline{\vec{u}_b|\vec{u}_b|} = u_0 \left\{ (u_{0x}\beta_2 + V_{bx}\beta_1)\,\vec{e}_x + (u_{0y}\beta_2 + V_{by}\beta_1)\,\vec{e}_y \right\} \tag{10.4.23}$$

Alternatively, we can introduce the tensor notation

$$V_{b\alpha} = (V_{bx},\ V_{by}) \quad ; \quad u_{0\alpha} = (u_{0x},\ u_{0y}) \quad ; \quad u_0 = \mid u_{0\alpha} \mid \tag{10.4.24}$$

Substituting into (10.4.23), that equation can then be written

$$\overline{u_{b\alpha} \mid u_{b\alpha} \mid} = u_0 \left\{ V_{b\alpha}\beta_1 + u_{0\alpha}\beta_2 \right\} \tag{10.4.25}$$

For the mean bottom shear stress this means

$$\boxed{\overline{\vec{\tau}_b} = \overline{\tau_{b\alpha}} = \frac{1}{2}\rho f_{cw} u_0 \left\{ V_{b\alpha}\beta_1 + u_{0\alpha}\beta_2 \right\}} \tag{10.4.26}$$

This result clearly shows that β_1 and β_2 represent the weight of the contributions of $V_{b\alpha}$ and $u_{0\alpha}$ to the mean shear stress. Since (10.4.26) represents a vectorial sum, $\overline{\tau_{b\alpha}}$ will in general not be in the direction of either $V_{b\alpha}$ or $u_{0\alpha}$ (or $u_{b\alpha}$).

The variation of β_1 and β_2 is shown in Figs. 10.4.3 and 10.4.4. Except for small V/u_0, the more important parameter for β_1 is V/u_0 whereas μ is of minor importance for β_1. β_2 on the other hand shows a significant variation with μ, but is largely independent of V/u_0 for that parameter larger than 1 (strong currents).

Since, however, most cases in the nearshore region will correspond to $V/u_0 < 1$ we conclude that both parameters are important for both β_1 and β_2.

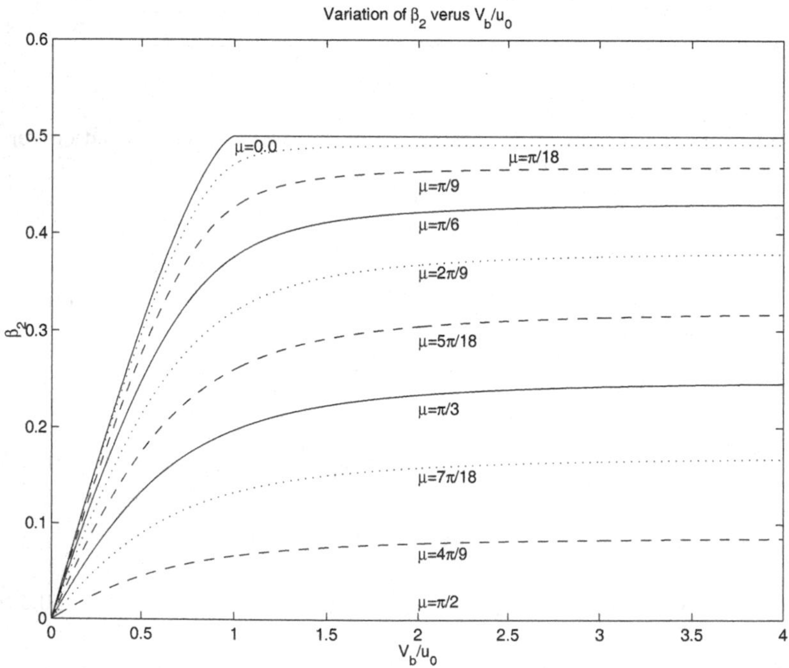

Fig. 10.4.4 The variation of β_2 versus V_b/u_0 for different values of μ.

10.4.4 *Special cases*

Of particular interest are the two special cases: weak currents, and strong currents relative to the oscillatory wave velocity.

Weak currents. $\frac{V}{u_0} \ll 1$

An obvious special case is that of a weak current relative to the waves. In that case we get (since $(\cos^2\theta)^{1/2} = |\cos\theta|$ which has the mean value $2/\pi$).

$$\frac{V}{u_0} \quad \text{small} \quad \Rightarrow \qquad \beta_1 = \overline{|\cos\theta| \left[1 + 2\frac{V_b}{u_0}\frac{\cos\mu}{\cos\theta} + O\left(\frac{V_b^2}{u_0^2}\right)\right]^{1/2}}$$

$$\sim \overline{|\cos\theta| \left[1 + \frac{V_b}{u_0}\frac{\cos\mu}{\cos\theta}\right]}$$

$$\beta_1 \sim \frac{2}{\pi} + O\left(\frac{V_b}{u_0}\right) \tag{10.4.27}$$

and

$$\beta_2 = \overline{\left\{\left[\cos^2\theta + 2\frac{V_b}{u_0}\cos\mu\cos\theta + O\left(\frac{V_b^2}{u_0^2}\right)\right]^{1/2}\cos\theta\right\}} \tag{10.4.28}$$

which gives

$$\begin{aligned} \beta_2 &\sim \overline{|\cos\theta|\cos\theta} + \overline{\tfrac{V_b}{u_0}\cos\mu\,|\cos\theta|} \\ \beta_2 &\sim 0 + \tfrac{2}{\pi}\,\tfrac{V_b}{u_0}\,\cos\mu = \tfrac{2}{\pi}\,\tfrac{V_k}{u_0} \end{aligned} \tag{10.4.29}$$

where $V_k \equiv V_b \cos\mu$ is the projection of the current onto the wave direction.

Thus, for a weak current (10.4.26) becomes

$$\bar{\tau}_{b\alpha} = \frac{1}{2}\rho f_{cw} u_0 \left\{\frac{2}{\pi}V_{b\alpha} + \frac{2}{\pi}\,\frac{V_k}{u_0}u_{b\alpha}\right\} \tag{10.4.30}$$

which can also be written

$$\bar{\tau}_{b\alpha} = \frac{1}{\pi}\rho f_{cw} u_0 \left\{V_{b\alpha} + \frac{V_k}{u_0}u_{b\alpha}\right\} \tag{10.4.31}$$

Clearly, $\bar{\tau}_{b\alpha}$ is not generally in the direction of $\vec{V}_b$. Also — as could be expected since $\overline{|\cos\phi|\cos\phi} = 0$ — **when** $(V_{b\alpha},\ V_k) \to 0$ **we get** $\overline{\tau_{b\alpha}} = 0$ **(pure waves).**

In the figures for β_1 and β_2, this comes out clearly:

$\beta_1 \sim$ const. $= 2/\pi = 0.6367$

$\beta_2 \sim$ linearly varying with V_b/u_0 with a slope proportional to $2/\pi \cos\mu$

The particle motion for this case is as shown Fig. 10.4.5

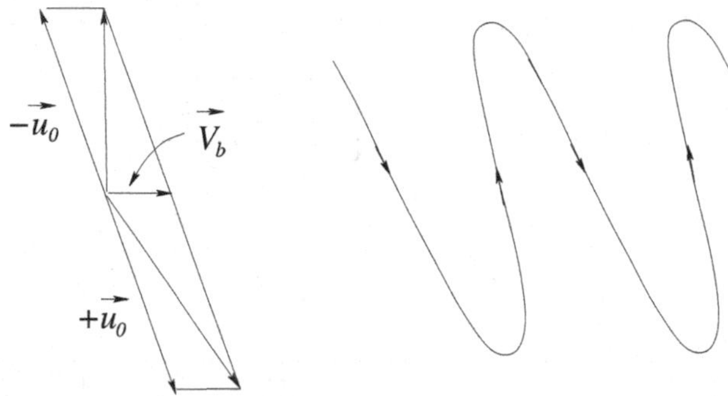

Fig. 10.4.5 The velocity variation and particle paths for a case with a weak current relative to the oscillatory velocity component (the wave motion).

Waves and currents perpendicular to each other, $\mu = \pi/2$. Of particular interest is the case of $\mu = \pi/2$ which means the current is perpendicular to the waves. We then get

$$V_k = 0 \tag{10.4.32}$$

or

$$\boxed{\vec{\bar{\tau}}_b = \bar{\tau}_{b\alpha} = \frac{1}{\pi}\rho f_{cw} u_0 V_{b\alpha}} \tag{10.4.33}$$

This result will be use in Section 11.7.3 for the case of longshore currents on a long straight coast with nearly perpendicular wave incidence. Note that here $\vec{\bar{\tau}}_b$ **is** in the direction of $\vec{V}_b$, and $\vec{\bar{\tau}}_b$ is proportional to the wave velocity amplitude u_0, even though the wave particle velocity is perpendicular to the current and to $\vec{\bar{\tau}}_b$.

Waves and currents parallel, $\mu = 0$. The other special example for a weak current is $\mu = 0$ which corresponds to current and waves parallel. Thus (10.4.31)

$$V_k = V_b \tag{10.4.34}$$

and since $V_k u_{b\alpha}/u_0 = V_{b\alpha}$ we get

$$\boxed{\bar{\tau}_{b\alpha} = \frac{2}{\pi}\rho f_{cw} u_0 V_{b\alpha} = \frac{2}{\pi}\rho f_{cw} u_0 \vec{V}_b} \tag{10.4.35}$$

Hence, again $\vec{\bar{\tau}}_b$ is in the direction of $\vec{V}_b$—in this case this could be expected because there are no velocities at all perpendicular to that direction.

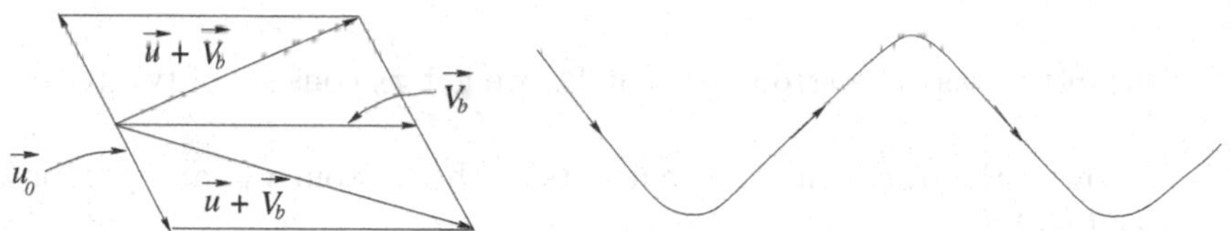

Fig. 10.4.6 The velocity variation and particle paths for a case with a strong current relative to the oscillatory velocity component (the wave motion).

Note also that the expressions for $\mu = 0$ and $\mu = \pi/2$ are similar except for the fact that for $\mu = 0$ the $\vec{\bar{\tau}}_b$ is twice the value it attains for $\mu = \pi/2$.

Strong currents. $\frac{V_b}{u_0} \gg 1$

When the current velocity V_b is much stronger than the wave motion, we essentially have a slight deviation from a uniform current with particle motion and path lines as shown in Fig. 10.4.6.

For $V_b/u_0 \gg 1$ we get

$$\beta_1 \simeq \frac{V_b}{u_0}\left[1+O\left(\frac{V_b}{u_0}\right)^{-1}\right]^{1/2} \simeq \frac{V_b}{u_0} \tag{10.4.36}$$

$$\begin{aligned}\beta_2 &= \frac{V_b}{u_o}\overline{\left[1+2\frac{\cos\mu\cos\theta}{V/u_0}+O\left(\frac{V_b}{u_0}\right)^{-2}\right]^{1/2}\cos\theta} \\ &\simeq \frac{V_b}{u_0}\overline{\left[1+\frac{\cos\mu\cos\theta}{V_b/u_0}\right]\cos\theta} \\ &= \frac{V_b}{u_0}\left[\overline{\cos\theta}+\frac{\cos\mu}{V_b/u_0}\overline{\cos^2\theta}\right]\end{aligned} \tag{10.4.37}$$

or

$$\beta_2 \simeq \frac{1}{2}\cos\mu \tag{10.4.38}$$

Thus for this special case we find from (10.4.26)

$$\vec{\overline{\tau}}_b = \overline{\tau}_{b\alpha} = \frac{1}{2}\rho f_{cw}\left\{V_b V_{b\alpha}+\frac{1}{2}u_0 u_{0\alpha}\cos\mu\right\} \tag{10.4.39}$$

Thus in the case of a strong current $\vec{V}_b$, we get $\vec{\overline{\tau}}_b$ consists of two terms

(1) A term in the direction of the currents at the bottom $\vec{V}_{b\alpha}$ and proportional to V_b^2.
(2) A term in the wave direction $u_{b\alpha}$, proportional to u_o^2 and $\cos\mu$.

Waves and currents perpendicular to one another, $\mu = \pi/2$. Hence if $\mu = \pi/2$ (waves and current perpendicular to each other), the term in the wave direction is zero (no effect of the waves). We therefore get

$$\boxed{\vec{\overline{\tau}}_b = \frac{1}{2}\rho f_{cw} V_b V_{b\alpha}} \tag{10.4.40}$$

which shows the mean stress in this case is in the direction of the current and equal to what we would expect for a current without waves, even though there is also a wave motion.

Waves and currents parallel, $\mu = 0$. If waves and currents are in the same direction ($\mu = 0$), we simply get

$$\boxed{\overline{\vec{\tau}_b} = \frac{1}{2}\rho f_{cw} V_b V_{b\alpha} + \frac{1}{4}\rho f_{cw} u_0 u_{0\alpha}} \tag{10.4.41}$$

For completeness we also note, that in the limit where there is no wave motion at all, ($u_0 = 0$), we get the result

$$\boxed{\overline{\vec{\tau}_b} = \frac{1}{2}\rho f_{cw} V_b V_{b\alpha} \quad \text{(currents only)}} \tag{10.4.42}$$

as one should expect.

10.5 References - Chapter 10

Cox, D. T., N. Kobayashi, and A. Okayasu (1995). Experimental and numerical modeling of surfzone hydrodynamics. Res. Rep. No. CACR-95-07, University of Delaware. 293 pp.

Fredsøe, J. and R. Deigaard (1992). Mechanics of coastal sediment transport. World Scientific, Singapore, 369 pp.

Grant, W. D. and O. S. Madsen (1979). Combined wave and current interaction with a rough bottom. J. Geophys. Res., **84**, C4, 1797 – 1808.

Grant, W. D. and O. S. Madsen (1986). The continental shelf bottom bounday layer. Ann Rev. Fluid Mech., **18**, 265 – 305.

Jonsson, I. G. (1966), Wave boundary layers and friction factors. 10th Int'l. Coast. Eng. Conf., Tokyo.

Jonsson, I. G. and N. A. Carlsen (1976). Experimental and theoretical investigations in an oscillatory rough turbulent boundary layer. J. Hydr. Res., **14**, 1, 45 – 60.

Kamphuis, J. W. (1975). Friction factor under oscillatory waves. ASCE J. Wtrway. Harbors Coast. Engrg. **101**, 2, 135 – 144.

Kemp, P. H. and R. R. Simons (1982). The interaction between waves and a turbulent currents: Waves propagating with the current. J. Fluid Mech., **116**, 227–250.

Kemp, P. H. and R. R. Simons (1983). The interaction of waves and a turbulent current: Waves propagating against the current. J. Fluid Mech., **130**, 73–89.

Liu, P. L.-F. and R. A. Dalrymple (1978). Bottom frictional stresses and longshore currents due to waves with large angles of incidence. J. Marine Res., **36**, 357 – 375.

Longuet-Higgins, M. S. (1953). Mass transport in water waves. Phil. Trans. Roy. Soc. London. **A, 245**, 535 – 581.

Longuet-Higgins, M. S. (1956). The mechanics of the boundary layer near the bottom in a progressive wave. ASCE Proc 6th Int. Conf. Coastal Eng. 184–193.

Nielsen, P. (1993). Coastal bottom boundary layers and sediment transport. World Scientific, Singapore, 324 pp.

Putrevu, U. and I. A. Svendsen (1991). Wave induced nearshore currents: A study of the forcing, mixing and stability characteristics. Res. Rep CACR-91-11, Center for Applied Coastal Research, University of Delaware, 242 pp.

Sleath, J. (1984). Sea bed mechanics. John Wiley and Son, New York.

Soulsby, R. L., L. Hamm, G. Klopman, D. Myrhaug, R. R. Simons, and G. P. Thomas (1993). Wave-current interaction within and outside the bottom boundary layer. Coastal Eng., **21**, 41–70.

Svendsen, I. A. and I. G. Jonsson (1976, 1980). (SJ76). Hydrodynamics of coastal regions. Den Private Ingeniør Fond, Technical University of Denmark.

Svendsen, I. A. and U. Putrevu, (1990). Nearshore circulation with 3-D profiles. Proc. 22 Int. Conf. Coastal Engrg., Delft.

Swart, D. H. (1974). Offshore sediment transport and equilibrium beach profiles. Delft Hydr. Lab. Publ 131, Delft Univ. Technol., Diss., Delft.

Trowbridge, J. and O. S. Madsen (1984). Turbulent boundary layers, Parts I and II, J. Geophys. Res., **89**, C5, 7989–8007.

Chapter 11

Nearshore Circulation

11.1 Introduction

Nearshore circulation is the term used for the complex nearshore currents generated by the short wave motion. These currents are determined by the depth integrated and time averaged equations of continuity and and momentum. In the classical approach the currents are assumed depth uniform. Since, however, it turns out that their depth variation is actually of significant importance for the way way in which the currents interact in the horizontal dimension we will avoid as long as possible in the derivations to make any assumptions about the variation over depth of the currents. However, after deriving the equations governing general depth varying currents the special case of depth uniform currents is analysed in detail. Various aspects of the depth variation of nearshore currents are then analysed in Sections 12.2 and 13.

The derivation of the general depth-integrated, time-averaged equations requires some care. The benefit, however, of such a derivation is a more clear understanding of which assumptions are involved and how the assumptions allow us to simplify the equations.

When we integrate the differential equations over the depth from the (fixed, sloping) bottom to the (moving) free surface, we will need to apply the relevant boundary conditions as well. Since we are talking about highly turbulent flows, there will of course be contributions from bottom friction, and similarly wind stresses (if any) on the surface will contribute to the stresses there. In all, this makes the boundary conditions somewhat more complicated than for simple potential flow wave motions. The derivation of those boundary conditions is shown in Section 11.3.

In the following, it is convenient to separate the vertical coordinate, z, from the two horizontal coordinates. Thus, where tensor notation is used, we will use a special version of the tensor notation. It is distinguished from the traditional tensor notation by the use of α, β instead of i, j. The relation is the following:

$$i, j = 1,\ 2\ (x, y) \qquad \text{becomes} \quad \alpha,\ \beta\ (1, 2) \tag{11.1.1}$$

$$i, j = 3(z) \qquad \text{becomes} \quad z\ ,\ u_i \quad \text{becomes} \quad w \tag{11.1.2}$$

As an example, $\partial u_i\, u_j / \partial x_i$ will be written

$$\frac{\partial u_i\, u_j}{\partial x_i} = \frac{\partial u_\alpha\, u_\beta}{\partial x_\alpha} + \frac{\partial w\, u_\beta}{\partial z} \quad \text{for } j = \beta = 1, 2 \tag{11.1.3}$$

$$\frac{\partial u_\alpha\, w}{\partial x_\alpha} + \frac{\partial w^2}{\partial z} \quad \text{for } j = 3 \tag{11.1.4}$$

Surface elevation and mean water surface, MWS

In this section, as in later sections on wave generated circulation, we cannot count on the local mean water level being horizontal so we now call it the mean water **surface MWS**. We therefore also cannot place it in the x, y-plane of the coordinate system which **is** horizontal. On the other hand we want to keep the name of η for the distance from the MWS to the instantaneous surface so that we still have $\overline{\eta} = 0$ in the wave motion. Consequently we must define a separate variable ζ which measures the instantaneous surface elevation from the horizontal x, y-plane. Then we have that the elevation of the MWS above the x, y-plane is $\overline{\zeta}$. At the same time we define the depth below $z = 0$ as h_0 which implies that the local mean water depth is

$$h = h_0 + \overline{\zeta} \tag{11.1.5}$$

These definitions are shown in Fig. 11.1.1.

Procedure for deriving the depth-integrated, wave-averaged equations

In brief terms the procedure we are going to follow to derive the depth integrated, wave averaged continuity and momentum equations consists of the following steps

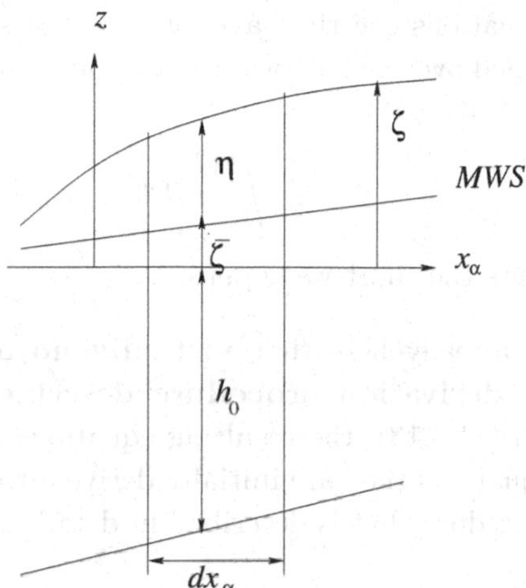

Fig. 11.1.1 Definition sketch for this chapter.

(1) The two horizontal components of the Reynolds equations are integrated from the bottom ($z = -h_0$) to the surface ($z = \zeta$)

(2) This yields terms of the form $\int \frac{\partial}{\partial t}, \int \frac{\partial}{\partial x_\alpha}$, etc., in the equations. Leibniz's rule is used to transform those terms into terms of the form $\frac{\partial}{\partial t} \int, \frac{\partial}{\partial x_\alpha} \int$, etc.

(3) The boundary terms in the equations left by those steps are terms evaluated at the bottom $-h_0$ and the free surface ζ. Invoking the boundary conditions at the bottom and the free surface essentially eliminates most of these boundary terms. Left is only the effect of wind stress at the surface, τ_α^S and bottom friction at the bottom, τ_α^B.

(4) The variation over a vertical of the pressure is derived by integrating the vertical momentum equation from the free surface to an arbitrary location in the fluid z. Usually hydrostatic pressure is assumed for the currents and also for the infragravity (IG)-waves discussed in Chapter14.

(5) The expression for the pressure obtained in the previous step is used to eliminate the pressure from the depth-integrated horizontal momentum equations.

(6) The horizontal velocity, u_α, is divided into a current part, V_α, and a wave-part, $u_{w\alpha}$, as

$$u_\alpha = V_\alpha + u_{w\alpha} \tag{11.1.6}$$

(7) The resulting equations are then averaged over a short-wave period. Quantities averaged over a short-wave period are hereafter denoted by an overbar

$$\overline{f} = \frac{1}{T} \int_t^{t+T} f dt \tag{11.1.7}$$

where T represents the short-wave period.

It is emphasized already here that **virtually no assumptions are introduced in the derivation procedure described above**. Hence in their general form (11.5.13), the resulting equations are as accurate as the Navier-Stokes equations they are initially derived from (see also Chapter 13). It is this procedure that is described in detail in this chapter.

11.2 Depth integrated conservation of mass

We first consider the conservation of mass for the column of fluid with dimensions $(dx_\alpha,\ h_o + \zeta)$ (see Fig. 11.1.1). If we define the instantaneous mass flux M_α as

$$M_\alpha = \int_{-h_o}^{\zeta} \rho u_\alpha \, dz \tag{11.2.1}$$

then simple continuity considerations show that the conservation of mass becomes

$$\frac{\partial(\rho_S \zeta)}{\partial t} + \frac{\partial M_\alpha}{\partial x_\alpha} = 0 \qquad \rho_S = \rho(z = \zeta) \tag{11.2.2}$$

where ρ_s is the fluid density at the surface. This equation simply says that the net difference between what flows into the column and what flows out is stored inside through change in the surface elevation.

In most of the analysis of nearshore flows, we will assume that the density ρ of the fluid is constant, although there will be regions — like at the turbulent front of a breaking wave — where the content of bubbles in the water may reduce the density of the air-water mixture.

With $\rho =$ constant, the conservation of mass is equivalent to a conservation of volume. Hence, we may define the volume flux equivalent to (11.2.1) as

$$Q_\alpha = \int_{-h_o}^{\zeta} u_\alpha \, dz \tag{11.2.3}$$

as in Section 3.3. The depth integrated continuity equation may then be written simply as

$$\frac{\partial \zeta}{\partial t} + \frac{\partial Q_\alpha}{\partial x_\alpha} = 0 \tag{11.2.4}$$

Although this result as shown above can be realized directly on an intuitive basis, it is useful already here to introduce the procedure of depth integrating the equations of motion in differential form. Hence in the following, we go through the derivation of (11.2.4) directly from the equation of continuity at a point.

We have

$$\frac{\partial u_\alpha}{\partial x_\alpha} + \frac{\partial w}{\partial z} = 0 \tag{11.2.5}$$

where u_α, w represent the total velocities, including turbulent fluctuation.

Formal integration of (11.2.5) over $(-h_o|\zeta)$ gives

$$\int_{-h_o}^{\zeta} \frac{\partial u_\alpha}{\partial x_\alpha}\, dz + w(\zeta) - w(-h_o) = 0 \tag{11.2.6}$$

For convenience we repeat the kinematic boundary condition (2.4.8) at the bottom

$$w(-h_o) = -u_\alpha \frac{\partial h_o}{\partial x_\alpha} \tag{11.2.7}$$

and (2.4.4) for the free surface

$$w(\zeta) = \frac{\partial \zeta}{\partial t} + u_\alpha \frac{\partial \zeta}{\partial x_\alpha}. \tag{11.2.8}$$

Substituting these two expressions into (11.2.6) then yields

$$\int_{-h_o}^{\zeta} \frac{\partial u_\alpha}{\partial x_\alpha}\, dz + \frac{\partial \zeta}{\partial t} + u_\alpha \frac{\partial \zeta}{\partial x_\alpha} + u_\alpha \frac{\partial h_o}{\partial x_\alpha} = 0 \tag{11.2.9}$$

Leibniz' rule of differentiation of integrals indicates that

$$\frac{\partial}{\partial x_\alpha} \int_{-h_o}^{\zeta} u_\alpha\, dz = \int_{-h_o}^{\zeta} \frac{\partial u_\alpha}{\partial x_\alpha}\, dz + u_\alpha(\zeta) \frac{\partial \zeta}{\partial x_\alpha} - u_\alpha(-h_0) \frac{\partial(-h_o)}{\partial x_\alpha} \tag{11.2.10}$$

which reduces (11.2.9) to

$$\frac{\partial \zeta}{\partial t} + \frac{\partial}{\partial x_\alpha} \int_{-h_o}^{\zeta} u_\alpha\, dz = 0 \tag{11.2.11}$$

or with (11.2.3)

$$\frac{\partial \zeta}{\partial t} + \frac{\partial Q_\alpha}{\partial x_\alpha} = 0 \tag{11.2.12}$$

which of course is the same as (11.2.4). It is noticed that this derivation is valid whether the bottom condition is a $u_\alpha = 0$ as in a boundary layer or a finite slip velocity is assumed at the bottom.

11.2.1 *Separation of waves and currents*

The next step in the process is the time averaging of (11.2.12). For that purpose, we divide the total particle velocity (u_α, w) into three parts: A time (or wave) averaged part (V_α, 0) which represents the "current,"[1] an oscillatory part ($u_{w,\alpha}$, w_w) representing the wave motion and a turbulent fluctuation (u'_α, w').

Thus, we have

$$u_\alpha = V_\alpha + u_{w\alpha} + u'_\alpha \tag{11.2.13}$$

$$w = 0 + w_w + w' \tag{11.2.14}$$

where we will assume that

$$\widehat{u'_\alpha}, \ \widehat{w'} = 0 \tag{11.2.15}$$

$$\overline{u_{w,\alpha}}, \ \overline{w_w} = 0 \tag{11.2.16}$$

and use $\widehat{\ }$ for ensemble or turbulent averaging, and — for time or wave averaging. Notice that strictly speaking (11.2.16) only applies below trough level (see later discussion).

Note also that the definition (11.2.13) implies that $\widehat{u_\alpha}$, the turbulent averaged total velocity is

$$\widehat{u_\alpha} = V_\alpha + u_{w\alpha} \tag{11.2.17}$$

i.e., $\widehat{u_\alpha}$ encompasses both the current and the oscillatory wave component of the instantaneous velocity.

---

[1] Assuming gently sloping bottom implies $W \cong 0$, but this is not a necesary simplification.

The time average of (11.2.12) becomes

$$\overline{\frac{\partial \zeta}{\partial t}} + \overline{\frac{\partial Q_\alpha}{\partial x_\alpha}} = 0 \tag{11.2.18}$$

or

$$\frac{\partial \bar{\zeta}}{\partial t} + \frac{\partial \overline{Q_\alpha}}{\partial x_\alpha} = 0 \tag{11.2.19}$$

Substituting (11.2.13) into (11.2.3) yields

$$Q_\alpha = \int_{-h_o}^{\zeta} (V_\alpha + u_{w,\alpha} + u'_\alpha)\, dz \tag{11.2.20}$$

which after ensemble average and time average yields

$$\overline{Q_\alpha} = \overline{\widehat{Q_\alpha}} = \overline{\int_{-h_o}^{\zeta} V_\alpha + u_{w\alpha} dz} \tag{11.2.21}$$

Comment: here we have assumed

$$\widehat{\int_{-h_o}^{\zeta} u'_\alpha dz} = \int_{-h_o}^{\zeta} \widehat{u'_\alpha}\, dz = 0 \tag{11.2.22}$$

which is equivalent to neglecting the turbulent fluctuations ζ' of the free surface. In the region of the turbulent front of the wave, ζ' is not quite negligible, however. Inclusion of ζ' would lead to acceptance of a term $\widehat{(\zeta' u'_\alpha)}$ in the equation, the value of which would depend on the correlation between fluctuations in the surface elevation and particle velocity. For various reasons, this term is suspected to be small but we do not have any measurements or models available in support of that estimate in the turbulent front of a breaking wave.

Introducing the surface elevation ζ_t at the wave trough level, and dropping the $\widehat{\ }$, we can now write

$$\overline{\widehat{Q_\alpha}} = \overline{Q_\alpha} = \overline{\int_{-h_o}^{\zeta} V_\alpha dz} + \overline{\int_{-h_o}^{\zeta_t} u_{w\alpha} dz} + \overline{\int_{\zeta_t}^{\zeta} u_{w,\alpha} dz} \tag{11.2.23}$$

which, by virtue of (11.2.16) means

$$\overline{Q_\alpha} = \overline{\int_{-h_o}^{\zeta} V_\alpha\, dz} + \overline{\int_{\zeta_t}^{\zeta} u_{w,\alpha}\, dz} \tag{11.2.24}$$

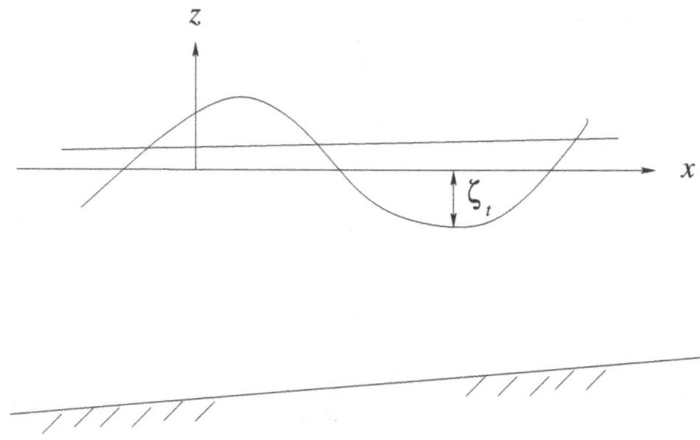

Fig. 11.2.1 Definition of ζ_t.

For the last integral we can introduce the definition

$$Q_{w\alpha} = \overline{\int_{\zeta_t}^{\zeta} u_{w\alpha} dz} \tag{11.2.25}$$

which means (11.2.21) can be written

$$\overline{Q_\alpha} = \int_{-h_o}^{\bar{\zeta}} V_\alpha \, dz + Q_{w\alpha} \tag{11.2.26}$$

The quantity $Q_{w\alpha}$ is the volume flux generated by the short wave motion. $Q_{w\alpha}$ was analysed for linear waves in Section 3.3, in Chapter 8 for second order Stokes waves, and in Section 9.6 for cnoidal waves. Substituting this into the continuity equation we then get

$$\boxed{\frac{\partial \bar{\zeta}}{\partial t} + \frac{\partial}{\partial x_\alpha} \left(\int_{-h_o}^{\bar{\zeta}} V_\alpha \, dz + Q_{w\alpha} \right) = 0} \tag{11.2.27}$$

which is the general form of the continuity equation.

A comment on the flow above wave trough level

As (11.2.3) indicates, for the calculation of $\overline{Q_\alpha}$ it is necessary to integrate the total velocity all the way to the instantaneous surface ζ. Thus we need, not surprisingly, to know the flow in the entire region between bottom and surface. When we average over the wave period and separate into the wave and current motion this become an integration to $\bar{\zeta}$. Thus (11.2.26) inevitably requires that we are able to define the current above wave trough level ζ_t. There is no unique way of doing this because the averaging over the wave period is not uniquely defined in that region where there is only water part of the time. However, the choice of how we do this will not affect the result for $\overline{Q}$ as long as we are consistent: what is not included in the current V_α above trough level must be included in the wave motion $u_{w\alpha}$.

Depth uniform current $V_\alpha = V_\alpha(x_\alpha, t)$

So far we have made no assumptions about the variation of the current component V_α over depth.

In the simple case, where the current component of the total velocity is uniform over depth, we simply get

$$\overline{Q_\alpha} = V_\alpha h + Q_{w\alpha} \tag{11.2.28}$$

which is the equivalent to (11.2.26) valid for a depth uniform current that is assumed extended to the MWS.

It is emphasized that V_α is the time-averaged velocity which will be measured below wave trough level by a current meter placed there.

11.3 Conditions at fixed and moving boundaries, II

Before embarking on the depth integration of the momentum equations we derive the boundary conditions for the turbulent flow with stresses tangential and normal to the free surface and the bottom.

11.3.1 *Kinematic conditions*

As mentioned in Section 2.4 the kinematic conditions remain the same as discussed there. Thus we have the free surface condition is

$$\frac{\partial \zeta}{\partial t} - w + u_\alpha \frac{\partial \zeta}{\partial x_\alpha} = 0 \quad \text{at} \;\; z = \zeta \tag{11.3.1}$$

and at the bottom we get

$$w + u_\alpha \frac{\partial h}{\partial x_\alpha} = 0 \quad \text{at} \quad z = -h(x_\alpha) \tag{11.3.2}$$

11.3.2 *Dynamic conditions*

Kinematic conditions relate the motion of particles in time — i.e., the velocity field — to the development in time of the position of the surface.

In contrast, dynamic conditions are **instantaneous conditions** expressing that the external stresses on a boundary surface must be balanced by equivalent internal stresses immediately inside the fluid.

The stress variations at a point

For general turbulent flows we need first to analyse the stresses at the surface. Since boundary surfaces usually form an angle to all coordinate axes, whereas the internal stresses in the fluid are expressed in terms of stresses σ_{ij} on the coordinate surfaces, we must first derive a relationship between stresses on a surface of arbitrary direction and stresses on the coordinate surfaces at the same point.

Let R_j be the total stress (i.e., force per unit area) on a surface with arbitrary normal vector $\overrightarrow{n} = n_i$. Letting $|n_i| = 1$ implies n_i are the direction cosines of $\overrightarrow{n}$. Fig. 11.3.1 shows the situation which we consider. This surface element will later be taken as part of the free surface or bottom.

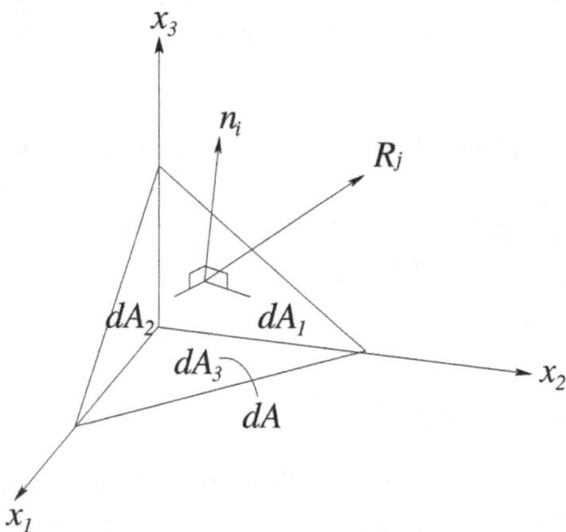

Fig. 11.3.1 The tetrahedron considered for analysing the stresses on an arbitrary surface with normal vector n_i.

The area of the boundary surface element is dA and it cuts off area sections $dA_i = (dA_1,\ dA_2,\ \ dA_3)$ on the three coordinate planes to form a tetrahedron. Since n_i are the direction cosines, we have

$$dA_i = n_i dA \tag{11.3.3}$$

On each of the coordinate-parallel area elements dA_i, we have total stress components σ_{ij}. Hence, the total force component in the j-direction from surface stresses on the tetrahedron is

$$R_j dA - \sigma_{ij} dA_i = R_j dA - \sigma_{1j} dA_1 - \sigma_{2j} dA_2 - \sigma_{3j} dA_3 \quad (j = 1, 2, 3) \tag{11.3.4}$$

where the minus signs reflect that the σ_{ij} stresses are opposing R_j.

The equation of momentum for the tetrahedron is then (with dV = the volume of the tetrahedron)

$$\rho dV \frac{du_j}{dt} = \rho g_j dV + R_j dA - \sigma_{ij} n_i dA \tag{11.3.5}$$

Here $dV \propto (dx_i)^3$ whereas $dA_i \propto (dx_i)^2$. Hence

$$\begin{aligned} \frac{dV}{dA_i} &\to 0 \\ dx_i &\to 0 \end{aligned} \tag{11.3.6}$$

so that volume forces in Eq. (11.3.5) are infinitesimal relative to surface forces and we simply get

$$\boxed{R_j = \sigma_{ij}\, n_i} \tag{11.3.7}$$

Fig. 11.3.2 shows a two dimensional version of the balance

If, for example, dA is a free surface, then $\overrightarrow{R}$ can be a combination of a (wind) shear stress and a surface pressure that together add up to the total stress $\overrightarrow{R}$ on the surface.

The dynamic condition in a general flow

To utilize Eq. (11.3.7) we need the normal vector at the surface point in question. Writing the equation for the surface as

$$F(x_\alpha,\ z,\ t) = z - \zeta(x_\alpha,\ t) \tag{11.3.8}$$

we have the normalized normal vector given by

$$n_i = \frac{\nabla F}{|\nabla F|} \tag{11.3.9}$$

where

$$\nabla = \frac{\partial}{\partial x_i} \qquad i = 1, 2, 3 \tag{11.3.10}$$

and

$$|\nabla F| = \left(1 + \left(\frac{\partial \zeta}{\partial x}\right)^2 + \left(\frac{\partial \zeta}{\partial y}\right)^2\right)^{1/2} \tag{11.3.11}$$

In x, y, z coordinates, this yields

$$n_i = \left(-\frac{\partial \zeta}{\partial x}\, ,\ -\frac{\partial \zeta}{\partial y}\, ,\ 1\right) \frac{1}{|\nabla F|} \tag{11.3.12}$$

We now assume the external stress on the boundary surface F is R_j^s where j refers to the j the component (see e.g. Fig. 11.3.2) and superscript s to the surface. We also assume that σ_{ij} at a point immediately inside the fluid can be written as $(-\widehat{\;p\;}\,\delta_{ij} + \tau_{ij})$ where τ_{ij} is the Reynolds stress (see Section 2.5). Thus, according to (11.3.7), we have

$$R_j^s = \sigma_{ij}\, n_i = \left(-\widehat{\;p\;}\,\delta_{ij} + \tau_{ij}\right) n_i \tag{11.3.13}$$

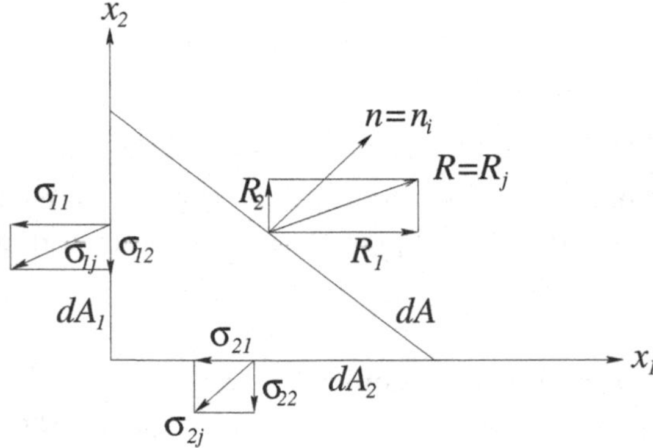

Fig. 11.3.2 The stresses on a two dimensional triangle.

In tensor form, distinguishing between α and z, we split σ_{ij} into two components: one on vertical surfaces (normal vector n_α) and one on horizontal surfaces (normal vector n_z):

$$\sigma_{ij} = \left(-\widehat{p}\,\delta_{\alpha j} + \tau_{\alpha j}\right) + \left(-\widehat{p}\,\delta_{zj} + \tau_{zj}\right) \tag{11.3.14}$$

which means (11.3.13) can be written as

$$R_j^s = \sigma_{ij}\, n_i = \underbrace{(-p\delta_{\alpha j} + \tau_{\alpha j})\; n_\alpha}_{\text{vert surface}} + \underbrace{\left(-\widehat{p}\,\delta_{zj} + \tau_{zj}\right)\, n_z}_{\text{hor. surface}} \tag{11.3.15}$$

Since we will later on need the horizontal and vertical components of R_j^s separately, we separate R_j^s in (11.3.15) into those components already now.

The horizontal component R_β^s of R_j^s ($\beta = 1{,}2$ corresponding to x or y) becomes, substituting (11.3.12) for n_i into (11.3.15)

$$\text{Hor. comp (tensor)}\;\; R_\beta^s = \left[\left(-\widehat{p}\,\delta_{\alpha\beta} + \tau_{\alpha\beta}\right)\left(-\frac{\partial\zeta}{\partial x_\alpha}\right) + \tau_{z\beta}\right]/|\nabla F|\;;\; z = \zeta \tag{11.3.16}$$

where $\tau_{\alpha\beta}$ and $\tau_{z\beta}$ are the horizontal components of the internal stresses in the fluid, and it has been used that $\delta_{z\beta} = 0$.

Comment: In (11.3.16) there is no $-\widehat{p}\,\delta_{z\beta}$-term because $\beta \sim x, y$ means $\delta_{z\beta} = 0$. In physical words: $-\widehat{p}\,\delta_{z\beta}$ would be the horizontal contribution to the stresses below the surface from the pressure on a surface with vertical normal, n_z, which is of course zero.

Hence, (11.3.16) indicates which internal horizontal stresses are generated by the horizontal component R_β^s of the external stress on the boundary F.

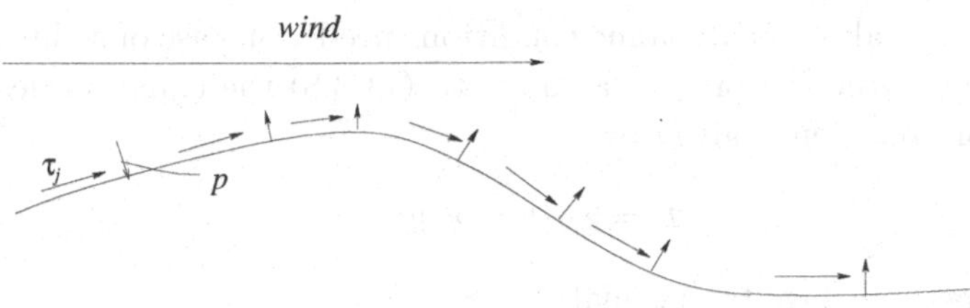

Fig. 11.3.3 The stresses on the free surface.

In x, y, z coordinates, this reads

$$\text{Hor. } x\text{-coord.} \quad R_x^s|\nabla F| = \widehat{p}\,\frac{\partial \zeta}{\partial x} - \tau_{xx}\frac{\partial \zeta}{\partial x} - \tau_{yx}\frac{\partial \zeta}{\partial y} + \tau_{zx} \tag{11.3.17}$$

$$\text{Hor. } y\text{-coord.} \quad R_y^s|\nabla F| = \widehat{p}\,\frac{\partial \zeta}{\partial y} - \tau_{xy}\frac{\partial \zeta}{\partial x} - \tau_{yy}\frac{\partial \zeta}{\partial y} + \tau_{zy} \tag{11.3.18}$$

Similarly, we get for the vertical component

$$\text{Vert. comp.} \quad R_z^s|\nabla F| = -\widehat{p} - \tau_{\alpha z}\frac{\partial \zeta}{\partial x_\alpha} + \tau_{zz} \tag{11.3.19}$$

or

$$\text{Vert } z\text{-comp.} \quad R_z^s|\nabla F| = -\widehat{p} - \tau_{xz}\frac{\partial \zeta}{\partial x} - \tau_{yz}\frac{\partial \zeta}{\partial y} + \tau_{zz} \tag{11.3.20}$$

Discussion

It may be recalled that when we talk about turbulent flows, the τ_{ij} are the Reynolds (or turbulent) stresses (see Section 2.5).

The forms (11.3.17), (11.3.18), (11.3.19), and (11.3.20) are convenient for the depth integration of the equations to be derived later. Usually, however, the external stress R_j^s is actually known by its normal component and its tangential component. Thus, e.g., in the case of strong winds blowing over waves this will create a pressure distribution and a shear stress distribution on the (sloping) surface.

The two components will vary quite differently along the surface. The shear stress will usually be largest on the upstream side of the wave where the pressure may be small and either positive or negative. At the rear side of the wave, however, the pressure is strongly negative and the shear stress small.

The bottom condition

Here we analyse the dynamic conditions needed in case of a slip velocity being used at the bottom. In analogy to (11.3.8) the equation describing the bottom position is given by

$$B = z_B + h(x, y) = 0 \tag{11.3.21}$$

We therefore have the normal

$$n_i = \left(\frac{\partial h}{\partial x}, \frac{\partial h}{\partial y}, 1\right) \frac{1}{|\nabla B|} \tag{11.3.22}$$

where

$$|\nabla B| = \left(1 + h_x^2 + h_y^2\right)^{1/2} \tag{11.3.23}$$

The horizontal component R_β^B of the total stress on the bottom can then be written (from Eq. (11.3.16))

$$R_\beta^B |\nabla B| = (-p\delta_{\alpha\beta} + \tau_{\alpha\beta}) \frac{\partial h}{\partial x_\alpha} + \tau_{z\beta} \tag{11.3.24}$$

Parallel to the bottom we have a shear stress τ_j^B (in addition to whatever pressure occurs there). It turns out to be convenient to rename the horizontal component of R_β^B as τ_β^B for which we then have from (11.3.24)

$$\tau_\beta^B |\nabla B| = \tau_{\alpha\beta} \frac{\partial h}{\partial x_\alpha} + \tau_{z\beta} \tag{11.3.25}$$

The reason for this separation of the pressure from the shear stress component of the total stresses will be discussed later. For the x and y components, this means

$$\tau_x^B |\nabla B| = \tau_{xx} \frac{\partial h}{\partial x} + \tau_{yx} \frac{\partial h}{\partial y} + \tau_{zx} \tag{11.3.26}$$

$$\tau_y^B |\nabla B| = \tau_{xy} \frac{\partial h}{\partial x} + \tau_{yy} \frac{\partial h}{\partial y} + \tau_{zy} \tag{11.3.27}$$

It is useful to realize that τ_j^B is the stress actually created on the (sloping) bottom by the fluid motion (see Fig. 11.3.4), and τ_x^B, τ_y^B are the x and y components of that shear stress. τ_j^B is the shear stress we can measure or compute by modelling the physical processes in the flow near the bottom. τ_{zx} and τ_{zy}, on the other hand, are the x and y stress components that appear in the x and y (horizontal) components of the momentum equation(s). Thus (11.3.26) and (11.3.27) make it possible to evaluate the contribution of the bottom shear stresses τ_j^B to the horizontal momentum equation(s).

Approximation for gently sloping bottom

The general equation (11.3.24) can be simplified if we assume gently sloping bottom corresponding to

$$\frac{\partial h}{\partial x}, \frac{\partial h}{\partial y} \ll 1 \tag{11.3.28}$$

we get,

$$n_i = (n_x,\ n_y,\ n_z) \approx (0,\ 0,\ 1) \tag{11.3.29}$$

Hence we get

$$\tau_x^B \sim \tau_{zx} \tag{11.3.30}$$

$$\tau_y^B \sim \tau_{zy} \tag{11.3.31}$$

which of course is the approximation frequently used in practical applications.

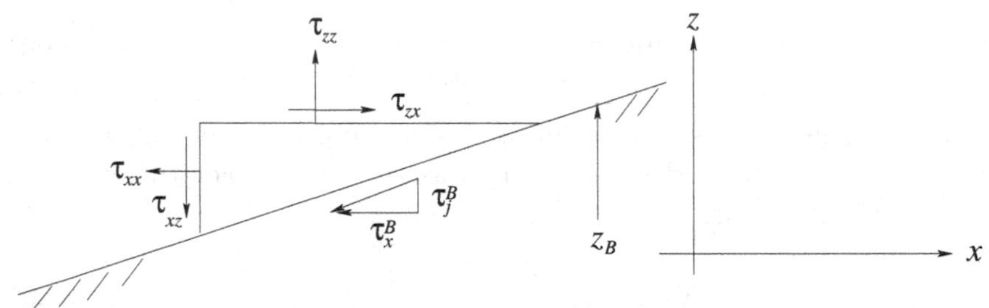

Fig. 11.3.4 The stresses on coordinate surfaces at the bottom.

11.4 Depth integrated momentum equation

In order to derive the depth integrated momentum equations, we start with the momentum part of the Reynolds equations in the form (2.5.28).

For simplicity, we also here from now on omit the $\frown$ that mark turbulent averaging except where it can be confusing not to do so. Thus we have

Horizontal components

$$\frac{\partial u_\beta}{\partial t} + \frac{\partial u_\alpha u_\beta}{\partial x_\alpha} + \frac{\partial u_\beta w}{\partial z} = -\frac{1}{\rho}\delta_{\alpha\beta}\frac{\partial p}{\partial x_\alpha} + \frac{1}{\rho}\left(\frac{\partial \tau_{\alpha\beta}}{\partial x_\alpha} + \frac{\partial \tau_{z\beta}}{\partial z}\right) \tag{11.4.1}$$

Notice that β indicates the component we look at so that $\beta = 1$ means the x component, $\beta = 2$ is the y component.

Vertical component

$$\frac{\partial w}{\partial t} + \frac{\partial u_\alpha w}{\partial x_\alpha} + \frac{\partial w^2}{\partial z} = -\frac{1}{\rho}\frac{\partial p}{\partial z} - g + \frac{1}{\rho}\left(\frac{\partial \tau_{\alpha z}}{\partial x_\alpha} + \frac{\partial \tau_{zz}}{\partial z}\right) \qquad (11.4.2)$$

11.4.1 *Integration of horizontal equations*

The first step is to integrate (11.4.1) over depth and use the Leibnitz rule to move the $\partial/\partial t$, $\partial/\partial x$ operations from under the integration sign to outside. We get term by term

$$\int_{-h_0}^{\zeta} \frac{\partial u_\beta}{\partial t} dz = \frac{\partial}{\partial t}\int_{-h_0}^{\zeta} u_\beta dz - u_\beta(\zeta)\frac{\partial \zeta}{\partial t} \qquad (11.4.3)$$

and

$$\int_{-h_0}^{\zeta} \frac{\partial u_\alpha u_\beta}{\partial x_\alpha} dz = \frac{\partial}{\partial x_\alpha}\int_{-h_0}^{\zeta} u_\alpha u_\beta dz - u_\alpha(\zeta)u_\beta(\zeta)\frac{\partial \zeta}{\partial x_\alpha} + u_\alpha(-h_0)u_\beta(-h_0)\frac{\partial(-h_0)}{\partial x_\alpha} \qquad (11.4.4)$$

and

$$\int_{-h_0}^{\zeta} \frac{\partial u_\beta w}{\partial z} dz = u_\beta(\zeta)w(\zeta) - u_\beta(-h_0)w(-h_0) \qquad (11.4.5)$$

The entire left-hand side of (11.4.1) then becomes

$$\begin{aligned} LHS = {} & \frac{\partial}{\partial t}\int_{-h_0}^{\zeta} u_\beta dz + \frac{\partial}{\partial x_\alpha}\int_{-h_0}^{\zeta} u_\alpha u_\beta dz \\ & + u_\beta(\zeta)\left(-\frac{\partial \zeta}{\partial t} - u_\alpha(\zeta)\frac{\partial \zeta}{\partial x_\alpha} + w(\zeta)\right) \\ & - u_\beta(-h_0)\left(u_\alpha(-h_0)\frac{\partial h_0}{\partial x_\alpha} + w(-h_0)\right) \end{aligned} \qquad (11.4.6)$$

Using the kinematic boundary conditions at the free surface (2.4.4) and at the bottom (2.4.8) we simply get

$$LHS = \frac{\partial}{\partial t}\int_{-h_0}^{\zeta} u_\beta dz + \frac{\partial}{\partial x_\alpha}\int_{-h_0}^{\zeta} u_\alpha u_\beta dz \qquad (11.4.7)$$

We see that all the boundary terms on the LHS vanish **exactly**.

For the right-hand side we get:

$$\rho \cdot RHS = \int_{-h_0}^{\zeta} \frac{\partial}{\partial x_\alpha} (-p\delta_{\alpha\beta} + \tau_{\alpha\beta})\, dz + \int_{-h_0}^{\zeta} \frac{\partial \tau_{z\beta}}{\partial z} dz$$

$$= \frac{\partial}{\partial x_\alpha} \int_{-h_0}^{\zeta} (-p\delta_{\alpha\beta} + \tau_{\alpha\beta})\, dz - \frac{\partial \zeta}{\partial x_\alpha} (-p\delta_{\alpha\beta} + \tau_{\alpha\beta})_\zeta$$

$$+ \frac{\partial(-h_0)}{\partial x_\alpha} \cdot (-p\delta_{\alpha\beta} + \tau_{\alpha\beta})_{-h_0} + (\tau_{z\beta})_\zeta - (\tau_{z\beta})_{-h_0} \quad (11.4.8)$$

Here we see (from (11.3.16)) that for the free surface terms in (11.4.8) we get

$$(\tau_{z\beta})_\zeta - (-p\delta_{\alpha\beta} + \tau_{\alpha\beta})_\zeta \frac{\partial \zeta}{\partial x_\alpha} = R^s_\beta \mid \nabla F \mid \quad (11.4.9)$$

For the bottom terms, we similarly get from (11.3.24):

$$-(\tau_{z\beta})_{-h_0} + (p\delta_{\alpha\beta} - \tau_{\alpha\beta})_{-h_0} \frac{\partial h_0}{\partial x_\alpha} = -R^B_\beta \mid \nabla B \mid + (p\delta_{\alpha\beta})_{-h_0} \frac{\partial h_0}{\partial x_\alpha}$$

$$\sim -\tau^B_\beta + p(-h_0)\delta_{\alpha\beta} \frac{\partial h_0}{\partial x_\alpha} \quad (11.4.10)$$

the latter by assuming the bottom sufficiently gently sloping to let $|\Delta B| = 1$ in the bottom stress term. Thus the entire right hand side becomes

$$\rho RHS = \frac{\partial}{\partial x_\alpha} \int_{-h_0}^{\zeta} (-p\delta_{\alpha\beta} + \tau_{\alpha\beta})\, dz + R^s_\beta \mid \nabla F \mid - \tau^B_\beta + p(-h_0)\delta_{\alpha\beta} \frac{\partial h_0}{\partial x_\alpha} \quad (11.4.11)$$

In (11.4.1), we notice that by the summation rule we have

$$\delta_{\alpha\beta} \frac{\partial h_0}{\partial x_\alpha} = \frac{\partial h_0}{\partial x_\beta} \quad (11.4.12)$$

Combining (11.4.7), (11.4.11) and (11.4.12) then yields

$$\rho \frac{\partial}{\partial t} \int_{-h_0}^{\zeta} u_\beta dz + \rho \frac{\partial}{\partial x_\alpha} \int_{-h_0}^{\zeta} u_\alpha u_\beta dz =$$

$$p(-h_0) \frac{\partial h_0}{\partial x_\beta} + \frac{\partial}{\partial x_\alpha} \int_{-h_0}^{\zeta} (-p\delta_{\alpha\beta} + \tau_{\alpha\beta})\, dz + R^s_\beta \mid \nabla F \mid - \tau^B_\beta \quad (11.4.13)$$

which is the <u>instantaneous</u> depth integrated horizontal momentum equation. We see that the RHS of (11.4.13) describes the sum of all horizontal

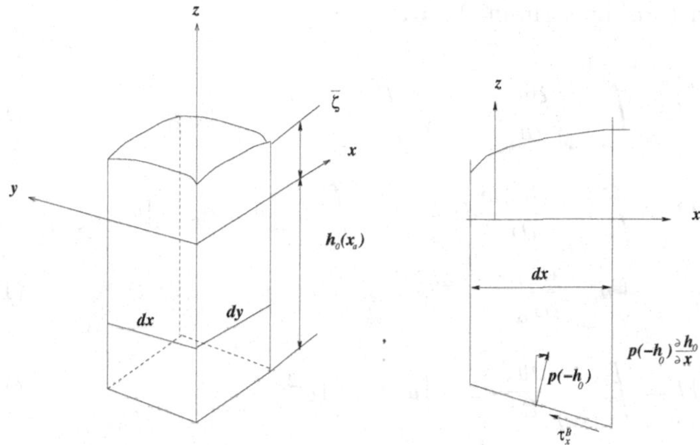

Fig. 11.4.1 Sketch of column (or control volume) considered. Left panel shows $3D$ elevation, right panel $2DV(x.z)$ only.

forces acting on a water column (or control volume) as shown in Fig. 11.4.1. The column has fixed vertical sides distance 1 apart in the x and y direction and follows the sloping bottom. The upper boundary of the control volume coincides with the free surface and follows its motion at any time.

In the following we will for simplicity assume that the surface slope is small enough to let $|\nabla F| \sim 1$ so that R_β^s represents the horizontal stress at the surface. This, however, is not a necessary asumption.

Note that we have continued to keep the term with the bottom pressure $p(-h_0)$ separate. The pressure force is normal to the bottom so it has the horizontal components $p(-h_0)\,(\partial h_0/\partial x,\ \partial h_0/\partial y)$ in the x and y directions, respectively.

11.4.2 *Integration of the vertical momentum equation*

For convenience, we repeat the vertical Reynolds equation:

$$\frac{\partial w}{\partial t} + \frac{\partial u_\alpha w}{\partial x_\alpha} + \frac{\partial w^2}{\partial z} = -\frac{1}{\rho}\frac{\partial}{\partial z}(p + \rho g z) + \frac{1}{\rho}\left(\frac{\partial \tau_{\alpha z}}{\partial x_\alpha} + \frac{\partial \tau_{zz}}{\partial z}\right) \tag{11.4.14}$$

I II III IV V VI

We now integrate this equation from a chosen level z_0 below the surface to the free surface $\zeta(x_\alpha,\ t)$. In order later to allow z_0 to take the value $-h_0(x_\alpha)$, we formally let z_0 be $z_0(x_\alpha)$.

We get term by term in (11.4.14):

$$I=\int_{z_0(x_\alpha)}^{\zeta}\frac{\partial w}{\partial t}dz=\frac{\partial}{\partial t}\int_{z_0(x_\alpha)}^{\zeta}wdz-[w]_\zeta\frac{\partial\zeta}{\partial t} \tag{11.4.15}$$

$$II=\int_{z_0(x_\alpha)}^{\zeta}\frac{\partial u_\alpha w}{\partial x_\alpha}dz=\frac{\partial}{\partial x_\alpha}\int_{z_0(x_\alpha)}^{\zeta}u_\alpha wdz-[u_\alpha w]_\zeta\frac{\partial\zeta}{\partial x_\alpha}$$
$$+u_\alpha w\frac{\partial z_0}{\partial x_\alpha} \tag{11.4.16}$$

$$III=\int_{z_0(x_\alpha)}^{\zeta}\frac{\partial w^2}{\partial z}dz=\left[w^2\right]_\zeta-\left[w^2\right]_{z_0} \tag{11.4.17}$$

Hence, the total LHS becomes:

$$LHS=\frac{\partial}{\partial t}\int_{z_0}^{\zeta}wdz+\frac{\partial}{\partial x_\alpha}\int_{z_0}^{\zeta}u_\alpha wdz-[w]_\zeta\left[\frac{\partial\zeta}{\partial t}+u_\alpha\frac{\partial\zeta}{\partial x_\alpha}-w\right]_\zeta$$
$$+[w]_{z_0}w\left[u_\alpha\frac{\partial z_0}{\partial x_\alpha}-w\right]_{z_0} \tag{11.4.18}$$

Here the surface terms in the bracket are zero by virtue of (11.3.1), so we get

$$LHS=\frac{\partial}{\partial t}\int_{z_0}^{\zeta}wdz+\frac{\partial}{\partial x_\alpha}\int_{z_0}^{\zeta}u_\alpha wdz-w\left[u_\alpha\frac{\partial z_0}{\partial x_\alpha}-w\right]_{z_0} \tag{11.4.19}$$

On the RHS, we have the following terms:

$$IV=-\int_{z_0}^{\zeta}\frac{\partial}{\partial z}(p+\rho gz)\,dz=-p(\zeta)+p(z_0)-\rho g(\zeta-z_0) \tag{11.4.20}$$

$$V=\int_{z_0}^{\zeta}\frac{\partial\tau_{\alpha z}}{\partial x_\alpha}dz=\frac{\partial}{\partial x_\alpha}\int_{z_0}^{\zeta}\tau_{\alpha z}dz-[\tau_{\alpha z}]_\zeta\frac{\partial\zeta}{\partial x_\alpha}+[\tau_{\alpha z}]_{z_0}\frac{\partial z_0}{\partial x_\alpha} \tag{11.4.21}$$

$$VI=\int_{z_0}^{\zeta}\frac{\partial\tau_{zz}}{\partial z}dz=[\tau_{zz}]_\zeta-[\tau_{zz}]_{z_0} \tag{11.4.22}$$

Hence, the total RHS becomes

$$\rho \cdot RHS = -p(\zeta) - \left(\tau_{\alpha z}\frac{\partial \zeta}{\partial x_\alpha}\right)_\zeta + (\tau_{zz})_\zeta + p(z_0) - \rho g(\zeta - z_0)$$
$$+\frac{\partial}{\partial x_\alpha}\int_z^\zeta \tau_{\alpha z} dz - \left[\tau_{zz} - \tau_{\alpha z}\frac{\partial z_0}{\partial x_\alpha}\right]_{z_0} \quad (11.4.23)$$

Combining LHS and RHS then gives the result for $p(z_0)$

$$p(z_0) = \rho g(\zeta - z_0) - \rho w\left[w - u_\alpha \frac{\partial z_0}{\partial x_\alpha}\right]_{z_0} + \frac{\partial}{\partial t}\int_{z_0}^\zeta \rho w dz + \frac{\partial}{\partial x_\alpha}\int_{z_0}^\zeta \rho u_\alpha w dz$$
$$+\left[\tau_{zz} - \tau_{\alpha z}\frac{\partial z_0}{\partial x_\alpha}\right]_{z_0} - \left[-p + \tau_{zz} - \tau_{\alpha z}\frac{\partial \zeta}{\partial x_\alpha}\right]_\zeta - \frac{\partial}{\partial x_\alpha}\int_{z_0}^\zeta \tau_{\alpha z} dz$$
$$(11.4.24)$$

In (11.4.24), the $[\ ]_\zeta$ bracket represents the vertical component of the stresses caused by the vertical force R_z^s on the free surface (see (11.3.19)). This force consists of the combination of the atmospheric pressure p_s and the wind stresses. As in the earlier parts of the derivation (for the horizontal components of the equations) we assumed that the vertical contribution from the wind stress is small in comparison to the atmospheric pressure which is constant. Thus we assume here for simplicity that

$$\left[-p + \tau_{zz} - \tau_{\alpha z}\frac{\partial \zeta}{\partial x_\alpha}\right]_\zeta = R_z^s \cong p^s = 0 \quad (11.4.25)$$

Again this is not a necessary condition. (11.4.24) then becomes

$$p(z_0) = \rho g(\zeta - z_0) - \rho w\left[w - u_\alpha \frac{\partial z_0}{\partial x_\alpha}\right]_{z_0} + \frac{\partial}{\partial t}\int_{z_0}^\zeta \rho w dz + \frac{\partial}{\partial x_\alpha}\int_{z_0}^\zeta \rho u_\alpha w dz$$
$$+\left[\tau_{zz} - \tau_{\alpha z}\frac{\partial z_0}{\partial x_\alpha}\right]_{z_0} - \frac{\partial}{\partial x_\alpha}\int_{z_0}^\zeta \tau_{\alpha z} dz \quad (11.4.26)$$

The pressure $p(z)$ at arbitrary level

From the expression (11.4.26) the pressure at arbitrary level z_0 can then be obtained by letting z_0 be independent of x_α, so that

$$\frac{\partial z_0}{\partial x_\alpha} = 0 \quad (11.4.27)$$

With $z_0 = z$ we then get

$$p(z) = \rho g(\zeta - z) - \rho w^2 + \tau_{zz} + \frac{\partial}{\partial t}\int_z^\zeta \rho w dz + \frac{\partial}{\partial x_\alpha}\int_z^\zeta (\rho u_\alpha w - \tau_{\alpha z})\, dz \tag{11.4.28}$$

We recall that τ_{ij} are the turbulent shear stresses defined as

$$\tau_{ij} = -\rho \widehat{u_i' u_j'} \tag{11.4.29}$$

Hence we can also write $p(z)$ as

$$\boxed{p(z) = \rho g(\zeta - z) - \rho(w^2 + \widehat{w'^2}) + \frac{\partial}{\partial t}\int_z^\zeta \rho w dz + \frac{\partial}{\partial x_\alpha}\int_z^\zeta \rho\left(u_\alpha w + \widehat{u'_\alpha w'}\right) dz} \tag{11.4.30}$$

I II III IV

where the terms are numbered to facilitate the discussion below.

For the purpose of the discussion below, we also calculate the value of the wave averaged pressure $\overline{p(z)}$. We get from (11.4.30)

$$\overline{p(z)} = \rho g\left(\bar{\zeta} - z\right) - \rho\left(\overline{w^2} + \overline{\widehat{w'^2}}\right) + \frac{\partial}{\partial x_\alpha}\overline{\int_{-z}^\zeta \rho\left(u_\alpha w + u'_\alpha w'\right) dz} \tag{11.4.31}$$

The pressure $p(-h_0)$ at the bottom

We next use (11.4.26) (with (11.4.25) inserted) to determine $p(-h_0)$ in the $p(-h_0)\partial h_0/\partial x_\beta$ term in (11.4.13). Thus in (11.4.26) we take

$$z_0 = -h_0 \;\; ; \;\; \frac{\partial z_0}{\partial x_\alpha} = -\frac{\partial h_0}{\partial x_\alpha} \tag{11.4.32}$$

We also introduce the bottom boundaary condition

$$u_\alpha \frac{\partial h_0}{\partial x_\alpha} - w = 0 \tag{11.4.33}$$

In dealing with the term

$$\tau_{zz} - \tau_{\alpha z}\frac{\partial z_0}{\partial x_\alpha}$$

at the bottom we use (11.4.29) for τ_{ij}. Hence, we have at $z_0 = -h_0$

$$\tau_{zz} + \tau_{\alpha z}\frac{\partial h_0}{\partial x_\alpha} = -\rho\left(\widehat{w'^2} + \widehat{w'u'_\alpha}\frac{\partial h_0}{\partial x_\alpha}\right)$$

$$= -\rho\,\widehat{w'\left(w' + u'_\alpha\frac{\partial h_0}{\partial x_\alpha}\right)} = 0 \qquad (11.4.34)$$

since (u'_α , w') must also satisfy (11.4.33).

Using all these results (11.4.26) at $z = -h_0$ becomes

$$\boxed{p(-h_0) = \rho g(\zeta + h_0) + \frac{\partial}{\partial t}\int_{-h_0}^{\zeta} \rho w dz + \frac{\partial}{\partial x_\alpha}\int_{-h_0}^{\zeta}\left(\rho u_\alpha w + \widehat{u'_\alpha w'}\right) dz} \qquad (11.4.35)$$

We see that in general the pressure at the bottom is **not** just equal to the weight of the water in the column above. Both the vertical acceleration in that column (the second term) and the vertical shear stresses along the vertical sides of the column (the last term) contribute to the instantaneous value of $p(-h_0)$.

Finally, it is illustrative also to do a time average over the short wave period of (11.4.35) to determine the mean pressure at the bottom. Invoking the periodicity of those waves, the result is

$$\overline{p(-h_0)} = \rho g\left(\bar{\zeta} + h_0\right) + \frac{\partial}{\partial x_\alpha}\overline{\int_{-h_0}^{\zeta}(\rho u_\alpha w - \tau_{\alpha z})\, dz} \qquad (11.4.36)$$

Negelecting the last term (see later discussion) then leads to the expression

$$\overline{p(-h_0)} \simeq \rho g(\bar{\zeta} + h_0) \qquad (11.4.37)$$

as one would intuitively expect.

Notice that the $\overline{w^2}$ and $\widehat{w'^2}$ terms in (11.4.31) vanish exactly at the bottom as (11.4.36) shows, not because the bottom slope $\partial h_0/\partial x_\alpha$ is small, but because (11.4.33) applies to both u_α, w and u'_α, w'.[2]

[2]Mei (1983) finds that it is necessary to require h_{0x} small to eliminate the bottom terms. The reason is he omits the $\partial z_0/\partial x_\alpha$ terms in (11.4.26). As the derivation above shows the results apply for arbitrary steep bottom slopes.

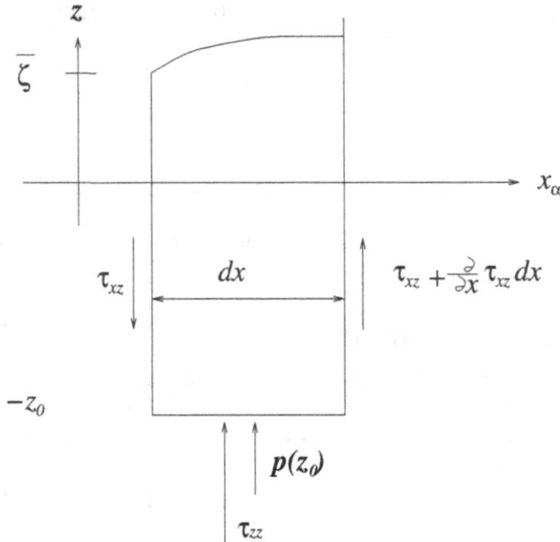

Fig. 11.4.2 The vertical stresses on a column of fluid reaching from the free surface to an arbitrary depth z. Note that for convenience the stresses in this figure have been named τ, though this τ is different from the $\tau_{\alpha\beta}$ used elsewhere. The definitions of the $\tau's$ in this figure are: $\tau_{xz} = \overline{\int_{-z}^{\zeta} (\rho u_\alpha w - \tau_{\alpha z})\, dz}$, $\tau_{zz} = -\rho \left(\overline{w^2} + \overline{\widehat{w'^2}} \right)$.

Interpretation of results of the pressure variation. The terms indicated by the numbers I through IV, which were shown beneath of (11.4.30), represent the following:

I. The term I is the hydrostatic pressure component. It represents the weight of the column of water (up to the instantaneous free surface $z = \zeta(x_\alpha, t)$) above the point $z = z_0$ which we consider.

II. represents the contribution to the flux of vertical momentum from the vertical fluid motion. This term can actually be written

$$\left(\widehat{w}\right)^2 + \widehat{w'^2} = w^2$$

where $w = \widehat{w} + w'$ is the total vertical velocity. In other words, the total vertical velocity of water flowing through the element of $z = z_0$ helps carrying the weight of the water above and therefore reduces the pressure $\widehat{p}(z_0)$.

III. represents the (integrated) vertical accelerations of the column above $z = z_0$. (This is in fact the only source of the dynamic pressure in linear waves.)

IV. is the net (due to the $\frac{\partial}{\partial x_\alpha}$) vertical force due to (Reynolds) shear stresses along the vertical sides of the column from organized motion $-\rho\left(\overbrace{u_\alpha}\,\overbrace{w}\right)$ and from turbulence $-\rho\left(\overbrace{u'_\alpha w'}\right)$, respectively.

Terms II, III, and IV together constitute the effect which the flow has on the hydrostatic pressure relative to the **instantaneous** water surface $z = \zeta$.

If the term IV differs from zero, it means that some of the weight of the column is transferred to the neighboring columns because of the water motion. In waves, this happens all the time throughout the wave period. (This effect is non-linear and is often neglected as e.g. in the weakly non-linear Boussinesq waves). However, if we consider (11.4.31) for the time averaged pressure $\overline{p(z)}$ we see that this expression can be written

$$\underbrace{\rho g(\bar{\zeta} - z)}_{\text{weight of water}} = \overline{p(z)} + \rho\left(\overline{\overbrace{w^2}} + \overline{\overbrace{w'^2}}\right) - \overline{\frac{\partial}{\partial x_\alpha}\int_z^\zeta \rho\,(u_\alpha w + u'_\alpha w')\,dz} \tag{11.4.38}$$

This can be interpreted as that in the average over a wave period, part of the weight of water $\rho g(\zeta - z)$ above a point z is carried by the pressure $\overline{p(z)}$ as we should expect. In addition some of the weight is carried by the normal momentum stress from the wave and turbulent oscillations $\left(\overline{\overbrace{w^2}} \text{ and } \overline{\overbrace{w'^2}} \text{ terms}\right)$. However, the expression also indicates that some weight is carried by the $\partial/\partial x_\alpha$ term, meaning neighboring water columns are helping carrying each other. Simple intuitive arguments suggest that over longer periods of time this is not possible in progressive waves. Each water column must very nearly carry itself. Thus it can be expected that in the wave averaged sense these terms are negligible.[3]

The time averaged dynamic pressure p_D is defined by

$$\overline{p} = \rho g(\bar{\zeta} - z) + \overline{p_D} \tag{11.4.39}$$

[3]Notice that in standing waves this is not the case.

Averaged over the wave period and omiting the $\frac{\partial}{\partial x_\alpha}$-terms this becomes

$$\overline{p_D} = \rho \left(\overline{w^2} + \overline{\widetilde{w'^2}} \right) \tag{11.4.40}$$

which shows that the time averaged dynamic pressure at z_0 comes from the vertical motion.

Finally, from (11.4.30), it is worth to notice that in a fluid in motion, the pressure is not completely independent of the direction of the fluid surface considered (since $p(z_0)$ in general depends on $\partial z_0/\partial x_\alpha$).

11.5 The nearshore circulation equations

11.5.1 *The time averaged momentum equation*

We are now ready to establish the various final forms of the depth integrated, time averaged momentum equations.

We first do the time averaging of (11.4.13), although this will be rather formal without also separating the horizontal velocities into wave and current components. We get

$$\overline{\rho \frac{\partial}{\partial t} \int_{-h_0}^{\zeta} u_\beta dz} + \overline{\rho \frac{\partial}{\partial x_\alpha} \int_{-h_0}^{\zeta} u_\alpha u_\beta dz} =$$

$$\overline{p(-h_0)} \frac{\partial h_0}{\partial x_\beta} + \frac{\partial}{\partial x_\alpha} \overline{\int_{-h_0}^{\zeta} -\delta_{\alpha\beta} p dz} + \frac{\partial}{\partial x_\alpha} \overline{\int_{-h_0}^{\zeta} \tau_{\alpha\beta} dz} + \overline{R_\beta^s} - \overline{\tau_\beta^B} \tag{11.5.1}$$

We then eliminate the pressure at the bottom by substituting (11.4.37) into (11.5.1) which yields

$$\overline{\rho \frac{\partial}{\partial t} \int_{-h_0}^{\zeta} u_\beta dz} + \overline{\rho \frac{\partial}{\partial x_\alpha} \int_{-h_0}^{\zeta} u_\alpha u_\beta dz} =$$

$$\rho g \left(\bar{\zeta} + h_0 \right) \frac{\partial h_0}{\partial x_\beta} + \frac{\partial}{\partial x_\alpha} \overline{\int_{-h_0}^{\zeta} -\delta_{\alpha\beta} p dz} + \frac{\partial}{\partial x_\alpha} \overline{\int_{-h_0}^{\zeta} \tau_{\alpha\beta} dz} + \overline{R_\beta^s} - \overline{\tau_\beta^B} \tag{11.5.2}$$

On the right hand side we add and subtract $\rho g h \partial\bar{\zeta}/\partial x_\beta$ which means the pressure term at the bottom can be written (with $h \equiv h_0 + \bar{\zeta}$)

$$\begin{aligned}\rho g\left(\bar{\zeta}+h_0\right)\frac{\partial h_0}{\partial x_\beta} &+ \rho g h\frac{\partial\bar{\zeta}}{\partial x_\beta} - \rho g h\frac{\partial\bar{\zeta}}{\partial x_\beta}\\ &= \rho g h\frac{\partial h}{\partial x_\beta} - \rho g h\frac{\partial\bar{\zeta}}{\partial x_\beta}\\ &= \delta_{\alpha\beta}\frac{\partial}{\partial x_\alpha}\left(\frac{1}{2}\rho g h^2\right) - \rho g h\frac{\partial\bar{\zeta}}{\partial x_\beta}\end{aligned} \tag{11.5.3}$$

Here we have also utilized that

$$\frac{\partial}{\partial x_\beta} = \delta_{\alpha\beta}\frac{\partial}{\partial x_\alpha} \tag{11.5.4}$$

The result is that (11.5.2) can be written

$$\begin{aligned}\rho\frac{\partial}{\partial t}\overline{\int_{-h_0}^{\zeta} u_\beta dz} + \rho\frac{\partial}{\partial x_\alpha}\overline{\int_{-h_0}^{\zeta} u_\alpha u_\beta dz} = -\rho g h\frac{\partial\bar{\zeta}}{\partial x_\beta}\\ -\frac{\partial}{\partial x_\alpha}\left[\overline{\int_{-h_0}^{\zeta}\delta_{\alpha\beta} p dz} - \delta_{\alpha\beta}\frac{1}{2}\rho g h^2\right] + \frac{\partial}{\partial x_\alpha}\overline{\int_{-h_0}^{\zeta}\tau_{\alpha\beta} dz} + \overline{R_\beta^s} - \overline{\tau_\beta^B}\end{aligned} \tag{11.5.5}$$

Separation of wave and current components

In order to obtain the momentum part of the circulation equations we need to to separate the total velocity (u_α, w) into a "current" and a short wave component the same way it was done for the continuity equation (see (6.3.13)). However, since the Reynolds equations we are using have already been averaged over the turbulence we only need to separate the waves from the currents. This is done by letting

$$u_\alpha(=\widehat{u_\alpha}) = V_\alpha + u_{w\alpha} \;\; ; \;\; w(=\widehat{w}) = w_w \tag{11.5.6}$$

Here $u_{w\alpha}$, w_w are the short wave components for which have

$$\overline{u_{w\alpha}}\,,\ \overline{w_w} = 0 \qquad \text{below wave trough level} \tag{11.5.7}$$

V_α is the current. If the short waves are irregular, we will expect the "current" to be varying with time. These time variations of the currents correspond to long wave components, and also to slow transient changes in the flows.

Taking the terms in (11.5.5) one by one we get:

$$\overline{\int_{-h_0}^{\zeta} u_\beta dz} \equiv \overline{Q_\beta} \tag{11.5.8}$$

according to (11.2.3). The second term becomes

$$\overline{\int_{-h_0}^{\zeta} u_\alpha u_\beta dz} = \overline{\int_{-h_0}^{\zeta} (u_{w\alpha} + V_\alpha)(u_{w\beta} + V_\beta)\, dz}$$

$$= \int_{-h_0}^{\bar{\zeta}} V_\alpha V_\beta dz + \overline{\int_{-h_0}^{\zeta} u_{w\alpha} u_{w\beta} dz} + \overline{\int_{\zeta_t}^{\zeta} (u_{w\alpha} V_\beta + u_{w\beta} V_\alpha)\, dz} \tag{11.5.9}$$

where we have used (11.5.7) in the last term. Hence, the left hand side of (11.5.2) becomes

$$LHS = \rho \left[\frac{\partial}{\partial t} \overline{Q_\beta} + \frac{\partial}{\partial x_\alpha} \int_{-h_0}^{\bar{\zeta}} V_\alpha V_\beta dz \right]$$

$$+ \rho \left[\frac{\partial}{\partial x_\alpha} \overline{\int_{\zeta_t}^{\zeta} (u_{w\alpha} V_\beta + u_{w\beta} V_\alpha]\, dz} + \frac{\partial}{\partial x_\alpha} \overline{\int_{-h_0}^{\zeta} u_{w\alpha} u_{w\beta} dz} \right] \tag{11.5.10}$$

Rearranging terms we may then write the total horizontal momentum equation as

$$\rho \frac{\partial \overline{Q_\beta}}{\partial t} + \rho \frac{\partial}{\partial x_\alpha} \int_{-h_0}^{\bar{\zeta}} V_\alpha V_\beta dz + \rho \frac{\partial}{\partial x_\alpha} \overline{\int_{\zeta_t}^{\zeta} (u_{w\alpha} V_\beta + u_{w\beta} V_\alpha)\, dz} =$$

$$- \frac{\partial}{\partial x_\alpha} \left[\overline{\int_{-h_0}^{\zeta} (\rho u_{w\alpha} u_{w\beta} + p\delta_{\alpha\beta}) dz} - \delta_{\alpha\beta} \frac{1}{2} \rho g h^2 \right]$$

$$+ \frac{\partial}{\partial x_\alpha} \overline{\int_{-h_0}^{\zeta} \tau_{\alpha\beta} dz} - \rho g h \frac{\partial \bar{\zeta}}{\partial x_\beta} + R_\beta^S - \tau_\beta^B \tag{11.5.11}$$

On the left hand side of (11.5.11) we now have the acceleration terms (local and convective) for the time varying current/long wave component of the motion. In this form of the equations the acceleration is expressed as the time rate of change of the momentum plus the gradient of the momentum flux. The third term represents the convective acceleration associated with the net mass flux Q_w in the wave (defined by (11.2.25)).

On the right hand side, the first term is seen to represent the effect of the momentum flux (again in terms of its gradient) due to the waves. This is the term we call the radiation stress for the waves which hence is defined as

$$S'_{\alpha\beta} \equiv \overline{\int_{-h_0}^{\zeta} (\rho u_{w\alpha} u_{w\beta} + \delta_{\alpha\beta} p)\, dz} - \delta_{\alpha\beta} \frac{1}{2} \rho g h^2 \tag{11.5.12}$$

The momentum equation (11.5.11) can then be written

$$\begin{aligned} &\rho \frac{\partial}{\partial t} \overline{Q_\beta} + \rho \frac{\partial}{\partial x_\alpha} \int_{-h_0}^{\bar{\zeta}} V_\alpha V_\beta dz + \rho \frac{\partial}{\partial x_\alpha} \overline{\int_{\zeta_t}^{\zeta} u_{w\alpha} V_\beta + u_{w\beta} V_\alpha dz} \\ &+ \rho g (\bar{\zeta} + h_0) \frac{\partial \bar{\zeta}}{\partial x_\beta} + \frac{\partial}{\partial x_\alpha} \left[S'_{\alpha\beta} - \overline{\int_{-h_0}^{\zeta} \tau_{\alpha\beta} dz} \right] \\ &- R_\beta^S + \tau_\beta^B = 0 \end{aligned} \tag{11.5.13}$$

This is the general momentum equation for nearshore circulation of currents with depth variation, expressed in terms of the total volume flux Q_α in the circulation.

In this form of the equation, we have placed together the $S_{\alpha\beta}$ which is the wave contribution and the $\tau_{\alpha\beta}$ which is the turbulent contribution to the momentum flux. This emphasizes the parallel mechanism behind these two terms, one caused by organized (wave) fluctuations, the other by disorganized (turbulent) fluctuations. In fact, in some texts this is further emphasized by using the same letter "S" for the two contributions

Wave radiation stress	$S_{\alpha\beta}$
Turbulent "radiation stress"	$S^t{}_{\alpha\beta} = -\overline{\int_{-h_0}^{\zeta} \tau_{\alpha\beta} dz}$

Notice the "−" in front of the $\tau_{\alpha\beta}$ term and the "+" in front of the $S_{\alpha\beta}$-term in (11.5.13). This indicates the difference between the + sign on the $u_\alpha u_\beta$ term in the definition (11.5.12) for $S_{\alpha\beta}$ and the − sign on the $u_\alpha u_\beta$ term in the definition (5.3.34) for $\tau_{\alpha\beta}$. This also implies that the sign convention for $S_{\alpha\beta}$ and $\tau_{\alpha\beta}$ is opposite, a point worth bearing in mind when checking direction of terms in the equations. The positive directions for the two terms are shown in Fig. 11.5.1.

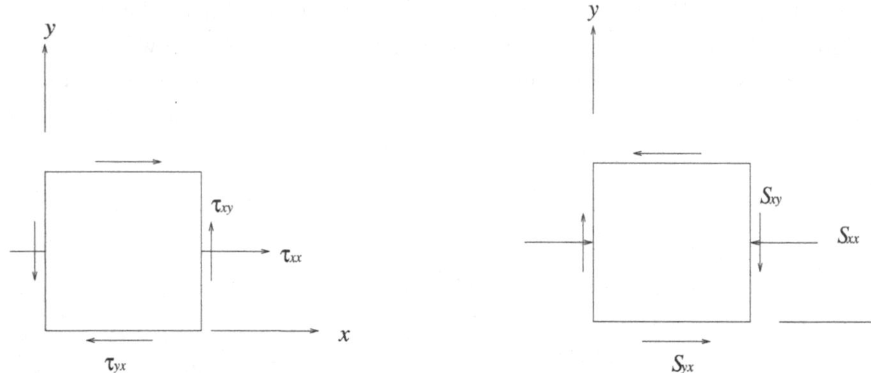

Fig. 11.5.1 Positive directions for $\tau_{\alpha\beta}$ and $S_{\alpha\beta}$.

11.5.2 *The equations for depth uniform currents*

Equation (11.5.13) allows the currents to vary over depth and in fact we know today that not only do nearshore currents normally do so but this depth variation is an important part of the mechanisms that control nearshore circulation. The depth variation of the currents result in a horizontal transfer of horizontal momentum which is of the same nature as the mixing caused by turbulent stresses. This effect is called **dispersive mixing**. It will be discussed later in Chapter 13.

However, making the assumption of depth uniform currents allows us to simplify (11.5.13) somewhat.

We therefore introduce the assumption that V_α, V_β are independent of z. The terms in (11.5.13) we can simplify using this assumption are the nonlinear current momentum terms. For depth uniform currents we have for the second term in (11.5.13):

$$\frac{\partial}{\partial x_\alpha} \int_{-h_0}^{\bar{\zeta}} V_\alpha V_\beta dz = \frac{\partial}{\partial x_\alpha} (V_\alpha V_\beta h) \tag{11.5.14}$$

Since the continuity equation for this case reads (see (11.2.28)) (letting $\overline{\widehat{Q_\alpha}} = \overline{Q_\alpha}$ for short)

$$\overline{Q_\alpha} = V_\alpha h + Q_{w\alpha} \tag{11.5.15}$$

(11.5.14) may be written

$$\frac{\partial}{\partial x_\alpha}\int_{-h_0}^{\bar{\zeta}} V_\alpha V_\beta dz = \frac{\partial}{\partial x_\alpha}\left(\left(\overline{Q_\alpha} - Q_{w\alpha}\right)V_\beta\right) \tag{11.5.16}$$

Remember that $\overline{Q_\alpha}$ is the total discharge including the wave mass flux $Q_{w\alpha}$, so that $\overline{Q_\alpha} - Q_{w\alpha}$ is the volume flux due to the current.

For the second current term in (11.5.13), we again assume the (constant) current profile can be extended to the mean water level as we did in Section 11.2 for the continuity equation. This implies we get

$$\begin{aligned}\frac{\partial}{\partial x_\alpha}\overline{\int_{\zeta_t}^{\zeta}\left(u_{w\alpha}V_\beta + u_{w\beta}V_\alpha\right)dz} &= \frac{\partial}{\partial x_\alpha}\left(V_\beta\overline{\int_{\zeta_t}^{\zeta} u_{w\alpha}dz} + V_\alpha\overline{\int_{\zeta_t}^{\zeta} u_{w\beta}dz}\right)\\ &= \frac{\partial}{\partial x_\alpha}\left(V_\beta Q_{w\alpha} + V_\alpha Q_{w\beta}\right)\end{aligned} \tag{11.5.17}$$

Hence, the two terms together may be written

$$\begin{aligned}&\frac{\partial}{\partial x_\alpha}\int_{-h_0}^{\bar{\zeta}} V_\alpha V_\beta dz + \overline{\int_{\zeta_t}^{\zeta}\left(u_{w\alpha}V_\beta + u_{w\beta}V_\alpha\right)dz}\\ &= \frac{\partial}{\partial x_\alpha}\left[\overline{Q_\alpha}V_\beta - Q_{w\alpha}V_\beta + V_\beta Q_{w\alpha} + V_\alpha Q_{w\beta}\right]\\ &= \frac{\partial}{\partial x_\alpha}\left[\overline{Q_\alpha}\frac{\overline{Q_\beta} - Q_{w\beta}}{h} + \frac{\overline{Q_\alpha} - Q_{w\alpha}}{h}Q_{w\beta}\right]\\ &= \frac{\partial}{\partial x_\alpha}\left[\frac{\overline{Q_\alpha Q_\beta}}{h} - \frac{Q_{w\alpha}Q_{w\beta}}{h}\right]\end{aligned} \tag{11.5.18}$$

Here the second term depends only on the wave motion since it is caused by interaction between mass flux components caused by the waves.

Therefore an alternative definition of the radiation stress that includes all terms generated by the waves is:

$$S_{\alpha\beta} = S'_{\alpha\beta} - \rho\frac{Q_{w\alpha}Q_{w\beta}}{h} \tag{11.5.19}$$

which is the definition used by Phillips (1966,77). The last term in (11.5.19) is strictly speaking of $O(H/L)^4$ and could therefore be considered very small, but this is not necessarily the case for nearbreaking waves or waves in the surf zone. It is often just neglected. However, it has a principal meaning which makes it desirable to keep it in the formula.

When substituting the result (11.5.18) into (11.5.13), and using the definition (11.5.19) for $S_{\alpha\beta}$, (11.5.13) takes the form

$$\boxed{\rho \frac{\partial \bar{Q}_\beta}{\partial t} + \frac{\partial}{\partial x_\alpha}\left(\rho \frac{\overline{Q_\alpha Q_\beta}}{h} + S_{\alpha\beta} - \overline{\int_{-h_0}^{\zeta} \tau_{\alpha\beta} dz}\right) = -\rho g h \frac{\partial \bar{\zeta}}{\partial x_\beta} + \overline{\tau_\beta^S} - \overline{\tau_\beta^B}} \tag{11.5.20}$$

valid for depth uniform currents only.

The equations for depth uniform currents were first derived by Longuet-Higgins and Stewart (1962, 1964). Those were also the publications where the term **Radiation Stress** was invented.[4] The form derived here is equivalent to the momentum equation derived by Phillips except for the turbulent stresses $\tau_{\alpha\beta}$, the horizontal components of the surface stress τ_β^S, and the bottom shear stress $\overline{\tau_\beta^B}$, all of which are neglected by Phillips. In nearshore circulation, however, these terms are of paramount importance as later analysis will show. Similar equations are also derived by Mei (1983).

It turns out, that the definition (11.5.12) introduced above looks equivalent to that used by Mei (1983). However, Mei has a different definition of the wave and current velocities. That definition is based on $\overline{\int_{-h_0}^{\zeta} u_{w\alpha} dz} = 0$ (versus our $\overline{u_\alpha} = 0$ at each point below wave trough), which, with depth uniform currents as here, causes the integral (11.5.17) to vanish altogether. Thus in Mei's case, the only contribution to the momentum equation explicitly from the current comes from the integral (11.5.14) whereas other current contributions are actually hidden in the definition of $S_{\alpha\beta}$ because in Mei's version $\overline{u_{w\alpha}} \neq 0$ below trough level.

Exercise 11.5-1

The purpose of this exercise is to analyse the difference between Phillips (1966, 1977) and Mei (1983).

In Phillips the wave motion is defined so that $\overline{u_{w\alpha,Phil}} = 0$ below wave trough level. The equaivalent current is called $V_{\alpha,Phil}$.

In Mei the wave motion is defined so that

$$\overline{\int_{-h_0}^{\zeta} u_{w\alpha,Mei}\, dz} = 0 \tag{11.5.21}$$

[4]It is interesting for historical reasons to mention that the concept of a time averaged momentum signature of the short waves - which is what the radiation stress is - was actually developed separately already in the late 1950'ties by H. Lundgren. The author of the present book learned about this when taking the advanced coastal engineering course from Lundgren already in 1959 at The Technical University of Denmark. Lundgren used the name **Wave Thrust** and published his results in Lundgren (1963).

Show that with the mass flux $Q_{w\alpha}$ defined as

$$Q_{w\alpha} = \overline{\int_{-h_0}^{\zeta} u_{w\alpha} dz} \tag{11.5.22}$$

we get

$$V_{\alpha,Phil} = \frac{\overline{Q_\alpha} - Q_{w\alpha,Phil}}{h} \tag{11.5.23}$$

Show similarly that (with the same definition of Q_w) we get

$$V_{\alpha,Mei} = \frac{\overline{Q_\alpha}}{h} \tag{11.5.24}$$

Thus we have

$$u_{w\alpha,Mei} = u_{w\alpha,Phil} - \frac{Q_w}{h} \tag{11.5.25}$$

The radiations stresses are also defined differently. We have

$$S_{\alpha\beta,Phil} = \overline{\int_{-h0}^{\zeta} \rho \left(u_{w\alpha,Phil} u_{w\beta,Phil} + \delta_{\alpha\beta}\, p\right) dz} - \delta_{\alpha\beta} \frac{1}{2} \rho g h^2 - \frac{Q_{w\alpha} Q_{w\beta}}{h} \tag{11.5.26}$$

which is exactly what the equations dictate, while Mei neglects the $Q_{w\alpha}Q_{w\beta}$-term (allegedly because it is of fourth order). Thus

$$S_{\alpha\beta,Mei} = \overline{\int_{-h0}^{\zeta} \rho \left(u_{w\alpha,Mei} u_{w\beta,Mei} + \delta_{\alpha\beta}\, p\right) dz} - \delta_{\alpha\beta} \frac{1}{2} \rho g h^2 \tag{11.5.27}$$

Show, introducing the respective definitions for the wave velocity components and the radiation stress, that in spite of the seeming differences we actually have

$$S_{\alpha\beta,Mei} = S_{\alpha\beta,Phil} \tag{11.5.28}$$

Thus Mei's omission of the $Q_{w\alpha}Q_{w\beta}$-term actually represents a correct definition of $S_{\alpha\beta}$, given his different definitions of the wave and current components, **not** an approximation to $O(kH)^4$.

11.6 Analysis of the radiation stress in two horizontal dimensions, 2DH

We now return to the definition (11.5.19) for the radiation stress $S_{\alpha\beta}$ (where we for the time being neglect the the last term which makes it equal to the expression for $S'_{\alpha\beta}$). We repeat here for convenience

$$S_{\alpha\beta} = \overline{\int_{-h_0}^{\zeta} (\rho u_{w\alpha} u_{w\beta} + \delta_{\alpha\beta} p)\, dz} - \delta_{\alpha\beta} \frac{1}{2} \rho g h^2 \tag{11.6.1}$$

The purpose is to use the results (11.4.28) for $p(z)$ to eliminate p from (11.6.1).

We have seen (see Section 3.3) that for simple sine waves in the x-direction, we can calculate the 1DH-version of the $u_\alpha u_\beta$-term from linear wave theory because we have

$$u_x u_x = c^2 \left(\frac{kH}{2} \frac{\cosh k(z+h)}{\sinh kh} \right)^2 \cos^2 \theta = O(kH^2) \tag{11.6.2}$$

and we then found (see (3.3.14)) that the momentum part S_m of the $S_{\alpha\beta}$ was given by

$$S_m = \frac{1}{16} \rho g H^2 (1+G) + 0(kH)^3 \tag{11.6.3}$$

This result will be qualitatively the same for any wave theory used for S_m.

For the $\int p dz$, however, we get, using linear theory:

$$\begin{aligned} \overline{\int_{-h_0}^{\zeta} p\, dz} &= \overline{\int_{-h_0}^{0} \rho g \left(-z + \frac{H}{2} \frac{\cosh k(z+h)}{\cosh kh} \right) \cos\theta dz} + \overline{\int_0^{\zeta} pdz} \\ &= \qquad 0 \qquad + \overline{\int_0^{\zeta} pdz} \end{aligned} \tag{11.6.4}$$

Whereas the second term on the RHS of (11.6.4) apparently is $O(H^2)$ as we should expect even when we use first order theory for p, it turns out that the zero value of the first term is not correct to $O(H^2)$. As mentioned in Section 3.3, this term actually would give the correct $O(H^2)$ contribution if we for p used Stokes second order theory instead of the simple linear theory used here.

To get the lowest order correct evaluation of $\int pdz$ directly from p will therefore require that second order approximations for p are used. This is not only inconvenient. It could also give the (erroneous) impression that the radiation stress of a first order wave cannot consistently be calculated by using only first order theory. Finally the definition (11.6.1) is exact and applies to any wave theory and we may not readily know a second order approximation for the wave theory we want to use to evaluate $S_{\alpha\beta}$.

In particular, the last two objections are quite serious, and it is very useful for our understanding of wave mechanics in general to see how this difficulty can be avoided by using the earlier results for the pressure variation along a vertical in the wave motion.

We therefore turn to considering the net contribution to the radiation stress from the pressure, which is $\overline{\int_{-h_0}^{\zeta} p dz} - \frac{1}{2}\rho g h^2$. Using (11.4.30) for p we get

$$
\begin{aligned}
&\overline{\int_{-h_0}^{\zeta} p dz} - \frac{1}{2}\rho g h^2 \\
&= \overline{\int_{-h_0}^{\zeta} \Big\{ \underbrace{\rho g (\zeta - z)}_{I} - \underbrace{\rho \left(\left(\widetilde{w}\right)^2 + \widetilde{w'^2} \right)}_{II} + \underbrace{\frac{\partial}{\partial t} \int_z^{\zeta} \rho \widetilde{w} dz'}_{III}} \\
&\quad \overline{+ \underbrace{\frac{\partial}{\partial x_\alpha} \int_z^{\zeta} \rho \left(\widetilde{u_\alpha} \widetilde{w} + \widetilde{u'_\alpha w'} \right) dz'}_{IV} \Big\} dz} - \frac{1}{2}\rho g h^2
\end{aligned}
\tag{11.6.5}
$$

Term by term we get

$$
I = \overline{-\frac{1}{2}\rho g (\zeta - z)^2 \Big]_{-h_0}^{\zeta}} = \frac{1}{2}\rho g \overline{(\zeta + h_0)^2} - \frac{1}{2}\rho g h^2 \tag{11.6.6}
$$

We leave II unchanged. For III we get, using Leibniz rule

$$
III = \overline{\frac{\partial}{\partial t} \int_{-h_0}^{\zeta} dz \int_z^{\zeta} \rho \widetilde{w} dz'} - \overline{\int_z^{\zeta} \rho \widetilde{w} dz' \,]_{z=\zeta} \frac{\partial \zeta}{\partial t}} \tag{11.6.7}
$$

Here the first term vanishes due to the periodicity $\left(\overline{\frac{\partial}{\partial t}} = 0\right)$. In the second term, the integral becomes zero where $z = \zeta$. Hence

$$
III = 0 \tag{11.6.8}
$$

$$
IV = \overline{\int_{-h_0}^{\zeta} \frac{\partial}{\partial x_\alpha} \int_z^{\zeta} \rho \left(\widetilde{u_\alpha} \widetilde{w} + \widetilde{u'_\alpha w'} \right) dz' dz} \tag{11.6.9}
$$

This term represents the effect on $S_{\alpha\beta}$ discussed earlier of one water column in average not being able to carry the weight of its neighbors. Since the wave averaged value of this contribution to the pressure was deemed negligible, it is consistent also to omit it in the expression for $S_{\alpha\beta}$.

The result for I can be combined with the hydrostatic pressure we subtract in (11.6.1). Using that $h = \bar{\zeta} + h_0$ we get:

$$\begin{aligned}\overline{(\zeta + h_0)^2} - h^2 &= \overline{(\zeta + h_0)^2} - \left(\bar{\zeta} + h_0\right)^2 = \overline{\zeta^2} + h_0^2 + 2h_0\bar{\zeta} - \bar{\zeta}^2 - h_0^2 - 2h_0\bar{\zeta} \\ &= \overline{\zeta^2} - \bar{\zeta}^2 = \overline{\zeta^2} + \bar{\zeta}^2 - 2\bar{\zeta}^2 = \overline{\zeta^2} + \bar{\zeta}^2 - 2\overline{\zeta\bar{\zeta}} \\ &= \overline{\left(\zeta - \bar{\zeta}\right)^2} = \overline{\eta^2} \qquad (11.6.10)\end{aligned}$$

Thus, we finally get

$$\int_{-h_0}^{\zeta} p\, dz - \frac{1}{2}\rho g h^2 = -\rho \int_{-h_0}^{\zeta} \left(\widetilde{w_w^2} + \widetilde{w'^2}\right) dz + \frac{1}{2}\rho g \overline{\eta^2} \qquad (11.6.11)$$

Substituting all these results back into (11.6.1), we eventually get

$$\boxed{S_{\alpha\beta} = \rho \overline{\int_{-h_0}^{\zeta} \left[u_{w\alpha}u_{w\beta} - \delta_{\alpha\beta}\left(w_w^2 + \widetilde{w'^2}\right)\right] dz} + \delta_{\alpha\beta}\frac{1}{2}\rho g \overline{\eta^2}} \qquad (11.6.12)$$

which is an expression for the radiation stress $S_{\alpha\beta}$ that, in the case of linear waves, can be evaluated by means of linear wave expressions for η, u_w and w_w. Notice that the w^2-terms represent the effect of the deviation from hydrostatic pressure, and that turbulence contributes to the pressure through the $\widetilde{w'^2}$ term. In practice, however, the effect of the turbulence is usually neglected. Also (11.6.12) was the expression used in Section 3.3 for the 1DH S_{xx} (without proof and without the $\widetilde{w'^2}$-term).

It is emphasized that (11.6.12) is exact[5] and applies for any wave theory. We see that all terms in (11.6.12) are quadratic in the flow quantities u_w, w_w and η. Hence, for any wave theory, the lowest order of approximation for $S_{\alpha\beta}$ can be determined by inserting into (11.6.12) a (consistent) lowest order of approximation for that wave theory.

[5] Apart from the contribution from term IV in (11.4.30), which however we showed is totally negligible.

11.6.1 $S_{\alpha\beta}$ expressed in terms of S_m and S_p

Omitting for convenience the index w for "wave" and $\widehat{\ }$ where possible, we can write (11.6.12) as

$$S_{\alpha\beta} = \overline{\int_{-h_0}^{\zeta} \left(\rho u_\alpha u_\beta - \delta_{\alpha\beta}\rho \left(w^2 + \widehat{w'^2} \right) \right) dz} + \frac{1}{2}\delta_{\alpha\beta}\rho g \overline{\eta^2} \qquad (11.6.13)$$

We now assume that the wave is locally plane; i.e., fronts are straight enough for the motion to be considered locally two dimensional, with particle velocities only in the vertical plane of the local wave orthogonal. This is normally an excellent approximation.

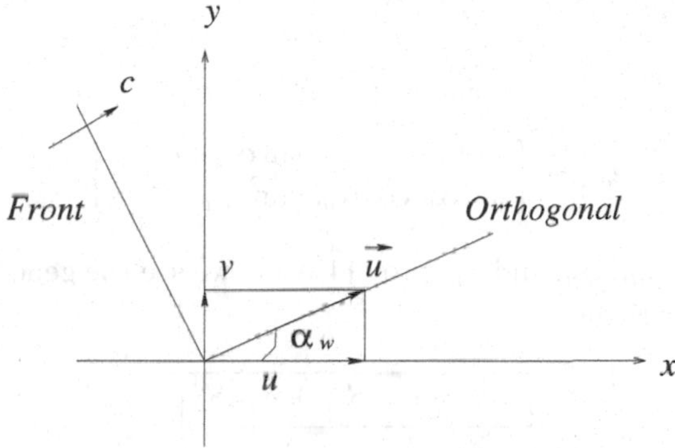

Fig. 11.6.1 Definition of the wave angle α_w. Note that this is only equal to the angle of incidence for the waves when the y-direction is tangent to the bottom contour through the point considered.

In that plane, velocities are

$$u^2 = (u_\alpha u_\alpha) \qquad (11.6.14)$$

Then the x and y components u_x and u_y are related to u by

$$u_x = |\,u\,| \cos\alpha_w \quad ; \quad u_y = |\,u\,| \sin\alpha_w \qquad (11.6.15)$$

where α_w is the angle of the wave orthogonal (or $\mathbf{k}$) to the x-axis as shown in Fig. 11.6.1. We may also define

$$S_m = \overline{\int_{-h_0}^{\zeta} \rho u^2 dz} \qquad (11.6.16)$$

$$S_p = \overline{-\int_{-h_0}^{\zeta} \rho \left(w^2 + \widetilde{w'^2} \right) dz} + \frac{1}{2}\rho g \overline{\eta^2} \qquad (11.6.17)$$

With these definitions we clearly have

$$\overline{\int_{-h_0}^{\zeta} \rho u_{w\alpha} u_{w\beta} dz} = e_{\alpha\beta} \overline{\int_{-h_0}^{\zeta} \rho u^2 dz} = S_m e_{\alpha\beta} \qquad (11.6.18)$$

where

$$e_{\alpha\beta} = \begin{Bmatrix} \cos^2 \alpha_w & \sin \alpha_w \cos \alpha_w \\ \sin \alpha_w \cos \alpha_w & \sin^2 \alpha_w \end{Bmatrix} \qquad (11.6.19)$$

Introducing $e_{\alpha\beta}$, S_m and S_p into (11.6.13) we see the general expression for $S_{\alpha\beta}$ can be written

$$\boxed{S_{\alpha\beta} = e_{\alpha\beta} S_m + \delta_{\alpha\beta} S_p} \qquad (11.6.20)$$

Thus, in terms of (x, y), we have the following $S_{\alpha\beta}$ components

$$S_{\alpha\beta} = \begin{Bmatrix} S_{xx} & S_{xy} \\ S_{yx} & S_{yy} \end{Bmatrix} \qquad (11.6.21)$$

Essentially, this expresses $S_{\alpha\beta}$, the radiation stress components in the α, β directions in terms of the two scalars S_m and S_p which represent the radiation stress on a vertical section of unit width perpendicular to the local wave direction. The convenience of this is clear since S_m, S_p would be the radiation stress values usually calculated from a given (plane) wave theory. In other words (11.6.20) indicates how $S_{\alpha\beta}$ varies with the direction of the waves.

It is also clear that the dimensionless radiation stress P_{xx} introduced in Chapter 3 can now be generalized to $P_{\alpha\beta}$ by writing

$$S_{\alpha\beta} = \rho g H^2 P_{\alpha\beta} \qquad (11.6.22)$$

where we apparently have

$$P_{\alpha\beta} = e_{\alpha\beta} \cdot \frac{S_m}{\rho g H^2} + \delta_{\alpha\beta} \frac{S_p}{\rho g H^2} \tag{11.6.23}$$

Sometimes it is convenient to describe $\sin \alpha_w$ and $\cos \alpha_w$ by the components of the wave number vector k_α. We have

$$k_\alpha = (k_x \ , \ k_y) = \mathbf{k} \tag{11.6.24}$$

and defining

$$k^2 = k_\alpha \ k_\alpha \tag{11.6.25}$$

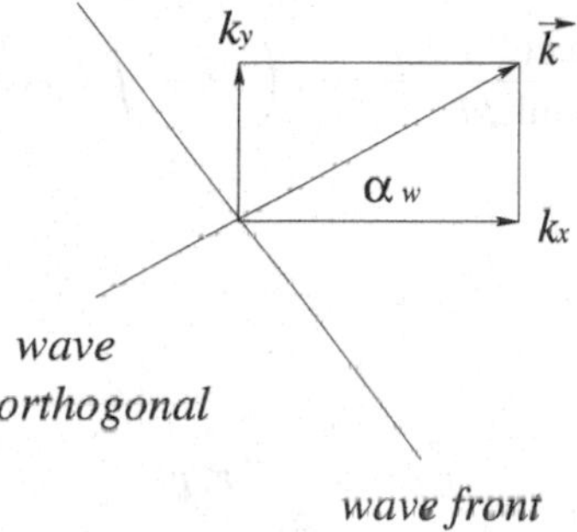

Fig. 11.6.2 The wave number vector **k**.

this means

$$\begin{aligned} k_x &= k \cos \alpha_w \\ k_y &= k \sin \alpha_w \end{aligned} \tag{11.6.26}$$

or

$$(\cos \alpha_w \ , \ \sin \alpha_w) = (k_x \ , \ k_y) \, / k \tag{11.6.27}$$

so that $e_{\alpha\beta}$ may be written

$$e_{\alpha\beta} = (k_\alpha \ k_\beta) \, / k^2 \tag{11.6.28}$$

Again all these results are exact and apply to any periodic wave motion.

11.6.2 *Radiation stress for linear waves in two horizontal dimensions (2DH)*

In Section 3.3 we derived the expression for the normal component S_{xx} of the radiation stress, provided the wave motion is described by linear wave theory. Using the results of the previous section, we can now directly extend these results to the general radiation stress for linear waves. The situation we consider is shown in Fig. 11.6.3.

For reference, it is worth first to recall the features of the locally plane linear wave motion we are talking about. It has a wave number vector $\mathbf{k} = k_\beta$ given by (11.6.24) with $k = 2\pi/L$ given by (11.6.25). We assume the wave height is H, phase velocity is c and frequency $\omega = 2\pi/T$. Then the velocity potential ϕ is represented by (invoking the locally-constant-depth assumption).

$$\phi = -\frac{H}{2}c\,\frac{\cosh k(z+h_0)}{\sinh kh}\,\sin\left(\omega t - \int k_\beta\, dx_\beta\right) \tag{11.6.29}$$

where

$$k_\beta\, dx_\beta = \mathbf{k}\cdot\mathbf{dx} \tag{11.6.30}$$

with

$$x_\beta = \mathbf{x} = (x, y) \tag{11.6.31}$$

being the horizontal position vector for the point considered (see Fig. 11.6.3). The phase velocity c is determined through the (constant depth) dispersion relationship

$$\omega^2 = gk \tanh kh \tag{11.6.32}$$

where $k = |\,k_\alpha\,|$ and

$$c = \omega/k \tag{11.6.33}$$

The horizontal velocities $u_\beta(= (u, v))$ are then obtained by differentiation of (11.6.29) which yields

$$u_\beta = \frac{H}{2}\,c\,k_\beta\,\frac{\cosh k(z+h_0)}{\sinh kh}\,\cos\left(\omega t - \int k_\beta\, dx_\beta\right) \tag{11.6.34}$$

which clearly is a vector in the direction of k_β, that is, it has x and y components that are obtained from (11.6.34) by inserting the x and y components of k_β.

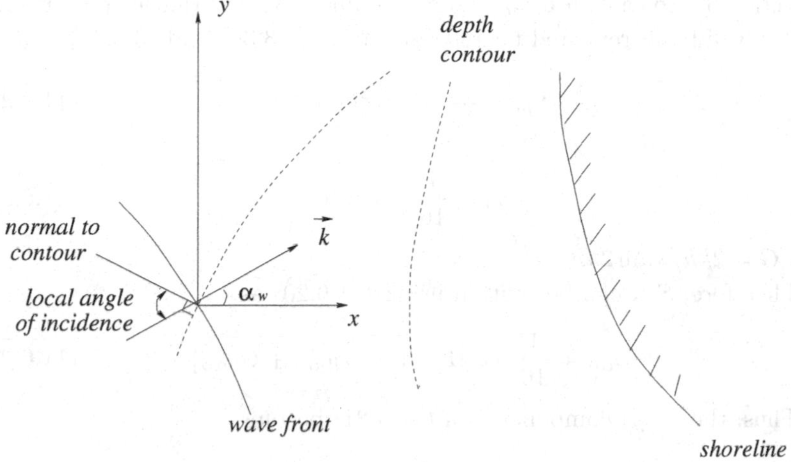

Fig. 11.6.3 Wave fronts, orthogonals and depth contours on an arbitrary topography. Note the anlge of incidence α is not equal to the wave angle α_w.

Similarly, the vertical velocity w becomes

$$w = \frac{H}{2} ck \frac{\sinh k(z + h_0)}{\sinh kh} \sin \left(\omega t - \int k_\beta dx_\beta \right) \tag{11.6.35}$$

From (11.6.34), however, we can also calculate $u = |u|$ which is the total horizontal velocity in the direction of k_β. Since by (11.6.28) $(k_\beta\, k_\beta)^{1/2} = k$ we have

$$u = (u_\beta\, u_\beta)^{1/2} = \frac{H}{2} ck \frac{\cosh k(z + h_0)}{\sinh kh} \cos \left(\omega t - \int k_\beta\, dx_\beta \right) \tag{11.6.36}$$

which of course is equivalent to the expression from 1DH linear wave theory except $\int k_\beta\, dx_\beta$ has replaced kx.

Comment: Essentially, all these results are equivalent to originally considering a plane wave in the direction of one of the coordinate axes (x, say) in an x, y, z-system, and then turn the x, y axes to form an angle with the wave fronts. This rotation about the z-axis of course does not change w which is reflected in the fact that w is not a function of k_β but of k only (since $\int k_\beta\, dx_\beta$ is a scalar that remains invariant at the rotation).

It is now straightforward to establish the expressions for $S_{\alpha\beta}$. We first notice that (11.6.16) and (11.6.17) are equivalent to the expressions used in Section 3.3 to calculate S_m and S_p for linear waves. Hence those results are still valid and repeated for reference from (3.3.30) and (3.3.35)

$$S_m = \frac{1}{16} \rho g H^2 (1 + G) \tag{11.6.37}$$

$$S_p = \frac{1}{16} \rho g H^2 G \tag{11.6.38}$$

with $G = 2kh/\sinh 2kh$.

Therefore, $S_{\alpha\beta}$ can be written using (11.6.20) as

$$S_{\alpha\beta} = \frac{1}{16} \rho g H^2 \left[(1 + G) e_{\alpha\beta} + G \delta_{\alpha\beta}\right] \tag{11.6.39}$$

Thus, the (x, y)-components in (11.6.21) become

$$S_{xx} = \frac{1}{16} \rho g H^2 \left[(1 + G) \cos^2 \alpha_w + G\right] \tag{11.6.40}$$

$$S_{xy} = S_{yx} = \frac{1}{16} \rho g H^2 \cos \alpha_w \sin \alpha_w (1 + G) \tag{11.6.41}$$

$$S_{yy} = \frac{1}{16} \rho g H^2 \left[(1 + G) \sin^2 \alpha_w + G\right] \tag{11.6.42}$$

Alternatively using (11.6.28) for $e_{\alpha\beta}$ we get

$$S_{\alpha\beta} = \frac{1}{16} \rho g H^2 \left[(1 + G) \frac{k_\alpha k_\beta}{k^2} + G \, \delta_{\alpha\beta}\right] \tag{11.6.43}$$

From (11.6.23) we see that the dimensionless radiation stress $P_{\alpha\beta}$ becomes

$$\begin{aligned} P_{\alpha\beta} &= \frac{1}{16} \left[(1 + G) e_{\alpha\beta} + G \delta_{\alpha\beta}\right] \\ &= \frac{1}{16} \left[(1 + G) \frac{k_\alpha k_\beta}{k^2} + G \, \delta_{\alpha\beta}\right] \end{aligned} \tag{11.6.44}$$

The function G is sometimes expressed in terms of the ratio of group velocity to phase velocity c_g/c (see (3.3.16)). Thus **for linear waves only** $S_{\alpha\beta}$ can also be written

$$S_{\alpha\beta} = \frac{E}{2} \left[\frac{2c_g}{c} \, e_{\alpha\beta} + \left(\frac{2c_g}{c} - 1\right) \delta_{\alpha\beta}\right] \tag{11.6.45}$$

$$= \frac{E}{2} \left[\frac{2c_g}{c} \, \frac{k_\alpha k_\beta}{k^2} + \left(\frac{2c_g}{c} - 1\right) \delta_{\alpha\beta}\right] \tag{11.6.46}$$

where $E = \frac{1}{8} \rho g H^2$.

For completeness, it is emphasized that the angle α_w in Fig. 11.6.3 is the angle between the wave front and the x-axis. As the figure indicates, this is *not* the "angle of incidence" for the wave in a refraction sense.

11.7 Examples on a long straight beach

11.7.1 *The momentum balance*

In order to illustrate the nature of the wave generated nearshore circulation it is instructive to consider two examples, which can be used to gain insight into how the equations work. They are both for a long straight beach. The first is the simple cross-shore momentum balance for normally incidence waves, the second analyzes the longshore current generated by obliqiely incident waves.

On a long, straight (or 2D) beach, depth contours are straight and parallel to the shoreline and all cross-shore profiles are equal. (A special 2D beach is the plane beach where the bottom slope is the same at all points.)

The traditional simplification (the validity of which we will examine further in a later chapter) is to consider only depth uniform currents. Then we can use the momentum equation in the form (11.5.20).

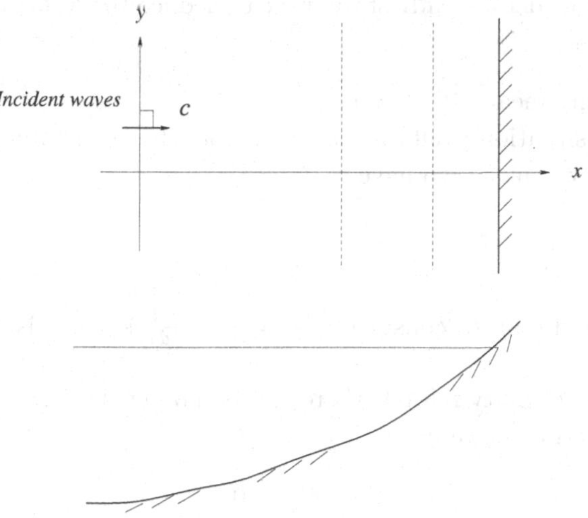

Fig. 11.7.1 A long straight (or "cylindrical") beach with normal incident waves.

The x-component becomes:

$$\rho\frac{\partial}{\partial t}\bar{Q}_x + \rho\frac{\partial}{\partial x_\beta}\left(\frac{\bar{Q_\beta}\,\bar{Q_x}}{h}\right)$$
$$= -\rho g(\bar{\zeta}+h_0)\frac{\partial\bar{\zeta}}{\partial x} + \frac{\partial}{\partial x_\beta}\left[-S_{\beta x} + \overline{\int_{-h_0}^{\zeta}\tau_{\beta x}dz}\right] - \overline{\tau_x^B} \quad (11.7.1)$$

and similarly for the y-component:

$$\rho\frac{\partial}{\partial t}\bar{Q}_y + \rho\frac{\partial}{\partial x_\beta}\left(\frac{\bar{Q_\beta}\,\bar{Q_y}}{h}\right)$$
$$= -\rho g(\bar{\zeta}+h_0)\frac{\partial\bar{\zeta}}{\partial y} + \frac{\partial}{\partial x_\beta}\left[-S_{\beta y} + \overline{\int_{-h_0}^{\zeta}\tau_{\beta y}dz}\right] - \overline{\tau_y^B} \quad (11.7.2)$$

In addition the depth integrated, time averaged continuity equation (11.2.19) is valid

$$\frac{\partial\bar{\zeta}}{\partial t} + \frac{\partial\overline{Q_\alpha}}{\partial x_\alpha} = 0 \quad (11.7.3)$$

In (11.7.1) and (11.7.2) the surface force terms R^s_β have been assumed equal to zero in accordance with the assumption that wind stresses are neglected.

When we introduce the above mentioned constraint of a 2D beach, and add that we consider a steady state, we can deduce the following about the flow:

(1) Steady state means $\overline{\partial/\partial t} = 0$
(2) Since the situation would be the same for all cross-shore profiles along the coast, we must also have

$$\frac{\partial}{\partial y} = 0 \quad (11.7.4)$$

i.e., we need only to consider $\frac{\partial}{\partial x_\beta} = \frac{\partial}{\partial x} = \frac{d}{dx}$ since x is now the only variable.
(3) Then for continuity reasons there can be no net-discharge in the cross-shore direction so we have

$$\bar{Q_x} = 0 \quad (11.7.5)$$

These simplifications will be used below in formulating the final version of the equations for the two examples.

First, however, it is recalled that in these equations the radiations stress and (hidden behind $\overline{Q_\alpha}$) the volume flux $Q_{w\alpha}$ for the waves are the wave averaged signatures of the short wave motion and these contributions represent the wave forcing for the currents and variation of the MWS $\bar{\zeta}$. These driving forces are assumed known and the unknowns in the equations are $\overline{Q_\alpha}$ and $\bar{\zeta}$.

11.7.2 The cross-shore momentum balance: Setdown and setup

We first consider the cross-shore momentum balance in the simple case of waves normal to the shore, i.e. $\alpha_w = 0$.

The simplifications listed above then reduce (11.7.1) to the following (with $h = h_0 + \bar{\zeta}$):

$$0 = -\rho g\, h \frac{d\bar{\zeta}}{dx} + \frac{d}{dx}\left(-S_{xx} + \overline{\int_{-h_0}^{\zeta} \tau_{xx} dz}\right) - \tau_{xb} \tag{11.7.6}$$

Comment: It is also worth to notice that we might think there is a longshore current Q_y. However, with normally incident waves we have

$$S_{\alpha x} = (S_{xx}\ ,\ S_{yx}) = (S_{xx}\ ,\ 0) \tag{11.7.7}$$

Hence there is no driving force for such a current in the y-direction and, as we will see later, this means the longshore bottom friction component which a current would cause would make it die out quickly.

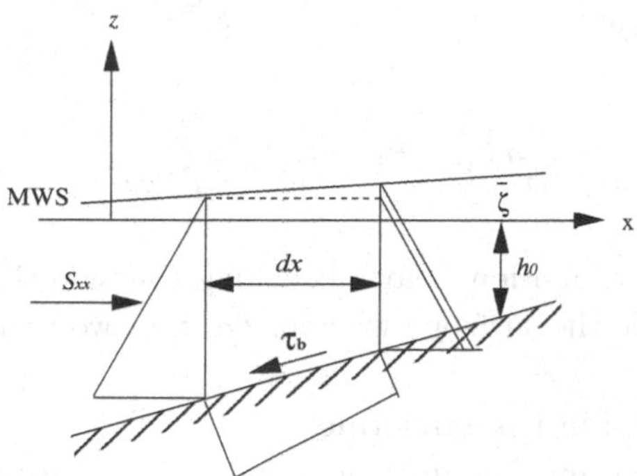

Fig. 11.7.2 The controle volume for the simple cross-shore momentum balance of equation (11.7.8). The triangles indicate the pressure forces.

This also means that all terms in the continuity equations vanish.

In (11.7.6) the τ_{xx}-term represents normal stresses on vertical surfaces x=const due to the turbulence (and viscosity). These terms are normally neglected because they are small in comparison to the other terms in (11.7.6), and we will do so here.

Hence the general Eq. (11.7.1) reduces to the expression

$$\boxed{\frac{dS_{xx}}{dx} = -\rho\, g\, h \frac{d\bar{\zeta}}{dx} - \overline{\tau_{xb}}} \qquad (11.7.8)$$

Fig. 11.7.2 illustrates this force balance. It is emphasized that with the present derivation (11.7.8) applies to any type of wave motion under the conditions listed above. This includes breaking waves.

Notice also that the assumption of no mean current in the cross-shore direction does *not* justify the assumption that the bottom friction $\overline{\tau_x^B}$ vanishes (as claimed e.g., by Mei, p. 463). As we will see later, there is a cross shore circulation, which cannot be described by the depth integrated equations considered here and the mean bottom friction is linked to the current at the bottom, not to the (zero) mean current. Here we will neglect $\overline{\tau_{xb}}$ because it turns out to be less than 5% of the other two terms in the equation.

We now turn to ways of assessing the terms in (11.7.8) inside the surf zone, in particular the radiation stress S_{xx}, and subsequently to integrating this equation.

Without loss of generality we can introduce (11.6.22) for S_{xx} into (11.7.8). We then get

$$\frac{d}{dx}(P_{xx} H^2) = -(h_0 + \bar{\zeta}) \frac{d\bar{\zeta}}{dx} \qquad (11.7.9)$$

Because the wave motion before breaking (outside the surfzone) is so different from inside the surfzone we consider the two regions separately.

The solution outside the surfzone

In order to determine the driving forces $dS_x x/dx$ we must use a wave model to determine the S_{xx}. On the sufficiently gently sloping bottom assumed here the locally constant depth assumption makes the linear wave theory a valid approximation outside the surfzone.

Disregarding losses from bottom friction etc. in the wave motion the energy flux $E_f = E_f(h, H)$ is conserved so that

$$\frac{d\,E_f}{dx} = 0 \tag{11.7.10}$$

In principle (11.7.10) is then solved to provide information about the wave height variation, and at each depth the value of P_{xx} is given by the expression (11.6.44)

$$P_{xx} = \frac{1}{16}\,(1 + 2G) \tag{11.7.11}$$

where $G = 2kh/\sinh\, 2kh$.

As can be seen (11.7.9) is a nonlinear equation in $\overline{\zeta}$ and hence we should expect it to be impossible to solve it analytically. This is not the case, however. The following method was developed by Longuet-Higgins and Stewart (1964).

Exercise 11.7-1

In order to integrate (11.7.9) consider the two compponents S_m and S_p of the radiation stress defined by (11.6.37) and (11.6.38). Then using the linear wave expressions and (11.7.10) first show that

$$\frac{d}{dx}(S_m c) = 0 \tag{11.7.12}$$

Then also show that

$$\frac{dc}{dh} = \frac{c}{h}\,\frac{G}{1+G} \tag{11.7.13}$$

Third show that (11.7.12) also implies that

$$\frac{dS_m}{dx} = -\frac{S_p}{h}\,\frac{dh}{dx} \tag{11.7.14}$$

Then substituting these results into (11.7.9) show that this modifies that equation to the form

$$\frac{d}{dx}\left(\frac{S_p}{h}\right) = -\rho g \frac{d\overline{\zeta}}{dx} \tag{11.7.15}$$

which can be integrated to give

$$\overline{\zeta} = -\frac{h^2}{16h}\,G + C \tag{11.7.16}$$

valid for the non-breaking waves outside the surfzone.

It is first noticed that for infinitely deep water this gives

$$\overline{\zeta} = C \tag{11.7.17}$$

Therefore, choosing $\overline{\zeta} = 0$ at deep water means $C = 0$. $\overline{\zeta} = 0$ at deep water implies that the horizontal x-axis (which represents the reference level $z = 0$) is now positioned at the deep water MWL. This is the most convenient choice and **this position is used everywhere in this text in the discussions of nearshore circulation**.

We also see that relative to deep water $\overline{\zeta}$ is < 0, meaning as the water depth decreases toward the breaking point the height of the MWL decreases and it reaches its lowest point at the breaking point. This is called the **setdown**.

Exercise 11.7-2

Verify the following statements: Waves in a storm may easily reach a height of $H = 5m$ on a depth of $h = 10m$ and with a wave period of $T = 8s$. This implies a deep water wave length of $L_0 = \frac{gT^2}{2\pi} \cong 100m$ which will correspond to $L = 71m$ on $h = 10m$. Thus we have $G = 0.62$ and we get $\underline{\bar{\zeta} = \text{-9.7 cm.}}$ Thus even in a storm with very large waves the setdown is only a few cm.

The solution inside the surfzone

The variations of the mean water surface MWS after breaking, i.e. the conditions inside the surfzone were first analysed by Bowen (1968).

As before the variation of H and P_{xx} must be assessed independently. Inside the surfzone we do not use sinewave theory. Instead, to circumvent the need for establishing a wave model to determine H, we assume that in the breaking zone H is a constant fraction γ of the local depth h, so we let

$$H = \gamma h \tag{11.7.18}$$

The value of P_{xx} essentially represents the shape of the wave motion in the widest sense (surface profile, velocity field, pressure variation), i.e., the entire phase motion. And that is information which we loose when we average over a wave period. For the present purpose we simply assume that

$$P_{xx} = \text{const.} = P \tag{11.7.19}$$

where we can use empirical information for P (see e.g. Section 5.5).

Substituting these simplifications into (11.7.9) that equation becomes

$$\gamma^2 P \frac{dh^2}{dx} = -(h_0 + \bar{\zeta}) \frac{d\bar{\zeta}}{dx} \tag{11.7.20}$$

and with $h = h_0 + \bar{\zeta}$, we get by solving with respect to $d\bar{\zeta}/dx$:

$$\frac{d\bar{\zeta}}{dx} = -\frac{2P\gamma^2}{1 + 2P\gamma^2} \frac{dh_0}{dx} \tag{11.7.21}$$

or

$$\bar{\zeta} = -A \int_{x_0}^{x} h_{ox} dx + \bar{\zeta}(x_0) \tag{11.7.22}$$

where h_{ox} is short for dh_0/dx and

$$A = \frac{2P\gamma^2}{1 + 2P\gamma^2} \tag{11.7.23}$$

This results in a water level which increases shoreward from the breaking point. It is called the wave **setup**.

Exercise 11.7-3

Typical values for the wave height-to-water depth-ratio γ and P in the surf zone are $\gamma = 0.6$; $P = 0.20$, respectively. Show that this results in

$$A = 0.126 \tag{11.7.24}$$

i.e.: The slope on the mean water level is 12.6% of the bottom slope.

Analysis of a large number of measurements on a coast (Longuet-Higgins and Stewart, 1963) show that the slope on the mean water level is of the order

$$\frac{d\bar{\zeta}}{dx} \sim -0.15 h_{ox} \tag{11.7.25}$$

though the scatter of the data naturally is substantial.

Plane beach, h_{ox} = const.

For a plane beach calculation of the integral in (11.7.22) is straight forward and yields

$$\int_{x_0}^{x} h_{0x} dx = h_0(x) - h_0(x_0) \qquad (11.7.26)$$

and hence

$$\bar{\zeta}(x) - \bar{\zeta}(x_0) = -\frac{2P\gamma^2}{1+2P\gamma^2}\left(h_0(x) - h_0(x_0)\right) \qquad (11.7.27)$$

If we have $x_0 = x_B$ at the breaking point (11.7.27) becomes

$$\bar{\zeta} - \bar{\zeta}_B = \frac{2P\gamma^2}{1+2P\gamma^2}(h_0(x_B) - h_0(x)) = A(h_0(x_B) - h_0(x)) \qquad (11.7.28)$$

that is, a linear variation of the MWS shoreward of the breaking point. We see that using the value of $A = 0.126$ found above this result means that at the point of the initial mean shoreline where $h_0 = 0$ we have a setup of $\bar{\zeta} = 0.126 h_B$, and with a breaker depth of $h_B = 5\ m$ as in the example earlier that means a setup of $\bar{\zeta} = 0.65m$ at the initial shoreline $x = x_s$. Thus the setup is typically much larger than the setdown.

Exercise 11.7-4

This substantial setup, however, moves the actual shoreline a substantial distance shoreward to position we can call x_r. Show that on a plane beach the distance $x_r - x_s$ from the shoreline defined by the MWL to the actual setup-defined shoreline can be written

$$x_r - x_s = \frac{A}{A + h_{0x}}(x_s - x_B) \qquad (11.7.29)$$

Show that if $A = 0.126$ as in the example above and $h_{0x} = -0.2$ this means the shoreline moves the distance $x_r - x_s \sim 1.7(x_s - x_B)$ shoreward, that is a distance which is 1.7 times the surfzone width. Similarly show that for $h_{0x} = -0.3$ the distance will be 0.72 times the surfzone width.

Clearly this is a very crude model. However, it contains essential elements of the real world such as that the set-up is much larger than the set-down before breaking.

Fig. 11.7.1 shows measurements of wave setdown and setup on a plane, but very steep, beach (1:12) by Bowen et al.(1968). They were some of the first lab measurements published showing the mean water variations on a beach. As (11.7.28) suggests, we should expect a constant slope of the mean water surface in this case and the experiments seem to confirm this. Later, more accurate and detailed experiments have shown, however, that this is an over-simplification originating from the two assumptions (11.7.19) and (11.7.18) which do not hold in general and in some cases are quite inaccurate.

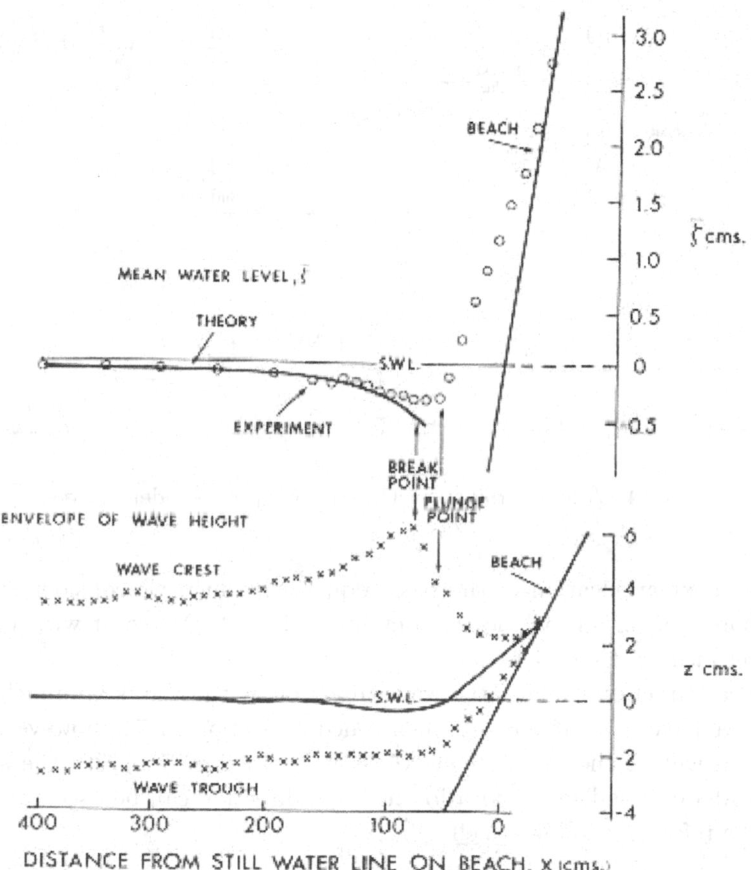

Fig. 11.7.3 Comparison of experiments with theory for set-down and set-up on a plane beach. Data: wave period = 1.14 s; deep water wave height $H_\alpha = 6.45$ cm; breaker height $H_b = 8.55$ cm; beach slope = 0.082 (from Bowen et al., 1968).

11.7.3 *Longshore currents*

Introductory analysis of the problem

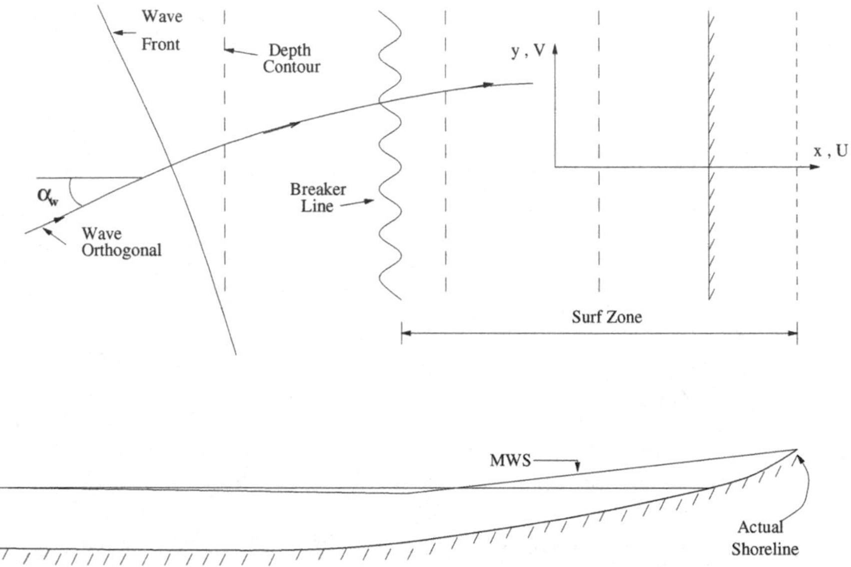

Fig. 11.7.4 A long straight beach with obliquely incident waves.

The next application of the basic equations which we consider is the longshore momentum balance on a cylindrical (or "2D") coast with waves incident at an angle α_w.

Since refraction changes the wave direction as the waves approach the shore, we have $\alpha_w = \alpha_w(x)$. As mentioned in Section 11.7.2, however, the situation will be the same at all points in the y direction with the same x-coordinate. The lack of variation in the y-direction can be used to show that the refraction follows Snell's law

$$\frac{\sin \alpha_w}{c} = \text{const.} \tag{11.7.30}$$

where c is the phase speed for the wave.

It is worth to recall that Snell's law is entirely based on the congruence of the wave orthogonals in the y-direction. Hence, Snell's law is valid for any wave type, linear or nonlinear, breaking or non-breaking, only provided they are regular with constant wave period T. This is often overlooked because Snell's law is often derived in the context of linear wave theory. As shown in Section 3.5.2 for pure wave motion Snell's law applies exactly on any cylindrical beach (i.e. longshore uniform topography), regardless of coastal profile and the shape of the waves. Similarly for waves with longshore currents Snell's law holds if we use the absolute phase velocity c_a.

Exercise 11.7-5

Show that for the situation with waves incident at an angle the cross-shore momentum balance given by (11.7.9) formally remains the same. The influence of the angle of incidence α_w shows up only in the value of P_{xx}.

Determine also that P-value assuming linear wave theory.

Introducing the assumption (11.7.4) and the assumption of stady state $(\overline{\partial/\partial t} = 0)$, the longshore (i.e., y) component of the momentum equation (11.7.2) can be written

$$\boxed{-\frac{d}{dx}S_{xy} + \frac{d}{dx}\overline{\int_{-h_0}^{\zeta} \tau_{xy}\, dz} - \tau_y^B = 0} \qquad (11.7.31)$$

i iii ii

We see that this equation says that on a long straight coast, the longshore momentum variation is a balance between three forces (or processes).

(1) The cross-shore rate of variation of the longshore radiation stress component S_{xy} — which acts as a driving force for currents.
(2) The bottom shear stress τ_y^B, which restrains the currents but which is zero unless there is a current.
(3) The rate of change of the turbulent shear stresses — which act as a distributing (or "dispersion") mechanism that transfers $\partial S_{xy}/\partial x$ driving forces in the cross-shore direction. This mechanism is also called **lateral mixing**.

The figure 11.7.5 shows the situation. Note that the arrows indicate the positive directions for each of the forces shown.

Hence, it is evident that in the longshore direction the momentum balance can only be achieved if there is a current (the "longshore current")

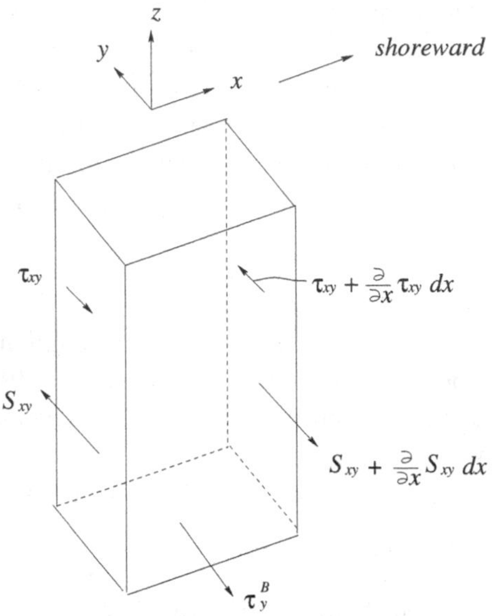

Fig. 11.7.5 The radiation and turbulent shear stresses acting in the longshore (y-direction) on a water column.

that creates a bottom shear stress, because there is no MWS-variation in the longshore direction to counteract the driving radiation stress component (as was the case for the cross-shore momentum balance).

Analysis of each of the terms in (11.7.31)

The equation (11.7.31) for the momentum balance in the y-direction was first solved by Thornton (1970) and later by Longuet-Higgins (1970a,b) (LH70 in the following). The following is a modified version of Longuet-Higgins' solution, using our nomenclature and the results from previous Chapters.

Though it cannot be seen in the present form (11.7.31) actually represents an equation that describes the longshore current. To expose that directly we need to replace the symbols for S_{xy}, τ_{xy} and τ_{yb} with expressions that relate these quantities to the waves and to the shear stresses generated by the current.

The value of $\partial S_{xy}/\partial x$ outside the breaking point

We first show the remarkable result that under fairly general conditions $\partial S_{xy}/\partial x = 0$ outside the breaking point.

In the direction of wave propagation, we have the following exact expression for the wave energy flux (see Section 3.3)

$$E_f = \overline{\int_{-h_0}^{\zeta} \left[p_D + \frac{1}{2}\rho\left(u_w^2 + w_w^2\right) \right] u_w dz} \tag{11.7.32}$$

In the general 3-D case, the vector property E_f can be written, using tensor notation

$$E_{f\beta} = \int_{-h_0}^{\zeta} \left[p_D + \frac{1}{2}\rho\left(u_{w\alpha}u_{w\alpha} + w_w^2\right) \right] u_\beta dz \tag{11.7.33}$$

For the waves with orthogonals at an angle α_w with the x direction this means

$$u_x = u\cos\alpha_w \quad ; \qquad u_y = u\sin\alpha_w \quad ; \qquad u^2 = u_\alpha u_\alpha \tag{11.7.34}$$

so that

$$E_{fx} = E_f \cos\alpha_w \tag{11.7.35}$$

Since the wave motion is a potential flow, the Bernoulli equation applies

$$\frac{\partial\phi}{\partial t} + \frac{p}{\rho} + z + \frac{1}{2}\left(u_{w\alpha}u_{w\alpha} + w_w^2\right) = \text{const} \tag{11.7.36}$$

or with $p_D = p - \rho g(b - z)$ introduced and the constant adjusted accordingly to zero this means

$$p_D + \frac{1}{2}\rho\left(u_{w\alpha}u_{w\alpha} + w_w^2\right) = -\rho\frac{\partial\phi}{\partial t} \tag{11.7.37}$$

which substituted into $E_{f\beta}$ yields

$$E_{f\beta} = -\overline{\int_{-h_0}^{\zeta} \rho\frac{\partial\phi}{\partial t} u_\beta dz} \tag{11.7.38}$$

We also assume that the waves are progressive and described by a velocity potential ϕ of the form

$$\phi(x, y, z, t) = \phi\left(z, \left(\omega t - \int k_x dx - k_y y\right)\right) \tag{11.7.39}$$

Therefore, we have

$$\frac{\partial \phi}{\partial t} = -\frac{\omega}{k_y}\frac{\partial \phi}{\partial y} = \frac{-c}{\sin \alpha_w}\frac{\partial \phi}{\partial y} \tag{11.7.40}$$

Substituting into (11.7.38) then leads to

$$E_{f\beta} = \overline{\frac{c}{\sin \alpha_w}\int_{-h_0}^{\zeta} \rho \frac{\partial \phi}{\partial y} u_\beta dz} \tag{11.7.41}$$

For the x-component and with $\frac{\partial \phi}{\partial y} = v$ this becomes

$$E_{fx} = \frac{c}{\sin \alpha_w}\int_{-h_0}^{\zeta} \rho uv dz = \frac{c}{\sin \alpha_w} S_{xy} \tag{11.7.42}$$

By Snell's law $c/\sin \alpha_w$ is constant on a cylindrical coast for any long-crested monochromatic wave, and since E_{fx} on a long straight coast with $\partial/\partial y = 0$ is constant too outside the surf zone, we get

$$\frac{\partial E_{fx}}{\partial x} = 0 = \frac{c}{\sin \alpha_w}\frac{\partial S_{xy}}{\partial x} \tag{11.7.43}$$

which shows that $\partial S_{xy}/\partial x = 0$ for non-breaking, progressive, periodic waves of any height and shape. Note that (11.7.43) does not assume linear wave theory.

Comparing with momentum equation (11.7.31) we see that this means that outside the breaking point there is no driving force for the longshore current even though the radiation stress is changing there too.

The radiation stress term

The solution we are talking about assumes that the wave motion can be described by the linear sine-wave theory both before and after breaking. Hence, we have from Section 11.6.2

$$S_{xy} = \frac{1}{16}\rho\, g\, H^2 \cos \alpha_w \sin \alpha_w (1 + G) \tag{11.7.44}$$

Outside the breaker line

Outside the breakerline we have already found that

$$\frac{\partial S_{xy}}{\partial x} = 0 \qquad x_B > x \tag{11.7.45}$$

In other words, and relating to the analysis above, this means that outside the breaker line there are no driving forces in the longshore direction.

Inside the breaker line

In this simplified analysis it is assumed that the waves are linear long waves so that

$$G = 1 \quad ; \quad c = \sqrt{gh} \tag{11.7.46}$$

We also assume that we are dealing with saturated breakers which means

$$H = \gamma h \tag{11.7.47}$$

Substituted into (11.7.44) this yields

$$S_{xy} = \frac{1}{8}\,\rho\, g\, \gamma^2\, h^2 \sin\alpha_w \cos\alpha_w \tag{11.7.48}$$

Here $h = h_0 + \overline{\zeta}$.

Using again Snell's law, and assuming α_w small enough to neglect the variation of $\cos\alpha_w$, (11.7.48) gives inside the surf zone

$$\frac{dS_{xy}}{dx} = \frac{5}{16}\,\rho\, g^{3/2}\, \gamma^2\, h^{3/2}\, h_x \left(\frac{\sin\alpha_w}{c}\right)_0 \tag{11.7.49}$$

where index $_0$ indicates a reference value of the Snell-const (e.g. deep water).

The bottom friction term

Clearly one of the crucial elements in the balance is the bottom friction term in (11.7.31). We are dealing with bottom friction due to a water motion which is a combination of waves and currents. This problem was analysed in Section 10.4.3. The conditions assumed in this case are that

(1) $\alpha_w \sim 0$, i.e., waves are, for the computation of bottom friction, normal incident on the beach.
(2) $V \ll u_0$, i.e., the current is weak relative to the wave particle motion.

Thus we have that the angle μ between the waves and the current is

$$\mu = \pi/2 \tag{11.7.50}$$

This corresponds to one of the cases analysed in Section 10.4.3 where it was found that $\overline{\tau_{yb}}$ is given by (10.4.33)

$$\overline{\tau_{yb}} \cong \frac{1}{\pi}\, \rho\, f\, u_0\, V \tag{11.7.51}$$

In the case of sinusoidal shallow water waves assumed here, we also have

$$u_0 = c\, \frac{H}{2h} = \sqrt{gh}\, \frac{H}{2h} = \frac{1}{2}\, \gamma \sqrt{gh} \tag{11.7.52}$$

so that

$$\overline{\tau_{yb}} \cong \frac{1}{2\pi}\, \rho\, f\, V\, \gamma \sqrt{gh} \tag{11.7.53}$$

which is a shear stress in the direction of the current.

The lateral mixing term

As mentioned, the turbulent shear stresses τ_{xy} in (11.7.31) represent a distributing mechanism which is often modelled by an eddy viscosity ν_t. Thus we assume

$$\tau_{xy} = \rho\, \nu_t\, \frac{dV}{dx} \tag{11.7.54}$$

which means that

$$\overline{\int_{-h_0}^{\zeta} \tau_{xy} dz} = \rho\, \nu_t \left(\bar{\zeta} + h_0\right) \frac{dV}{dx} \tag{11.7.55}$$

By dimensional considerations

$$\nu_t \propto U L \tag{11.7.56}$$

where U and L are a characteristic velocity and a characteristic length, respectively. Longuet-Higgins chose

$$U \sim (gh)^{1/2}\,, \quad L \sim x \tag{11.7.57}$$

whence we get

$$\nu_t = N\, x\, \sqrt{gh} \tag{11.7.58}$$

where N is supposed to be nearly constant. This assumption will be discussed later in Section 12.2 and Chapter 13 and found to to be less realistic than a choice based on $L \sim h$.

It is convenient to non-dimensionalize the eddy viscosity by $h\sqrt{gh}$ so we get in dimensionless form

$$\frac{\nu_t}{h\sqrt{gh}} = \frac{N}{h_x} \tag{11.7.59}$$

Inserted into (11.7.55) together with (11.7.63), this yields

$$\overline{\int_{-h_0}^{\zeta} \tau_{xy}\, dz} = \rho\, N\, g^{1/2}\, h_x^{3/2}\, x^{5/2}\, \frac{dV}{dx'} \tag{11.7.60}$$

11.7.4 *Longshore current solution for a plane beach*

From here on we will restrict the considerations to a plane bottom (h_{ox} = const) because in that case we can find an analytical solution to the problem, which helps the understanding.

Change of coordinate x to include setup at the shoreline

At the shoreline we found there is a significnt setup which means $x = 0$ is not the actual shoreline. However, since both the data mentioned earlier and the simple theory for $\bar{\zeta}$ inside the surfzone shows that on a plane bottom we have

$$\bar{\zeta}_x \approx C\, h_{ox} \qquad C = \text{const} \tag{11.7.61}$$

we may include the effect of set-up in the solution without further complication (although this was not done by Longuet-Higgins) by a simple change in the definition of x. Hence, we let (with x' the distance from the actual shoreline (see Fig. 11.7.6))

$$h = h_x\, x' \qquad h = h_0 + \bar{\zeta} \tag{11.7.62}$$

and use (11.7.21) to find

$$h_x = \frac{h_{ox}}{1 + 2P\gamma^2} \tag{11.7.63}$$

which can be used together with (11.7.53) to get

$$\tau_y^B \equiv \overline{\tau_{yb}} \simeq \frac{1}{2\pi}\, \rho\, f\, \gamma (g h_x\, x')^{1/2} V \tag{11.7.64}$$

Note that: $V > 0 \rightarrow \overline{\tau_{yb}} > 0$ which according to the sign convention means τ is <u>opposing</u> V. That has already been built into (11.7.31).

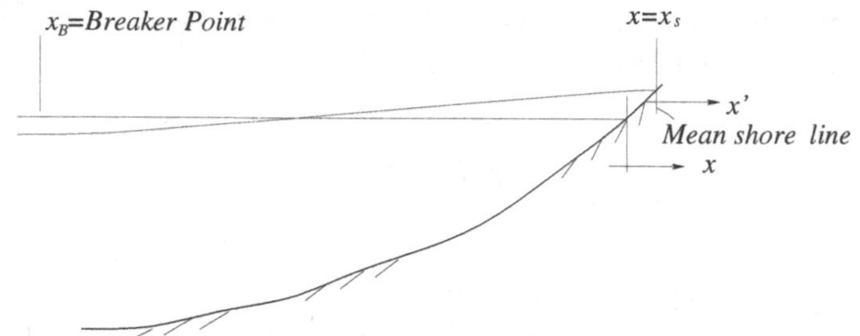

Fig. 11.7.6 Definition of the x and x' coordinates.

Substituting (11.7.49), (11.7.64) and (11.7.60) into (11.7.31) the entire equation may then be written (with $(\sin\alpha_w/c)_0 = C_0$)

$$N\,h_x^{3/2}\,\frac{d}{dx'}\left(x'^{5/2}\,\frac{dV}{dx'}\right) - \frac{f}{2\pi}\,\gamma\,h^{1/2}\,V$$
$$= \begin{cases} 0 & -\infty < x' < x'_B \\ \frac{5}{16}\,g\,\gamma^2\,h^{3/2}\,h_x\,C_0 & x_B < x' < 0 \end{cases} \tag{11.7.65}$$

where the RHS represents the forcing dS_{xy}/dx. There are two regions for the equation because outside the breaking point the forcing is zero.

The solution procedure

It is convenient to change the independent variable in (11.7.65) from x' to h. Then (11.7.65) can be written

$$P_N\frac{d}{dh}\left(h^{5/2}\frac{dV}{dh}\right) - h^{1/2}V = \begin{cases} 0 & h_B < h < \infty \\ -V_0h^{1/2} & 0 < h < h_B \end{cases} \tag{11.7.66}$$

where we have defined

$$P_N = -2\pi\frac{Nh_x}{\gamma f} \tag{11.7.67}$$

and

$$V_0 = -\frac{5}{8}\pi gh\frac{h_x\gamma}{f}\,C_0 \tag{11.7.68}$$

By writing (11.7.66) in the form

$$P_N h^2 \frac{d^2V}{dh^2} + \frac{5}{2} P_N h \frac{dV}{dh} - V = \begin{cases} 0 \\ -V_0 \end{cases} \tag{11.7.69}$$

it is seen that (11.7.66) is an "equidimensional" equation (see e.g., Greenberg, 1988 p. 899), (also called a (Cauchy-)Euler equation).

Special case: No lateral mixing

We first consider the special case of $N = 0$ i.e. $P_N = 0$, which corresponds to neglecting the turbulent or lateral mixing. Then (11.7.66) reduces to

$$V(\nu_t = 0) \equiv \begin{cases} 0 & h_B < h < \infty \\ V_0(h) & 0 < h < h_B \end{cases} \tag{11.7.70}$$

Fig. 11.7.7 shows the solution (11.7.70) which has a discontinuity at the breaker line and no longshore current at all outside the breaker line.

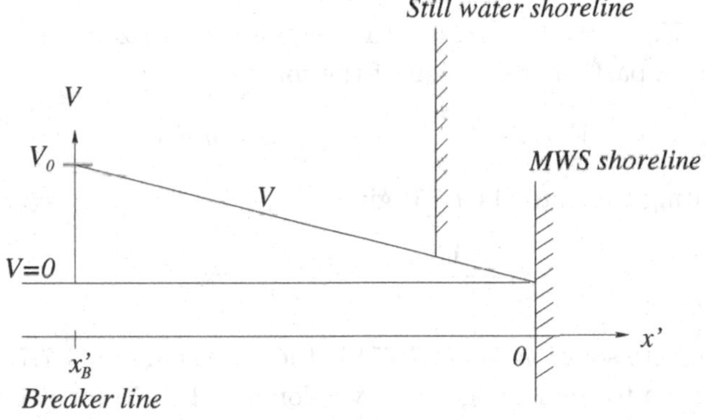

Fig. 11.7.7 The cross-shore variation of the longshore current if lateral mixing is neglected.

Solution with turbulent mixing

The solution is obtained using the boundary condition

$$V = 0 \quad \text{at} \quad x' = 0 \quad \text{(shoreline)} \tag{11.7.71}$$

and the matching condition that the velocity and shear stress are the continuous at the breaker line, i.e.,

$$V(x'_B+) = V(x'_B-) \tag{11.7.72}$$

$$\left(\frac{dV}{dx'}\right)_{x'_B+} = \left(\frac{dV}{dx'}\right)_{x'_B-} \tag{11.7.73}$$

The general solution to (11.7.68) can be found by the transformation of the independent variable h given by

$$h' = \frac{h}{h_0} = e^z \tag{11.7.74}$$

where h_0 is a reference depth. This changes (11.7.69) into an equation with constant coefficients.

Substituting (11.7.74) into (11.7.69) we get

$$P_N \frac{d^2V}{dz^2} + \frac{3}{2} P_N \frac{dV}{dz} - V = \begin{cases} 0 \\ -A_0 h_0 e^z \end{cases} \tag{11.7.75}$$

where $A_0 = V_0/h$. We first realize that inside the surf zone ($x_B < x' < 0$) (11.7.75) has a particular solution of the form

$$V = Ae^z (= Ah') \qquad (h_B < h < 0) \tag{11.7.76}$$

Substituting this into (11.7.75) gives

$$A = \frac{A_0 h_0}{1 - \frac{5}{2} P_N} \qquad P_N \neq \frac{2}{5} \tag{11.7.77}$$

The complete solution to (11.7.75) is then the sum of (11.7.76) and the general solution to the homogeneous version of (11.7.75) (or (11.7.70))

$$\frac{d^2V}{dz^2} + \frac{3}{2} \frac{dV}{dz} - \frac{V}{P_N} = 0 \tag{11.7.78}$$

The solution to this equation has the form

$$V = B_1 e^{p_1 z} + B_2 e^{p_2 z} \tag{11.7.79}$$

where p_1, p_2 are solutions to the characteristic equation

$$p^2 + \frac{3}{2} p - \frac{1}{P_N} = 0 \tag{11.7.80}$$

Solving this equation for p gives the two values

$$\left.\begin{aligned} p_1 &= -\tfrac{3}{4} + \left(\tfrac{9}{16} + \tfrac{1}{P_N}\right)^{1/2} \\ p_2 &= -\tfrac{3}{4} - \left(\tfrac{9}{16} + \tfrac{1}{P_N}\right)^{1/2} \end{aligned}\right\} \tag{11.7.81}$$

Substituting back the inverse of (11.7.74) then gives the homogeneous solution for V

$$V = B_1 h'^{p_1} + B_2 h'^{p_2}$$

Since $p_1 > 0$ and $p_2 < 0$ we only get bounded solutions in the two intervals if we choose $B_1 = 0$ outside the surf zone and $B_2 = 0$ inside so that the complete solution to (11.7.66) (or 11.7.70) becomes

$$V = \left\{\begin{matrix} B_1 h'^{p_1} + Ah' & 0 < h' < h'_B \\ B_2 h'^{p_2} & h'_B < h' < \infty \end{matrix}\right\} \tag{11.7.82}$$

The boundary conditions at $h = h_B$ are then satisfied by taking

$$B_1 = \frac{p_2 - 1}{p_1 - p_2} A \qquad B_2 = \frac{p_1 - 1}{p_2 - p_1} A \tag{11.7.83}$$

In (11.7.77) we had to exclude the case $P_N = 2/5$ which we see from (11.7.81) corresponds to

$$\begin{aligned} p_1 &= 1 \quad , \qquad V = B_1 h' + Ah' \\ p_2 &= -\frac{5}{2} \end{aligned} \tag{11.7.84}$$

Therefore for $P_N = 2/5$ the particular solution $const \cdot h'$ is also a solution to the homogeneous equation and hence cannot be used as a particular solution. An alternative to (11.7.76) is then

$$V = A\, z\, e^z \; (= Ah' \ln h') \tag{11.7.85}$$

which turns out to give $A = -\frac{5}{7}$ and the complete solution for $P_N = 2/5$ becomes

$$V = \left\{\begin{matrix} \frac{10}{49} h' - \frac{5}{7} h' \ln h' & 0 < h' < h'_B \\ \\ \frac{10}{49} h'^{-5/2} & h'_B < h' < \infty \end{matrix}\right\} \tag{11.7.86}$$

where the constants have again been chosen so that the solution satisfies the conditions (11.7.72) and (11.7.73). It also turns out that when $P_N \rightarrow$

2/5 then the solution given by (11.7.82) approaches the solution given by (11.7.86).

Fig. 11.7.8 shows the variation of the solution for different values of P_N, which is the only other parameter of the problem in addition to V_0. Clearly, the $P_N \neq 0$ solution is radically different from the solution with no lateral mixing in Fig. 11.7.7, first of all in that the longshore current is nonzero also in a region outside the breaker line.

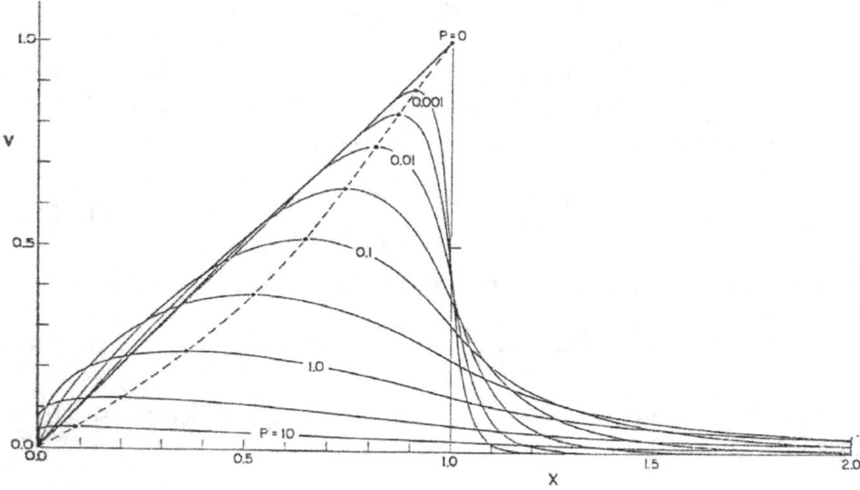

Fig. 11.7.8 Theoretical form of the longshore current V/V_b as a function of $X = x/x_b$ and the lateral mixing parameter $P = P_N$ (from Longuet-Higgins, 1970b).

The best fit value of P_N

Fig. 11.7.9 from Longuet-Higgins (1970b) shows one of the comparisons with laboratory measurements by Galvin and Eagleson (1965). It is seen that most of the eperimental results lie in the range of P_N-values for the solution described above of 0.1–0.4. We later (Chapter 13) want to compare these results with a case with a bottom slope of 1 : 35 and therefore we need the value of $C = N/h_x$ in (11.7.59). We see that for $\gamma = H/h = 0.6$, $f = 0.01$ and $h_x = 1 : 35$ we get that

$$\begin{aligned} P_N &= 0.1 \qquad \text{corresponds to } C \sim 0.45 \\ P_N &= 0.4 \qquad \text{corresponds to } C \sim 1.80 \end{aligned} \tag{11.7.87}$$

with a central fit corresponding to perhaps $P_N = 0.25$ or $C \sim 1.1$.

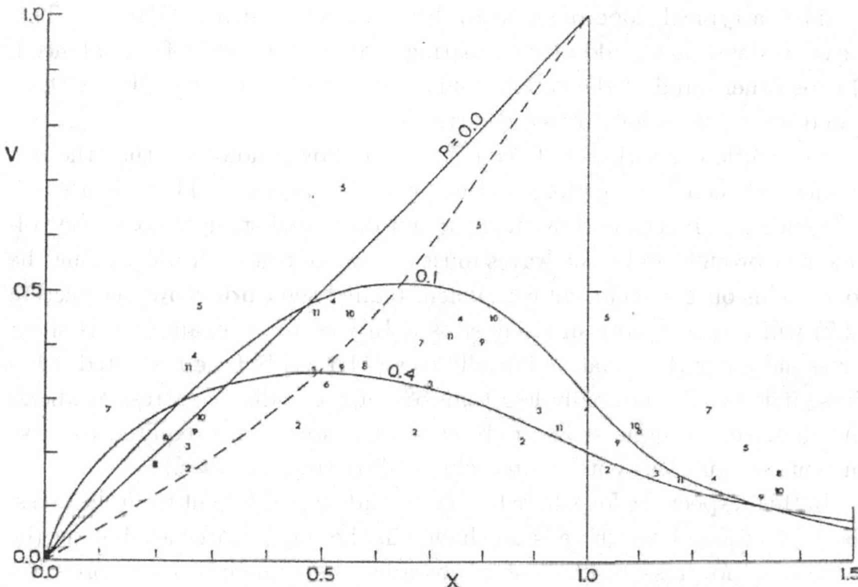

Fig. 11.7.9 Comparison between theoretical prediction of longshore current profile and laboratory measurements (from Longuet-Higgins, 1970b).

11.7.5 *Discussion of the examples*

The two examples analysed in this chapter form canonical cases in that they contain some of the most basic and important elements of nearshore circulation.

In the first example, the cross-shore momentum balance, we find that the change in radiation stress - in particular the spectacular decline from maximum at the breaking point to zero at the shoreline - represent forces, that to hydrodynamical standards actually are quite large.

However, except for a small bottom friction the force from the decreasing radiation stress is balanced entirely by the pressure force represented by the increase in mean water level. As we saw in the example the slope on the mean water level would be $10-15\%$ of the bottom slope. If that is $1/30$ the slope on the mean water surface will be of the order 3×10^{-3}. This is roughly

30 times a typical slope required to drive a river, which is $O(10^{-4})$. This means a slope that could drive a roaring stream if it were left unbalanced. On the other hand, if the two forces balanced each other completely there would be no forces left to create currents.

As mentioned earlier a closer inspection shows, however, that the difference between the two forces is not completely zero. There is a small difference which is caused by the fact that on a long straight coast the volume flux brought in by the waves must also go out at each point along the coast. This outgoing current component (called the **undertow**, see Section 12.2) will create a bottom shear stress. In the above example this shear stress did enter the equation initially (see (11.7.8)) but we neglected it because it is small - normally less than 5% of the radiation stress gradient. (As the river example shows, only very small forces are required to drive currents because currents create only small bottom stresses.)

In this respect the longshore balance is radically different from the cross-shore situation. Here there is no change in the mean water level along the shore (i.e. no pressure gradient) to balance the change in longshore radiation stress component S_{xy}. As (11.7.31) shows the longshore component of the radiation stress is essentially balanced by the bottom shear stress, because the lateral mixing only distributes the forcing in the cross-shore direction, it does not balance it in an overall sense. Only the bottom friction can do that.

However, that longshore component of the driving radiation stress S_{xy} is also much smaller than the cross-shore component, because the incident angle α_w of the waves relative to the normal to the shore is assumed small. Hence there is again only a relatively small amount of forcing available to generate currents - in this case the longshore component.

Thus we see that the forcing available for driving currents depends on how the radiation stress gradients and the pressure gradients develop. In more general nearshore flow situations the forcing available for driving currents can be expressed by the **Forcing Residual** R_α which is defined as the vectorial sum of the two gradients

$$R_\alpha = -\rho g h \frac{\partial \bar{\zeta}}{\partial x_\alpha} - \frac{\partial S_{\alpha\beta}}{\partial x_\beta} \tag{11.7.88}$$

The use of R_α for analysis of nearshore flows with complicated horizontal variation as is often the case on natural beaches can facilitate the understanding of what drives the currents.

A consequence of the fact that the magnitude of the cross-shore radiation stress gradients often are much larger than the longshore gradients leads to a possible source of longshore current forcing. If the depth variations are such that the setup generated at two neighboring cross-shore profiles is even moderately different we can have a situation where there is a (perhaps small) longshore difference in the setup at the two positions. This corresponds to a longshore gradient in $\overline{\zeta}$ which can be a noticeable forcing in comparison to the other (small) longshore forces. Therefore, as the expression for R_α shows, this small gradient can drive a longshore current, which can be strong locally. Thus in more complex situations on natural beaches it is often not a good approximation to neglect the first term in R_α as we are able to do on a long straight coast. One can say: a coast only needs to deviate very little from long and straight for the longshore variations to be important. This also applies to the incoming waves. This can easily result in longshore variations caused by (small) offshore depth variations that create refraction of the waves resulting in focusing or spreading of the waves near the shore.

It is also worth here to emphasize that the two examples above follow the classical approach of depth uniform currents. Not only do we know that the currents are not depth uniform, but the results of more detailed analysis suggests that the $\int UV$ and the $\int u_w V$-terms in (11.5.13), which we would get without the simplification of depth uniform current, would in fact have been important terms in (11.7.2) had we initially used (11.5.13) instead of (11.5.20). This is discussed in a continuation of the example in Section 13.5

11.8 Wave drivers

In the time averaging process used to derive the nearshore circulation equations we loose the detailed information about the short wave motion. What is left is the time averaged signature of the wave motion in the form of the short wave mass flux $Q_{w\alpha}$ and radiation stress $S_{\alpha\beta}$.

In order to solve the nearshore circulation equations (11.2.27) and (11.5.13) or (11.5.20) we therefore need to determine $Q_{w\alpha}$ and $S_{\alpha\beta}$ by separately solving the equations of a wave model that provides information about $Q_{w\alpha}$ and $S_{\alpha\beta}$. Because the variations of $Q_{w\alpha}$ and $S_{\alpha\beta}$ are driving the nearshore circulations a wave model used in that capacity is also called a **wave driver** for the circulation equations.

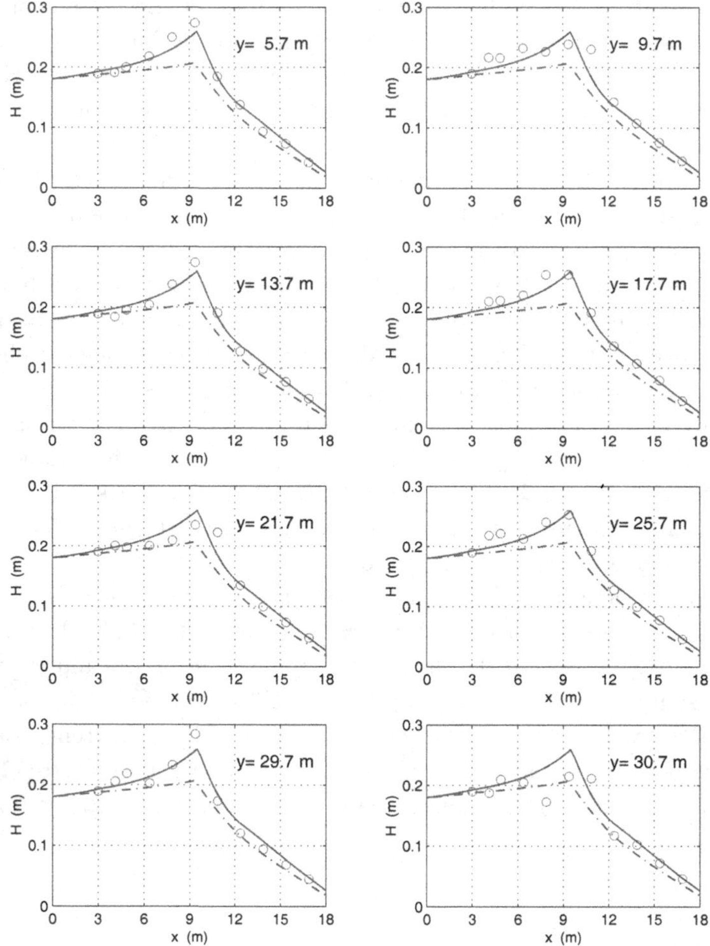

Fig. 11.8.1 Comparisons of wave height (H) between experimental data (○) from Hamilton and Ebersole and the cnoidal-bore model (—), the sine wave model with a roller (–·) (Svendsen et al., 2003).

To avoid that the wave driver becomes a major part of the total computational effort for the circulation problems wave drivers need to be reasonably simple. Typical examples of wave models that are suitable as wave drivers are the models described in Chapter 6 such as

- refraction models
- wave action equation models

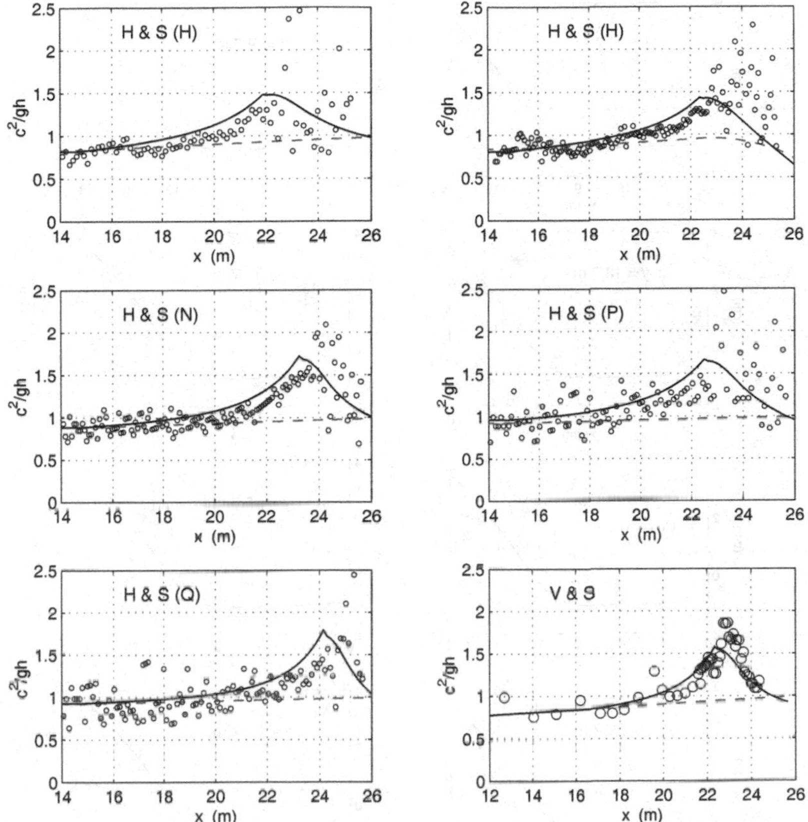

Fig. 11.8.2 Comparisons of c^2/gh between the experimental data (○) from Hansen and Svendsen, and Svendsen and Veeramony and the cnoidal-bore model (—), and the sine wave model (–·) (Svendsen et al., 2003).

- kinematic wave models
- mild Slope Equation (MSE) models
- parabolic wave models

These are all in some way wave averaged models from which the driving parameters can be determined fairly readily.

On the other hand Boussinesq type models are not suited because the computational time needed to solve those equations is one or more orders of magnitude larger than it takes to solve the circulation equations. Furthermore such models will already generate some of the circulation themselves.

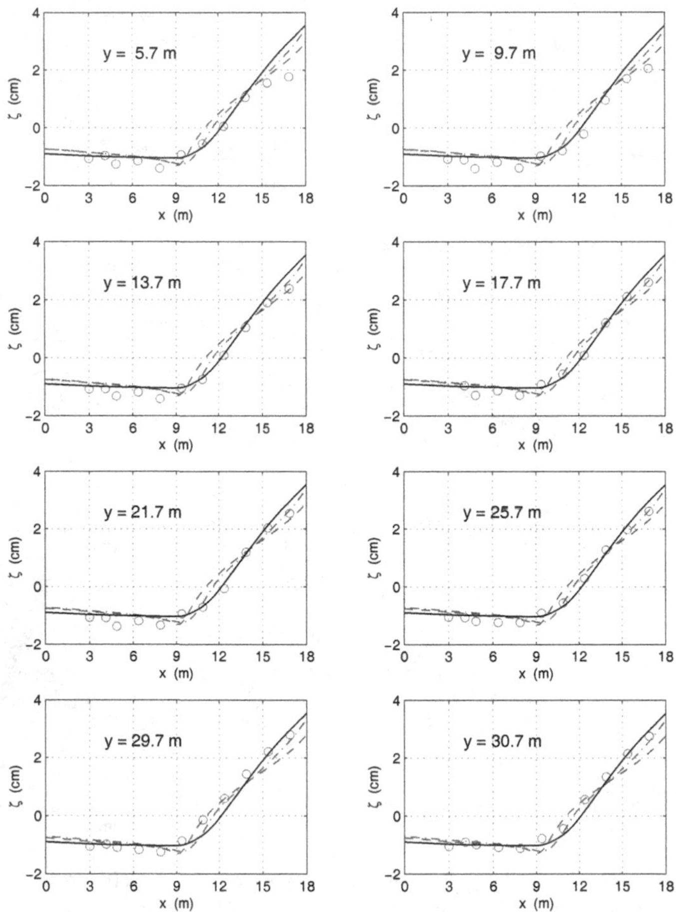

Fig. 11.8.3 Comparisons of the mean water level ($\bar{\zeta}$) between experimental data (○) from Hamilton and Ebersole and the SC with the cnoidal-bore wave driver (—), with the sine wave driver with a roller (–·) and with the sine wave driver without a roller (– –) (Svendsen et al., 2003).

The accuracy of wave driver models

As described in the past chapters the above mentioned wave averaged models are all, one way or the other, based on linear (or sinusoidal) wave theory, which is only a first approximation for the wave motion. It is therefore important to assess how accurate the predictions of $Q_{w\alpha}$ and $S_{\alpha\beta}$ provided by linear wave theory actually are.

A recent investigation (Svendsen et al., 2003) shows comparison of a sinusoidal wave driver model with a wide selection of laboratory measurements. Particular emphasis is on the extensive 3D experiments on a long straight coast conducted at the Large scale Sediment Transport Facility (LSTF) at the US-Army CHL-laboratory in Vicksburg, but many other published laboratory data are used as well. For illustration the comparisons also include a wave driver built on a combination of cnoidal waves (outside the surfzone) and a bore-type model using Hansens (1990) empirical results for the shape factors (inside the surfzone). Fig. 11.8.1 shows a direct comparison of the wave height predictions with the LSTF-data (Hamilton and Ebersole, 2001). The eight panels are for eight cross-shore transects in the wave basin. Similar comparison with published data from a range of wave flume (i.e. 1DH) experiments show similar results: The model based on sinusoidal waves underestimate the increase of the wave height toward breaking. We can see the reason for this if we consider the expresssion for the energy dissipation

$$E_f = \rho g c H^2 B \tag{11.8.1}$$

where B is the dimensionless shape factor. As discussed in Section 5.4 B for sine waves is larger than for real waves. Thus for constant E_f a too large B means a too small H. Consistent with this it turns out that the cnoidal wave model comes closer to predicting the decrease in B that corresponds to the actual increasing peakedness of the waves approaching breaking.

In the comparisons the breaking point is fixed at the point where the waves break in the experiments so the surfzone wave heights reflect the errors before breaking. Because all models of this type have wave heights tending to zero at the shoreline the absolute errors decrease shoreward.

However, this is not the full story as we can see by considering Fig. 11.8.2 which shows the phase velocities c^2/gh. It is clear that the sinusoidal theory predicts c^2/gh almost constant meaning c is decreasing rapidly shoreward while the measurements (and the cnoidal model) shows an increase in c^2/gh toward breaking. Considering (11.8.1) a too small c (sine waves) would tend to increase the growth of H. So this effect counteracts the other error in the sine wave theory. As Fig. 11.8.1 shows, however, the bottomline is that failure to predict the changes in B dominates over the errors in phase velocity and H is not increasing fast enough.

Inside the surfzone the scatter in the measuremnts is substantial, but the behaviour of the two models considered is analogous to the variation of the wave height.

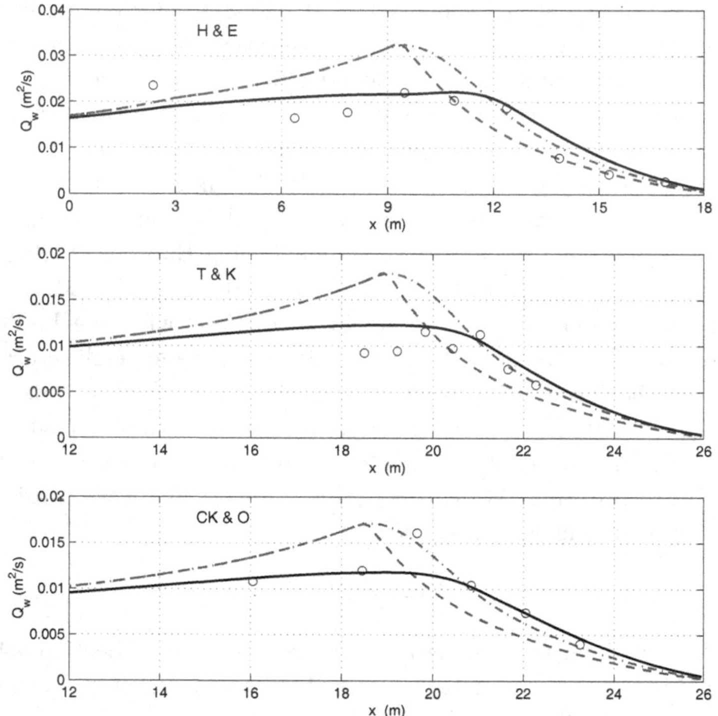

Fig. 11.8.4 Comparisons of wave volume flux Q_w between experimental data (o) from Hamilton and Ebersole, Ting and Kirby, and Cox et al. and the cnoidal-bore model (—), the sine wave model with a roller and without a roller(– –) (Svendsen et al., 2003).

Turning to the (1DH for simplicity) radiation stress S_{xx} this can be written

$$S_{xx} = \rho g H^2 P_{xx} \tag{11.8.2}$$

With H too small one would expect the radiation stress to increase too little toward breaking. Unfortunately it is not possible to measure S_{xx} directly. However, we saw in Section 11.7 that (if we disregard the small bottom friction) the gradient dS_{xx}/dx is directly linked to the gradient on the mean water surface $d\overline{\zeta}/dx$ (see (11.7.8)), which is easy to measure with great accuracy. Therefore the comparisons in Fig. 11.8.3 of the measured and computed variations in mean water surface $\overline{\zeta}$ are direct indicators of the accuracy of the radiation stress, and, surprisingly, the sinusoidal theory seems to predict the radiation stresses quite well both before and after breaking. The only possible explanation is that the too large value of P_{xx} in sinusoidal waves compensates fully for the too small values of H.

We finally consider the volume balance where the wave generated volume flux Q_{wx} is the important parameter. The expression found for the volume flux can be written

$$Q_{wx} = \frac{g}{c} H^2 B_Q \tag{11.8.3}$$

where B_Q is the dimensionless shape factor. Again, we cannot measure Q_{wx} directly. However, in a 1DH setting of a wave flume or a 2DH situation with no longshore variation as in the LSTF experiment the cross-shore undertow current balances the Q_{wx} (See Section 12.2), and the undertow can be measured with good accuracy. The resulting values for Q_{wx} are shown in Fig. 11.8.4 and we find that the sinusoidal waves have too large volume flux relative to the measurements as they approach breaking. Again this is likely to be because the sinusoidal waves are too bulky (B_Q too large) and c is too small. And in this case of the volume balance there are no compensating factors that can eliminate the errors.

So in conclusion we find that for wave drivers based on sinusoidal wave motion

- we can expect the radiation stresses, which is one of the important driving forces for nearshore circulation, to be surprisingly well represented.
- in contrast the volume fluxes will be predicted as too large.

The consequence would be that longshore currents and cross-shore setup will be predicted relatively well, while currents dominated by the volume flux, such as the undertow, will be predicted too large.

11.9 Conditions along open boundaries

11.9.1 *Introduction about open boundaries*

Many of the boundary conditions in the horizontal direction differ fundamentally in nature from the boundary conditions in the vertical direction. Both the bottom and the free surface are actual physical boundaries and so is the shoreline, and the conditions are set by the physical conditions at those boundaries.

In contrast, except for the shoreline, there are no natural horizontal boundaries for the computations in a limited coastal domain. In fact the ideal would be to stretch the computations to the entire ocean.[6] That, however, would lead to computational conditions where the detailing required in the nearshore would be diffficult to achieve.

Therefore practical computations are always done for a computational domain that is limited by a boundary at some distance offshore and some crsoss-shore boundaries that usually are nearly perpendicular to the shoreline at some location up- and downstream of the region of the coast, which is of immediate interest.

The problem then is that the boundary conditions we want to specify along those **open** boundaries are intended to simulate the effect inside the computational domain from what happens **outside**. In principle that can only be specified either by making assumptions about what is happening outside (as for example a certain incident short wave motion along the offshore boundary, or the conditions upstream and downstream along the coast are a certain type of beach, often a long straight beach with the same incident wave motion), or running a larger scale, coarser model that covers a much larger area than we are considering and using the information from that model to specify the conditions along the open boundaries of our model to nest it into the coarser model. The latter method is increasingly done in engineering applications.

It is inherent, however, in the problem that specifying the boundary conditions along an open boundary implies making assumptions about what happens outside the computational domain, that is assumptions about conditions which are in principle unknown.

Because the open boundaries are imaginary except in the computations, we also want to make them as transparent as possible, which includes allowing waves and currents from inside to propagate out of the computational domain. Those issues are discussed briefly in the following.

[6]That would, however, require a different set of equations, that among other things should include the Coriolis acceleration, which we have omitted because it only contributes significantly in larger scale problems than we are normally dealing with on a coast. It would also be necessary to consider density variations due to temperature and salinity differences in the ocean. Finally some consideration of the depth variation of the currents would be required, and the simple analytical treatment introduced in Section 13 would have to be extended to include in particular vertical density variations even on the continental shelf.

11.9.2 *Absorbing-generating boundary conditions*

On the types of coasts we consider here the incoming short wave motion is generally dissipated by breaking on the beach. Therefore along the offshore boundary for the computational domain we can generally specify a short wave motion motion corresponding to an incoming wave motion. For the nearshore circulation equations the incoming wave motion is represented by its radiation stress and its volume flux. The parameters for that motion needs to be specified through the assumptions about the short wave motion.

However, as will be discussed in more detail in Chapter 14, an irregular wave train will carry with it a long wave motion (termed an **infragravity or IG-wave**) which is generated by the time and space variation of the radiation stresses of the short waves. In principle this component of the incoming wave motion which is a variation of the mean water surface $\overline{\zeta}$, can also be specified along the open boundaries, but those long waves will generally **not** be destroyed by breaking. Instead they are reflected from the beach and head back out to sea. It is necessary to design the conditions at the open boundary in such a way that it allows the reflected IG-waves to freely pass through the boundary. That leads to a boundary condition that generates the incoming variations of the radiation stress and volume flux for the computations, generates the incoming IG-waves, and absorbs the outgoing IG-waves. It is therefore called an **absorbing-generating boundary condition**.

A brief literature review

The problem of absorbing (sometimes called radiating or non-reflective) boundary conditions has been discussed in the literature since Sommerfeld about a hundred years ago suggested the 1D-condition at each point of the boundary for the wave particle velocity u normal to the boundary

$$\left(\frac{\partial}{\partial t} + c_0 \frac{\partial}{\partial n}\right) u = 0 \tag{11.9.1}$$

where c_0 is the local linear wave speed. This condition (often called a Sommerfeld condition, see Sommerfeld (1964)) essentially says that the wave is a progressive wave that propagates in the offshore directed normal direction n without change of form, which is exactly the condition satisfied if we have no reflection from the boundary. It allows update to the next timestep of the value of u at the boundary in the same way the circulation equations are solved. For a thorough review of the subject see Givoli (1991).

The condition becomes inaccurate, however, if the wave approaches the boundary at an angle. An alternative condition was derived by Engquist and Majda (1977). That however, is a global condition that requires the entire time history along the entire boundary to update the values of u at a point of the boundary, which is highly impractical in a computational setting. Instead various modifications of the Sommerfeld condition have been suggested which are centered around a chosen angle of incidence θ_n relative to the local normal to the boundary. Those conditions can be shown (Higdon, 1986, 1987) to be on the form

$$\left(\frac{\partial}{\partial t}+\frac{c_0}{\cos\,\theta_n}\frac{\partial}{\partial n}\right)^m u=0 \tag{11.9.2}$$

where m is the chosen order of accuracy. This condition of course has the disadvantage that the wave angle θ_n must be known in advance. This is usually not the case for the nearshore problems we are considering, and the accuracy of the condition decreases if θ_n is not the same as the actual wave angle with the normal.

The boundary condition briefly outlined below was developed by van Dongeren and Svendsen (1997) (VS97 for short) as an extension to 2DH of the 1DH condition given by Kobayashi et al. (1987). See also Verboom et al. (1981).

Formulation of the problem

Thus we can set up the following requirements for a local absorbing-generating boundary condition:

The region outside the computaional domain can influence the motion inside only through the incident (long) waves and through the currents along the boundaries. They both need to be specified/determined otherwise because they represent the influence from the outside which we do not model.

(Long) waves must be allowed to freely propagate out of the computational domain, that is with minimal reflection from the boundary.

The equations considered are depth-integrated, shortwave averaged continuity and momentum equations which form the circulation equations. When solved numericaly these equations provide updates in time and space of the dependent variables ζ, Q_α. The objective is to develop a boundary condition that does the same along the open boundary. If we place the open boundaries carefully (e.g. in a region with constant depth and no short wave breaking) the local forcing will be weak near these boundaries. This means that the dominating terms in the continuity and momentum equations near those boundaries are the terms corresponding to the nonlinear shallow water (NSW) equations. In matrix form they read:

$$\frac{\partial}{\partial t}\begin{bmatrix}\bar{u}'\\ \bar{v}'\\ h\end{bmatrix} + \begin{bmatrix}\bar{u}' & 0 & g\\ 0 & \bar{u}' & 0\\ h & 0 & \bar{u}'\end{bmatrix}\frac{\partial}{\partial x'}\begin{bmatrix}\bar{u}'\\ \bar{v}'\\ h\end{bmatrix} + \begin{bmatrix}\bar{v}' & 0 & 0\\ 0 & \bar{v}' & g\\ 0 & h & \bar{v}'\end{bmatrix}\frac{\partial}{\partial y'}\begin{bmatrix}\bar{u}'\\ \bar{v}'\\ h\end{bmatrix}$$
$$= \begin{bmatrix} g\,\partial h_o/\partial x' + f_x\\ g\,\partial h_o/\partial y' + f_y\\ 0\end{bmatrix} \tag{11.9.3}$$

Here h is the total water depth $h = h_o + \bar{\zeta}$, h_o is the still water depth and $\bar{\zeta}$ is the shortwave averaged surface elevation. $\bar{u}'$ and $\bar{v}'$ are the depth averaged and shortwave averaged velocities in the x' and y' directions, respectively, which at a general point is at an angle to the normal to the boundary. Usually $\bar{\zeta}$ will also include a steady set-down or set-up component which we disregard here for simplicity. f represents all the local forcing terms for the motion, which comprises the radiation stress gradients, and the bottom and wind shear stresses. In the quasi-3D case described in Chapter 13 f also includes the current-current and current-wave integrals (originating from the non-uniform variation of the velocities over depth).

The solution of the nearshore equations in the domain provides the **total** values of the velocities and surface elevations for the IG-waves, but no direct information about in which direction they are propagating. With both incident and reflected IG-waves present at the same time at the boundary points it is necessasy to divide the total signal for velocities and surface elevations into incoming and outgoing components so that the outgoing components can be allowed to pass freely. This is done by writing the NSW-equations on characteristic form.

To simplify the derivation we assume that locally the coordinate system is x, y with the x-axis normal to the boundary and pointing inwards into the computational domain. We first notice that the continuity equation can be written

$$\frac{\partial(h+\bar{\zeta})}{\partial t} + \frac{\partial \bar{u}(h+\bar{\zeta})}{\partial x} + \frac{\partial \bar{v}(h+\bar{\zeta})}{\partial y} = 0 \tag{11.9.4}$$

and since

$$c^2 = g(h + \overline{\zeta}) \tag{11.9.5}$$

this can also be changed to

$$\frac{\partial c^2}{\partial t} + \frac{\partial}{\partial x}(\bar{u}c^2) + \frac{\partial}{\partial y}(\bar{v}c^2) = 0 \tag{11.9.6}$$

or

$$\frac{\partial(2c)}{\partial t} + c\frac{\partial \bar{u}}{\partial x} + \bar{u}\frac{\partial(2c)}{\partial x} + \bar{v}\frac{\partial(2c)}{\partial y} + c\frac{\partial \bar{v}}{\partial y} = 0 \tag{11.9.7}$$

Considering the x-momentum equation this can be written

$$\frac{\partial \bar{u}}{\partial t} + \bar{u}\frac{\partial \bar{u}}{\partial x} + \bar{v}\frac{\partial \bar{u}}{\partial y} + g\frac{\partial(h + \overline{\zeta})}{\partial x} = g\frac{\partial h_0}{\partial x} + f_x \tag{11.9.8}$$

which, again by means of (11.9.5), becomes

$$\frac{\partial \bar{u}}{\partial t} + \bar{u}\frac{\partial \bar{u}}{\partial x} + c\frac{\partial(2c)}{\partial x} + \bar{v}\frac{\partial \bar{u}}{\partial y} = g\frac{\partial h_0}{\partial x} + f_x \tag{11.9.9}$$

Adding and subtracting (11.9.6) and (11.9.9) then gives the two equations

$$\frac{\partial(\bar{u} + 2c)}{\partial t} + (\bar{u} + c)\frac{\partial(\bar{u} + 2c)}{\partial x} + \bar{v}\frac{\partial(\bar{u} + 2c)}{\partial y} + c\frac{\partial \bar{v}}{\partial y} = g\frac{\partial h_0}{\partial x} + f_x \tag{11.9.10}$$

$$\frac{\partial(\bar{u} - 2c)}{\partial t} + (\bar{u} - c)\frac{\partial(\bar{u} - 2c)}{\partial x} + \bar{v}\frac{\partial(\bar{u} - 2c)}{\partial y} - c\frac{\partial \bar{v}}{\partial y} = g\frac{\partial h_0}{\partial x} + f_x \tag{11.9.11}$$

which, along with the y-equation written as

$$\frac{\partial \bar{v}}{\partial t} + \bar{u}\frac{\partial \bar{v}}{\partial x} + \bar{v}\frac{\partial \bar{v}}{\partial y} = -g\frac{\partial \zeta}{\partial y} \tag{11.9.12}$$

constitute the equations on characteristic form.

We introduce the Riemann-variables β^+, β^-, and γ defined as

$$\beta^- = \bar{u} + 2c = \frac{Q_x}{(h_o + \bar{\zeta})} + 2\sqrt{g(h_o + \bar{\zeta})} \quad (11.9.13)$$

$$\beta^- = \bar{u} - 2c = \frac{Q_x}{(h_o + \bar{\zeta})} - 2\sqrt{g(h_o + \bar{\zeta})} \quad (11.9.14)$$

$$\gamma = \bar{v} = \frac{Q_y}{(h_o + \bar{\zeta})} \quad (11.9.15)$$

where Q_x, Q_y are the total fluxes in the x, y direction, respectively, and $\bar{u}, \bar{v}$ the depth averaged velocities.

Substituting into (11.9.10) and (11.9.11) the equations in characteristic form then simplify to:

$$\frac{\partial \beta^-}{\partial t} + (\bar{u} - c)\frac{\partial \beta^-}{\partial x} + \bar{v}\frac{\partial \beta^-}{\partial y} = c\frac{\partial \bar{v}}{\partial y} + g\frac{\partial h_o}{\partial x} + F_{\beta^-} \quad (11.9.16)$$

$$\frac{\partial \beta^+}{\partial t} + (\bar{u} + c)\frac{\partial \beta^+}{\partial x} + \bar{v}\frac{\partial \beta^+}{\partial y} = -c\frac{\partial \bar{v}}{\partial y} + g\frac{\partial h_o}{\partial x} + F_{\beta^+} \quad (11.9.17)$$

$$\frac{\partial \gamma}{\partial t} + \bar{u}\frac{\partial \gamma}{\partial x} + \bar{v}\frac{\partial \gamma}{\partial y} = -g\frac{\partial \bar{\zeta}}{\partial y} + F_\gamma \quad (11.9.18)$$

It is noticed that the γ-equation is the y-momentum equation itself.

As the definition sketch Fig. 11.9.1 shows, β^+ propagates along a characteristic in the positive x direction, β^- in the negative x direction and γ in the y direction. The forcing terms F_{β^+}, F_{β^-} and F_γ originate from the f-terms in (11.9.3). These terms imply that β^+, β^- and γ vary along their characteristics and hence they are variables rather than invariants.

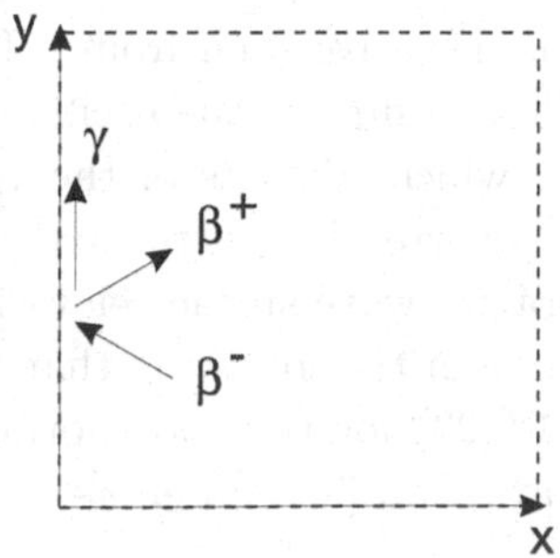

Fig. 11.9.1 Definition sketch of the characteristics.

During the computation we will at time step n know the value of $\bar{\zeta}^n$ and (Q_x^n, Q_y^n) at interior as well as at boundary points. The incoming IG-wave motion is specified along the boundaries through specification of $(Q_{x,i}, Q_{y,i})$, which represent the x and y components of the flux of the incident wave. The numerical integration of (11.9.3) will then provide the values of the total $\bar{\zeta}, Q_x, Q_y$ at time step $n+1$, and the problem is to determine how large a fraction of this represents the outgoing wave at $n+1$.

Assuming linear superposition of the incoming wave (subscripted i in the following) and the outgoing wave (subscripted r), we can write

$$Q_x = Q_{x,i} + Q_{x,r} \tag{11.9.19}$$

$$\bar{\zeta} = \bar{\zeta}_i + \bar{\zeta}_r \tag{11.9.20}$$

Without further approximation the outgoing Riemann-variable (11.9.13) can then be rewritten as

$$\frac{\beta^-}{c_o} = \frac{Q_{x,i}}{c_o h_o}\left(1+\frac{\bar{\zeta}_i+\bar{\zeta}_r}{h_o}\right)^{-1} + \frac{Q_{x,r}}{c_o h_o}\left(1+\frac{\bar{\zeta}_i+\bar{\zeta}_r}{h_o}\right)^{-1} - 2\sqrt{1+\frac{\bar{\zeta}_i+\bar{\zeta}_r}{h_o}} \tag{11.9.21}$$

where

$$c_o = \sqrt{g h_o} \tag{11.9.22}$$

In order to solve this for $Q_{x,r}$ we need to eliminate the surface elevations. For waves of constant form Q is related to the surface elevation by the (exact) equation (see VS97)

$$Q = c_a(\bar{\zeta} - \bar{\bar{\zeta}}) + \bar{\bar{Q}} \tag{11.9.23}$$

where c_a is the celerity of the wave seen from a fixed coordinate system and $c_a \bar{\zeta}$ represents the volume flux in the oscillatory part of the motion. $\bar{\bar{Q}}$ is the net volume flux, which consists of the volume flux, Q_w, in the infragravity waves and the "current". $\bar{\bar{\zeta}}$ is the average over the infragravity wave period of the infragravity wave surface elevation $\bar{\zeta}$.

For simplicity we assume in the following that $\bar{\bar{\zeta}}$ as well as $\bar{\bar{Q}}$ are zero. Then $c_a = c$ and Eq. (11.9.23) for the shorenormal x-components of the incoming and outgoing waves can be written as

$$Q_{x,i} = c\,\bar{\zeta}_i \cos\theta_i \tag{11.9.24}$$

$$Q_{x,r} = -c\,\bar{\zeta}_r \cos\theta_r \tag{11.9.25}$$

where θ_i and θ_r are defined as the angles between the normal to the boundary and the incoming and outgoing waves in the range $[-\frac{\pi}{2}, \frac{\pi}{2}]$, respectively. Eq. (11.9.21) then becomes

$$\frac{\beta^-}{c_o} = \frac{Q_{x,i}}{c_o h_o}\left(1 + \frac{Q_{x,i}}{h_o c\cos\theta_i} - \frac{Q_{x,r}}{h_o c\cos\theta_r}\right)^{-1} + \frac{Q_{x,r}}{c_o h_o}\left(1 + \frac{Q_{x,i}}{h_o c\cos\theta_i} - \frac{Q_{x,r}}{h_o c\cos\theta_r}\right)^{-1} - 2\left(1 + \frac{Q_{x,i}}{h_o c\cos\theta_i} - \frac{Q_{x,r}}{h_o c\cos\theta_r}\right)^{\frac{1}{2}} \tag{11.9.26}$$

To this point no real approximations have been made (except $\overline{\overline{\zeta}} = 0$ and $\overline{\overline{Q}} = 0$). However, we can expect that the volume fluxes are small in comparison to $h_O c$. We therefore expand this expression with respect to $Q_x/h_o c_o$ and to first order we get

$$\frac{\beta^-}{c_o} + 2 = c_1\, Q'_{x,i} + b_1\, Q'_{x,r} + O\left(\frac{Q_x}{c_o h_o}\right)^2 \tag{11.9.27}$$

where for convenience we have defined

$$Q'_{x,i} \equiv \frac{Q_{x,i}}{c_o\, h_o} \tag{11.9.28}$$

$$Q'_{x,r} \equiv \frac{Q_{x,r}}{c_o\, h_o} \tag{11.9.29}$$

$$b_1 \equiv \frac{(\cos\theta_r + 1)}{\cos\theta_r} \tag{11.9.30}$$

$$c_1 \equiv \frac{(\cos\theta_i - 1)}{\cos\theta_i} \tag{11.9.31}$$

We solve this with respect to $Q'_{x,r}$ which yields

$$\boxed{Q'_{x,r} = \frac{1}{b_1}\left(\frac{\beta^-}{c_o} + 2 - c_1\, Q'_{x,i}\right) + O\left(\frac{Q_x}{c_o h_o}\right)^2} \tag{11.9.32}$$

which is the lowest order approximation for $Q_{x,r}$ in terms of quantities that we can consider known at the boundary.

Using (11.9.15) $Q'_{y,r}$, can be determined as

$$Q'_{y,r} = \frac{\gamma\,(h_o + \bar{\zeta}) - Q_{y,i}}{c_o\, h_o} \tag{11.9.33}$$

This means that in (11.9.32) the unknown θ_r can be determined by realizing that

$$\theta_r = \arctan\left(\frac{Q'_{y,r}}{Q'_{x,r}}\right) \tag{11.9.34}$$

which gives

$$\boxed{\theta_r = \arctan\left(\frac{\gamma\,(h_o + \bar{\zeta}) - Q_{y,i}}{c_o\, h_o\, Q'_{x,r}}\right)} \tag{11.9.35}$$

Here $Q_{y,i}$ is specified and γ is determined by integration of the last of the characteristic equations (11.9.18). Thus we have determined the outgoing volume flux $Q_{x,r}$ and its angle of incidence θ_r relative to the normal of the boundary.

It turns out, that for larger amplitude waves the expansion (11.9.32) is one of the most significant error sources. VS97 therefore extend the expansion of (11.9.26) to second order, which yields a quadratic equation in $Q'_{x,r}$. For further details about this step the reader is referred to the paper.

In these expressions β^- is the Riemann-variable, which is updated to the next time level by (11.9.16) at the $x = 0$ boundary. (The value of β^- in the interior points, which is needed to calculate $\frac{\partial \beta^-}{\partial x}$ in (11.9.16), is constructed from Q_x and $\bar{\zeta}$ using (11.9.13).) From (11.9.32) and (11.9.34) we can find the unknowns $Q'_{x,r}$ and θ_r iteratively. With the incoming wave known through specification, the boundary value of total flux Q_x can then be determined at the next time stepand $\bar{\zeta}$ frp, the continuity equation. This concludes the upgrading along the boundary to time step $n+1$.

It should be emphasized that boundary condition (11.9.32) is derived for the $x = 0$ boundary and that it can readily be generalized for any other boundary that is normal to a coordinate axis. The boundary condition can also be generalized to boundaries that are not normal to a coordinate axis by rotating the coordinate system in the solution presented here. For brevity, this derivation is omitted here.

Absorption errors

The errors relative to complete absorption of the outgoing wave have been determined using tests with suitably designed domains. A least square error ϵ_2 averaged over the test domain can then be defined as the error on the amplitude of the wave reflected from the absorbing boundary divided by the amplitude of the outgoing wave coming toward the boundary from inside the domain.

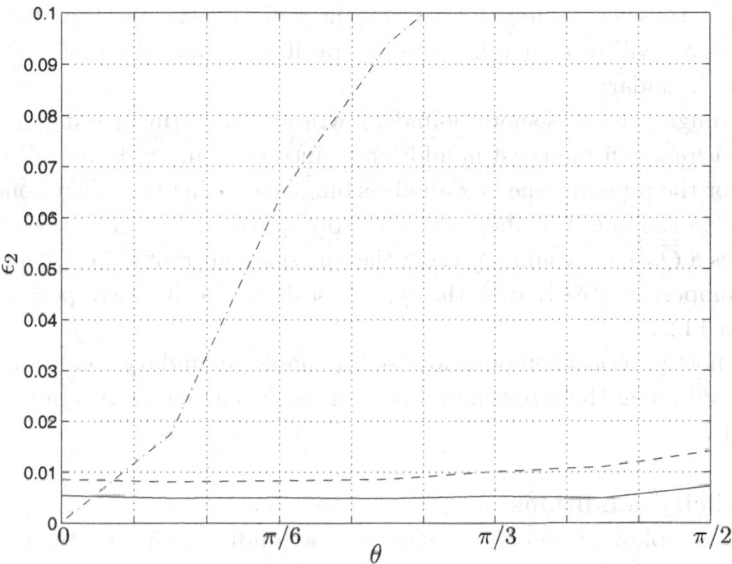

Fig. 11.9.2 The relative errors using the linear approximation (11.9.32) (- - -), the second order approximation (—), and the Sommerfeld condition (11.9.1) (- . - . -) (from VS97).

Fig. 11.9.2 shows the error ϵ_2 as a function of the angle of incidence θ of the wave at the boundary. Curves are shown both for the linear solution outlined above and for the second order approximation. For comparison is also shown the error caused by the Sommerfeld condition (11.9.1). We see that the absorbing boundary condition gives errors that are unifom in magnitude for all angles of incidence up to 90° and even the linear solution gives errors that hardly exceed 1%. In contrast the error from the Sommerfeld condition grows rapidly with increasing angles of incidence and reaches 10% at an angle of about 45°. The more general radiation condition (11.9.2) will grow slower but also reach high values for larger angles of incidence.

11.9.3 *Boundary conditions along cross-shore boundaries*

In coastal regions the computational domain will typically, in addition to the offshore boundary, be bounded by some cross-shore boundaries limiting the domain along the coast up- and downstream of the area of interest.

Along those cross-shore boundaries the radiation stresses from incoming short waves will of course have to be specified as was the case along the offshore boundary.

Through the cross-shore boundary will also flow (in or out) the longshore currents on the coast in addtion to incoming and reflected IG-waves. Thus for the total volume flux an absorbing-generating boundary condition similar to the one described above is approproppriate, but in a version that includes a $\overline{\overline{Q}}$. This would represent the longshore current if we assume that flow component steady over the typical scale of the IG-wave periods (see Section 14.2).

In most practical computations so far simpler boundary conditions have been used along the cross-shore boundaries. In particular two options are possible.

Periodicity conditions

The simplest option to implement is a condition that states that the condition at the left and right cross-shore boundaries are the same at the same time and at corresponding points in the cross-shore direction. Thus if we consider the simple case of a long straight coast with cross-shore boundaries a y_{left} and y_{right} we would have

$$Q(x, y_{left}, t) = Q(x, y_{right}, t) \tag{11.9.36}$$

and the surface elevations determined from the continuity equation.

However, on a coast with longshore varying topography the words "corresponding points in the cross-shore direction" cannot simply be identified as points with the same x-coordinate. In the first place if we talk about a simple cartesian x, y, z-coordinate system the shoreline is not at the same x-position for different points along the shore. And even at points at the same distance offshore the depth will in general be different at the left and right boundaries.

There is not a "correct" solution to this problem. A frequently used approach is to extend the coast artificially at both ends beyond the intended domain by adding regions in which the cross-shore depth profiles are allowed to gradually approach a common average cross-shore profile which is then used at both ends, usually in a periodicity condition. This, however, does not mean that the volume flux at "corresponding points" are the same, because the cross-shore variation of Q_y is determined by the conditions over quite a distance "upstream" (i.e. against the longshore current) of the boundary.

Prescribed volume flux

The second option used is to assume that the coast "upstream" of the upstream boundary is long and straight and use this to calculate the cross-shore distribution of the longshore current at the upstream boundary by solving the equations for a long straight coast (Section 11.7) with the chosen cross-shore profile. This essentially means that $Q_{y,i}$ is prescribed as we assumed for the offshore boundary. The downstream boundary condition can then be treated as a freee boundary where $Q_{y,out}$ is determined by the internal solution in combination with $\partial/\partial y = 0$ along the boundary.

Discussion

As mentioned none of these options can solve the problem that by formulating the "upstream" conditions we try to estimate (guess!) the distribution along the boundary of the incoming currents created in the region outside the domain we model (unless those conditions are determined in another larger scale computation). (In general terms the word "upstream" should be taken to mean all boundary sections where the volume flux is into the domain. This can in principle also be parts of the offshore boundary and even parts of a "downstream" boundary with e.g. a return flow). The question therefore arizes: how do these errors influence the flow inside the computational domain?

Some insight into this was established by Chen and Svendsen (2003). They performed numerical experiments on a long straight coast where the true solution is known in which errors were imposed along the upstream boundary (i.e. $Q_{y,i}$ specified). At the downstream boundary the same distribution was specified by a periodicity condition. The results show that the upstream errors spread inside the domain so that the local errors become smaller but the total volume deficit of course remaines the same. However, the picture was complicated by the fact that when specifying for example an excessive total volume flux at the cross-shore boundary that was not distributed at the boundary in the cross-shore direction the same way as in a uniform longshore current with that volume flux. Then the flow inside created recirculation flows that entrained extra fluid from the offshore, thus increasing the longshore flux inside the domain. When the flow reached the downstream exit, where the cross-shore distribution of the volume flux was forced to be the same as at the upstream bounday (periodicity imposed), a strong local disturbance emerged that over a very short distance (as in the runout to a drain in a bathtub) recirculated the extra volume flux entrained, and redistributed the longshore velocity profile to the prescribed profile.

For decreased volume flow other disturbances occurred. In general, however, the conclusion is that errors in volume flux specified along cross-shore boundaries do disturb the internal flow more than one might expect.

11.10 References - Chapter 11

Bowen, A. J. (1969). The generation of longshore currents on a plane beach. J. Geophys. Res. **74**, 5479 - 5490.

Bowen, A. J., D. L. Inman and V. P. Simmons (1968). Wave 'set down' and set-up. J. Geophys. Res., **73**, 2569 - 2577.

Chen, Q. and I. A. Svendsen (2003). Effects of cross-shore boundary condition errors in nearshore circulation modeling. Coast. Engrg., **48**, 243–256.

Engquist, B. and A. Majda (1977). Absorbing boundary conditions for the numerical simulations of waves. Math. Comp., **31**, 139, 629–651.

Galvin, C. J. and P. S. Eagleson (1965). Experimental study of longshore currents on a plane beach. U.S. Army Corps of ewng. Res. Center, Tech. Mem. 10, 1-80.

Givoli, D. (1991). Non-reflective boundary conditions. J. Comp. Phys., **94**, 1–29.

Greenberg, M. D. (1988). Advanced engineering mathematics. Prentice Hall, Englewood Cliffs, New Jersey 07632. pp 946.

Hamilton, D. G. and B. A. Ebersole (2001). Establishing uniform longshore currents in a large-scale sediment transport facility. *Coastal Engineering*, 42, 199-218.

Hansen, J. B. (1990). Periodic waves in the surf zone: Analysis of experimental data. *Coast. Eng.*, **14**, 14–41.

Hansen, J. B. and I. A. Svendsen (1979). Regular waves in shoaling water experimental data. Inst. Hydrodyn. Hydraul. Eng. (ISVA), Tech. Univ. Denmark, Lyngby, Ser. Pap., 21.

Higdon, R. L. (1986). Absorbing boundary conditions for difference approximations to the multi-dimensional wave equation. Math. Comp., **47**, 176, 437–459.

Higdon, R. L. (1987). Numerical absorbing boundary conditions for the wave equation. Math. Comp., **49**, 179, 65–90.

Kobayashi, N., A. K. Otta and I. Roy (1987). Wave reflection and runup on rough beaches. ASCE J. Waterw. Port, Coast. and Oc. Engrg., **113**, 3, 282–298.

Longuet-Higgins, M. S. (1970a,b). Longshore currents generated by obliquely incident sea waves, J. Geophys. Res., **75**, 6778–6789 and 6790–6801.

Longuet-Higgins, M. S. and R. W. Stewart (1962). Radiation stress and mass transport in gravity waves with application to 'surf-beats'. J. Fluid Mech., **8**, 565 – 583.

Longuet-Higgins, M. S. and R. W. Stewart (1963). A note on wave set-up. J. Marine Research, **21**, 4–10.

Longuet-Higgins, M. S. and R. W. Stewart (1964). Radiation stress in water waves, a physical discussion with application. Deep Sea Research, **11**, 529 – 563.

Lundgren, H. (1963). Wave thrust and wave energy level. IAHR Proc. 10th Int. Congr. Assoc. Hydr. Res. London, **1**, Paper 1.20, 147–151.

Mei, C. C. (1983). The applied dynamics of ocean surface waves. John Wiley and Sons, Inc., 740pp.

Phillips, O. M. (1966, 1977). Dynamics of the upper ocean. Cambridge University Press.

Sommerfeld, A. (1964). Lectures on theoretical physics. Academic Press Inc. San Diego.

Svendsen, I. A. and J. Veeramony (2001). Wave breaking in wave groups. *J. of Wtrwy., Port, Coast., and Oc. Engrg.*, ASCE, 127(4), 200-212.

Svendsen, I. A., W. Qin and B. A. Ebersole (2003). Modelling waves and currents at the LSTF and other laboratory facilities. Coast. Engrg., **50**, 19–45.

Thornton, E. B. (1970). Variation of longshore current across the surf zone, Proc., 12th Coast. Engrg. Conf., ASCE, Washington, D.C., 291–308.

Van Dongeren, A. R. and I. A. Svendsen (1997). Absorbing-generating boundary conditions for shallow water models. ASCE J. Waterwaves, Port, Coastal, and Ocean Engrg., **123**, 6, 303–313.

Verboom, G. K., G. S. Stelling, and M. J. Officier (1981). Boundary conditions for the shallow water equations. In: Engineering applictions of computational hydraulics (eds. M. B. Abbott and J. A. Cunge), Pitman Publishing, Ltd., London. 230 - 262.

Chapter 12

Cross-Shore Circulation and Undertow

12.1 The vertical variation of currents

12.1.1 *Introduction*

One of the most serious problems in coastal engineering is the disastrous erosion of beaches that often occurs during heavy storms. In a few hours, large amounts of material is removed from the beach and at first deposited as bars at some distance from the shoreline.

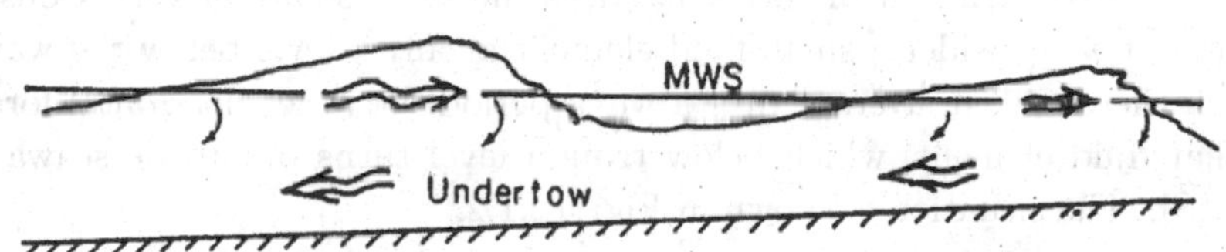

Fig. 12.1.1 The undertow current (from Svendsen, 1984).

Though it has not yet been finally proved, this process is likely to be linked to the strong seaward oriented current flowing in the lower part of the water column under breaking waves. This current, called the **undertow**, is fed by the water volume brought shoreward by the volume flux of the breakers. In the two-dimensional situation on a long straight beach shown in Fig. 12.1.1, the two fluxes are equal. In cases with a horizontal circulation, the undertow will be superimposed on a net current. Inside the surf zone the amount of water carried shoreward in the roller of the breaking waves is further increasing the volume flux in such a wave relative to the ordinary volume flux in the waves, as found in Chapter 5.

As found earlier (see Section 11.7.2), in the longshore uniform case the cross-shore depth integrated, time averaged momentum balance reads

$$\frac{\partial S_{xx}}{\partial x} = -\rho g \left(h_0 + \bar{\zeta}\right) \frac{\partial \bar{\zeta}}{\partial x} - \bar{\tau}_b \tag{12.1.1}$$

where $\bar{\zeta}$ is the set-up, τ_b the mean bottom shear stress, S_{xx} the radiation stress.

In the strictly 2D circulation flow shown in Fig. 12.1.1 $\bar{\tau}_b$ normally is a small term ($O(5\%)$ of the other two terms). Thus, the momentum balance (12.1.1) essentially means that, averaged over depth, the radiation stress gradient is balanced by the force from the slope on the MWS.

However, S_{xx} is defined as (see Sect. 6.6)

$$S_{xx} = \overline{\int_{-h_0}^{\zeta} \rho \left(u_w^2 - w_w^2\right) dz} + \frac{1}{2} \rho g \overline{\eta^2} \tag{12.1.2}$$

which shows that S_{xx} is composed of contributions from $\overline{u_w^2}$ and w_w^2 (which in principle vary over depth) and a contribution from $\overline{\eta^2}$ which originates between trough and crest of the waves. Thus the S_{xx}-term varies over the depth.

The $\partial \bar{\zeta}/\partial x$ term, on the other hand, is the same at all z-levels. Consequently, if we consider a small fluid element at any z-level below the wave trough there will, in average over a wave period, be a net horizontal force on that fluid element, which below trough level turns out to be seaward oriented. The situation is shown in Fig. 12.1.2.

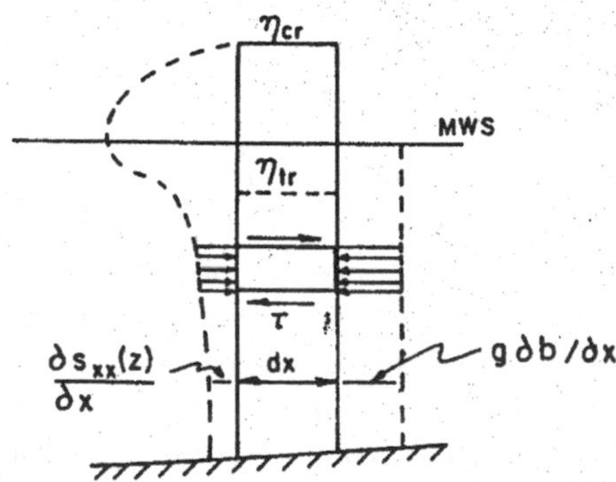

Fig. 12.1.2 Stresses in undertow (from Svendsen, 1984).

In order to properly analyze the phenomenon of cross-shore circulation, we need to go back to the original equations—before we started integrating those equations over depth—and analyze what they imply at each point in the vertical when we consider the combination of a turbulent wave motion and a current, which we know is what we have in the surf zone.

Although at first we are only looking for the cross-shore variation, let us take a more general approach and first derive the equations for the full 3D circulation, since this will not add much to the complexity of the equations at this point.

12.1.2 *The governing equations for the variation over depth of the 3D currents*

The equations used to describe the vertical distribution of the mean flow in the surf zone are the Reynolds' equations averaged over a wave period.

The instantaneous Reynolds' equations read (see Section 2.5)

$$x: \quad \frac{\partial \widehat{u}}{\partial t} + \frac{\partial \widehat{u^2}}{\partial x} + \frac{\partial \widehat{uv}}{\partial y} + \frac{\partial \widehat{uw}}{\partial z}$$

$$= -\frac{1}{\rho}\frac{\partial p}{\partial x} - \frac{\partial \widehat{u'^2}}{\partial x} - \frac{\partial \widehat{u'v'}}{\partial y} - \frac{\partial \widehat{u'w'}}{\partial z} \tag{12.1.3}$$

$$y: \quad \frac{\partial \widehat{v}}{\partial t} + \frac{\partial \widehat{uv}}{\partial x} + \frac{\partial \widehat{v^2}}{\partial y} + \frac{\partial \widehat{vw}}{\partial z}$$

$$= -\frac{1}{\rho}\frac{\partial p}{\partial y} - \frac{\partial \widehat{u'v'}}{\partial x} - \frac{\partial \widehat{v'^2}}{\partial y} - \frac{\partial \widehat{v'w'}}{\partial z} \tag{12.1.4}$$

$$z: \quad \frac{\partial \widehat{w}}{\partial t} + \frac{\partial \widehat{uw}}{\partial x} + \frac{\partial \widehat{uw}}{\partial y} + \frac{\partial \widehat{w^2}}{\partial z}$$

$$= -\frac{1}{\rho}\frac{\partial p}{\partial y} - g - \frac{\partial \widehat{u'w'}}{\partial x} - \frac{\partial \widehat{v'w'}}{\partial y} - \frac{\partial \widehat{w'^2}}{\partial z} \tag{12.1.5}$$

where, as before, $\widehat{\ }$ means turbulent averaging, and where

$$\left(\widehat{u}, \widehat{v}\right) = (u_w, v_w) + (U, V) \quad ; \quad \widehat{w} = w_w \tag{12.1.6}$$

divides the total velocity (u, v, w) into a wave and a current component. The currents will in general be functions of the vertical possition z and of time t so that $U, V = U(x, y, z, t), V(x, y, z, t)$, and we need to keep that option open now.

We have chosen u_{wi} so that

$$\overline{u_{wi}} = 0 \qquad \text{below trough level} \tag{12.1.7}$$

and the wave motion is periodic, soin consequence, we have

$$\overline{\frac{\partial u_{wi}}{\partial t}} = 0 \tag{12.1.8}$$

When we introduce $(u_w,\ v_w)$ and $U,\ V$, we notice that

$$\frac{\partial \overline{u_i u_j}}{\partial x_j} = \frac{\partial \overline{(u_{wi} + U_i)\,(u_{wj} + U_j)}}{\partial x_j} = \frac{\partial \overline{u_{wi} u_{wj}}}{\partial x_j} + \frac{\partial U_i U_j}{\partial x_j} \tag{12.1.9}$$

where the overbar as usual means average over a wave period.

The current U_i, however, may in the general case vary over a time scale which is much longer than the wave period.

Time averaging over a wave period then yields for the two horizontal equations

$$x: \quad \frac{\partial U}{\partial t} + \frac{\partial U^2}{\partial x} + \frac{\partial UV}{\partial y} + \frac{\partial \overline{u_w^2}}{\partial x} + \frac{\partial \overline{u_w v_w}}{\partial y} + \frac{\partial \overline{u_w w_w}}{\partial z} = -\frac{1}{\rho}\frac{\partial \overline{\widehat{p}}}{\partial x} - \frac{\partial \overline{\widehat{u'^2}}}{\partial x} - \frac{\partial \overline{\widehat{u'v'}}}{\partial y} - \frac{\partial \overline{\widehat{u'w'}}}{\partial z} \tag{12.1.10}$$

$$y: \quad \frac{\partial V}{\partial t} + \frac{\partial UV}{\partial x} + \frac{\partial V^2}{\partial y} + \frac{\partial \overline{u_w v_w}}{\partial x} + \frac{\partial \overline{v_w^2}}{\partial y} + \frac{\partial \overline{v_w w_w}}{\partial z} = -\frac{1}{\rho}\frac{\partial \overline{\widehat{p}}}{\partial x} - \frac{\partial \overline{\widehat{u'v'}}}{\partial x} - \frac{\partial \overline{\widehat{v'^2}}}{\partial y} - \frac{\partial \overline{\widehat{v'w'}}}{\partial z} \tag{12.1.11}$$

The vertical z-component of the momentum equation is integrated from the surface to an arbitrary level z from which we get an expression for $p(z)$ (see (11.4.30)). As in Chapter 11 we again assume hydrostatic pressure for the currents. After time averaging, this leads to

$$\overline{\overline{p}} = \rho g\left(\bar{\zeta} - z\right) - \rho\left(\overline{w_w^2} + \overline{\widehat{w'^2}}\right) \tag{12.1.12}$$

which substituted into (12.1.10) and (12.1.11) yields

$$\frac{\partial U}{\partial t} + \frac{\partial U^2}{\partial x} + \frac{\partial UV}{\partial y} + \frac{\partial \left(\overline{u_w^2} - \overline{w_w^2}\right)}{\partial x} + \frac{\partial \overline{u_w v_w}}{\partial y} + \frac{\partial \overline{u_w w_w}}{\partial z} =$$

$$-g\frac{\partial \bar{\zeta}}{\partial x} - \frac{\partial (\overline{\widehat{u'^2}} - \overline{\widehat{w'^2}})}{\partial x} - \frac{\partial \widehat{\overline{u'v'}}}{\partial y} - \frac{\partial \widehat{\overline{u'w'}}}{\partial z} \qquad (12.1.13)$$

and

$$\frac{\partial V}{\partial t} + \frac{\partial UV}{\partial x} + \frac{\partial V^2}{\partial y} + \frac{\partial \left(\overline{v_w^2} - \overline{w_w^2}\right)}{\partial y} + \frac{\partial \overline{u_w v_w}}{\partial x} + \frac{\partial \overline{v_w w_w}}{\partial z} =$$

$$-g\frac{\partial \bar{\zeta}}{\partial y} - \frac{\partial \widehat{\overline{u'v'}}}{\partial x} - \frac{\partial (\overline{\widehat{v'^2}} - \overline{\widehat{w'^2}})}{\partial y} - \frac{\partial \widehat{\overline{v'w'}}}{\partial z} \qquad (12.1.14)$$

Eqs. (12.1.13) and (12.1.14) are the ensemble and time averaged momentum equations governing the variation of the current velocity in the nearshore at any point in the vertical, which is below wave trough level.

In (12.1.13) and (12.1.14) the $\partial U^2/\partial x$ and $\partial V^2/\partial y$ represent the momentum flux from the cross-shore and longshore currents respectively. The UV-terms represent coupling between the two current components. In the 2D situation of pure undertow on a long straight coast (shore normal wave incidence), these terms are zero because $\partial/\partial y = 0$ and $V = 0$.

In the general 3D case with obliquely incident waves, cross-shore circulation and longshore currents, these terms have traditionally been neglected. However, even with longshore uniformity in both topography (long straight coast) and flow ($\partial/\partial y = 0$), there is a $\partial UV/\partial x$-term which turns out to be crucial for a correct analysis of the current pattern. This will be discussed further in Chapter 13.

However, in the following we concentrate on the simple 2D case with shore normal waves so all the UV-terms are zero.

The $\overline{u_w^2} - \overline{w_w^2}$ and $\overline{v_w^2} - \overline{w_w^2}$ terms are the local contributions to the radiation stress (compare to (12.1.2)) and so are the $u_w w_w$-terms.

The $\partial \overline{u_w w_w}/\partial z$ and $\partial \overline{v_w w_w}/\partial z$ term represents the Reynolds-type stresses due to various factors that disturb the wave motion, such as the depth variation, the bottom boundary layer, and the effect of the wave breaking. It was found in Section 5.5 for 1D waves that between the bottom boundary layer and the surface the $\partial \overline{u_w w_w}/\partial z$-term is small in comparison with the turbulent shear stresses $\widehat{u'w'}$ (see Section 5.6). We will therefore neglect it above the boundary layer, but in the bottom boundary layer we found that the term is determined by the boundary layer solution and is responsible for generating the steady streaming (see Section 10.1.5). AS a consequence we leave those terms in the equations and assume they are known as the other wave averaged terms.

On the right-hand side, the $\partial\bar{\zeta}/\partial x$ and $\partial\bar{\zeta}/\partial y$-terms are clearly the pressure gradients due to the sloping MWS.

The rest of the terms in (12.1.13) and (12.1.14) are turbulent Reynolds' stresses.

12.2 The cross-shore circulation, undertow

12.2.1 *Formulation of the 2-D problem and general solution*

The purpose is first to study the cross-shore circulation $U(z)$ in the vertical plane. In the momentum-equations 12.1.13 and 12.1.14 derived above that means concentrating on the x-component 12.1.13 and assuming that the situation is the same at all positions along the shore; i.e., assume

$$\frac{\overline{\partial}}{\partial y} = 0 \tag{12.2.1}$$

We further consider a case of steady currents so that we are looking at a situation similar to that found in a wave flume. This means that the volume flux Q_w in the incoming waves is compensated by a by a return flow below wave trough level (see Section 3.3) with depth averaged velocity U_m given by

$$U_m = -\frac{Q_w}{h} \tag{12.2.2}$$

Here Q_w is assumed known (see e.g. Chapter 5).

We also assume that we can neglect the turbulent normal stresses $\widehat{u'^2}$ since in most of the water column these terms were found by Stive and Wind (1982) to be relatively small in comparison with $\partial S_{xx}/\partial x$ in the depth integrated equations.

The cross-shore momentum equation (12.1.10) can then be written

$$-\frac{\partial}{\partial z}\left(\overline{\widetilde{u'w'}}\right) = \frac{\partial}{\partial x}\left(\overline{u_w^2} - \overline{w_w^2} + g\bar{\zeta} + U^2\right) + \frac{\partial}{\partial z}\overline{u_w w_w} \tag{12.2.3}$$

In the shore normal direction we then find (as earlier for the depth integrated equation) that inside the surf zone the pressure term $\partial\bar{\zeta}/\partial x$ dominates relative to the U^2-term. We will account for the U^2 term in an approximate way.

For the turbulent Reynolds-stresses on the left-hand side of (12.2.3) we use an eddy viscosity ν_t to connect those stresses to the cross-shore circulation velocity U. This means we introduce

$$-\overline{\widetilde{u'w'}} = \nu_t \frac{\partial U}{\partial z} \tag{12.2.4}$$

which substituted into (12.2.3) yields:

$$\frac{\partial}{\partial z}\left(\nu_t \frac{\partial U}{\partial z}\right) = \frac{\partial}{\partial x}\left(\overline{u_w^2} - \overline{w_w^2} + g\bar{\zeta} + U^2\right) + \frac{\partial}{\partial z}\overline{u_w w_w} \tag{12.2.5}$$

In order to solve this equation, we assume as usual that the short wave driver used for the circulation equations gives $\overline{u_w^2}$ and $\overline{w_w^2}$ (which are the components of the radiation stress), that the $\overline{u_w w_w}$-term is either known or neglected, and that ν_t is known.

The U^2-term is the effect of the undertow on itself, and in this case of 1DH cross-shore circulation it is small but non-negligible. We approximate it with the depth averaged value U_m given by (12.2.2). Thus we can write (12.2.5) as

$$\frac{\partial}{\partial z}\left(\nu_t \frac{\partial U}{\partial z}\right) = \alpha_1(x, z) \tag{12.2.6}$$

where $\alpha_1(x, z)$ is defined as

$$\alpha_1(x, z) = \frac{\partial}{\partial x}\left(\overline{u_w^2} - \overline{w_w^2} + g\bar{\zeta} + U_m^2\right) + \frac{\partial}{\partial z}\overline{u_w w_w} \tag{12.2.7}$$

We see that $\alpha_1(x, z)$ consists of quantities that are determined by the wave driver in addition to the $g\bar{\zeta}$-term which is determined by the solution to the depth integrated circulation equations. Therefore $\alpha_1(x, z)$ can be assumed known in this context. The general solution to (12.2.6) can then be determined by direct integration on both sides of the equation.

For convenience we introduce the new vertical coordinate ξ, which is the distance from the bottom, defined by

$$\xi = z + h \tag{12.2.8}$$

because it makes the values of the integration constants simpler. Without further assumptions, we can then integrate (12.2.6) directly twice and get for $U(z)$

$$U(z) = \int \frac{1}{\nu_t} \int \alpha_1(x, \xi) d\xi d\xi + A_1(x) \int \frac{d\xi}{\nu_t} + A_2(x) \tag{12.2.9}$$

where $A_1(x)$, $A_2(x)$ are the (arbitrary) integration functions we must determine from the boundary conditions.

12.2.2 *Boundary conditions*

We need two boundary conditions to determine A_1 and A_2.

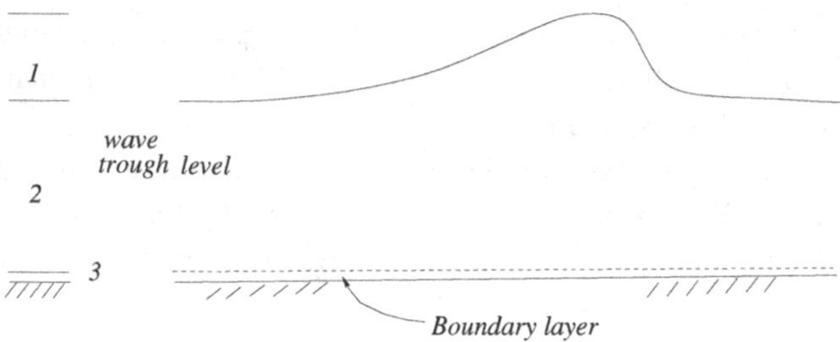

Fig. 12.2.1 The three layer concept.

A particular point is associated with the fact mentioned earlier that the current cannot be identified by ordinary averaging above wave trough level of the short waves where there is only water part of the time. Hence the momentum equation (12.2.6) is strictly speaking not valid above that level.

We by-pass this problem by solving (12.2.6) below trough level only and extending the analytical solution up to the mean water level $\overline{\zeta}$. To clarify this we introduce the formal concept of a **three layer model** (see Fig. 12.2.1).

The three layers are described as follows

Layer 1: The region above the level of the wave trough (the upper region).

Layer 2: The region below the level of the wave trough but above the top of the bottom boundary layer (the center region).

Layer 3: The bottom boundary layer.

The solution given by (12.2.9) is valid in both layer 2 and layer 3. All that is needed is to specify ν_t corresponding to the conditions in these two regions. However, because the oscillatory bottom boundary layer is very thin we will first focus on the solution for region 2, the central layer from the top of the oscillatory boundary layer to the wave trough level and seek the necessary boundary conditions to detemine the integration constants in the solution for that layer.

There are several ways of prescribing the necessary conditions. We first describe the approach that is particularly convenient when the vertical profiles are analyzed in association with the solution of the nearshore circulation equations described in Chapter 11, and then describe alternative boundary conditions.

When solving the circulation equations in the horizontal plane we obtain information about the total volume flux, $\overline{Q}$ and the vertical position of the MWS, $\overline{\zeta}$ at all grid points in the computational domain and at each time step. Therefore, whatever boundary conditions we specify for the vertical profile equations the parameters involved must be calculated from $\overline{Q}$ and $\overline{\zeta}$ along with the information about the short wave forcing provided by the wave driver. This is readily possible if we use the following two boundary conditions for the 2D vertical case. In Chapter 13 this is generalized to 3D.

I. The first boundary condition

The first is a boundary condition at the bottom. In the presence of the undertow we essentially have a bottom boundary layer which is a combination of a wave and a current motion.

In the derivation we utilize the usual concept that the boundary layer is thin enough to only disturb the flow above by exerting a (mean) shear stress on that flow at the bottom. Hence, at the interface between layer 2 and 3 in Fig. 12.2.1 we assume that the flow in layer 2 has a velocity distribution with a finite bottom value, the slip velocity U_b (see Fig. 12.2.2).

U_b is then connected to the (mean) shear stress $\overline{\tau_b}$ at the bottom by the expression found in Section 10.4.3.

In terms of the results in that chapter the case we are dealing with here has the angle $\mu = 0$ between the wave and current motion, and the current

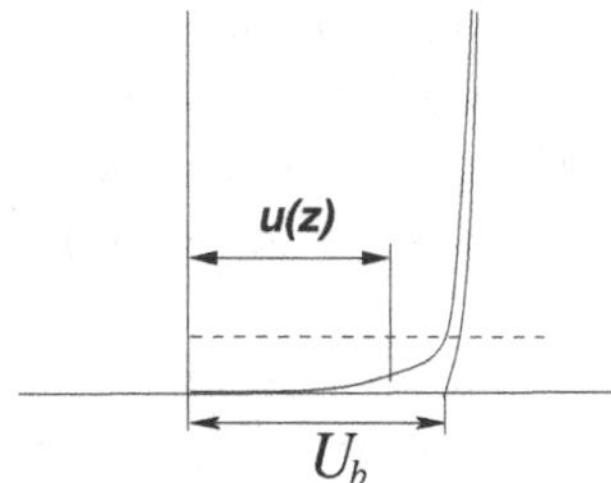

Fig. 12.2.2 The slip velocity U_b at the bottom.

velocity $U \ll u_o$. Hence $\overline{\tau_b}$ is given by (10.4.35).

$$\overline{\tau_{bx}} = \frac{2}{\pi} \rho f_{cw} u_o U_b \tag{12.2.10}$$

Since the mean shear stress varies very little over the boundary layer this shear stress must also be the shear stress at the "upper limit" of the boundary layer in the $U(z)$ distribution at the bottom of layer 2. Hence, we have the matching condition

$$\overline{\tau_{bx}} = \rho \nu_t \left(\frac{\partial U}{\partial \xi} \right)_b \tag{12.2.11}$$

Eliminating $\overline{\tau_{bx}}$ between (12.2.10) and (12.2.11) then yields the mixed boundary condition at the bottom

$$\boxed{\left(\frac{\partial U}{\partial \xi} \right)_b - \frac{2}{\pi} \frac{f_{cw}}{\nu_t} u_o U_b = 0} \tag{12.2.12}$$

(Svendsen and Putrevu, 1990).

II. The second boundary condition

The second condition is specified as a **continuity** condition for the cross-shore flow that replaces a boundary condition. It expresses that the total (cross-shore) flow equals $\overline{Q_x}$ (=0). Because there is no net volume flux in the x-direction (which is what we assume here) then the depth averaged velocity U_m satisfies the relation

$$U_m = \frac{1}{h} \int_{-h_o}^{\overline{\zeta}} U dz = -\frac{Q_w}{h} \tag{12.2.13}$$

It may be recalled from the note in Section 11.2 that the integration has to be carried all the way to $\overline{\zeta}$ meaning we have to define the current above trough level. Here we assume the analytical expressions found later are continued to $\overline{\zeta}$.

Eq. (12.2.13) is a boundary condition that can readily be specified from the information provided by the nearshore circulation equations.

Ia. Alternative boundary condition specifying a surface shear stress

An alternative boundary condition is to specify the value of the short wave generated shear stress that acts at wave trough level (Stive and Wind, 1986). This surface stress represents the part of the radiation stress generated above trough level. If we consider a control volume covering the region above the trough level $z = \zeta_t$ then we get that the shear stress $\tau(\zeta_t)$ at trough level must satisfy the horizontal momentum balance

$$\frac{\partial S_{xx,s}}{\partial x} + \rho g(\overline{\zeta} - \zeta_t)\frac{\partial \overline{\zeta}}{\partial x} + \tau(\zeta_t) = 0 \tag{12.2.14}$$

The radiation stress component $S_{xx,s}$ in this expression is given by

$$S_{xx,s} = \overline{\int_{\zeta_t}^{\zeta} \rho(u_w^2 - w_w^2)dz} + \frac{1}{2}\rho g\overline{\eta^2} \tag{12.2.15}$$

Stive and Wind uses this condition together with boundary condition II described above. The undertow profiles obtained from this approach using measured input for $\partial\overline{\zeta}/\partial x$ fit measurements as well as other methods. Since the τ_s-condition in essence is a condition for the derivative of U the integration constant must be detrermined so that the velocity profiles determined this way have the correct $\overline{Q}$.

12.2.3 *Solution for the undertow profiles with depth uniform ν_t and α_1*

We see that (12.2.9) is valid for arbitrary variations of both ν_t and α_1. For simplicity, however, we will (following Svendsen and Hansen, 1988) consider the simplified version of (12.2.9) that corresponds to assuming $\overline{u_w^2}$ constant over depth below trough level. For justification, see Fig. 5.5.14 which show measurements of $\overline{u_w^2}$ for a large number of experiments. It also turns out that $\overline{w_w^2} \ll \overline{u_w^2}$ so that we can neglect $\overline{w_w^2}$ (or assume $\overline{w_w^2} \sim C \cdot \overline{u_w^2}$, where $C \sim 0.05$). Then $\alpha_1(x, z) = \alpha_1(x)$.

For constant ν_t (12.2.9) simplifies to

$$U(\xi) = \frac{1}{2}\frac{\alpha_1}{\nu_t}\xi^2 + \frac{A_1}{\nu_t}\xi + A_2 \tag{12.2.16}$$

We see that

$$A_2 = U_b \tag{12.2.17}$$

and by (12.2.11)

$$\left(\frac{\partial U}{\partial \xi}\right)_b = \frac{\overline{\tau_b}}{\rho \nu_t} = \frac{A_1}{\nu_t} \tag{12.2.18}$$

or

$$A_1 = \frac{\overline{\tau_b}}{\rho} = \frac{2}{\pi} f_{cw} u_o U_b \tag{12.2.19}$$

the latter by virtue of (12.2.10). Thus $U(z)$ expressed in terms of U_b becomes

$$\boxed{U(\xi) = \frac{1}{2}\;\frac{\alpha_1}{\nu_t}\xi^2 + \frac{2}{\pi}\;\frac{f_{cw}}{\nu_t}u_o U_b \xi + U_b} \tag{12.2.20}$$

U_b is then determined by means of the second boundary condition (12.2.13).

Alternatively we can write $U(\xi)$ in terms of $\overline{\tau_b}$ as

$$U(\xi) = \frac{1}{2}\;\frac{\alpha_1}{\nu_t}\xi^2 + \frac{\overline{\tau_b}}{\rho \nu_t}\xi + U_b \tag{12.2.21}$$

Exercise 12.2-1 Show that using (12.2.13) we get for U_b

$$U_b' = \left(U_m' - \frac{1}{6}A\right)(1+R)^{-1} \tag{12.2.22}$$

where

$$U_b' = \frac{U_b}{\sqrt{gh}} \quad ; \quad U_m' = \frac{U_m}{\sqrt{gh}}$$
$$A = \frac{\alpha_1 h^2}{\nu_t \sqrt{gh}} \quad ; \quad R = \frac{f_{cw} u_o h}{\pi \nu_t} \tag{12.2.23}$$

Therefore we can write the solution (12.2.20) in dimensionless form by dividing with $\sqrt{gh}$. We get

$$U'(\xi) = \frac{U(\xi)}{\sqrt{gh}} = \frac{1}{2}A\left(\frac{\xi}{h}\right)^2 + 2\,R\,U_b'\,\frac{\xi}{h} + U_b' \qquad (12.2.24)$$

We see that $U'(z)$ depends on two parameters:

- A is a measure of the ratio between α_1 — the forcing that generates the flow (see the introduction) — and ν_t — the viscosity that indicates the capability of the flow to carry the shear stresses.
- R is the ratio between bottom friction factor f_{cw} and the eddy viscosity ν_t. As (12.2.22) shows, R represents the change in U_b due to the bottom shear stress. It may be expected that R is small as the following example shows.

12.2.4 *Discussion of results and comparison with measurements*

In order to evaluate the results for the undertow profile from (12.2.20) (or (12.2.24)), the driving forces α_1 given by (12.2.7) are required. The wave driver will provide information about $\overline{u_w^2}$ and $\overline{w_w^2}$. Furthermore, the gradient $\partial\bar{\zeta}/\partial x$ of the MWS is an important part of α_1, which must be determined from the solution of the circulation equations, here the momentum equation (12.1.1).

In the simple 1DH cross-shore case considered here the direct solution of the momentum equation can be avoided by realizing that $\partial\bar{\zeta}/\partial x$ can be eliminated from the problem by substituting $\partial\bar{\zeta}/\partial x$ from (12.1.1) into α_1. The result is

$$\alpha_1(x,z) = \frac{\partial}{\partial x}\left(\overline{u_w^2} - \overline{w_w^2} + U_m^2\right) + \frac{\partial}{\partial z}\overline{u_w w_w} - \frac{1}{\rho h}\frac{\partial}{\partial x}S_{xx} - \frac{\tau_b}{\rho h} \qquad (12.2.25)$$

However, $\bar{\zeta}$ will still be needed to determine the local water depth $h = h_0 + \bar{\zeta}$.

If we accept that $\overline{\tau_b}$ is negligible then (12.2.25) expresses the driving force in terms that can be calculated from information about the wave motion only i.e. from the wave driver.

As we can see from (12.2.25) the forcing for the undertow can also be interpreted as being equal to the difference between the local contribution to the radiation stress - represented by the $\overline{u_w^2} - \overline{w_w^2} + U_m^2$-term and the $\overline{u_w w_w}$-term - and the depth averaged total radiation stress represented by the S_{xx}-term. This forcing is only slightly modified by the mean bottom friction τ_b.

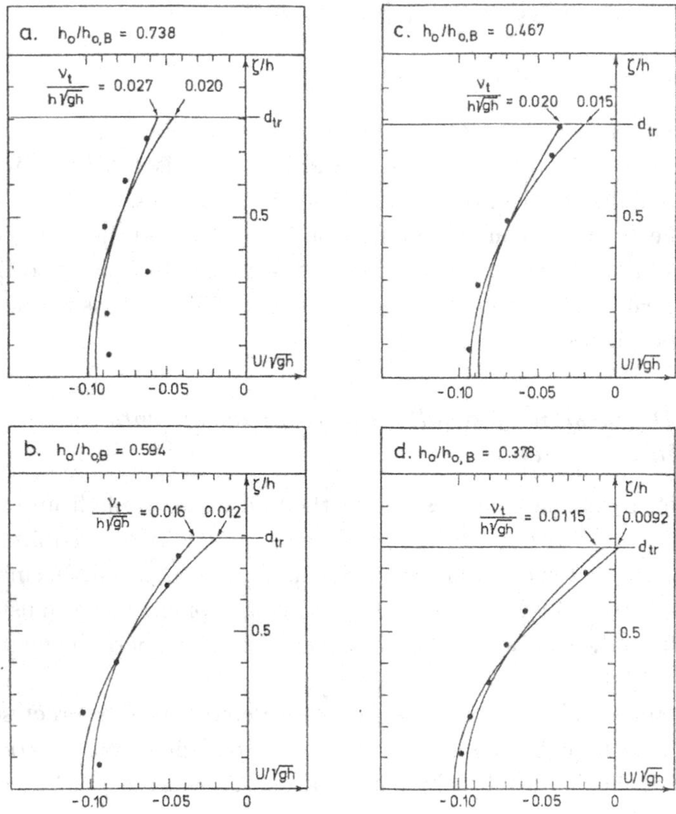

Fig. 12.2.3 Measured and computed undertow velocities at four locations in the surfzone from (12.2.20), (12.2.22) and (12.2.23) for different values of ν_t. Slope 1:35. It is seen that the maximum value of $U \sim 0.05 - 0.10\sqrt{gh}$.

Evaluation of the eddy viscosity

A frequently used form for ν_t is

$$\nu_t = \ell\sqrt{q} \tag{12.2.26}$$

where ℓ is a characteristic length scale for the turbulence, and $q = \frac{1}{2}\left(\widetilde{u'^2} + \widetilde{v'^2} + \widetilde{w'^2}\right)$ is the turbulent kinetic energy.

The major part of the turbulence is created by the wave breaking which suggests that $\ell \propto h$ and $\sqrt{q} \propto c \sim \sqrt{gh}$. This leads to an expression of the form

$$\nu_t = Ch\sqrt{gh} \tag{12.2.27}$$

where C is a proportionality factor. In this form ν_t is constant over depth. This, however, is not necessary in the general solution (12.2.9).

Thus the solution requires specification of the constant C (or its variation over depth if (12.2.9) is used) in the expression for the eddy viscosity ν_t. ν_t can be determined by comparing computed undertow profiles to experimental results for $U(z)$ and adjust ν_t to obtain the best possible agreement.

As an example such comparisons were done by Svendsen et al. (1987). Fig. 12.2.3 shows comparison of the profiles at four different positions in the surfzone on a plane laboratory beach.[1] In each figure panel is shown the eveluation of (12.2.20) for different values of C. We see that the shape of the indertow profile is quite sensitive to the value of ν_t, which means that the best fit gives a relatively accurate estimate of ν_t.

We also see form the figure that actually the best fit for C is not quite constant accross the surfzone, the variation being in the range of $C = 0.01 - 0.02$.

The shear stress variation and the bottom shear stress τ_b

The bottom shear stress can also be evaluated from the solution for $U(\xi)$. Using (12.2.21) we have for the shear stress variation over the column

$$\overline{\tau} = \rho\nu_t\frac{\partial U}{\partial z} = \rho\alpha_t\xi + \overline{\tau_b} \tag{12.2.28}$$

Svendsen and Hansen (1988) analyzed the bottom shear stress from model simulations. They considered the patching solution of Svendsen et al. (1987) with two different, but depth uniform values, of ν_t, one small value inside the bottom boundary layer one much larger in layer 2 above the bottom boundary layer. They gave the variation of $\overline{\tau_b}$ with a number of parameters including the level at which the patching between the two layers was made. The values of $\overline{\tau_b}$ corresponding to the physically realistic patching level of $\zeta_0 = 2.5\delta_w$ (where δ_w is the boundary layer thickness defined by equation (10.3.15)) are the most realistic so here we focus on those results. They show shear stresses of the order $0.05 - 0.10\ \rho\alpha_1 d_{tr}$. As

[1] Fig. 12.2.3 was originally intended as Fig. 4 in Svendsen et al. (1987) but was accidentally replaced with a wrong figure.

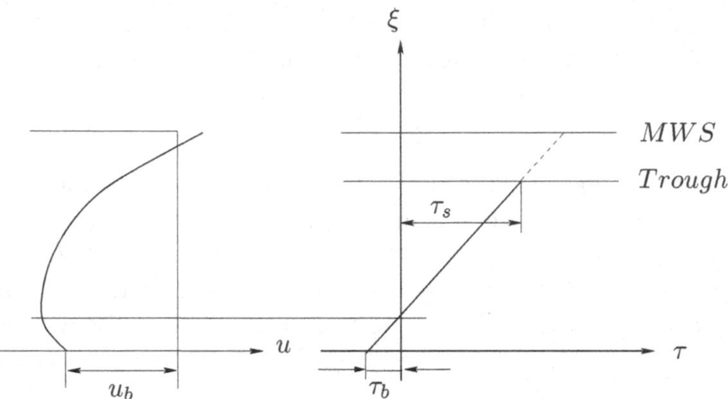

Fig. 12.2.4 The shear stress distibution for depth uniform α_1 and the relation to the velocity profile.

(12.2.28) shows $\rho\alpha_1 d_{tr}$ is a relevant measure in comparison to the variation of $\overline{\tau(\xi)}$ in layer 2, because $\rho\alpha_1 d_{tr}$ is the total (depth integrated) driving force for the undertow below trough level. From (12.2.28) we see this means that the $\overline{\tau}$ varies linearly over depth and the total variation $\overline{\tau(d_{tr})} - \overline{\tau_b}$ is simply $\rho\alpha_1 d_{tr}$. Thus the value of $0.05 - 0.10$ means the bottom shear stress is on a small fraction of the total variation. This also means $\overline{\tau_b} \ll \overline{\tau_s}$.

Going back to the undertow velocity that then confirms the impression from the figures 12.2.3 that the maximum velocity - which occurs at the level where $\overline{\tau} = 0$ - is found close to the bottom. Fig. 12.2.4 shows the relation between the shear stress and the velocity profile.

Their results also showed that the value of $\overline{\tau_b}$ depends quite strongly on the bottom roughness parameter, k_N which is consistent with the results for turbulent boundary layers.

Finally, since the total forcing $\rho\alpha_1 d_{tr}$ over the column for the undertow is much less than than either of the two large terms in (12.1.1) it follows that the value of $\overline{\tau_b}$ in that equation is really very small as indicated at several places before.

12.2.5 *Solutions including the effect of the boundary layer*

Depth varying eddy viscosity

As mentioned the general solution (12.2.9) is valid in both layer 2 and 3 and for an arbitrary distribution of the eddy viscosity. However, when ν_t is constant over depth we cannot satisfy the condition of zero velocity at the bottom, which is what leads to the assumption of a slip velocity there, and the boundary layer is only included through the bottom shear stress it generates. This is the situation expressed by boundary condition II.

A simple way to model the velocity distribution including the boundary layer is to specify a variation of the eddy viscosity which realistically describes the conditions in the boundary layer (layer 3 in Fig. 12.2.1) as well as in the central layer 2.

Simple considerations suggest that near the bottom the turbulence is restricted so that both q and ℓ in (12.2.27) may be smaller there. In the oscillatory boundary layer the wave (and current) motion creates its own turbulence, and there are strong indications that the boundary layer turbulence is much weaker than the breaker turbulence in layer 2, which means ν_t becomes much smaller near the bottom, and tends to 0 at the bottom. Hence, we would expect the nature of the ν_t-variation is as shown in Fig. 12.2.5.

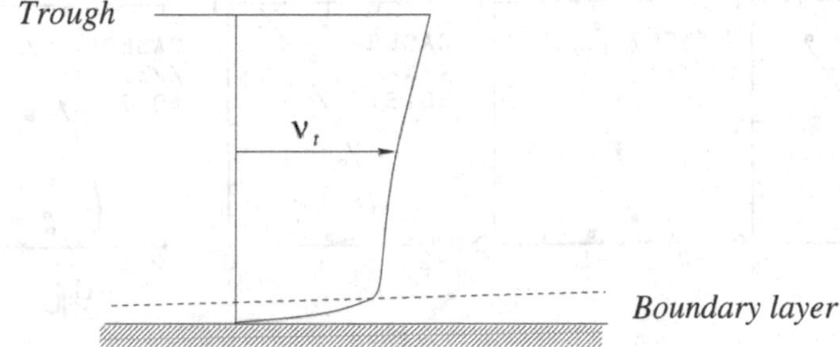

Fig. 12.2.5 Estimated variation of ν_t.

In the literature, the variation of ν_t has been included in various ways. In particular, if $\nu_t \to 0$ at the bottom, then the solution (12.2.9) allows $U = 0$ at $\xi = 0$.

An example is the eddy viscosity distribution (Okayasu et al., 1988) given by

$$\nu_t = C\xi\sqrt{gd_c} \tag{12.2.29}$$

This ν_t varies linearly upwards from $\nu_t = 0$ at the bottom. The value at the surface differs slightly from (12.2.27) by using a velocity based on d_c

which is the depth under the wave crest. Okayasu makes C depend on the bottom slope by setting

$$C = 0.30\, h_x \tag{12.2.30}$$

but combines this with an empirical expression for the driving force α_1 which is also proportional to h_x so that α_1/ν_t in the first term of (12.2.9) is again independent of h_x and has the form

$$\frac{\alpha_1}{\nu_t} = C_1 \frac{\sqrt{g d_c}}{d_t} \frac{1}{\xi} \tag{12.2.31}$$

where d_t is the depth at wave trough.

Substituted into (12.2.9) and using the bottom condition that $U = 0$ at $\xi = 0$, the following solution can be obtained

$$U(\xi) = U_m - 0.017c \ln \xi/d_t + 0.15c\xi/d_t - 0.10\, c \tag{12.2.32}$$

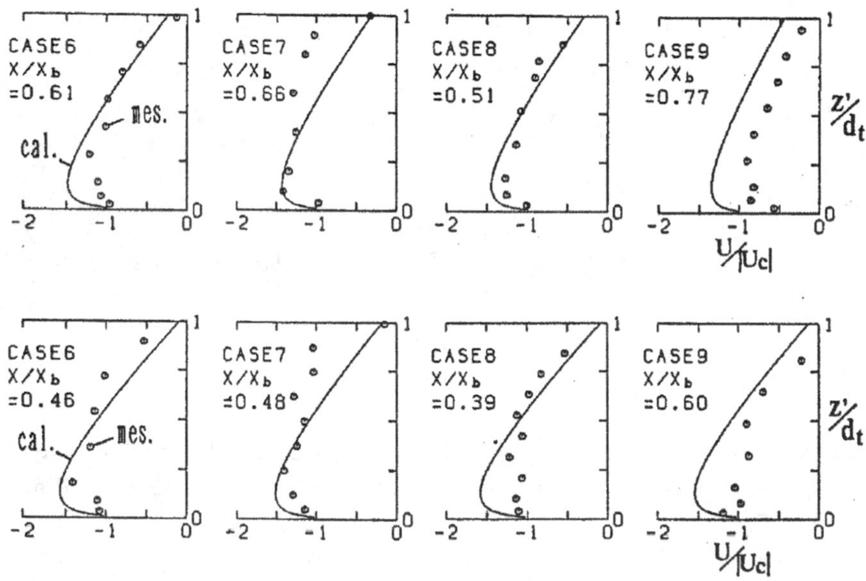

Fig. 12.2.6 Comparison between the measured and the calculated undertow profiles for a 1/30 slope (from Okayasu et al. (1988)).

As an example Fig. 12.2.6 shows comparison with measurements of U/U_m for a bottom slope of 1/30. We see that this model predicts the variation $U(\xi)$ close to the bottom reasonably well but is less accurate in the upper part of the water column.

What is important to notice is that for a bottom slope in the range of $1/20 - 1/30$ found in most experiments for comparisons the ν_t-expression (12.2.29) gives values of $\nu_t = 0.01 - 0.015\ h\ \sqrt{gh}$ at the surface where the values in this model are largest. This is in the same range as found for $\nu_t = C\ h\sqrt{gh}$. This observation is essential for the discussion in Chapter 13.

The effect of the steady streaming u_s in the boundary layer

Another approach is to consider a matched problem of two layers (layer 2 and 3) with vastly different ν_t: The breaker dominated layer 2 above the bottom boundary layer with $\nu_t \sim C_1 h\sqrt{gh}$ where $C_1 = O(10^{-2})$, and at the bottom the boundary layer with $\nu_t \sim C_2 h\sqrt{gh}$ as a region with very low eddy viscosity with C_2 typically $O(10^{-4})$ (Svendsen et al. 1987, Svendsen and Hansen, 1988). In the latter case, the $\partial\overline{u_w w_w}/\partial z$ terms were included in the bottom boundary layer, and the effect of the steady streaming analyzed.

The following mechanisms are at play for the averaged motion:

- In purely oscillatory boundary layers, we found a shoreward oriented steady streaming (see Chapter 10) driven by a $u_w w_w$ variation inside the boundary layer.
- However, measurements in the surfzone show that even close to the bottom boundary layer we have strong **seaward** oriented currents.
- We must expect that the seaward oriented undertow in layer 2 interacts with the generation of the steady streaming generated in layer 3. To get realistic results this interaction must obviously be properly accounted for.

Svendsen et al. (1987) gives a solution based on patching the two solutions in layer 2 and layer 3 using continuity in velocities and shear stresses. Based on the results for turbulent boundary layers described in chapter 10 the eddy vicosity in the boundary layer is $O(10^{-2})$ times the viscosity of $10^{-2}h\sqrt{gh}$ found as typical between wave trough and bottom. Svendsen and Hansen (1988) derive a simplified version that treats the boundary layer as a black box while still including the effect of the shear stress reponsible for the steady streaming.

The results show that even inside the boundary layer the undertow has a tendency to be stronger and hence override the steady streaming.

12.2.6 *Undertow outside the surfzone*

Outside the surfzone we still have a volume flux in the waves and therefore also and undertow flow, but because the Q_w is much smaller the velocities in the undertow are also much smaller than inside the surfzone. However, it also turns out that because of the weaker forcing the steady streaming in the boundary layer is no longer negligible.

Results by Putrevu and Svendsen (1993) shown in Fig. 12.2.7 for the theoretical profiles both inside and outside the surfzone illustrate the dramatical difference between to two regions. Their solution is based on the patching of the solutions for the boundary layer and layer 2 above between wave trough and the boundary layer. In the figure the ratio between the two eddy viscositie is $\nu_{tg}/\nu_t = 100$. Fig. 12.2.8 shows comparisons between theoretical results and laboratory measurements that confirm that outside the surfzone the velocity profile is almost linear over depth with the largest values near the surface.

12.2.7 *Conclusions*

The analysis shows that for the cross-shore 2DV conditions on a long straight coast or in a wave flume there is a cross-shore circulation caused by the volume flux generated by the short wave motion. As shown in Chapter 5 this volume flux is particularly strong in the surf zone. In such simpel 2DV cases the return current (the "undertow") can be analysed and predicted with good accuracy provided we are able to obtain the necessary input for the calculations. The necessary input is

- an appropriate model for the short wave motion (the wave driver) from which we can determine the variation of the radiation stress S_{xx} and the short wave volume flux Q_w.
- an appropriate model for the depth avaraged flow variables $\bar{\zeta}$ and Q_α from solving the circulation equations (Chapter 11),
- and the eddy viscosity distribution in the region.

More importantly, however, the analysis shows that the undertow in the 2-D flow under breaking waves is far from depth uniform.

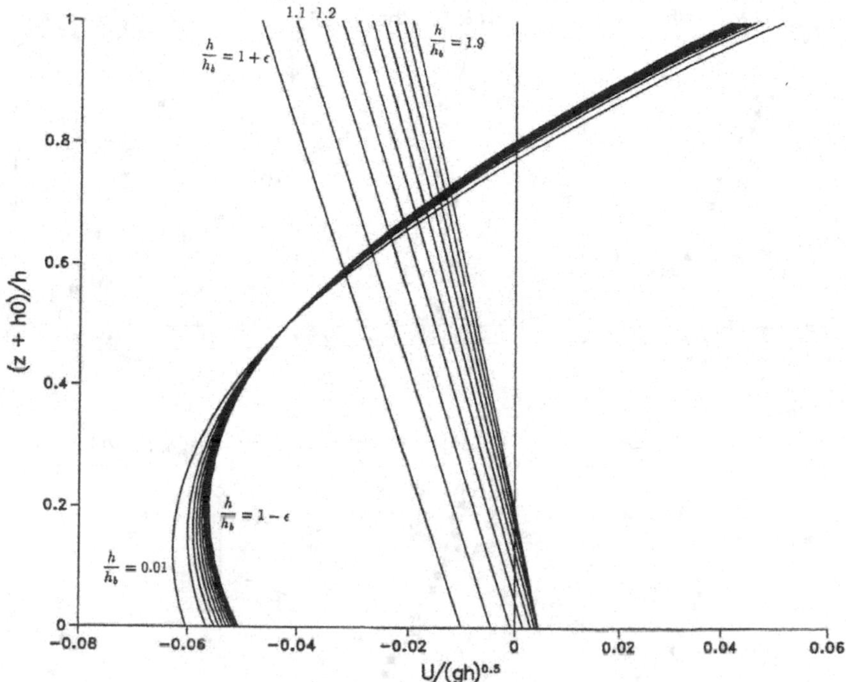

Fig. 12.2.7 Undertow profiles inside and outside the surfzone for $\nu_{tg}/\nu_t = 100$ and for different values of h/h_b, h_b being the brealing depth (from Putrevu and Svendsen, 1993).

The analysis of the undertow essentially outlines the approach that can be used for a more general analysis of the 3-dimensional variation of wave generated currents in nearshore regions discussed in the Chapter 13.

12.3 References - Chapter 12

Hansen, J. B. and I. A. Svendsen (1986). Experimental investigation of the wave and current motion over a longshore bar. ASCE Proc. 20th Int. Conf. Coastal Enrgr., 544–556.

Hansen, J. B. and I. A. Svendsen (1987). Proc IUTAM Symposium on Wave Breaking, Tokyo, Japan.

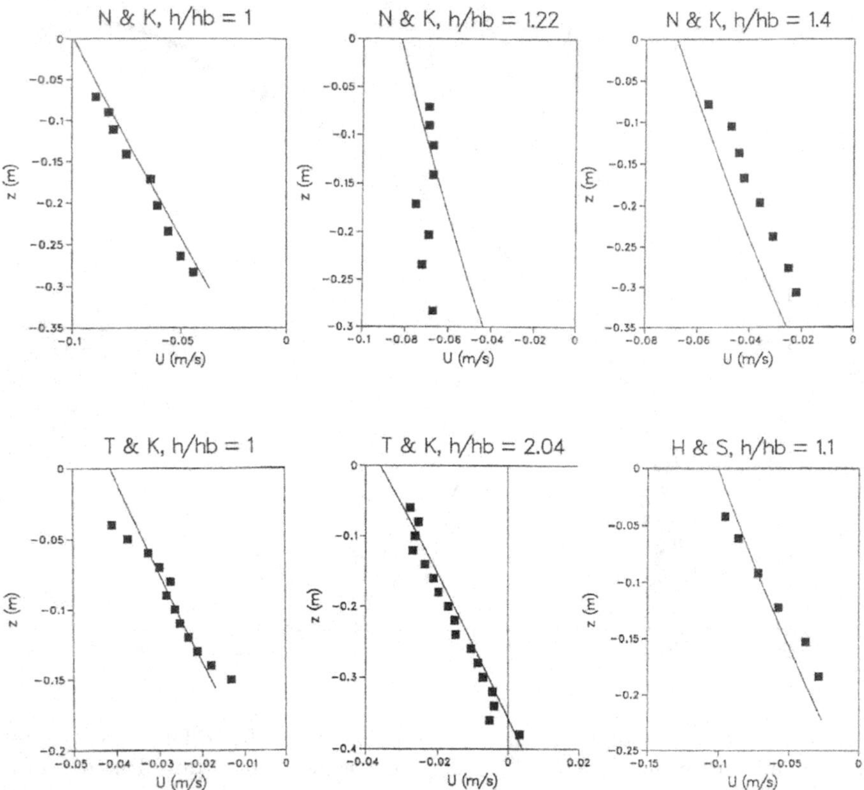

Fig. 12.2.8 Comparison of theoretical predictions with experimental observations of undertow profiles outside the surfzone. N & K refers to Nadaoka and Kondoh (1982), T & K to Ting and Kirby (1994, 1995), and H & S to Hansen and Svendsen (1986) (adapted from Putrevu and Svendsen, 1993).

Nadaoka, K. and T. Kondoh (1982). Laboratory measurements of velocity field structure in the surfzone by LDV. Coastal Eng. Japan, **25**, 125—145.

Okayasu, A., T. Shibayama, and K. Horikawa (1988). Vertical variation of undertow in the surf-zone. ASCE Proc 21st Int. Conf. Coastal Engrg., 478–491.

Putrevu, U. and I. A. Svendsen (1993). Vertical structure of the undertow outside the surf zone. J. Geophys. Res., **98**, C12, 22707–22716.

Stive, M. J. F. and H. G. Wind (1986). Cross-shore mean flow in the surf zone. Coast, Engrg. **10**, 325–340.

Svendsen, I. A. Mass flux and undertow in a surf zone (1984). Coastal Engrg., 8, 4, pp. 347–366, Closure, **10**, 4, 299–307.

Svendsen, I. A., H. A. Schäffer, and J. B. Hansen (1987). The interaction between undertow and the boundary layer flow on a beach. J. Geophys. Res., **92**, C. 11, pp. 11845–11856.

Svendsen, I. A. and J. Buhr Hansen (1988). Cross-shore currents in surf-zone modelling. Coastal Engineering, **12**, 1, 23–42.

Ting, F. C. K. and J. T. Kirby (1994). Observation of undertow and turbulence in a laboratory surfzone. Coastal Engrg., **24**, 51 - 80.

Ting, F. C. K. and J. T. Kirby (1995). Dynamics of surf-zone turbulence in a strong plunging breaker. Coastal Engineering, **24**, 177–204.

Chapter 13

Quasi-3D Nearshore Circulation Models

13.1 Introduction

In Section 11.7.3 we saw that for longshore currents on a straight beach the lateral mixing coefficient could be written

$$\nu_t = C_x \, h\sqrt{gh} \tag{13.1.1}$$

where index x refers to mixing in the cross-shore direction , and it was found that a C_x-value of $0.45 - 1.80$ is required to obtain cross-shore variations of the longshore current that resembled the measured variations. The order of magnitude of this number has been verified repeatedly in the literature.

On the other hand the vertical variation of the cross-shore undertow profile investigated in Section 12.2 used a ν_t, which can be written as

$$\nu_t = C_z \, h\sqrt{gh} \tag{13.1.2}$$

requires a C_z-value of only $0.01-0.02$, at most, to achieve similarly reallistic comparisons to measured undertow velocities. Again this number has been verified in the litrature by several authors using different models.

Thus, there is an order of magnitude difference between the ν_t required to predict the measured cross shore variation of longshore currents and the measured vertical variation of undertow, respectively.

It is obvious that such a discrepancy is disturbing from a physical point of view. A natural hypothesis is to assume that the mixing is caused by the turbulence which is abundant due to the wave breaking. We may then express ν_t as

$$\nu_t = \ell\sqrt{k} \tag{13.1.3}$$

where k is the turbulent kinetic energy and ℓ a characteristic length scale for the turbulence. Based on measurements, Svendsen (1987) estimated $k^{1/2} = O(0.05\sqrt{gh})$, a result which was based on many laboratory experiments by independent authors. With $C_z = 0.01$, this then corresponds to $\ell_z = O(0.2h)$ which seems reasonable. In contrast, $C_x \sim 0.45$–1.80 will correspond to $\ell_x = O(9-36h)$. Although there clearly could be a difference between the vertical and horizontal structure of the turbulence generated by the wave breaking, a factor of 45–180 between the respective mixing coefficients seems unlikely.

In fact, it would be a more reasonable assumption that the characteristic length scale ℓ in the horizontal and vertical directions are similar (so that $\nu_{tx} = \nu_{tz} = \nu_t$ or $C_x = C_z$).

As an illustration, Fig. 13.1.1 shows calculations of the longshore current distribution $V(x)$ with ν_{tx} based on $C_x = 0.01$ and 0.35, respectively. The value of $C_x = 0.35$ gives a cross-shore distribution compares well with the measurements used by Longuet-Higgins (1970a,b) and the very detailed and accurate laboratory measurements by Visser (1984). For simplicity, the analytical solution derived in Section 11.7.3 has been used in the computation.

Svendsen and Putrevu (1994) (SP94) analysed this problem for the simple case of the longshore current on a long straight beach and found that the discrepancy is caused by the omission of the vertical variation of the cross- and longshore currents. This variation represents mechanisms that cause far more lateral mixing than even turbulence generated by the wave breaking in the surf zone. This mixing mechanism has been termed **dispersive mixing** because it disperses horizontal momentum. In fact in the longshore uniform case it was found that the mixing caused by the turbulence only accounts for approximately 5% while the dispersive mixing accounts for 95% or more of the total lateral mixing. Hence they found that the smaller values for the eddy viscosity required to correctly predict the undertow were close to the real values and when used together with the mixing caused by the vertical current profile variation the lateral mixing would also be large enough to fit the measured longshore current data.

The 3-dimensional nature of the current profiles has been known for a long time and it follows directly from the strong vertical variation of the undertow. In comparison the longshore currents vary only a little over depth. However, when cross- and longshore velocities are added vectorily the result is a spiral-like profile as shown in Fig. 13.1.2 (Svendsen and Lorenz, 1989).

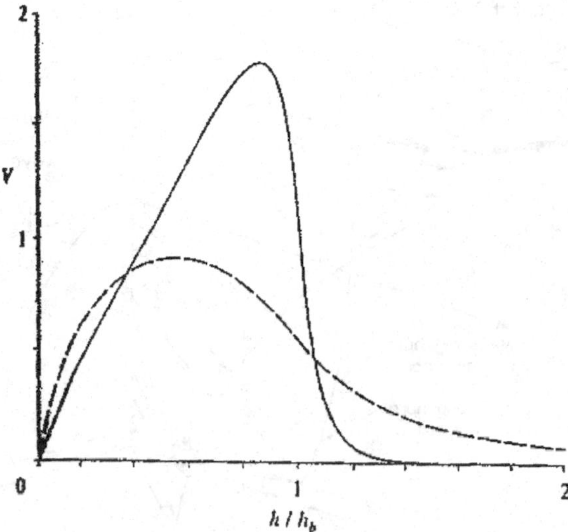

Fig. 13.1.1 Cross-shore variation of longshore current according to the analytical model derived in Section 11.7.3 for $C_x = 0.01$ and 0.35. From Svendsen and Putrevu (1994).

The analysis of this mechanism has since been extended to the general 3-dimensional situation of arbitrary topography and current variation by Putrevu and Svendsen (1999), and Haas and Svendsen (2000).

The conclusion is that physically realistic models for general nearshore circulation phenomena should include the effect of the 3D current profiles.

Discussion

The possibility exists that turbulence with a much larger length scale than the the turbulence generated by the wave breaking could play a role in the total lateral mixing. Such turbulence occurs on real beaches due to shear instability of the longshore currents (see Section 14.2). This was discussed and dismissed by SP94 because it is unlikely such turbulence could develop over the short longshore lengths of Vissers laboratory experiments used by SP94 for illustrating the effect of the dispersive mixing. Later model computaions with large scale conditions similar to natural beaches described in Section 14.2 show, however, that such large scale turbulence does have a significant lateral mixing effect. The remakable finding is that when this occurs the dispersive mixing due to the 3-dimensional current

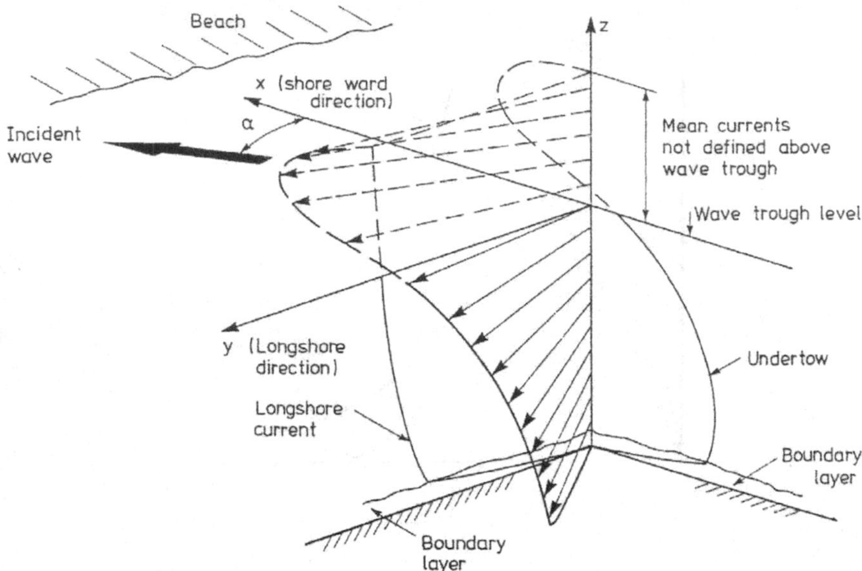

Fig. 13.1.2 The depth variation of the combined cross- and longshore velocity profiles in the surfzone (from Svendsen and Lorenz, 1989).

profiles will automatically be reduced similarly. For discussion of this see Section 14.2.6.

Quasi-3D models

On the other hand, fully 3D models would be an order of magnitude more demanding computationally than the 2DH models represented by the circulation equations. This has lead to the introduction of quasi-3D models in which the 3D effects of the currents are represented in the 2DH circulation equations by coefficients the same way the effect of the velocity profile in a hydraulic flow is represented by a momentum correction factor in the 1DH momentum equation. The equivalent in the quasi-3D equations is an array of coefficients that are determined analytically by solving the local momentum equations for the vertical distribution of the current, from which the coefficients are then calculated. In this way the quasi-3D models essentially solve 2DH equations and therefore are numerically almost as fast as traditional 2DH models although they include the depth integrated effecs of the 3D current profiles.

In the following we briefly outline the derivation the equations for the generalized case of the Quasi-3D circulation equations. The resulting equations are also the governing equations used in nearshore circulation model SHORECIRC (SC) which was first developed by Van Dongeren et al. (1994).

13.2 Governing equations

In many ways this chapter presents generalizations of the results derived and discussed in Chapters 11 and 12.2 for special cases.

The idea of a circulation model which includes 3D current profiles was probably first introduced by DeVriend and Stive (1987) and by Sanchez-Arcilla et al. (1990). Svendsen and Putrevu (1994) showed that the 3D current profiles is not just a matter of adding cosmetic details to the circulation models. As mentioned above they found the 3-dimensionality of the currents stongly influences the horizontal exchange of horizontal momentum and thereby the lateral mixing.

The derivation of the depth integrated equations for the flow was given in Chapter 11 and lead to the continuity equation (11.2.27) and the momentum equation (11.5.13) for depth varying currents. However, it also gave the form of the equations for currents constant over depth.

13.2.1 *Time-averaged depth-integrated equations*

For reference we repeat here the continuity equation given as

$$\frac{\partial \overline{\zeta}}{\partial t} + \frac{\partial \overline{Q_\alpha}}{\partial x_\alpha} = 0 \tag{13.2.1}$$

where $\overline{Q_\alpha}$ as usual is the total volume flux including the volume flux Q_w in the short wave motion.

Similarly the momentum equation for arbitrarily varying currents was given as

$$\frac{\partial \overline{Q_\alpha}}{\partial t} + \frac{\partial}{\partial x_\beta} \overline{\int_{-h_o}^{\zeta} V_\alpha V_\beta dz} + \frac{\partial}{\partial x_\beta} \overline{\int_{\zeta_t}^{\zeta} u_{w\alpha} V_\beta + u_{w\beta} V_\alpha dz} =$$
$$-gh \frac{\partial \overline{\zeta}}{\partial x_\alpha} - \frac{1}{\rho} \frac{\partial}{\partial x_\beta} \left[S'_{\alpha\beta} - \overline{\int_{-h_o}^{\zeta} \tau_{\alpha\beta} dz} \right] + \frac{\tau_\alpha^S}{\rho} - \frac{\tau_\alpha^B}{\rho} \tag{13.2.2}$$

with the radiation stress $S'_{\alpha\beta}$ defined as

$$S'_{\alpha\beta} \equiv \overline{\int_{-h_o}^{\zeta} p\delta_{\alpha\beta} + \rho u_{w\alpha} u_{w\beta}\, dz} - \delta_{\alpha\beta}\frac{1}{2}\rho g h^2 \tag{13.2.3}$$

13.2.2 *Choices for splitting the current*

In (13.2.2) the momentum equation is written in terms of the total current velocity, V_α and a short-wave velocity, $u_{w\alpha}$. However, it is convenient to further divide the current velocity by splitting it into a depth uniform and a depth varying component as follows,

$$V_\alpha(z) = V_{m\alpha} + V_{d\alpha}(z) \tag{13.2.4}$$

where we define the depth uniform component by

$$V_{m\alpha} \equiv \frac{\overline{Q_\alpha} - Q_{w\alpha}}{h} \tag{13.2.5}$$

Thus the depth uniform component of the current only represents the true current part and does not include the wave volume flux which originates between crest and trough.

The combination of (13.2.4) and (13.2.5) implies that

$$\overline{\int_{-h_o}^{\zeta} V_{d\alpha}\, dz} = 0 \tag{13.2.6}$$

Substituting (13.2.4) into (13.2.2) yields, after some algebra

$$\begin{aligned}
&\frac{\partial \overline{Q_\alpha}}{\partial t} + \frac{\partial}{\partial x_\alpha}\left(\frac{\overline{Q_\alpha}\,\overline{Q_\beta}}{h}\right) + \frac{\partial}{\partial x_\beta}\overline{\int_{-h_o}^{\zeta} V_{d\alpha} V_{d\beta} dz} \\
&+\frac{\partial}{\partial x_\beta}\overline{\int_{\zeta_t}^{\zeta} u_{w\alpha} V_{d\beta} + u_{w\beta} V_{d\alpha} dz} = \\
&-gh\frac{\partial \bar{\zeta}}{\partial x_\alpha} - \frac{1}{\rho}\frac{\partial}{\partial x_\beta}\left[S_{\alpha\beta} - \overline{\int_{-h_o}^{\zeta} \tau_{\alpha\beta} dz}\right] + \frac{\tau_\alpha^S}{\rho} - \frac{\tau_\alpha^B}{\rho}
\end{aligned} \tag{13.2.7}$$

where a modified radiation stress similar to the definition used by Phillips (1966,1977) is defined as,

$$S_{\alpha\beta} = S'_{\alpha\beta} - \frac{Q_{w\alpha} Q_{w\beta}}{h} \tag{13.2.8}$$

Because of (13.2.6), $V_{d\alpha}$ vanishes for the depth uniform current case. This means that for depth uniform currents the integral terms in (13.2.7) simply vanish as well. In this case the contribution from the integral terms for the depth uniform currents are already included in the modified radiation stress as evident in (13.2.8). Thus the depth uniform equations are obtained directly from (13.2.7) by simply omitting all the integral terms. These terms therefore clearly represent the effect of the depth variation of the currents, and the objective becomes to calculate those terms. We see that they are nonlinear current-current and current-wave interaction terms.

We notice that, when we have solved the depth integrated, time averaged equations for $\overline{Q}, \overline{\zeta}$, V_m can be determined directly from (13.2.5) because Q_w is known from the wave driver.

Exercise 13.2-1

Derive (13.2.8). Notice that this is not quite trivial!

13.3 Solution for the vertical velocity profiles

The solution for the vertical profiles is essentially a generalization of the solution presented in Section 12.2 for the 1-DH undertow.

The momentum equations given by (13.2.7) are all in terms of depth-averaged properties except for the integral terms. In order to evaluate the integral terms we must derive a local (non depth-integrated) momentum equation for the depth varying part $V_{d\alpha}$ of the currents.

We begin with the local horizontal momentum balance which is rewritten here for convenience,

$$\frac{\partial u_\alpha}{\partial t} + \frac{\partial u_\alpha u_\beta}{\partial x_\beta} + \frac{\partial u_\alpha w}{\partial z} = -\frac{1}{\rho}\frac{\partial p}{\partial x_\alpha} + \frac{1}{\rho}\left(\frac{\partial \tau_{\alpha\beta}}{\partial x_\beta} + \frac{\partial \tau_{\alpha z}}{\partial z}\right) \tag{13.3.1}$$

In this we introduce the velocity separation into a wave and a current part, time-average the equation and assume the short-wave-averaged pressure is hydrostatic

$$\overline{p} = \rho g(\overline{\zeta} - z) \tag{13.3.2}$$

This gives after some algebra

$$\frac{\partial V_\alpha}{\partial t} + \frac{\partial}{\partial x_\beta}\left(V_\alpha V_\beta + \overline{u_{w\alpha} u_{w\beta}}\right) + \frac{\partial}{\partial z}\left(V_\alpha W + \overline{u_{w\alpha} w_w}\right)$$
$$= -\frac{1}{\rho}\frac{\partial \overline{\zeta}}{\partial x_\alpha} + \frac{1}{\rho}\left(\frac{\partial \overline{\tau_{\alpha\beta}}}{\partial x_\beta} + \frac{\partial \overline{\tau_{\alpha z}}}{\partial z}\right) \tag{13.3.3}$$

By using the local continuity equation written as

$$\frac{\partial V_\alpha}{\partial x_\alpha} + \frac{\partial W}{\partial z} = 0 \tag{13.3.4}$$

we rewrite (13.3.3) as

$$\frac{\partial V_\alpha}{\partial t} + V_\beta \frac{\partial V_\alpha}{\partial x_\beta} + W\frac{\partial V_\alpha}{\partial z} + \frac{\partial}{\partial x_\alpha}(\overline{u_{w\alpha} u_{w\beta}}) + \frac{\partial}{\partial z}(\overline{u_{w\alpha} w_w}) =$$
$$-\frac{1}{\rho}\frac{\partial \overline{\zeta}}{\partial x_\alpha} + \frac{1}{\rho}\left(\frac{\partial \overline{\tau_{\alpha\beta}}}{\partial x_\beta} + \frac{\partial \overline{\tau_{\alpha z}}}{\partial z}\right) \tag{13.3.5}$$

It may be noticed that we no longer neglect the vertical component W of the current.

Further we introduce the split for the current given by (13.2.4) and express the turbulent stresses using an eddy viscosity model,

$$\overline{\tau_{\alpha\beta}} = \rho(\nu_t + \nu_s)\left(\frac{\partial V_\beta}{\partial x_\alpha} + \frac{\partial V_\alpha}{\partial x_\beta}\right) \tag{13.3.6}$$

$$\overline{\tau_{\alpha z}} = \rho(\nu_t + \nu_s)\left(\frac{\partial V_{d\alpha}}{\partial z} + \frac{\partial W}{\partial x_\alpha}\right) \tag{13.3.7}$$

with ν_t being the the eddy viscosity representing the turbulence created by the bottom friction and the short-wave breaking, and ν_s being the eddy viscosity created by the shear in the flow. This results in

$$\frac{\partial V_{d\alpha}}{\partial t} - \frac{\partial}{\partial z}\left((\nu_t + \nu_s)\frac{\partial V_{d\alpha}}{\partial z}\right)$$
$$= -\left(\frac{\partial V_{m\alpha}}{\partial t} + V_{m\beta}\frac{\partial V_{m\alpha}}{\partial x_\beta} + g\frac{\partial \overline{\zeta}}{\partial x_\alpha} + f_\alpha\right)$$
$$-\left(V_{m\beta}\frac{\partial V_{d\alpha}}{\partial x_\beta} + V_{d\beta}\frac{\partial V_{m\alpha}}{\partial x_\beta} + V_{d\beta}\frac{\partial V_{d\alpha}}{\partial x_\beta} + W\frac{\partial V_{d\alpha}}{\partial z}\right)$$
$$+\frac{\partial}{\partial x_\alpha}\left((\nu_t + \nu_s)\left(\frac{\partial V_\alpha}{\partial x_\beta} + \frac{\partial V_\beta}{\partial x_\alpha}\right)\right) + \frac{\partial}{\partial z}\left((\nu_t + \nu_s)\frac{\partial W}{\partial x_\alpha}\right) \tag{13.3.8}$$

Here we have defined f_α, which is the local contribution to the radiation stress, as

$$f_\alpha = \frac{\partial}{\partial x_\alpha}(\overline{u_{w\alpha}u_{w\beta}}) + \frac{\partial}{\partial z}(\overline{u_{w\alpha}w_w}). \tag{13.3.9}$$

Using the definition for $V_{m\alpha}$ (13.2.5) in (13.3.8) yields

$$\begin{aligned}
&\frac{\partial V_{d\alpha}}{\partial t} - \frac{\partial}{\partial z}\left((\nu_t + \nu_s)\frac{\partial V_{d\alpha}}{\partial z}\right) \\
&\quad = -\left(\frac{\partial \frac{\overline{Q}_\alpha}{h}}{\partial t} + \frac{\overline{Q}_\beta}{h}\frac{\partial \frac{\overline{Q}_\alpha}{h}}{\partial x_\beta} + g\frac{\partial \overline{\zeta}}{\partial x_\alpha} + f_\alpha\right) \\
&\quad\quad - \left(\frac{\overline{Q}_\beta}{h}\frac{\partial V_{d\alpha}}{\partial x_\beta} + V_{d\beta}\frac{\partial \frac{\overline{Q}_\alpha}{h}}{\partial x_\beta} + V_{d\beta}\frac{\partial V_{d\alpha}}{\partial x_\beta} + W\frac{\partial V_{d\alpha}}{\partial z}\right) \\
&\quad\quad + \left(\frac{\partial \frac{Q_{w\alpha}}{h}}{\partial t} + \frac{\overline{Q}_\beta}{h}\frac{\partial \frac{Q_{w\alpha}}{h}}{\partial x_\beta} + \frac{Q_{w\beta}}{h}\frac{\partial \frac{\overline{Q}_\alpha}{h}}{\partial x_\beta} - \frac{Q_{w\beta}}{h}\frac{\partial \frac{Q_{w\alpha}}{h}}{\partial x_\beta} + \frac{Q_{w\beta}}{h}\frac{\partial V_{d\alpha}}{\partial x_\beta}\right. \\
&\quad\quad \left. + V_{d\beta}\frac{\partial \frac{Q_{w\alpha}}{h}}{\partial x_\beta}\right) + \frac{\partial}{\partial x_\alpha}\left((\nu_t + \nu_s)\left(\frac{\partial V_\alpha}{\partial x_\beta} + \frac{\partial V_\beta}{\partial x_\alpha}\right)\right) \\
&\quad\quad + \frac{\partial}{\partial z}\left((\nu_t + \nu_s)\frac{\partial W}{\partial x_\alpha}\right)
\end{aligned} \tag{13.3.10}$$

This is the equation we need to solve for $V_{d\alpha}$. To do that $V_{d\alpha}$ is divided into two parts by the definition

$$V_{d\alpha} = V_{d\alpha}^{(0)} + V_{d\alpha}^{(1)} \tag{13.3.11}$$

The detailed derivation of the principal results is shown in Appendix 13A. A general version is in Putrevu and Svendsen (1999) (PS99) who uses a different way of splitting the current into a depth uniform and a depth varying part. Therefore their results differ slightly from this presentation. The slightly simplified form used in the present version of the model can be found in Haas and Svendsen (2000) (HS2000) and some parts of the derivation are quoted from that report.

The results (13.A.12) and (13.A.15) found in the Appendix 13A for $V_{d\alpha}^{(0)}$ and $V_{d\alpha}^{(1)}$ can be evaluated analytically for depth uniform eddy viscosity and $F_\alpha^{(0)}$. As can be seen from the coefficients in section 13.4 below a further simplification is possible: only the $V_{d\alpha}^{(0)}$-component of the velocity is required to account for the final values of the current-current and current-wave interaction terms. The effect of $V_{d\alpha}^{(1)}$ can be accounted for analytically too. For details about the transformation leading to this, reference is made to (PS99), Appendix B.

We therefore focus on the expression for $V_{d\alpha}^{(0)}$. When evaluating (13.A.12) the vertical variation of $V_{d\alpha}^{(0)}$ can then be expressed in the form (HS2000)

$$V_{d\alpha}^{(0)} = d_{1\alpha}\xi^2 + e_{1\alpha}\xi + f_{1\alpha} + f_{2\alpha} \tag{13.3.12}$$

where

$$\xi = z + h_0 \tag{13.3.13}$$

and

$$d_{1\alpha} = -\frac{F_\alpha^{(0)}}{2\nu_t} \tag{13.3.14}$$

$$e_{1\alpha} = \frac{\tau_\alpha^B}{\rho\nu_t} \tag{13.3.15}$$

$$f_{1\alpha} = -\frac{h}{2}\frac{\tau_\alpha^B}{\rho(\nu_t + \nu_s)} \tag{13.3.16}$$

$$f_{2\alpha} = \frac{h^2 F_\alpha^{(0)}}{6(\nu_t + \nu_s)} \tag{13.3.17}$$

with F_α given by

$$F_\alpha^{(0)} = \left\{\frac{1}{\rho h}\frac{\partial S'_{\alpha\beta}}{\partial x_\beta} - \frac{\tau_\alpha^S}{\rho h} + \frac{\tau_\alpha^B}{\rho h} - f_\alpha\right\} \tag{13.3.18}$$

and f_α given by (13.3.9)

In addition, $V_{d\alpha}^{(1)}$ is given by

$$V_{d\alpha}^{(1)} = V_{d\alpha}^{(1,a)} + V_{d\alpha}^{(1,b)} + V_{d\alpha}^{(1,w)} + V_{d\alpha}^{(1,c)} \tag{13.3.19}$$

with

$$V_{d\alpha}^{(1,a)} = V_{d\alpha}^{(1,a,4)}\xi^4 + V_{d\alpha}^{(1,a,3)}\xi^3 + V_{d\alpha}^{(1,a,2)}\xi^2 \tag{13.3.20}$$

$$V_{d\alpha}^{(1,b)} = V_{d\alpha}^{(1,b,4)}\xi^4 + V_{d\alpha}^{(1,b,3)}\xi^3 + V_{d\alpha}^{(1,b,2)}\xi^2 \tag{13.3.21}$$

$$V_{d\alpha}^{(1,w)} = V_{d\alpha}^{(1,w,4)}\xi^4 + V_{d\alpha}^{(1,w,3)}\xi^3 + V_{d\alpha}^{(1,w,2)}\xi^2 \tag{13.3.22}$$

and

$$
\begin{aligned}
V_{d\alpha}^{(1,c)} = & \left[V_{d\alpha}^{(1,a,4)} + V_{d\alpha}^{(1,b,4)} + V_{d\alpha}^{(1,w,4)}\right]\frac{h^4}{5} \\
& \left[V_{d\alpha}^{(1,a,3)} + V_{d\alpha}^{(1,b,3)} + V_{d\alpha}^{(1,w,3)}\right]\frac{h^3}{4} \\
& \left[V_{d\alpha}^{(1,a,2)} + V_{d\alpha}^{(1,b,2)} + V_{d\alpha}^{(1,w,2)}\right]\frac{h^2}{3}
\end{aligned} \tag{13.3.23}
$$

For further details see Haas and Svendsen (2000).

13.4 Calculation of the integral terms: The dispersive mixing coefficients

From the current velocity profiles the current-current and current-wave interaction terms in (13.2.7) can be rewritten in terms of a set of coefficients $M_{\alpha\beta}$, $D_{\alpha\gamma}$, $B_{\alpha\beta}$, and $A_{\alpha\beta\gamma}$ which are the 3D dispersion coefficients. After substantial derivations (see HS2000) the integral terms can then be written in terms of those coefficients by introducing the following approximations

$$
\overline{\int_{-h_o}^{\zeta} V_{d\alpha}\, dz} \approx \int_{-h_o}^{\bar{\zeta}} V_{d\alpha}\, dz \tag{13.4.1}
$$

$$
\begin{aligned}
& \overline{\textstyle\int_{-h_o}^{\zeta} V_{d\alpha} V_{d\beta} dz} + \overline{\int_{\zeta_t}^{\zeta} u_{w\alpha} V_{d\beta} + u_{w\beta} V_{d\alpha} dz} \approx \\
& \qquad \int_{-h_o}^{\bar{\zeta}} V_{d\alpha} V_{d\beta} dz + V_{d\beta}(\bar{\zeta}) Q_{w\alpha} + V_{d\alpha}(\bar{\zeta}) Q_{w\beta}
\end{aligned} \tag{13.4.2}
$$

This essentially assumes that $V_{d\alpha}$ remains constant between ζ_t and ζ. In the first place it is not clear how we should define $V_{d\alpha}$ above trough level. However, when the surface wind stresses are neglected this is not a bad approximation.

We then get

$$
\begin{aligned}
& \int_{-h_o}^{\bar{\zeta}} V_{d\alpha} V_{d\beta} dz + V_{d\beta}(\bar{\zeta}) Q_{w\alpha} + V_{d\alpha}(\bar{\zeta}) Q_{w\beta} \\
& \quad = M_{\alpha\beta} + A_{\alpha\beta\gamma} \frac{\overline{Q}_\gamma}{h} - h \left(D_{\gamma\beta} \frac{\partial \frac{\overline{Q}_\alpha}{h}}{\partial x_\gamma} + D_{\gamma\alpha} \frac{\partial \frac{\overline{Q}_\beta}{h}}{\partial x_\gamma} + B_{\alpha\beta} \frac{\partial \frac{\overline{Q}_\gamma}{h}}{\partial x_\gamma} \right)
\end{aligned} \tag{13.4.3}
$$

The 3D-dispersive mixing coefficients in (13.4.3) are defined by[1]

$$
\begin{aligned}
A_{\alpha\beta\gamma} = -\Bigg\{ & \int_{-h_o}^{\bar\zeta} \frac{1}{(\nu_t + \nu_s)} \left(\int_{-h_o}^{z} \frac{\partial V_{d\alpha}^{(0)}}{\partial x_\gamma} - \frac{\partial \frac{Q_{w\alpha}}{h}}{\partial x_\gamma} - \frac{\partial h_o}{\partial x_\gamma} \frac{\partial V_{d\alpha}^{(0)}}{\partial z} \right) \\
& \times \left(\int_{-h_o}^{z} V_{d\beta}^{(0)} - \frac{Q_{w\beta}}{h} dz' \right) dz \\
& + \int_{-h_o}^{\bar\zeta} \frac{1}{(\nu_t + \nu_s)} \left(\int_{-h_o}^{z} \frac{\partial V_{d\beta}^{(0)}}{\partial x_\gamma} - \frac{\partial \frac{Q_{w\beta}}{h}}{\partial x_\gamma} - \frac{\partial h_o}{\partial x_\gamma} \frac{\partial V_{d\beta}^{(0)}}{\partial z} \right) \\
& \times \left(\int_{-h_o}^{z} V_{d\alpha}^{(0)} - \frac{Q_{w\alpha}}{h} dz' \right) dz \Bigg\}
\end{aligned}
\tag{13.4.4}
$$

$$
\begin{aligned}
B_{\alpha\beta} = -\frac{1}{h} \Bigg\{ & \int_{-h_o}^{\bar\zeta} \frac{1}{(\nu_t + \nu_s)} \left(\int_{-h_o}^{z} (h_o + z') \frac{\partial V_{d\alpha}^{(0)}}{\partial z} dz' \right) \\
& \times \left(\int_{-h_o}^{z} V_{d\beta}^{(0)} - \frac{Q_{w\beta}}{h} dz' \right) dz \\
& + \int_{-h_o}^{\bar\zeta} \frac{1}{(\nu_t + \nu_s)} \left(\int_{-h_o}^{z} (h_o + z') \frac{\partial V_{d\beta}^{(0)}}{\partial z} dz' \right) \\
& \times \left(\int_{-h_o}^{z} V_{d\alpha}^{(0)} - \frac{Q_{w\alpha}}{h} dz' \right) dz \Bigg\}
\end{aligned}
\tag{13.4.5}
$$

$$
\begin{aligned}
D_{\alpha\beta} = \frac{1}{h} & \int_{-h_o}^{\bar\zeta} \frac{1}{(\nu_t + \nu_s)} \left(\int_{-h_o}^{z} V_{d\alpha}^{(0)} - \frac{Q_{w\alpha}}{h} dz' \right) \\
& \times \left(\int_{-h_o}^{z} V_{d\beta}^{(0)} - \frac{Q_{w\beta}}{h} dz' \right) dz
\end{aligned}
\tag{13.4.6}
$$

$$
M_{\alpha\beta} = \int_{-h_o}^{\bar\zeta} V_{d\alpha}^{(0)} V_{d\beta}^{(0)} dz + V_{d\beta}^{(0)}(\bar\zeta) Q_{w\alpha} + V_{d\alpha}^{(0)}(\bar\zeta) Q_{w\beta} \tag{13.4.7}
$$

Here the last coefficient $M_{\alpha\beta}$ represents a simplification of the full coefficient that comes out of the derivations which is justified in (HS2000).

[1] It is emphasized again that although these formulas only seem to contain $V_{d\alpha}^{(0)}$ the full effect of $V_{d\alpha}^{(1)}$ in (13.3.11) is included in the formulas because in the above expressions for the dispersive mixing coefficients the contributions involving $V_{d\alpha}^{(1)}$ have been expressed analytically in terms of $V_{d\alpha}^{(0)}$. See Putrevu and Svendsen (1999), Appendix B.

Alternative form of the coefficient $D_{\alpha\beta}$

It is interesting to notice that the form (13.4.6) shown here for the $D_{\alpha\beta}$-coefficient is in a form which is convenient for analytical and numerical computations. However, its resemblance with the mixing coefficient in turbulent pipe flow found by Taylor (1954), and with the coefficients found by Fischer (1978) currents can better be realized if the coefficient is transformed the following way. We define the quantities p and q by

$$p = \frac{1}{h}\int_{-h_o}^{z} V_{d\beta}^{(0)} - \frac{Q_{w\beta}}{h} dz' \; ; \qquad q = \int_{z}^{\overline{\zeta}} \frac{1}{(\nu_t + \nu_s)} \int_{-h_o}^{z} V_{d\alpha}^{(0)} - \frac{Q_{w\alpha}}{h} dz' \tag{13.4.8}$$

By these definitions we see that

$$\frac{\partial q}{\partial z} = -\frac{1}{(\nu_t + \nu_s)} \int_{-h_o}^{z} V_{d\alpha}^{(0)} - \frac{Q_{w\alpha}}{h} dz' \tag{13.4.9}$$

Thus we can write $D_{\alpha\beta}$ as

$$\begin{aligned} D_{\alpha\beta} &= \frac{1}{h}\int_{-h_o}^{\overline{\zeta}} \frac{1}{(\nu_t + \nu_s)} \left(\int_{-h_o}^{z} V_{d\alpha}^{(0)} - \frac{Q_{w\alpha}}{h} dz'\right) \left(\int_{-h_o}^{z} V_{d\beta}^{(0)} - \frac{Q_{w\beta}}{h} dz'\right) dz \\ &= -\int_{-h_o}^{\overline{\zeta}} \frac{\partial q}{\partial z} \, p \, dz = -\int_{-h_o}^{\overline{\zeta}} p \, dq \\ &= -[pq]_{-h_o}^{\zeta} + \int_{-h_o}^{\overline{\zeta}} q \, dp \end{aligned} \tag{13.4.10}$$

Here the boundary terms vanish because $p(-h_o) = 0$ and $q(\overline{\zeta}) = 0$. We therefore get for the coefficient

$$D_{\alpha\beta} = \frac{1}{h}\int_{-h_o}^{\overline{\zeta}} \left(V_{d\alpha}^{(0)} - \frac{Q_{w\alpha}}{h}\right) \int_{z}^{\overline{\zeta}} \frac{1}{(\nu_t + \nu_s)} dz' \int_{-h_o}^{z'} \left(V_{d\alpha}^{(0)} - \frac{Q_{w\alpha}}{h}\right) dz \tag{13.4.11}$$

which esentially is the form in which the above mentioned references by Taylor (1954) and Fischer (1978) gave the results.

13.4.1 *Final form of the basic equations*

The final depth-averaged horizontal momentum equation is then

$$\frac{\partial \overline{Q_\alpha}}{\partial t} + \frac{\partial}{\partial x_\alpha}\left(\frac{\overline{Q_\alpha Q_\beta}}{h} + A_{\alpha\beta\gamma}\frac{\overline{Q}_\gamma}{h}\right) + \frac{1}{\rho}\frac{\partial}{\partial x_\beta}\left[S'_{\alpha\beta} + \rho M_{\alpha\beta}\right]$$
$$+ gh\frac{\partial \overline{\zeta}}{\partial x_\alpha} + \frac{\tau_\alpha^B}{\rho} - \frac{\partial}{\partial x_\beta}\left[\overline{\int_{-h_o}^{\zeta} \tau_{\alpha\beta} dz} + h\left(D_{\gamma\beta}\frac{\partial \frac{\overline{Q}_\alpha}{h}}{\partial x_\gamma} + D_{\gamma\alpha}\frac{\partial \frac{\overline{Q}_\beta}{h}}{\partial x_\gamma}\right)\right]$$
$$- \frac{\partial}{\partial x_\beta}\left(hB_{\alpha\beta}\frac{\partial \frac{\overline{Q}_\gamma}{h}}{\partial x_\gamma}\right) = 0 \tag{13.4.12}$$

We see that together with the continuity equation this forms a system with surface elevation $\overline{\zeta}$ and the total volume flux components Q_x and Q_y as dependent unknowns. When written out in x,y-coordinates the results are as follows

The continuity equation

$$\frac{\partial \overline{\zeta}}{\partial t} + \frac{\partial Q_x}{\partial x} + \frac{\partial Q_y}{\partial y} = 0 \tag{13.4.13}$$

and

The x-momentum equation

$$\frac{\partial Q_x}{\partial t} + \frac{\partial}{\partial x}\left(\frac{Q_x^2}{h} + M_{xx}\right) + \frac{\partial}{\partial y}\left(\frac{Q_x Q_y}{h} + M_{xy}\right)$$
$$- \frac{\partial}{\partial x}h\left[(2D_{xx} + B_{xx})\frac{\partial}{\partial x}\left(\frac{Q_x}{h}\right) + 2D_{xy}\frac{\partial}{\partial y}\left(\frac{Q_x}{h}\right) + B_{xx}\frac{\partial}{\partial y}\left(\frac{Q_y}{h}\right)\right]$$
$$- \frac{\partial}{\partial y}h\left[(D_{xy} + B_{xy})\frac{\partial}{\partial x}\left(\frac{Q_x}{h}\right) + D_{yy}\frac{\partial}{\partial y}\left(\frac{Q_x}{h}\right) + D_{xx}\frac{\partial}{\partial x}\left(\frac{Q_y}{h}\right)\right.$$
$$\left. + (D_{xy} + B_{xy})\frac{\partial}{\partial y}\left(\frac{Q_y}{h}\right)\right]$$
$$+ \frac{\partial}{\partial x}\left[A_{xxx}\frac{Q_x}{h} + A_{xxy}\frac{Q_y}{h}\right] + \frac{\partial}{\partial y}\left[A_{xyx}\frac{Q_x}{h} + A_{xyy}\frac{Q_y}{h}\right]$$
$$= -gh\frac{\partial \overline{\zeta}}{\partial x} - \frac{1}{\rho}\left(\frac{\partial S_{xx}}{\partial x} + \frac{\partial S_{xy}}{\partial y}\right) + \frac{1}{\rho}\left(\frac{\partial}{\partial x}\overline{\int_{-h_0}^{\zeta} \tau_{xx} dz} + \frac{\partial}{\partial y}\overline{\int_{-h_0}^{\zeta} \tau_{xy} dz}\right)$$
$$+ \frac{\tau_x^S - \tau_x^B}{\rho} \tag{13.4.14}$$

The y-momentum equation

$$\frac{\partial Q_y}{\partial t} + \frac{\partial}{\partial x}\left(\frac{Q_x Q_y}{h} + M_{xy}\right) + \frac{\partial}{\partial y}\left(\frac{Q_y^2}{h} + M_{yy}\right)$$
$$-\frac{\partial}{\partial x} h \left[(D_{xy} + B_{xy})\frac{\partial}{\partial x}\left(\frac{Q_x}{h}\right) + D_{yy}\frac{\partial}{\partial y}\left(\frac{Q_x}{h}\right) + D_{xx}\frac{\partial}{\partial x}\left(\frac{Q_y}{h}\right)\right.$$
$$\left.+(D_{xy} + B_{xy})\frac{\partial}{\partial y}\left(\frac{Q_y}{h}\right)\right]$$
$$-\frac{\partial}{\partial y} h \left[B_{yy}\frac{\partial}{\partial x}\left(\frac{Q_x}{h}\right) + 2D_{xy}\frac{\partial}{\partial x}\left(\frac{Q_y}{h}\right) + (2D_{yy} + B_{yy})\frac{\partial}{\partial y}\left(\frac{Q_y}{h}\right)\right]$$
$$+\frac{\partial}{\partial x}\left[A_{xyx}\frac{Q_x}{h} + A_{xyy}\frac{Q_y}{h}\right] + \frac{\partial}{\partial y}\left[A_{yyx}\frac{Q_x}{h} + A_{yyy}\frac{Q_y}{h}\right]$$
$$= -gh\frac{\partial \bar{\zeta}}{\partial y} - \frac{1}{\rho}\left(\frac{\partial S_{xy}}{\partial x} + \frac{\partial S_{yy}}{\partial y}\right) + \frac{1}{\rho}\left(\frac{\partial}{\partial x}\overline{\int_{-h_0}^{\zeta} \tau_{xy} dz} + \frac{\partial}{\partial y}\overline{\int_{-h_0}^{\zeta} \tau_{yy} dz}\right)$$
$$+\frac{\tau_y^S - \tau_y^B}{\rho} \tag{13.4.15}$$

In the above, $\tau_\alpha^B, \tau_\alpha^S$, are the bottom shear stresses, the surface (wind) shear stresses, and $S_{\alpha\beta}, \tau_{\alpha\beta}$ are the radiation stresses and the turbulent Reynold's stresses, respectively.

The equations (13.4.13), (13.4.14) and (13.4.15) are the equations solved by the SHORECIRC model.

13.5 Example: Longshore currents on a long straight coast

The nature of the quasi-3D equations is perhaps best illustrated by considering the simple example of cross- and longshore currents on a long straight coast. This was the examples that were analysed in detail for depth uniform currents in Section 11.7.3. Comparison of the former results with the quasi-3D results will help explain the mechanisms involved in the dispersive mixing.

Thus we first need to extract the form of the general equations for this case. As before we consider a steady situation and, on a long straight coast, there are no variations in the y-direction and no net volume flux in the x-direction, so

$$\frac{\partial}{\partial t} = 0 \ ; \qquad \frac{\partial}{\partial y} = 0 \ ; \qquad \overline{Q}_x = 0 \tag{13.5.1}$$

As in the case with depth uniform currents we also disregard for simplicity the surface stresses τ_α^S The cross-shore **x-momentum** equation then reduces to

$$\frac{\partial M_{xx}}{\partial x} + \frac{\partial}{\partial x}\left(A_{xxy}\frac{\overline{Q}_y}{h}\right) =$$

$$-gh\,\frac{\partial\overline{\zeta}}{\partial x} - \frac{1}{\rho}\frac{\partial}{\partial x}\left(S_{xx} - \overline{\int_{-h_0}^{\zeta}\tau_{xx}\,dz}\right) - \frac{\tau_x^B}{\rho} \tag{13.5.2}$$

Inspection of (13.4.4) shows that for a long straight coast $A_{xxy} = 0$. which is analysed further below. As before we also disregard the turbulent normal stresses which are negligeable. We then simply get

$$\boxed{\frac{\partial M_{xx}}{\partial x} = -gh\,\frac{\partial\overline{\zeta}}{\partial x} - \frac{1}{\rho}\frac{\partial S_{xx}}{\partial x} - \frac{\tau_x^B}{\rho}} \tag{13.5.3}$$

For the longshore **y-component** of the momentum equation we similarly get by introducing the assumptions (13.5.1)

$$\frac{\partial M_{xy}}{\partial x} - \frac{\partial}{\partial x}\left(h\,D_{xx}\frac{\partial}{\partial x}\left(\frac{\overline{Q}_y}{h}\right)\right) + \frac{\partial}{\partial x}\left(A_{xyy}\frac{\overline{Q}_y}{h}\right) =$$

$$-gh\,\frac{\partial\overline{\zeta}}{\partial x} - \frac{1}{\rho}\frac{\partial}{\partial x}\left(S_{xy} - \overline{\int_{-h_0}^{\zeta}\tau_{xy}\,dz}\right) - \frac{\tau_y^B}{\rho} \tag{13.5.4}$$

where inspection of the difinitions for the coefficinets show that

$$M_{xy} = \int_{-h_o}^{\overline{\zeta}} V_{dx}V_{dy}dz + V_{dy}(\overline{\zeta})Q_{wx} + V_{dx}(\overline{\zeta})Q_{wy} \tag{13.5.5}$$

and $A_{xyy} = 0$. Also, for D_{xx} we get

$$D_{xx} = \frac{1}{h}\int_{-h_o}^{\overline{\zeta}} \frac{1}{(\nu_t + \nu_s)}\left(\int_{-h_o}^{z} V_{dx} - \frac{Q_w}{h}dz'\right)^2 dz \tag{13.5.6}$$

At this point we also introduce the expression for τ_{xy}^B

$$\tau_{xy}^B = \rho\nu_t\,\frac{\partial}{\partial x}\left(\frac{Q_y}{h}\right) \tag{13.5.7}$$

The final form of the longshore component of the momentum equation can then be written after integration of the τ_{xy}-term

$$\boxed{\frac{\partial M_{xy}}{\partial x} - \frac{\partial}{\partial x}\left[(D_{xx} + \nu_t)h\frac{\partial}{\partial x}\left(\frac{\overline{Q}_y}{h}\right)\right] + \frac{1}{\rho}\frac{\partial}{\partial x}S_{xy} + \frac{\tau_y^B}{\rho} = 0} \qquad (13.5.8)$$

Discussion of the results for a long straight beach

When we compare the cross-shore momentum equation (13.5.2) with the equivalent equation (11.7.8) in Section 11.7.2 for the cross-shore momentum balance, we see that the quasi-3D version of the equation contains the additional M_{xx}-term which represents the momentum correction factor for the cross-shore velocity profile in which the total volume flux is $\overline{Q_x} = 0$. Otherwize the equation is the same as for depth uniform currents.

Similarly when we compare the longshore momentum equation (13.5.8) with the equivalent equation (11.7.31) in Section 11.7.3 we see that the quasi-3D version of the equation contains an additional M_{xy}-term and, more importantly, a D_{xx}-term.

The $M_{\alpha\beta}$ is directly similar to the hydraulic momentum correction factor and is for this case given as

$$M_{xx} = \int_{-h_0}^{\overline{\zeta}} V_{dx}^{(0)2} dz + 2V_{dx}^{(0)}(\overline{\zeta})Q_{wx} \qquad (13.5.9)$$

which with the above simplifications becomes

$$M_{xx} \sim 2V_{dx}(0)(\overline{\zeta})Q_{wx} \qquad (13.5.10)$$

At $z = \overline{\zeta}$ the undertow velocity is close to zero meaning $V_{dx}^{(0)}(\overline{\zeta}) \sim -V_{mx} = Q_{wx}/h$ so that we get

$$M_{xx} \sim 2\frac{Q_{wx}^2}{h} \qquad (13.5.11)$$

In comparison with the other terms in the cross-shore equation this is a smalll contribution.

In the y-direction we get from (13.5.5)

$$M_{xy} = V_{dy}(\overline{\zeta})Q_{wx} + V_{dx}(\overline{\zeta})Q_{wy} \qquad (13.5.12)$$

and since the longshore current is almost uniform over depth we have $V_{dy}(\overline{\zeta}) \sim 0$ and with a small angle of incidence for the waves (as is often the case) we may also expect $Q_{wy} \sim 0$ so that we can conclude that M_{xy} is a very small contribution.

The D_{xx}-term is the important result here. In (13.5.8) D_{xx} obviously has the same role as the lateral mixing coefficient ν_t so that its function is to enhance the lateral mixing, which we found is too weak if we let ν_t correspond to the value it has in undertow. Thus it beome of paramount importance what the value of D_{xx} is because we found that we need a factor of the order 20 on ν_t to fit the measurements.

In order to get a simple feeling for the magnitude of the additional terms in the equations let us further disregard the vertical variation of the current represented by V_{dx} and V_{dy}. That means **approximating the undertow with a depth uniform return current** balancing the Q_w and the longshore current profile with its depth average. In the cross-shore direction the current variation over depth then consists of the (depth uniform) return flow V_{mx} and the (forward oriented) volume flux Q_w in the short waves. In the longshore direction the profile is given by a depth uniform longshore current V_{my} and the y-component of the short wave volume flux Q_{wy}.

The D_{xx} is more complicated to calculate than the M_{xx}-term. Since V_{dx} has been assumed smaller than the Q_w/h we may get a first approximation to D_{xx} by neglecting V_{dx} in (13.5.6). We also neglect the background eddy viscosity ν_s. The expression for D_{xx} then reduces to

$$D_{xx} = \frac{1}{h} \int_{-h_o}^{\bar{\zeta}} \frac{1}{\nu_t} \left(\int_{-h_o}^{z} \frac{Q_w}{h} dz' \right)^2 dz \tag{13.5.13}$$

which can be integrated directly to give

$$D_{xx} = \frac{1}{3} \frac{Q_w^2}{\nu_t} \tag{13.5.14}$$

which means that the ratio between the two lateral mixing coefficients is

$$\frac{D_{xx}}{\nu_t} = \frac{1}{3} \frac{Q_w^2}{\nu_t^2} \tag{13.5.15}$$

In Exercise 5.6-6 in Section 5.6 we found that a reasonable value for Q_{wx} would be

$$Q_{wx} = 0.066 h \sqrt{gh} \tag{13.5.16}$$

which combined with $\nu_t = 0.01 h \sqrt{gh}$ shows that

$$\frac{D_{xx}}{\nu_t} \sim 14.5 \tag{13.5.17}$$

The approximations made in the above evaluation of (13.5.6) means that the actual D_{xx} will be somewhat larger. Numerical evaluations show that a value of $D_{xx}/\nu_t = 20$ is a realistic estimate for the conditions considered here. Thus we see that the D_{xx}-term in the longshore momentum equation **does** provide an additional lateral mixing that is far stronger than that provided by the eddy viscosity which we can defend to use on the basis of the undertow comparisons shown in Section 12.2. In physical terms it means the even the strong turbulence generated by the wave breaking only provides about 5% of the lateral mixing.

13.6 Applications and further developments of quasi-3D modelling

The quasi-3D models have been applied to a number of nearshore phenomena and also developed further. A few examples are briefly reveiwed here.

13.6.1 *The start-up of a longshore current*

This hypothetical example was first analysed by Van Dongeren et al. (1994). It considers the temporal development of the 3-D flow on a long straight beach where the water is initially at rest. At $t = 0$, however, the forcing from obliquely incident regular waves is started without any transition. The numerical experiment then shows the development in time of cross- and longshore currents and of the setup.

The temporal variation shows principal features that can be expected to be found on any coast when the wave motion is varying.

In the analysis the time scale T_b used to describe the results is given by

$$T_b = \frac{L_b}{\sqrt{gh_b}} \tag{13.6.1}$$

where L_b is the cross-shore width of the surfzone, h_b the depth at the breaking point of the waves. Thus T_b is about half the time it takes a shallow water wave to propagate from the breaking point to the shoreline on a plane beach.

The first major change occurs in the setup which is almost fully established after only $2T_b$. The equivalent shoreward velocity (averaged over the depth) represents the volume flux required to establish the setup. It already reaches its maximum after just $1T_b$, and at just $4T_b$ there is virtually no net coss-shore flow left.

On the other hand, the growth of the longshore currents reveals that even at $2T_b$ the longshore current has hardliy started flowing, It is not until $16 - 25\ T_b$ that the longshore currents velocity has grown to near its steady value. Thus it is clear that variations in the longshore currents occur much more slowly (at a time scale $O(20T_b)$) than the the variations in setup which has a time scale of only $O(2T_b)$

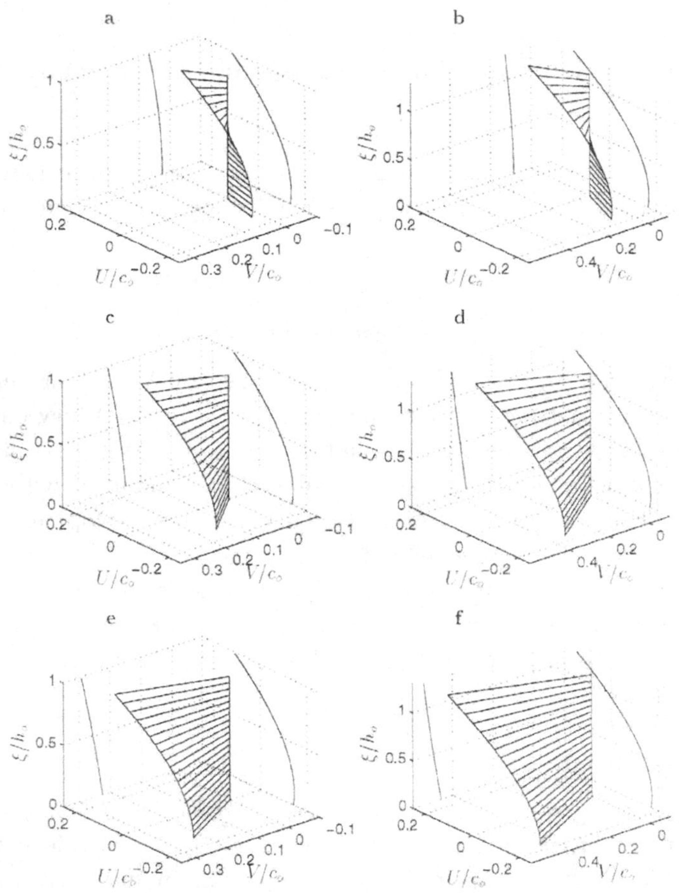

Fig. 13.6.1 Development of the velocity profiles at the breaking point (left column set of panels) and at a point well into the surfzone ($h/h_b = 0.35$) at three different times: $t = 6T_b$ (top row), $20T_b$ (center) and $48T_b$ (bottom) (Van Dongeren et al. 1994).

The main reason for the difference is - as discussed in Section 11.7 - that the forcing of the longshore current is much weaker thatn the cross-shore forcing.

For a natural beach the variations in wave heights associated with wave groupiness accur on a time scale of typically $3-5$ times the mean wave period and with a wave period of perhaps $T_b/2$ (depending on the steepness of the beach) this means that the setup can readily follow the groupiness variations in wave height whereas the changes generated in the longshore currents due to such wave height variations will be minimal.

The uneven growth rates of the cross-shore and longshore flows give raise to similar changes in the vertical current profiles. Fig. 13.6.1 shows the velocity profiles at the breaking point (left column set of panels) and at a point well into the surfzone ($h/h_b = 0.35$) at three different times: $t = 6T_b$ (top row), $20T_b$ (center) and $48T_b$ (bottom). At $6T_b$, when there is no more net cross-shore volume flux the undertow is fully establlished while the longshore current is only just beginning to flow. So the velocities are mainly crosshore (though the net cross-shore volume flux is close to zero). At $20T_b$ the longshore current is partly established and at $48T_b$ the situation has become steady and the velcity is mainly longshore.

13.6.2 *Rip currents*

Rip currents are highly localized cross-shore currents the flow seaward usually through channels that cut through a longshore bar located so the waves break over the bar. They are among the most dangerous phenomena that occur on sandy beaches. It has been reported by the American Lifesaving Association that 80% of all rescues by lifeguards at surf beaches are the result of swimmers being caught in rip currents (Haas et al., 2003), and every year many swimmers drown becasue rapidly flowing rip currents in a very short time carry them far seawards to deeper water.

Extensive testing of the quasi-3D SHORECIRC model has included analysis of the flow in rip currents on a barred beach. The work has also included comprehensive laboratory measurements. The results have been reported in a large number of publications, and the experimental results were ducumented by Haller et al. (2000, 2002), and by Haas and Svendsen (2000). The quasi-3D modelling has been applied to the problem and the combination of experimental and numerical modelling has revealed that rip currents are surprisingly complicated flows. Here we can only describe a

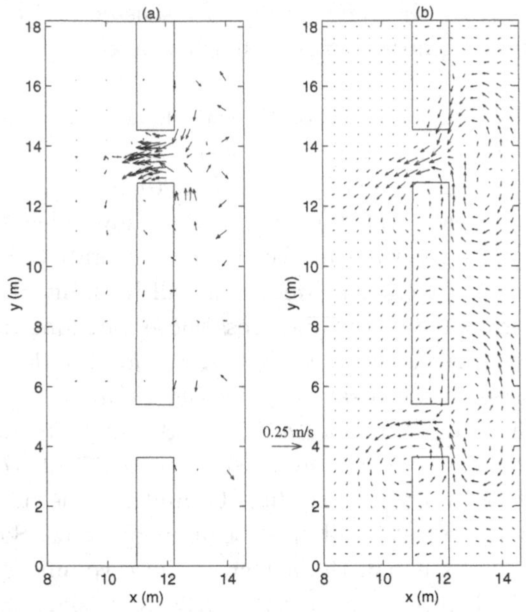

Fig. 13.6.2 Time-averaged below-trough velocity from (a) experimental data (Haller et al., 1997) and (b) the SC simulation (Haas and Svendsen, 2000)

few of the principal results. In addition to showing how complex the rip currents are they also illustrate the versatility of the quasi-3D modelling approach.

The experiments were carried out in a wave basin with shore normal wave incidence of regular waves. Thus the forcin gof the flow was steady in time. A plane slope was superimposed with a longshore bar with two openigs that functioned as rip channels. Fig. 13.6.2 shows the outline and also shows longtime averaged velocity vectors, to the left the measured values, to the right quasi-3D simulations. Though there are differences between the measured and computed results similarities are clear. In particular the tendency of the upper rip (which is the one covered in most detail in the measurements) to veer off to the left (downwards in the fig). The analysis showed that this was due to small irregularities in the intended longshore uniform topography. The computed results in the fig were made with a carefully surveyed version of the actual topography in the basin.

It was also found that both the experimental and the computed flows were highly unstable. An analysis of the linear stability of the rip as a jet was conducted by Haller and Dalrymple (2001). Quasi-3D computations shown in Fig. 13.6.3 indicate that flow is fluctuating dramatically over a period of time that corresponds to approximately 500 s laboratory time. It appears that large scale vortices are generated and convected seaward from the bar by the rip current. Unfortunately it is not possible with the limited number of velocity meters available for the experiments to illustrate this flow field but the time series recorded at the individual gages showed radical time variations on a similar time scale that fully consisted with the computed pattern. It is also mentioned that the bottom friction coefficient f plays a role in how the flow develops.

The velocities also showed variations over depth that meant the vertical velocity profiles were twisting and turning with velocities at the surface sometimes being opposite or perpendicular to the velocities near the bottom. Those variations are by far the strongest some distance offshore of the rip channel. Fig. 13.6.4 shows a sequence velocity vectors and their cross- and longshore projections measured at three different water depths. The times of the panels range over a period of just 70 s in the laboratory timescale.

In comparison the measurements taken at the centerline of the channel between the longshore bars shown in Fig. 13.6.5 fluctuate much less in even over the longer period of 110 s covered.

For comparison the entire velocity field obtained by quivalent computations are shown in a snapshot in Fig. 13.6.6. We see that qualitatively the picture is similar to the measured results but a direct comparison between the measured and the computed results is difficult to obtain because the flow is clearly turbulent in nature so the exact experimantal conditions cannot be reproduced from repetition to repetition.

On sandy beaches rip currents are ubquitous and these investigations reveal how complex rip current flows are. However, there is still much to be learned about the occurence and development rip current systems and the role they play in the morhological evolution of sandy beaches.

13.6.3 *Curvilinear version of the SHORECIRC model*

Recently the quasi-3D SHORECIRC model has been developed for non-orthogonal curvilinear coordinates (Shi et al., 2002, 2003). This highly facilitates the possibilities for fitting the coordinate grid to irregular boundaries. After extensive testing of the scheme against canonical cases where the exact solution is known the paper shows application of the model to

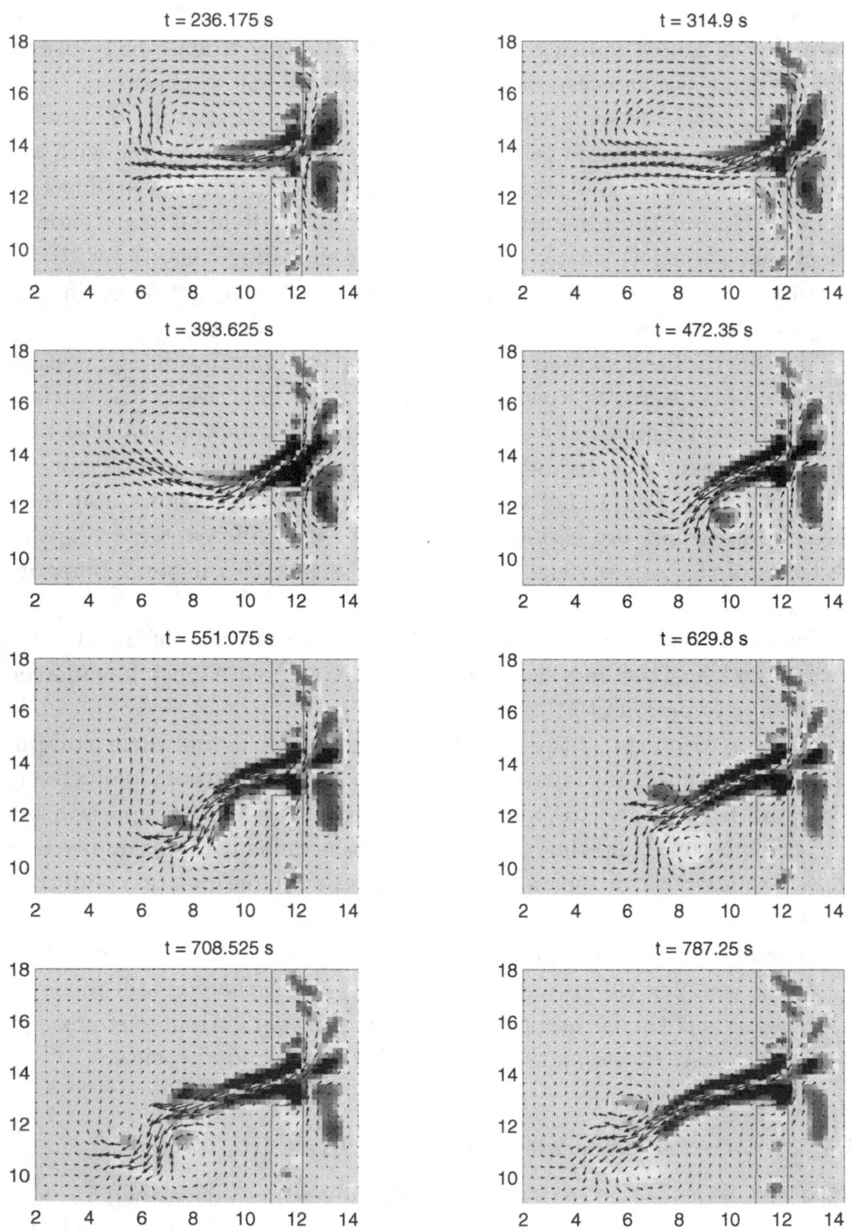

Fig. 13.6.3 A series of instantaneous snapshots of vorticity and velocity vectors from the SC simulation. Only an excerpt of the entire computational domain is shown (Haas and Svendsen, 2000).

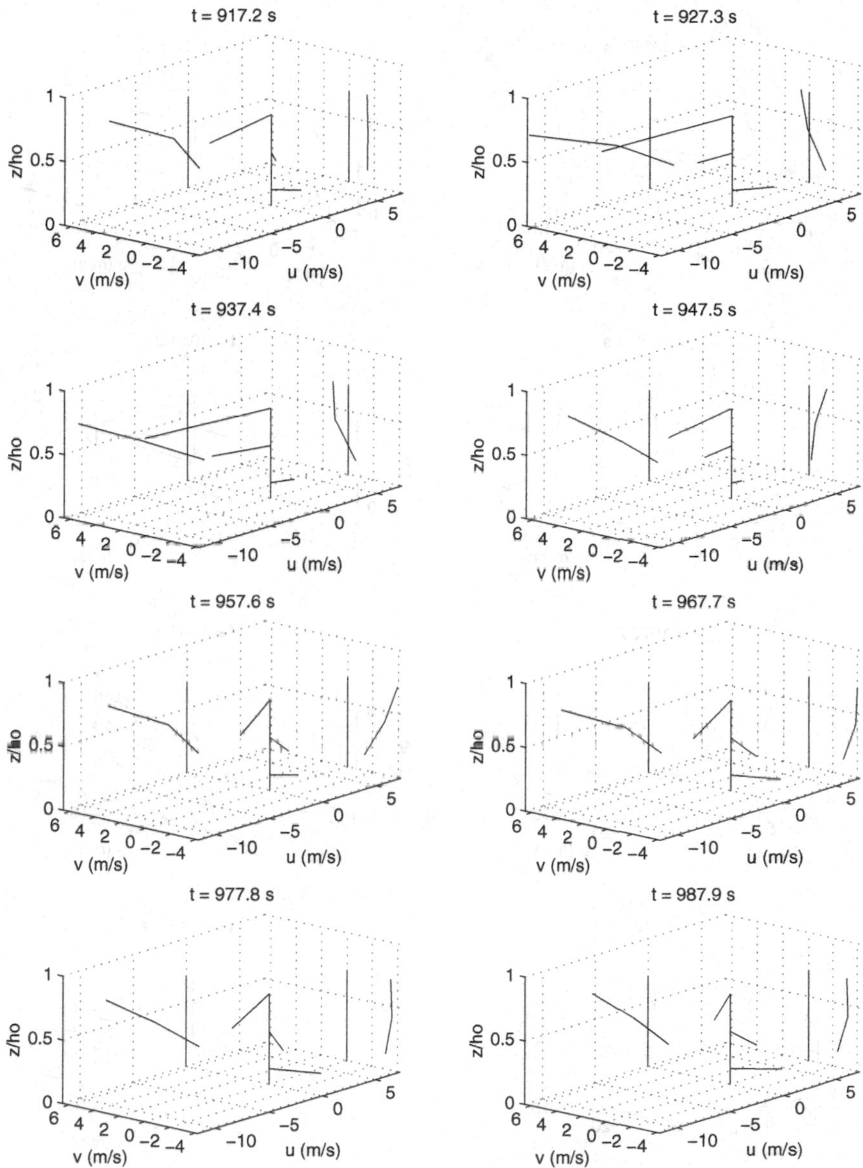

Fig. 13.6.4 Snapshots of the measured velocity vectors with projections of the cross-shore and longshore currents 2 m offshore of the channel ($x = 9$, $y = 13.6$ m in Fig. 13.6.2) (Haas and Svendsen, 2002).

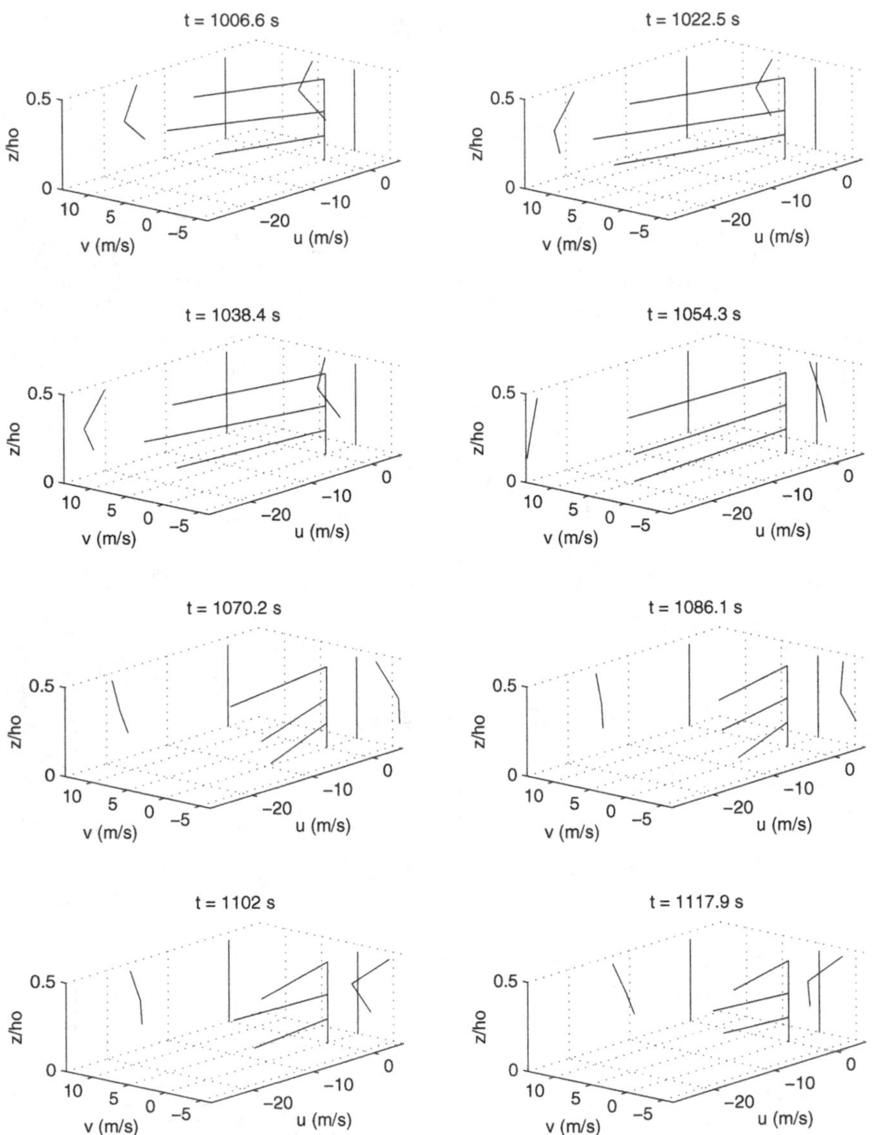

Fig. 13.6.5 Snapshots of the velocity vectors with projections of the cross-shore and longshore currents in the channel ($x = 11.75$, $y = 13.6$ m) (Haas and Svendsen, 2002).

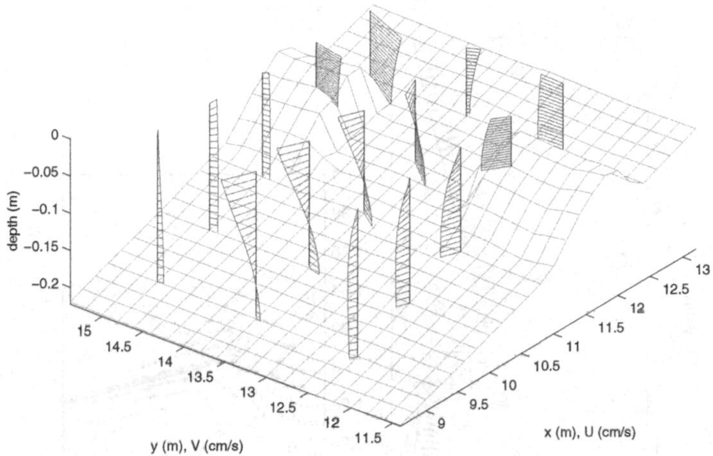

Fig. 13.6.6 An instantaneous snapshot at $t = 771$ s of the 3D variation of V_α from the SC simulations with the real topography. The middle row of profiles is along the edge of the rip (Haas and Svendsen, 2000).

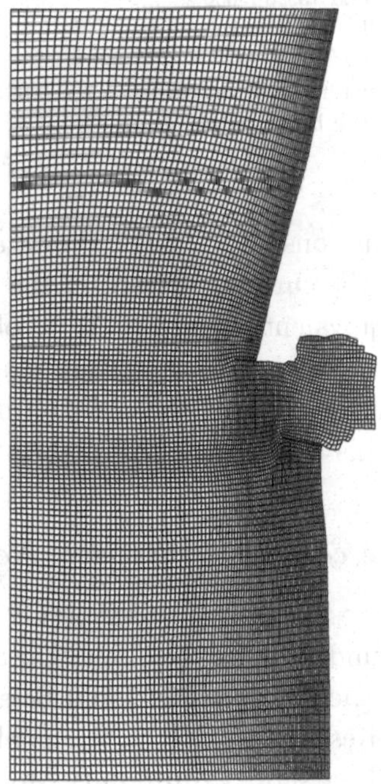

Fig. 13.6.7 The non-orthogonal curvilinear grid used for the computations in Fig. 13.6.8 (from Shi et al. 2003).

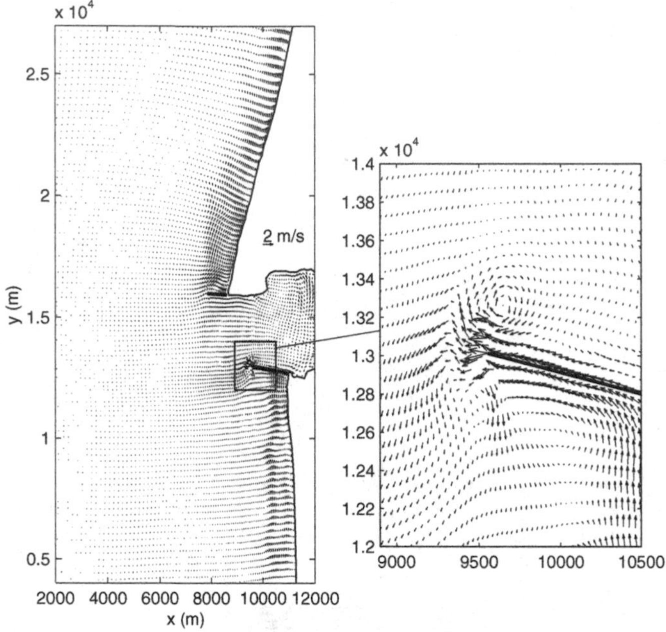

Fig. 13.6.8 The computed current field around Grey's harbour. The insert shows the flow details around the tip of the breakwater (from Shi et al. 2003).

several more realistic situations. Fig 13.6.7 shows an example of the grid generated for calculating the current pattern in the neighbohood of Grey's harbor California. The equivalent current patterns obtained from the model are shown in Fig. 13.6.8. The left part of this figure gives a view of the entire computational domain of approximately 10 times 23 km and the right panel shows the flow details around the tip of the southern breakwater.

13.6.4 *The nearshore community model, NearCoM*

In 1999 a large 5-year research project involving researchers from many universities in USA was funded by the National Oceanographic Partnership Program (NOPP) which includes the National Science Foundation (NSF) and the Office of Naval Research (ONR). The goal of the project was to develop a comprehensive Nearshore Community Model for nearshore hydrodynamic and sediment processes. This model, which has now been called **The Nearshore Community Model** or **NearCoM** will become available at the end of 2004 at the NOPP home page at the University of Delaware Center for Applied Coastal Research website at www.coastal.udel.edu. In parallel with oceanographic models like the Priceton Ocean Model (POM) this is an open-source model which is availble to the community free of charge.

To quote from the website: "The community model predicts waves, currents, sediment transport and bathymetric change in the nearshore ocean, between the shoreline and about 10 m water depth."

The centerpiece of the system is a program, master.f, which is used to couple individual modules, including

- wave driver
- Circulation module
- Morphology module

Components of the system are accessible (as they become available) from the NOPP page or from the link given at the website. The initial configured test system will include REF/DIF 1 as the wave driver, SHORECIRC as the circulation module and a sediment transport calculation based on the Bagnold-Bailard-Bowen formulation. Many of the programs that have previously been distributed as free standing models are configured as subroutine calls within the larger modeling system.

13.7 References - Chapter 13

DeVriend, H. J. and M. J. F. Stive (1987). Quasi-3D modelling of nershore currents. Coastal Engrg. **11**, 565–601.

Fischer, H. B. (1978). On the tensor form of the bulk dispersion coefficient in a bounded skewed shear flow. *Journal of Geophysical Research*, 83, pp. 2373–2375.

Haller, M. C., R. A. Dalrymple and I. A. Svendsen (1997). Rip channels and nearshore circulation: Experiments. Proc. ASCE costak Dynamics conf., 594–603.

Haller, M. C. and R. A. Dalrymple (2001). Rip current instabilities. J. Fluid Mech., **433**, 161–192.

Haller, M. C., R. A. Dalrymple and I. A. Svendsen (2002). Experiments on rip currents and nearshore circulation. Univ of Delaware, Center for Applied Coastal Res. Rep. CACR-00-04.

Haller, M. C., R. A. Dalrymple and I. A. Svendsen (2002). Experimental study of nearshore dynamics on a barred beach with rip currents. J. Geophys. Res. **107**, C6, 10.1029/2001JC000955.

Haas, K. A. and I. A. Svendsen (2000). Three-dimensional modeling of rip-current systems. Univ of Delaware, Center for Applied Coastal Res. Rep. CACR-00-06.

Haas, K. A. and I. A. Svendsen (2002). Laboratory measurements of the vertical structure of rip currents. J. Geophys. Res., **107**, 3047, doi:10.1029/2001JC000911.

Longuet-Higgins, M. S. (1970a,b). Longshore currents generated by obliquely incident sea waves, 1 and 2, *J. Geophys. Res.*, **75**, 6778–6789 and 6790–6801.

Phillips, O. M. (1966, 1977). Dynamics of the upper ocean. Cambridge University Press.

Putrevu, U. and I. A. Svendsen, (1991). Wave induced nearshore currents: A study of the forcing, mixing and stability characteristics. Univ. of Delaware, Center for Applied Coastal Res. Rep. CACR-91-11.

Putrevu, U. and I. A. Svendsen, (1999). Three-dimensional dispersion of momentum in wave-induced nearshore currents. European Journal of Mechanics-B/Fluids, 409–427.

Sanchez-Arcilla, A., F. Collado, M. Lemos and F. Rivero (1990). Another quasi-3D model for surfzone flows. ASCE Proc. 23rd Int. Conf. Coastal Engrg., Chap. 215, 2811–2824.

Shi, F., I. A. Svendsen, J. T. Kirby and J. McKee Smith (2003). A curvilinear version of a quasi=3D nearshore circulation model. Coast. Engrg., **49**, 99-124.

Shi, F., I. A. Svendsen and J. T. Kirby (2002). Curvilinear modeling of nearshore circulation at Grays Habour, ASCE Proc. 28th Intl. Conf. Coastal Engrng., 810-822.

Svendsen, I. A. (1987). Analysis of surfzone turbulence. J. Geophys. Res. **92**, C5, 5115–5124.

Svendsen, I. A. and R. S. Lorenz (1989). Velocities in combined undertow and longshore currents. *Coastal Engineering*, 13, pp. 55–79.

Svendsen, I. A. and U. Putrevu (1994). Nearshore mixing and dispersion. Proc. Roy. Soc. Lond. A **445**, 561–576.

Taylor, G. I. (1954). The dispersion of matter in a turbulent flow through a pipe. *Proceedings of the Royal Society of London.* Series A, 219, pp. 446–468.

Van Dongeren, A. R., F. E. Sancho, I. A Svendsen and U. Putrevu (1994). Quasi 3-D modelling of infragravity waves. ASCE Proc. 24th Int. Conf. Coastal Engrg.

Visser, P. J. (1984). A mathematical model of uniform longshore currents and comparison with laboratory data. Communications on Hydraulics. *Report 84-2*, Department of Civil Engineering, Delft University of Technology, 151 pp.

Appendix

Appendix 13A: Solution of the equation for the vertical current profiles

The pricipal solution for the vertical profiles of the currents starts with equation (13.3.10).

We first elliminate the first parenthesis on the left hand side of (13.3.10) the following way. Using the depth integrated continuity equation (Chapter 11.2) the depth-integrated momentum equation, (13.2.7) is rewritten as

$$\frac{\partial \frac{\overline{Q}_\alpha}{h}}{\partial t} + \frac{\overline{Q}_\beta}{h}\frac{\partial \frac{\overline{Q}_\alpha}{h}}{\partial x_\beta} + g\frac{\partial \overline{\zeta}}{\partial x_\alpha} = -\frac{1}{\rho h}\frac{\partial}{\partial x_\beta}\left[S'_{\alpha\beta} - \overline{\int_{-h_o}^{\zeta} \tau_{\alpha\beta} dz}\right] + \frac{\tau_\alpha^S}{\rho h} - \frac{\tau_\alpha^B}{\rho h}$$
$$-\frac{1}{h}\frac{\partial}{\partial x_\beta}\overline{\int_{-h_o}^{\zeta} V_{d\alpha}V_{d\beta} dz} - \frac{1}{h}\frac{\partial}{\partial x_\beta}\overline{\int_{\zeta_t}^{\zeta} u_{w\alpha}V_{d\beta} + u_{w\beta}V_{d\alpha} dz} \tag{13.A.1}$$

which is then used to rewrite (13.3.10) as

$$\frac{\partial V_{d\alpha}}{\partial t} - \frac{\partial}{\partial z}\left((\nu_t + \nu_s)\frac{\partial V_{d\alpha}}{\partial z}\right)$$
$$= \left\{\frac{1}{\rho h}\frac{\partial S'_{\alpha\beta}}{\partial x_\beta} - \frac{\tau_\alpha^S}{\rho h} + \frac{\tau_\alpha^B}{\rho h} - f_\alpha\right\}$$
$$-\left\{\left(\frac{\overline{Q}_\beta}{h}\frac{\partial V_{d\alpha}}{\partial x_\beta} + V_{d\beta}\frac{\partial \frac{\overline{Q}_\alpha}{h}}{\partial x_\beta} + V_{d\beta}\frac{\partial V_{d\alpha}}{\partial x_\beta} + W\frac{\partial V_{d\alpha}}{\partial z}\right) - \left(\frac{\partial \frac{Q_{w\alpha}}{h}}{\partial t}\right.\right.$$
$$\left. + \frac{\overline{Q}_\beta}{h}\frac{\partial \frac{Q_{w\alpha}}{h}}{\partial x_\beta} + \frac{Q_{w\beta}}{h}\frac{\partial \frac{\overline{Q}_\alpha}{h}}{\partial x_\beta} - \frac{Q_{w\beta}}{h}\frac{\partial \frac{Q_{w\alpha}}{h}}{\partial x_\beta} + \frac{Q_{w\beta}}{h}\frac{\partial V_{d\alpha}}{\partial x_\beta} + V_{d\beta}\frac{\partial \frac{Q_{w\alpha}}{h}}{\partial x_\beta}\right)$$

$$- \frac{1}{h}\frac{\partial}{\partial x_\beta}\overline{\int_{-h_o}^{\zeta} V_{d\alpha}V_{d\beta}dz} - \frac{1}{h}\frac{\partial}{\partial x_\beta}\overline{\int_{\zeta_t}^{\zeta} u_{w\alpha}V_{d\beta} + u_{w\beta}V_{d\alpha}dz}\Bigg\}$$
$$+ \left\{\frac{\partial}{\partial x_\alpha}\left((\nu_t+\nu_s)\left(\frac{\partial V_\alpha}{\partial x_\beta} + \frac{\partial V_\beta}{\partial x_\alpha}\right)\right) + \frac{\partial}{\partial z}\left((\nu_t+\nu_s)\frac{\partial W}{\partial x_\alpha}\right)\right\} \quad (13.A.2)$$

The previous equation governs the vertical structure of $V_{d\alpha}$. Solving (13.A.2) requires boundary conditions for the current $V_{d\alpha}$. We match the shear stress at the bottom and specify no net flow per (13.2.6) which is written as

$$(\nu_t+\nu_s)\frac{\partial V_{d\alpha}}{\partial z}\bigg|_{z=-h_o} = \frac{\tau_\alpha^B}{\rho}, \qquad \int_{-h_o}^{\overline{\zeta}} V_{d\alpha} = 0. \quad (13.A.3)$$

Equations (13.A.2) and (13.A.3) are similar to the equations governing $V_{1\alpha}$ (19) and (20) in PS99. The terms in the third line of (13.A.2) are new. In addition, the depth integral boundary condition in (13.A.3) is equal to zero whereas in PS99 (20) it is equal to $-Q_{w\alpha}$.

We solve these equations for the vertical variation of $V_{d\alpha}$ by assuming time variations slow enough to allow us to neglect $\frac{\partial}{\partial t}$. This results in

$$-\frac{\partial}{\partial z}\left((\nu_t+\nu_s)\frac{\partial V_{d\alpha}}{\partial z}\right) = F_\alpha^{(0)} + F_\alpha^{(1)} + F_\alpha^{(2)} \quad (13.A.4)$$

where the forcing on the right hand side is given by

$$F_\alpha^{(0)} = \left\{\frac{1}{\rho h}\frac{\partial S'_{\alpha\beta}}{\partial x_\beta} - \frac{\tau_\alpha^S}{\rho h} + \frac{\tau_\alpha^B}{\rho h} - f_\alpha\right\} \quad (13.A.5)$$

$$F_\alpha^{(1)} = -\Bigg\{\left(\frac{\overline{Q}_\beta}{h}\frac{\partial V_{d\alpha}}{\partial x_\beta} + V_{d\beta}\frac{\partial \frac{\overline{Q}_\alpha}{h}}{\partial x_\beta} + V_{d\beta}\frac{\partial V_{d\alpha}}{\partial x_\beta} + W\frac{\partial V_{d\alpha}}{\partial z}\right)$$
$$-\left(\frac{\overline{Q}_\beta}{h}\frac{\partial \frac{Q_{w\alpha}}{h}}{\partial x_\beta} + \frac{Q_{w\beta}}{h}\frac{\partial \frac{\overline{Q}_\alpha}{h}}{\partial x_\beta} - \frac{Q_{w\beta}}{h}\frac{\partial \frac{Q_{w\alpha}}{h}}{\partial x_\beta} + \frac{Q_{w\beta}}{h}\frac{\partial V_{d\alpha}}{\partial x_\beta} + V_{d\beta}\frac{\partial \frac{Q_{w\alpha}}{h}}{\partial x_\beta}\right)$$
$$- \frac{1}{h}\frac{\partial}{\partial x_\beta}\overline{\int_{-h_o}^{\zeta} V_{d\alpha}V_{d\beta}dz} - \frac{1}{h}\frac{\partial}{\partial x_\beta}\overline{\int_{\zeta_t}^{\zeta} u_{w\alpha}V_{d\beta} + u_{w\beta}V_{d\alpha}dz}\Bigg\}$$
$$(13.A.6)$$

$$F_\alpha^{(2)} = \left\{\frac{\partial}{\partial x_\alpha}\left((\nu_t+\nu_s)\left(\frac{\partial V_\alpha}{\partial x_\beta} + \frac{\partial V_\beta}{\partial x_\alpha}\right)\right) + \frac{\partial}{\partial z}\left((\nu_t+\nu_s)\frac{\partial W}{\partial x_\alpha}\right)\right\}$$
$$(13.A.7)$$

A rigorous discussion about the scaling of the problem is given by PS99 and the details are omitted here. It suffices here to say that the relative size of the forcing is as follows

$$F_\alpha^{(0)} \gg F_\alpha^{(1)} \gg F_\alpha^{(2)} \tag{13.A.8}$$

This allows us to solve (13.A.4) using a perturbation method where we utilize

$$V_{d\alpha} = V_{d\alpha}^{(0)} + V_{d\alpha}^{(1)} + \ldots \tag{13.A.9}$$

The equation governing $V_{d\alpha}^{(0)}$ is then

$$-\frac{\partial}{\partial z}\left((\nu_t + \nu_s)\frac{\partial V_{d\alpha}^{(0)}}{\partial z}\right) = F_\alpha^{(0)} \tag{13.A.10}$$

with the boundary conditions

$$(\nu_t + \nu_s)\frac{\partial V_{d\alpha}^{(0)}}{\partial z}\bigg|_{z=-h_o} = \frac{\tau_\alpha^B}{\rho}, \qquad \int_{-h_o}^{\overline{\zeta}} V_{d\alpha}^{(0)} = 0. \tag{13.A.11}$$

The solution method for $V_{d\alpha}^{(0)}$ is the same as shown by PS99 for $V_{1\alpha}^{(0)}$. It gives the following result

$$\begin{aligned}
V_{d\alpha}^{(0)} = {} & \frac{\tau_\alpha^B}{\rho}\left(\int_{-h_o}^{z}\frac{dz'}{(\nu_t+\nu_s)} - \frac{1}{h}\int_{-h_o}^{\overline{\zeta}}\int_{-h_o}^{z}\frac{dz'}{(\nu_t+\nu_s)}dz\right) \\
& - \int_{-h_o}^{z}\frac{1}{(\nu_t+\nu_s)}\int_{-h_o}^{z'} F_\alpha^{(0)}\,dz'dz \\
& + \frac{1}{h}\int_{-h_o}^{\overline{\zeta}}\int_{-h_o}^{z}\frac{1}{(\nu_t+\nu_s)}\int_{-h_o}^{z'} F_\alpha^{(0)}\,dz''dz'dz
\end{aligned} \tag{13.A.12}$$

This equation is the same as the steady solution for $V_{1\alpha}^{(0)}$ given by PS99 (34) without the $-\frac{Q_{w\alpha}}{h}$ term.

Similarly, the equation governing $V_{d\alpha}^{(1)}$ is

$$-\frac{\partial}{\partial z}\left((\nu_t + \nu_s)\frac{\partial V_{d\alpha}^{(1)}}{\partial z}\right) = F_\alpha^{(1)} \tag{13.A.13}$$

with the boundary conditions

$$(\nu_t + \nu_s)\frac{\partial V_{d\alpha}^{(1)}}{\partial z}\bigg|_{z=-h_o} = 0, \qquad \int_{-h_o}^{\overline{\zeta}} V_{d\alpha}^{(1)} = 0 \tag{13.A.14}$$

where we substitute $V_{d\alpha}^{(0)}$ for $V_{d\alpha}$ in the equation for $F_{\alpha}^{(1)}$. The solution method is the same as for $V_{d\alpha}^{(0)}$ with the following result

$$V_{d\alpha}^{(1)} = -\int_{-h_o}^{z} \frac{1}{(\nu_t + \nu_s)} \int_{-h_o}^{z'} F_{\alpha}^{(1)} dz' dz$$
$$+\frac{1}{h} \int_{-h_o}^{\overline{\zeta}} \int_{-h_o}^{z} \frac{1}{(\nu_t + \nu_s)} \int_{-h_o}^{z'} F_{\alpha}^{(1)} dz'' dz' dz \quad (13.A.15)$$

The equation (13.A.12) and (13.A.15) are the solutions for the depth varying part of the vurrents currents in terms of the depth-integrated properties.

Chapter 14

Other Nearshore Flow Phenomena

14.1 Infragravity waves

14.1.1 *Introduction*

Long waves (or "infra-gravity" waves, or IG-waves for short) are waves with significantly longer periods than the peak frequency of the incident wave spectrum. Field measurements show that such waves occur very frequently on many beaches. The measurements show that close to the shore - well inside the surfzone - the major part of the wave energy is often concentrated in spectral components with wave periods much longer than the dominating periods (generally in the range 20-300s) of the incoming wave motion. Therefore it is natural to suspect that some mechanism is present in the very nearshore region that transforms short wave energy into longer wave energy.

Different mechanisms have been considered for their generation.

One mechanism is the height variation of incoming waves which Munk (1949) suggested create shoreward mass transport under high wave groups that, at the break point where wave groups are destroyed, transfers into long waves. The long waves are reflected from the shore and become free waves called **surf beat**. Tucker (1950) found negative correlation between incident wave amplitude and low frequency motion. This observation was later by Longuet-Higgins and Stewart (1962) shown to agree with the forced **set-down waves** under wave groups that is a consequence of the radiation stress variations. This mechanism is analysed in a section below.

Symonds et al. (1982) pointed out that varying wave heights causes the break point of the waves to vary with height. This means varying the point where the surfzone starts and therefore the start of the strong decrease in radiation stress and the generation of setup starts. This generating

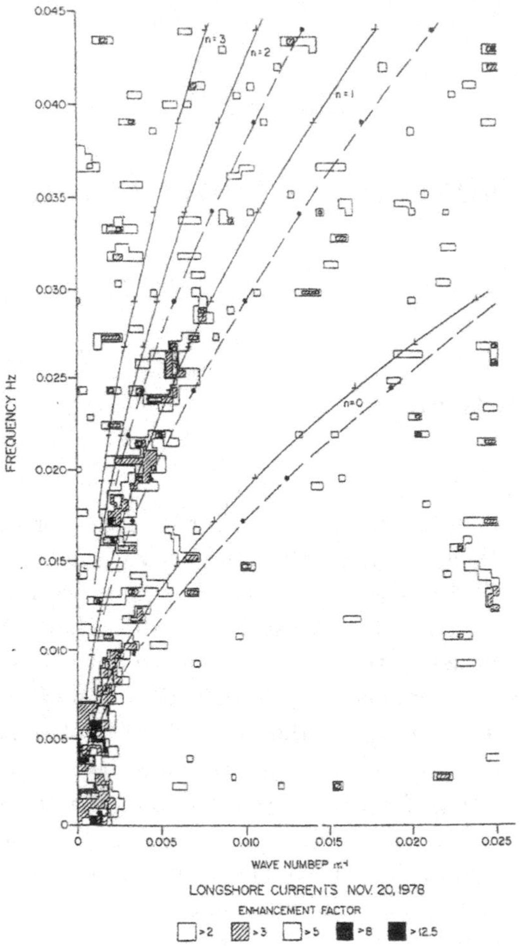

Fig. 14.1.1 Frequency-longshore wave number diagram showing spectral energy at each point in frequency-wave number space (from Huntley et al. 1981).

mechanism has turned out to create much stronger forcing than the nonlinear resonant interaction described below. It leads to the idea (which we will use as basis for analysis of long waves in this Chapter) that a time varying set-down/setup, and the (time varying) currents associated with it, can be considered a (long) wave in itself. This has been pursued further by Schäffer and Svendsen (1988), Lo (1988), List (1992) and many others thereafter.

Another proposed mechanism is nonlinear interaction between shorter waves: Gallagher (1971) analysed interaction between various spectral components in the incident wave train suggesting that longshore variation of the wave height in wave groups forces the low frequency wave motion. Bowen and Guza (1978) found that even a monochromatic wave train may transfer energy into longer waves by non-linear resonant interaction with the long waves generated, and in Bowen and Guza (1978) discussed whether the resonant wave generation versus nonresonant motion at surf beat frequencies could be responsible for edge wave motion. Since it has turned out that variations in wave breaking (variations in break point as well as in height) generate much stronger forcing of IG waves that type of mechanism has been a dominating line of research in recent years.

The special type of infragravity waves called **edge waves** has been known as a mathematical solution to the linear equations for a plane beach since Stokes (1846) (see Lamb, 1945, p 446). Eckart (1951) provided an extensive analysis of the shallow water solutions for what is actually several modes of edge waves and Ursell (1952) gave the general solution for arbitrary depths. However, verification of their existence in nature remained uncertain for many years though many publications showed measurements indicating that the abundance of infragravity wave energy could be interpreted as signs of edge wave activity. Inman and Bowen (1967) may have been the first to associate the large energy in the surf zone with progressive edge waves. Probably one of the most important breakthroughs came when field experiments made it possible to measure the longshore wave number of the infragravity wave motion. That made it possible to verify that some of the long wave energy originated from waves following the edge wave dispersion relation, and it became clear that edge waves are actually ubiquitous in the nearshore environment.

Fig. 14.1.1 shows an early example from Huntley et al. (1981) of a frequency-longshore wave number diagram. In the diagram the dispersion relations for the lowest mode edge waves appear as curves (numbered $n = 0, 1, 2, 3$ in the figure). The local the slope of the curves represent the group velocity $d\omega/dk$, the ratio of the coordinates ω/k at each point of the curves corresponds to the phase speed of the waves. The small framed and tainted areas indicate frequency-wave number combinations for which spectral energy was identified in the experiment in question. The darker the more energy. Though perhaps not as conclusive as later examples of similar plots the concentration of energy along the curves clearly is consistent with the presence of edge wave motion on the beach. It may be mentioned that in the diagram ordinary storm waves would appear as lines that almost cling to the vertical axis.

In this section we first derive the proper equations for long, infragravity wave motion from the basic circulation equations. This will provide a generalized wave equation that clearly demonstrates how short, time varying waves can act as forcing that generates the long waves. We then analyse the simplest case of free long waves without considering where and how they were generated. Mathematically that turns out to correspond to solving a homogeneous equation, while generation of the long waves discussed in the following subsection is governed by an inhomogenous wave equation.

14.1.2 *Basic equations for infragravity waves*

The fact that we assume the long waves to be "long" in the sense of classical long wave theory, $h/L \ll 1$, leads in the first approximation (that of small waves) to the conclusion that in those waves we have hydrostatic pressure

$$p = \rho g \overline{\zeta} \tag{14.1.1}$$

where $\overline{\zeta} = \overline{\zeta}(x_\alpha, t)$ is the surface elevation of the long wave (setdown-setup). If we also as a first approximation neglect the bottom friction and other effects causing a vertical variation of the horizontal particle velocity (= the "current") then V_α will be depth uniform, and therefore we can also consider the depth averaged velocities V_α in such waves.

As already indicated, however, those are exactly the conditions we have built into our depth integrated, wave averaged equations for depth uniform nearshore currents derived earlier.

Hence, we turn back to those equations which are essentially conservation of mass and momentum.

Continuity equation

$$\frac{\partial \overline{\zeta}}{\partial t} + \frac{\partial Q_\alpha}{\partial x_\alpha} \tag{14.1.2}$$

where as usual for depth uniform currents

$$Q_\alpha = V_\alpha h + Q_{w,\alpha} \tag{14.1.3}$$

is the sum of the current discharge $V_\alpha h$ and the volume flux $Q_{w,\alpha}$ in the short wave motion.

We also have that

$$h = h_0 + \overline{\zeta} \tag{14.1.4}$$

Momentum equation

The momentum equation for depth uniform currents are given as (11.5.20)

$$\rho \frac{\partial Q_\alpha}{\partial t} + \frac{\partial}{\partial_{x_\beta}} \left(\rho \frac{Q_\alpha Q_\beta}{h} + S'_{\alpha\beta} - \overline{\int_{-h_0}^{\zeta} \tau_{\alpha\beta} dz} \right) + \rho g h \frac{\partial \overline{\zeta}}{\partial x_\alpha} + \tau_\alpha^s - \tau_\alpha^B = 0 \tag{14.1.5}$$

In order to concentrate our attention on the long wave generation we will neglect here all the effects in (14.1.5) that are not strictly necessary for the generation and propagation of such waves. The point here is that neglecting those effects does not eliminate the phenomena we want to study.

Thus we disregard

- Turbulent shear stresses $\tau_{\alpha\beta}$ (resposible e.g. for lateral mixing).
- Bottom friction τ_α^B.
- Wind stress τ_α^S.

The result is a reduced version of (14.1.5) that actually corresponds to "potential flow theory":

$$\frac{\partial Q_\alpha}{\partial t} + \frac{\partial}{\partial x_\beta} \left(\frac{Q_\alpha Q_\beta}{h} \right) + \frac{1}{\rho} \frac{\partial S_{\alpha\beta}}{\partial x_\beta} + g h \frac{\partial \overline{\zeta}}{\partial x_\alpha} = 0 \tag{14.1.6}$$

Notice that the effect of the short waves is still represented by the $\frac{\partial S_{\alpha\beta}}{\partial x_\beta}$-term in (14.1.6).

We plan to assume the long waves are small, and so are then the particle velocitites V_α in those waves. Therefore also $Q_\alpha \simeq V_\alpha h$ is small, which implies we can neglect the $Q_\alpha Q_\beta$-terms, and replace h with h_0. In all, this amounts to linearizing the equation. The final version of the momentum equation then becomes

$$\frac{\partial Q_\alpha}{\partial t} + g h_0 \frac{\partial \overline{\zeta}}{\partial x_\alpha} = -\frac{1}{\rho} \frac{\partial S_{\alpha\beta}}{\partial x_\beta} \tag{14.1.7}$$

We can now eliminate Q_α between (14.1.2) and (14.1.7) by cross differentiation. This gives

$$\boxed{\frac{\partial^2 \overline{\zeta}}{\partial t^2} - \frac{\partial}{\partial x_\alpha}\left(gh_0 \frac{\partial \overline{\zeta}}{\partial x_\alpha}\right) = \frac{1}{\rho}\frac{\partial^2 S_{\alpha\beta}}{\partial x_\alpha \partial x_\beta}} \tag{14.1.8}$$

which represents a generalized version of the traditional long wave mild slope equation (MSE).

The coordinate form of the general equation (14.1.8) is

$$\frac{\partial^2 \overline{\zeta}}{\partial t^2} - \frac{\partial}{\partial x}\left(gh_0 \frac{\partial \overline{\zeta}}{\partial x}\right) - \frac{\partial}{\partial y}\left(gh_0 \frac{\partial \overline{\zeta}}{\partial y}\right) = \frac{1}{\rho}\left(\frac{\partial^2 S_{xx}}{\partial x^2} + 2\frac{\partial^2 S_{xy}}{\partial x \partial y} + \frac{\partial^2 S_{yy}}{\partial y^2}\right) \tag{14.1.9}$$

For long waves the traditional MSE reads:

$$\frac{\partial^2 \overline{\zeta}}{\partial t^2} - \frac{\partial}{\partial x_\alpha}\left(gh_0 \frac{\partial \overline{\zeta}}{\partial x_\alpha}\right) = 0 \tag{14.1.10}$$

The difference is that (14.1.8) contains the forcing term coming from the variation of $S_{\alpha\beta}$ in the short waves. Thus where the ordinary (long wave version of the) mild slope equation (14.1.10) describes the propagation over a varying depth $h_0(x_\alpha)$ of (long) waves already in existence, Eq. (14.1.8) also includes the possible generation of such waves by the $S_{\alpha\beta}$ variation. The short waves, it is recalled, are assumed known, also in terms of how they propagate (refract, diffract) over the same varying topography $h_0(x_\alpha)$.

14.1.3 *Homogeneous solutions – Free edge waves*

Eq. (14.1.8) is essentially an inhomogeneous equation. As with all such systems, it is useful first to consider which solutions the homogeneous version of the equation has for the geometry of interest here. Such solutions would indicate how free long waves—waves not forced by the short waves—would behave and propagate.

Analytical solutions for (14.1.10) turn out to be possible for a long plane beach with $h_0 = h_x x$. Fig. 14.1.2 illustrates the situation considered. For that topography we now seek solutions to this equation which have the form

$$\overline{\zeta} = \eta(x) e^{i(k_y y \pm \omega t)} \tag{14.1.11}$$

where both + and $-\omega t$ are possible but where we for simplicity look at $-\omega t$ only. We see that these solutions are waves with a sinusoidal shape

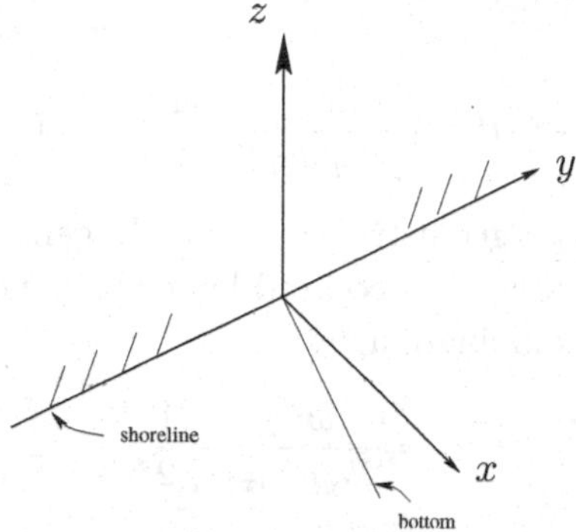

Fig. 14.1.2 The long straight beach assumed for the analytical solution. The cross-shore coordinate x is pointed seawards and the longshore coordinate is y.

corresponding to $Re[e^{i(k_y y \pm \omega t)}]$ that propagate along the shore in the y-direction. The amplitude varies in the cross-shore direction as $\eta(x)$ which is to be determined by the solution to the equation.

Substituting into (14.1.10) we get the terms

$$(gh_0\overline{\zeta}_x)_x = gh_{0x}\frac{\eta'}{\eta}\overline{\zeta} + gh_0\frac{\eta''}{\eta}\overline{\zeta} \tag{14.1.12}$$

and

$$(gh_0\overline{\zeta}_y)_y = -gh_0k_y^2\overline{\zeta} \tag{14.1.13}$$

and for the entire equation

$$gh_0\eta'' + gh_{0x}\eta' + (\omega^2 - gh_0k_y^2)\eta = 0 \tag{14.1.14}$$

Since on a plane beach $h_0 = h_{0x}x$ this can also be written

$$x\eta'' + \eta' + \left(\frac{\omega^2}{gh_{0x}} - k_y^2 x\right)\eta = 0 \tag{14.1.15}$$

The variation of $\overline{\zeta(x)}$ in the cross-shore direction turns out to be related to e^{-x}. It turns out that the substitution

$$\eta = a_n e^{-\xi/2} f(\xi) \qquad \text{where} \qquad \xi = 2k_y x \tag{14.1.16}$$

transforms Eq. (14.1.15) into a standard equation. Substituting (14.1.16) into (14.1.15) yields

$$\xi f'' + (1-\xi) f' + \left(\frac{\omega^2}{2k_y g h_{0x}} - \frac{1}{2} \right) f = 0 \tag{14.1.17}$$

which is a confluent hypergeometric equation. In general, this equation has solutions that tend to ∞ as $x \to \infty$ (and hence have no physical relevance) unless the parameter combination

$$n = \frac{\omega^2}{2k_y g h_{0x}} - \frac{1}{2} \tag{14.1.18}$$

is a non-negative integer. Thus, we get the constraint for meaningful solutions by solving (14.1.18) with respect to ω^2:

$$\boxed{\omega^2 = g k_y (2n+1) h_x = g k_y (2n+1) \tan \beta} \tag{14.1.19}$$

where β is the beach slope angle. This is the shallow water dispersion relation for these waves, which are called **Edge Waves** because their energy turns out to be concentrated near the shore (the "edge" of the water).

As the values of ω and k_y satisfying (14.1.19) corresponds to n being a non-negative integer, the permitted solutions are for (14.1.17) on the form

$$\xi f'' + (1-\xi) f' + n f = 0 \qquad n = 0,\ 1,\ 2, \ldots \tag{14.1.20}$$

Those solutions (eigenfunctions) are (proportional to) Laguerre polynomials which are given by

$$\begin{aligned} f(\xi) = L_n(\xi) = \frac{(-1)^n}{n!} \Bigg[& \xi^n - \frac{n^2}{1!} \xi^{n-1} + \frac{n^2(n-1)^2}{2!} \xi^{n-2} \\ & - \frac{n^2(n-1)^2(n-2)^2)}{3!} \xi^{n-3} + \cdots + \frac{n^2(n-1)^2 \cdots 2^2}{(n-1)!} \xi^1 \Bigg] + 1 \end{aligned} \tag{14.1.21}$$

See Abramowitz and Stegun (1964). Evaluation for the first values of n gives

$$\begin{array}{lll} n=0 & L_0 = 1 & ; \eta = e^{-k_y x} \\ n=1 & L_1 = 1-\xi & ; \eta = e^{-k_y x}(1-2k_y x) \\ n=2 & L_2 = 1 - 2\xi + \frac{1}{2}\xi^2 & ; \eta = e^{-k_y x}(1 - 4k_y + 2k_y^2 x^2) \end{array} \tag{14.1.22}$$

Thus we have

$$\bar{\zeta} = e^{-k_y x}\, L_n(2k_y x)\, e^{i(k_y y \pm \omega t)} \tag{14.1.23}$$

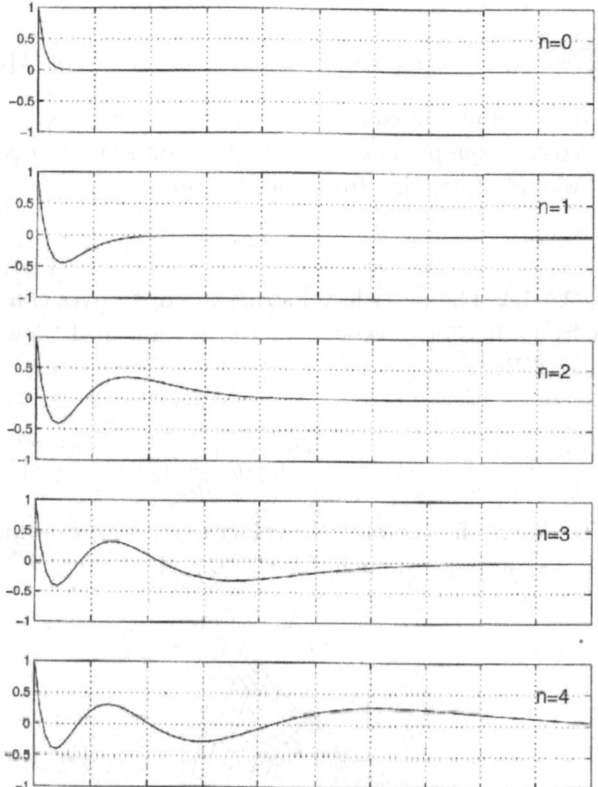

Fig. 14.1.3 Edge wave profiles. Surface elevation versus x.

These solutions all tend to zero as $x \to \infty$. From (14.1.19), we see that the larger n the smaller k_y for given ω (and h_x). Hence, the higher the mode n, the slower the offshore attenuation.

Figure 14.1.3 shows the cross-shore variation of the lowest order edge wave modes. We see that the 0th order mode has no zero-crossings in the cross-shore direction, the first order has one, the second two, etc.

In Fig. 14.1.4 is shown a snapshot of the surface elevations in a $0th$-order edge wave. This wave could propagate in both directions along the shore, and the picture could even be for a standing edge wave.

Exercise 14.1-1

Show by adding two edge waves traveling in opposite direction that the surface elevation of a standing edge wave can be written (e.g.) as

$$\overline{\zeta} = 2\, a_n e^{-k_y x}\, L_n(2k_y x)\, \cos\, k_y x\, \cos\, \omega t \tag{14.1.24}$$

It is noted that standing edge waves can emerge when edge waves are enclosed for example between two promontories on a coast or two breakwaters protruding from the shoreline.

Exercise 14.1-2 The particle velocities in edge waves can be determined from the linear version of the nonlinear shallow water equations (NSW)

$$\frac{\partial u}{\partial t} + g\,\frac{\partial \overline{\zeta}}{\partial x} = 0\,; \qquad \frac{\partial v}{\partial t} + g\,\frac{\partial \overline{\zeta}}{\partial y} = 0 \tag{14.1.25}$$

Use this to determine the particle velocities. As an example substitute the Laguerre polynomial for $n = 0$ and $n = 1$.

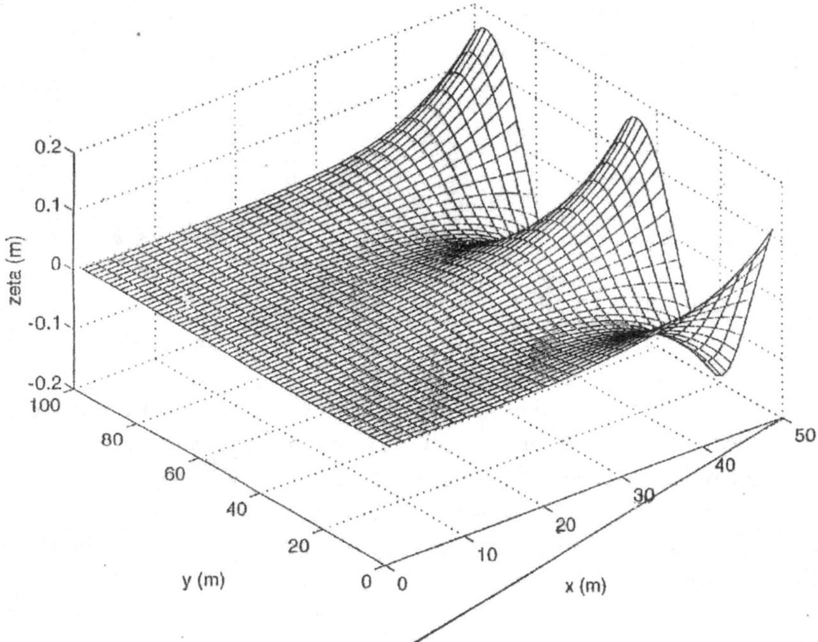

Fig. 14.1.4 A snapshot of a $0th$-order edge wave.

Recalling (14.1.11), it is realized that edge waves are sinusoidal along $x =$ constant lines in the longshore direction and travel in that direction at the speeds

$$c = \frac{\omega}{k_y} = \sqrt{\frac{g}{k_y}(2n+1)\tan\beta} \tag{14.1.26}$$

We see that since $\beta \ll 1$, these waves will generally travel much slower than ordinary deep water waves with the same wave length which have

$$c = \sqrt{\frac{g}{k_y}} \tag{14.1.27}$$

This solution is due to Eckart (1951). However it only represents the shallow water approximation. In particular for larger n this approximation starts to deviate from the general edge wave solution which our equations do not provide because we have assumed hydrostatic pressure in our long waves. The complete solution referred to was found by Ursell (1952). It has the slightly different dispersion relation

$$\omega^2 = gk_y \sin[(2n+1)\beta] \; ; \quad \text{which is only valid for:} \quad \left((2n+1)\beta < \frac{\pi}{2}\right) \tag{14.1.28}$$

This means that there is a limited number of possible edge wave modes. A comparison between the cross-shore profiles of the shallow water and the general solutions, respectively, is shown in Fig. 14.1.5 for two different beach slopes, 0.10 and 0.20. We see that the differences between the two solutions are negligible for the lowest order modes but start showing up for $n = 2$ and 3. This is because the higher order modes also reach farther out from the shoreline where the depths become larger. This is also the reason why the differences are more pronounced for the steeper slope of $h_x = 0.20$.

Trapped and leaky IG-wave modes

Finally, it is useful to analyze the propagation pattern of the lowest order edge waves.

Since there is no energy dissipation associated with these waves, they can be thought of as fully reflected from the shoreline. The propagation pattern becomes particularly clear if we look at a wave orthogonal.

For $n = 0$, all orthogonals are parallel with and the fronts perpendicular to the shoreline. For $n > 0$, however, the solution given by (14.1.11) with $\eta(x)$ according to (14.1.21) or (14.1.22) represent a situation where waves reflected from the shoreline are refracted to an angle of incidence $= 90°$

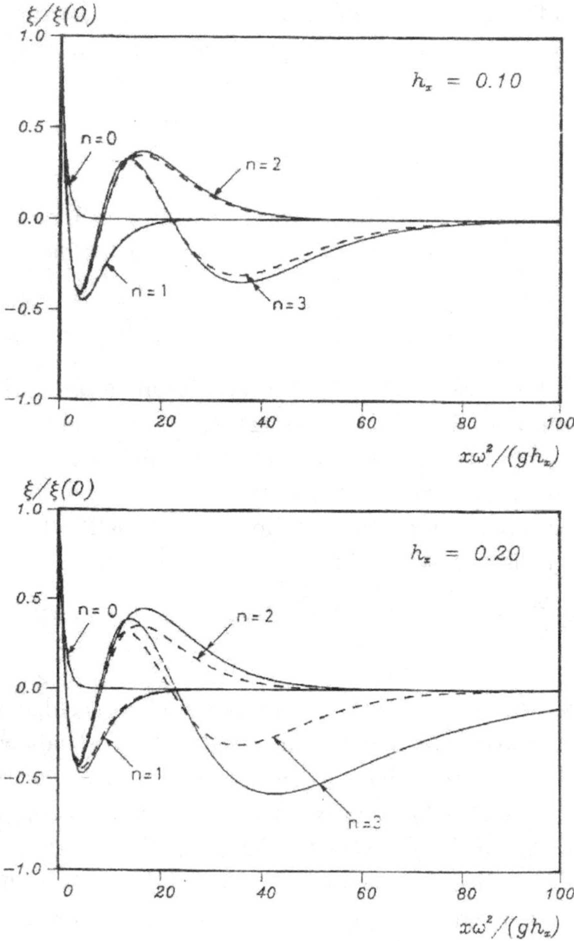

Fig. 14.1.5 Cross-shore variations of edge-wave surface elevation amplitudes for beach slopes $h_x = 0.1$ and 0.2. Comparison between Ursell's general solution ξ_{Ur} (eq. 5) (normalized by its value at the shoreline) (—) and Eckart's shallow-water approximation ξ_{Ec} (eq. 11) (- - -) for n = 1,2,3. For the dominant mode $n = 0$ these coincide (from Schäffer and Jonsson, 1992).

and then refracted back to the shore, reflected again, and so forth. Fig. 14.1.6 (Schäffer and Jonsson, 1992) shows some examples for a long straight beach. Waves of this type are also called **trapped waves**. The offshore line in the figure indicating the farthest the waves reach is called a **caustic**

curve. The nature of the wave motion around caustics is complicated. An extensive discussion is given in Peregrine (1976).

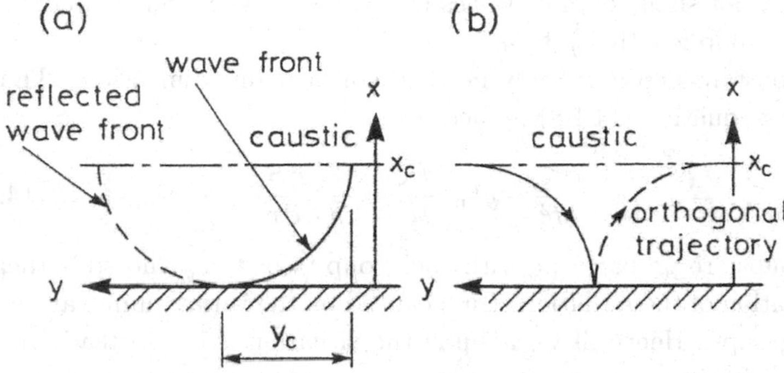

Fig. 14.1.6 Sketch of trapped higher order edge wave motion. The waves are propagating to the right in the figure. Left panel shows the wave fronts, right panel the wave othogonals (from Schäffer and Jonsson, 1992).

For IG-waves in general there is another type of motion of which the surf beat is an example. Those are waves for which the refraction, as the wave moves away from the coast, does not reach the angle of 90° at any depth. Therefore such waves leave the coast and move out into the open ocean where they can travel long distances before they reach another coast. They are called **leaky mode waves**. A typical example are set-down waves generated by incoming short waves or waves that are generated as leaky waves elsewhere and which arrive to the beach from offshore. On a long straight beach the reciprocity of the refraction process will always ensure that incoming IG-waves also leave the beach again as leaky mode waves (unless they break on the shore). On curved shorelines like an island, however, such waves can become trapped. Similarly on curved beaches edge waves can turn into leaky waves.

14.1.4 *IG wave generation*

This subject has been thoroughly researched and an abundance of literature is available. For brevity we can only outline a few of the principal questions and ideas and with much regret have to refer the reader to the individual publications, of which only a few are quoted here.

Set-down waves

We first consider short wave groups propagating in a region with constant depth h_0. It is assumed that the depth is large enough that the short waves are not shallow water waves ($kh = O(1)$), while h_0 is small enough to make group length $L_g \gg h_0$.

In constant depth we only need to consider one dimension. Thus the governing equation (14.1.8) reduces to

$$\frac{\partial^2 \overline{\zeta}}{\partial t^2} - gh_0 \frac{\partial^2 \overline{\zeta}}{\partial x^2} = \frac{1}{\rho} \frac{\partial^2 S_{xx}}{\partial x^2} \tag{14.1.29}$$

The groups are propagating with the group velocity c_g and so is therefore the variation of the radiation stress caused by the variation in wave heights in the groups. Hence, if we assume the situation is quasi-steady in time, variations in the mean surface elevation $\overline{\zeta}$ must also travel with speed c_g so we have

$$\overline{\zeta} = \overline{\zeta}(x - c_g t) \tag{14.1.30}$$

which implies that

$$\frac{\partial \overline{\zeta}}{\partial t} = -c_g \frac{\partial \overline{\zeta}}{\partial x} \tag{14.1.31}$$

Substituted into (14.1.29) this yields

$$(c_g^2 - gh_0) \frac{\partial^2 \overline{\zeta}}{\partial x^2} = \frac{1}{\rho} \frac{\partial^2 S_{xx}}{\partial x^2} \tag{14.1.32}$$

which, using the assumption of steady conditions and space uniformity to eliminate the integration functions, can be integrated twice to give

$$\boxed{\overline{\zeta} = -\frac{S_{xx}(x,t)}{\rho(gh_0 - c_g^2)}} \tag{14.1.33}$$

(see Longuet-Higgins and Stewart, 1964). We see that this solution is exactly in anti-phase with the radiation stress variation, so that the set-down is largest where the wave groups have their maximum, and the long wave motion is bound to the wave groups. Notice that as written this expression for $\overline{\zeta}$ has a mean value $\overline{\overline{\zeta}} \neq 0$.

It is noticed that as long as the short waves are long enough that $c_g < \sqrt{gh_0}$ this expression is finite. However, if the length of the short waves, which create the groups become shallow water waves with $L \gg h_0$ then $c_g \to \sqrt{gh}$ and the $\overline{\zeta}$ given by (14.1.33) will be very large. This means that the solution (14.1.33) cannot readily be used as the waves approach the shoreline. On a sloping bottom this solution would essentially be a geometrical optics approach which assumes locally constant depth. This issue is discussed more closely later.

Generation of IG-waves: The forcing

In the special case of shore normal wave incidence (14.1.5) also reduces to the one dimensional wave equation

$$\frac{\partial^2 \overline{\zeta}}{\partial t^2} - \frac{\partial}{\partial x}\left(gh_0 \frac{\partial \overline{\zeta}}{\partial x}\right) = \frac{1}{\rho}\frac{\partial^2 S_{xx}}{\partial x^2} \tag{14.1.34}$$

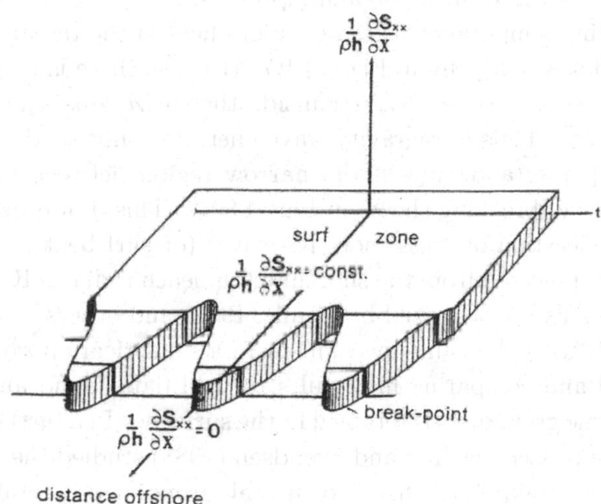

Fig. 14.1.7 Schematic illustration of the variation of the radiation stress with a time varying break point (from Symonds et al., 1982).

As mentioned in the introduction to this chapter, one of the essential problems related to infragravity waves is the nature of the forcing, and as (14.1.8) shows the forcing comes from changes in the radiation stress. It was shown in Chapter 5 that the overwhelming changes in radiation stress are due to the wave breaking and therefore occurs in the surfzone. However, it is also clear that simple uniform waves with initially constant wave height and period do not produce significant second order space-derivatives of S_{xx} as (14.1.8) or (14.1.34) require for IG wave generation. As the example above showed that requires wave motion varying in time (and therefore in space) as in propagating wave groups or bichromatic waves.

The strongest change in the gradients of S_{xx} occurs at the start of breaking, and in wave groups the breaking point changes position due to the fact that the waves of different height break at diferent depths. The effect of changing break point was analysed by Symonds et al. (1982). The wave height variation is simplified as partial groupiness in which the wave amplitude is assumed to vary sinusoidally around a mean value with a given amplitude of variation. The waves are shallow water waves. It is further assumed that the breaking occurs at a fixed value of the breaker index $\gamma = H/h$, so that after breaking all waves have the same height $H = \gamma h$ at a given depth. This also means that the groupiness of the incoming waves is completely destroyed in the breaking process.

The resulting temporal changes in the gradient of the radiation stresses are illustrated schematically in Fig. 14.1.7. Because there is no groupiness left inside the surfzone the $\partial S_{xx}/\partial x$ inside the surfzone is simply proportional to $(\gamma\, h_x)^2$. Thus infragravity wave energy is only produced outside the surfzone per se and only in the narrow region between the two extreme positions of breaking shown in Fig. 14.1.7. This, however, does lead to (strong) generation of cross-shore IG-waves (or surf beat). Since these waves are fully reflected from the shoreline the beach radiates IG-waves out to the open sea as hypothesized by Munk (1949) and others.

List (1992) used the same mechanism in a numerical approach with a similar model and comparing his results to field data found among other things that some groupiness is retained in the surfzone. In a further development of this approach Schäffer and Svendsen (1988) studied the alternative case where the breakpoint is fixed so that all groupiness is retained inside the surfzone. This approach had already been pursued by Foda and Mei (1981) using a multiple scale expansion. The assumption that groupiness is maintained inside the surfzone means that IG-wave energy is generated throughout the surfzone. This again creates strong cross-shore IG-motions. Schäffer (1993) generalized the approach to 2DH generation of IG-waves. The breaking was also generalized to an arbitrary combination of the two extreme breaking hypotheses by introducing a parameter κ which takes the value of 0 for a fixed breaking point and has the value of 1 for a time varying breaking point with no groupiness iside the surfzone. Fig. 14.1.8 shows a schematic illustration of the wave height variation in the two extreme cases of $\kappa = 0$ and 1. A possibility that shows up in the analysis of laboratory data is that κ can be > 1 which correponds to an inversion of the groupiness so that smaller waves before breaking starts become the larger waves inside the surfzone. This was also found in laboratory experiments with wave groups by Svendsen and Veeramony (2001).

Notice that Fig. 14.1.8 also shows in the bottom panel that the analysis starts from a constant depth region which can be at any depth. This ensures that at the outer boundary of the analysis we can count on actually having the constant depth solution for set-down waves given by (14.1.33) as boundary condition for the IG wave motion.

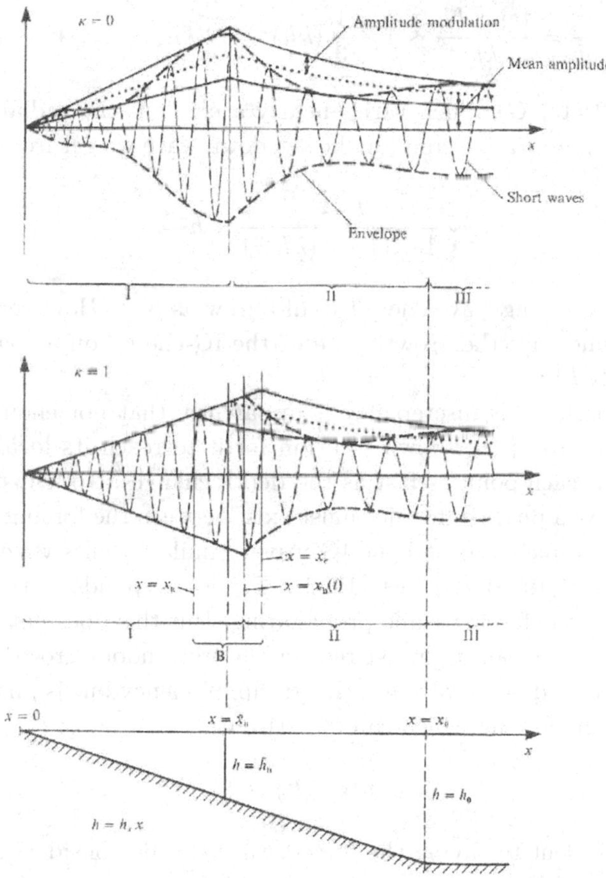

Fig. 14.1.8 Schematic illustration of the variation of the wave heights in the two extreme cases of $\kappa = 0$ and $\kappa = 1$. The bottom panel also shows the constant depth shelf used as offshore boundary for the analysis (from Schäffer, 1993).

Generation of IG-waves: energy transfer into IG-waves

As IG waves are generated from the forcing from short waves, energy must be transfered from the short wave to the long waves,[1] and a question becomes how this happens and how fast the long waves grow.

As an introductory example we consider the constant depth bound set-down wave solution given by (14.1.33) applied to a gently sloping bottom. To simplify we would expect that sufficiently close to the shore the short waves would be close to shoaling following the Green's law $\sim h^{-1/4}$, and the phase velocity c would satisfy

$$\frac{c^2}{gh} = \frac{\tanh kh}{kh} \sim 1 - \frac{1}{3}(kh)^2 + O(kh)^4 \tag{14.1.35}$$

and $c_g = 1/2c(1+G)$ which varies nearly a c. Thus the radiation stress increases as $S_{xx} \sim H^{-1/2}$ and for the set-down wave we get from (14.1.33)

$$\overline{\zeta} \sim \frac{h^{-\frac{1}{2}}}{1-(1-\frac{1}{3}(kh)^2)} \sim h^{-\frac{5}{2}} \tag{14.1.36}$$

which means the long wave energy would grow as h^{-5}. However, Elgar et al. (1992) found that the growth rate of the IG-energy on a coast is only approximately $h^{1.1}$.

The reason for this discrepancy is apparently that not enough energy is transferred into the set-down solution for it to reach its local constant depth value at each point as fast as the depth changes. To transfer energy into an IG wave a phase difference must exist between the forcing (the wave group in the simple case) and the IG-wave. Similar results were found by Elgar and Guza (1985) and List (1992). This corresponds to the IG-wave lacking behind the forcing while propagating. But this does not guarantee that enough energy can be transfered for the tremendous growth of h^{-5}.

Thus the forced total wave has the group frequency but is phase shifted ϕ relative to the forcing which can be written

$$\overline{\zeta} = a\ \cos(\theta_g - \phi) \tag{14.1.37}$$

which is equivalent to saying the forced wave can be considered a sum of a bound wave locked to the forcing moving with the phase θ_g and a free wave. This is often the interpretation used.

[1]Usually the resulting loss of energy from the short waves is not considered which implicitly assumes this to be small in comparison to the total short wave energy.

The forcing in the form of $\partial^2 S_{xx}/\partial x^2$ is propagating shorewards only and dissipated with the short waves. Throughout the nearshore (both before and after breaking) the total incoming IG wave (bound plus free) may attain any phase relatively to the forcing. Therefore energy transfer form the short to the incoming long waves can be both negative (growth of IG energy) and positive (loss of IG energy) depending on the phase difference.

This conjecture was confirmed by van Dongeren and Svendsen (1997) solving numerically the quasi-3D circulation equations for an incoming bichromatic wave group. In order to explicitly study the energy transfer to the IG waves they used the general energy equation for the current motion given by (4.4.19) which in 1D and linearized form becomes

$$\frac{\partial E}{\partial t} + \frac{\partial E_{fx}}{\partial x} + \frac{Q_x}{h_0}\frac{\partial S_{xx}}{\partial x} = 0 \tag{14.1.38}$$

where E is the long wave energy, $E_{fx} = \rho g \overline{\eta} Q_x$ is the energy flux and the third term represents the work done by the short wave radiation stress on the long waves. This is the term that represents the energy transfer. The expression is averaged over the IG wave period and the long wave motion is further divided into the incoming and outgoing parts of the total motion.

For a particular selection of parameters in the computations they found that the incoming part of the long wave (essentially the set-down wave generated on a shelf at the boundary of the computation) did not even grow as fast as the Green's law of $h^{-1/4}$.

On the other hand, once the IG waves have been reflected from the shoreline and move seaward they pass through the forcing pattern so quickly that virtually no energy transfer takes place. Van Dongeren and Svendsen even found in another example that the reflection coefficient R (defined as the ratio of the amplitude of the outgoing wave over the incoming wave) can be as large as 100, indicating large generation of IG energy in the nearshore.

Recently Janssen et al. (2003) analysed this mechanism analytically for arbitrary waves and gave an explicit solution for the phase shift for a bichromatic wave field. In confirmation of the above qualitative explanation they found that the phase shifts occur already before the breaking starts, and experimental data are confirming these results. These findings have been further analysed and discussed by Battjes et al. (2004).

Other results

As mentioned the literature about this subject is very extensive and it is beyond the scope of the present text to cover all the results properly.

Here we only quote a few papers addressing some principal issues.

Edge waves on longshore currents

Edge waves on longshore currents have been analysized mostly under the assumption of current velocities that are much smaller than the propagation speed of the edge waves. Examples are Howd et al. (1992) and Falqués and Iranzo (1992).

In both cases the analysis is restricted to the shallow water version of the equations. They use different methods to analyse the problem and Howd et al. find that the effect of a weak current is equivalent to the solution to the ordinary edge wave using a modified depth given by

$$h'(x) = \frac{h(x)}{(1 - \frac{V(x)}{c})^2} \tag{14.1.39}$$

The solution can be used for in principle arbitrary beach profiles on a long straight beach. As usual for waves on currents it is found that opposing currents will increase the longshore wave number k_y, while a following current decreases k_y. The currents also strongly modify the cross-shore edge wave profiles and velocities with the nodal structure shifting in the cross-shore direction.

Falqués and Iranzo (1992) also solve the problem on a long straight beach for an arbitrary beach profile and for arbitrary cross-shore variations. Their method is to consider the periodic problem that emerge by assuming longshore periodicity. Then the linear shallow water equations can be subject to Fourier transformation in the longshore direction which reduces them to a set of ordinary differential equations in the cross-shore direction. Those equations are then solved using orthogonal Chebyshev polynomials. This approach has several advantages including computational efficiency and potential high accuracy. The solution includes both the dispersion relation and the wave profiles.

Resonant generation of edge waves

The initial growth of edge wave has been studied in a complex wave environment described by a wave spectrum (Lippmann et al., 1997). The forcing is determined by considering the interaction of pairs of progressive shallow water waves that approach the beach obliquely and generate the radiation stress gradients that constitute the forcing with frequencies and wave numbers equal to the difference between the two components corresponding to the groupiness. A varying break point is assumed so that inside the surf zone all groupiness is destroyed and no forcing generated there. Only frequency-wave number combinations that satisfy the edge wave dispersion relation are considered so that the edge waves show resonant growth. In establising a linearized evolution equation for the growth rate of the edge wave amplitudes a_n the cross-shore variation is eliminated by a cross-shore integration. This essentially utilizes that edge waves is a complete orthogonal function set. Finally introduction of a weak frictional damping rate makes it possible to determine equilibrium amplitudes for the edge wave components.

14.2 Shear instabilities of longshore currents

14.2.1 *Introduction*

The phenomenon which is today most frequently refered to as shear waves or shear instabilities were not identified until 1989 when Oltman-Shay, Howd and Birkemeier discovered that the data from one of the large field experiments at the Army's Field Research Facility at Duck, North Carolina, the SUPERDUCK experiment, showed charateristics that did not fit any known wave theory. The signal analysis showed that some of the oscillations in the measured data corresponded to wavy fluctuations that propagated in the longshore direction at a speed much smaller than edge waves.

In this section we review the findings of Oltman-Shay et al. (1989) and then give an account of the first attempt to explain the observations theoretically, the analysis by Bowen and Holman (1989). Since then intensive research has been conducted leading to much better insight into this phenmenon. This includes analytical analysis of situations more complex and realistic than the first results, and, at least as exciting, extensive numerical computations showing how the initial linear instabilities develop over long time. For this, use has been made of advanced numerical models based on the circulation equations developed in Chapters 11 and 13. As will be discussed these results have far reaching implications for our understanding of nearshore phenomena.

14.2.2 *The discovery of shear waves*

The topography of the beach on which the measurements of the SUPERDUCK experiment were taken is shown in Fig. 14.2.1. The beach had a clear longshore bar-system. The figure also shows the locations of the sensors primarily used in the analysis. The longshore array placed mainly

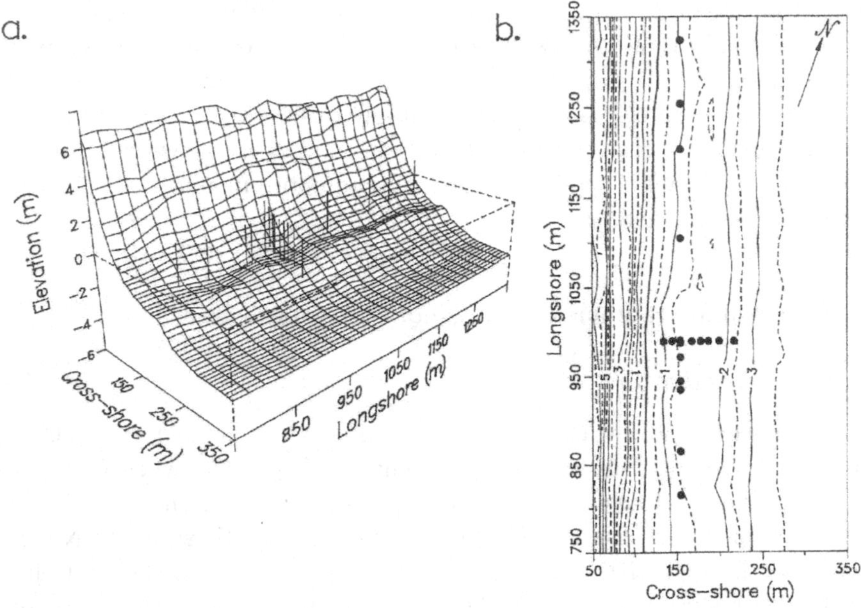

Fig. 14.2.1 Nearshore topography and positions of the sensors used in the analysis for the shear waves (from Oltman-Shay et al., 1989).

in the trough behind the bar consisted of electromagmetic velocity sensors, the cross-shore array of pressure sensors and accoustic altimeters. There was an offshore array of sensors to monitor incoming wave characteristics.

The time series from the instruments showed slow variations much larger that expected for a reasonably steady wave situation. An example of a time series of longshore (v), cross-shore (u) velocities and surface displacements on a particular date are shown in Fig. 14.2.2. The longshore velocity measurements (panel a in the figure) clearly illustrate the strong variations. The amplitude of the velocity variations is of the order $20cm/s$ and the period in the range of 100 to 1000s. The cross-shore velocity (panel b) and to some extent the surface elevation (panel c) also show these oscillations though to a less pronounced extent.

All the sensors in the longshore array were used for analysis of the statistical features of the in-phase and quadrature components of the cross-spectra to obtain longshore wavenumber and direction of propagation for the disturbances. It was found that unidirectional progressive waves with longshore wave lengths of 300m and 200m and with frequncies of 0.0017 Hz

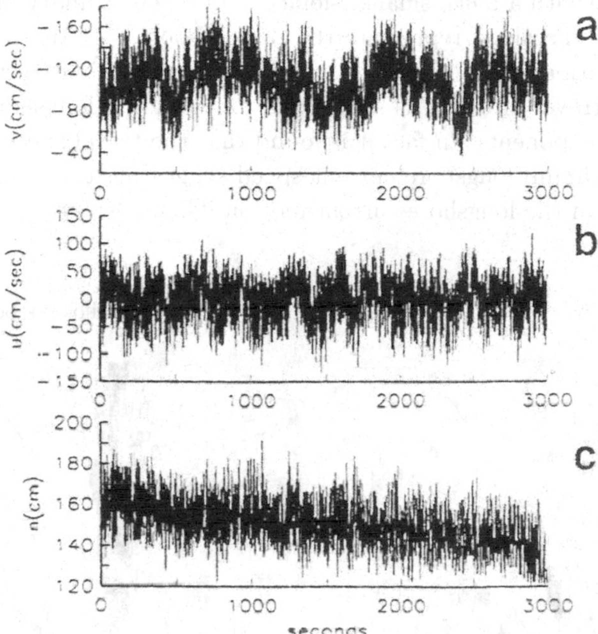

Fig. 14.2.2 Time series measurements of (a) longshore velocities v, (b) cross-shore velocities u, and (c) surface elevations at a position in the surfzone (from Oltman-Shay et al., 1989).

and 0.0037 Hz, respectively, were present on the morning of October 15. Such waves are an order of magnitude shorter than the shortest (0th-order) edge waves on the beach so the authors ruled out that the fluctuations could be edge wave related.

The availabililty of the longshore wave number for spectral components in the time series makes it possible to construct wavenumber frequency-spectra for both longshore and cross-shore velocities. An example is shown in Fig. 14.2.3. In this figure the rectangular boxes mark the location of variance peaks in the spectrum for the respective velocity, and the shading indicates the energy density within that box. In the figure is also marked the dispersion curves for the lovest order infragravity edge waves (0th, 1st, and 2nd order), and the boundary between trapped and leaky wave modes.

The detail to observe in this figure is that in addition to substantial amounts of energy in the area of edge wave dispersion there is in each of the two panels a significant amount of spectral energy concentrated along a straight line with a much smaller slope than for the ordinary infragravity waves. Since the slope represents the phase velocity of the motion this means fluctuations that travel at a much lower speed than even the edge waves. The travel direction is southward for both the longshore and the cross-shore components. In fact it is found that the travel speed is smaller than the maximum longshore current speed suggesting that this could be a fluctuation in the longshore current motion itself.

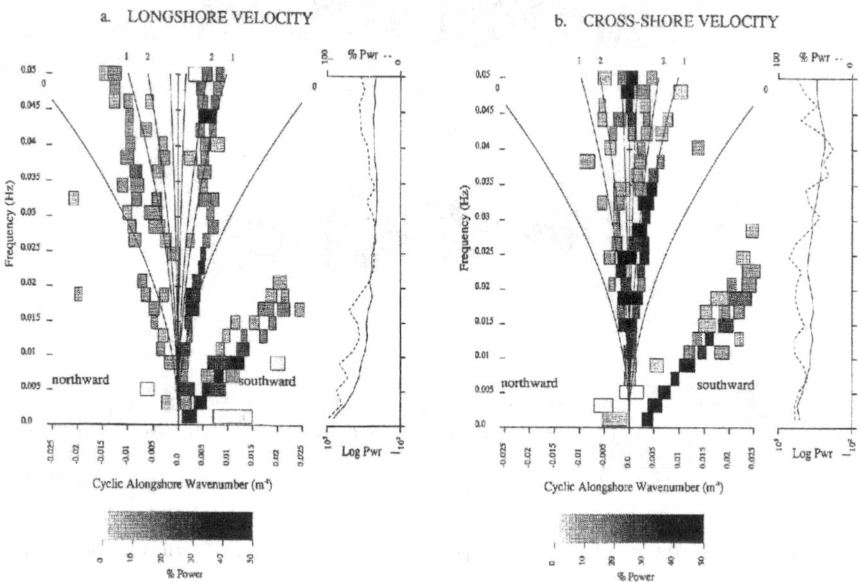

Fig. 14.2.3 Wavenumber-frequency spectra of (a) longshore velocity, and (b) cross-shore velocity in the surfzone for the October sata set (from Oltman-Shay et al., 1989).

14.2.3 *Derivation of the basic equations*

Simultaneously with the discovery by Oltman-Shay et al. (1989) that some of the very low frequency fluctuations behaved entirely differently than ordinary edge waves an analysis was presented by Bowen and Holman (1989) suggesting that the observed fluctuations had characteristics that were consistent with the characteristics found in stability analysis of the longshore currents. The initial analysis was made for very simplified conditions but were later extended by others to more realistic conditions and essentially confirmed.

Here we first show how to derive the equations analysed by Bowen and Holman illustrating the assumptions and approximations needed to get from the basic nearshore circulation equations, and then briefly describe the principle of the stability analysis.

In the first approximation it suffices to consider the equations for depth uniform currents. Thus we start with the continuity and momentum equations on the form derived in Chapter 11

$$\frac{\partial \bar{\zeta}}{\partial t} + \frac{\partial Q_\alpha}{\partial x_\alpha} = 0 \tag{14.2.1}$$

$$\frac{\partial Q_\alpha}{\partial t} + \frac{\partial}{\partial x_\beta}\left(\frac{Q_\alpha Q_\beta}{h} + \frac{S_{\alpha}}{\rho} - \frac{1}{\rho}\overline{\int_{-h_0}^{\zeta} \tau_{\alpha\beta}\, dz}\right) + gh\,\frac{\partial \bar{\zeta}}{\partial x_\alpha} + \frac{\tau_\alpha^S - \tau_\alpha^B}{\rho} = 0 \tag{14.2.2}$$

where for simplicity we have skipped the overbar on the volume flux Q_α.

It is convenient here to introduce he depth averaged velocity

$$\tilde{V_\alpha} = \frac{Q_\alpha}{h} \tag{14.2.3}$$

and

$$\tau_{\alpha\beta}^t = \overline{\int_{-h_0}^{\zeta} \tau_{\alpha\beta}\, dz} \tag{14.2.4}$$

which gives

$$\frac{\partial \bar{\zeta}}{\partial t} + \frac{\partial}{\partial x_\alpha}\tilde{V}_\alpha h = 0 \tag{14.2.5}$$

$$\frac{\partial}{\partial t}(\tilde{V}_\alpha h) + \frac{\partial}{\partial x_\beta}\left(\tilde{V}_\alpha \tilde{V}_\beta h + \frac{1}{\rho}\left(S_\alpha - \tau_{\alpha\beta}^t\right)\right) + gh\,\frac{\partial \bar{\zeta}}{\partial x_\alpha} + \frac{\tau_\alpha^S - \tau_\alpha^B}{\rho} = 0 \tag{14.2.6}$$

It is further assumed that the variations of the surface elevation $\bar{\zeta}$ are so small that they have little influence on the acceleration terms so that we can write the momentum equation as

$$h\frac{\partial}{\partial t}\tilde{V}_\alpha + h\frac{\partial}{\partial x_\beta}\tilde{V}_\alpha \tilde{V}_\beta + \frac{1}{\rho}\frac{\partial}{\partial x_\beta}\left(S_\alpha - \tau_{\alpha\beta}^t\right) + gh\,\frac{\partial \bar{\zeta}}{\partial x_\alpha} + \frac{\tau_\alpha^S - \tau_\alpha^B}{\rho} = 0 \tag{14.2.7}$$

The same assumption used in the continuity equation gives

$$\frac{\partial \tilde{V}_\alpha}{\partial x_\alpha} = 0 \tag{14.2.8}$$

This is often referred to as a **rigid lid assumption.**[2] Hence we have

$$\frac{\partial}{\partial x_\beta} \tilde{V}_\alpha \tilde{V}_\beta = \tilde{V}_\alpha \frac{\partial \tilde{V}_\beta}{\partial x_\beta} + \tilde{V}_\beta \frac{\partial \tilde{V}_\alpha}{\partial x_\beta} = \tilde{V}_\beta \frac{\partial \tilde{V}_\alpha}{\partial x_\beta} \tag{14.2.9}$$

so that (14.2.7) becomes

$$\frac{\partial}{\partial t} \tilde{V}_\alpha + \tilde{V}_\beta \frac{\partial \tilde{V}_\alpha}{\partial x_\beta} + \frac{1}{\rho h} \frac{\partial}{\partial x_\beta} \left(S_\alpha - \tau^t_{\alpha\beta} \right) + g \frac{\partial \overline{\zeta}}{\partial x_\alpha} + \frac{\tau^S_\alpha - \tau^B_\alpha}{\rho h} = 0 \tag{14.2.10}$$

If we now introduce the coordinate components of the velocity we have

$$\tilde{V}_\alpha = (u, v) \tag{14.2.11}$$

and writing the derivatives $\partial/\partial t, \partial/\partial x, \partial/\partial y$ as indices we get the following set of equations

$$u_t + uu_x + vu_y = -g\overline{\zeta}_x - \frac{1}{\rho h}(S_{xx,x} + S_{yx,y} - \tau_{bx} - \tau^t_{xx.x} - \tau^t_{yx,y}) \tag{14.2.12}$$

$$v_t + uv_x + vv_y = -g\overline{\zeta}_y - \frac{1}{\rho h}(S_{xy,x} + S_{yy,y} - \tau_{by} - \tau^t_{xy,x} - \tau^t_{yy,y}) \tag{14.2.13}$$

At this point we introduce that the flow is a combination of a steady long-shore current and cross-shore setup plus a time varying fluctuation (u_1, v_1) that may be stable or unstable. This is expressed as

$$u = 0 + u_1(x, y, t) \tag{14.2.14}$$

$$v = V(x) + v_1(x, y, t) \tag{14.2.15}$$

$$\overline{\zeta} = \zeta_0(x) + \zeta_1(x, y, t) \tag{14.2.16}$$

where we assume that $(u_1, v_1) \ll V$, and $\zeta_1 \ll h$. This implies that

$$\tau^t_{\alpha\beta} = \tau^t_{0\alpha\beta}(x) \;\; ; \quad \tau_{bx}, \tau_{by} = \tau_{0bx}(x), \tau_{0by}(x) \tag{14.2.17}$$

We also assume that the smallness of (u_1, v_1) implies that effects which the fluctuations in the currents have on the short wave motion are so weak that

[2] This term can be misleading because it still allows the pressure to be constant at the free surface which a real rigid lid would not.

the radiation stress components are not affected. That means the forcing of the longshore current is not changing, so that

$$S_{\alpha\beta} = S_{0\alpha\beta}(x) \tag{14.2.18}$$

Substituting all this into (14.2.12) and (14.2.14) then gives

$$u_{1,t} + u_1 u_{1,x} + (V + v_1)u_{1,y} = \\ -g(\zeta_{0,x} + \zeta 1, x) - \frac{1}{\rho h}\left(S_{0xx,x} + \tau_{bx} - \tau^t_{0xx,x}\right) \tag{14.2.19}$$

$$v_{1,t} + u_1(V + v_1)_y + (V + v_1)v_{1,y} = -g\zeta_{1,y} - \frac{1}{\rho h}(S_{0xy,x} + \tau_{by} - \tau^t_{0xy,x}) \tag{14.2.20}$$

From these equations we get the **zeroth order eqations** are

$$S_{0xx,x} = -\rho g h \zeta_{0,x} - \tau_{bx} + \tau^t_{xx,x} \tag{14.2.21}$$
$$S_{0xy,x} = -\tau_{by} + \tau^t_{xy,x} \tag{14.2.22}$$

in the x and y directions, respectively. Here the $\tau^t_{xx,x}$-term represents the cross-shore turbulent normal stresses which we normally neglect as small. We then see that to zeroth order the basic situation is the usual steady longshore current and setup on a long straight coast analysed in Section 11.7.

For the **first order equations** we get similarly

$$u_{1,t} + V u_{1,y} = -g\zeta_{1,x} \tag{14.2.23}$$
$$v_{1,t} + u_1 V_{,x} + V v_{1,y} = -g\zeta_{1,y} \tag{14.2.24}$$

Exercise 14.2-1

Show that by substituting the perturbation expansions (14.2.15) and (14.2.16) into the continuity equation we get

$$\zeta_{0,t} = 0 \qquad \text{zeroth-order} \tag{14.2.25}$$
$$\zeta_{1,t} + (hu_1)_x + (hv_1)_y = 0 \qquad \text{1st-order} \tag{14.2.26}$$

Thus if we neglect the fluctuations ζ_1 of the surface caused by the velocity fllucutuations (which is very reasonable) then we get

$$(hu_1)_x + (hv_1)_y = 0 \qquad \text{or} \qquad \nabla \cdot (h\mathbf{u}) = 0 \tag{14.2.27}$$

It is worth to emphasize that the neglect of the surface variation in these fluctuations is not a trivial matter because it underlines the fact that, as pointed out already by Oltman-Shay et al. (1989), the shear instabilities we are studying here are not surface waves per se.

These are essentially the equations for the shear wave instabilities analysed by Bowen and Holman (1989).

As the derivation above shows the basic assumptions underlying their equations is that the shear instability is a fluctuation on a steady longshore current on a long straight coast. The fluctuations are assumed so small that they do not disturb the wave motion that generates and maintains the longshore current. Finally, as (14.2.22) shows, it is assumed that the fluctuations generate no fluctuations in the bottom shear stresses or in the lateral mixing. The latter of those imply that the fluctuations are not damped: the constant bottom shear stresses are balanced in the zeroth order equations by the steady radiation stress forcing component and lateral mixing term $\tau^t_{xy,x}$. This assumption is of cause not realistic and was abandoned in later investigations.

We also see that the resulting equations are linear in the unknown first order variables u_1, v_1, ζ_1.

14.2.4 *Stability analysis of the equations*

The next step is to eliminate ζ_1 from (14.2.23) and (14.2.24). By cross-differentiation and subtraction the result becomes

$$\left(\frac{\partial}{\partial t} + V\frac{\partial}{\partial y}\right)(v_{1.x} - u_{1,y}) = -(u_1 V_x)_x - v_{1,y} V_x \tag{14.2.28}$$

As (14.2.27) shows the volume flux $h\mathbf{u}$ is now divergence free. Therefore we can introduce a stream function for the volume flux defined so that

$$u_1 h = -\psi_y \quad ; \quad v_1 h = \psi_x \tag{14.2.29}$$

With this (14.2.28) can be written

$$\left(\frac{\partial}{\partial t} + V\frac{\partial}{\partial y}\right)\left(\left(\frac{\psi_x}{h}\right)_x + \frac{\psi_{yy}}{h}\right) = \psi_y \left(\frac{V_x}{h}\right) \tag{14.2.30}$$

It is this equation for which we now analyse the possibility that the disturbance described by ψ could be an unstable oscillation propagating along the shore. Mathematically this is expressed as

$$\psi = Re\left(\phi(x)\ e^{i(\omega t - ky)}\right) \qquad (14.2.31)$$

This means that (dropping Re and setting $\omega t - ky = \theta$) (14.2.30) can be written

$$\omega\left(\frac{\phi_{xx}}{h} - \frac{\phi_x h_x}{h^2}\right) - \frac{k^2}{h}\omega\phi - k\left(\frac{\phi_{xx}}{h} - \frac{\phi_x h_x}{h^2}\right)V + \frac{k^3}{h}V\phi = k\phi\left(\frac{V_x}{h}\right)_x \qquad (14.2.32)$$

After introducing the definition

$$c = \frac{\omega}{k} \qquad (14.2.33)$$

where c is the propagation speed of the fluctuations this may be further modified to the form

$$\boxed{(c - V)\left(\phi_{xx} - k^2\phi - \frac{\phi_x h_x}{h}\right) - \phi h\left(\frac{V_x}{h}\right)_x = 0} \qquad (14.2.34)$$

Exercise 14.2-2

Derive equations (14.2.32) and (14.2.34)

This is an eigenvalue problem for the unknown x-variation ϕ of the stream function, and c is the (unknown) eigenvalue. This means that solutions for ϕ may only exist for certain values of c. c may be complex so that

$$c = c_{re} - ic_{im} \qquad (14.2.35)$$

which means that the corresponding frequency ω is complex

$$\omega = \omega_{re} - i\omega_{im} \qquad (14.2.36)$$

Substituting this, (14.2.31) then gives

$$\psi = Re\left(e^{+\omega_{im}t}\ \phi(x)e^{i(\omega_{re}t - ky)}\right) \qquad (14.2.37)$$

Thus if $\omega_{im} > 0$ then we get a ψ-solution that grows exponentially with time. That means an **unstable** solution and the growth rate is proportional to ω_{im}.

Solution of (14.2.34) for a simple case

To illustrate the nature of the problem we briefly review the solution given by Bowen and Holman (1989) for the very simple distribution of the longshore currents velocity $V(x)$ shown in Fig. 14.2.4. With x_0 the seaward limit of the assumed longshore current domain we have the maximum current velocity at δx_0.

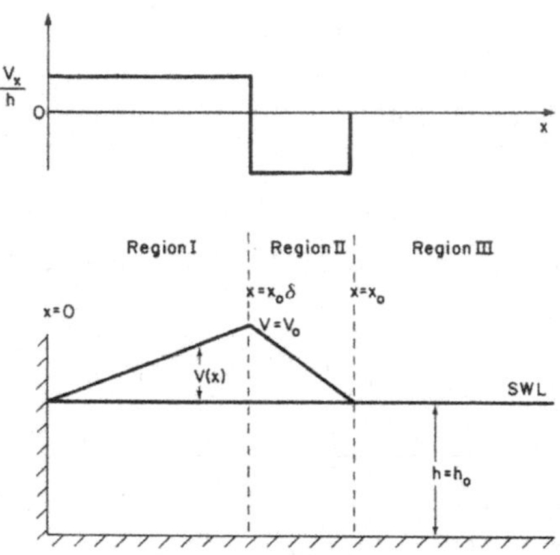

Fig. 14.2.4 The longshore current distribution used in the example (from Bowen and Holman, 1989).

We see that in this case $h_x = 0$ everywhere, and for this case (14.2.34) reduces to the equation

$$(c - V)\left(\phi_{xx} - k^2\phi\right) - \phi V_{xx} = 0 \qquad (14.2.38)$$

which is called the Rayleigh equation.[3]

Before discussing the solution we notice that the boundary conditions for $\phi(x)$ are

$$\phi = 0 \qquad \text{at} \qquad x = 0 \qquad (14.2.39)$$

$$\phi \to 0 \qquad \text{at deep water} \qquad x \to \infty \qquad (14.2.40)$$

It is also clear that in (14.2.34) we cannot have $c = V(x)$ because c is not a function of x. Therefore $c - V \neq 0$ and we can divide by $c - V$. Thus the

[3]For further details about hydrodynamic stability theory reference is made to Drazin and Reid (1982).

equation we need to solve is

$$(\phi_{xx} - k^2\phi) - \phi V_{xx} = 0 \tag{14.2.41}$$

and for the special velocity profile shown in Fig. 14.2.1 this reduces to an equation of the form

$$\phi_{xx} - k^2\phi = 0 \tag{14.2.42}$$

Exercise 14.2-3

Show that for the conditions described in Fig. 14.2.1 equation (14.2.41) reduces to (14.2.42).

Thus the problem is reduced to solving this equation in the three regions shown in Fig. 14.2.4 and determining the integration constants so that we get a continuous solution including the solution for ζ (which also represents the pressure variation). This is in principle a straightforward problem but leads to quite complicated algebraic manipulations. A few details are given in Bowen and Holman and the task concludes in finding that the frequency ω must satisfy a quadratic equation of the form

$$\omega^2 + a_1\omega + a_0 = 0 \tag{14.2.43}$$

where a_1, a_0 are complicated functions of the parameters x_0, δ, V_0 that describe the velocity distribution assumed for V, and the wave number k in the assumed solution (14.2.31). The conclusion is then that for each value of k the equation will have two roots, which can be either both real or complex conjugate, depending on the value of a_0 and a_1. The unstable cases that grow exponentially are those with complex conjugate roots.

Fig. 14.2.5 shows the range of solutions for a case with chosen V_0, δ, x_0. The numbers are in line with the field situation described by Oltman-Shay et al. The top panel in the figure shows the real comoponent of ω, the bottom panel the imaginary part. The abscissa is $L^{-1} = k/2\pi$. In the two outer intervals for L_1 there are two real solutions for ω and hence the imaginary part is zero. In the center interval however, $\omega_{im} \neq 0$. The philosophy is then that the solution with the largest positive value of ω_{im} will grow faster than any other and therefore be the one that after some time dominates and provides the characteristics of the instabilites that occur. We see that the most unstable frequencies have $\omega_{re} = O(0.0012-0.0015)$ which

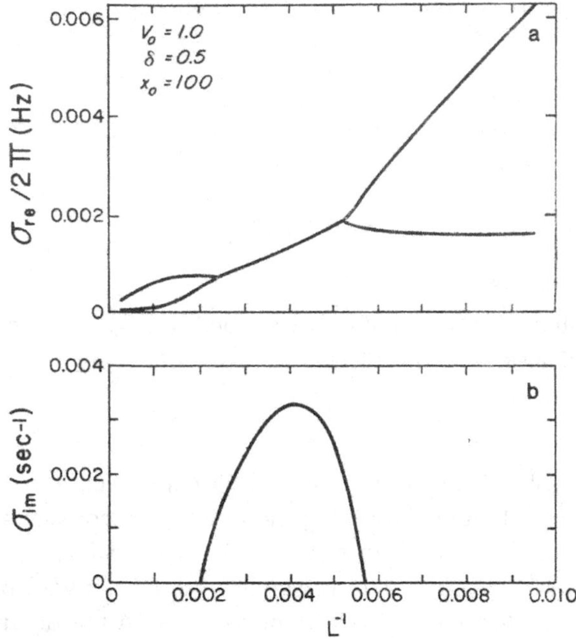

Fig. 14.2.5 The solution for ω for the parameter values shown in the diagram. The top panel gives the real value ω_{re} (named σ_{re} in the figure), the bottom panel ω_{im}, (σ_{im}) (from Bowen and Holman, 1989).

is in the range of the observed shear wave instabilities in the SUPERDUCK data analysed by Oltman-Shay et al.

Once we have found the most unstable k-value we can go back into the solution and determine how it varies with x, y. The solution does not provide any information about the magnitude because it is growing exponentially, only the relative amplitudes. Fig. 14.2.6 shows an example of the distibution of the fluctuation velocities. This is an instantaneous snapshot, the pattern propagates along the shore with velocity $c = \omega/k$.

14.2.5 *Further analyses of the initial instability*

The discovery that longshore currents in some field data showed slow fluctuations that could be interpreted as shear instabilities set off extensive research into the theoretical nature of this mechanism. As shown above the first analysis was for the very simplified bottom topography and cross-shore

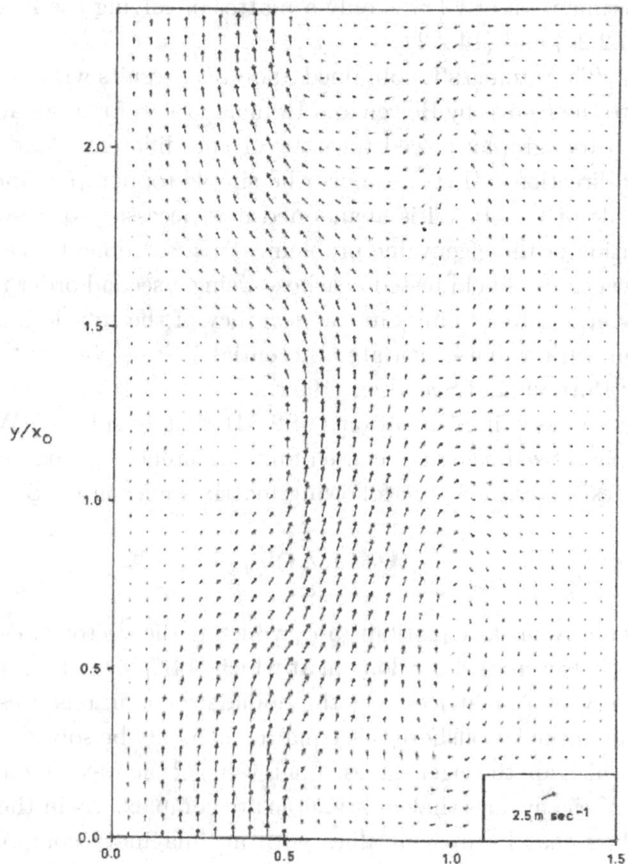

Fig. 14.2.6 The distribution of the total velocities (steady longshore current plus fluctuations) for the most unstabble k-value in the case described in Fig. 14.2.2 (from Bowen and Holman, 1989).

distribution of the longshore current shown in Fig. 14.2.4, with no frictional damping or wave-current interaction. Here we briefly describe a few of the many contributions expanding the simplistic analysis described above. We mainly focus on the primary problem formulation and the solution techniques, rather than on the details of the derivations.

It may also be mentioned that a condition for the following solutions is that the cross-shore profile $V(x)$ of the longshore current is preset. This is not a necessary condition since - as the analysis earlier in this section shows - determination of $V(x)$ is only a matter of solving the zeroth order equatins (14.2.21) and (14.2.22)

Dodd and Thornton (1990) obtained analytical results with characteristics similar to the results by Bowen and Holman (1989) for a biplanar beach (i.e. a beach topography pieced together of two different plane slopes in the seaward direction). However, once the simple topography and current velocity profile in Fig. 14.2.4 is abandoned it is necessary to resort to numerical solution of the eigenvalue problem. This was done by Dodd et al. (1992) for two cases of field tested beaches. Using a second order numerical scheme they encountered limits in the accuracy of the solution. A similar approach but with a more accurate differential scheme correct to $O(\Delta x^4)$ was used by Putrevu and Svendsen (1992).

In all those cases it is equation (14.2.34) that is solved. When this equation is discretized using the chosen finite difference approximations for the derivatives we arrive at the following matrix equation for ϕ

$$[\mathbf{A}]\phi_i = c[\mathbf{B}]\phi_i \tag{14.2.44}$$

This represents N linear equations in ϕ_i which is the vector of ϕ-values at the N points in the cross-shore domain at which (14.2.34) is discretized, and $\mathbf{A}$ and $\mathbf{B}$ are N by N matrices. For the boundary conditions prescribed at the shore and in the far offshore this equation can only be solved for certain values of c, which are the eigenvalues. Each value of the wave number vector k will yield N eigenvalues that in principle are complex. As in the example above ω-values that have eigenvalues with an imaginary component > 0 represents unstable wave numbers, and it is assumed that when subjected to a whole spectrum of infinitesimal disturbances the eigenvalue with the largest imaginary component will dominate the instability of that wave number. Similarly, if for some k all eigenvalues are real then disturbances with that k are stable.

Putrevu and Svendsen analysed three different cross-shore profiles: a plane beach, an equilibrium beach and barred beach and found substantial sensitivity to the topography. In particular for a barred beach they found more than one unstable mode for some k-values: one is a strong instability that develops on the outer side of the bar, the second, much weaker, is found close to the shoreline. Fig. 14.2.7 shows characteristics of the velocity field. The classical criterion for instability is a socalled Rayleigh condition which requires that the longshore velocity distribution $V(x)$ must have a shape so that V_x/h has a minimum (i.e. $(V_x/h)_x = 0$) somewhere in the cross-shore

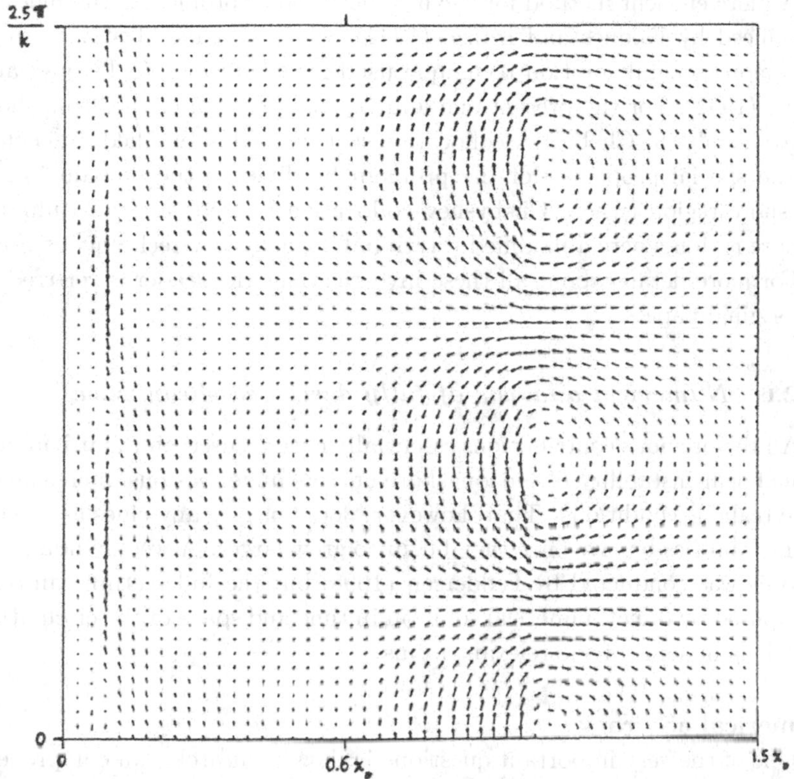

Fig. 14.2.7 A plot of the distribution of the shear wave induced valocity field for a barred beach with breaking point at $x_0 = 1$, and the bar positioned at $x_0 = 0.6$. In this case there are two unstable modes at the chosen wave number (which corresponds to $kx_0 = 3.5$): one, the strongest, at the outer side of the bar, the other close to the shoreline (from Putrevu and Svendsen, 1992).

domain (Bowen and Holman, 1989). Putrevu and Svendsen show that, if we define V_s as the velocity at the point where $(V_x/h)_x = 0$, then in fact it is also necessary that $(V_x/h)_x(V - V_s) < 0$ in some interval (Fjortofts theorem). This is typically satisfied on the seaward side of the barcrest.

Thus each version of (14.2.44) represents the solution of (14.2.34) for one value of k, which means the solution to the entire eigenvalue problem requires solution of (14.2.44) for a large number of k-values. Computationally this is quite demanding (Dodd et al. 1992 found that $N = 100$ was about what was practically possible for their computer capacity).

A more efficient method for solving the stability problem is the method introduced by Falqués and Iranzo (1994), which is a modified version of their more general solution technique used for edge waves in Falqués and Iranzo (1992). For the present one dimensional problem of the cross-shore variation only the Chebyshev collocation scheme used is modified to account for the special properties of the problem by dividing the domain into a nearshore region (where Chebyshev collocation is used) and an (infinite) offshore region where a modified ("rational") Chebyshev technique is used.

For more details about all these investigations the reader is referred to the original papers.

14.2.6 *Numerical analysis of fully developed shear waves*

All the previous contributions essentially have focused on the **initiation** of the linear instability of the longshore curents under various, increasingly more realistic conditions. That, however, does not give any clues as to how the instabilities behave once they begin to grow larger. A weakly nonlinear analysis was conducted by Feddersen (1998) but the full picture can only be exposed by direct numerical solution in time and space of the circulation equations in some of the relevant forms.

Numerical accuracy

One of the very important questions is: how accurately can comprehensive numerical models predict the long term temporal development of so complicated flows. In order to assess that Fig. 14.2.8 shows the development over the first 12 hours of the surface elevation and the two horizontal, depth averaged velocity components at a point in a flow which shows shear instability (Zhao et al. 2003). The figure shows the results from two different models, the 2DH version of the SHORECIRC (SC for short), and the 2DH model by Özkan-Haller and Kirby (1999). The two models represent essentially the same mathematical formulations of depth averaged 2DH flow with the same bottom friction, the same lateral mixing, etc. The difference in the two analyses is entirely in the numerical schemes. The SC model is based on the numerical predictor-corrector method developed by Wei and Kirby (1995) with 4th order finite difference accuracy for the spatial derivatives, using a third order predictor step and a fourth order corrector step. The Özkan-Haller-Kirby analysis uses a numerical scheme similar to the scheme used by Falqués and Iranzo (1992) for edge wave motion. It uses spectral (Fourier) decomposition in the longshore direction and a Chebyshev collocation for the cross-shore variation.

Considering the complexity of the flow problem (the computational domain is approximately $500-800m \times 2500-3000m$ in the cross- and longshore directions, respectively) and the time over which the computations stretch (12 hrs in beach time) the similarities between the two simulations are really remarkable. There are differences in particular in the initial growth rate of the shear wave instabilities. That can for example be due to small differences in the amplitudes of the disturbance used to seed the instabilities. However, once developed, the two timeseries look very alike. The similarities range from roughly the same period and magnitude of the dominating fluctuations and the slow variation of those amplitudes, to the temporal shape of those fluctuations, which perhaps is particularly characteristic in the longshore velocity. It may be mentioned that the computations by Allen et al. (1996) made with a lower order difference scheme show similar patterns but in a comparison with the two models above comes out rather differently (Sancho and Svendsen, 1997).

Thus Fig. 14.2.8 is interpreted as a very strong illustration of the accuracy and reliability of the numerical schemes available today. The conclusion is, as one could hope, that the accuracy (or the error sources) of the computations relative to the nature we try to represent lies in the mathematical formulation of the models including the approximations and empirical formulas that it has been necessary to introduce such as the eddy viscosity, and the friction factor for the bottom shear stress, but also in the short wave forcing terms. The numerical errors do not seem to dominate the results.

The long term development of the shear waves

When we look closer at the computations we find that as the shear waves grow, so does the vorticity they represent. The solutions available all assume that in the longshore direction the flow conditions are periodic so that the flow patterns that leave the domain in the downstream end are re-entered in the upstream end of the computational domain. Over time this is equivalent to following the flow as it develops downstream over longshore distances many times the longshore extent of the computational domain.

After some time the picture typically looks as shown in Fig. 14.2.9 (Özkan-Haller and Kirby, 1999). The figure shows the intensity of the vorticity rather than the velocities. Unfortunately the grey scale does not give all the details available in the original color photo. We see that the

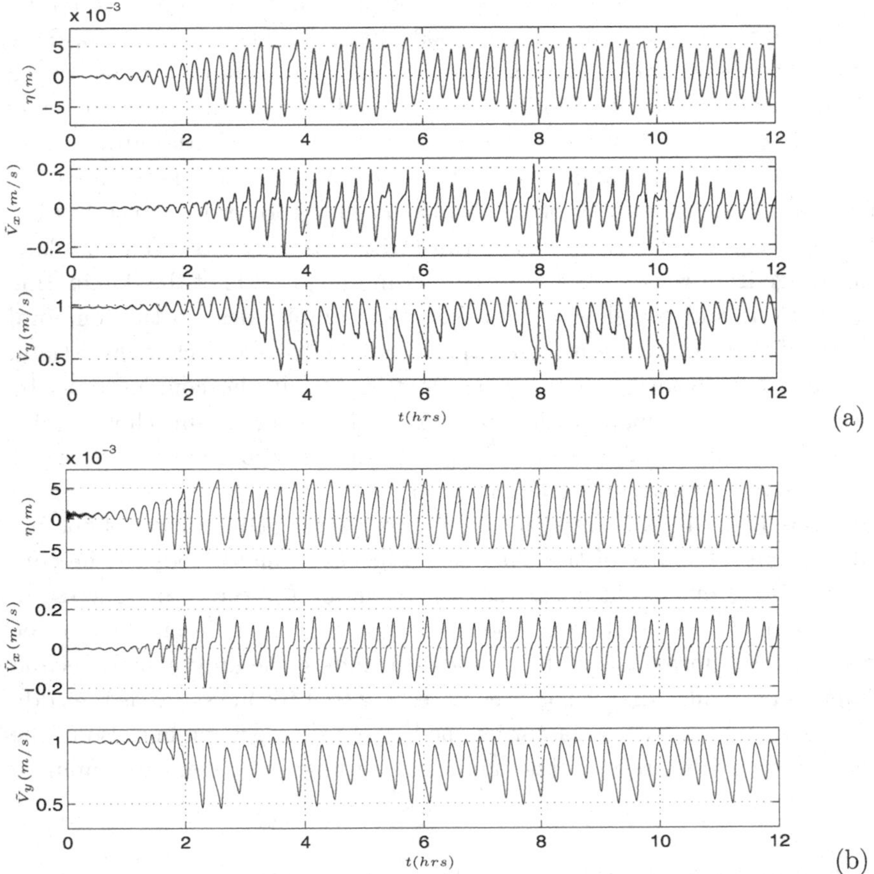

Fig. 14.2.8 Comparison of 2D models with two different numerical schemes. (a) the 2D version of the SHORECIRC model. (b) The analysis of Özkan-Haller and Kirby (1999) (from Zhao et al., 2003).

nature of a shear wave undulation found in the linear instability analysis has given way to what appears to be a vorticity distribution with a series of individual, but interacting vortices, moving largely with the local longshore current velocity. This is the picture found in all such long-term simulations withvorticity intensities varying from case to case.

The three panels also show that the intensity of the vorticity depends strongly on the parameter M which is a measure of the lateral mixing imposed in the equations.

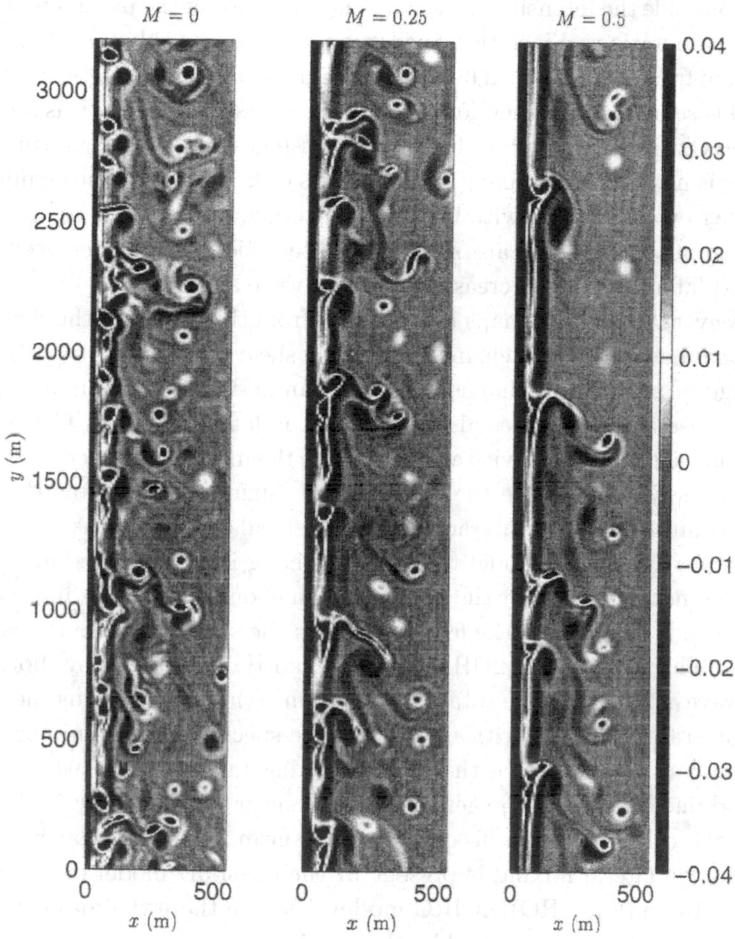

Fig. 14.2.9 Contour plots of vorticity after 5 hours of model simulations with three different values of the constant in the lateral mixing eddy viscosity (from Özkan-Haller and Kirby, 1999).

Overall effect of the shear waves on the mean flow

It is instructive to consider the relations between the vorticity distributions and the various mechnisms involved in creating the mean flow, which in the cases studied is primarily a longshore uniform longshore current.

On one side the intensity of the vorticity depends on the parameters used in the computations. Since the shear wave motion is weakly forced changes in bottom friction and lateral mixing give significant changes in the vorticity level. These parameters therefore need to be assessed carfully. (It is recalled that the Özkan-Haller and Kirby analysis assumes depth uniform currents, so there is no dispersive mixing.) The results clearly indicate the significant decrease in vorticity level with increasing lateral mixing (factor M) in these 2DH computations. We are seeing an interaction where increased pre-specified lateral mixing decreases the shear wave activity.

However, as might perhaps be expected from the nature of the flow, the additional horizontal particle motions in the shear waves cause a horizontal exchange of momentum that is similar to an added lateral mixing. Thus the presense of shear waves also increases the lateral mixing. Conversely decreasing shear wave activity also decreases the momentum exchange from the shear waves and hence the total lateral mixing. This means that the two mechanisms tend to interact and counter-balance each other.

However, in a 2DH model the lateral mixing is pre-specified at a level that does not change with the solution. Thus one side of the interaction mechanism is eliminated: the feed back from the shear waves to the lateral mixing. Therefore in the 2DH case of Özkan-Haller and Kirby, both the shear wave activity and the total lateral mixing (the specified plus the shear wave generated) depend critically on the pre-specified lateral mixing. In a way one can say that once the lateral mixing (and other constants) are specified one can expect the general level of shear wave activity is also set.

On the other hand this feed back mechanism from the shear wave activity to the lateral mixing is present in the quasi-3D model described in Chapter 13 and the SHORECIRC model, because there the major part of the lateral mixing is determined by the nonlinear current-current and wave-current interactions due to the vertical variation of the current velocities, that is by the flow itself and hence by the shear wave activity.

Zhao et al. (2003) conducted simulations with the quasi-3D model with and without the dispersive mixing. The dispersive mixing mechanisms of course also suppresses the shear wave activity, and because it is controled by the flow itself, not pre-specified, it allows the shear wave activity to influence the lateral mixing. The overall reductions in the shear wave energy for the domain when dispersive mixing is included are found to be a factor ranging from 1.6 early in the computations to $4-5$ after close to 3 hours simulation time. The quasi-3D computations also show the shear wave energy is more confined to the region closer nearshore.

In a longshore uniform case discussed here a change in lateral mixing shows up as a change in the cross-shore distribution of the longshore current velocity. This seems to be one of the major results of both Özkan-Haller and Kirby (1999) and of Zhao et al. (2003) where the effect has been clearly demonstrated. It is not clear at this point if there is a well defined equilibrium or balance between the two mechanisms the are responsible for the cross-shore current profile and, if so, what characterizes such an equilibrium. Further research into these mechanisms would be interesting.

The quoted papers further contain analysis of various other flow related features such as the interaction between vortices, the propagation paths of the vortices, influence of a barred topography, and Zhao et al. (2003) also analyse the energetics of the vorticity (the socalled enstrophy) along with the energy transfer to and from shear wave vortices and mean flow, and the vertical distribution of the vorticity. For these and other topics the reader is referred to the papers.

Concluding remarks

In all there is a growing understanding of how the different mechanisms interact in the complicated nearshore flows. The examples quoted above were still idealized in the sense that only flows over longshore uniform topography were analysed, but this is only to clarify the already complicated mechanisms. However, as the latest literatue shows the modelling techniques based on the equations described in Chapters 11 and 13 are fully capable of handling general topographies, and with the wave drivers that provide time series information about volume flux and radiation stress variations in irregular waves then the circulation models are capable of also predicting the resulting time vaying currents and IG-wave motions.

14.3 References - Chapter 14

Abramowitz, M. and I. A. Stegun (1964). Handbook of mathematical functions. Dover Publications, New York.

Allen, J. S., P. A. Newberger, and R. A. Holman (1996). Nonlinear shear instabilities of longshore currents on planar beaches. J. Fluid Mech., **310**, 181–213.

Battjes, J. A., H. J. Bakkenes, T. T. Janssen and A. R. van Dongeren (2004) "Shoaling of subharmonic waves". J. Geoph. Res., 109(C2), C02009, doi:10.1029/2003JC001863.

Bowen, A. J. and R. T. Guza (1978). Edge waves and surf beat. J. Geophys. Res., **83**, 1913–1920.

Bowen, A. J. and R. A. Holman, (1989) Shear instabilities of the mean longshore current, 1. Theory, J. Geophys. Res., **94**, 18,023-18,030.

Dodd, N. and E. B. Thornton (1990). Growth and energetics of shear waves in the nearshore. J. Geophys. Res., **95**, C9, 16075–16083.

Dodd, N., J. Oltman-Shay, and E. B. Thornton (1992). Shear instabilities in the longshore current: A comparion of observation and theory. J. Phys. Ocean., **22**, 1, 62–82.

Drazin, P. G. and W. H. Reid (1982). Hydrodynamic stability. Cambridge Univ. Press, New York.

Eckart, C. (1951). Surface waves on water of variable depth, Wave Rep. 100, Scripps Inst. Oceanogr., Univ Calf., La Jolla.

Elgar, S. and R. T. Guza (1985). Observations of bispectra in shoaling waves. J. Fluid Mech., **161**, 425–448.

Elgar, S., T. H. C. Herbers, M. Okino, J. Oltman-Shay, and R. T. Guza (1992). Observations of ingragravity waves. J. Geophys. Res., **97**, 15573–15577.

Falqués, A. and V. Iranzo (1992). Edge waves on a longshore shear flow. Phys. Fluids, A, **10**, 2169–2190.

Falqués, A. and V. Iranzo (1994). Numerical simulation of vorticity waves in the nearshore. J. Geophys. Res., **99**, C1, 825–841.

Feddersen, F. (1998). Weakly nonlinar shear waves. J. Fluid Mech., **372**, 71–91.

Foda, M. A. and C. C. Mei (1981). Nonlinear excitation of long trapped waves by groups on short swells. J. Fluid Mech., **111**, 319–345.

Gallagher, B. (1971). Generation of surf beat by nonlinear wave interactions. J. Fluid Mech. **49**, 1–29.

Guza, R. T. and A. J. Bowen (1975). The resonant instabilities of long waves obliquely incident on a beach. J. Geophys. Res., **80**, 4529–4534.

Howd, P. A., A. J. Bowen, and R. A. Holman (1992). Edge waves in the presence of longshore currents. J. Geophys. Res., **97**, C7, 11357–11371.

Huntley, D. A., R. T. Guza, and E. B. Thornton (1981). Field observations of surf beat 1. Progressive edge waves. J. Geophys. Res. **86**, C7, 6451–6466.

Inman, D. L. and A. J. Bowen (1967). Spectra of breaking waves. Ann. Rev. of Fluid Mech. **8**, 275-310.

Janssen, T. T., J. A. Battjes, and A. R. van Dongeren (2003). Long waves induced by short wave groups over a sloping bottom. J. Geophys. Res., **108**, C8, 3252.

Lippmann, T. C., R. A. Holman, and A. J. Bowen (1997). Generation of edge waves in shallow water. J. Geophys. Res., **102**, C4, 8663–8679.

Lamb, H. (1945). Hydrodynamics. Dover publications, New York.

List, J. H. (1992). A model for the generation of two-dimensional surf beat. J. Geophys. Res., **97**, C7, 5623-5635.

Lo, J.-M. (1988). Dynamic wave setup. ASCE Proc. 21st Int. Conf. Coastal Engrg., 999-1010.

Longuet-Higgins, M. S. and R. W. Stewart (1962). Radiation stress and mass transport in gravity waves with application to 'surf-beats'. J. Fluid Mech., **8**, 565 – 583.

Longuet-Higgins, M. S. and R. W. Stewart (1964). Radiation stress in water waves, a physical discussion with application. Deep Sea Research, **11**, 529-563.

Munk, W. H. (1949). Surf beats. EOS Trans. AGU **30**, 6, 849-854.

Oltman-Shay, J., P. A. Howd and W. A. Birkemeier (1989). Shear instabilities of the mean longshore currents 2. Field observations. J. Geophys. Res., **94**, C12, 18031 - 18042.

Peregrine, D. H. (1976). Interaction of waves and currents. *Adv. Appl. Mech.*, **16**, 9-117.

Putrevu, U. and I. A. Svendsen (1992). Shear instability of longshore currents: A numerical study. J. Geophys. Res., **97**, C5, 7283–7303.

Sancho, F. E. P. and I. A. Svendsen (1998). Unsteady nearshore currents on longshore varying topographies, Center for Applied Coastal Research, Rep. No. CACR-97-10, 320 pp.

Schäffer, H. A. (1993). Infragravity waves induced by short wave groups. J. Fluid Mech. **247**, 551–588.

Schäffer, H. A. and I. A. Svendsen (1988). Surf beat generation on a mild-slope beach. ASCE Proc 21st Int. Conf. Coastal Engrg., 1058–1072.

Schäffer, H. A. and I. G. Jonsson (1992). Edge waves revisited. Coastal Eng., **16**, 349–368.

Stokes, G. G. (1846). Report on recent researches in hydrodynamics. Brit. Ass. rep. (see also Papers, 1880, vol 1, p 167).

Symonds, G., D. A. Huntley, and A. J. Bowen (1982). Two-dimensional surf beat: Long wave generation by a time-varying break point. J. Geophys. Res., **87**, C12, 9499–9908.

Svendsen, I. A. and J. Veeramony, (2001), Experiments with groups of breaking waves: Short wave motion. ASCE J. Waterway, Port, Coastal and Ocean Engrg., **127**, 4, 200-212.

Tucker, M. J. (1950). Surf beats: Sea waves of 1 to 5 min. period. Proc. Roy. Soc. Lond., A **202**, 565-573.

Ursell, F. (1952). Edge waves on a sloping beach. Proc. Roy. Soc. Lond., A, **214**, 79–97.

Van Dongeren, A. R. and I. A. Svendsen (1997). Quasi-3D modeling of nearshore hydrodynamics. Center for Applied Coastal Reseach, Rep. no CACR-97-04, 243 pp.

Wei, G. and J. T. Kirby (1995). A time dependent numerical code for extended Boussinesq equations. ASCE J. WPCOE, **120**, 251 - 261.

Zhao, Q., I. A. Svendsen and K. Haas (2003). Three-dimensional effects in shear waves. J. of Geophys. Res., **108** (C8) art. no.-3270.

Özkan-Haller, H. T. and J. T. Kirby (1999). Nonlinear evolution of shear instabilities of the longshore current: A comparison of observations and computations. J. Geophys. Res., **104**, C11, 25953-25984.

Author Index

Subject Index

Printed in the United States
By Bookmasters